# System Dynamics and Response

**S. Graham Kelly**
The University of Akron

CENGAGE
Learning™

Australia • Brazil • Japan • Korea • Mexico • Singapore • Spain • United Kingdom • United States

**System Dynamics and Response,
First Edition**

S. Graham Kelly

Associate Vice President and Editorial
    Director: Evelyn Veitch

Publisher: Chris Carson

Developmental Editors: Kamilah Reid Burrell/
    Hilda Gowans

Permissions Coordinator: Vicki Gould

Production Services: RPK Editorial Services

Copy Editor: Pat Daly

Proofreader: Erin Wagner

Indexer: Shelly Gerger-Knechtl

Production Manager: Renate McCloy

Creative Director: Angela Cluer

Interior Design: Carmela Pereira

Cover Design: Andrew Adams

Compositor: International Typesetting and
    Composition

Cover Image Credit: © Alan Schein/zefa/
    Corbis

MATLAB and SIMULINK are registered
trademarks of The MathWorks, 3 Apple Hill
Drive, Natick, MA 01760.

For product information and technology assistance, contact us at
**Cengage Learning Customer & Sales Support, 1-800-354-9706**

For permission to use material from this text or product,
submit all requests online at **www.cengage.com/permissions**
Further permissions questions can be e-mailed to
**permissionrequest@cengage.com**

Library of Congress Control Number: 2006904572

ISBN-13: 978-0-534-54930-5

ISBN-10: 0-534-54930-6

**Cengage Learning**
5191 Natorp Boulevard
Mason, OH 45040
USA

Cengage Learning is a leading provider of customized learning solutions
with office locations around the globe, including Singapore, the United
Kingdom, Australia, Mexico, Brazil, and Japan. Locate your local office at
**www.cengage.com/global**

Cengage Learning products are represented in Canada by
Nelson Education, Ltd.

To learn more about Cangage Learning, visit **www.cengage.com**

Purchase any of our products at your local college store or at our
preferred online store **www.cengagebrain.com**

Printed and bound in the United States of America
3 4 5  16

To Seala and Graham

# Contents

Preface    xiv

## 1    Introduction    1

1.1    Dynamic Systems    1
    1.1.2  Control Systems    3
1.2    Dimensions and Units    3
1.3    Mathematical Modeling of Dynamic Systems    9
1.4    System Response    12
1.5    Linearization of Differential Equations    15
1.6    Unit Impulse Function and Unit Step Function    20
    1.6.1  Unit Impulse Function    21
    1.6.2  Unit Step Function    23
1.7    Stability    25
1.8    MATLAB    27
1.9    Scope of Study    43
    1.9.1  Model Formulation    43
    1.9.2  Mathematical Solutions    44
    1.9.3  System Response    44
    1.9.4  Introduction to Control Systems    44
1.10    Summary    45
    1.10.1 Chapter Highlights    45
    1.10.2 Important Equations    45
Problems    46

## 2    Mechanical Systems    50

2.1    Inertia Elements    51
    2.1.1  Particles    51
    2.1.2  Rigid Bodies    52

**2.1.3** Deformable Bodies   58

**2.1.4** Degrees of Freedom   59

**2.2**   Springs   61

**2.2.1** Force-Displacement Relations   62

**2.2.2** Combinations of Springs   64

**2.2.3** Static Deflections   67

**2.3**   Friction Elements   68

**2.3.1** Viscous Damping   68

**2.3.2** Coulomb Damping   72

**2.3.3** Hysteretic Damping   74

**2.4**   Mechanical System Input   74

**2.4.1** External Forces and Torques   75

**2.4.2** Impulsive Forces   78

**2.4.3** Step Forces   78

**2.4.4** Periodic Forces   79

**2.4.5** Motion Input   80

**2.5**   Free-Body Diagrams   81

**2.6**   Newton's Laws   87

**2.6.1** Particles   87

**2.6.2** Rigid Body Motion   90

**2.6.3** Pure Rotational Motion about a Fixed Axis of
         Rotation   91

**2.6.4** Planar Motion of a Rigid Body   93

**2.6.5** Three-Dimensional Motion of Rigid Bodies   95

**2.6.6** D'Alembert's Principle   95

**2.6.6.1** Particles   95

**2.6.6.2** Rigid Bodies Undergoing Planar Motion   95

**2.7**   Single-Degree-of-Freedom Systems   101

**2.8**   Multidegree-of-Freedom Systems   103

**2.9**   Energy Methods   104

**2.9.1** Principle of Work and Energy   104

**2.9.2** Equivalent Systems   105

**2.9.3** Energy Storage   110

**2.9.4** Lagrange's Equations for Multidegree-of-Freedom
         Systems   111

**2.9.5** States and Order   112

**2.10**   Further Examples   114

**2.11**   Summary   123

**2.11.1** Modeling Methods   123

**2.11.2** Chapter Highlights   124

**2.11.3** Important Equations   125

Problems   127

3    **Electrical Systems**   133

**3.1**    Charge, Current, Voltage, and Power   134
**3.2**    Circuit Components   136
    **3.2.1** Resistors   136
    **3.2.2** Capacitors   138
    **3.2.3** Inductors   141
    **3.2.4** Voltage and Current Sources   143
    **3.2.5** Operational Amplifiers   143
    **3.2.6** Electric Circuits and Mechanical Systems   143
**3.3**    Kirchoff's Laws   144
**3.4**    Circuit Reduction   146
    **3.4.1** Series and Parallel Components   147
    **3.4.2** Series Combinations   147
    **3.4.3** Parallel Combinations   148
**3.5**    Modeling of Electric Circuits   151
**3.6**    Mechanical Systems Analogies   156
    **3.6.1** Energy Principles   156
    **3.6.2** Single-Loop Circuits with Voltage Sources   157
    **3.6.3** Single-Loop Circuits with Current Sources   159
    **3.6.4** Multiple-Loop Circuits   161
    **3.6.5** Mechanical Systems with Motion Input   163
    **3.6.6** States   165
**3.7**    Operational Amplifiers   166
**3.8**    Electromechanical Systems   172
    **3.8.1** Magnetic Fields   172
    **3.8.2** General Theory   173
    **3.8.3** Direct Current (dc) Servomotors   176
    **3.8.4** Microelectromechanical Systems (MEMSs) and Nanoelectromechanical Systems (NEMSs)   179
**3.9**    Further Examples   182
**3.10**   Summary   189
    **3.10.1** Mathematical Modeling of Electrical Systems   189
    **3.10.2** Other Chapter Highlights   190
    **3.10.3** Important Equations   190
Problems   191

## 4    Fluid, Thermal, and Chemical Systems    198

**4.1**    Introduction    198
**4.2**    Control Volume Analysis    200
    **4.2.1**    Conservation of Mass    201
    **4.2.2**    Energy Equation    202
    **4.2.3**    Bernoulli's Equation    204
**4.3**    Pipe Flow    207
    **4.3.1**    Losses    208
    **4.3.2**    Orifices    213
    **4.3.3**    Compressible Flows    214
**4.4**    Modeling of Liquid-Level Systems    215
**4.5**    Pneumatic and Hydraulic Systems    222
    **4.5.1**    Pneumatic Systems    223
    **4.5.2**    Hydraulic Systems    226
**4.6**    Thermal Systems    232
**4.7**    Chemical and Biological Systems    240
    **4.7.1**    Continuous Stirred Tank Reactors (CSTRs)    240
    **4.7.2**    Biological Systems    244
**4.8**    Further Examples    245
**4.9**    Summary    261
    **4.9.1**    Mathematical Modeling of Transport Systems    261
    **4.9.2**    Chapter Highlights    261
    **4.9.3**    Important Equations    262
Problems    264

## 5    Laplace Transforms    269

**5.1**    Definition and Existence    271
**5.2**    Determination of Transform Pairs    272
    **5.2.1**    Direct Integration    272
    **5.2.2**    Use of MATLAB    276
**5.3**    Laplace Transform Properties    278
**5.4**    Inversion of Transforms    289
    **5.4.1**    Use of Tables and Properties    290
    **5.4.2**    Partial Fraction Decompositions    292
        **5.4.2.1**    Real Distinct Poles    294
        **5.4.2.2**    Complex Poles    295
        **5.4.2.3**    Repeated Poles    297
        **5.4.2.4**    Brute Force Methods    299

5.4.3 Inversion of Transforms of Periodic Functions   300

5.4.4 Use of MATLAB   301

5.5   Laplace Transform Solution of Differential Equations   307

5.5.1 Systems with One Dependent Variable   307

5.5.2 Systems of Differential Equations   310

5.5.3 Integrodifferential Equations   313

5.5.4 Use of MATLAB   315

5.6   Further Examples   317

5.7   Summary   331

5.7.1 Mathematical Solutions for Response of Dynamic Systems   331

5.7.2 Important Equations   332

Problems   332

# 6    Transient Analysis and Time Domain Response   336

6.1   Transfer Functions   336

6.1.1 Definition and Determination   336

6.1.2 Multiple Inputs and Multiple Outputs   340

6.1.3 System Order   342

6.2   Transient Response Specification   342

6.2.1 Free Response   343

6.2.2 Impulsive Response   343

6.2.3 Step Response   343

6.2.4 Ramp Response   345

6.2.5 Convolution Integral   345

6.2.6 Transient System Response Using MATLAB   346

6.3   Stability Analysis   349

6.3.1 General Theory   349

6.3.2 Routh's Method   353

6.3.3 Relative Stability   355

6.3.4 An Introduction to Root-Locus Analysis   357

6.4   First-Order Systems   361

6.4.1 Free Response   362

6.4.2 Impulsive Response   365

6.4.3 Step Response   366

6.4.4 Ramp Response   367

6.5   Second-Order Systems   368

6.5.1 Free Response   370

6.5.2 Impulsive Response   379

**6.5.3** Step Response   383

**6.5.4** General Transient Response   391

**6.6**   Higher-Order Systems   394

**6.6.1** General Case   394

**6.6.2** Multidegree-of-Freedom Mechanical Systems   399

**6.6.2.1** Transfer Functions   399

**6.6.2.2** Undamped Systems   400

**6.7**   Systems with Time Delay   403

**6.8**   Further Examples   408

**6.9**   Summary   426

**6.9.1** Chapter Highlights   426

**6.9.2** Important equations   428

Problems   430

## 7   Frequency Response   435

**7.1**   Undamped Second-Order Systems   436

**7.2**   Sinusoidal Transfer Function   440

**7.3**   Graphical Representation of the Frequency Response   448

**7.3.1** Frequency Response Curves   448

**7.3.2** Bode Diagrams   450

**7.3.2.1** Construction and Asymptotes   450

**7.3.2.2** Products of Transfer Functions   451

**7.3.2.3** Bode Diagrams for Common Transfer Functions   453

**7.3.2.4** Bode Diagram Parameters   461

**7.3.3** Nyquist Diagrams   462

**7.3.4** Use of MATLAB to Develop Bode Plots and Nyquist Diagrams   463

**7.4**   First-Order Systems   467

**7.5**   Second-Order Systems   472

**7.5.1** One-Degree-of-Freedom Mechanical System   472

**7.5.2** Motion Input   478

**7.5.3** Filters   482

**7.6**   Higher-Order Systems   485

**7.6.1** Dynamic Vibration Absorbers   486

**7.6.2** Higher-Order Filters   489

**7.7**   Response Due to Periodic Input   493

**7.8**   Further Examples   497

**7.9**   Summary   510

**7.9.1** Chapter Highlights   510

**7.9.2** Important Equations   511

Problems   512

---

**8   Feedback Control Systems**   515

**8.1** Block Diagrams   515

**8.1.1** Block Diagram Algebra   516

**8.1.2** Block Diagram Modeling of Dynamic Systems   522

**8.2** Using SIMULINK in Block Diagram Modeling   525

**8.3** Feedback Control   533

**8.3.1** Proportional Control   534

**8.3.2** Integral Control and Proportional Plus Integral (PI) Control   535

**8.3.4** Derivative Control and Proportional Plus Derivative (PD) Control   536

**8.3.5** Proportional Plus Integral Plus Derivative (PID) Control   536

**8.3.6** Error and Offset   538

**8.3.7** Response Due to Unit Step Input   538

**8.4** Feedback Control of First-Order Plants   539

**8.5** Control of Second-Order Plants   551

**8.6** Control System Design   562

**8.6.1** Design Using Root-Locus Diagrams   563

**8.6.2** Ziegler-Nichols Tuning Rules   573

**8.7** Further Examples   576

**8.8** Summary   594

**8.8.1** Chapter Highlights   594

**8.8.2** Important Equations   595

Problems   597

---

**9   State-Space Methods**   601

**9.1** An Example in the State Space   601

**9.2** General State-Space Modeling   607

**9.2.1** Basic Concepts   607

**9.2.2** Multidegree-of-Freedom Mechanical Systems   615

**9.3** State-Space Solutions for Free Response   620

**9.3.1** Laplace Transform Solution   620

**9.3.2** Exponential Solution   621

**9.3.3** General Description of Free Response   624

**9.4** State-Space Analysis of Response Due to Inputs   627

**9.4.1** Laplace Transform Solution   627

**9.4.2** Numerical Solutions   631

**9.4.3** Use of MATLAB Program ode45.m   636

**9.5** Relationship Between Transfer Functions and State-Space Models   638

**9.6** MATLAB and SIMULINK Modeling in the State-Space   641

**9.6.1** MATLAB   641

**9.6.2** SIMULINK   649

**9.7** Nonlinear Systems and Systems with Variable Coefficients   656

**9.8** Further Examples   663

**9.9** Summary   669

**9.9.1** Chapter Highlights   669

**9.9.2** Important Equations   670

Problems   670

---

**A    Complex Algebra**   673

---

**B    Matrix Algebra**   676

**B.1** Definitions   676

**B.2** Matrix Arithmetic   677

**B.3** Determinants   677

**B.4** Matrix Inverse   678

**B.5** Systems of Equations   679

**B.6** Cramer's Rule   680

**B.7** Eigenvalues and Eigenvectors   680

---

**C    MATLAB**   683

**C.1** MATLAB Basics   684

**C.2** Plotting and Annotating Graphs   688

**C.3** Programming commands   690

**C.3.1** Input and Output   691

**C.3.2** Conditional Statements   692

**C.3.3** Looping   694

**C.3.4** User-Defined Functions   695

C.4    Symbolic Math Toolbox    696
C.6    Control System Toolbox    696

# D    Construction of Root-Locus Diagrams    698

D.1    Definitions    698
D.2    Angle and Magnitude Criteria    799
D.3    Construction Guidelines    700
D.4    Summary of Construction Steps    703
D.5    Examples    704

# References    709

# Index    711

# Preface

Engineering systems are becoming increasingly more interdisciplinary. Many modern systems contain both mechanical and electrical components. Engineers designing dynamic systems often work in interdisciplinary teams. If only for the sake of communication within the team, a mechanical engineer must have some knowledge of the modeling and response of electrical systems and an electrical engineer must be familiar with the modeling and response of mechanical systems. All engineers should have general facility with the modeling of dynamic systems and determining their response. The objective of this book is to provide engineers with the framework for a general understanding of system dynamics and response.

The prerequisites for pursuing this study are knowledge of calculus of a single variable, including ordinary differential equations, and a course in engineering physics that includes mechanics, electromagnetism, and thermal systems. Some advanced topics require additional mathematics such as multivariable calculus. It is suggested that readers have a familiarity with concepts studied in the core engineering subjects of statics and dynamics with a familiarity of particle dynamics and dynamics of rigid bodies undergoing planar motion. Previous or concurrent study of chemistry, circuit analysis, fluid mechanics, and thermodynamics, while not essential, will enhance understanding of some topics and examples.

The text focuses on the modeling of dynamic systems and the determination and description of system response. The text also introduces the concepts of feedback control systems, but these are not the focus of the text but is presented as an application of a topic in which knowledge of response of dynamic systems is required. After an introductory chapter the study material is presented in four distinct parts.

Mathematical modeling of dynamic systems is studied in Chapters 2–4. Mechanical systems are the focus of Chapter 2. Electrical systems including electromechanical systems are studied in Chapter 3. Chapter 4 introduces the modeling of systems involving transport processes such as fluid, thermal, chemical, and biological systems. The modeling technique is common for all systems, but the details of its implementation for specific systems vary. System components and basic laws are studied as a prelude to development of mathematical models of each system. Numerous examples are presented.

The mathematical solution of the differential equations and integral differential equations obtained during the modeling process is studied in Chapter 5. The Laplace transform method is the chosen method of solution.

Chapters 6 and 7 provide the third distinct part of the study, the response of dynamic systems. The concept of transfer functions, functions in the Laplace domain which contain information about the system dynamics, is introduced at the beginning of Chapter 6 and is then used in Chapter 6 to study the transient response of dynamic systems and in Chapter 7 to study their steady-state response. Frequency response, including its graphical representation, is also studied in Chapter 7. Applications to mechanical systems with motion input, vibration absorbers, and electronic filters are developed.

The book concludes with an introduction to feedback control systems and their analysis presented in Chapters 8 and 9. The concepts of system response are used to analyze feedback control systems in Chapter 8. State-space analysis of dynamic systems, used in modern control system design, is presented in Chapter 9.

The text frequently uses MATLAB as a tool for the determination of the response of dynamic systems. MATLAB is used as a computational tool, a programming tool, and a graphical tool. Specific applications of functions from the Symbolic Toolbox and the Control System Toolbox are presented. Appendix C provides a short introduction to MATLAB.

SIMULINK, a MATLAB-based simulation and modeling tool, is used in Chapters 8 and 9. SIMULINK allows the development of models using either the transfer function approach or the state-space method.

All MATLAB programs and SIMULINK models developed in the book as well as extensions of these programs are available online at the Cengage website at www.cengage.com/engineering.

The author wishes to thank the editorial staff at Thomson publishing for developing this work. Included in this acknowledgement are Bill Stenquist, who recruited this project, Chris Carson who continued to lead the development of the project, and Developmental Editors Kamilah Reid Burrell and Hilda Gowans. The author gratefully acknowledges excellent suggestions made by the following reviewers: Mohammad H. Elahinia of Virginia Polytechnic Institute and State University, Brandon W. Gordon of Concordia University, Frank Kelso of the University of Minnesota, Nilanjan Sarkar of Vanderbilt University, Kim A. Stetson of the University of Minnesota, Jiong Tang of the University of Connecticut, and David C. Zimmerman of the University of Houston. Comments and suggestions from students who have taken the System Dynamics and Response class at The University of Akron have been appreciated. Finally, the author acknowledges the contributions of his wife Seala Fletcher-Kelly and his son Graham Kelly. Not only did they provide valuable support, they assisted in the development of the manuscript.

*S. Graham Kelly*

# 1

# Introduction

## 1.1 DYNAMIC SYSTEMS

A model for many physical processes is illustrated by Figure 1.1(a). An **input** is provided to a **system** that delivers an **output**. The system is the set of **components** that act together developing the output from the input. The components are individual units within the system. The system is **static** if the output is dependent only on the instantaneous input. The system is **dynamic** when the output is a function of the history of the input. The output from the system for a given system input is called the **system response**.

This book is concerned with dynamic systems and their response. The problem defining the relationship between input, the system components, and the system response can be posed in two ways:

1. Given the input and the system components, determine the system response.
2. Given the input and the desired output, determine a set of system components that can be used to achieve the desired output.

The first problem is the main focus of this book. A thorough knowledge of dynamic system response is necessary to study the second problem, which is related to the design and synthesis of dynamic systems.

The output of a dynamic system is dependent on the history of the input, which may be time dependent. Thus the output varies with time. Time is an independent variable in the study of dynamic system response and is designated by $t$. The dynamic response of a system is obtained when the system output, represented by, say, the dependent variable $x$, is determined as a function of time $x(t)$.

**Mathematical modeling** is the process through which the dynamic response of a system is obtained. Mathematical modeling, described in Section 1.3, leads to the development of mathematical equations that describe the behavior of the system. The behavior of a dynamic system is usually governed by a differential equation, an integral equation, an integrodifferential equation, or a set of differential equations in which time is

1

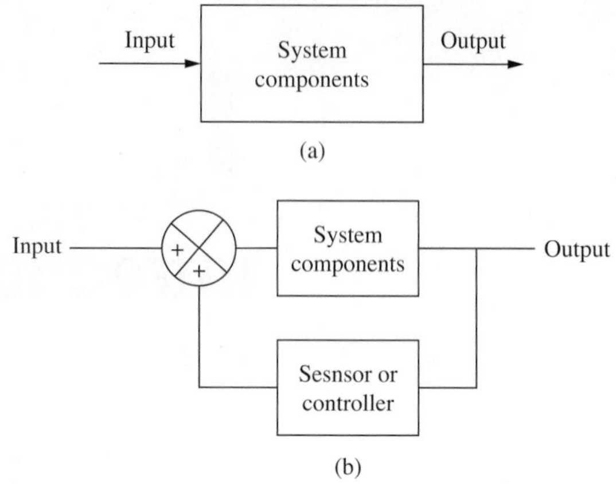

| Input | System components | Output |

(a)

Input ── ⊗(+, +) ── System components ── Output

Sesnsor or controller

(b)

**FIG. 1.1** (a) System model of a physical process; (b) schematic of feedback control system.

the independent variable. The dependent variables represent the system outputs. An assumption made in this book is that all system components are discrete and have lumped properties. This assumption implies that time is the only independent variable in the system modeling. Since the dependent variables are functions of a single independent variable, the derived equations are ordinary differential equations.

A differential equation is **linear** if it contains only linear functions of the dependent variables and their time derivatives. A **nonlinear** differential equation contains nonlinear functions of the dependent variables or their time derivatives. Equations (1.1) are examples of linear differential equations while Equations (1.2) are examples of nonlinear differential equations

$$\ddot{x} + 4\dot{x} + 3x = 2 \sin(5t) \tag{1.1a}$$

$$\ddot{x} + \sin(t)x = 0 \tag{1.1b}$$

$$\ddot{x} + 2\dot{x} + 3x + 4x^3 = 0 \tag{1.2a}$$

$$\ddot{x} + \sin(x) = 2 \sin(5t) \tag{1.2b}$$

The **dot notation** is used in Equations (1.1) and (1.2); a dot above a dependent variable represents differentiation of the variable with respect to time. The term $\ddot{x}$ is read "x double dot" and signifies two derivatives of $x$ with respect to time. Dot notation is used throughout this book.

A system whose mathematical model involves only linear differential equations is called a **linear system**. A system whose mathematical model contains nonlinear differential equations is called a **nonlinear system**. Linear systems are the focus of this book. All systems are inherently nonlinear. Aerodynamic drag and sliding friction lead to nonlinear terms in differential equations governing the response of mechanical systems. Equations used to specify the behavior of system components such as springs and resistors are linearized by neglecting nonlinear terms. Assumptions are often made to eliminate nonlinear terms from the differential equations, thus approximating nonlinear systems using linear equations. Such assumptions include neglecting nonlinear effects such as aerodynamic drag, assuming linear behavior of system components such as springs and resistors, or postulating system operation in a range where nonlinear effects are small. In other cases when a nonlinear differential equation is obtained, it is linearized using the techniques described in Section 1.5.

A system whose input is known at all values of time is a **continuous time system**. The response of a continuous time system is governed by differential equations. A system whose input is known only at discrete times, but at regularly spaced intervals, is called a **discrete time system**. Discrete time systems are governed by difference equations.

### 1.1.2 Control Systems

Dynamic systems are designed such that their response is predictable when subject to defined input. However, input to a system may change unpredictably. For example, an output variable for a heating and air conditioning system is the temperature of a room it is designed to service. An input to the system is the temperature to which the room is to be heated or cooled, the thermostat setting. When the input is changed, the system responds dynamically to change the room temperature. This is an example of a predictable change in input. However, for the heating and air conditioning system to function, it must also be able to sense the room temperature and respond to changes in room temperature caused by external sources when the thermostat setting is constant. The heating and air conditioning system must be designed so that the current room temperature is a system input, and the system must be able to respond to the unpredictable changes caused by external sources.

The heating and air conditioning system just described must be designed with a sensor to determine the room temperature and compare it to the thermostat temperature. When the two temperatures are different, the system must respond so that the output temperature dynamically approaches the input temperature.

A system that senses its output and responds to a difference between input and output is called a **feedback control system**. Feedback control systems are designed to provide stable responses to unpredictable changes in system input. A simple model for a feedback control system is illustrated in Figure 1.1(b). This system is closed loop in that the output from the system components is fed to a sensor whose output is compared with the input, thus closing the loop. The closed-loop response differs from that of the open loop.

## 1.2 DIMENSIONS AND UNITS

A **dimension** is a representation of how a physical variable is expressed quantitatively. The seven basic dimensions are mass, length, time, temperature, electric current, luminous intensity, and amount of a substance in moles. The dimensions of every physical variable can be expressed in terms of these basic dimensions using either the definition of the variable or a physical law relating it to other variables. The dimensions of a

**TABLE 1.1 BASIC DIMENSIONS**

| Quantity | M-L-T dimensions | F-L-T dimensions | SI unit | English unit | Conversion |
|---|---|---|---|---|---|
| Mass | $[M]$ | $[FT^2L^{-1}]$* | kilogram (kg) | slug* | 1 kg = 0.00685 slug |
| Length | $[L]$ | $[L]$ | meter (m) | foot (ft) | 1 m = 3.28 ft |
| Time | $[T]$ | $[T]$ | second (s) | second (s) | |
| Electric current | $[i]$ | $[i]$ | ampere (A) | ampere (A) | |
| Number of moles | [mol] | [mol] | gram mole (gmol) | pound mole (lbmol) | |
| Temperature | $\Theta$ | $\Theta$ | kelvin (K) | rankine (R) | 1 K = 1.8 R |
| Luminous intensity | $[I]$ | $[I]$ | candela (cd) | candela (cd) | |
| Force | $[MLT^{-2}]$* | $[F]$ | newton (N)* | pound (lb) | 1 N = 0.225 lb |

* Derived dimension or unit

quantity are expressed using square brackets. For example, $[M]$ represents a variable whose dimensions are mass, and $[MT^{-2}]$ represents the dimensions of a variable whose dimensions are mass per time squared. Symbols and other information for the basic dimensions are found in Table 1.1.

The dimensions of any physical variable can be derived in terms of the basic dimensions. A quantity such as a displacement has the basic dimension of length. Velocity is defined as the time rate of change of displacement; thus its dimensions are $[LT^{-1}]$. Similarly, the dimensions of acceleration are obtained as $[LT^{-2}]$. The dimensions of force, $[MLT^{-2}]$, are obtained using a physical law, Newton's second law, which states that force equals mass times acceleration.

The set of basic dimensions is called the *M-L-T* system. An alternate formulation of the set of basic dimensions, called the *F-L-T* system, uses force, $[F]$, as a basic dimension rather than mass. In the *F-L-T* system the dimensions of mass are derived as $[FT^2L^{-1}]$. The *F-L-T* system can be used for any physical system but is often used for physical systems in which mass is not a system parameter.

---

**Example 1.1**

The force, $F$, developed on the piston in a piston-cylinder viscous damper is proportional to the velocity, $v$, of the piston. The equation commonly used to state this proportionality is

$$F = cv \tag{a}$$

Determine the dimensions of $c$, called the viscous damping coefficient. Use both the *M-L-T* system and the *F-L-T* system.

**Solution**

Solving Equation (a) for $c$ leads to

$$c = \frac{F}{v} \tag{b}$$

Thus the dimensions of $c$ are equal to the dimensions of force divided by the dimensions of velocity. Using the *M-L-T* system

$$\text{dimensions of } c = \frac{[MLT^{-2}]}{[LT^{-1}]} \tag{c}$$
$$= [MT^{-1}]$$

Using the *F-L-T* system the dimensions of the damping coefficient are

$$\text{dimensions of } c = \frac{[F]}{[LT^{-1}]} \tag{d}$$
$$= [FTL^{-1}]$$

Even though Equations (c) and (d) are equivalent, Equation (d) provides a better sense of the physical meaning of the damping coefficient.

The basic dimension used for electrical systems is that of electric current. Current results form the motion of charged particles. If $i$ represents electric current and $q$ represents electric charge, then

$$i = \frac{dq}{dt} \tag{1.3}$$

Thus electric charge has dimensions of $[iT]$. The electric potential represents the amount of work required to move an electric charge through an electric field. Usually given the algebraic symbol $v$, electric potential has the dimensions of work divided by electric charge. In terms of the basic dimensions using the *F-L-T*, system electric potential has dimensions of $[FLi^{-1}T^{-1}]$.

**Example 1.2**

The relation between voltage and current in a circuit component called an inductor is

$$v = L\frac{di}{dt} \tag{a}$$

Determine the dimensions of $L$, the inductance using the *F-L-T* system of dimensions.

**Solution**

Solving Equation (a) for $L$ leads to

$$L = \frac{v}{\left(\dfrac{di}{dt}\right)} \tag{b}$$

Thus

$$\text{dimensions of } L = \frac{[FLi^{-1}T^{-1}]}{[iT^{-1}]} \tag{c}$$
$$= [FLi^{-2}]$$

The dimensions for physical quantities used in this book are listed in Table 1.2.

Any equation derived from physical principles satisfies the **principle of dimensional homogeneity** which states that every additive term in the equation has equal dimensions. Its inverse can be stated in terms of a popular cliché, "You can't add apples and oranges." A corollary to the principle of dimensional homogeneity is that in mathematical expressions, arguments of transcendental functions must be dimensionless.

---

### Example 1.3

The integrodifferential equation derived in Example 3.10 to model the response of a series $LRC$ circuit is

$$L\frac{di}{dt} + Ri + \frac{1}{C}\int_0^t i\,dt = v(t) \tag{a}$$

where $i(t)$ is the current in the circuit and $v(t)$ is an electric potential.

**(a)**  Show that Equation (a) satisfies the principle of dimensional homogeneity.
**(b)**  If $v(t) = V_0 \sin(\omega t)$ where $t$ is time, what are the dimensions of $\omega$?

**Solution**

(a) The dimensions of each of the parameters are given in Table 1.2. Since the left-hand side of the equation is an electric potential, for the principle of dimensional homogeneity to be satisfied, every term in the equation must also have dimensions of electric potential, $[FLi^{-1}T^{-1}]$. Example 1.2 shows that the first term has the same dimensions as electric potential. Checking the remaining terms

$$\text{dimensions of } Ri = [FLi^{-2}T^{-1}][i]$$
$$= [FLi^{-1}T^{-1}] \tag{b}$$

$$\text{dimensions of } \frac{1}{C}\int_0^t i\,dt = \frac{[i][T]}{[i^2T^2F^{-1}L^{-1}]}$$
$$= [FLi^{-1}T^{-1}] \tag{c}$$

The results in Equations (b) and (c) verify that all terms in Equation (a) have the same dimensions.
(b) Since the argument of the transcendental function $\sin(\omega t)$ must be dimensionless, the dimensions of $\omega$ are $[T^{-1}]$.

---

A **system of units** is a system in which numerical values for basic dimensions are defined. A benchmark quantity for each basic dimension is designated, and all variables with that dimension are referenced to the benchmark quantity.

The **English system** is an $F$-$L$-$T$ system that uses pound as the fundamental unit of force, foot as the fundamental unit of length, second as

**TABLE 1.2 DERIVED DIMENSIONS AND UNITS**

| Quantity (symbol) | M-L-T Dimensions | F-L-T Dimensions | SI units | English units |
|---|---|---|---|---|
| Acceleration ($a$) | $[LT^{-2}]$ | $[LT^{-2}]$ | m/s$^2$ | ft/s$^2$ |
| Angular acceleration ($\alpha$) | $[T^{-2}]$ | $[T^{-2}]$ | r/s$^2$ | r/s$^2$ |
| Angular velocity ($\omega$) | $[T^{-1}]$ | $[T^{-1}]$ | r/s | r/s |
| Area ($A$) | $[L^2]$ | $[L^2]$ | m$^2$ | ft$^2$ |
| Capacitance ($C$) | $[i^2T^4M^{-1}L^{-2}]$ | $[i^2T^2F^{-1}L^{-1}]$ | farad (F) = C/V | F |
| Damping coefficient ($c$) | $[MT^{-1}]$ | $[FTL^{-1}]$ | N·s/m | lb·s/ft |
| Density ($\rho$) | $[ML^{-3}]$ | $[FT^2L^{-4}]$ | kg/m$^3$ | slug/m$^3$ |
| Dynamic viscosity ($\mu$) | $[ML^{-1}T^{-1}]$ | $[FTL^{-2}]$ | N·s/m$^2$ | lb·s/ft$^2$ |
| Electric charge ($q$) | $[iT]$ | $[iT]$ | coulomb (C) = A·s | C |
| Electric potential ($v$) | $[ML^2i^{-1}T^{-3}]$ | $[FLi^{-1}T^{-1}]$ | volt (V) = J/C | V |
| Energy ($E$) | $[ML^2T^{-2}]$ | $[FL]$ | joule (J) = N·m | lb·ft |
| Film coefficient ($h$) | $[MT^{-3}\Theta^{-1}]$ | $[FL^{-1}T^{-1}\Theta^{-1}]$ | J/(m$^2$·K·s) | lb/(ft·R·s) |
| Frequency ($f$) | $[T^{-1}]$ | $[T^{-1}]$ | hertz (Hz) = cycles/s | Hz |
| Frequency ($\omega$) | $[T^{-1}]$ | $[T^{-1}]$ | r/s | r/s |
| Gas constant ($R$) | $[L^2T^{-2}\Theta^{-1}]$ | $[L^2T^{-2}\Theta^{-1}]$ | J/(kg·K) | ft·lb/(lbm·R) |
| Head ($h$) | $[L]$ | $[L]$ | m | ft |
| Impulse ($I$) | $[MLT^{-1}]$ | $[FT]$ | N·s | lb·s |
| Inductance ($L$) | $[ML^2i^{-2}T^{-2}]$ | $[FLi^{-2}]$ | henry (H) = Wb/A | H |
| Magnetic flux ($\psi$) | $[ML^2i^{-1}T^{-2}]$ | $[FLi^{-1}]$ | weber (Wb) = V·s | Wb |
| Mass flow rate ($\dot{m}$) | $[MT^{-1}]$ | $[FTL^{-1}]$ | kg/s | slug/s |
| Moment of inertia ($I$) | $[ML^2]$ | $[FLT^2]$ | kg·m$^2$ | slug·ft$^2$ |
| Mutual inductance ($M$) | $[ML^2i^{-2}T^{-2}]$ | $[FLi^{-2}]$ | H | H |
| Pipe roughness ($\varepsilon$) | $[L]$ | $[L]$ | m | ft |
| Power ($P$) | $[ML^2T^{-3}]$ | $[FLT^{-1}]$ | watt (W) = J/s | lb·ft/s |
| Pressure ($p$) | $[ML^{-1}T^{-2}]$ | $[FL^{-2}]$ | pascal (Pa) = N/m$^2$ | lb/ft$^2$ |
| Rate of reaction ($k$) | $[(mol)L^{-3}T^{-1}]$ | $[(mol)L^{-3}T^{-1}]$ | gmol/(m$^3$·s) | lbmol/(ft$^3$·s) |
| Resistance ($R$) | $[ML^2T^{-3}i^{-2}]$ | $[FLT^{-1}i^{-2}]$ | ohm ($\Omega$) = V/A | $\Omega$ |
| Species concentration ($C$) | $[(mol)L^{-3}]$ | $[(mol)L^{-3}]$ | gmol/m$^3$ | lbmol/ft$^3$ |
| Specific energy ($e$) | $[ML^{-1}T^{-2}]$ | $[FL^{-2}]$ | J/m$^3$ | lb·ft/m$^3$ |
| Specific heat ($c_p$) | $[L^2T^{-2}\Theta^{-1}]$ | $[L^2T^{-2}\Theta^{-1}]$ | J/(kg·K) | ft·lb/(lbm·R) |
| Stiffness ($k$) | $[MT^{-2}]$ | $[FL^{-1}]$ | N/m | lb/ft |
| Thermal conductivity ($k$) | $[MLT^{-3}\Theta^{-1}]$ | $[FT^{-1}\Theta^{-1}]$ | J/(m·K·s) | lb/(s·R) |
| Torsional damping coefficient ($c_t$) | $[ML^2T^{-1}]$ | $[FLT]$ | N·m·s/r | lb·ft·s/r |
| Torsional stiffness ($k_t$) | $[ML^2T^{-2}]$ | $[FL]$ | N·m/r | lb·ft/r |
| Velocity ($v$) | $[LT^{-1}]$ | $[LT^{-1}]$ | m/s | ft/s |
| Volume ($V$) | $[L^3]$ | $[L^3]$ | m$^3$ | ft$^3$ |
| Volume flow rate ($Q$) | $[L^3T^{-1}]$ | $[L^3T^{-1}]$ | m$^3$/s | ft$^3$/s |

the fundamental unit of time, and degrees Rankine as the fundamental unit of temperature. The **SI (System International)** system of units is an *M-L-T* system that uses the kilogram as the fundamental unit of mass, the meter as the fundamental unit of length, the second as the fundamental unit of time, and degrees Kelvin as the fundamental unit of temperature. Both systems use amperes as the fundamental unit of current and candelas as the fundamental unit of luminous intensity. Both systems use the mole as the basic unit of chemical matter. One mole of a substance is the mass of Avogadro's number ($6.022 \times 10^{23}$) of fundamental molecular units (atoms, molecules, ions) of the substance. The SI system uses the gram-mole (gmol), which is the mass in grams of Avogadro's number of molecules, whereas the English system uses the pound-mole (lbmol), which is the mass of the same number of molecules, but measured in pound-mass. These units are summarized in Table 1.1 with conversions between the two systems. In the English system the unit of mass, called a slug, is a derived unit defined by

$$1 \text{ slug} = 1 \text{ lb·s}^2/\text{ft} \tag{1.4}$$

As noted, an alternate unit of mass is the pound mass (lbm) which is the mass of a substance that exerts a gravitational force of one pound. In the SI system the unit of force, call a Newton, is a derived unit defined by

$$1 \text{ newton (N)} = 1 \text{ kg·m/s}^2 \tag{1.5}$$

Derived units for common variables are given in Table 1.2 for both the English and the SI systems. Often derived units are given names, especially in the SI system. For example the unit for electric potential is called the volt and is defined as

$$1 \text{ volt(V)} = 1 \frac{\text{joule(J)}}{\text{coulomb(C)}} = 1 \text{ N·m/A·s} \tag{1.6}$$

It is common practice to use scientific notation to express an answer when using the English system. For example a displacement may be expressed as $x = 1.35 \times 10^{-5}$ ft. In the SI system it is common practice to use prefixes representing powers of 10 rather than scientific notation. For example a numerical value for energy may be expressed as $E = 35.6$ MJ. Table 1.3 lists the prefixes used in the SI system. In either system, unless otherwise needed, three significant digits are used to express numerical values.

Angular measure is inherently dimensionless. However, there are several units in which angular measure is expressed. The measure of an angle in **radians**, the fundamental unit of angular measure, is illustrated in Figure 1.2. The angle between the two line segments of unit length, measured in radians, is equal to the length of the arc subtended from the line segments. There are $2\pi$ radians in the arc of a full circle, the circumference of a circle. A protractor is used to measure angles in **degrees**, with 360 degrees in a circle. Thus

$$2\pi \text{ radians (r)} = 360 \text{ degrees (°)} \tag{1.7}$$

**TABLE 1.3 SI PREFIXES**

| Prefix | Symbol | Factor |
|--------|--------|--------|
| Terra  | T      | $10^{12}$ |
| Giga   | G      | $10^{9}$ |
| Mega   | M      | $10^{6}$ |
| Kilo   | k      | $10^{3}$ |
| Centi  | c      | $10^{-2}$ |
| Milli  | m      | $10^{-3}$ |
| Micro  | $\mu$  | $10^{-6}$ |
| Nano   | n      | $10^{-9}$ |

**FIG. 1.2** Angular measure in radians is equal to the arc length of a sector of a circle of radius 1 formed by two radii making an angle $\theta$.

The units of **cycles** and **revolutions** measure the number of times a full circle has been executed. Thus

$$1 \text{ cycle} = 2\pi \, r \qquad (1.8)$$

If an angular measure is determined using degrees, cycles, or revolutions, it must be converted to radians using Equation (1.7) or (1.8) before it is used in calculations.

**Frequency** measures the rate at which a cyclic motion is executed. Frequency is often measured in hertz (Hz) defined by

$$1 \text{ Hz} = 1 \text{ cycle/s} \qquad (1.9)$$

Frequency in hertz is usually given the symbol $f$. However, in calculations the frequency in r/s, $\omega$, must be used. The conversion is

$$\omega = 2\pi f \qquad (1.10)$$

Since angular measure is inherently dimensionless frequency has dimensions of $[T^{-1}]$.

# 1.3 MATHEMATICAL MODELING OF DYNAMIC SYSTEMS

Mathematical modeling is a process through which the output of a dynamic system, as illustrated in Figure 1.1, is determined given the system input. Mathematical modeling of a system can be used to achieve one of three objectives. **(1) System analysis** is used to determine the output for a specific system. The system components and their parameters are defined. The system may be analyzed for a variety of system inputs. **(2) System design** is used to determine the system components and their parameters such that a specific system output is achieved. **(3)** There is a subtle difference between system design and **system synthesis**, which is the determination of system components and their parameters to achieve a specific performance for a variety of system inputs. System synthesis is used in the analysis and design of control systems.

The procedure for the mathematical modeling of dynamic systems is outlined in this section. The steps are not necessarily followed in the order presented. For example steps 2–4 may be done simultaneously or step 4 may be performed before step 2 or 3. All steps are not necessary for some systems. For example, step 6, determining the system's initial conditions, is not necessary when only a steady-state response is required.

*Step 1:* Define the system to be modeled. Identify the input to the system and what will constitute output. For example an electromechanical system may consist of a circuit designed to operate a motor, which in turn provides power to turn a shaft that operates a turbine. This system consists of several subsystems. The output from one subsystem is the input for another. Figure 1.3(a) illustrates subsystem that are uncoupled, whereas Figure 1.3(b) illustrates subsystems that are coupled. In Figure 1.3(b) there

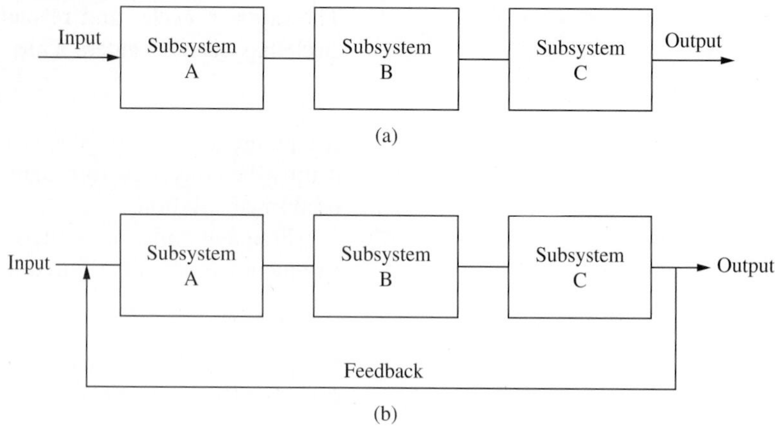

FIG. **1.3** (a) The system is composed of uncoupled subsystems; (b) feedback from subsystem C into subsystem A couples the subsystems.

is **feedback** from subsystem C into subsystem A; the output from subsystem C affects the input to subsystem A. Each subsystem in Figure 1.3(a) could be modeled as a system. The feedback in the system of Figure 1.3(b) requires that the modeling include the entire system.

*Step 2:* The assumptions under which the modeling occurs must be identified and stated. **Implicit assumptions** are those made for almost any system. They are taken for granted and rarely stated. Indeed these assumptions should be explicitly stated when they are not applicable. Examples of implicit assumptions are that the earth is an inertial reference frame, that a continuum model can be used for all matter in the system, that no nuclear reactions are occurring in the system, and that the acceleration due to gravity is a constant. **Explicit assumptions** must be stated and later verified. Examples of explicit assumptions are displacements are small, specific laws for system components apply, and friction effects are negligible.

Assumptions are made to simplify the modeling. A mathematical model is an approximation of a true physical system. If it were possible to model a physical system exactly, a mathematical analysis would be most likely impossible. Thus approximations, in the form of assumptions, must be made. These approximations reduce the accuracy of the predicted response. Some assumptions are made so that the modeling is not needlessly complicated by including negligible effects. Other assumptions are made to linearize a system. Because all systems are inherently nonlinear, some linearizing assumptions are essential. The validity of assumptions can be determined upon completion of the modeling.

This book considers mostly linear systems whose response is governed by linear ordinary differential equations. Thus all nonlinear effects are assumed to be negligible. In addition all system components are assumed to be discrete and their properties are lumped. This assumption implies that variables have no variation with spatial coordinates. Specific assumptions leading to this result include that all bodies in mechanical systems are rigid, all springs and viscous dampers are massless, and the resistance of a wire can be modeled by a single resistor in a circuit. The discrete, lumped parameter system is implicit throughout this study.

*Step 3:* System components are identified and their behavior quantified. Relations defining the behavior of system components are the result of constitutive equations (equations defining the behavior of materials and specifying their properties) and equations of state (equations defining the physical state in which the component exists). Hooke's law relating stress and strain for an elastic solid is an example of a constitutive equation, while the ideal gas law is an example of an equation of state.

For example the force-displacement relationships for springs are quantified in mechanical systems. The equations for the voltage drop across a resistor and stored energy in the electric field of a capacitor are defined for an electrical system. Appropriate equations for friction losses in pipes are established for hydraulic systems. Laws for heat transfer mechanisms are established for thermal systems. Rate of reaction equations are determined for components in chemical systems.

*Step 4:* **Variables** and **parameters** are defined. The independent variable in a dynamic system is time. The dependent variables in a dynamic system are related to the system output. Parameters are properties of system components that are assumed to remain constant as the dependent variables change. Examples of parameters include geometric quantities such as length and area, stiffnesses of springs, properties of electric circuit components such as resistance and capacitance, rates of reactions in chemical systems, and properties such as density viscosity, and specific heat for fluid and thermal systems.

*Step 5:* Applicable physical laws are applied resulting in an equation or a set of equations to solve for the dependent variables. The applicable physical laws vary depending on the nature of the system. Newton's second law is the basic law applied for a mechanical system. Kirchoff's circuit laws, derived from Maxwell's equations, are applied in modeling discrete electric circuits. Conservation of mass and conservation of energy are often applied to model thermal and chemical systems.

*Step 6:* **Initial conditions**, conditions that define the state of the system at an initial value of time, usually at $t = 0$, are often specified.

*Step 7:* Mathematical analysis is performed to determine the time-dependent solution for the dependent variables. The Laplace transform method is used in this book as the tool for solving differential equations derived in step 5 subject to the initial conditions formulated in step 6.

*Step 8:* The system output is determined from the mathematical solution obtained in step 7. The output is analyzed to attain the objectives of the modeling process.

*Step 9:* The model is validated. Model validation may include checking the validity of the assumptions, comparing numerical results or system performance to benchmark situations, or simulating the system in a laboratory.

Often, especially in design and synthesis, the modeling process is iterative. For example, in a design situation, the system components are not defined in advance. Also, the process may proceed with proposed components. If the desired objective is not achieved, the components may be changed.

## 1.4 SYSTEM RESPONSE

The response of a system depends on many factors, including system components and how the system is modeled. It has been established that this book deals with the dynamics of lumped parameter linear systems modeled by ordinary differential equations. The **order** of the system is a key factor in understanding the dynamics of such a system. The term "order" can be defined in a variety of ways including a definition relating to energy. A definition of order is presented in Chapter 6, which relates order to a mathematical property of a system, called its transfer function. For now it suffices to define the term as the order of the derivatives appearing in the system's mathematical model. In this context, the order of a system is the highest order derivative appearing in the mathematical model of a system when it is expressed as a single differential equation. A **first-order system** is modeled by a first-order differential equation. A **second-order system** is modeled by a second-order differential equation. When the modeling of a system results in an integrodifferential equation, the order of the system is obtained by differentiating the equation to eliminate the integral. A **higher-order system** is modeled by a set of differential equations; a system modeled by three second-order differential equations is a sixth-order system.

The **free response** of a system is its response due to nonzero initial conditions and occurs in the absence of any other system input. A **forced response** occurs when the system is subject to a nonzero input for $t > 0$. When subject to a nonzero input, a linear system also has a free response caused by the sudden change in conditions. For linear systems a **general response** is the sum of the forced response and the free response.

The **transient response** of a system refers to either its free response or to the system response shortly after input is changed. The **steady-state response** is the system's response after a long period of time. In many systems the transient response decays and is insignificant after some time. In such a case the steady-state response is the forced response. If the input is periodic, the steady-state response is periodic. When the energy added to the system through the input is continually increased, the system may not reach a steady state. The steady-state response of a linear system, when it exists, is independent of initial conditions. The mass-spring-viscous damper system of Figure 1.4(a) is at rest in equilibrium when an external periodic force is applied. Mathematical modeling of the system leads to the dynamic response of Figure 1.4(b), which clearly shows an initial transient response that decays leaving the steady-state response.

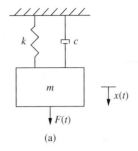

(a)

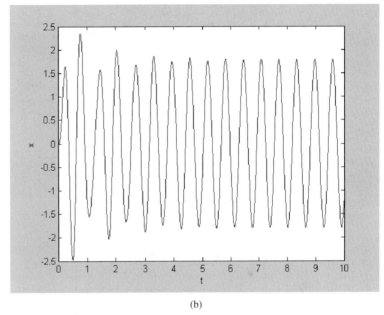

(b)

**FIG. 1.4**  (a) A mechanical system is initially at rest in equilibrium, when an external periodic force is applied. (b) The response of a mechanical system at rest in equilibrium when suddenly subject to a periodic input begins with a transient response, which eventually decays leaving only the steady-state response.

Figure 1.5(a) illustrates the free response of a typical second-order system. Figure 1.5(b) illustrates the general response of a typical first-order system due to a step input. Transient system response is studied in Chapter 6. Figure 1.5(c) illustrates the steady-state response of a typical second-order system due to a periodic input. Steady-state responses are studied in Chapter 7.

**Equilibrium** is the balance achieved between competing forces. The term is often used to describe the state of a system when system variables do not change with time. A mechanical system is in static equilibrium when the resultant of external forces is zero. When current does not flow through as circuit, the circuit is in equilibrium. When the total heat transfer to or from a thermal system is zero, the system is in equilibrium. A chemical system is in equilibrium when the concentrations of reaction components are constant. In each of these systems the equilibrium is disturbed when external conditions change. A switch may be closed in a circuit allowing current to flow, additional reactants may be added to a chemical system, or the ambient temperature may change in a thermal system. A transient response occurs when an equilibrium state is disturbed.

In system dynamics terms, input to the system disturbs an equilibrium position. Mathematical modeling of the system results in the prediction of the system response due to the input. Thus mathematical models of systems are built around predicting changes from equilibrium.

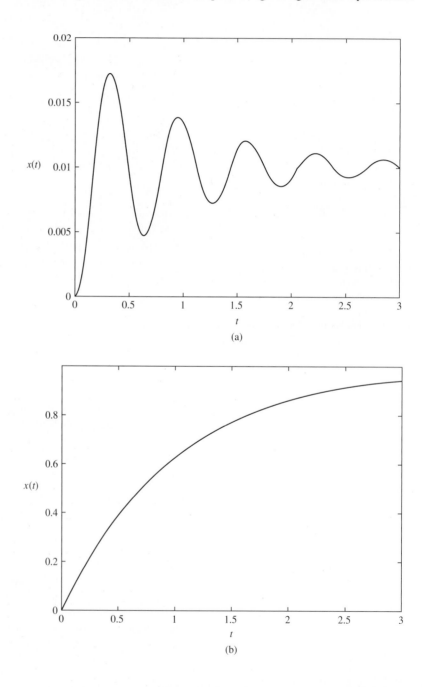

**FIG. 1.5** (a) Typical of free response of a second-order system; (b) typical response of a first-order system due to a step input; (c) typical steady-state response of a second-order system due to a periodic input.

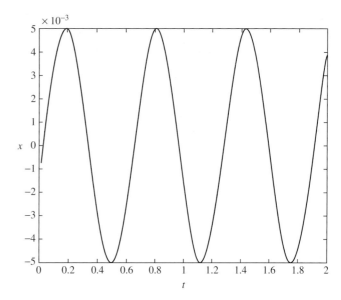

**FIG. 1.5** (*Continued*)

For this reason, dependent variables used in system analysis are often measured from the system's equilibrium position.

## 1.5 LINEARIZATION OF DIFFERENTIAL EQUATIONS

This book considers the modeling and response of linear systems, or systems governed by linear differential equations. As noted in Section 1.3 assumptions are often made to render systems linear. Even after the application of such assumptions, the behavior of some systems is governed by nonlinear differential equations. When appropriate, the equations may be **linearized** using mathematical methods to approximate the nonlinear equation by a linear equation.

As discussed in Section 1.4, changes in system input may disturb an established equilibrium or a steady state. The dependent variables used in the mathematical modeling of a physical system are measured with respect to an equilibrium position or from a defined steady-state position. The system response is a measure of the change in the dependent variable from this position. If the modeling results in a nonlinear differential equation, a linear approximation can often by obtained by expanding the nonlinear terms in a binominal or Taylor series expansion about the equilibrium position or the steady state.

Consider a differential equation of the form

$$\frac{d^2x}{dt^2} + f(x) = g(t) \tag{1.11}$$

Suppose $x = 0$ corresponds to the equilibrium position of the system. If $f(x)$ is continuous and continuously differentiable at $x = 0$, then it has a Taylor series expansion of the form

$$f(x) = f(0) + x\frac{df}{dx}(0) + \frac{x^2}{2}\frac{d^2f}{dx^2}(0) + \cdots + \frac{x^n}{n!}\frac{d^nf}{dx^n}(0) + \cdots \qquad (1.12)$$

In most cases, because $x$ is measured from the system's equilibrium position, $f(0) = 0$. In this case substituting Equation (1.12) into Equation (1.11) leads to

$$\frac{d^2x}{dt^2} + x\frac{df}{dx}(0) + \frac{x^2}{2}\frac{d^2f}{dx^2} + \cdots + \frac{x^n}{n!}\frac{d^nf}{dx^n} + \cdots = g(t) \qquad (1.13)$$

If $x$ is small, then $x^2 \ll x$, and higher-order terms in the Taylor series expansion for $f(x)$ are negligible in comparison to the linear term. An approximate linear differential equation is obtained by truncating the Taylor series expansion for $f(x)$ after the linear term leading to

$$\frac{d^2x}{dt^2} + \left[\frac{df}{dx}(0)\right]x = g(t) \qquad (1.14)$$

Equation (1.14) is an approximation to Equation (1.11) under the assumption of small $x$. After the solution is obtained, the assumption should be checked to validate the linear approximation.

---

**Example 1.4**

The differential equation governing the motion of a compound pendulum is derived in Example 2.15 as

$$\frac{d^2\theta}{dt^2} + \frac{3g}{2L}\sin\theta = 0 \qquad (a)$$

where $\theta$ is the counterclockwise angular displacement of the system measured from the system's vertical equilibrium position. Linearize Equation (a).

**Solution**

The McLaurin series expansion for $f(\theta) = \sin\theta$ about $\theta = 0$ is

$$\sin\theta = 0 - \frac{1}{6}\theta^3 + \frac{1}{120}\theta^5 - \cdots \qquad (b)$$

Truncating Equation (b) after the linear term leads to the approximation

$$\sin\theta \approx \theta \qquad (c)$$

Using Equation (c) in Equation (a) leads to the linear differential equation

$$\frac{d^2\theta}{dt^2} + \frac{3g}{2L}\theta = 0 \qquad (d)$$

Equation (d) is a linear approximation to Equation (a).

**TABLE 1.4 ACCURACY OF SMALL ANGLE APPROXIMATION**

| $\theta$ (degrees) | $\theta$ (rad) | $\sin \theta$ | % error |
|---|---|---|---|
| 0 | 0 | 0 | 0 |
| 1 | 0.017453 | 0.017452 | 0.005077 |
| 3 | 0.052359 | 0.052335 | 0.045706 |
| 4 | 0.069812 | 0.069756 | 0.081276 |
| 5 | 0.087266 | 0.087155 | 0.127034 |
| 7 | 0.122172 | 0.121868 | 0.2492 |
| 10 | 0.174531 | 0.173646 | 0.509495 |
| 13 | 0.22689 | 0.224949 | 0.863169 |
| 15 | 0.261797 | 0.258816 | 1.151492 |
| 18 | 0.314156 | 0.309014 | 1.664039 |
| 20 | 0.349062 | 0.342017 | 2.059983 |
| 25 | 0.436328 | 0.422614 | 3.244952 |
| 30 | 0.523593 | 0.499995 | 4.719654 |
| 35 | 0.610859 | 0.573571 | 6.500963 |
| 40 | 0.698124 | 0.642782 | 8.609822 |
| 45 | 0.78539 | 0.707101 | 11.07183 |
| 50 | 0.872656 | 0.766039 | 13.91796 |

The approximation defined by Equation (c) of Example 1.4 is called the **small angle approximation** and is often used to linearize differential equations whose dependent variable is an angular displacement. As shown in Table 1.4, the error in using Equation (c) to approximate $\sin \theta$ is only 2.1 percent for $\theta = 20°$ and 11.1 percent for $\theta = 45°$.

As an alternative to linearizing a derived nonlinear differential equation, the small angle approximation can be applied before the application of appropriate conservation laws. Displacements are identified in terms of the dependent variable at an arbitrary instant using the small angle approximation. For example the change in length of a spring may be $a \sin \theta$, but when the small angle approximation is applied it is written simply as $a \theta$. The a priori use of the small angle assumption often leads to the derivation of a linear differential equation.

The appropriate approximations used for trigonometric functions when the small angle approximation is used are

$$\sin \theta \approx \theta \qquad (1.15a)$$

$$\cos \theta \approx 1 \qquad (1.15b)$$

$$\tan \theta \approx \theta \qquad (1.15c)$$

$$1 - \cos \theta \approx \frac{1}{2}\theta^2 \qquad (1.15d)$$

Consider a system that has a steady-state response $x_s(t)$ due to a system input $f_s(t)$. The system input is changed to $F(t)$ and the resulting system response is $X(t)$. However, mathematical modeling of the system with this change in input leads to a nonlinear differential equation for $X(t)$. This system is linearized by considering the **perturbation variables**, $x(t)$ and $f(t)$ defined by

$$X(t) = x_s(t) + x(t) \tag{1.16a}$$
$$F(t) = f_s(t) + f(t) \tag{1.16b}$$

The perturbation variables measure changes from steady-state conditions. Equations (1.16) are substituted into the governing differential equations. Nonlinear terms involving $X(t)$ are linearized assuming small $x(t)$ and using an appropriate binomial or Taylor series expansion truncated after the linear term. The steady-state relation between $x_s(t)$ and $f_s(t)$ is used to simplify the equation. Perturbation variables become the dependent variables used in the system.

Suppose a nonlinearity is of the form of $X^n$. Use of Equation (1.16a) leads to

$$X^n = (x_s + x)^n$$
$$= x_s^n \left(1 + \frac{x}{x_s}\right)^n \tag{1.17}$$

Recall the binomial expansion

$$(1 + a)^n = 1 + na + \frac{1}{2!}n(n-1)a^2$$
$$+ \frac{1}{3!}n(n-1)(n-2)a^3 + \cdots \qquad |a| < 1 \tag{1.18}$$

The perturbation variable should be much smaller than the steady-state variable, thus $\left|\frac{x}{x_s}\right| \ll 1$ and Equation (1.18) is used in Equation (1.17) to give

$$X^n = x_s^n \left[1 + n\frac{x}{x_s} + \frac{1}{2!}n(n-1)\left(\frac{x}{x_s}\right)^2 + \cdots\right] \tag{1.19}$$

Equation (1.19) is linearized by truncating after the linear term leading to

$$X^n \approx x_s^n + nx_s^{n-1}x \tag{1.20}$$

If the nonlinearity is not in the form of a power of the dependent variable, a Taylor series expansion is used to linearize the term. Suppose the nonlinear term is of the form $g(X)$. Then a Taylor series expansion for $g(X)$ about $X = x_s$ is

$$g(X) = g(x_s) + \left[\frac{dg}{dX}\right]_{X=x_s} x + \frac{1}{2!}\left[\frac{d^2g}{dX^2}\right]_{X=x_s} x^2 + \cdots \tag{1.21}$$

Truncating Equation (1.21) after the linear term leads to

$$g(X) = g(x_s) + \left[\frac{dg}{dX}\right]_{X=x_s} x \tag{1.22}$$

## Example 1.5

The differential equation governing the level of a fluid in a reservoir is

$$A\frac{dH}{dt} + K\sqrt{H} = Q \tag{a}$$

where $H$ is the liquid level in the reservoir, $A$ is the area of the reservoir, $Q$ is the input flow rate, and $K$ is a constant. For a constant input flow rate $q_s$, the steady-state liquid level is $h_s$ where

$$K\sqrt{h_s} = q_s \tag{b}$$

Suppose the flow rate is perturbed from the steady state such that

$$Q(t) = q_s + q(t) \tag{c}$$

If $h(t)$ is the perturbation of the liquid level, then

$$H(t) = h_s + h(t) \tag{d}$$

Substitute Equations (b)–(d) into Equation (a), and linearize the system to obtain a linear differential equation for $h(t)$.

### Solution

Substituting Equations (b) and (c) into Equation (a) leads to

$$\frac{d}{dt}[h_s + h(t)] + K\sqrt{h_s + h(t)} = q_s + q(t) \tag{e}$$

Since $h_s$ is a constant, $\frac{dh_s}{dt} = 0$. Use of Equation (1.20) with $n = 1/2$ leads to

$$\frac{dh}{dt} + K\left(\sqrt{h_s} + \frac{h}{2\sqrt{h_s}}\right) = q_s + q(t) \tag{f}$$

Equation (b) is used to simplify Equation (f) to

$$\frac{dh}{dt} + \frac{K}{2\sqrt{h_s}}h = q(t) \tag{g}$$

## Example 1.6

In the mathematical modeling of nonisothermal systems with chemical reactions, the Arrhenius law is used to specify the dependence of the rate of reaction $k$ on absolute temperature $T$

$$k = \alpha e^{-\frac{E}{RT}} \tag{a}$$

where $\alpha$ is a constant of proportionality, $E$ is the activation energy, and $R$ is the gas constant. Suppose the temperature is perturbed from a steady-state temperature of $T_s$ by a perturbation $T_p$. Determine a linear relationship between the perturbation in the rate of reaction $k_p$ and the perturbation temperature.

**Solution**

The rate of reaction at the steady state is

$$k_s = \alpha e^{-\frac{E}{RT_s}} \tag{b}$$

Let $k_p$ be the perturbation in the reaction rate. Using perturbation variables, Equation (a) is written as

$$k_s + k_p = \alpha e^{-\frac{E}{R(T_s + T_p)}} \tag{c}$$

Equation (1.22) is used to linearize the exponential function

$$e^{-\frac{E}{R(T_s + T_p)}} = e^{-\frac{E}{RT_s}} + \left[\frac{d}{dT}\left(e^{-\frac{E}{RT}}\right)\right]_{T=T_s} T_p \tag{d}$$

Noting that

$$\frac{d}{dT}\left(e^{-\frac{E}{RT}}\right) = \frac{E}{RT^2} e^{-\frac{E}{RT}} \tag{e}$$

Equation (d) becomes

$$e^{-\frac{E}{R(T_s + T_p)}} = e^{-\frac{E}{RT_s}} + \frac{E}{RT_s^2} e^{-\frac{E}{RT_s}} T_p \tag{f}$$

Using Equations (b) and (f) in Equation (c) leads to

$$k_p = \alpha \frac{E}{RT_s^2} e^{-\frac{E}{RT_s}} T_p$$

$$= \frac{k_s E}{RT_s^2} T_p \tag{g}$$

## 1.6  UNIT IMPULSE FUNCTION AND UNIT STEP FUNCTION

Nature is continuous; that is, physical systems occurring in nature have continuous properties and their responses are continuous functions of time. True discontinuities rarely occur in nature. However, for convenience, some phenomena are often modeled using discontinuous functions. Consider a golfer about to strike a golf ball, as illustrated in Figure 1.6(a). The ball is at rest on the tee while the club head is approaching and about to strike the ball with some velocity. It appears that the ball instantaneously is given a velocity and begins its trajectory; that is, it appears that the velocity of the ball is discontinuous in time. Immediately before the ball is struck, it has a velocity of zero, and immediately after it is stuck, it has a finite velocity, as illustrated in

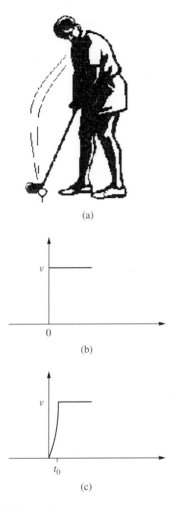

(a)

(b)

(c)

**FIG. 1.6** (a) Before the golfer strikes ball, its velocity is zero, but immediately after being struck the ball appears to have a velocity, $v$; (b) the discontinuous model of velocity of ball; (c) the actual time-dependent velocity of the ball.

Figure 1.6(b). In reality the velocity of the ball is continuous; the club is in contact with the ball for a finite time, after which the ball has attained its initial velocity. During this very short time the force applied to the ball is very large causing a large acceleration and leading to a finite velocity after contact. The actual time-dependent velocity may be as in Figure 1.6(c) where $t_0$ is a very small time.

A very large force applied over a very short interval of time is called an impulsive force, and its application results in the application of an impulse. When the club is in contact with the ball it is applying an impulse to the ball. The modeling of the ball's subsequent motion is dependent on knowledge of the initial velocity imparted to it. Thus it is necessary to mathematically model impulsive forces.

Figures 1.7 and 1.8 illustrate systems in which it is convenient to use discontinuous models for system input. The inlet flow rate into the tank of Figure 1.7(a) is a constant $Q_1$, when an upstream valve is opened allowing more flow into the tank. The actual time-dependent inlet flow rate may be given by Figure 1.7(b), whereas a useful discontinuous approximation is given in Figure 1.7(c). When the switch in the circuit of Figure 1.8 is closed, a finite time is required for the battery to reach its full potential. The true potential supplied by the battery may be given by Figure 1.8(b), whereas a useful discontinuous approximation is given in Figure 1.8(c).

These examples show that it is necessary and useful to approximate continuous input with discontinuous approximations. To this end, mathematical functions that model these discontinuities are necessary.

### 1.6.1  Unit Impulse Function

The mathematical description of the function illustrated in Figure 1.9 is

$$F(t;a) = \begin{cases} 0 & t < -\frac{a}{2} \\ \frac{1}{a} & -\frac{a}{2} < t < \frac{a}{2} \\ 0 & t > \frac{a}{2} \end{cases} \tag{1.23}$$

Independent of the value of $a$, the function defined by Equation (1.23) has the property

$$\int_{-\infty}^{\infty} F(t;a)dt = 1 \tag{1.24}$$

The function $\delta(t)$ is defined by

$$\delta(t) = \lim_{a \to 0} F(t;a) \tag{1.25}$$

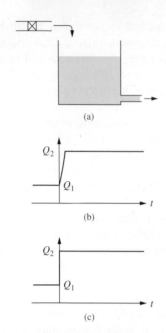

Substitution of Equation (1.23) into Equation (1.25) leads to

$$\delta(t) = \begin{cases} 0 & t \neq 0 \\ \infty & t = 0 \end{cases} \tag{1.26}$$

Integrating Equation (1.25) leads to

$$\int_{-\infty}^{\infty} \delta(t)dt = \int_{-\infty}^{\infty} \lim_{a \to 0} F(t; a)dt \tag{1.27}$$

Interchanging the order of the limit process and the integration in Equation (1.27) gives

$$\int_{-\infty}^{\infty} \delta(t)dt = \lim_{a \to 0} \int_{-\infty}^{\infty} F(t; a)dt \tag{1.28}$$

Using Equation (1.25) in Equation (1.28) yields

$$\int_{-\infty}^{\infty} \delta(t)dt = 1 \tag{1.29}$$

**FIG. 1.7** (a) Single tank system is at a steady state when are upstream valve is opened, increasing flow rate into the tank; (b) the actual time history of the inlet flow rate; (c) the discontinuous model of the inlet flow rate.

The function defined by Equations (1.26) and (1.29) is called the **unit impulse function** or the **Dirac delta function**. The unit impulse function is zero everywhere except at $t = 0$, where it is infinite. However, a definite integral of the unit impulse function over a region containing the impulse is finite and equal to one.

The unit impulse function is used to model system input that is applied instantaneously. In reality the input is applied over a very short interval. Instantaneous input leads to an instantaneous change in a physical variable. Recall that an impulse applied to a mechanical system leads to an instantaneous change in velocity. The unit impulse function provides a mathematical representation of the force whose instantaneous application results in a unit impulse. A force whose instantaneous application results in an impulse of magnitude $I$ is modeled mathematically as

$$F(t) = I\delta(t) \tag{1.30}$$

The mathematical representation of the force whose instantaneous application at time $t_0$ results in an impulse of magnitude $I$ is modeled as

$$F(t) = I\delta(t - t_0) \tag{1.31}$$

The unit impulse function can be used to model the impulsive force applied to the golf ball as it is being struck by the club head. Indeed

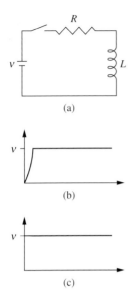

FIG. **1.8** (a) When the switch in the LR circuit is closed at $t = 0$, a battery provides potential to the circuit; (b) the actual potential supplied to the circuit; (c) the discontinuous model of potential.

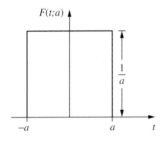

FIG. **1.9** The unit impulse function: $\lim\limits_{a\to 0} F(t; a) = \delta(t)$.

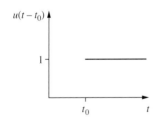

FIG. **1.10** Definition of unit step function, $u(t - t_0)$.

it can be used to model impulsive forces applied to many mechanical systems.

### 1.6.2 Unit Step Function

Consider the function defined as

$$u(t - t_0) = \int_{-\infty}^{t} \delta(\tau - t_0)d\tau \qquad (1.32)$$

If $t < t_0$, then the integrand is identically zero everywhere over the range of integration and the value of the integral is zero. If $t > t_0$, then the range of integration includes the time where the impulse is applied; thus the value of the integral is one. In summary

$$u(t - t_0) = \begin{cases} 0 & t < t_0 \\ 1 & t > t_0 \end{cases} \qquad (1.33)$$

The function $u(t - t_0)$ defined in Equation (1.33) and illustrated in Figure 1.10 is called the **unit step function**. The unit step function is zero when its argument is negative and one when its argument is positive. The unit step function is used to model input that is discontinuous at time $t_0$. For example if a switch in an electrical circuit of Figure 1.8 is closed at $t = t_0$ leading to a battery of potential $V$ being connected to the circuit, then the voltage supplied to the circuit through this source is mathematically modeled as $Vu(t - t_0)$. The mathematical model for the time-dependent flow rate into the tank of Figure 1.7 when the upstream valve is opened at $t = 0$ is $Q(t) = Q_1 + (Q_2 - Q_1)u(t)$.

Differentiation of Equation (1.33) with respect to time leads to

$$\frac{d}{dt}[u(t - t_0)] = \delta(t - t_0) \qquad (1.34)$$

Thus the unit impulse function is the derivative of the unit step function. Two useful integral formulas are

$$\int_{0}^{t} \delta(\tau - t_0)F(\tau)d\tau = F(t_0)u(t - t_0) \qquad (1.35)$$

$$\int_{0}^{t} F(\tau)u(\tau - a)d\tau = u(t - a)\int_{a}^{t} F(\tau)d\tau \qquad (1.36)$$

**Example 1.7**

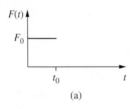

(a)

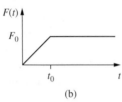

(b)

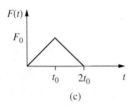

(c)

**FIG. 1.11** Time-dependent forces for Example 1.7: (a) step excitation removed at time $t_0$; (b) step excitation with rise time $t_0$; (c) triangular pulse excitation.

Develop a mathematical model for each of the forces of Figure 1.11.

**Solution**

Each of the forces can be written as a linear superposition of forces, illustrated in Figure 1.12. Each force in the superposition has a mathematical representation that is written using the unit step function.

(a) The force if Figure 1.11(a) is that of a step excitation applied at $t = 0$ and removed at $t = t_0$. The mathematical representation of the force is obtained using Figure 1.12(a) as

$$F(t) = F_0 u(t) - F_0 u(t - t_0) \qquad (a)$$

$F_0$ ⎯⎯⎯⎯ $- F_0$ ⎯⎯⎯⎯

$F_0 u(t)$ $\qquad$ $F_0 u(t - t_0)$

(a)

$\dfrac{F_0 t}{t_0} u(t)$ $\;- F_0\;$ $\dfrac{F_0 t}{t_0} u(t - t_0)$ $\;+ F_0\;$ $F_0 u(t - t_0)$

(b)

$\dfrac{F_0 t}{t_0} u(t)$ $\;- F_0\;$ $\dfrac{F_0 t}{t_0} u(t - t_0)$

$+ F_0$ $\dfrac{2F_0 - F_0 t}{t_0} u(t - t_0)$ $\;- F_0\;$ $\dfrac{2F_0 - F_0 t}{t_0} u(t - 2t_0)$

(c)

**FIG. 1.12** Superpositions used to develop mathematical model for forces of Example 1.7.

(b) The force of Figure 1.11(b) represents a step excitation, but with a finite rise time of $t_0$. The appropriate superposition for this force is shown in Figure 1.12(b), leading to

$$F(t) = F_0 \frac{t}{t_0} u(t) - F_0 \frac{t}{t_0} u(t - t_0) + F_0 u(t - t_0)$$

$$= \frac{F_0}{t_0} [tu(t) - (t - t_0)u(t - t_0)] \tag{b}$$

(c) The superposition for the triangular pulse is illustrated in Figure 1.12(c). The mathematical model for this pulse is

$$F(t) = F_0 \frac{t}{t_0} u(t) - F_0 \frac{t}{t_0} u(t - t_0) + \left(2F_0 - F_0 \frac{t}{t_0}\right) u(t - t_0)$$

$$- \left(2F_0 - F_0 \frac{t}{t_0}\right) u(t - 2t_0)$$

$$= \frac{F_0}{t_0} [tu(t) - 2(t - t_0)u(t - t_0) + (t - 2t_0)u(t - 2t_0)] \tag{c}$$

The **unit ramp function** is defined by

$$r(t) = \int_0^t u(t)dt$$

$$= tu(t) \tag{1.37}$$

The delayed unit ramp function is

$$r(t - t_0) = (t - t_0)u(t - t_0) \tag{1.38}$$

The unit ramp function is illustrated in Figure 1.13.

The mathematical form of the triangular pulse of Example 1.7(c) can be written in terms of ramp functions as

$$F(t) = \frac{F_0}{t_0} [r(t) - 2r(t - t_0) + r(t - 2t_0)] \tag{1.39}$$

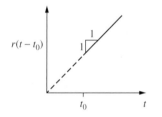

**FIG. 1.13** Delayed unit ramp function.

## 1.7 STABILITY

An equilibrium position may be **stable**, **unstable**, or **neutrally stable**. When a stable equilibrium position is disturbed, a transient response, if developed, will die out when the disturbance is removed and the system will return to the equilibrium position. The transient response of a system about an unstable equilibrium position will continue when the disturbance is removed and equilibrium will not be reestablished. The response of a linear system about a **neutrally stable** equilibrium position is such that the system will establish a new equilibrium position when the disturbance is removed.

FIG. **1.14** (a) Stable equilibrium position;
(b) unstablable equilibrium position;
(c) neutrally stable equilibrium position.

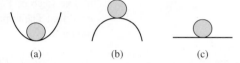

(a)                    (b)                    (c)

A famous example illustrating the stability of equilibrium positions is illustrated in Figure 1.14. In each case the sphere is at rest in equilibrium. If the sphere of Figure 1.14(a) is displaced slightly from this equilibrium position, the action of friction causes the sphere to eventually return to its original equilibrium position; thus the system is stable. When the sphere in Figure 1.14(b) is displaced from equilibrium, it indefinitely continues to move farther away from equilibrium; the system is unstable. If the sphere of Figure 1.14(c) is displaced from equilibrium, it remains in its displaced position; the equilibrium position is neutrally stable.

In this study the stability of a dynamic system about an equilibrium position is determined from the response of the system when a unit impulse is applied. This response is called the system's **impulsive response**. If, after application of a unit impulse, the system approaches its equilibrium position in the steady state, then the system is stable. If no steady state exists for the impulsive response, the system is unstable. If, in the steady state, after the application of a unit impulse the response is bounded but does not approach the original equilibrium position, the system is neutrally stable.

---

## Example 1.8

An actuator is often used in feedback control systems to modify transient system response. The system response is dependent on parameters of the actuator, called gains. Consider a feedback system in which an actuator is used such that the time-dependent impulsive responses of the third-order system for different values of the actuator gain are

**(a)** $x(t) = 0.03e^{-0.25t} + 0.05e^{0.15t} \sin(120t + 0.5)$
**(b)** $x(t) = 0.03e^{-0.25t} + 0.05e^{-0.15t} \sin(120t + 0.5)$
**(c)** $x(t) = 0.03e^{-0.25t} + 0.05 \sin(120t + 0.5)$

Discuss the stability of the system when each actuator is used.

### Solution

(a) Due to the presence of the exponential function with a positive exponent, the impulsive response of this system grows without bound, and a steady state does not exist. Thus this system is unstable.
(b) Since both terms in the impulsive response contain exponential functions with negative exponents, the impulsive response approaches zero for large $t$ and the system is stable.
(c) The steady-state impulsive response is $0.05 \sin(120t + 0.5)$, which is bounded but does not approach zero. Thus the system is neutrally stable.

The forced response of a system is unstable if it grows without bound. Instabilities in the forced response occur due to the nature of the system input as well as system parameters. Resonance, explained in Section 7.2, is an example of an instability that occurs in the forced response of an undamped dynamic system when the frequency of a periodic input is equal to a certain value, which depends on system parameters. Instabilities frequently occur in forced responses of nonlinear systems. An example is the "pull-in" instability in many microelectric mechanical (MEMS) devices, as discussed in Section 3.8. However, instabilities in the forced responses of linear systems are the exception rather than the rule.

## 1.8 MATLAB

MATLAB is a high-performance language with many applications in engineering. The capabilities of MATLAB relevant to this study are computation, programming, and graphics. Advanced capabilities of MATLAB include data acquisition, data analysis, modeling, prototyping, and building interfaces. MATLAB is a command-based language in that it acts on commands provided interactively or from a script.

MATLAB has all the features of any scientific programming language. It accepts formatted input and provides formatted output; variables are declared through arithmetic assignment statements that follow a standard hierarchy of operation; decision-making statements allow for development of structured programs; and subroutines may be written that allow for modular programming.

There are many significant differences between MATLAB and programming languages such as FORTRAN and C. The name MATLAB is a contraction of Matrix Laboratory. MATLAB treats each piece of data as a matrix. A scalar is treated and stored as a one-by-one matrix. However, unlike FORTRAN, MATLAB does not require data stored in matrices to be declared as such and dimensioned. MATLAB is free formatted and less syntax driven than most programming languages.

MATLAB can be used interactively, as in using a calculator or worksheet, or it can be used by writing a script, storing the script in a text file with an .m extension (called an M-file) and executing the M-file. Solutions to problems using MATLAB are presented in this book as M-files. Some M-files are listed as figures; in other cases, only the results obtained through using M-files are presented in the solution to some problems. All M-files used in the solution of examples are available on the Web site for the book.

Special collections of M-files, called toolboxes, are available for use with MATLAB. The toolboxes extend the capability of MATLAB for a variety of applications. The Symbolic Toolbox allows for symbolic computation, that is, representation of data in terms of symbolic variables. SIMULINK is an add-on that allows system simulation by building systems from components and using block diagrams. The Control System Toolbox contains many M-files appropriate for computing and plotting dynamic system response. Use of the Symbolic Toolbox is illustrated in Chapter 5 in the Laplace transform solution of differential equations. Simulations of mechanical and fluid systems using SIMULINK are presented in Chapters 8 and 9. Programs from the Control Systems Toolbox are used in Chapters 5–9.

MATLAB is best learned through continual usage. The language is introduced in this section. A compendium of commands used in this text is presented in Appendix C. Each M-file presented it is annotated with comment statements so that the meaning of each line is clear. The comment statements serve as a tutorial for the M-file. For further elaboration the reader is encouraged to read one of the many books published on using MATLAB in the solution of engineering problems. Also, MATLAB's help function provides valuable assistance with syntax. A major goal of this book is to provide the reader with an understanding of the modeling and response of dynamic systems. MATLAB is a valuable tool used to achieve this goal.

The following examples illustrate MATLAB's capabilities in computation, programming, and graphics that will be used throughout this text. Each example includes comment statements that begin with the percent (%) symbol.

**Example 1.9**

The steady-state response $x(t)$ of a second-order system of mass $m$, natural frequency $\omega$, and damping ratio $\zeta$ due to a sinusoidal input of the form $F_0 \sin(\omega t)$ is obtained in Chapter 7 as

$$x(t) = X \sin(\omega t - \phi) \tag{a}$$

where the steady-state amplitude is

$$X = \frac{F_0}{m\omega_n^2} \frac{1}{\sqrt{(1 - r^2)^2 + (2\zeta r)^2}} \tag{b}$$

the steady-state phase is

$$\phi = \tan^{-1}\left(\frac{2\zeta r}{1 - r^2}\right) \tag{c}$$

and the frequency ratio is defined as

$$r = \frac{\omega}{\omega_n} \tag{d}$$

A special case occurs when $\zeta = 0$ and $r = 1$. In this case a condition called resonance occurs and the response that grows without bound is given by

$$x(t) = \frac{F_0}{2m\omega_n} t \cos(\omega_n t) \tag{e}$$

Develop a MATLAB M-file that, when executed, will

(a) input values of all parameters;
(b) calculate the frequency ratio and check to see if resonance occurs;
(c) if resonance occurs, determine and plot the response given by Equation (e);
(d) if resonance does not occur, determine and plot the response given by Equations (a)–(c).

### Solution

The script of the MATLAB M-file is provided in Figure 1.15. The MATLAB screen obtained from execution of the M-file and the resulting plot file is shown in Figure 1.16 for the case in which the response is given by Equation (a). Figure 1.17 shows the same when resonance occurs, and the response is given by Equation (e).

```
% Example1_9.m
% m file for Example 1.9
%
disp('Steady-state response of second-order mechanical system')
disp('due to harmonic input of F(t)=F_0sin(omega*t)')
disp('Enter system parameters')
%
% Input for this m file is provided interactively when running the script.
% Text contained in single quotes, ' ' ,is printed on the screen.
% The following statements prompt the user to input parameters
% and assigns a numerical value to the stated variable.
%
mass   = input('Enter mass in kg  m= ');
omegan = input('Enter natural frequency in rad/s  omega_n= ');
zeta   = input('Enter dimensionless damping ratio   zeta= ');
omega  = input('Enter frequency in rad/s  omega= ');
F0     = input('Enter magnitude of input force in N   F_0= ');
%
% Calculate and display frequency ratio
%
disp('Dimensionless frequency ratio')
r=omega/omegan;
disp(r)
%
% Resonance is a special case which occurs when zeta=0 and r=1
%
if zeta==0 & r==1                        % Beginning of if loop A
   disp('Undamped system with input frequency equal to natural frequency')
   disp('Resonance occurs and steady state not attained')
%
% When resonance occurs the response is calculated using Eq.(e) of Example
% 1.9. The response is calculated for 0<t<tf
%
   tf=15*pi/omega;
%
% The values of time at which the response is calculated is stored in the
% vector t(i). The corresponding displacements at these times are stored in
% the vector x(i)
```

**FIG. 1.15** Script of file Example1_9.m used for solution of Example 1.9.

```
%
   for i=1:501;                              % Beginning of loop to assign
     t(i)=tf/500*(i-1);                      % values of t and calculate
     c1=-F0/(2*mass*omegan);                 % corresponding x
     x(i)=c1*t(i)*cos(omegan*t(i));
   end                            % Specifies end of for loop
%
% Plot graph of system response
%
   plot(t,x)
%
% Annotation of graph
%
   xlabel('time (s)')                        % Prints label on x axis
   ylabel('x (m)')                           % Prints label on y axis
   title('Resonance response for undamped system')      % Prints title on graph
%
% Set up strings of text to identify, on the graph, values of system parameters
%
   str2(1)={['F(t)=',num2str(F0),'sin(',num2str(omega),'t) N']};
   str2(2)={['m=',num2str(mass),' kg']};
   str2(3)={['\omega_n=',num2str(omegan),' rad/s']};
   str2(4)={['\zeta=',num2str(zeta)]};
   str2(5)={['x(t)=',num2str(c1),'tcos(',num2str(omegan),'t) m']};
   xmax=max(x);
%
% The text will be printed on the graph beginning at the point
% corresponding to t=0.05*tf, and x=0.8*xmax
%
   text(0.05*tf,0.8*xmax,str2)
%
% If resonance occurs the above is the last statement that will be executed
% Program control is shifted to the End statement closing this loop. The
% statements following the Else are executed only when resonance does not occur
%
else
   tf=6*pi/omega;
%
% The steady-state amplitude is calculated using Equation (b) of Example 1.9
%
   amp=F0/(mass*omegan^2)/((1-r^2)^2+(2*zeta*r)^2)^0.5;
   disp('steady-state amplitude in m')
   disp(amp)
   disp('phase angle in rad')
%
% When zeta=0 the phase angle is either zero or pi, depending on the sign
% of r.
%
   if zeta==0
     if r<1
       phi=0;
     else
       phi=pi;
     end
```

**FIG. 1.15** (*Continued*)

```
% If zeta>0 then the phase angle is calculated using Equation (c) of Example 1.9
   else
     if r==1
        phi=pi/2;
     else
        phi=atan(2*zeta*r/(1-r^2));
     end
   end
   disp(phi)
%
% The steady-state response is calculated using Equation (a) of Example 1.9
% The For loop sets up a vector of times at which the response is
% determined and a vector of the response at these times
%
   for i=1:501
     t(i)=tf/500*(i-1);
     x(i)=amp*sin(omega*t(i)-phi);
   end
%
% Plot of response with t on horizontal axis and x on vertical axis
  plot(t,x)
%
%
% Provide title and labels for the graph
%
   xlabel('time (s)')
   ylabel('x (m)')
   title('Steady-state response of second-order system')
%
% Defining text strings to annotate graph with values of system parameters
%
   str2(1)={['F(t)=',num2str(F0),'sin(',num2str(omega),'t) N']};
   str2(2)={['m=',num2str(mass),' kg']};
   str2(3)={['\omega_n=',num2str(omegan),' rad/s']};
   str2(4)={['\zeta=',num2str(zeta)]};
%
% The case when the phase angle is negative is treated as a special case such that only
% one sign is printed
%
   if phi<0
     str2(4)={['x(t)=',num2str(amp),'sin(',num2str(omega),'t+',...
          num2str(-phi),') m']};
   else
     str2(5)={['x(t)=',num2str(amp),'sin(',num2str(omega),'t-',...
          num2str(phi),') m']};
   end
%
% Placement of text on graph
%
  C=min(x);
  D=max(x);
  xp=C+0.2*(D-C);
  text(0.05*tf,xp,str2)
end
```

**FIG. 1.15** *(Continued)*

```
>> Example9
Steady-state response of second-order mechanical system
due to harmonic input of F(t)=F_0sin(omega*t)
Enter system parameters
Enter mass in kg  m= 10
Enter natural frequency in rad/s  omega_n= 100
Enter dimensionless damping ratio   zeta= 0.25
Enter frequency in rad/s  omega= 150
Enter magnitude of input force in N   F_0= 50
Dimensionless frequency ratio
   1.5000

steady-state amplitude in m
 3.4300e-004

phase angle in rad
 -0.5404

>>
```

(a)

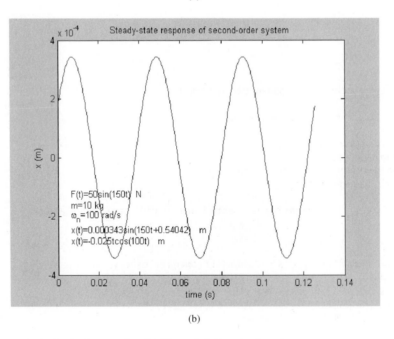

(b)

**FIG. 1.16** (a) MATLAB screen obtained when running file Example1_9.m for the nonresonant case; (b) MATLAB-generated plot for system response.

The M-file is annotated with comment statements to illustrate the syntax of certain statements and the flow of the program. The following additional notes are made:

- The logic in the program is controlled using if loops. A logical statement follows the word if. If the statement is true, the succeeding statements

```
>> Example9
Steady-state response of second-order mechanical system
due to harmonic input of F(t)=F_0sin(omega*t)
Enter system parameters
Enter mass in kg  m= 10
Enter natural frequency in rad/s  omega_n= 100
Enter dimensionless damping ratio   zeta= 0
Enter frequency in rad/s  omega= 100
Enter magnitude of input force in N   F_0= 50
Dimensionless frequency ratio
   1

Undamped system with input frequency equal to natural frequency
Resonance occurs and steady-state not attained
>>
```

(a)

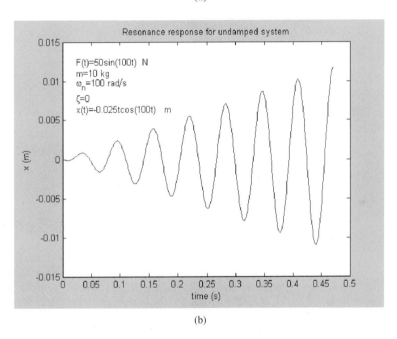

(b)

**FIG. 1.17** (a) MATLAB screen obtained when running file Example1_9.m for the resonant case; (b) MATLAB-generated plot for the response of the system with resonance.

are executed. When the else statement is reached, the control is shifted to the End statement for the loop. If the statement is false, control is shifted to the Else statement and the statements following the word else are executed.

- In the logical sentence within the first If statement, the two equals signs ($==$) are a relational operator used to test equality, and the ampersand (&) is a logical operator meaning "and". Other relational operators that may be used in logical sentences are testing less than ($<$), testing less than or equal to ($<=$), testing greater than ($>$), testing greater than or equal to ($>=$), and testing not equal to ($-=$). The logical operator meaning "or" is a vertical bar (|).

- The hierarchy of operations, from highest to lowest in arithmetic assignment statements for scalars ($1 \times 1$ matrix), is

  **(i)** parentheses ( );
  **(ii)** power (^);
  **(iii)** multiplication (*) and division (/);
  **(iv)** addition(+) and subtraction (−).

  In the case of operations at equal level in the hierarchy, evaluation proceeds from left to right. Since MATLAB stores all data in matrices, additional levels are used for matrix operations such as matrix transpose.
- The title, xlabel, ylabel, and text statements are used to annotate the graphs from the execution of an M-file. It is often easier to annotate a graph using the mouse and visible menus when viewing the graph.
- Unless otherwise specified, the value of a variable is printed after the execution of an assignment statement. The use of the semicolon (;) at the end of the statement suppresses the printing.
- An M-file is executed from the MATLAB command prompt by typing its name.

---

## Example 1.10

In Chapter 3 Kirchoff's Current Law and Kirchoff's Voltage Law are applied to derive equations whose solutions lead to currents in multiloop circuits. Such a circuit analysis of the three-loop circuit of Figure 1.18 leads to the following equations, formulated in a matrix form

$$
\begin{bmatrix} R_1 + R_2 + R_3 & -R_2 & 0 \\ -R_2 & R_2 + R_4 + R_5 & -R_5 \\ 0 & -R_5 & R_5 + R_6 + R_7 \end{bmatrix} \begin{bmatrix} i_1 \\ i_2 \\ i_3 \end{bmatrix} = \begin{bmatrix} v_1 \\ 0 \\ v_2 \end{bmatrix} \quad \text{(a)}
$$

Write a MATLAB M-file that inputs the resistances and dc voltages and that uses matrix methods to solve for the currents. Use the program to solve for the currents when $R_1 = 200\Omega$, $R_2 = 100\Omega$, $R_3 = 400\Omega$, $R_4 = 200\Omega$, $R_5 = 600\Omega$, $R_6 = 300\Omega$, $R_7 = 100\Omega$, $v_1 = 12V$, and $v_2 = 20V$.

### Solution

The file Example10.m developed to solve this problem is shown in Figure 1.19. The output generated when the file is executed using the stated input values is shown in Figure 1.20. Note the following about the programming in this M-file:

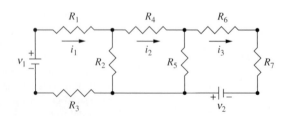

**FIG. 1.18** Circuit for Example 1.10.

```
% Example1_10.m
% Example 1.10
% m file inputs parameter values, sets up matrices, computes currents,
% and outputs values
%
disp('Example 1.10')
disp('Referring to Figure 1.18, input values of resistances, R(i), i=1,2,...,7, each in ohms')
%
% Input resistances
%
for i=1:7
  str={['Please input R(',num2str(i),') ']};
  disp(str)
  R(i)=input('>> ');
%
% Each resistance must have a positive numerical value
%
  while R(i)<=0
    disp('Invalid entry, resistance must be positive')
    str={['Please reeenter R(',num2str(i),') ']};
    disp(str)
    R(i)=input(' ');
  end
end
disp('The coefficient matrix of resistances is ')
%
% Develop coefficient matrix
%
A=[R(1)+R(2)+R(3) -R(2) 0;-R(2) R(2)+R(4)+R(5) -R(5);0 -R(5) R(5)+R(6)+R(7)]
%
% Input dc voltages
%
disp('Referring to Figure 1.18, input values of dc voltages in V ')
for i=1:2
  str={['Please input v(',num2str(i),')']};
  disp(str)
  v(i)=input('');
end
disp('The right-hand side vector is')
%
% Develop right-hand side vector
%
y=[v(1);0;v(2)]
%
% The solution vector is obtained by multiplying the inverse of the
% coefficient matrix by the right-hand side vector
%
I=A^-1*y;
%
% Output results
%
disp('As defined in Figure 1.18 the currents are ')
for i=1:3
  str={['i(',num2str(i),')= ',num2str(I(i)),' A']};
```

**FIG. 1.19** Script of file Example1_10.m that provides solution of Example 1.10.

```
   disp(str)
end
%
% End of m file
%
```

**FIG. 1.19**  (*Continued*)

```
>> Example10
Example 1.10
Referring to Figure 1.18, input values of resistances, R(i), i=1,2,...,7, each in ohms
  'Please input R(1) '

>> 200
  'Please input R(2) '

>> 100
  'Please input R(3) '

>> 400
  'Please input R(4) '

>> 200
  'Please input R(5) '

>> 600
  'Please input R(6) '

>> 300
  'Please input R(7) '

>> 100
The coefficient matrix of resistances is

A =

      700     -100        0
     -100      900     -600
        0     -600     1000
Referring to Figure 1.18, input values of dc voltages in V
  'Please input v(1)'

12
  'Please input v(2)'

20
The right-hand side vector is

y =

   12
    0
   20
```

**FIG. 1.20**  MATLAB output for Example 1.10.

As defined in Figure 1.18 the currents are
  'i(1)= 0.02087 A'

  'i(2)= 0.026087 A'

  'i(3)= 0.035652 A'

\>\>

**FIG. 1.20**  (*Continued*)

---

- The syntax for defining a matrix in MATLAB is to enclose the elements of the matrix in square brackets with entry by rows. Elements in the same row but in adjacent columns are separated by spaces. A semicolon (;) is used to separate rows.
- Contrary to programming languages such as FORTRAN and BASIC, the size of a matrix is not declared before defining the matrix. MATLAB checks for compatibility in performing arithmetic operations using matrices.
- A while loop is used to check the validity of the entered resistances. If an entered resistance is less than zero, the user is requested to re-enter the resistance. The program will continue to request re-entry of the data until a valid value is entered.
- The use of string variables is not the only method to develop the text to be displayed when requesting input data. The command num2str(i) converts the numerical value of the variable $i$ into a text string for the purpose of display.

---

### Example 1.11

A transfer function, $G(s)$, a function introduced in Chapter 6, is used to determine dynamic system response. The Nyquist diagram, introduced in Chapter 7, is a diagram used in the analysis of system response and control system design. The Nyquist diagram is a plot of the real part of the $G(j\omega)$ on the horizontal axis and the imaginary part of $G(j\omega)$ on the vertical axis where $\omega$ is a parameter that varies such that $-\infty < \omega < \infty$. MATLAB has a command for developing the Nyquist plot from the definition of the transfer function. However, it is instructive to develop an M-file that, given the transfer function, develops and plots the Bode diagram. The script is to be developed so that it can be executed without modification for any transfer function. The transfer function is to be provided in a separate file called trans.m. Use the file to develop Nyquist diagrams for (a)

$$G(s) = \frac{2s + 10}{s^2 + 2s + 10}$$

and (b)

$$\frac{2}{(s + 2)(s + 5)}$$

### Solution

The MATALB script Example1_11.m is shown in Figure 1.21. Execution of the script requires the presence of a function subprogram in the trans.m file.

```
%Example 1.11
% Program to develop Nyquist diagram for arbitrary transfer function
%
% The variable i is used as an index for setting up the vectors to store the
% real and imaginary parts of the transfer function
%
i=0
%
% The variable omega is used as a parameter in developing the Nyquist
% diagram. The smallest value of omega is -200, the largest value is 200.
% omega is incremented by 0.1 on each successive pass through the "for"
% loop
%
for omega=-200:.1:200
  i=i+1;
%
% The value of G is the value transferred from function subprogram trans
% when the input to trans is j*omega where j is the square root of -1
%
  G=trans(omega*j);
%
% Calculation of real and imaginary parts of G
%
  x(i)=real(G);
  y(i)=imag(G);
end
% Plot of Nyquist diagram. The values of the vector x are plotted on the
% horizontal axis while the values of y are plotted on the vertical axis.
%
plot(x,y)
%
% Annotating graph
%
title('Nyquist diagram')
xlabel('Re[G(j\omega)]')
ylabel('Im(G(j\omega)]')
%
% End of Example1_11.m
```

**FIG. 1.21**  MATLAB script Example1_11.m developed for Example 1.11.

The appropriate scripts for trans.m used to provide the transfer function to Example1_11.m are shown in Figure 1.22. The Nyqusit diagrams generated from the execution of Example11.m using these files for trans.m are in Figure 1.23. Note the following regarding these files:

- The file uses $-200 \leq \omega \leq 200$ for development of the Nyquist diagram.
- MATALB recognizes either $i$ or $j$ to represent $\sqrt{-1}$.
- MATLAB performs algebra using complex variables without requiring declaring them as complex. The MATLAB functions real and imag evaluate the real and imaginary parts of a complex number respectively.

```
% Required function for Example 1.11
% Name of function subprogram is trans
%
function t=trans(s)
%
% s is the variable that is being transferred from the calling program to
% be used in calculations in function trans
%
% t is the name of the variable which will be transferred to the calling
% program
%
% Calculation of transfer function
t=(2*s+10)/(s^2+2*s+10);
%
% End of function subprogram
```

(a)

```
% Required function for Example 1.11
% Name of function subprogram is trans
%
function t=trans(s)
%
% s is the variable that is being transferred from the calling program to
% be used in calculations in function trans
%
% t is the name of the variable which will be transferred to the calling
% program
%
% Calculation of transfer function
t=2/(s+2)/(s+5);
%
% End of function subprogram
```

(b)

**FIG. 1.22** Function subprogram trans.m required for execution of MATLAB script Example1_11.m. The function subprogram provides the transfer functions for Example 1.11: (a) $G(s) = (2s + 10)/(s^2 + 2s + 10)$ and (b) $G(s) = 2/[(s + 2)(s + 5)]$.

- The program uses a counting index, $i$, to correlate the values of $\text{Re}(G)$ and $\text{Im}(G)$.
- The syntax used for the function subprogram trans.m is explained in the comments statements in Figure 1.22. When MATALB encounters a statement such as G=trans(omega*j), it searches for the file trans.m. The first statement in trans.m must be the function statement defining what is transferred between the calling file and the function subprogram. In mathematical terms, in this example, this statement is comparable to G=trans(s) where trans is the name of the function, s is the independent variable, and G is the dependent variable. The argument list for a function statement may contain

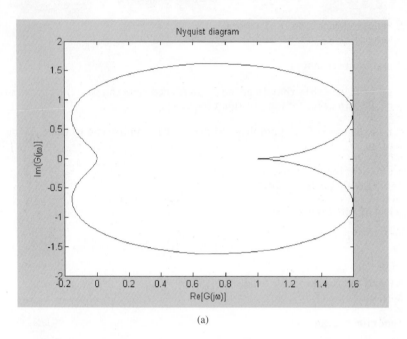

(a)

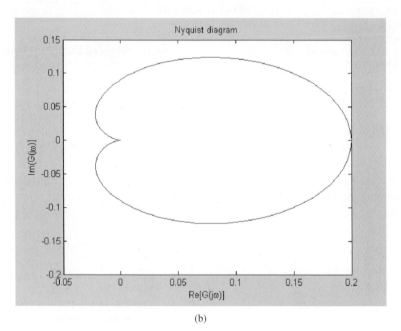

**FIG. 1.23** Nyquist diagrams obtained from execution of MATLAB script Example1_11.m with
(a) $G(s) = (2s + 10)/(s^2 + 2s + 10)$ and
(b) $G(s) = 2/(s + 2)(s + 5)$.

(b)

more than one independent variable whose values are specified from the calling program and more than one dependent variable whose values are calculated in the function subprogram. For example the statement function $[x,y] = h(a,b,c)$ indicates that $x = f(a, b, c)$ and $y = g(a, b, c)$ and that $a$, $b$, and $c$ are specified from the calling program, and $x$ and $y$ are calculated in the function subprogram.

## Example 1.12

It is often instructive to compare the response of a system for several values of a system parameter and thus be able to plot several responses on the same set of axes. Equation (7.82) presents a function used in the frequency response of mechanical systems

$$T = \sqrt{\frac{1 + (2\zeta r)^2}{(1 - r)^2 + (2\zeta r)^2}} \tag{a}$$

It is instructive to plot $T$ vs. $r$ for several values of the parameter $\zeta$. Write a MATLAB M-file that plots on the same set of axes $T$ vs. $r$ for $0 < r < 3.5$ for $\zeta = 0.1, 0.25, 0.4, 0.6,$ and $0.8$.

### Solution

The script for Example1_12.m is given in Figure 1.24, while the plot developed from the execution of Example1_12.m is given in Figure 1.25. Note the following regarding this program:

- This program uses an alternate method of defining and working with vectors. The statement r = 0:.03:3.5 establishes the vector $r$ and specifies that $r(1) = 0, r(2) = 0.03, r(3) = 0.06, \ldots, r(116) = 3.45, r(117) = 3.48$. In general the statement $r = a : b : c$ establishes the vector $r$ such that $r(1) = a, r(i) = r(i-1) + b$ and continues until $r(k) > c$.
- The vectors **T1**, **T2**, **T3**, **T4**, and **T5** are established using the defined values of $r$. The operations (.\*, ./, and .^) in these equations use individual values of $r$ in calculating individual values of the vector **T**. The operations without the preceding period (.) refer to matrix operations. The preceding period indicates that these are scalar operations performed for each value of $r$.
- When executed, the file produces a figure with five plots on the same set of axes. The extra arguments in the plot statement refer to the line style. Other arguments for line width and color can be used. These are listed by using MATLAB's Help function.

The preceding programs illustrate most of MATALB's programming concepts, operations, and plotting functions that are used in the remainder of the book. Appendix C provides a list of important commands and their explanations. The symbolic capabilities of MATLAB are introduced in Chapter 5 and use of the Controls Toolbox in Chapters 6–9.

```
% Example 1.12
% Development of plots of T vs. r for different values of zeta
%
% Input values of zeta
%
disp('Enter five damping ratios for which plots are to be developed')
for i=1:5
   str=['zeta(',num2str(i),')= '];
   disp(str)
   zeta(i)=input('');
%
% Negative values of damping ratio are invalid entries
%
   while zeta(i)<0
      disp('The damping ratio must be non-negative')
      str=['Please reenter zeta(',num2str(i),')'];
      disp(str)
      zeta(i)=input('');
   end
end
%
% Setting up vector of r values
%
r=0:.03:3.5;
%
% Calculation of vectors of T values corresponding to defined values of r
%
T1=((1.+(2*zeta(1)*r).^2)./((1-r.^2).^2+(2*zeta(1)*r).^2)).^0.5;
T2=((1+(2*zeta(2).*r).^2)./((1-r.^2).^2+(2*zeta(2).*r).^2)).^0.5;
T3=((1+(2*zeta(3).*r).^2)./((1-r.^2).^2+(2*zeta(3).*r).^2)).^0.5;
T4=((1+(2*zeta(4).*r).^2)./((1-r.^2).^2+(2*zeta(4).*r).^2)).^0.5;
T5=((1+(2*zeta(5).*r).^2)./((1-r.^2).^2+(2*zeta(5).*r).^2)).^0.5;
%
% Plot of T vs. r for various values of zeta
%
plot(r,T1,'-',r,T2,'.',r,T3,'-.',r,T4,'--',r,T5,'.')
%
% Annotating graph
%
title('T vs r')
xlabel('r')
ylabel('T')
%
% Defining string variables to be used in legend
%
str1=['\zeta=',num2str(zeta(1))];
str2=['\zeta=',num2str(zeta(2))];
str3=['\zeta=',num2str(zeta(3))];
str4=['\zeta=',num2str(zeta(4))];
str5=['\zeta=',num2str(zeta(5))];
legend(str1,str2,str3,str4,str5)
%
% End of Example1_12.m
```

**FIG. 1.24**  Script of MATLAB file Example1_12.m.

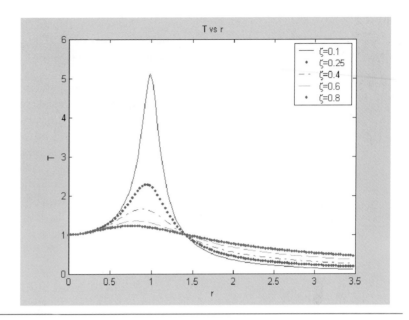

**FIG. 1.25** Plot generated from execution of Example1_12.m.

## 1.9 SCOPE OF STUDY

This study is concerned with the modeling and response of dynamic systems. The study is presented in four distinct parts.

### 1.9.1 Model Formulation

Model formulation, the development of an appropriate mathematical model for a dynamic system, is discussed in Chapters 2–4. Given a dynamic system, basic assumptions are made to simplify the model without compromising its integrity. Appropriate diagrams defining the system at an arbitrary instant are drawn. Basic conservation laws, constitutive equations, and equations of state are applied to formulate a mathematical model. The resulting model is a differential equation, an integrodifferential equation, or a set of coupled differential and/or integrodifferential equations.

Mathematical models for mechanical systems are formulated in Chapter 2. Free-body diagrams of the system at an arbitrary instant are drawn. An appropriate form of Newton's second law (conservation of momentum) is applied leading to the differential equations governing the dynamic response of systems with one and multiple degrees of freedom. An energy method provides an alternate method of deriving the governing equations.

Mathematical models for electrical and electromechanical systems are formulated in Chapter 3. Kirchoff's laws are applied to circuit diagrams drawn of discrete electric circuits leading to differential and integrodifferential equations. Circuits containing active elements such as operational amplifiers are considered. Analogies between mechanical and electrical systems are developed. The techniques studied in Chapters 2 and 3 are used to develop models for electromechanical and microelectric mechanical (MEMS) systems.

Formulation of mathematical models whose systems involve transport processes are considered in Chapter 4. Applications are made to fluid, thermal, and chemical systems and biological systems. The concept of the control volume is introduced, and control volume forms of conservation of mass and conservation of energy (first law of thermodynamics) are developed. Incompressible and compressible flows in pipes and orifices are considered. Applications are developed for liquid level problems, hydraulic servomotors, and pneumatic systems. Heat transfer mechanisms are considered in modeling dynamic thermal systems. Applications are developed for systems in which chemical reactions occur and simple biological models are considered.

### 1.9.2  Mathematical Solutions

The Laplace transform method is developed in Chapter 5 as a tool to solve linear differential equations and integrodifferential equations with constant coefficients. Laplace transform solutions are developed for models derived in Chapters 2–4.

### 1.9.3  System Response

System response is the focus of Chapters 6 and 7. Laplace transforms are used to introduce transfer functions in Chapter 6. The transfer function is then used to study system response. Transient responses of first-order, second-order, and higher-order systems are studied in Chapter 6. The free response of each system is considered as well as the response due to an impulsive input and a unit step input. The transient response is characterized in terms of system parameters.

System response due to periodic input is developed in Chapter 7. The sinusoidal transfer function is used to develop the steady-state response. Frequency response, the dependency of steady-state response on the frequency of periodic input, is studied. Graphical methods are introduced to study frequency response. Applications are made to electrical and mechanical systems with emphasis on vibration isolation and electronic filter design.

### 1.9.4  Introduction to Control Systems

An introduction to feedback control systems and their effect on the response of dynamic systems is considered in Chapter 8. The focus in this chapter is on the types of control systems and on the analysis and response of systems with feedback control.

Modern control systems are analyzed using the state-space methods introduced in Chapter 9. The governing equations are transformed into the state space, and matrix methods are used to determine the system response.

MATLAB is used as an analysis tool in Chapters 5–9. Examples are presented using MATLAB to perform numerical computations and graphically illustrate system response. SIMULINK is used in Chapters 8 and 9 to build and numerically evaluate models.

Each subsequent chapter has a section titled "Further Examples," which illustrate in some detail the concepts of the current and preceding chapters. Each chapter concludes with a summary of the important points in the chapter and a recap of the most important equations from the chapter. Problems for solution are presented after each chapter.

## 1.10 SUMMARY

### 1.10.1 Chapter Highlights

The important points of Chapter 1 are as follows:

- A system is a set of components working together to achieve a desired output from a specified input. A system is dynamic if its output is a function of the history of the input.
- A dimension is a representation of how a physical quantity is expressed quantitatively. Every numerical representation of a physical quantity must be accompanied by appropriate units.
- The principle of dimensional homogeneity requires that all additive terms in an equation have the same dimensions and that all arguments of transcendental functions are dimensionless.
- Mathematical modeling is a process by which the output of a system is determined knowing the system input. A formal process is established and applied throughout this book.
- All physical systems are inherently nonlinear. Assumptions are made to approximate systems using linear differential equations. A formal linearization process may be used if certain assumptions are valid.
- The unit impulse function is the mathematical model of a force whose instantaneous application results in the application of a unit impulse.
- The unit step function is used to mathematically model system input that has finite changes at discrete time.
- The stability of a system about an equilibrium position is defined in this book in relation to the impulsive response of the system.
- MATLAB is a valuable tool that can be used to aid in the numerical determination and graphical representation of system response.

### 1.10.2 Important Equations

- Definition of unit impulse function

$$\delta(t) = \begin{cases} 0 & t \neq 0 \\ \infty & t = 0 \end{cases} \tag{1.26}$$

$$\int_{-\infty}^{\infty} \delta(t)dt = 1 \qquad (1.29)$$

- Definition of unit step function

$$u(t - t_0) = \begin{cases} 0 & t < t_0 \\ 1 & t > t_0 \end{cases} \qquad (1.33)$$

- Definition of unit ramp function

$$r(t - t_0) = (t - t_0)u(t - t_0) \qquad (1.38)$$

## PROBLEMS

**1.1** Newton's law of cooling is used to calculate the rate at which heat is transferred by convection to a solid body at a temperature $T$ from a surrounding fluid at a temperature $T_\infty$. The formula is

$$\dot{Q} = hA(T - T_\infty) \qquad (a)$$

where $\dot{Q}$ is the rate of heat transfer (rate of energy transfer), $A$ is the area over which heat is transferred by convection, and $h$ is the heat transfer coefficient (also called the film coefficient). Use Equation (a) to determine the basic dimensions of $h$. Suggest appropriate units for $h$ using the English system and the SI system.

**1.2** The Reynolds number is used as a measure of the ratio of inertia forces to the friction forces in the flow of a fluid in a circular pipe. The Reynolds number (Re) is defined as

$$\text{Re} = \frac{\rho V D}{\mu} \qquad (a)$$

where $\rho$ is the mass density of the fluid, $V$ is the average velocity of the flow, is the diameter of the pipe, and $\mu$ is the dynamic viscosity of the fluid. Show that the Reynolds number is dimensionless.

**1.3** The relationship between voltage and current in a capacitor is $i = C\frac{dv}{dt}$. Use this relation to determine the basic dimensions of the capacitance $C$.

**1.4** The natural frequency $\omega_n$ of a mechanical system of mass $m$ and stiffness $k$ is calculated by $\omega_n = \sqrt{k/m}$. (a) What are the dimensions of $\omega_n$? (b) A mass-spring system of mass 100 g has a natural frequency of 20 Hz. What is the stiffness of the system?

**1.5** Carbon naontubes are new materials that consist of carbon atoms and are of significant interest because of their good conductivity properties, light weight, and high strength. (a) The density of a carbon nanotube is 1,300 kg/m³. A carbon nanotube is a tube of the diameter of 1 carbon atom, $d = 0.68$ nm. Determine the mass of a carbon nanotube whose length is 80 nm. (b) The elastic modulus of a carbon nanotube is $E = 1.1$ TPa. What is its elastic modulus in pounds per square inch? (c) If modeled as a fixed-fixed beam, the fundamental frequency of the nanotube is

$$\omega = 22.37\sqrt{\frac{EI}{\rho A L^4}}$$

where $I = \frac{\pi}{4}r^4$ is the cross-sectional moment of inertia of the beam and $A$ is its cross-sectional area. Calculate the fundamental frequency in hertz of a nanotube of length 80 nm.

**1.6** During 24 hours of operation, a motor expends 900 kW·hr of energy. Noting that a horsepower (hp) is a unit of power such that 1 hp = 550 ft·lb/s, determine the power delivered by the motor in horsepower.

**1.7** The density of water is 1.94 slugs/ft³. Express the density of water in kilograms per cubic meter (kg/m³).

**1.8** A train is traveling at a speed of 180 km/hr and is decelerating at 6 m/s². Assuming uniform deceleration, how far will the train travel, in miles, before it stops?

**1.9** The differential equation governing the angular velocity $\omega$ of a shaft of mass moment of inertia $J$, acted on by a torque $T$ and attached to bearings of torsional damping coefficient $c_t$ is

$$J\frac{d\omega}{dt} + c_t\omega = T \qquad (a)$$

Use Equation (a) to determine the appropriate dimensions of $c_t$.

**1.10** In an armature controlled dc servomotor, the torque applied to the armature due to a magnetic coupling field is given by

$$T = K_a i_a i_f \qquad \text{(a)}$$

and the back electromotive force (voltage) generated as the shaft rotates through the magnetic field is

$$v = K_v i_f \omega \qquad \text{(b)}$$

where $i_a$ is the current in the armature circuit, $i_f$ is the current in the field circuit, and $\omega$ is the angular velocity of the shaft. Show that the constants $K_a$ and $K_f$ have the same dimensions. Referring to Table 1.2, what physical quantity has the same dimensions as these constants?

**1.11** The rate of heat transfer by radiation from one body to another is given by

$$\dot{Q} = \sigma A \varepsilon \left( T^4 - T_b^4 \right) \qquad \text{(a)}$$

where $A$ is the area of heat transfer, $\varepsilon$ is the dimensionless emissivity of the body, $T$ is the temperature of the body, $T_b$ is the temperature of the radiating body, and the Stefan-Boltzmann constant is

$$\sigma = 1.73 \times 10^{-7}\,\text{Btu/ft}^2 \cdot \text{hr} \cdot \text{R}^4 = 5.67 \times 10^{-8}\,\text{W/m}^2 \cdot \text{K}^4 \quad \text{(b)}$$

(a) What are the basic dimensions of $\dot{Q}$ ? (b) The differential equation governing the transient temperature in a body due to radiation heat transfer is

$$\rho c \frac{dT}{dt} + \sigma \varepsilon \left( T^4 - T_b^4 \right) = 0 \qquad \text{(c)}$$

where $\rho$ is the mass density of the body and $c$ is its specific heat. The body is in a steady state with a uniform temperature defined by a temperature $T_s$ when the temperature of the radiating body is suddenly changed from $T_{bs}$ to $T_{b1}$. Let $T_1$ be perturbation in temperature from the steady state such that the temperature $T$ is

$$T = T_s + T_1(t) \qquad \text{(d)}$$

Substitute Equation (d) into Equation (c), use the binomial theorem to expand the nonlinear term, and derive a linearized differential equation to solve for $T_1(t)$.

**1.12** A nonlinear differential equation governing the motion of a mechanical system is

$$\frac{1}{3} mL^2 \ddot{\theta} + \frac{1}{4} c L^2 \dot{\theta} + kL^2 \sin \theta \cos \theta = 0 \qquad \text{(a)}$$

Assuming small $\theta$ derive a linearized differential equation for the system.

**1.13** The differential equation governing the motion of the system of Figure P1.13 is

$$\frac{1}{3} mL^2 \ddot{\theta} + L\ddot{y} \sin \theta + L\ddot{x} \cos \theta + mg \frac{L}{2} \sin \theta = 0 \qquad \text{(a)}$$

where $x(t)$ and $y(t)$ are known functions of time. Derive a linearized equation governing the motion of the system for small $\theta$.

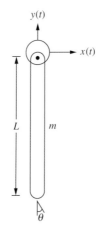

**FIG. P1.13**

**1.14** A chemical reaction that converts moles of a reactant A into moles of a substance B occurs in a continuously stirred tank reactor. The reaction is exothermic, and heat is given off during the reaction. The reactor is cooled by adding heat at a rate $\dot{Q}$ to the reactor. The differential equations governing the concentration of the reactant $C_A$ and the temperature in the reactor $T$ are

$$V\frac{dC_A}{dt} + \left( q + \alpha V e^{-E/(RT)} \right) C_A = q C_{Ai} \qquad \text{(a)}$$

$$\rho q c_p T_i - \rho q c_p T - \dot{Q} + \lambda V \alpha e^{-E/(RT)} C_A = \rho V c_p \frac{dT}{dt} \quad \text{(b)}$$

where $V$ is the volume of the reactor, $q$ is the flow rate of the reactant into the reactor, $\lambda$ is the heat of reaction, $E$ is the activation energy of the reaction, $R$ is the gas constant, $\alpha$ is a constant related to the rate of reaction, $\rho$ is the mass density of the inlet stream, $c_p$ is the specific heat of the mixture, $C_{Ai}$ is the concentration of the reactant in the inlet stream, and $T_i$ is the temperature of the inlet stream. The steady-state conditions $C_{As}$ and $T_s$ are determined from Equations (a) and (b) by setting $\frac{dC_A}{dt} = \frac{dT}{dt} = 0$. The reactor is operating at steady state when it is subject to a sudden change in the flow rate of the inlet stream defined so that the flow rate is

$$q = q_s + q_p(t) \qquad \text{(c)}$$

where $q_s$ is the flow rate at steady state and $q_p$ is the perturbation in the flow rate. The sudden change in flow rate leads to transient behavior in the reactor with time-dependent changes in reactant concentration and temperature defined by

$$C_A = C_{As} + C_{Ap}(t) \tag{d}$$
$$T = T_s + T_p(t) \tag{e}$$

Derive a set of linearized equations that could be solved to obtain the perturbations in reactant concentration and temperature.

**1.15** The differential equation governing the transient temperature in a structure is

$$c_p \frac{dT}{dt} + \frac{1}{R}T = \frac{1}{R}T_\infty \tag{a}$$

where $c_p$ is the specific heat of the structure, $R$ is its thermal resistance, and $T_\infty$ is the ambient temperature of the surrounding medium. The specific heat of a material is a function of temperature. However, in many thermal problems the changes in temperature are small and the specific heat not greatly affected. It is common to assume a constant value for $c_p$, perhaps evaluated at $T_\infty$. Consider a situation in which the ambient temperature has a sudden large perturbation $T_{\infty p}(t)$ from its steady-state value of $T_{\infty s}$, resulting in a transient temperature in the structure defined by $T = T_s + T_p(t)$. The temperature-dependent form of the specific heat for the structure is

$$c_p(T) = A_1 + A_2 T^{1.5} + A_3 T^{2.6} \tag{b}$$

(a) Use binominal expansions to linearize Equation (b), assuming the perturbation temperature is small compared to the steady-state temperature. (b) Derive a linearized mathematical model for the perturbation temperature.

**1.16** An air spring consists of a piston moving in a cylinder of compressed air. When the piston is in equilibrium, the height of the column of air is $h$ and exerts a pressure $p_0$ on the face of the piston. The area of the face is $A$. When the piston moves a distance $x$ downward, as illustrated in Figure P1.16,

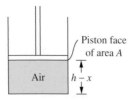

**FIG. P1.16**

the air is further compressed. The process is assumed to be isentropic so that the air pressure is related to its density by

$$p = C\rho^\gamma \tag{a}$$

where $C$ is a constant and $\gamma$ is the ratio of specific heats. (a) Derive a relation between the force on the piston and the distance the piston moves into the column of air $x$. Hint: The mass of the air in the cylinder is constant. (b) Linearize the relationship of part(a) so that the force acting on the face of the piston is approximated as

$$F = p_0 A + kx \tag{b}$$

Specify $k$.

**1.17** The time-dependent voltage of Figure P1.17 is supplied by a source in a series LRC circuit. Write a unified mathematical representation of the voltage provided by the source using unit step functions.

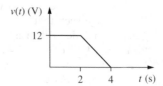

**FIG. P1.17**

**1.18** A mechanical system is subject to the time-dependent force of Figure P1.18. Develop a unified mathematical representation of the force using unit step functions.

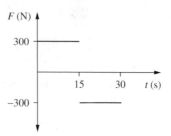

**FIG. P1.18**

**1.19** A cam and follower system has been designed to provide a periodic displacement to a valve in a mechanical system. Figure P1.19 illustrates the displacement provided over one period. (a) Use unit step functions to write a unified mathematical representation for the displacement over one period. (b) The displacement provided by the cam and follower system is periodic of period 0.5 s. Generalize the results of part (a) to provide a unified mathematical representation of the displacement for all time. Hint: Use an

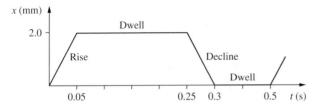

**FIG. P1.19**

infinite series representation of the displacement with the index of summation representing one period.

**1.20** A particle of mass 4 kg is at rest when it is subject to an impulsive force that imparts a total impulse of 12 N·s to the mass. Knowing that Newton's second law can be written as $F = m\frac{dv}{dt}$, determine the velocity of the particle after application of the impulse.

**1.21** No current is flowing through a circuit when it is subject to an impulsive voltage such that the total impulse is 20 V·s. Knowing that the relation between voltage and current in an inductor is $v = L\frac{di}{dt}$ where L = 0.4 H is the inductance, determine the current through the inductor immediately after application of the impulse.

**1.22** A machine is subject to three impulses: an impulse of magnitude 100 N·s applied at $t = 0$, an impulse of magnitude 150 N·s applied at $t = 2.5$ s, and an impulse of magnitude 50 N·s applied at $t = 3.8$ s. What is the mathematical formulation for the force applied to the machine?

**1.23** The steady-state current in a series LRC circuit due to a sinusoidal voltage source is

$$i(t) = \frac{V_0\omega}{L} \frac{1}{\sqrt{(\omega^2 - \omega_n^2)^2 + (2\zeta\omega\omega_n)^2}}$$

$$\times \sin\left[\omega t + \tan^{-1}\left(\frac{\omega^2 - \omega_n^2}{2\zeta\omega\omega_n}\right)\right] \quad (a)$$

where the circuit's natural frequency is $\omega_n = \frac{1}{\sqrt{LC}}$ and damping ratio is $\zeta = \frac{R}{2}\sqrt{\frac{C}{L}}$. Write a MATLAB M-file that does all of the following: (a) inputs the values of resistance (R) in ohms, capacitance (C) in farads, inductance (L) in henrys,

input frequency ($\omega$) in rad/s, and input amplitude ($V_0$) in volts; (b) evaluates $i(t)$ for 200 discrete values of $t$ in the interval $0 \leq t \leq \frac{10\pi}{\omega}$; (c) plots the response with an annotated graph.

**1.24** Write a MATLAB M-file that (a) inputs the elements of two 5×5 matrices **A** and **B**; (b) calculates **A**+**B**; (c) calculates **AB**; (c) calculates $\mathbf{A}^{-1}$; (d) calculates det(**A**); and (e) determines the eigenvalues and eigenvectors of **A** (see Appendix B) using the MATLAB function eigs.

**1.25** The function

$$\Lambda = \frac{r^2}{\sqrt{(1 - r^2)^2 + (2\zeta r)^2}} \quad (a)$$

is used in the analysis of the steady-state response of certain mechanical systems. Write a MATLAB M-file that plots, on the same axes, $\Lambda$ vs. $r$ for five different values of $\zeta$. The plot should be annotated.

**1.26** The step response of an underdamped mechanical system of natural frequency $\omega_n$ and damping ratio $\zeta$ is

$$x(t) = \frac{1}{\omega_n^2}\left[1 - e^{-j\omega_n t}\cos(\omega_d t) + \frac{\zeta\omega_n}{\omega_d}e^{-\zeta\omega_n t}\sin(\omega_d t)\right] \quad (a)$$

where $\omega_d = \omega_n\sqrt{1 - \zeta^2}$. Write a MATLAB M-file that (a) inputs the values of natural frequency and damping ratio and (b) evaluates and plots the step response on an annotated graph. The response should be evaluated for at least five cycles, $t = 5\left(\frac{2\pi}{\omega_d}\right)$.

**1.27** The perturbation in the liquid level in a tank of area $A$ and hydraulic resistance $R$ due to a flow rate perturbation $q$ is

$$h(t) = qR\left(1 - e^{-t/(RA)}\right) \quad (a)$$

(a) If $A$ is in square meters (m²) and $t$ is measured in seconds (s), what are the units of $R$?
(b) It is noted that $\lim_{t\to\infty} h(t) = qR$, which is its steady-state value. Write a MATLAB program that, given $q$, $R$, and $A$, calculates an $h(t)$ until it is within 0.1 % of the steady-state value.
(c) Use the MATALB program to plot $h(t)$.

# 2

# Mechanical Systems

Mechanical system components include inertia elements (elements with mass and kinetic energy), stiffness elements (flexible elements that store energy), and damping elements (elements that dissipate energy). External sources of force or motion provide input to the system. Dependent variables used in the modeling of mechanical systems are displacements of particles within the system or angular displacements. Application of conservation laws leads to differential equations whose solutions provide time-dependent descriptions of the motion of the system's components.

The automobile in Figure 2.1(a) is a familiar mechanical system. To simplify the modeling of the system, assume that when traveling on a straight bumpy road the body of the vehicle has displacements in two directions. Let $x$ be the displacement, measured from a reference position, of the vehicle in the direction along the road and let $y$ be the displacement, measured from an established equilibrium position, in the direction perpendicular to the average road surface. The vehicle's drive train provides an external force that propels the car along the road. The body of the automobile is assumed to be rigid and acts as the system's inertia element. The vehicle has a suspension system which protects the body from large accelerations caused by vertical changes in road contour. There are a variety of suspension elements, but many are modeled by a spring in parallel with a viscous damper. The flexible elements that store energy are the suspension system and the tires. Energy dissipation is provided in the vertical direction by the suspension system and in the horizontal direction by application of the brakes. A simplified model of the automobile is shown in Figure 2.1(b).

Implicit assumptions used throughout this chapter include

- the Earth is an inertial reference frame;
- particles are moving at speeds such that Newtonian mechanics applies;
- gravity is the only force field present;
- the acceleration due to gravity $g$ is constant and equal to $9.81 \text{ m/s}^2$ or $32.2 \text{ ft/s}^2$;
- force-displacement relations for springs are linear;

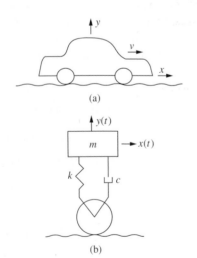

**FIG. 2.1** (a) The automobile as a complex mechanical system; (b) a simplified model of vehicle motion when it is traveling on a straight bumpy road.

- unless otherwise specified, viscous damping is the only dissipative mechanism considered (specifically all forms of aerodynamic drag, friction at pin supports, belt friction, pulley friction, and internal damping are assumed to be negligible);
- all system parameters are constant.

Explicit assumptions will be stated as needed. Assumptions such as the small angle assumption are often required to linearize nonlinear differential equations obtained through the mathematical modeling process.

## 2.1 INERTIA ELEMENTS

An inertia element is a component of a mechanical system with finite mass. The mass of a **particle** is concentrated at a single point. A **rigid body** has a fixed distribution of mass about its center of gravity. The mass of a **deformable body** is constant, but its distribution may change relative to a fixed point on the body.

### 2.1.1 Particles

Consider a particle of mass $m$, instantaneously located at a point $P$ in a three-dimensional Cartesian coordinate system, as illustrated in Figure 2.2. The particle is located relative to the origin of a fixed coordinate system by the **position vector**

$$\mathbf{r} = x\mathbf{i} + y\mathbf{j} + z\mathbf{k} \tag{2.1}$$

where $x$, $y$, and $z$ are the coordinates of $P$ in the Cartesian system. As the particle moves, the location of $P$ changes; thus $x$, $y$, and $z$ and the position vector $\mathbf{r}$ are functions of time.

The **velocity vector** is defined as the time rate of change of the position vector

$$\mathbf{v} = \frac{d\mathbf{r}}{dt} = \frac{dx}{dt}\mathbf{i} + \frac{dy}{dt}\mathbf{j} + \frac{dz}{dt}\mathbf{k} \tag{2.2}$$

Using the shorthand dot notation described in Section 1.1, $\mathbf{v}$ is written as

$$\mathbf{v} = \dot{\mathbf{r}} = \dot{x}\mathbf{i} + \dot{y}\mathbf{j} + \dot{z}\mathbf{k} \tag{2.3}$$

The **acceleration vector** is defined as the time rate of change of the velocity vector

$$\mathbf{a} = \dot{\mathbf{v}} = \mathbf{r} \tag{2.4}$$

$$\mathbf{a} = \ddot{x}\mathbf{i} + \ddot{y}\mathbf{j} + \ddot{z}\mathbf{k} \tag{2.5}$$

The energy associated with the motion of a particle, its **kinetic energy**, is

$$T = \frac{1}{2}m\mathbf{v}\cdot\mathbf{v} \tag{2.6}$$

$$T = \frac{1}{2}m\left(\dot{x}^2 + \dot{y}^2 + \dot{z}^2\right) \tag{2.7}$$

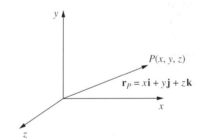

**FIG. 2.2** Position vector of a particle moving in a Cartesian system.

A particle's **linear momentum** is defined as

$$\mathbf{L} = m\mathbf{v} \qquad (2.8)$$

## 2.1.2 Rigid Bodies

The mass of a rigid body is distributed throughout the body, but the distance between two particles on the body is fixed as the body moves in space. Consider particles $A$ and $B$ on the rigid body of Figure 2.3. Let $\mathbf{r}_A$ and $\mathbf{r}_B$ be the position vectors between the origin and particles $A$ and $B$ respectively. Let $\mathbf{r}_{B/A}$ be the vector between $A$ and $B$ and directed toward $B$ and the position vector of particle $B$ relative to particle $A$. As the body moves, the direction of $\mathbf{r}_{B/A}$ may change, but its magnitude is constant.

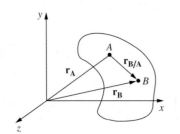

**FIG. 2.3** As the rigid body moves $\mathbf{r}_{B/A}$ may change direction, but its magnitude is constant.

Application of the triangle law of vector addition gives

$$\mathbf{r}_B = \mathbf{r}_A + \mathbf{r}_{B/A} \qquad (2.9)$$

Differentiation of Equation (2.9) with respect to time leads to

$$\mathbf{v}_B = \mathbf{v}_A + \mathbf{v}_{B/A} \qquad (2.10)$$

Rigid body motion is a superposition of translation and rotation. An observer fixed to the body at $A$ only views particle $B$ as rotating about an axis through $A$. The **angular velocity** vector $\omega$ for the body is

$$\omega = \omega\mathbf{e} \qquad (2.11)$$

where $\omega$ is the rate at which the rigid body is rotating about an axis defined by the unit vector $\mathbf{e}$. The angular velocity is positive when the rotation is counterclockwise about the axis of rotation, viewed from the positive axis. The angular velocity vector is a property for the rigid body that may change with time; in general both $\omega$ and $\mathbf{e}$ may change with time. Equation (2.10) is written using the angular velocity vector as

$$\mathbf{v}_B = \mathbf{v}_A + \omega \times \mathbf{r}_{B/A} \qquad (2.12)$$

Differentiating Equation (2.12) with respect to time leads to the relative acceleration equation

$$\mathbf{a}_B = \mathbf{a}_A + \alpha \times \mathbf{r}_{B/A} + \omega \times (\omega \times \mathbf{r}_{B/A}) \qquad (2.13)$$

where the angular acceleration vector is

$$\alpha = \frac{d\omega}{dt} \qquad (2.14)$$

A rigid body is undergoing **planar motion** if the path of each point lies within a plane and the axis of rotation is perpendicular to the plane.

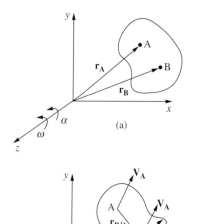

(a)

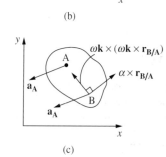

(b)

(c)

**FIG. 2.4** (a) The rigid body is undergoing planar motion. Particles A and B travel in the $x$-$y$ plane. The body rotates about the $z$ axis with an angular velocity $\omega$ and angular acceleration $\alpha$. (b) Illustration of the relative velocity equation. (c) Illustration of the relative acceleration equation.

An observer fixed at $A$ on a rigid body undergoing planar motion views particle $B$ moving in a circular path whose center is at $A$. The relative velocity and acceleration between any two particles $A$ and $B$ on a rigid body undergoing planar motion are given by

$$\mathbf{v}_{B/A} = \left|\mathbf{r}_{B/A}\right|\omega\mathbf{i_t} \tag{2.15}$$

$$\mathbf{a}_{B/A} = \left|\mathbf{r}_{B/A}\right|\alpha\mathbf{i_t} - \left|\mathbf{r}_{B/A}\right|\omega^2\mathbf{i_n} \tag{2.16}$$

where $\mathbf{i_t}$ and $\mathbf{i_n}$ are unit vectors tangent and normal to the circular path. These unit vectors and the components of relative velocity and relative acceleration are illustrated in Figure 2.4.

The **center of mass** of a rigid body is defined as the particle whose position vector is given by

$$\bar{\mathbf{r}} = \frac{1}{m}\int_m \mathbf{r}\,dm \tag{2.17}$$

or in component form

$$\bar{x} = \frac{1}{m}\int_m x\,dm \quad \bar{y} = \frac{1}{m}\int_m y\,dm \quad \bar{z} = \frac{1}{m}\int_m z\,dm \tag{2.18}$$

The position vector for the center of mass of a rigid body may change with time, however the distance between the center of mass and every other particle on the rigid body is constant.

An overbar is used to denote quantities referenced to the mass center. The velocity of the mass center is denoted by $\bar{\mathbf{v}}$ while the acceleration of the mass center is denoted by $\bar{\mathbf{a}}$.

The **moment of inertia** of a rigid body about an axis is a measure of how the body's mass is distributed about that axis. The larger the moment of inertia the more mass is concentrated away from the axis. The definition of moment of inertia about an axis $AA$ through a particle is

$$I_{AA} = \int_m r^2\,dm \tag{2.19}$$

where $r$ is the distance from the axis to a particle of mass $dm$. An **inertia tensor** can be developed at every point in the body

$$\mathbf{I} = \begin{pmatrix} I_{xx} & I_{xy} & I_{xz} \\ I_{yx} & I_{yy} & I_{yz} \\ I_{zx} & I_{zy} & I_{zz} \end{pmatrix} \tag{2.20}$$

For example the moment of inertia of a body about the $x$-axis is

$$I_{xx} = \int_m (y^2 + z^2)\,dm \tag{2.21}$$

The off-diagonal elements of the inertia tensor are called **products of inertia**. For example

$$I_{xy} = \int_m xy\,dm \tag{2.22}$$

The $x$, $y$, and $z$ axes used to define the inertia tensor are mutually perpendicular and form a right-handed coordinate system. The components of the inertia tensor depend on the orientation of these axes. A set of axes for which all products of inertia are zero is called a set of principal axes. An axis of symmetry is a principal axis.

If $x$ and $y$ are any set of orthogonal axes in a plane with an origin at a point $O$, then the sum of the moments of inertia about these axes is a constant regardless of the orientation of the axes. This sum is called the **polar moment of inertia** in this plane about $O$,

$$J_O = I_{xx} + I_{yy} \tag{2.23}$$

The polar moment of inertia is a useful quantity for rigid bodies undergoing planar motion or in pure rotation about a fixed axis.

The components of the inertia tensor depend on the location of the particle for which the tensor is calculated. Centroidal moments of inertia and products of inertia are calculated for axes passing through the center of mass and are denoted by an overbar. The **Parallel Axis Theorem** relates moments of inertia calculated about a centoridal axis and a parallel axis passing through another particle. For example

$$I_{xx} = \bar{I}_{xx} + md_x^2 \tag{2.24}$$

where $\bar{I}_{xx}$ is the moment of inertia about an $x$ axis passing through the center of mass of a rigid body of mass $m$, and $I_{xx}$ is the moment of inertia about an axis parallel to the $x$ axis that passes through a particle such that the distance between the two parallel axes is $d_x$. Equation (2.24) shows that the moment of inertia is smallest through an axis passing through the center of mass. The corresponding form of the Parallel Axis Theorem for a product of inertia is

$$I_{xy} = \bar{I}_{xy} + m\bar{x}\bar{y} \tag{2.25}$$

where $\bar{x}$ and $\bar{y}$ are the $x$ and $y$ coordinates of the of the $x$-$y$ axes relative to the centroidal axes. Moments of inertia of several rigid bodies about principal axes though the body's center of mass are given in Table 2.1.

The mass center of a rigid body in planar motion travels in a plane, say the $x$-$y$ plane. The body rotates about the $z$ axis. Without loss of generality the subscript referring to the axis about which the moment of inertia is taken for bodies undergoing planar motion will be dropped. Thus $\bar{I}$ refers to the mass moment of inertia of the body about an axis parallel to the axis of rotation passing through the center of mass.

**TABLE 2.1  MOMENTS OF INERTIA OF THREE-DIMENSIONAL BODIES**

| Body | General Shape | Centroidal Moments of Inertia |
|---|---|---|
| General shape | | $\bar{I}_x = \int (y^2 + z^2)\, dm$ <br> $\bar{I}_y = \int (x^2 + z^2)\, dm$ <br> $\bar{I}_z = \int (x^2 + y^2)\, dm$ |
| Slender rod | | $\bar{I}_x \approx 0$ <br> $\bar{I}_y = \dfrac{1}{12} mL^2$ <br> $\bar{I}_z = \dfrac{1}{12} mL^2$ |
| Thin disk | | $\bar{I}_x = \dfrac{1}{2} mr^2$ <br> $\bar{I}_y = \dfrac{1}{4} mr^2$ <br> $\bar{I}_z = \dfrac{1}{4} mr^2$ |
| Thin plate | | $\bar{I}_x = \dfrac{1}{12} m(w^2 + h^2)$ <br> $\bar{I}_y = \dfrac{1}{12} mw^2$ <br> $\bar{I}_z = \dfrac{1}{12} mh^2$ |
| Circular cylinder | | $\bar{I}_x = \dfrac{1}{2} mr^2$ <br> $\bar{I}_y = \dfrac{1}{12} m(3r^2 + L^2)$ <br> $\bar{I}_z = \dfrac{1}{12} m(3r^2 + L^2)$ |
| Sphere | | $\bar{I}_x = \dfrac{2}{5} mr^2$ <br> $\bar{I}_y = \dfrac{2}{5} mr^2$ <br> $\bar{I}_z = \dfrac{2}{5} mr^2$ |

## Example 2.1

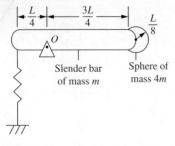

**FIG. 2.5** The system of Example 2.1.

A slender bar of mass $m$ and length $L$ is pinned as shown in Figure 2.5. A sphere of mass 4m and radius $L/8$ is attached to the end of the bar. The bar rotates about an axis through $O$, at the pin support. Determine the mass moment of inertia of the assembly about the axis of rotation.

### Solution

The mass moment of inertia of the slender bar about its centroidal axis is obtained using Table 2.1 as

$$\bar{I}_b = \frac{1}{12}mL^2 \tag{a}$$

Application of the Parallel Axis Theorem between the centroidal axis and the axis of rotation, which is a distance $L/4$ from the centroidal, axis leads to

$$I_{O_b} = \bar{I}_b + m\left(\frac{L}{4}\right)^2$$

$$= \frac{1}{3}mL^2 \tag{b}$$

The mass moment of inertia of the sphere about its centroidal axis is determined using Table 2.1 as

$$\bar{I}_s = \frac{2}{5}(4m)\left(\frac{L}{8}\right)^2$$

$$= \frac{1}{40}mL^2 \tag{c}$$

The parallel axis theorem is used to determine the mass moment of inertia of the sphere about an axis through $O$

$$I_{O_s} = \frac{1}{40}mL^2 + (4m)\left(\frac{3}{4}L\right)^2$$

$$= \frac{37}{40}mL^2 \tag{d}$$

The moment of inertia of the assembly about an axis through $O$ is

$$I_O = I_{O_b} + I_{O_s}$$

$$= \frac{1}{3}mL^2 + \frac{37}{40}mL^2$$

$$= \frac{151}{120}mL^2 \tag{e}$$

The total kinetic energy of a rigid body is

$$T = \frac{1}{2}\int_m \mathbf{v} \cdot \mathbf{v}\,dm \tag{2.26}$$

For a rigid body undergoing planar motion evaluation of Equation (2.26) leads to

$$T = \frac{1}{2}m\bar{v}^2 + \frac{1}{2}\bar{I}\omega^2 \qquad (2.27)$$

where $\bar{v}$ is the magnitude of the velocity of the body's center of mass.

If the rigid body is rotating about a fixed axis at $O$ a distance $d$ from the center of mass, the velocity of the mass center is $\bar{v} = d\omega$, which when substituted into Equation (2.27) leads to

$$T = \frac{1}{2}\left(\bar{I} + md^2\right)\omega^2 \qquad (2.28)$$

Use of the Parallel Axis Theorem, Equation (2.24), in Equation (2.28) leads to

$$T = \frac{1}{2}I_O\omega^2 \qquad (2.29)$$

Equation (2.29) is applicable only when a rigid body undergoing planar motion is rotating about a fixed axis.

The linear momentum of a rigid body undergoing planar motion is

$$\mathbf{L} = m\bar{\mathbf{v}} \qquad (2.30)$$

The angular momentum of a rigid body about an axis through its mass center is

$$\bar{\mathbf{H}} = \bar{I}\omega \qquad (2.31)$$

The angular momentum of a rigid body about an axis $O$ is

$$\mathbf{H_O} = \bar{\mathbf{H}} + \mathbf{r}_{O/G} \times \mathbf{L} \qquad (2.32)$$

If $O$ is a fixed axis of rotation

$$\mathbf{H_O} = I_O\omega \qquad (2.33)$$

## Example 2.2

At the instant shown, the massless collar of Figure 2.6 has a velocity of 20 m/s to the right. Slender bar $AB$ has mass 2.8 kg, while bar $BC$ has mass 2.1 kg. Determine the kinetic energy of the system at this instant.

### Solution

The law of sines is used to determine the angle that bar $BC$ makes with the horizontal at this instant

$$\frac{\sin 30°}{\ell_{BC}} = \frac{\sin\theta}{\ell_{AB}} \qquad (a)$$

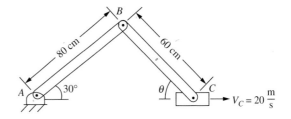

**FIG. 2.6** System of Example 2.2.

leading to $\theta = 42.1°$. The relative velocity equation, Equation (2.12), is used on bar $AB$ between $A$ and $B$

$$\mathbf{v_B} = \mathbf{v_A} + \boldsymbol{\omega_{AB}} \times \mathbf{r_{B/A}}$$

$$\mathbf{v_B} = \mathbf{0} + \omega_{AB}\mathbf{k} \times 0.8(\cos 30°\mathbf{i} + \sin 30°\mathbf{j})$$

$$= -0.4\omega_{AB}\mathbf{i} + 0.693\omega_{AB}\mathbf{j} \tag{b}$$

The relative velocity equation applied on bar $BC$ between $B$ and $C$ is

$$\mathbf{v_C} = \mathbf{v_B} + \boldsymbol{\omega_{BC}} \times \mathbf{r_{C/B}} \tag{c}$$

Substitution of Equation (b) into Equation (c) leads to

$$(20\mathbf{i})\text{m/s} = -0.4\omega_{AB}\mathbf{i} + 0.693\omega_{AB}\mathbf{j} + \omega_{BC}\mathbf{k} \times 0.6(\cos 42.1°\mathbf{i} - \sin 42.1°\mathbf{j})$$

$$(20\mathbf{i})\text{m/s} = (-0.4\omega_{AB} + 0.402\omega_{BC})\mathbf{i} + (0.693\omega_{AB} + 0.445\omega_{BC})\mathbf{j} \tag{d}$$

The component forms of Equation (d) are

$$-0.4\omega_{AB} + 0.402\omega_{BC} = 20 \tag{e}$$

$$0.693\omega_{AB} + 0.445\omega_{BC} = 0 \tag{f}$$

The solution of Equations (e) and (f) is $\omega_{AB} = -19.5\frac{r}{s}$ and $\omega_{BC} = 30.4\frac{r}{s}$. Since bar $AB$ rotates about $A$, the velocity of its mass center is

$$\bar{v}_{AB} = \left| \frac{\ell_{AB}}{2} \omega_{AB} \right| \tag{g}$$

which is evaluated yielding $\bar{v}_{AB} = 7.80\frac{m}{s}$. The relative velocity equation, Equation (2.12), is used to determine the velocity of the mass center of bar BC

$$\bar{\mathbf{v}}_{BC} = \omega_{AB}\mathbf{k} \times \ell_{AB}(\cos 30°\mathbf{i} + \sin 30°\mathbf{j}) + \omega_{BC}\mathbf{k} \times \frac{\ell_{BC}}{2}(\cos 42.1°\mathbf{i} - \sin 42.1°\mathbf{j})$$

$$= (13.9\mathbf{i} - 6.74\mathbf{j}) \text{ m/s} \tag{h}$$

The magnitude of the velocity of the mass center of bar $BC$ is $\bar{v}_{BC} = 15.45$ m/s.

The kinetic energy of the system is obtained using Equation (2.27) as

$$T = \tfrac{1}{2}m_{AB}\bar{v}_{AB}^2 + \tfrac{1}{2}\bar{I}_{AB}\omega_{AB}^2 + \tfrac{1}{2}m_{BC}\bar{v}_{BC}^2 + \tfrac{1}{2}\bar{I}_{BC}\omega_{BC}^2 \tag{i}$$

Substituting calculated values in Equation (i) gives

$$T = \tfrac{1}{2}\Big[(2.8\,kg)(7.80\,\text{m/s})^2 + \tfrac{1}{12}(2.8\,\text{kg})(0.8\,\text{m})^2(-19.5\,\text{r/s})^2$$

$$+ (2.1\,\text{kg})(15.5\,\text{m/s})^2 + \tfrac{1}{12}(2.1\,\text{kg})(0.6\,\text{m})^2(30.4\,\text{r/s})^2\Big] = 394.8\text{ N·m} \tag{j}$$

Since bar $AB$ rotates about a fixed axis through $A$, Equation (2.29) may be used in lieu of Equation (2.27) to calculate its kinetic energy. However, since the axis of rotation for bar $BC$ is not fixed, Equation (2.29) cannot be used to determine its kinetic energy.

## 2.1.3 Deformable Bodies

A deformable body loses its original shape when external forces or moments are applied. The elastic bar of Figure 2.7(a) has a change in

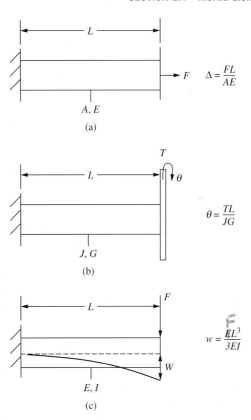

**FIG. 2.7** Examples of deformable bodies:
(a) The elastic bar increases in length and
decreases in cross section when an axial
load is applied. (b) The elastic shaft
twists when a torque is applied.
(c) Application of a transverse load to the
end of the beam leads to a deflected
shape.

length when a longitudinal force is applied to its free end. The shaft of
Figure 2.7(b) twists from its original shape when a torque is applied at
its end. The beam of Figure 2.7(c) has a transverse deflection when a vertical load is applied. In each of theses cases the relative position of particles on the body change when external forces or moments are applied.
The distance between particles on a deformable body changes when
forces are applied; the change in distance is dependent on the magnitude
and direction of the forces and where they are applied.

Assumptions are often made to simplify the analysis of the motion
of a deformable body. A common assumption used in the analysis of
the bar of Figure 2.7(a) is that all particles in a plane perpendicular to
the axis of the bar have the same displacement. Particles in a cross section of the shaft made by a plane perpendicular to the axis of the shaft of
Figure 2.7(b) are assumed to remain on the same radius as twisting
occurs. Planar sections of the beam of Figure 2.7(c) are assumed to
remain plane during deformation.

### 2.1.4  Degrees of Freedom

The particle of Figure 2.2 may move independently in each of the three
coordinate directions. However, its motion may be limited by constraints.

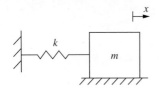

**FIG. 2.8** Since the block is constrained to move only in the $x$ direction, the system has one degree of freedom.

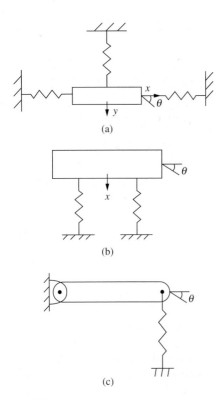

**FIG. 2.9** Rigid bodies undergoing planar motion with (a) three degrees of freedom, (b) two degrees of freedom, (c) one degree of freedom.

For example the particle in the system of Figure 2.8 is constrained to move along the surface and in the horizontal direction. The number of independent directions in which a particle may move is the number of **degrees of freedom** used in the analysis of the particle's motion. An unconstrained particle has three degrees of freedom. The number of degrees of freedom necessary for the analysis of the motion of a particle is

$$n_p = 3 - k \qquad (2.34)$$

where $k$ is the number of constraints.

The mass center of a rigid body may move independently in each of the three coordinate directions. The rigid body may also be free to rotate independently about each of the coordinate axes. Thus a rigid body without any constraints has six degrees of freedom.

The mass center of a rigid body undergoing planar motion is constrained to move in a plane and may rotate only about an axis perpendicular to the plane. Thus a rigid body undergoing planar motion has at most three degrees of freedom. The system of Figure 2.9(a) has three degrees of freedom. The system of Figure 2.9(b) is constrained from side-to-side motion, but the displacement of its mass center is independent of the angular rotation of the bar and the system has two degrees of freedom. The pin support provides an additional constraint to the motion of the system of Figure 2.9(c), which only has one degree of freedom.

The number of degrees of freedom for a system composed of multiple particles or rigid bodies is the sum of the number of degrees of freedom for each body in the system minus the number of constraints added through the connections between the bodies. The system of Figure 2.10(a) has three degrees of freedom: (1) the vertical displacement of the mass center, bar $AB$ (2) the angular rotation of bar $AB$, and (3) the angular rotation of bar $CD$. However, the system of Figure 2.10(b) has only two degrees of freedom because the rigid connection between bar $AB$ and bar $CD$ adds a constraint: the displacement of particle $B$ must equal the displacement of particle $C$. A system composed of a finite number of particles and rigid bodies is called a **discrete system**.

A $n$-degree-of-freedom discrete system requires $n$ kinematically independent coordinates to describe the motion of the system. These coordinates are functions of time and serve as the dependent variables in the system modeling. Whereas the number of degrees of freedom necessary to model the system is a unique number, the choice of the corresponding dependent variables is not unique.

The displacement of any particle along the neutral axis of the beam of Figure 2.7(c) is independent of the displacement of every other particle on the neutral axis. Deformable bodies such as this beam have an infinite number of degrees of freedom. Such a system is called a **continuous system** or a **distributed parameter system**.

Two independent variables are necessary for modeling the motion of the beam of Figure 2.7(c). Let $x$ be a spatial variable that measures the distance along the neutral axis of the beam from its left support. The

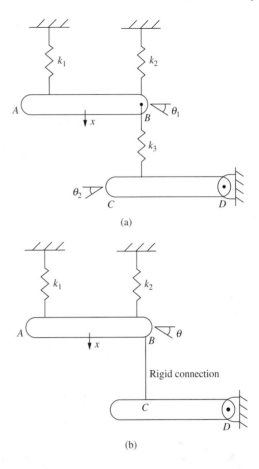

**FIG. 2.10** (a) Three-degree-of-freedom system composed of two rigid bodies; (b) two degrees of freedom.

dependent variable, the transverse displacement, is a function of both $x$ and time $t$. Thus, $w = w(x, t)$.

## 2.2 SPRINGS

A spring is a flexible link between two particles in a mechanical system. The length of a spring increases when it is subject to a tensile force. The length of a spring decreases when it is subject to a compressive force. A spring is, by definition, a deformable body. In this book, the mass of a spring is assumed to be small such that its kinetic energy is negligible in comparison to the kinetic energy of the system's inertia elements. Thus a spring is assumed to be massless.

A force, applied to a spring, does work when its application leads to a change in the length of the spring and energy is stored in the spring in form of strain energy. In an ideal system the process is reversible and all the energy stored is recovered when the force is removed. The system is conservative and the strain energy stored in the spring is a form of potential energy. In a real system, energy is released due to the breaking of bonds and the motion of dislocations. Springs in this study are assumed to be ideal.

### 2.2.1 Force-Displacement Relations

The length of a spring in an equilibrium state in the absence of external forces is called its **unstretched length**. Since the spring is flexible, the application of an external force results in a change in length of the spring. The general force-displacement relationship for the spring shown schematically in Figure 2.11 is of the form

$$F = f(x) \tag{2.35}$$

where $x$ is the change in length of the spring from its unstretched length, which occurs due to application of the force $F$. The specific form of the function in Equation (2.35) depends on the spring's geometry and properties of the material from which the spring is fabricated.

The state $x = 0$ occurs when the spring is unstretched. Assuming $f(x)$ is continuous and infinitely differentiable at $x = 0$, then it has a Taylor Shies expansion of the form

$$f(x) = f(0) + \left[\frac{df}{dx}(0)\right]x + \frac{1}{2}\left[\frac{d^2f}{dx^2}(0)\right]x^2 + \frac{1}{6}\left[\frac{d^3f}{dx^3}(0)\right]x^3 + \cdots$$
$$+ \frac{1}{n!}\left[\frac{d^nf}{dx^n}(0)\right]x^n + \cdots \tag{2.36}$$

When the spring is unstretched, $F = 0$; thus $f(0) = 0$. Many springs have the same properties in compression as in tension. That is, if a tensile force $F$ is required to stretch the spring a distance $x$, then a compressive force of $F$ is required to compress (reduce the length of) the spring a distance $x$. Using positive values for tensile forces and increases in length of the spring and negative values for compressive forces and decreases in length, the force-displacement relation must satisfy

$$F(-x) = -F(x) \tag{2.37}$$

Since Equation (2.37) is satisfied by odd powers of $x$ and not satisfied by even powers of $x$, Equation (2.36) reduces to

$$f(x) = k_1 x + k_3 x^3 + k_5 x^5 + \cdots \tag{2.38}$$

where

$$k_n = \frac{1}{n!}\frac{d^nf}{dx^n}(0) \tag{2.39}$$

Assumptions are often made that allow the nonlinear terms in Equation (2.38) to be neglected in comparison to $k_1 x$, thus linearizing the force-displacement relationship. Examples of such assumptions are that the spring is made of a linearly elastic material and stresses are below

**FIG. 2.11** Application of a tensile force $F$ increases the length of a spring from its unstretched length.

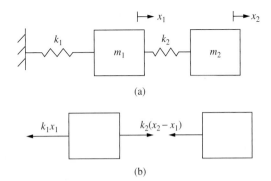

**FIG. 2.12** (a) Two-degree-of-freedom system with blocks connected by linear springs; (b) free-body diagrams illustrating the effects of springs on blocks.

the yield stress or that the change in length of the spring is small. When linearized Equation (2.38) is written as

$$F = kx \qquad (2.40)$$

where

$$k = \frac{df}{dx}(0) \qquad (2.41)$$

is called the spring stiffness.

The value of $x$ used in Equation (2.40) is the change in length of the spring, measured from its unstretched length. Consider the system of Figure 2.12(a). The displacements of the blocks of mass $m_1$ and $m_2$, measured from a position in which both springs are unstretched, are $x_1$ and $x_2$ respectively. Since the spring of stiffness $k_1$ is fixed at one end and attached to the block of mass $m_1$ at its other end, its change in length is $x_1$ and the force in the spring is $F_1 = k_1 x_1$. When the force from the spring acting on the block is illustrated on a free-body diagram of the block, Figure 2.12(b), it is shown acting away from the block. For a positive $x_1$ the spring has increased in length and is pulling on the block. If $x_1$ is negative the spring has decreased in length and is pushing against the block. The left end of the spring of stiffness $k_2$ has displaced a distance $x_1$ while its right end has a displacement of $x_2$. Thus its change in length is $x_2 - x_1$, which is the relative displacement between the ends of the spring. From Equation (2.40) the force developed in the spring is $F_2 = k_2(x_2 - x_1)$. When $x_2 > x_1$ the spring's length has increased and the spring is pulling on the blocks, as illustrated in Figure 2.12(b).

The force in a linear spring is conservative; the work done by the spring force as the spring deflects is dependent on only the spring's initial and final configurations and is expressed as a difference in potential energies

$$W_{1 \to 2} = V_1 - V_2 \qquad (2.42)$$

where the potential energy function for a linear spring is

$$V = \tfrac{1}{2}kx^2 \qquad (2.43)$$

A torsional spring is a flexible element that has an angular displacement between its ends when a moment or toque is applied. A linear torsional spring has a moment-angular displacement relation of the form

$$M = k_t\theta \tag{2.44}$$

where $M$ is the applied moment, $\theta$ is the relative angular displacement between the ends of the spring, and $k_t$ is the torsional stiffness, which has typical units of N-m/rad.

## 2.2.2 Combinations of Springs

Each of the mechanical systems of Figure 2.13 contains a combination of linear springs. Modeling of these systems is simplified when the combination of springs is replaced by a single spring of an equivalent stiffness, as in Figure 2.14. The model system of Figure 2.14 is equivalent to one of the systems of Figure 2.13 if both systems have the same displacement when they are acted on by identical forces. When the mass in the model system is displaced a distance $x$ from the system's equilibrium position, the force developed in the equivalent spring is

$$F = k_{eq}x \tag{2.45}$$

When the mass in the system of Figure 2.13(a) is displaced from equilibrium, the change in length of each of the springs in the combination is the same. The resultant force acting on the mass from the springs is the sum of the forces in the individual springs. These springs are said to be in **parallel**. A combination of springs is a parallel combination if all the springs have the same displacement and the resultant force from the springs in the combination is the sum of the individual spring forces.

Let $x$ be the displacement from equilibrium of the block of system in Figure 2.13(a). Since each spring in a parallel combination of $n$ springs has the same displacement, the resultant force acing on the block from this combination is

$$F = \left(\sum_{i=1}^{n} k_i\right)x \tag{2.46}$$

Equating Equation (2.46) with Equation (2.45) leads to the equivalent stiffness of a parallel combination of springs as

$$k_{eq} = \sum_{i=1}^{n} k_i \tag{2.47}$$

When the mass in the system of Figure 2.13(b) is displaced from equilibrium, the same force is developed in each spring in the combination. The change in length of each spring may be different, but the total change in length of the combination of springs, which is the sum of the changes in length of the individual springs, must equal the displacement of the mass. These springs are said to be in **series**. A combination of

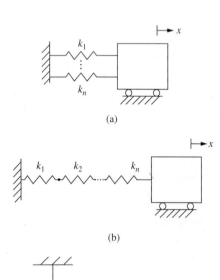

(a)

(b)

(c)

**FIG. 2.13** Combinations of springs: (a) springs in parallel; (b) springs in series; (c) springs that are neither in parallel nor in series.

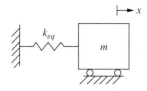

**FIG. 2.14** Equivalent model system for a system with a combination of springs.

springs is in series if each spring has the same force and the total change in the length of the combination is the sum of the changes in the lengths of the individual springs.

If $x$ is the displacement of the block of Figure 2.13(b), then the force developed in the $i$th spring in the combination is

$$F = k_i x_i \tag{2.48}$$

where $x_i$ is the change in length of the $i$th spring in the series combination. The total change in length of the springs is

$$x = \sum_{i=1}^{n} x_i \tag{2.49}$$

Solving Equation (2.48) for $x_i$ and then substituting into Equation (2.49) leads to

$$F = \left( \frac{1}{\sum_{i=1}^{n} \frac{1}{k_i}} \right) x \tag{2.50}$$

Comparing Equation (2.50) with Equation (2.45) leads to the equivalent stiffness of a series combination of springs as

$$k_{eq} = \frac{1}{\sum_{i=1}^{n} \frac{1}{k_i}} \tag{2.51}$$

Any combination of springs in a linear one-degree-of-freedom system can be replaced by a single spring of equivalent stiffness, even if they do not form a parallel combination or a series combination. Two combinations of springs acting on a system are equivalent if, when the two systems are given identical displacements, the potential energy of each of the combinations is the same. The potential energy of an equivalent spring connected to a mechanical system at a particle whose displacement from equilibrium is $x$ is

$$V = \tfrac{1}{2} k_{eq} x^2 \tag{2.52}$$

**Example 2.3**

Determine the equivalent stiffness of the combinations of springs for each of the systems of Figure 2.15.

**Solution**

(a) The springs acting on the left of the block of Figure 2.15(a) are in parallel. Their equivalent stiffness is calculated using Equation (2.47) as

$$k_{eq_a} = 2 \times 10^5 \, \text{N/m} + 4 \times 10^5 \, \text{N/m} = 6 \times 10^5 \, \text{N/m} \tag{a}$$

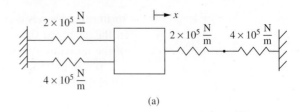

(a)

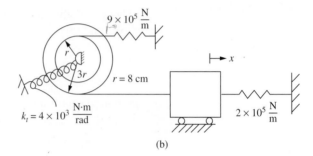

(b)

**FIG. 2.15** Systems for Example 2.3.

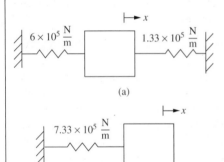

(a)

(b)

**FIG. 2.16** Steps in the reduction of the combination of springs in the system of Figure 2.15(a).

The springs acting on the right of the block of Figure 2.15(a) are in series. Their equivalent stiffness is calculated using Equation (2.51) as

$$k_{eq_b} = \frac{1}{\dfrac{1}{2 \times 10^5 \text{N/m}} + \dfrac{1}{4 \times 10^5 \text{N/m}}} = 1.33 \times 10^5 \text{ N/m} \qquad (b)$$

The results are summarized in Figure 2.16(a). When the block of Figure 2.16(a) is given a displacement $x$, the spring on the left of the block increases in length a distance $x$ while the spring on the right decreases in length by $x$. The forces from the springs acting on the block are in the same direction and their resultant is their sum. Thus these springs are in parallel with an equivalent stiffness determined using Equation (2.47) of

$$k_{eq_c} = 6 \times 10^5 \text{ N/m} + 1.33 \times 10^5 \text{ N/m} = 7.33 \times 10^5 \text{ N/m} \qquad (c)$$

The equivalent system is illustrated in Figure 2.16(b).

(b) Let $x$ be the displacement of the block of the system of Figure 2.15(b), let $\theta$ be the counterclockwise angular rotation of the disk, and let $y$ be the increase in length of the spring connected to the inner radius of the disk. The total potential energy developed in the springs at an arbitrary instant is

$$V = \tfrac{1}{2}k_1 x^2 + \tfrac{1}{2}k_2 y^2 + \tfrac{1}{2}k_t \theta^2 \qquad (d)$$

Assuming no slip between the cables and the disk, the coordinates are related by

$$x = 3r\theta \qquad y = r\theta = x/3 \qquad (e)$$

Substitution of Equation (e) into Equation (d) leads to

$$V = \frac{1}{2}k_1 x^2 + \frac{1}{2}k_2 \left(\frac{x}{3}\right)^2 + \frac{1}{2}k_t \left(\frac{x}{r}\right)^2$$

$$= \frac{1}{2}\left(k_1 + \frac{1}{9}k_2 + \frac{1}{r^2}k_t\right)x^2 \qquad (f)$$

Comparison of Equation (f) with Equation (2.52) reveals

$$k_{eq} = k_1 + \frac{1}{9}k_1 + \frac{1}{r^2}k_t \qquad \text{(g)}$$

Substitution of the given values into Equation (g) gives

$$k_{eq} = 2 \times 10^5 \,\text{N/m} + \tfrac{1}{9}(9 \times 10^5 \,\text{N/m}) + \frac{1}{(0.1\text{m})^2}\left(4 \times 10^3 \,\text{N·m/r}\right)$$

$$= 7 \times 10^5 \,\text{N/m} \qquad \text{(h)}$$

### 2.2.3 Static Deflections

The system of Figure 2.17(a) is in static equilibrium. Application of the equilibrium condition that $\sum F = 0$ to the free-body diagram of Figure 2.17(b) leads to

$$mg - F_s = 0 \qquad (2.53)$$

where $F_s$ is the force developed in the spring to maintain equilibrium. Using the force-displacement relationship for a linear spring, Equation (2.40), the static spring force can be written as

$$F_s = k\mathbf{\Delta}_s \qquad (2.54)$$

The **static deflection** of the spring $\mathbf{\Delta}_s$ is obtained by substituting Equation (2.54) into Equation (2.53) and rearranging, leading to

$$\mathbf{\Delta}_s = \frac{mg}{k} \qquad (2.55)$$

The static deflection of a spring in a system is the change in the length of the spring from its unstretched length when the system is in static equilibrium without the application of external forces. Since external forces are not present, a static spring force is often necessary to balance gravity forces in order to maintain equilibrium. If the dependent variables used in the system modeling are measured from the system's static equilibrium position, then a change in the dependent variable measures the change in the length of the spring from its equilibrium position, not from its unstretched length. If at some instant $x$ is the downward displacement of the mass of Figure 2.17(a), measured from the system's static equilibrium position, then the force developed in the spring at this instant is

$$F_s = k(x + \mathbf{\Delta}_x) \qquad (2.56)$$

The potential energy in the spring at this instant is

$$V = \frac{1}{2}k(x + \mathbf{\Delta}_s)^2 \qquad (2.57)$$

It is not possible to determine the static deflections solely from the equilibrium equations for a system that is statically indeterminate, such as the system of Figure 2.18(a). Application of the equilibrium condition,

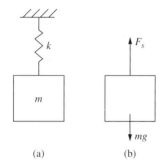

(a)                    (b)

**FIG. 2.17**  (a) When the system is in equilibrium position, a static force is developed in the spring to balance the gravity force of the block; (b) free-body diagram of the equilibrium position.

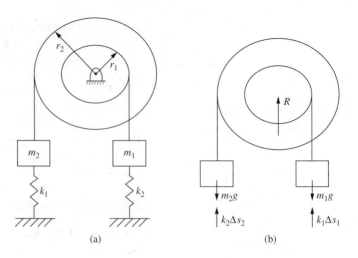

**FIG. 2.18** (a) Statically indeterminate system in which static deflections cannot be determined only by applying equilibrium conditions; (b) free-body diagram of the system in its equilibrium position.

$\sum M_O = 0$, to the free-body diagram of Figure 2.18(b) leads to

$$m_1 g r_1 - k_1 \Delta_{s1} r_1 - m_2 g r_2 + k_2 \Delta_{s2} r_2 = 0 \tag{2.58}$$

A second equation, developed from geometric considerations, is necessary to solve for the static deflections. However, the development of an equilibrium condition like Equation (2.58) is sufficient for purposes of modeling the dynamics and response of a linear system.

## 2.3 FRICTION ELEMENTS

Friction elements are components of a mechanical system that dissipate energy. Friction may be inherent in the system's inertia and energy storage elements, it may result from contact between the system and its surroundings, or it may be added to the system through a system component to control the system response.

The dissipation of energy from a mechanical system is called **damping**, and devices that dissipate energy are called dampers. Damping mechanisms dissipate energy by converting stored energy into heat, which is then transferred from a mechanical system to its surroundings. Thus, as opposed to inertia and stiffness elements, friction elements do not store energy but dissipate energy.

### 2.3.1 Viscous Damping

**Viscous damping** refers to any form of damping in which the damping force is proportional to the velocity of a system component

$$F = cv \tag{2.59}$$

where $c$ is the **damping coefficient**, which has dimensions of $[FTL^{-1}]$. Viscous damping leads to linear terms in differential equations governing the motion of a system. The system response is easy to obtain and study when viscous damping is present. Thus viscous damping is often introduced into systems through mechanical components to control the system response.

A **dashpot** is a mechanical device used to provide viscous damping. Dashpots use viscous fluids to develop forces proportional to the velocity of moving components. A common dashpot configuration, shown in Figure 2.19, is a piston moving in a cylinder of oil. As the piston moves in the cylinder with a velocity $v$, its motion is resisted by the fluid in the cylinder. This resistance leads to a pressure difference between the two surfaces of the piston, which in turn leads to a resisting force acting on the piston. If the piston is moving with a velocity $v$ and the cylinder is stationary, then the force acting on the piston is of the form of Equation (2.59). If the cylinder has a velocity $v_c$, then the resisting force is proportional to the velocity of the piston relative to the cylinder

$$F = c(v - v_c) \tag{2.60}$$

A **torsional viscous damper** develops a resisting moment that is proportional to the angular velocity, $\omega$, of a rotating component

$$M = c_t\omega \tag{2.61}$$

where $c_t$ is the **torsional viscous damping coefficient**, which has dimensions of $[FTL]$ and typical units of N·m·s/rad. The common torsional damper configuration illustrated in Figure 2.20 consists of a thin disk attached to the end of a shaft rotating at an angular speed $\omega$. As the disk rotates in a dish of viscous oil, a shear stress distribution develops on the face of the disk. The resultant of the shear stress distribution is a resisting moment that satisfies Equation (2.61).

**Viscoelastic materials** exhibit behaviors of both elastic materials and viscous fluids. Rubber is a viscoelastic material used in many mechanical systems. It is used, for example, in engine mounts or vibration isolators for both its elastic behavior as well as its damping properties. A common model for rubber, called a Kelvin model, is that of a spring in parallel with a viscous damper, as illustrated in Figure 2.21. The Kelvin model was developed to model the stress-strain behavior of the viscoelastic material, but it can also be used in modeling mechanical systems. The total force developed in a material modeled using a Kelvin model is

$$F = kx + c\dot{x} \tag{2.62}$$

Viscous dampers can be placed in combinations, as illustrated by the systems in Figure 2.22. For the purpose of modeling a mechanical system, a combination of viscous dampers can be replaced by a single damper with an equivalent damping coefficient, as in the system in Figure 2.23. When the mass in the system of Figure 2.23 has a velocity $\dot{x}$, the force developed in the viscous damper is

$$F = c_{eq}\dot{x} \tag{2.63}$$

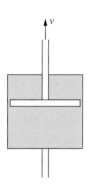

**FIG. 2.19** A piston-cylinder model of a viscous damper in which the piston moves in a cylinder filled with oil.

**FIG. 2.20** Model of a torsional viscous damper in which the disk rotates with angular velocity $\omega$ in a dish of viscous liquid.

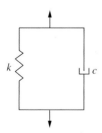

**FIG. 2.21** The Kelvin model for a viscoelastic material such as rubber consists of a spring in parallel with a viscous damper.

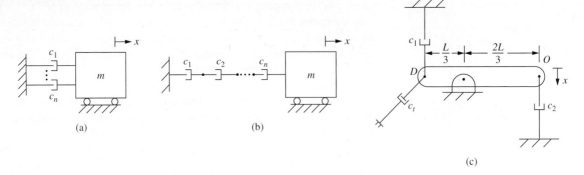

**FIG. 2.22** Combinations of viscous dampers: (a) dampers in parallel; (b) dampers in series; (c) dampers neither in parallel nor in series.

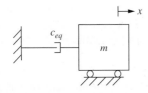

**FIG. 2.23** A combination of viscous dampers can, for modeling purposes, be replaced by a damper with the equivalent damping coefficient.

The viscous dampers of the system of Figure 2.22(a) are in parallel. The velocity of the piston in each damper is the same, and the resultant force acting on the block is the sum of the forces in the parallel dampers. The equivalent viscous damping coefficient for parallel viscous dampers is

$$c_{eq} = \sum_{i=1}^{n} c_i \qquad (2.64)$$

The viscous dampers of the system of Figure 2.22(b) are in series. The force developed in each viscous damper is the same, and the velocity of the block is the sum of the relative velocities between the piston and cylinder of each viscous damper. The equivalent viscous damping coefficient for a series combination of viscous dampers is

$$c_{eq} = \frac{1}{\sum_{i=1}^{n} \frac{1}{c_i}} \qquad (2.65)$$

The energy dissipated by a viscous damper is equal to the work done by the viscous damping force. If the piston of the viscous damper moves between two positions described by $x_1$ and $x_2$ then the work done by the viscous damping force in the system of Figure 2.23 is

$$W_{1 \rightarrow 2} = -\int_{x_1}^{x_2} c_{eq}\dot{x}dx \qquad (2.66)$$

The work done by the viscous damping force is negative because the damping force is always in the opposite direction of the motion.

The work done by the moment in a torsional viscous damper of damping coefficient $c_t$ as it rotates between angles $\theta_1$ and $\theta_2$ is

$$W = -\int_{\theta_1}^{\theta_2} c_t \dot{\theta}d\theta \qquad (2.67)$$

## Example 2.4

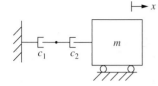

FIG. **2.24**  Two viscous dampers in series of Example 2.4.

Show that the equivalent viscous damping coefficient for two viscous dampers in series, shown in Figure 2.24, is

$$c_{eq} = \frac{c_1 c_2}{c_1 + c_2} \tag{a}$$

### Solution

Let $\dot{x}$ be the velocity of the block, which is also the velocity of the piston of the damper of damping coefficient $c_2$. Let $\dot{x}_1$ be the velocity of the piston of the damper of damping coefficient $c_1$. The forces developed in the viscous dampers are

$$F_1 = c_1 \dot{x}_1 \tag{b}$$
$$F_2 = c_2(\dot{x}_2 - \dot{x}_1) \tag{c}$$

The forces developed in viscous dampers in series are equal; thus Equations (b) and (c) lead to

$$c_1 \dot{x}_1 = c_2(\dot{x}_2 - \dot{x}_1) \tag{d}$$

Equation (d) is rearranged to yield

$$\dot{x}_1 = \frac{c_2}{c_1 + c_2} \dot{x}_2 \tag{e}$$

Substitution of Equation (e) into Equation (b) leads to

$$F = \frac{c_1 c_2}{c_1 + c_2} \dot{x} \tag{f}$$

which, when compared with Equation (a), gives the desired result.

## Example 2.5

Replace the combination of viscous dampers in the system of Figure 2.22(c) by a viscous damper of an equivalent viscous damping coefficient attached to the system at $O$.

### Solution

Let $x_A$ be the displacement of the left end of the bar, $x_B$ the displacement of the right end of the bar, and $\theta$ the clockwise rotation of the bar, all measured from the system's equilibrium position. The work done by the viscous damping forces as the bar rotates between two positions is determined using Equations (2.66) and (2.67) as

$$W_{1 \to 2} = -\int_{\theta_1}^{\theta_2} c_t \dot{\theta} d\theta - \int_{x_{A1}}^{x_{A2}} c_1 \dot{x}_A dx_A - \int_{x_{B1}}^{x_{B2}} c_2 \dot{x}_B dx_B \tag{a}$$

The equivalent viscous damper is to be placed at $O$; thus it is necessary to relate $x_A$ and $\theta$ to $x_B$. Assuming small displacements, these relations are

$$x_B = -2x_A \tag{b}$$

$$\theta = \frac{3}{2L} x_B \tag{c}$$

Substituting Equations (b) and (c) into Equation (a) leads to

$$W_{1 \to 2} = -\int_{x_{B1}}^{x_{B2}} c_t \left( \frac{3}{2L} \dot{x}_B \right) d\left( \frac{3}{2L} x_B \right) - \int_{x_{B1}}^{x_{B2}} c_1 \left( -\frac{1}{2} \dot{x}_B \right) d\left( -\frac{1}{2} x_B \right) - \int_{x_{B1}}^{x_{B2}} c_2 \dot{x}_B dx_B \tag{d}$$

Equation (d) is simplified to

$$W_{1 \to 2} = -\int_{x_{B1}}^{x_{B2}} \left( \frac{9}{4L^2} c_t + \frac{1}{4} c_1 + c_2 \right) dx_B \tag{f}$$

Comparison of Equation (f) with Equation (2.66) leads to

$$c_{eq} = \frac{9}{4L^2} c_t + \frac{1}{4} c_1 + c_2 \tag{g}$$

## 2.3.2 Coulomb Damping

The reaction force between two surfaces has two components: (1) a normal force, which is perpendicular to the plane of contact between the surfaces, and (2) a friction force, which lies within the plane of contact. Coulomb's friction law states that the magnitude of the friction force is directly proportional to the magnitude of the normal force. When there is no relative motion between the bodies, the constant of proportionality is called the **static coefficient of friction** $\mu_s$. One body will move relative to the other when an externally applied force exceeds the static friction force. When relative motion exists, the constant of proportionality between the normal force and the friction force decreases and is called the **kinetic coefficient of friction** $\mu_k$. The dynamic friction force is normal to the plane of contact and opposite to the direction of the motion, as illustrated in Figure 2.25. The force required to impart the relative motion is greater than the force required to maintain the motion as $\mu_s \geq \mu_k$.

Coulomb's friction law for sliding contact between two bodies is

$$F = \mu_k N \tag{2.68}$$

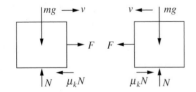

**FIG. 2.25** The friction force between two surfaces moving relative to one another is proportional to the normal force and opposes the direction of motion.

The kinetic coefficient of friction is determined empirically. The sliding between two surfaces results in energy dissipation from the mechanical system in the form of heat transfer to the surroundings. The energy dissipated by friction between two surfaces in contact moving relative to each other is called **Coulomb damping**. The amount of energy dissipated through

Coulomb damping is equal to the work done by the friction force

$$\Delta E = W \tag{2.69}$$

where $\Delta E$ is the change in energy of the mechanical system and $W$ is the work done by the friction force. Since the friction force always opposes the direction of motion, its work is negative and the amount of energy stored in the system decreases. If the distance traveled by the point of application of the friction force is $x$, then the work done by the friction force is

$$W = -\mu_k N x \tag{2.70}$$

Coulomb damping also occurs when a belt or cable slides over a drum such as rotor or a pulley, as illustrated in Figure 2.26. In the absence of friction the tension in both sides of the cable is the same. The resultant of the friction as the belt slides over the drum is a moment about the center of the cable opposite the direction of the motion. The tensions are related by

$$T_2 = T_1 e^{\mu \beta} \tag{2.71}$$

where $\mu$ is the coefficient of belt friction and $\beta$ is the angle of contact between the belt and the drum.

A friction force is developed between surface and a rolling body, as illustrated in Figure 2.27. If the disk is rolling without sliding, then there is no relative velocity between the surface and its point of contact with the disk. Application of the relative velocity equation, Equation (2.12), between the mass center of the disk and the point of contact leads to

$$\bar{v} = r\omega \tag{2.72}$$

When the disk is rolling without sliding, the friction force developed is less than the maximum available friction force

$$F < \mu_k N \tag{2.73}$$

Since there is no relative motion between the two bodies the friction force does no work and no energy is dissipated by friction.

The disk will roll and slide when there is a relative velocity between the surface and its point of contact with the disk. In this case the friction force is its maximum value

$$F = \mu_k N \tag{2.74}$$

When sliding occurs, Equation (2.72) is not applicable, and kinematics cannot be used to determine a relation between the velocity of the center of the disk and the angular velocity of the disk.

Coulomb damping is present in all systems. However, when included in the modeling of mechanical systems, Coulomb damping leads to non-linear terms in the differential equations governing the motion of the system. These terms result from the friction force having a constant absolute value, but always opposing the direction of motion. In contrast, the force in a viscous damper is proportional to the velocity and leads to linear terms in the governing differential equations.

**FIG. 2.26** Belt friction occurs due to the sliding of a belt relative to a fixed drum, As a result of belt friction $T_2 > T_1$ when the belt moves in the direction of $T_2$.

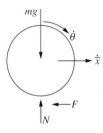

**FIG. 2.27** Illustration of rolling friction. If the disk rolls without sliding, then $\dot{x} = r\dot{\theta}$ and $F < \mu_k N$. If the disk rolls and slides, then $F = \mu_k N$ and no kinematic relation exists between $\dot{x}$ and $\theta$.

Since Coulomb damping leads to nonlinear terms in differential equations it is not included in the modeling of mechanical systems in this book. When viscous damping is added to a system, the effects of Coulomb damping are overshadowed by the effects of viscous damping. However, Coulomb damping is always present and leads to system behavior that is not predicted using viscous damping, such as cessation of motion.

### 2.3.3 Hysteretic Damping

In modeling a mechanical system with a linear spring, it is usually assumed that the spring's force-displacement relation, Equation (2.40), does not change with time. That is, at any time when a force $F$ is developed in the spring, the spring has a change in length of $x$ from its unstretched length, regardless of the history of the applied force. Equation (2.40) is assumed to apply while a force is being applied to the spring as well as when the spring is being unloaded.

Springs are made of engineering materials such as steel, in which imperfections exist. As the spring is loaded bonds break between molecules, releasing energy. These and other nonconservative effects due to real material behavior lead to the true force-displacement curve of Figure 2.28 for one cycle of completely reversed loading, which illustrates a **hysteresis loop**. The area enclosed by the hysteresis loop is the energy dissipated during the cycle due to nonideal material behavior. This dissipation of energy in a mechanical system is called **hysteretic damping**.

Hysteretic damping is present in all systems with elastic components, but it is quantified only empirically. Differential equations derived assuming hysteretic damping are often nonlinear. In some cases the behavior of a system with hysteretic damping is similar to the behavior of a system with viscous damping. In these cases hysteretic damping is modeled using an equivalent viscous damping coefficient. The effects of hysteretic damping in systems with viscous damping are often overshadowed by the effects of viscous damping.

In this book it is assumed that all elastic materials are ideal and the effects of hysteretic damping are not included.

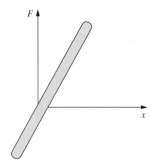

**FIG. 2.28** Force-displacement curve of a real engineering material during one cycle of completely reversed loading. The area enclosed by the hystersis loop is the energy dissipated.

## 2.4  MECHANICAL SYSTEM INPUT

External forces or externally imposed motions provide input for mechanical systems. For example a torque applied to a shaft with rotors and gears is developed by an electric motor, as discussed in Chapter 3. The operation of reciprocating machines produces inertia forces arising from the rotation of unbalanced components that act as external excitation to the machine. Hydraulic servomotors, discussed in Chapter 4, act as actuators to deliver

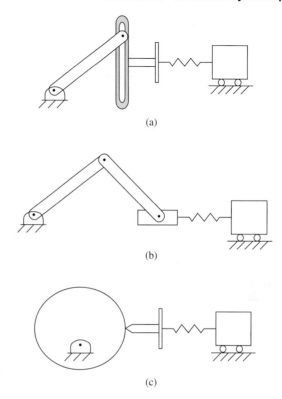

(a)

(b)

(c)

**FIG. 2.29** Mechanisms that impart a prescribed motion to a system: (a) Scotch yoke; (b) slider-crank; (c) cam and follower.

motion to mechanical systems. The Scotch yoke [Figure 2.29(a)], the slider-crank mechanism [Figure 2.29(b)], and cam and follower systems [Figure 2.29(c)] are mechanisms specially designed to deliver specific motions to system components.

A force is a vector quantity in that it requires a magnitude and a direction of application for its complete specification. A force, like any vector, can be resolved into scalar components. Since the directions of the application of forces are constant for all problems in this book, scalar equations are developed using force components.

### 2.4.1 External Forces and Torques

Figure 2.30 illustrates a mechanical system subject to a time-dependent external force $F(t)$, which may be provided by an external actuator. The work done by the force as its point of application moves between two positions described by the coordinates $x_1$ and $x_2$ is

$$W_{1 \to 2} = \int_{x_1}^{x_2} F(t)\,dx \qquad (2.75)$$

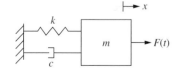

**FIG. 2.30** A force actuator provides a time-dependent force to a mechanical system.

An alternate expression for the work is obtained by noting that $dx = \dot{x}dt$, which when used in Equation (2.75) leads to

$$W_{1 \rightarrow 2} = \int_{t_1}^{t_2} F(t)\dot{x}dt \tag{2.76}$$

The work done by the force on the system is converted into kinetic and potential (or stored) energy.

The work done by an applied moment or torque $M$ as the component to which it is applied rotates between two positions described by $\theta_1$ and $\theta_2$ is

$$W_{1 \rightarrow 2} = \int_{\theta_1}^{\theta_2} M(t)d\theta \tag{2.77}$$

Noting that $d\theta = \dot{\theta}dt$, Equation (2.77) can be rewritten as

$$W_{1 \rightarrow 2} = \int_{\theta_1}^{\theta_2} M(t)\dot{\theta}dt \tag{2.78}$$

The power delivered to the system by an external force is

$$P = \frac{dW}{dt} \tag{2.79}$$

Use of Equation (2.76) in Equation (2.79) where the upper limit of integration is an arbitrary time leads to

$$P = F(t)\dot{x} \tag{2.80}$$

Using Equation (2.77) the power delivered by an externally applied moment is

$$P = M(t)\dot{\theta} \tag{2.81}$$

For the purposes of modeling a mechanical system, two forces applied to a mechanical system are considered to be equivalent if they deliver the same power to the system at all times. For example a force $F$ applied to the end of the bar of Figure 2.31 delivers a power equal to $FL\dot{\theta}$, which is the same power as a clockwise moment of magnitude $M = FL$. Thus, for purposes of dynamic system modeling, the two are equivalent. The dynamic response of the system is the same whether the force or the moment is applied.

The assumption that the response of the system is the same whether the force $F$ or the moment $M = FL$ is applied is true only for the dynamic response of the bar. The time-dependent reation at the pin support does depend upon whether the force or moment is applied. The same is true when other assumptions of equivalencies are used. The equivalency applies only to the motion of the system and not necessarily to constraint forces such as reations.

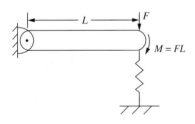

**FIG. 2.31** The power delivered to the system by the force $F$ applied at the end of the bar is the same as the power delivered by a moment of magnitude $FL$.

## Example 2.6

A motor delivers a torque $T = 2000$ N·m to drive shaft $AB$ of the system of Figure 2.32 at 1,000 rpm. The system has a gear reduction system with a gear reduction ratio of 2.5 to 1. (a) What is the power delivered to the system by the torque? (b) What is the equivalent torque that would be applied to shaft $CD$ such that the same power is delivered to the system and shaft $AB$ rotates at 1,000 rpm?

### Solution

(a) The power delivered to the system is calculated using Equation (2.81) as

$$P = T\omega_{AB} = (2,000 \text{ N·m})(1,000 \text{ rev/m})(2\pi \text{ r/rev})(1 \text{ m/60 s}) = 209.4 \text{ kW} \quad \text{(a)}$$

(b) The points of contact between two meshing gears have the same velocity. Thus

$$r_1\omega_{AB} = r_2\omega_{CD} \quad \text{(b)}$$

$$\omega_{CD} = \frac{r_1}{r_2}\omega_{AB} \quad \text{(c)}$$

The radius of a gear is proportional to its number of teeth. The gear ratio for two meshing gears is the ratio of the number of teeth for the two gears. Thus

$$\frac{r_1}{r_2} = \frac{n_1}{n_2} = 2.5 \quad \text{(d)}$$

Using Equation (d) in Equation (c) leads to

$$\omega_{CD} = 2.5\omega_{AB} = 2,500 \text{ rpm} \quad \text{(e)}$$

Let $T_{AB}$ be the torque applied to shaft CD needed to deliver the same power to the system as the 2,000 N·m torque applied to shaft $AB$, then

$$T_{AB}\omega_{AB} = T_{CD}\omega_{CD} \quad \text{(f)}$$

$$T_{CD} = \frac{\omega_{AB}}{\omega_{CD}}T_{AB} \quad \text{(g)}$$

Thus for the angular velocity of shaft $AB$ to be the same in both cases, the torque applied to shaft $AB$ is

$$T_{CD} = \left(\frac{1,000 \text{ rpm}}{2,500 \text{ rpm}}\right)(2000 \text{ N·m}) = 800 \text{ N·m} \quad \text{(h)}$$

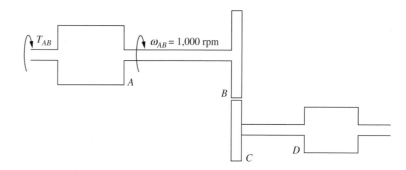

**FIG. 2.32** System of Example 2.6.

Note that this calculation provides only an equivalent torque for modeling purposes. In actuality a larger motor is required to drive a shaft at 2,500 rpm than at 1,000 rpm. The gear reduction system allows a shaft to develop a larger angular velocity with lower input power.

### 2.4.2 Impulsive Forces

The **impulse** due to a force applied from time equal to zero to an arbitrary time $t$ is defined as

$$I = \int_0^t F(t)dt \tag{2.82}$$

An **impulsive force** is a large force applied to a system over a short interval such that the response of the system is affected only by the total impulse imparted by the force, not by the time history of the force.

An impulsive force is often modeled as a force of very large but undetermined magnitude applied instantaneously such that the impulse applied by the force is finite. In this context the mathematical representation of an impulsive force $F(t)$ is

$$F(t) = I\delta(t - t_0) \tag{2.83}$$

where $I$ is the total impulse imparted due to the force, $\delta(t)$ is the unit impulse function as described in Section 1.6.1, and $t_0$ is the time at which the force is applied.

The application of an impulsive force to a system leads to an instantaneous change in the system's velocity. An impulse of magnitude $I$ applied to a particle of mass $m$ leads to an instantaneous change in the particle's velocity of

$$\Delta v = \frac{I}{m} \tag{2.84}$$

Although the velocity of the particle is discontinuous at the time of application of the impulse, its displacement is continuous.

### 2.4.3 Step Forces

A **step force** is a force that has a finite but nonzero change at some time. A step force is discontinuous at the time of the step change. The mathematical representation of the step force illustrated in Figure 2.33 is

$$F(t) = F_0 u(t - t_0) \tag{2.85}$$

where $u(t)$ is the unit step function defined in Section 1.6.2 and $t_0$ is the time at which the step change occurs. Since $u(t)$ is zero when its argument is negative and has a value of 1 when its argument is positive, $F(t)$ as evaluated using Equation (2.85) has a value of 0 for $t < t_0$ and a value of $F_0$ for $t > t_0$.

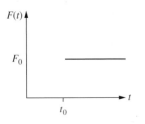

**FIG. 2.33** A step force applied to a system at time $t_0$ has the mathematical representation $F_0 u(t - t_0)$.

It is physically impossible to apply a force instantaneously. The application of a force occurs over a finite time. A step force is often used to model a force applied to a particle at rest when suddenly subject to a force that builds up quickly to a constant value. The displacement and velocity of a particle are continuous when subjected to a step force, but its acceleration has an instantaneous change.

### 2.4.4 Periodic Forces

A force $F(t)$ is **periodic** if there is a value $T$ such that

$$F(t + T) = F(t) \tag{2.86}$$

for all $t$. The value $T$ is called the **period** of the force. Periodic forces are often developed during the continuous operation of machines. The most familiar examples of periodic functions are the trigonometric functions $\sin(\omega t)$ and $\cos(\omega t)$. The sine and cosine functions repeat when their arguments increase by $2\pi$. The functions $\sin(\omega t)$ and $\cos(\omega t)$ are periodic of period

$$T = \frac{2\pi}{\omega} \tag{2.87}$$

where $\omega$ is the frequency of the force and has typical units of rad/s.

Figure 2.34 provides examples of other periodic forces. The **fundamental frequency** of a periodic force is defined as

$$\omega = \frac{2\pi}{T} \tag{2.88}$$

Any periodic force has a **Fourier series representation**, which is an infinite series of trigonometric terms with frequencies that are integer multiples of the fundamental frequency.

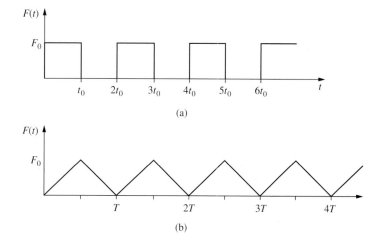

**FIG. 2.34** Examples of periodic functions: (a) square wave; (b) triangular wave.

A periodic force is often referred to as a **harmonic excitation**. A force that is represented as

$$F(t) = F_0 \sin(\omega t + \psi) \tag{2.89}$$

is a single-frequency harmonic excitation with amplitude $F_0$, frequency $\omega$, and phase angle $\psi$. The phase angle represents the shift from a purely sinusoidal excitation. A trigonometric identity can be used to write Equation (2.89) as

$$F(t) = F_0 \cos\psi \sin(\omega t) + F_0 \sin\psi \cos(\omega t) \tag{2.90}$$

### 2.4.5 Motion Input

Motion input occurs when a system particle is subject to a prescribed displacement, velocity, or acceleration. If one of these kinematic quantities is specified, the kinematic relations in Equations (2.3) and (2.4) may be used to determine the other two if necessary.

---

**Example 2.7**

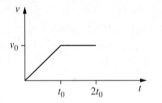

**FIG. 2.35** Velocity input for Example 2.7.

The velocity of a particle in a mechanical system is prescribed as in Figure 2.35. Determine the time-dependent displacement and acceleration of the particle.

**Solution**

The mathematical representation of the velocity, written using unit step functions is

$$v(t) = v_0\left[\frac{t}{t_0}u(t) + \left(1 - \frac{t}{t_0}\right)u(t - t_0) - u(t - 2t_0)\right]$$

The particle displacement is obtained by

$$x(t) = \int_0^t v(\tau)d\tau$$

$$= v_0 \int_0^t \left[\frac{\tau}{t_0}u(\tau) + \left(1 - \frac{\tau}{t_0}\right)u(\tau - t_0) - u(\tau - 2t_0)\right]d\tau \tag{a}$$

Equation (1.36) is used to evaluate the integrals in Equation(a), leading to

$$x(t) = v_0\left[u(t)\int_0^t \frac{\tau}{t_0}d\tau + u(t - t_0)\int_{t_0}^t \left(1 - \frac{\tau}{t_0}\right)d\tau - u(t - 2t_0)\int_{2t_0}^t d\tau\right]$$

$$= v_0\left[u(t)\frac{\tau^2}{2t_0}\Big|_{\tau=0}^{\tau=t} + u(t - t_0)\left(\tau - \frac{\tau^2}{2t_0}\right)\Big|_{\tau=t_0}^{\tau=t} - u(t - 2t_0)\tau\Big|_{\tau=2t_0}^{\tau=t}\right]$$

$$= v_0\left[\frac{t^2}{2t_0}u(t) + \left(t - \frac{t^2}{2t_0} - \frac{t_0}{2}\right)u(t - t_0) - (t - 2t_0)u(t - 2t_0)\right] \tag{b}$$

The particle's acceleration is obtained from

$$
\begin{aligned}
a &= \frac{dv}{dt} \\
&= v_0 \left\{ u(t) \frac{d}{dt}\left(\frac{t}{t_0}\right) + \frac{t}{t_0}\frac{d}{dt}[u(t)] + u(t - t_0)\frac{d}{dt}\left(1 - \frac{t}{t_0}\right) \right. \\
&\quad \left. + \left(1 - \frac{t}{t_0}\right)\frac{d}{dt}[u(t - t_0)] - \frac{d}{dt}[u(t - 2t_0)] \right\}
\end{aligned}
\tag{c}
$$

Noting from Equation (1.34) that $\frac{d}{dt}[u(t - t_0)] = \delta(t - t_0)$, Equation (c) becomes

$$
a(t) = v_0 \left[ \frac{1}{t_0}u(t) + \frac{t}{t_0}\delta(t) - \frac{1}{t_0}u(t - t_0) + \left(1 - \frac{t}{t_0}\right)\delta(t - t_0) - \delta(t - 2t_0) \right]
\tag{d}
$$

Since $\delta(t) = 0$ for all $t$ except $t = 0$, then $t\delta(t) = 0$ for all $t$. Similarly

$$
\left(1 - \frac{t}{t_0}\right)\delta(t - t_0) = 0
$$

for all $t$. Thus Equation (d) becomes

$$
a(t) = \frac{v_0}{t_0}[u(t) - u(t - t_0)] - v_0\delta(t - 2t_0)
\tag{e}
$$

The presence of the unit impulse function in Equation (e) is a result of the impulse that must be applied to cause a discrete change in velocity at $t = 2t_0$.

## 2.5 FREE-BODY DIAGRAMS

An essential step in the mathematical modeling of mechanical systems is the application of appropriate conservation laws to the entire system or its individual components. The application of a conservation law to a system component is facilitated by drawing a **free-body diagram (FBD)** of the component in which it is abstracted from its surroundings and the effects of the surroundings are illustrated on the diagram as forces and moments. The forces illustrated on a free-body diagram are drawn in the appropriate directions and labeled with appropriate magnitudes. The forces illustrated on a free-body diagram are of two types: body forces and surface forces.

A **body force** results from the presence of an external force field such as a gravitational field or an electromagnetic field. A body of mass $m$ is subject to a force from the earth's gravitational field, called the **gravity force**, which is the weight of the body

$$
W = mg
\tag{2.93}
$$

where $g$ is the acceleration due to gravity. On the surface of the earth

$$
g = 32.2 \text{ ft/s}^2 \text{ or } 9.81 \text{ m/s}^2
\tag{2.94}
$$

The gravity force is drawn on the free-body diagram with its point of application at the center of gravity and directed toward the center of the earth, usually the vertical direction. If the $y$ coordinate in a Cartesian system is the vertical coordinate, the vector representation of the gravity force is

$$\mathbf{F} = -mg\mathbf{j} \qquad (2.95)$$

**Surface forces** show the effects of contact between the body and its surroundings. Examples of surface forces include reactions at supports, forces due to sliding friction, forces exerted by springs, and externally applied forces. Surface forces act only on the surface of a body whereas a body force may be applied at an internal point.

Modeling of a mechanical system requires the identification of time-dependent coordinates that describe the motion of the system. Free-body diagrams are drawn at an arbitrary time and thus for arbitrary values of the dependent coordinates. Forces and moments are drawn on the free-body diagram in directions consistent with the chosen positive directions of the coordinates. The magnitudes of the forces are identified in terms of the coordinates, where appropriate.

The **resultant** of the forces acting on a free-body diagram is the vector sum of the forces. The **resultant moment** about an axis is the sum of the moments about the axis of the forces acting on the free-body diagram.

---

### Example 2.8

Draw free-body diagrams for each of the systems shown in Figure 2.36. The free-body diagrams are to be drawn at an arbitrary instant. Label the forces in terms of the indicated coordinate. All coordinates are measured from the system's static equilibrium position.

**Solution**

(a) The free-body diagram for the system of Figure 2.36(a) is drawn in Figure 2.37(a). Gravity is the only body force. The surface forces include the force exerted by the spring on the block, the force from the viscous damper, the reaction at the surface, and the external force. When the block's displacement is positive, the spring is stretched and pulls on the block. When the block's displacement is negative, the spring is compressed and pushes against the block. Labeling the spring force as $kx$ and drawing it so that it acts against the block is consistent with both positive and negative values of $x$.

At an arbitrary instant, the velocity of the block is $\dot{x}$ and thus the force acting on the block from the viscous damper is $c\dot{x}$. Since the viscous damping force always acts in the opposite direction of the velocity, it is drawn to the left on the free-body diagram, consistent with a positive velocity. This is also consistent with a negative velocity since the damping force is labeled $c\dot{x}$; when the velocity is negative the damping force is actually acting to the right, again opposing the direction of motion.

The reaction force between the block and the surface has two components: (1) a force perpendicular to the surface, called the normal force, and (2) a force

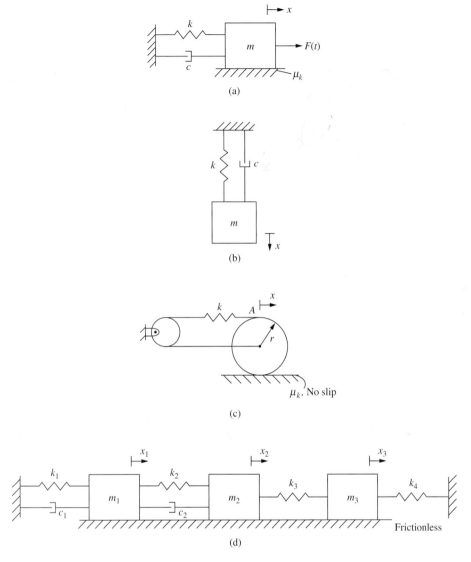

FIG. **2.36** Systems for Example 2.8.

opposite the direction of motion, the friction force. However, unlike the spring force and viscous damping force, the direction of the force from Coulomb friction does not take care of itself when drawn and labeled on a free-body diagram. Two separate free-body diagrams are required, one for a positive $\dot{x}$ [Figure 2.37(a)(i)] and one for a negative velocity [Figure 2.37(a)(ii)].

(b) The free-body diagram for the system of Figure 2.36(b), drawn at an arbitrary instant, is shown in Figure 2.37(b). The system is similar to that of Figure 2.36(a) except that the mass is suspended and its motion is in the vertical direction. As illustrated in Section 2.2.2, when the system is in equilibrium the spring is stretched with a static deflection $\Delta_{st} = \frac{mg}{k}$. The dependent variable $x$ is measured

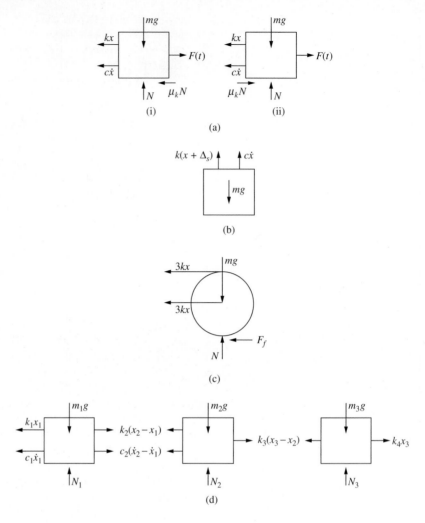

**FIG. 2.37** Free-body diagrams at an arbitrary instant for the systems of Example 2.8.

from the system's equilibrium position. Thus, at an arbitrary instant the change in length of the spring from its unstretched length is $x + \Delta_{st}$.

The resultant of the forces acting on the body is

$$\sum F = -mg + c\dot{x} + k(x + \Delta_{st})$$

$$= -mg + c\dot{x} + kx + k\left(\frac{mg}{k}\right)$$

$$= c\dot{x} + kx \tag{a}$$

When the equilibrium condition is applied, the static spring force cancels the gravity force. Since the resultant of the static spring force and the gravity force is zero the free-body diagram of the system could have been drawn without including either.

(c) The force developed in the spring is its stiffness times its change in length. One end of the spring is attached to the mass center of the disk, which has a displacement $x$ to the right. The other end of the spring is attached to point $A$. When the center of the disk has a displacement $x$, point $A$ has rotated through an angle $\theta$ and the end of the spring attached to point $A$ has increased in length by $x + r\theta$. Thus the total change in length of the spring is $2x + r\theta$. If the disk rolls without slip then $x = r\theta$ and the change in length of the spring is $3x$. At any instant the disk has a single point of contact with the surface. The reaction at the surface has two components: a normal force and a friction force. Even when the disk rolls without slip, a friction force is developed but is less than the maximum available friction force. The free-body diagram for this system assuming no slip is drawn in Figure 2.37(c).

(d) Modeling of this three-degree-of-freedom system requires free-body diagrams of each mass as drawn in Figure 2.37(d). The force developed in a spring is its stiffness times its change in length. Consider the spring of stiffness $k_2$, which links the particles of mass $m_1$ and $m_2$. The right end of the spring is attached to the block of mass $m_2$, which has a displacement $x_2$. The left end of the spring is attached to the block of mass $m_1$, which has a displacement $x_1$. The change in length of the spring is the difference between the displacements of its two ends, $x_2 - x_1$. If $x_2 > x_1$ then the spring's length has increased from its unstretched length and the spring is in tension. Thus if the spring force is labeled as $k_2(x_2 - x_1)$ it must be drawn directed away from each of the particles on their free-body diagrams.

Systems such as the one in Figure 2.36(b) include springs that are stretched or compressed when the system is in equilibrium. The dependent variable in Example 2.8(b) is measured from the system's equilibrium position. The spring force at an arbitrary instant is the sum of the static spring force $(k\Delta_{st})$ and the spring force developed due to the displacement from equilibrium $(kx)$. The resultant of the static spring force and the gravity force is zero, and thus it is not necessary to include either when applying conservation laws to the system.

The result in the example can be generalized for any system. If the system has springs that are at a length different from their unstretched lengths when the system is in equilibrium, then an equilibrium condition exists between the static spring forces and the gravity forces that cause the static forces. The equilibrium condition is derived from requiring either that the resultant of the static forces and the gravity forces equals zero or that the resultant moment due to these forces is zero. If the system's dependent variables are measured from the system's equilibrium position, then, when the conservation laws are applied to the system at an arbitrary instant, the application of the equilibrium condition leads to the static spring forces and the gravity forces canceling from the resulting equation.

Therefore it is not necessary to include static spring forces or gravity forces that lead to static spring forces in free-body diagrams when mathematically modeling a linear system and the dependent variables are measured from the system's equilibrium position. These forces must be included if the system is nonlinear or the dependent variable is measured from a position other than the static equilibrium position. The forces must also be included if a different conservation law is to be applied than the one used to develop the equilibrium condition. For example the equilibrium condition between the static spring force and gravity for the system of Figure 2.38 is obtained by summing moments about an axis

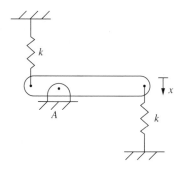

**FIG. 2.38** Static spring forces cancel with gravity when deriving the differential equation governing the motion of the system, but not when determining the reaction at $A$.

through $A$. When the moment equation about an axis through $A$ at an arbitrary instant is applied to derive the differential equation governing the motion of the system, the forces cancel. However, these forces do not cancel when the summation of forces is applied at an arbitrary instant to determine the reaction force at $A$.

## Example 2.9

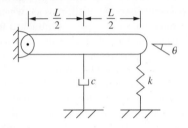

FIG. **2.39** Systems for Example 2.9.

Consider the system of Figure 2.39. (a) Determine $\theta_{st}$, the angular rotation of the bar, measured clockwise from the horizontal, when the system is in equilibrium. (b) Draw a free-body diagram of the system at an arbitrary instant including gravity and the static forces in the spring. (c) Show that the resultant moment about an axis through $A$ is independent of the gravity force and $\theta_{st}$. (d) Redraw the free-body diagram not including the gravity force and the static force in the spring. (e) What restrictions are placed on the use of this free-body diagram?

### Solution

(a) The free-body diagram of the static equilibrium position, assuming small angular displacements, is illustrated in Figure 2.40(a). Assuming the spring force is vertical is consistent with the small angle assumption. Summing moments, assuming clockwise positive, about the pin support leads to

$$\sum M_A = 0$$
$$mg(L/2) - kL\theta_{st}(L) = 0$$
$$\theta_{st} = \frac{mg}{2kL} \qquad (a)$$

(b) Let $\theta$ represent the clockwise angular displacement of the bar, measured from the system's equilibrium position. A free-body diagram of the system at an arbitrary

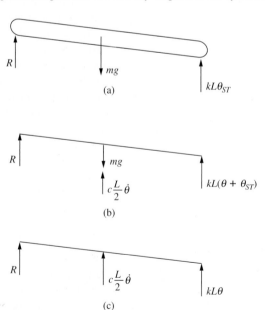

FIG. **2.40** (a) Free-body diagram of static equilibrium position of system of Example 2.9; (b) free-body diagram at an arbitrary instant, including gravity and static spring force; (c) modified free-body diagram in which gravity and static spring force are not included.

instant, assuming small angular displacements, is illustrated in Figure 2.40(b). The bar is rotating about an axis fixed at $A$; thus the velocity of the center of the bar is $\frac{L}{2}\dot{\theta}$ and the force form the viscous damper is $c\frac{L}{2}\dot{\theta}$. Assuming small $\theta$ the total change in length of the spring from its unstretched length is $L(\theta + \theta_{st})$.

(c) Summing moments, assuming positive clockwise, at an arbitrary instant about an axis at $A$ gives

$$\sum M_A = mg(L/2) - c\frac{L}{2}\dot{\theta}(L/2) - kL(\theta + \theta_{st})(L) \qquad \text{(b)}$$

Substitution of Equation (a) into Equation (b) leads to

$$\sum M_A = -c\frac{L^2}{4}\dot{\theta} - kL^2\theta \qquad \text{(c)}$$

(d) In view of Equation (c) the resultant moment about $A$ is independent of the gravity force and the static spring force. Thus for the purposes of deriving a mathematical model for the system, a modified free-body diagram can be used in which neither force is included. This useful free-body diagram is illustrated in Figure 2.40(c).

(e) The reaction at the pin support $A$ is determined by summing the forces acting on the bar. However, whereas the resultant moment of the gravity force and spring force about $A$ is zero, their resultant itself is not zero and thus must be included when summing forces.

## 2.6 NEWTON'S LAWS

Newton's three laws of motion provide a foundation for the modeling of mechanical systems. Newton's first law states that a body at rest or in motion with a constant velocity will continue to remain at rest or move with a constant velocity in the absence of external forces. Newton's first law is the basis of static analysis. Newton's third law, often stated as "every action has an equal and opposite reaction," is the basis on which surface forces are drawn on free-body diagrams. For example, consider the system of Figure 2.36(a). When the block of mass $m$ is displaced a distance $x$ to the right from equilibrium, the spring has an increase in length of $x$ from its unstretched length. The change in length is a result of a force $F = kx$ pulling on the spring by the block. Newton's third law implies that the spring exerts a force $F = kx$ on the block in the opposite direction of the force from the block on the spring. This force is illustrated on the free-body diagram of Figure 2.37(a).

Newton's second law, the subject of this section, is the basic conservation law used in the mathematical modeling of mechanical systems.

### 2.6.1 Particles

**Newton's second law** applied to a particle states that the resultant force acting on a particle is equal to the time rate of change of linear momentum of the particle. If the mass of the particle is constant then the time rate of change of linear momentum is equal to the mass of the particle times its acceleration. In this case Newton's second law is written as

$$\sum \mathbf{F} = m\mathbf{a} \tag{2.96}$$

where $\sum \mathbf{F}$ is the sum of all external forces acting on the particle, or the resultant of the external forces. The component forms of Newton's second law are

$$\sum F_x = m\ddot{x} \tag{2.97a}$$

$$\sum F_y = m\ddot{y} \tag{2.97b}$$

$$\sum F_z = m\ddot{z} \tag{2.97c}$$

Newton's second law is a basic law of nature. It cannot be derived from any law more basic and it can be proven only by empirical observation. The Principle of Work and Energy is obtained by integrating Newton's law spatially. The Principle of Impulse and Momentum is obtained by integrating Newton's law over time. While these principles are not independent of Newton's second law, they are often applied as an alternative to the direct application of it in the modeling of a mechanical system.

Mathematical modeling of a mechanical system requires the application of Newton's second law to the free-body diagram of a system component drawn at an arbitrary instant, for an arbitrary value of the chosen dependent variable. The external forces acting on the particle are labeled in terms of the dependent variable. The particle acceleration is often expressed in terms of time derivatives of the dependent variable. The application of Newton's second law to the free-body diagram leads to a differential equation whose solution is the time-dependent behavior of the system.

---

**Example 2.10**

Use Newton's law to derive the differential equations governing the motion of (a) the system of Figure 2.36(a) assuming $\mu_k = 0$ and (b) the system of Figure 2.36(b).

**Solution**

The free-body diagrams for these systems drawn at an arbitrary instant are illustrated Figures 2.37(a) and 2.37(b).

(a) The acceleration of the particle at an arbitrary instant is $\ddot{x}$. The component form of Newton's second law in the $x$ direction, Equation (2.97a) is applied to the free-body diagram of Figure 2.37(a). Taking positive forces acting to the right leads to

$$-kx - c\dot{x} + F(t) = m\ddot{x} \tag{a}$$

Equation (a) is rearranged so that all the terms involving $x$ are on the left-hand side leading to

$$m\ddot{x} + c\dot{x} + kx = F(t) \tag{b}$$

Equation (b) is a second-order linear ordinary differential equation whose solution describes the motion of the system.

(b) The application of Newton's second law in the $x$ direction with the positive direction downward to the free-body diagram of Figure 2.37(b) leads to

$$-mg - k(x + \Delta_S) - c\dot{x} + F(t) = m\ddot{x} \tag{c}$$

The equilibrium condition for this system is

$$mg + k\Delta_s = 0 \tag{d}$$

Using Equation (d) in Equation (c) and rearranging leads to

$$m\ddot{x} + c\dot{x} + kx = F(t) \tag{e}$$

Note that Equation (e) is identical to Equation (b). This reinforces the discussion in Section 2.5 regarding the use of the equilibrium condition in mathematical modeling of a linear system. When the equilibrium condition is applied, the static spring forces and gravity cancel with one another in the differential equation. Thus, for purposes of deriving a mathematical model for a linear system, static spring forces and the gravity forces leading to static spring forces do not need to be included in the analysis. From this point in this book, these forces will not be included on free-body diagrams.

## Example 2.11

A projectile of mass $m$ is fired at $t = 0$ with a velocity $v_0$ at an angle $\alpha$ with the horizontal. Determine a mathematical model for the motion of the projectile.

### Solution

Assuming no air resistance or other wind effects, the projectile travels in a plane defined by the vertical direction and the direction of the initial velocity. Let $y$ be the vertical coordinate and let $x$ be a coordinate in the plane of motion whose axis is perpendicular to the $y$ axis. The particle is fired from the point $(0, 0)$ at $t = 0$, as illustrated in Figure 2.41(a). The particle is constrained to move in the $x$-$y$ plane, thus its motion has two degrees of freedom. A free-body diagram of the particle at an arbitrary instant is shown in Figure 2.41(b). When friction is neglected, gravity is the only external force acting on the particle.

Application of the component forms of Newton's second law to the free-body diagram of the particle leads to

$$\sum F_x = ma_x \tag{a}$$
$$0 = m\ddot{x} \tag{b}$$

and

$$\sum F_y = ma_y \tag{c}$$
$$-mg = m\ddot{y} \tag{d}$$

Equations (b) and (d) are summarized as

$$\ddot{x} = 0 \tag{e}$$
$$\ddot{y} = -g \tag{f}$$

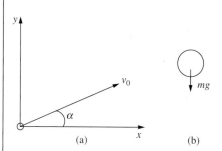

(a)   $x$   (b)

**FIG. 2.41** (a) The coordinate system for the projectile motion problem of Example 2.11; (b) the free-body diagram of the projectile.

Equations (e) and (f) form the mathematical model for projectile motion. The mathematical model is completed by determining the initial conditions that $x(t)$ and $y(t)$ must satisfy. Since the particle was fired from the origin of the coordinate system at $t = 0$,

$$x(0) = 0 \tag{g}$$
$$y(0) = 0 \tag{h}$$

The initial velocity is resolved into $x$ and $y$ components

$$\dot{x}(0) = v_0 \cos\alpha \tag{i}$$
$$\dot{y}(0) = v_0 \sin\alpha \tag{j}$$

Integration of Equations (e) and (f) twice with respect to time leads to

$$x(t) = C_1 t + C_2 \tag{k}$$
$$y(t) = -\tfrac{1}{2}gt^2 + C_3 t + C_4 \tag{l}$$

The constants of integration, $C_1, C_2, C_3,$ and $C_4$ are determined from application of the initial conditions, Equations (g)–(h), leading to

$$x(t) = (v_0 \cos\alpha)t \tag{m}$$
$$y(t) = -\tfrac{1}{2}gt^2 + (v_0 \sin\alpha)t \tag{n}$$

Equations (m) and (n) are the projectile motion equations.

Equations (m) and (n) can be combined to determine the equation for the path traveled by the particle in the $x$-$y$ plane as

$$y = -\frac{g\,s^2}{2v_0^2}x^2 + (\tan\alpha)x \tag{o}$$

Equation (o) is the equation of a parabola opening downward and passing through the origin. The vertex of the parabola corresponds to the maximum height of the projectile.

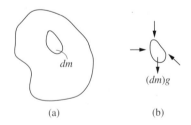

(a)                      (b)

**FIG. 2.42** (a) A rigid body composed of an infinite number of particles of infinitesimal mass; (b) the free-body diagram of a particle of mass $dm$ illustrates the internal forces between particles.

### 2.6.2 Rigid Body Motion

Newton's law, as written in Equation (2.96) applies to a single particle. A rigid body is a system with an infinite number of particles. A form of Newton's law for a rigid body may be derived by first applying the law to an arbitrary particle on the rigid body of mass $dm$, as illustrated in Figure 2.42. The surface forces shown acting on a particle are internal forces from the surrounding particles. When equations for adjacent particles are added together, the internal forces cancel as a result of Newton's third law. When the equations for all particles comprising the rigid body are added together, the summation becomes an integration over the mass of the body resulting in

$$\sum \mathbf{F} = m\bar{\mathbf{a}} \tag{2.98}$$

where $\bar{\mathbf{a}}$ is the acceleration of the mass center of the rigid body, $m$ is its total mass, and $\sum \mathbf{F}$ is the resultant of the external forces as shown on a free-body diagram of the rigid body.

A general rigid-body motion is a combination of rotation and translation. The rotational motion of a rigid body leads to the necessity for applying of a moment equation when deriving a mathematical model for the motion of the body. A moment equation for a rigid body is derived using a process similar to that used to derive the force equation. The result is

$$\sum \mathbf{M_G} = \dot{\mathbf{H}}_{\mathbf{G}} \tag{2.99}$$

where $\sum \mathbf{M_G}$ is the resultant moment about the mass center of all external forces and moments and $\dot{\mathbf{H}}_{\mathbf{G}}$ is the time rate of change of angular momentum of the rigid body about its mass center.

## 2.6.3 Pure Rotational Motion About a Fixed Axis of Rotation

The angular momentum about its mass center for a rigid body undergoing planar motion is given in Equation (2.31) as

$$H_G = \bar{I}\omega \tag{2.31}$$

where $\bar{I}$ is the body's mass moment of inertia about its centroidal axis and $\omega$ is the angular velocity about the axis of rotation. The angular momentum is in the direction of the axis of rotation. Equation (2.31) also applies for a symmetrical body undergoing pure rotation (no translation) about a fixed axis parallel to an axis of symmetry. The time rate of change of the angular momentum, determined using Equation (2.31) is

$$\dot{H}_G = \bar{I}\alpha \tag{2.100}$$

Thus, Equation (2.99) becomes

$$\sum M_G = \bar{I}\alpha \tag{2.101}$$

It is usually convenient to apply the moment equation about the axis of rotation. When the axis of rotation is not a centroidal axis, but parallel to a centroidal axis through a point $O$, the Parallel Axis Theorem is used to rewrite Equation (2.101) as

$$\sum M_O = I_O\alpha \tag{2.102}$$

where $I_O$ is the moment of inertia of the rigid body about an axis through $O$.

## Example 2.12

A rotor of moment of inertia $I$ is rotating about its centroidal axis at a constant angular speed $\omega_0$ when a constant braking torque $T$ is applied. Determine a mathematical model for the system and determine the time required to stop the rotor.

### Solution

Application of Equation (2.101) to the free-body diagram of Figure 2.43(b) gives

$$\sum M_G = I\alpha \qquad \text{(a)}$$

$$-T = I\alpha \qquad \text{(b)}$$

Noting that $\alpha = \dot{\omega}$ leads to

$$\dot{\omega} = -\frac{T}{I} \qquad \text{(c)}$$

The torque is applied at $t = 0$ when

$$\omega(0) = \omega_0 \qquad \text{(d)}$$

Integration of Equation (c) gives

$$\omega(t) = -\frac{T}{I}t + C \qquad \text{(e)}$$

Application of the initial condition, Equation (d), leads to $C = \omega_0$ and

$$\omega(t) = -\frac{T}{I}t + \omega_0 \qquad \text{(f)}$$

Setting $\omega = 0$ in Equation (f) to determine the time at which the rotor will stop leads to

$$t = \frac{I\omega_0}{T} \qquad \text{(g)}$$

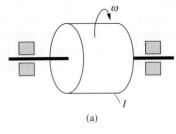

(a)

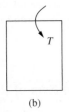

(b)

**FIG. 2.43** (a) The rotor has constant speed when a braking $T$ is applied; (b) the free-body diagram of the rotor.

## Example 2.13

A rotor of moment of inertia $I$ is connected to a torsional spring of stiffness $k_t$ and a torsional viscous damper of damping coefficient $c_t$, as shown in Figure 2.44(a). Let $\theta$ be the counterclockwise angular displacement of the rotor, measured from the system's equilibrium position. Derive a mathematical model for the system.

### Solution

Application of Equation (2.101) to the free-body diagram of Figure 2.44(b) leads to

$$\sum M_G = I\alpha \qquad \text{(a)}$$

$$-k_t\theta - c_t\dot{\theta} = I\ddot{\theta} \qquad \text{(b)}$$

Rearranging Equation (b) gives

$$I\ddot{\theta} + c_t\dot{\theta} + k_t\theta = 0 \qquad \text{(c)}$$

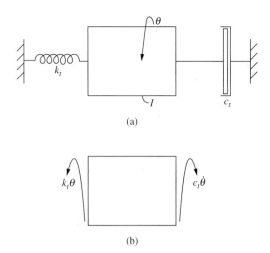

FIG. **2.44**  (a) The system of Example 2.13; (b) the free-body diagram of the system at an arbitrary instant.

### 2.6.4 Planar Motion of a Rigid Body

The applicable form of the force equation for a rigid body undergoing planar motion is Equation (2.98). In addition, because the axis of rotation is always perpendicular to the plane of motion, the moment Equation (2.101) also applies. Furthermore, if the axis of rotation is fixed, then Equation (2.102) applies. If the planar motion is constrained then a kinematic relationship exists between $\bar{a}$ and $\alpha$.

---

**Example 2.14**

Determine a mathematical model for the system of Figure 2.45(a). Let $\theta$ be the clockwise angular displacement of the bar, measured from the system's equilibrium position. Assume small $\theta$.

**Solution**

The free-body diagram of the system at an arbitrary instant is shown in Figure 2.45(b). Use of the small angle assumption linearizes the system. In addition, $\theta$ is measured from the system's equilibrium position, so the static force in the spring and the gravity force of the bar cancel each other when the differential equation is derived. Thus these forces are not drawn on the free-body diagram. For a positive value of $\theta$, the displacement of point $B$ is $(2L/3)\theta$ downward and the velocity of point $A$ is $(L/3)\dot{\theta}$ upward. The forces from the spring and viscous damper are drawn consistently on the free-body diagram.

The moment of inertia of a slender bar about an axis through its mass center is given in Table 2.1 as $\bar{I} = \frac{1}{12}mL^2$. Noting that the distance between the mass center and $O$ is $L/6$, the Parallel Axis Theorem is used to determine $I_O$

$$I_O = \frac{1}{12}mL^2 + m\left(\frac{L}{6}\right)^2 = \frac{1}{9}mL^2 \tag{a}$$

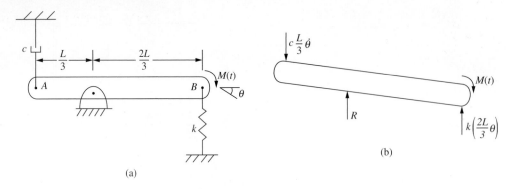

(a)

(b)

**FIG. 2.45** (a) The system of Example 2.14; (b) the free-body diagram at an arbitrary instant.

Applying Equation (2.102) to the free-body diagram of Figure 2.45(b) to sum moments about $O$ leads to

$$\sum M_O = I_O \alpha \qquad (b)$$

$$M(t) - \left(c\frac{L}{3}\dot{\theta}\right)\left(\frac{L}{3}\right) - \left(k\frac{2L}{3}\theta\right)\left(\frac{2L}{3}\right) = \frac{1}{9}mL^2\ddot{\theta} \qquad (c)$$

Rearranging Equation (d) results in

$$\frac{1}{9}mL^2\ddot{\theta} + \frac{1}{9}cL^2\dot{\theta} + \frac{4}{9}kL^2\theta = M(t) \qquad (d)$$

## Example 2.15

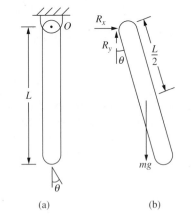

(a)                (b)

**FIG. 2.46** (a) The compound pendulum of Example 2.15; (b) the free-body diagram of the compound pendulum at an arbitrary instant.

Determine a linear mathematical model for the compound pendulum of Figure 2.46(a). Let $\theta$ be the counterclockwise angular rotation of the bar from the system's equilibrium position. Assume small $\theta$.

### Solution

The free-body diagram of the pendulum at an arbitrary instant is shown in Figure 2.46(b). The moment of inertia of the bar about $O$ is obtained using Table 2.1 and the Parallel Axis Theorem as

$$I_O = \frac{1}{12}mL^2 + m\left(\frac{L}{2}\right)^2 = \frac{1}{3}mL^2 \qquad (a)$$

Summing moments about $O$ using Equation (2.102) gives

$$-mg\left(\frac{L}{2}\sin\theta\right) = \frac{1}{3}mL^2\ddot{\theta} \qquad (b)$$

For small $\theta$, $\sin \theta \approx \theta$. Using this approximation, Equation (b) is approximated by

$$\frac{1}{3}mL^2\ddot{\theta} + \frac{1}{2}mgL\theta = 0 \tag{c}$$

Dividing by the coefficient of the $\ddot{\theta}$ term leads to

$$\ddot{\theta} + \frac{3g}{2L}\theta = 0 \tag{d}$$

### 2.6.5  Three-Dimensional Motion of Rigid Bodies

The position vector of the mass center of a rigid body has three components, each of which may vary independently with time. A rigid body may have angular velocities about three independent axes of rotation. Thus a rigid body may have up to six degrees of freedom and it may be necessary to develop up to six equations to model the motion of the system. The component forms of Equation (2.98) can be used to develop three independent equations. The angular momentum equation, Equation (2.99) is used to develop the following equations, called Euler's equations

$$\sum M_x = \bar{I}_x \dot{\omega}_x - (\bar{I}_y - \bar{I}_z)\omega_y\omega_z \tag{2.103a}$$

$$\sum M_y = \bar{I}_y \dot{\omega}_y - (\bar{I}_z - \bar{I}_x)\omega_z\omega_x \tag{2.103b}$$

$$\sum M_z = \bar{I}_z \dot{\omega}_z - (\bar{I}_x - \bar{I}_y)\omega_x\omega_y \tag{2.103c}$$

Euler's equations, as written in Equations (2.103), are for a set of principal axes located at the mass center of the rigid body.

### 2.6.6  D'Alembert's Principle

**2.6.6.1  Particles**  Newton's second law applied to a particle, Equation (2.96), can be rewritten as

$$\sum \mathbf{F} - m\mathbf{a} = \mathbf{0} \tag{2.104}$$

Defining the inertia force by $m\mathbf{a}$, Equation (2.104) states that the sum of the external forces acting on the particle minus the inertia force is equal to zero. That is, the resultant of the external forces and the negative of the inertia force is the zero vector. This reformulation of Newton's second law converts the dynamics problem into a statics problem and is called D'Alembert's prinicple.

**2.6.6.2  Rigid Bodies Undergoing Planar Motion**  D'Alembert's principle for a particle is a trivial reformulation of Newton's second law. It is more substantial and useful when developed for a rigid body. For this case the principles of equivalent force systems are used to develop an alternate formulation of the equations of motion.

The conservation laws that may be applied to rigid bodies undergoing planar motion are summarized by Equations (2.98) and (2.101). The application of these equations, which can be applied to derive the differential equations governing the motion of any rigid body undergoing planar motion, are illustrated in the free-body diagrams of Figure 2.47. Figure 2.47(a) illustrates the system of external forces acting on a rigid body at an arbitrary instant. Figure 2.47(b) defines the system of effective forces, a force equal to $m\bar{a}$ that is applied at the center of mass of the body and a couple equal to $\bar{I}\alpha$. From Equations (2.98) and (2.101) it is clear that the resultant forces of the two systems are the same and the moment about the mass center for each of the systems is the same. Thus the system of external forces and the system of effective forces are equivalent force systems. This is a statement of **D'Alembert's principle**: at any instant of time, the system of external forces acting on a rigid body undergoing planar motion is equivalent to the body's system of effective forces. The system of effective forces is defined as a force equal to $m\bar{a}$ applied at the mass center and a couple equal to $\bar{I}\alpha$.

The properties of equivalent force systems lead to the following statements of D'Alembert's principle for planar motion

$$\left(\sum \mathbf{F}\right)_{ext} = \left(\sum \mathbf{F}\right)_{eff} \tag{2.105}$$

$$\left(\sum M_A\right)_{ext} = \left(\sum M_A\right)_{eff} \tag{2.106}$$

The moments for Equation (2.106) are taken about an axis perpendicular to the plane of motion through $A$, where $A$ is any point in the plane.

D'Alembert's principle provides an alternative to the direct application of Equations (2.98) and (2.101). If $A$ is a fixed axis of rotation then the Parallel Axis Theorem can be used to reconcile Equation (2.102) with Equation (2.106). Two problems in which the application of D'Alembert's principle is more convenient and provides more flexibility than the direct application of Equations (2.98) and (2.101) or (2.102) are illustrated in Figures 2.48 and 2.49. The axis of rotation for bar $BC$ of the

**FIG. 2.47** The free-body diagrams of a rigid body undergoing planar motion at an arbitrary instant: (a) the system of external forces; (b) the system of effective forces. D'Alembert's principle states that the system of external forces is equivalent to the system of effective forces.

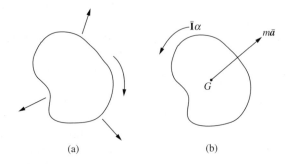

(a)                    (b)

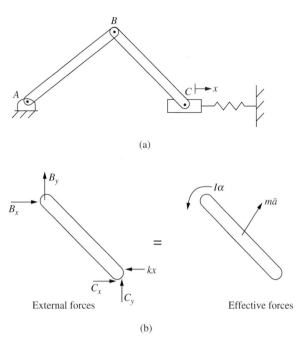

**FIG. 2.48** (a) The axis of rotation of bar *BC* changes as it rotates; (b) the free-body diagrams for the application of D'Alembert's principle to bar *BC*.

slider-crank mechanism of Figure 2.48(a) is not fixed; it changes as the bar rotates. Thus Equation (2.102) is not applicable. Free-body diagrams of external and effective forces for bar *BC* are shown in Figure 2.48(b). Principles of rigid-body kinematics must be applied to relate the acceleration of the mass center of the bar and its angular acceleration to the displacement of the collar. Note that gravity and the static deflection of the spring have not been included on the free-body diagram of the external forces because they cancel one another in the governing differential equation. D'Alembert's principle may be applied to this set of free-body diagrams.

The system of Figure 2.49(a) is composed of three bodies. The application of Newton's law requires drawing free-body diagrams and writing Equation (2.98) for each of the blocks and Equation (2.101) for the disk. In a drawing of the individual free-body diagrams, the tensions in the cables connecting the blocks to the disks are identified as unknown quantities. Algebra is used to eliminate the tensions from the equations to derive a single differential equation to solve for $x(t)$. Only one set of free-body diagrams is necessary for application of D'Alembert's principle, as illustrated in Figure 2.49(b). Kinematics is used to relate the accelerations of the various bodies (This is also necessary when using the direct application of Newton's laws). A single differential equation is derived by applying Equation (2.106) about the pin support of the disk.

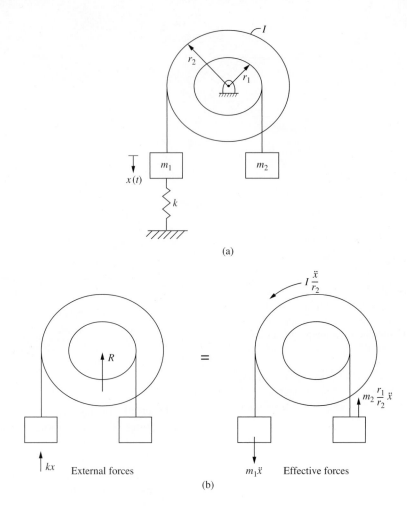

**FIG. 2.49** (a) Three free-body diagrams and the algebraic elimination of the tensions in the cables are necessary when directly applying Newton's law to derive the governing differential equation; (b) the free-body diagrams for the application of D'Alembert's principle.

External forces    =    Effective forces

(b)

---

### Example 2.16

Use D'Alembert's principle to derive the differential equation governing the motion of the system of Figure 2.50(a) using $\theta$ as the dependent variable. Assume small $\theta$.

**Solution**

Derivation of the differential equation governing the motion of the system of Figure 2.50(a) requires writing a moment equation about an axis through $A$. The direct application of Equation (2.101) requires determining the center of mass of the assembly, then the moment of inertia about the centroidal axis. Free-body diagrams showing the external and effective forces for the system at an arbitrary instant are shown in Figure 2.50(b). The small angle assumption is

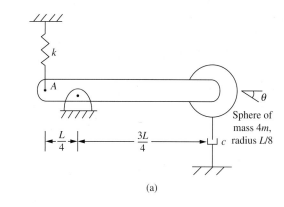

(a)

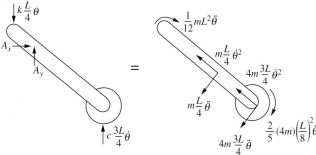

External forces

=

Effective forces

(b)

**FIG. 2.50**  (a) The system for Example 2.16; (b) the free-body diagram for the application of D'Alembert's principle.

used in labeling the external forces. Gravity and the static force in the spring are not included on the free-body diagram of external forces because they cancel from the resulting equation when the equilibrium condition is applied.

Application of

$$\left(\sum M_A\right)_{ext} = \left(\sum M_A\right)_{eff} \tag{a}$$

to the free-body diagram of Figure 2.50(b) leads to

$$-k\left(\frac{L}{4}\theta\right)\left(\frac{L}{4}\right) - c\left(\frac{3L}{4}\dot\theta\right)\left(\frac{3L}{4}\right) = m\left(\frac{L}{4}\ddot\theta\right)\left(\frac{L}{4}\right) + 4m\left(\frac{3L}{4}\ddot\theta\right)\left(\frac{3L}{4}\right)$$
$$+ \frac{1}{12}mL^2\ddot\theta + \frac{2}{5}(4m)\left(\frac{L}{8}\right)^2\ddot\theta \tag{b}$$

Equation (b) is simplified and rearranged, resulting in

$$\frac{351}{240}mL^2\ddot\theta + \frac{9}{16}cL^2\dot\theta + \frac{1}{16}kL^2\theta = 0 \tag{c}$$

## Example 2.17

Determine a mathematical model for the system of Figure 2.51 using $x$ as the dependent variable.

### Solution

Free-body diagrams of the system at an arbitrary instant are shown in Figure 2.51(b). If the block of mass $m_1$ moves down a distance $x$ from the system's equilibrium position, then the disk rotates counterclockwise through an angle $\theta = x/r$, and the block of mass $m_2$ moves up a distance $3x$. Since $x$ is measured from the system's equilibrium position, static spring forces and gravity forces are not included on the free-body diagrams.

D'Alembert's principle is applied in the form

$$\left( \sum M_A \right)_{ext} = \left( \sum M_A \right)_{eff} \tag{a}$$

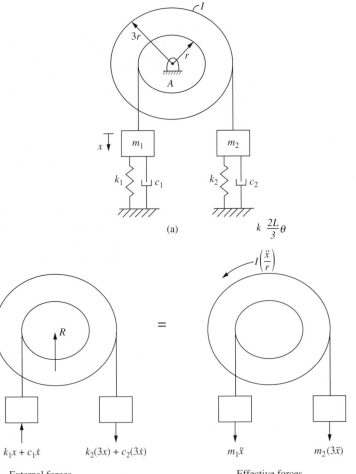

Application of Equation (a) to the free-body diagrams using counterclockwise moments as positive yields

$$-(k_1 x + c_1 \dot{x})(r) - (3k_2 x + 3c_2 \dot{x})(3r) = (m_1 \ddot{x})r + (3m_2 \ddot{x})(3r) + I(\ddot{x}/r) \quad \text{(b)}$$

Rearrangement of Equation (b) leads to

$$\left(m_1 r + 9m_2 r + \frac{I}{r}\right)\ddot{x} + (c_1 r + 9c_2 r)\dot{x} + (k_1 r + 9k_2 r)x = 0 \quad \text{(c)}$$

Equation (c) governs the motion of the system when displaced from its equilibrium position.

## 2.7  SINGLE-DEGREE-OF-FREEDOM SYSTEMS

Modeling of the motion of a single-degree-of-freedom (one-degree-of-freedom) system requires one time-dependent variable. The modeling process results in a differential equation whose solution provides the time-dependent behavior of the system. The systems of Examples 2.9, 2.13, 2.14, 2.15, 2.16, and 2.17 are single-degree-of-freedom systems. The differential equation for each of these systems is of the form

$$\tilde{m}\ddot{x} + \tilde{c}\dot{x} + \tilde{k}x = F(t) \tag{2.107}$$

The differential governing the motion of any linear one-degree-of-freedom system is of the form of Equation (2.107). The term $\tilde{m}\ddot{x}$ results from the inertia system's inertia forces, $\tilde{c}\dot{x}$ results from viscous damping forces, $\tilde{k}x$ results from forces in the system's mechanisms for potential energy storage, and $F(t)$ results from externally applied forces.

If $F(t) = 0$, the solution of Equation (2.107) is called the system's free response, which is a result of energy present in the system at $t = 0$. A nonzero initial velocity is a result of kinetic energy in the system at $t = 0$. Since $x$ is assumed to be measured from the system's equilibrium condition, a nonzero initial displacement occurs when potential energy is stored in the system at $t = 0$. The initial energies determine the values of $x(0)$ and $\dot{x}(0)$, the initial conditions necessary to determine a unique solution of Equation (2.106). If $F(t) \neq 0$ then the solution of Equation (2.106) is a forced response.

When divided through by $\tilde{m}$, Equation (2.107) becomes

$$\ddot{x} + \frac{\tilde{c}}{\tilde{m}}\dot{x} + \frac{\tilde{k}}{\tilde{m}}x = \frac{1}{\tilde{m}}F(t) \tag{2.108}$$

The system's **natural frequency** $\omega_n$ is calculated as

$$\omega_n = \sqrt{\frac{\tilde{k}}{\tilde{m}}} \tag{2.109}$$

The system's **viscous damping ratio** $\zeta$ is calculated from

$$2\zeta\omega_n = \frac{\tilde{c}}{\tilde{m}} \tag{2.110}$$

Equation (2.108) is rewritten using the natural frequency and viscous damping ratio as

$$\ddot{x} + 2\zeta\omega_n\dot{x} + \omega_n^2 x = \frac{1}{\tilde{m}}F(t) \tag{2.111}$$

Equation (2.111) is called the **standard form of the differential equation** governing the motion of a linear single-degree-of-freedom system. The natural frequency and viscous damping ratio are system properties. The natural frequency is dependent on the system's inertia and stiffness characteristics. The viscous damping ratio is dependent on the system's inertia, stiffness, and viscous damping characteristics. Their role in system response is developed in Chapters 6 and 7.

The forms or numerical values of the natural frequency and damping ratio for a single-degree-of-freedom system can be determined directly from the differential equation modeling the system. Once the differential equation is derived, it can be written in the standard form of Equation (2.111), from which the natural frequency and viscous damping ratio can be determined. The natural frequency determined directly from the differential equation has dimensions of r/s. The viscous damping ratio is dimensionless.

---

**Example 2.18**

Numerical values for the parameters of the system of Example 2.14 are $m = 10$ kg, $k = 2 \times 10^5$ N/m, $c = 120$ N·s/m, and $L = 2.4$ m. The moment is a single-frequency harmonic excitation of amplitude 200 N·m, frequency 100 rad/s, and phase 30°. (a) Determine the system's natural frequency. (b) Determine the system's viscous damping ratio. (c) Write the differential equation governing the angular displacement of the bar in the standard form of Equation (2.110), using the numerical values obtained in parts (a) and (b).

**Solution**

Equation (d) of Example 2.14 is divided by $mL^2/9$ leading to

$$\ddot{\theta} + \frac{c}{m}\dot{\theta} + \frac{4k}{m}\theta = \frac{9}{mL^2}M(t) \tag{a}$$

(a) The natural frequency is obtained by comparing Equation (a) with Equation (2.111) as

$$\omega_n = \sqrt{4k/m}$$

$$= \sqrt{\frac{4(2 \times 10^5 \text{ N/m})}{10 \text{ kg}}} = 282.8 \text{ r/s} \tag{b}$$

(b) Comparison of Equation (a) with Equation (2.111) leads to

$$2\zeta\omega_n = \frac{c}{m} \qquad (c)$$

Equation (c) is rearranged to give

$$\zeta = \frac{c}{2m\omega_n}$$

$$= \frac{120\,\text{N}\cdot\text{s/m}}{2(10\,\text{kg})(282.8\,\text{r/s})} = 0.0212 \qquad (d)$$

(c) From the information given, the time-dependent form of $M(t)$ is

$$M(t) = 200\sin\left(100t + \frac{\pi}{6}\right)\,\text{N}\cdot\text{m} \qquad (e)$$

Substitution of Equations (b), (d), and (e) in Equations (a) and (2.111) leads to

$$\ddot{\theta} + 2(0.0212)(282.8)\dot{\theta} + (282.8)^2\theta = \frac{9}{(10)(2.4)^2}\left[200\sin\left(100t + \frac{\pi}{6}\right)\right] \qquad (f)$$

$$\ddot{\theta} + 12\dot{\theta} + 80000\theta = 31.25\sin\left(100t + \frac{\pi}{6}\right) \qquad (g)$$

## 2.8 MULTIDEGREE-OF-FREEDOM SYSTEMS

The application of Newton's law to the modeling of multidegree-of-freedom systems is similar to that of one-degree-of-freedom systems. Free-body diagrams are drawn of system components at an arbitrary instant and the appropriate forms of Newton's law applied. The differential equations for linear multidegree-of-freedom systems are summarized in a matrix form as

$$\mathbf{M\ddot{x} + C\dot{x} + Kx = F} \qquad (2.112)$$

where $\mathbf{x}$ is a $n \times 1$ column vector whose elements are the chosen dependent variables, $\mathbf{M}$ is a $n \times n$ mass matrix, $\mathbf{C}$ is a $n \times n$ damping matrix, $\mathbf{K}$ is a $n \times n$ stiffness matrix, and $\mathbf{F}$ is a $n \times 1$ force vector. A dot above a vector represents differentiation of all elements of the vector with respect to time.

### Example 2.19

Derive the differential equations modeling the three-degree-of-freedom system shown in Figure 2.52(a). Dependent variables are measured from the system's equilibrium position. Write the differential equations in matrix form.

**Solution**
Free-body diagrams of each mass are shown in Figure 2.52(b). The static spring forces cancel the gravity forces, and thus neither is included in the free-body diagrams. The differential equations are derived by applying Newton's second law to each of the blocks. Forces are summed assuming positive downward.

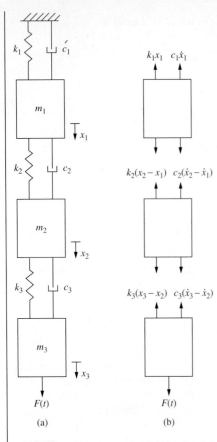

*Block 1*

$$\sum F = m_1\ddot{x}_1 \qquad \text{(a)}$$

$$-k_1x_1 - c_1\dot{x}_1 + k_2(x_2 - x_1) + c_2(\dot{x}_2 - \dot{x}_1) = m_1\ddot{x}_1 \qquad \text{(b)}$$

*Block 2*

$$\sum F = m_2\ddot{x}_2 \qquad \text{(c)}$$

$$-k_2(x_2 - x_1) - c_2(\dot{x}_2 - \dot{x}_1) + k_3(x_3 - x_2) + c_3(\dot{x}_3 - \dot{x}_2) = m_2\ddot{x}_2 \qquad \text{(d)}$$

*Block 3*

$$\sum F = m_3\ddot{x}_3 \qquad \text{(e)}$$

$$-k_3(x_3 - x_2) - c_3(\dot{x}_3 - \dot{x}_2) + F(t) = m_3\ddot{x}_3 \qquad \text{(f)}$$

Simplifying Equations (b), (d), and (f) leads to

$$m_1\ddot{x}_1 + (c_1 + c_2)\dot{x}_1 - c_2\dot{x}_2 + (k_1 + k_2)x_1 - k_2x_2 = 0 \qquad \text{(g)}$$

$$m_2\ddot{x}_2 - c_2\dot{x}_1 + (c_2 + c_3)\dot{x}_2 - c_3\dot{x}_3 - k_2x_2 + (k_2 + k_3)x_2 - k_3x_3 = 0 \qquad \text{(h)}$$

$$m_3\ddot{x}_3 - c_3\dot{x}_2 + c_3\dot{x}_3 - k_3x_2 + k_3x_3 = F(t) \qquad \text{(i)}$$

Equations (g)–(i) are summarized in matrix form as

$$\begin{bmatrix} m_1 & 0 & 0 \\ 0 & m_2 & 0 \\ 0 & 0 & m_3 \end{bmatrix} \begin{bmatrix} \ddot{x}_1 \\ \ddot{x}_2 \\ \ddot{x}_3 \end{bmatrix} + \begin{bmatrix} c_1 + c_2 & -c_2 & 0 \\ -c_2 & c_2 + c_3 & -c_3 \\ 0 & -c_3 & c_3 \end{bmatrix} \begin{bmatrix} \dot{x}_1 \\ \dot{x}_2 \\ \dot{x}_3 \end{bmatrix}$$

$$+ \begin{bmatrix} k_1 + k_2 & -k_2 & 0 \\ -k_2 & k_2 + k_3 & -k_3 \\ 0 & -k_3 & k_3 \end{bmatrix} \begin{bmatrix} x_1 \\ x_2 \\ x_3 \end{bmatrix} = \begin{bmatrix} 0 \\ 0 \\ F(t) \end{bmatrix} \qquad \text{(j)}$$

**FIG. 2.52** (a) The system of Example 2.19; (b) the free-body diagrams at an arbitrary instant.

## 2.9 ENERGY METHODS

Energy methods provide a convenient alternative to the direct application of Newton's laws in the derivation of differential equations modeling the motion of a mechanical system and are usually preferred for the derivation of equations governing the motion of multidegree-of-freedom systems. The application of energy methods does not require the drawing of free-body diagrams and is often less confusing than the direct application of Newton's laws in that it is not necessary to identify the directions of spring forces to determine the potential energy of the system. Energy methods can be applied to a system of particles and rigid bodies.

### 2.9.1 Principle of Work and Energy

The Principle of Work and Energy is derived by taking the dot product of Newton's second law with a differential displacement vector and both sides of the resulting equation integrated between two points in the path of motion of the system, represented in the final equation by subscripts 1 and 2. The general form of the Principle of Work and Energy is

$$T_1 + W_{1 \to 2} = T_2 \tag{2.113}$$

where $T$ is the kinetic energy of the system and $W_{1 \to 2}$ is the work done by all external forces as the system moves between positions 1 and position 2.

The work done by a conservative force is independent of the path of motion and dependent only on the initial and final positions of the system. Gravity and linear spring forces are examples of conservative forces. Viscous damping, Coulomb damping, and externally applied forces are examples of nonconservative forces. For any conservative force a potential energy function $V$ can be defined so that its work is expressed as a difference in potential energies

$$(W_{1 \to 2})_c = V_1 - V_2 \tag{2.114}$$

The potential energy of a system at a given instant is dependent on the system parameters and the instantaneous position of the system.

Most systems are subject to a combination of conservative and nonconservative forces. In such a case the total work is represented as

$$W_{1 \to 2} = V_1 - V_2 + (W_{1 \to 2})_{NC} \tag{2.115}$$

where $(W_{1 \to 2})_{NC}$ is the work done by all nonconservative forces. Using Equation (2.115), the Principle of Work and Energy, Equation (2.113) can be rewritten as

$$T_1 + V_1 + (W_{1 \to 2})_{NC} = T_2 + V_2 \tag{2.116}$$

The total energy in a system is the sum of the kinetic and potential energies. If the system is conservative (that is, all forces are conservative), then Equation (2.116) shows that the total energy is a constant and the system satisfies the Principle of Conservation of Energy

$$T + V = C \tag{2.117}$$

## 2.9.2 Equivalent Systems

The Principle of Work and Energy is used to develop the equivalent systems method, which can be directly applied to derive the differential equation of a linear one-degree-of-freedom system. Let $x$ be the chosen dependent variable for the system. The kinetic energy of a rigid body in planar motion is given by Equation (2.27). Since the system is linear and has only one degree of freedom, both $v$ and $\omega$ are proportional to $\dot{x}$. Thus the kinetic energy of the rigid body is of the form

$$T = \frac{1}{2} m_{eq} \dot{x}^2 \tag{2.118}$$

where $m_{eq}$ is a constant dependent on the inertia properties of the system. It is shown in Section 2.2 that a combination of springs can be replaced by a single spring of equivalent stiffness $k_{eq}$ such that the potential energy of the combination is

$$V = \frac{1}{2}k_{eq}x^2 \tag{2.119}$$

It is shown in Section 2.3 that a combination of viscous dampers can be replaced by a single viscous damper of damping coefficient $c_{eq}$ so that the work done by the viscous dampers is

$$W = -\int_{x_1}^{x_2} c_{eq}\dot{x}dx \tag{2.120}$$

Define position 1 as the system's initial position and position 2 as the position of the system at an arbitrary instant. Applying the Principle of Work and Energy, Equation (2.116), and using Equations (2.118), (2.119), and (2.120) leads to

$$T_1 + V_1 - \int_{x_{1dx}}^{x_2} c_{eq}\dot{x} = \frac{1}{2}m_{eq}\dot{x}^2 + \frac{1}{2}k_{eq}x^2 \tag{2.121}$$

Differentiating Equation (2.121) with respect to time, noting that $T_1 + V_1$ is a constant gives

$$\frac{1}{2}m_{eq}\frac{d\dot{x}^2}{dt} + \frac{1}{2}k_{eq}\frac{dx^2}{dt} + \frac{d}{dt}\left(\int_{x_1}^{x_2} c_{eq}\dot{x}dx\right) = 0 \tag{2.122}$$

The following algebraic relations are noted:

$$\frac{d\dot{x}^2}{dt} = 2\dot{x}\frac{d\dot{x}}{dt} = 2\dot{x}\ddot{x} \tag{2.123a}$$

$$\frac{dx^2}{dt} = 2x\frac{dx}{dt} = 2x\dot{x} \tag{2.123b}$$

$$\frac{d}{dt}\left(\int_{x_1}^{x} \dot{x}dx\right) = \frac{d}{dt}\left(\int_{t_1}^{t} \dot{x}\frac{dx}{dt}dt\right) = \frac{d}{dt}\left(\int_{t_1}^{t} \dot{x}^2 dt\right) = \dot{x}^2 \tag{2.123c}$$

Using Equations (2.123) in Equation (2.122) leads to

$$\dot{x}\left(m_{eq}\ddot{x} + c_{eq}\dot{x} + k_{eq}x\right) = 0 \tag{2.124}$$

Equation (2.124) implies that either $\dot{x} = 0$ for all $t$ or

$$m_{eq}\ddot{x} + c_{eq}\dot{x} + k_{eq}x = 0 \tag{2.125}$$

The corresponding form of Equation (2.125) for a rotational system is

$$I_{eq}\ddot{\theta} + c_{t,eq}\dot{\theta} + k_{t,eq}\theta = 0 \tag{2.126}$$

where

$$T = \frac{1}{2} I_{eq} \dot{\theta}^2 \tag{2.127}$$

$$V = \frac{1}{2} k_{t,eq} \theta^2 \tag{2.128}$$

$$W_{1 \to 2} = -\int_{\theta_1}^{\theta} c_{t,eq} \dot{\theta} d\theta \tag{2.129}$$

Any linear one-degree-of-freedom system with inertia, stiffness, and/or viscous damping elements can be modeled using either Equation (2.125) or Equation (2.126). The coefficients $m_{eq}$ and $I_{eq}$ can be determined directly from the system's kinetic energy at an arbitrary instant, Equation (2.118) or Equation (2.127). The coefficients $k_{eq}$ and $k_{t,eq}$ can be determined directly from the system's potential energy at an arbitrary instant, Equation (2.119) or Equation (2.128). If the system has viscous damping components, the coefficients $c_{eq}$ and $c_{t,eq}$ can be determined from the work done by the viscous damping forces, Equation (2.120) or Equation (2.129).

If the system is subject to external forces, Equations (2.125) and (2.126) are modified as

$$m_{eq} \ddot{x} + c_{eq} \dot{x} + k_{eq} x = F_{eq}(t) \tag{2.130}$$

$$I_{eq} \ddot{\theta} + c_{t,eq} \dot{\theta} + k_{t,eq} \theta = M_{eq}(t) \tag{2.131}$$

where $F_{eq}(t)$ or $M_{eq}(t)$ is obtained by equating the power delivered by the external forces to the power requirement for an equivalent force.

As is the case when applying Newton's second law to model a mechanical system, the static forces developed in springs and the gravity forces leading to static deflections cancel one another if the system is linear and the dependent variable is measured from the system's equilibrium position. In determining the potential energy developed in a spring, the value of $x$ used in Equation (2.43) is the change in length of the spring measured from the system's equilibrium position. The potential energy due to gravity forces that induce static deflections is not included in the potential energy calculation.

---

## Example 2.20

Use the energy method to derive the differential equation for the system of Figure 2.53 using $x$, measured from the system's equilibrium position, as the dependent variable.

### Solution

Let $\theta$ be the clockwise angular rotation of the pulley and $y$ the displacement of the mass center of the disk, both measured from the system's equilibrium

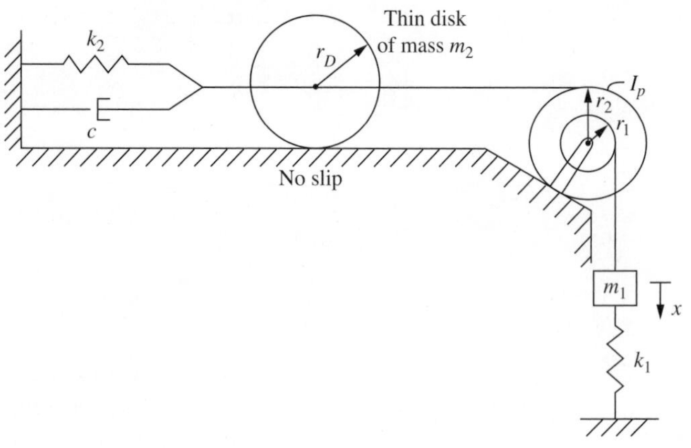

**FIG. 2.53** The system of Example 2.20 illustrating use of the energy method to derive the equation of motion for this one-degree-of-freedom system.

position. These are related to $x$ by

$$\theta = \frac{x}{r_1} \tag{a}$$

$$y = \frac{r_2}{r_1}x \tag{b}$$

The total kinetic energy of the system is the sum of the kinetic energy in each of the bodies

$$T = \tfrac{1}{2}m_1\dot{x}^2 + \tfrac{1}{2}I_p\dot{\theta}^2 + \tfrac{1}{2}m_2\dot{y}^2 + \tfrac{1}{2}(\tfrac{1}{2}m_2r_D^2)\omega^2 \tag{c}$$

where the moment of inertia of the disk is obtained using Table 2.1, $r_D$ is radius of the disk, and $\omega$ is its angular velocity. Since the disk rolls without slip

$$\omega = \frac{\dot{y}}{r_D} \tag{d}$$

Substituting Equations (a), (b), and (d) into Equation (c) leads to

$$T = \frac{1}{2}\left( m_1 + \frac{I_p}{r_1^2} + \frac{3}{2}\frac{r_2^2}{r_1^2}m_2 \right)\dot{x}^2 \tag{e}$$

Comparison of Equation (e) with Equation (2.118) gives

$$m_{eq} = m_1 + \frac{I_p}{r_1^2} + \frac{3}{2}\left(\frac{r_2}{r_1}\right)^2 m_2 \tag{f}$$

The total potential energy of the system is

$$V = \frac{1}{2}k_1x^2 + \frac{1}{2}k_2y^2 = \frac{1}{2}\left( k_1 + \frac{r_2^2}{r_1^2}k_2 \right)x^2 \tag{g}$$

Comparison of Equation (g) with Equation (2.119) leads to

$$k_{eq} = k_1 + \left(\frac{r_2}{r_1}\right)^2 k_2 \tag{h}$$

The work done by the viscous damper as the system moves between its initial position and an arbitrary position is

$$W_{1 \to 2} = -\int_{y_1}^{y} c\dot{y}\,dy = -\int_{x_1}^{x} c\left(\frac{r_2}{r_1}\dot{x}\right) d\left(\frac{r_2}{r_1}x\right) = -\int_{x_1}^{x} c\left(\frac{r_2}{r_1}\right)^2 \dot{x}\,dx \qquad \text{(i)}$$

Comparison of Equation (i) with Equation (2.120) leads to

$$c_{eq} = \left(\frac{r_2}{r_1}\right)^2 c \qquad \text{(j)}$$

The differential equation is obtained by using Equations (f), (h), and (j) in Equation (2.130), leading to

$$\left[m_1 + \frac{I_p}{r_1^2} + \frac{3}{2}\left(\frac{r_2}{r_1}\right)^2 m_2\right]\ddot{x} + \left(\frac{r_2}{r_1}\right)^2 c\dot{x} + \left[k_1 + \left(\frac{r_1}{r_2}\right)^2 k_2\right]x = 0 \qquad \text{(k)}$$

---

## Example 2.21

A gear reduction system is used to increase the angular velocity from an input motor as illustrated in Figure 2.54. The angular velocity of the output rotor is $\omega$. The rotors are mounted on identical shafts, each of which has a torsional viscous damping coefficient $c_t$. The input shaft is subject to a torque $T$. Derive a mathematical model for the system.

### Solution

Let $\omega_1$ be the angular velocity of the input shaft. The meshing equation for the gears is

$$n_1\omega_1 = n_2\omega \qquad \text{(a)}$$

$$\omega_1 = \frac{n_2}{n_1}\omega \qquad \text{(b)}$$

The total kinetic energy of the system is

$$T = \tfrac{1}{2}J_{r1}\omega_1^2 + \tfrac{1}{2}J_{G1}\omega_1^2 + \tfrac{1}{2}J_{r2}\omega^2 + \tfrac{1}{2}J_{G2}\omega^2 \qquad \text{(c)}$$

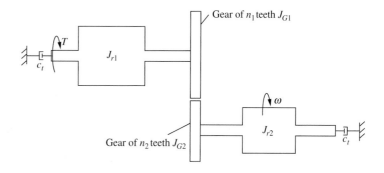

**FIG. 2.54** The system of Example 2.21. The energy method is used to derive a mathematical model of the gear reduction system.

Using Equation (b) in Equation (c) leads to

$$T = \frac{1}{2}\left[(J_{r1} + J_{G1})\left(\frac{n_2}{n_1}\right)^2 + J_{r2} + J_{G2}\right]\omega^2 \tag{d}$$

Comparison of Equation (d) with Equation (2.127) leads to

$$I_{eq} = (J_{r1} + J_{G1})\left(\frac{n_2}{n_1}\right)^2 + J_{r2} + J_{G2} \tag{e}$$

The work done by the torsional viscous dampers is

$$W_{1 \to 2} = -\int_{\theta_{1,0}}^{\theta_1} c_t \dot{\theta}_1 d\theta_1 - \int_{\theta_0}^{\theta} c_t \dot{\theta} d\theta = -\int_{\theta_0}^{\theta} c_t\left(\frac{n_2}{n_1}\dot{\theta}\right)d\left(\frac{n_2}{n_1}\theta\right) - \int_{\theta_0}^{\theta} c_t \dot{\theta} d\theta$$

$$W_{1 \to 2} = -\int_{\theta_0}^{\theta}\left[1 + \left(\frac{n_2}{n_1}\right)^2\right] c_t \dot{\theta} d\theta \tag{f}$$

Comparison of Equation (f) to Equation (2.129) leads to

$$c_{t,eq} = \left[1 + \left(\frac{n_2}{n_1}\right)^2\right] c_t \tag{g}$$

The torque is applied to the input shaft. An equivalent torque applied to the output shaft is obtained by equating the power developed by the torque to the power developed by a torque if applied to the output shaft

$$T\omega_1 = T_{eq}\omega$$

$$T_{eq} = \frac{\omega_1}{\omega} T$$

$$= \frac{n_2}{n_1} T \tag{h}$$

The differential equation modeling the system is obtained by substituting Equations (e), (g), and (h) into Equation (2.131), resulting in

$$\left[(J_{r1} + J_{G1})\left(\frac{n_2}{n_1}\right)^2 + J_{r2} + J_{G2}\right]\ddot{\theta} + \left[1 + \left(\frac{n_2}{n_1}\right)^2\right] c_t \dot{\theta} = \left(\frac{n_2}{n_1}\right) T \tag{i}$$

## 2.9.3 Energy Storage

Equation (2.117) shows that in the absence of nonconservative forces the total energy (the sum of the kinetic energy and the potential energy) remains constant. A decrease in kinetic energy is compensated by an increase in potential energy and vice versa. The term "energy" in this context refers to stored energy or energy that can be converted from kinetic to potential or vice versa.

The kinetic energy of a system consisting of a single particle of mass $m$ moving with a velocity $\dot{x}$ is $T = \frac{1}{2}m\dot{x}^2$. The initial velocity of the particle is determined by factors external to the system. However the mass is constant. In this context the mass can be viewed as a measure of the system's capacity for the storage of kinetic energy. Similarly $m_{eq}$, a system's equivalent mass, is the total capacity of the system to store kinetic energy.

The potential energy of a single spring of change in length $x$ is $V = \frac{1}{2}kx^2$. The initial displacement of the spring is determined by factors external to the system, but the stiffness is constant. The spring stiffness is a measure of its capacity to store potential energy. Similarly $k_{eq}$, a system's equivalent stiffness, is the total capacity of the system to store potential energy.

Equation (2.129) shows that the work done by the viscous damping forces is negative. Thus Equation (2.116) shows that when viscous damping is present the total energy is continually decreasing. A viscous damper does not store energy; it dissipates stored energy. The damping coefficient is a measure of the viscous damper's capacity to dissipate energy.

### 2.9.4 Lagrange's Equations for Multidegree-of-Freedom Systems

The energy method can be extended to multidegree-of-freedom systems by using a set of equations called Lagrange's equations. The development of Lagrange's equations and their use for nonconservative systems is beyond the scope of this book. Lagrange's equations can be used for nonlinear as well as linear systems. However, only their use for linear systems is considered in this book.

Consider an $n$-degree-of-freedom system with dependent variables $x_1, x_2, \ldots, x_n$ chosen for modeling of the system. The kinetic energy of a linear system at an arbitrary time can be developed in the form of

$$T = T(\dot{x}_1, \dot{x}_2, \ldots, \dot{x}_n) \tag{2.132}$$

The potential energy of a linear system at an arbitrary instant can be developed in the form of

$$V = V(x_1, x_2, \ldots, x_n) \tag{2.133}$$

The lagrangian $L$ is defined as the difference between the kinetic and potential energies

$$L = T - V \tag{2.134}$$

Thus the lagrangian is a function of the dependent variables and their first time derivatives.

Lagrange's equations for a conservative linear system are

$$\frac{d}{dt}\left(\frac{\partial L}{\partial \dot{x}_i}\right) - \frac{\partial L}{\partial x_i} = 0 \qquad i = 1, 2, \ldots, n \tag{2.135}$$

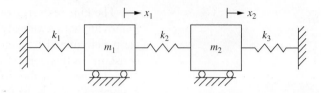

**FIG. 2.55** The two-degree-of-freedom system of Example 2.22.

The application of Lagrange's equations for $i = 1, 2, \ldots, n$ leads to $n$ equations that when simplified become the differential equations governing the motion of the $n$-degree-of-freedom system.

## Example 2.22

Use Lagrange's equations to derive the differential equations governing the motion of the two-degree-of-freedom system of Figure 2.55.

### Solution

The kinetic energy of the system at an arbitrary instant is

$$T = \tfrac{1}{2}m_1\dot{x}_1^2 + \tfrac{1}{2}m_2\dot{x}_2^2 \tag{a}$$

The potential energy of the system at an arbitrary instant is

$$V = \tfrac{1}{2}k_1x_1^2 + \tfrac{1}{2}k_2(x_2 - x_1)^2 + \tfrac{1}{2}k_3x_3^2 \tag{b}$$

Equations (a) and (b) are used in Equation (2.135) to determine the lagrangian

$$L = \tfrac{1}{2}m_1\dot{x}_1^2 + \tfrac{1}{2}m_2\dot{x}_2^2 - \left(\tfrac{1}{2}k_1x_1^2 + \tfrac{1}{2}k_2(x_2 - x_1)^2 + \tfrac{1}{2}k_3x_3^2\right) \tag{c}$$

Using Equation (c) in Lagrange's equations, Equation (2.135) leads to

$i = 1$:

$$\frac{d}{dt}\left(\frac{\partial L}{\partial \dot{x}_1}\right) - \frac{\partial L}{\partial x_1} = 0 \tag{d}$$

$$\frac{d}{dt}(m_1\dot{x}_1) - [-k_1x_1 - k_2(x_2 - x_1)(-1)] = 0 \tag{e}$$

$$m_1\ddot{x}_1 + (k_1 + k_2)x_1 - k_2x_2 = 0 \tag{f}$$

$i = 2$:

$$\frac{d}{dt}\left(\frac{\partial L}{\partial \dot{x}_2}\right) - \frac{\partial L}{\partial x_2} = 0 \tag{g}$$

$$\frac{d}{dt}(m_2\ddot{x}_2) - [-k_2(x_2 - x_1) - k_3x_2] = 0 \tag{h}$$

$$m_2\ddot{x}_2 - k_2x_1 + (k_2 + k_3)x_2 = 0 \tag{i}$$

## 2.9.5 States and Order

The lagrangian, defined by Equation (2.134), is viewed as a function of $2n$ independent variables, $n$ displacements, and $n$ time derivatives of displacements. Each variable is explicitly present in the lagrangian due to independent energy storage components. The displacements arise due to

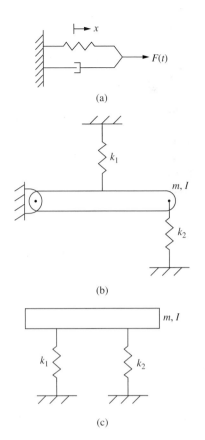

**FIG. 2.56** The systems of Example 2.23: (a) a first-order system; (b) a second-order system; (c) a fourth-order system.

potential energy stored in springs, and their time derivatives are present due to kinetic energy stored in the inertia elements.

One method of formulation and analysis, used in Chapter 9, is called state-space analysis. There are several methods for defining the terms "state" and "state variables." The clearest physical definition, which can be used for all the systems discussed, is made using energy methods. Simply put, a state is associated with each independent energy storage component. State variables can be identified from the energy storage elements, displacements, and their time derivatives. There are four states in Example 2.22. Two states are due to the two masses, which can store kinetic energy independently. Even though there are three springs, only two are necessary for independent energy storage. This is seen from the potential energy formulation, Equation (b) of Example 2.22, in which only two independent variables are necessary to determine the potential energy.

For a mechanical system it is possible to determine the states and the state variables from the system's lagrangian. However, the definition of state variables is not unique and the use of the lagrangian is only one method of identifying states and state variables.

Another term that can be defined using an energy definition is the "order" of a system. A definition in Chapter 1 defined order in terms of the highest order derivative appearing in the governing differential equation. A definition in Chapter 6 defines order in terms of the order of the polynomial in the denominator of the system's transfer function. These are both consistent with the definition of order as the total number of states in the system (or the total number of independent energy storage components).

The use of energy to define states leads to individual situations that require more discussion than is given here. However, it does allow a consistent definition of state for every physical system considered.

**Example 2.23**

Determine the states and the order of the system for each of the systems in Figure 2.56.

**Solution**

(a) The spring is the only energy storage element in the system of Figure 2.56(a). Since there is no inertia element in the system, kinetic energy cannot be stored. This system has one state and is of the first order.

(b) The bar in the system of Figure 2.56(b) stores kinetic energy as it rotates about the pin support. Each of the springs stores potential energy, but they are not independent energy storage elements because they can be replaced by a single spring of an equivalent stiffness. This system has two states: one associated with the kinetic energy storage of the bar and one associated with the potential energy storage of the spring. It is of the second order.

(c) Even though there is only one rigid body in the system of Figure 2.56(c) the rotational kinetic energy of the bar is independent of its translational kinetic energy. Thus there are two independent elements for storing kinetic energy. Since the springs cannot be replaced by a single spring of equivalent stiffness, they are independent potential energy storage elements. Thus this system has four states: two associated with kinetic energy storage and two associated with potential energy storage. The system is of the fourth order.

If the bar of Figure 2.56(c) is attached to three springs, the system still has only two independent potential energy storage elements. The system's lagrangian is a function of only two independent displacements and their time derivatives. The three springs could be replaced by a pair of springs of equivalents stiffnesses so that the potential energy function is the same.

## 2.10 FURTHER EXAMPLES

The examples in this section illustrate how the principles developed in previous sections are synthesized into the development of a mathematical model for a mechanical system. In addition to the implicit assumptions listed at the beginning of this chapter, the following assumptions are used throughout the examples:

- All springs are linear and massless.
- All viscous dampers are massless.
- Viscous damping is the only form of friction.
- Small displacements are assumed where appropriate to linearize the differential equations.
- All dependent variables are measured from the system's equilibrium position.
- Rigid bodies are undergoing planar motion.

Since all systems are linear and all coordinates are measured from the system's equilibrium position, the static forces developed in springs cancel the gravity forces that cause the static deflections. Thus neither are included in the free-body diagrams or the expressions of potential energy.

Two methods have been developed for the mathematical modeling of mechanical systems. The free-body diagram method requires the application of Newton's second law to a free-body diagram, drawn at an arbitrary instant, illustrating the external forces acting on the system. Problems involving rotational motion also require the application of a moment equation. For rigid bodies undergoing planar motion, an alternate formulation using D'Alembert's principle is developed in which the system of external forces is equivalent to the system of effective forces, defined as a force of $m\bar{\mathbf{a}}$ applied at the mass center and a couple of $\bar{I}\alpha$.

The energy method is based on determining the equivalent mass, stiffness, and viscous damping coefficient for an equivalent mass spring–viscous damper model. The equivalent mass and stiffness are determined from expressions for the kinetic and potential energies written at an arbitrary instant. The equivalent viscous damping coefficient is obtained from an expression for the work done by viscous damping forces.

## Example 2.24

Consider the system of Figure 2.57(a). The thin disk of mass $m_1$ and radius $r$ is attached to a shaft of torsional stiffness $k_t$. (a) Determine a mathematical model for the system using $\theta$ as the dependent variable. (b) Determine the system's natural frequency and damping ratio using the given values.

### Solution

(a) Any of the methods can be used to model the system. The polar moment of inertia of the disk obtained from Table 2.1 is $J = (1/2)m_1 r^2$. Assuming no slip between the disk and the cable, the velocity of a particle on the circumference of the disk is $r\dot{\theta}$ and the displacement of the suspended mass is $y = r\theta$.

*Direct application of Newton's second law:* Free-body diagrams of the disk and the suspended mass are shown in Figure 2.57(b). Applying Newton's law to

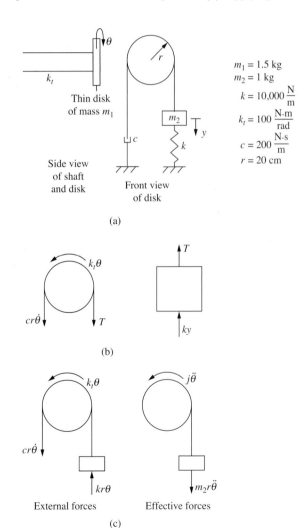

$m_1 = 1.5$ kg
$m_2 = 1$ kg
$k = 10{,}000 \dfrac{\text{N}}{\text{m}}$
$k_t = 100 \dfrac{\text{N·m}}{\text{rad}}$
$c = 200 \dfrac{\text{N·s}}{\text{m}}$
$r = 20$ cm

**FIG. 2.57** (a) The system of Example 2.24. The thin disk is attached to the end of a shaft. A mass is suspended from the disk, and a viscous damper is attached to the outer circumference of the disk. (b) the free-body diagrams of the disk and the suspended mass at an arbitrary instant; (c) the free-body diagrams of disk and suspended mass at an arbitraty instant showing external and effective forces.

the suspended mass using positive forces downward

$$\sum F = m_2\ddot{y} \tag{a}$$

$$-T - ky = m_2\ddot{y} \tag{b}$$

$$T = -m_2\ddot{y} - ky \tag{c}$$

Since the disk is undergoing pure rotational motion, Equation (2.102) is used with positive moments clockwise

$$\sum M_G = J\ddot{\theta} \tag{d}$$

$$-k_t\theta - \left(cr\dot{\theta}\right)r + Tr = \frac{1}{2}m_1 r^2\ddot{\theta} \tag{e}$$

Substituting Equation (c) into Equation (e) and noting that $y = r\theta$ leads to

$$-k_t\theta - cr^2\dot{\theta} - m_2 r^2\ddot{\theta} - kr^2\theta = \frac{1}{2}m_1 r^2\ddot{\theta} \tag{f}$$

$$\left(\frac{1}{2}m_1 r^2 + m_2 r^2\right)\ddot{\theta} + cr^2\dot{\theta} + (k_t + kr^2)\theta = 0 \tag{g}$$

*D'Alembert's principle*: Free-body diagrams showing the external forces and the effective forces acting on the system composed of the disk and the suspended mass are shown in Figure 2.57(c). Application of D'Alembert's principle in the form of Equation (2.106) leads to

$$\left(\sum M_G\right)_{ext} = \left(\sum M_G\right)_{eff} \tag{h}$$

$$-k_t\theta - \left(cr\dot{\theta}\right)r - (kr\theta)r = \tfrac{1}{2}m_1 r^2\ddot{\theta} + \left(m_2 r\ddot{\theta}\right)r \tag{i}$$

$$\left(\tfrac{1}{2}m_1 r^2 + m_2 r^2\right)\ddot{\theta} + cr^2\dot{\theta} + \left(k_t + kr^2\right)\theta = 0 \tag{j}$$

*Energy method*: The kinetic energy of the system at an arbitrary instant is

$$T = \tfrac{1}{2}J\dot{\theta}^2 + \tfrac{1}{2}m_2\left(r\dot{\theta}\right)^2 = \tfrac{1}{2}(\tfrac{1}{2}m_1 r^2 + m_2 r^2)\dot{\theta}^2 \tag{k}$$

Comparison of Equation (k) with Equation (2.127) leads to

$$I_{eq} = \tfrac{1}{2}m_1 r^2 + m_2 r^2 \tag{l}$$

The potential energy of the system at an arbitrary instant is

$$V = \tfrac{1}{2}k_t\theta^2 + \tfrac{1}{2}k(r\theta)^2 = \tfrac{1}{2}\left(k_t + kr^2\right)\theta^2 \tag{m}$$

Comparison of Equation (m) with Equation (2.128) leads to

$$k_{t,eq} = k_t + kr^2 \tag{n}$$

The work done by the viscous damper between two arbitrary positions is

$$W_{1\to 2} = -\int_{\theta_1}^{\theta_2} c\left(r\dot{\theta}\right)d(r\theta) = -\int_{\theta_1}^{\theta_2} cr^2\dot{\theta}d\theta \tag{o}$$

Comparison between Equation (o) and Equation (2.129) leads to

$$c_{t,eq} = cr^2 \tag{p}$$

When Equations (l), (n), and (p) are used in Equation (2.131), Equation (g) is obtained. (b) Equation (g) is put in the standard form of Equation (2.111) by dividing by its leading coefficient, yielding

$$\ddot{\theta} + \frac{c}{\frac{1}{2}m_1 + m_2}\dot{\theta} + \frac{k_t + kr^2}{(\frac{1}{2}m_1 + m_2)r^2}\theta = 0 \tag{q}$$

Comparison of Equation (q) with Equation (2.111) yields the natural frequency

$$\omega_n = \sqrt{\frac{k_t + kr^2}{(\frac{1}{2}m_1 + m_2)r^2}}$$

$$= \sqrt{\frac{100\,\text{N·m/rad} + (10,000\,\text{N/m})(0.2\,\text{m})^2}{[\frac{1}{2}(1.5\,\text{kg}) + 1\,\text{kg}](0.2\,\text{m})^2}} = 84.5\,\text{r/s} \tag{r}$$

and viscous damping ratio

$$\zeta = \frac{c}{2\omega_n(\frac{1}{2}m_1 + m_2)}$$

$$= \frac{200\,\text{N·s/m}}{2(84.5\,\text{r/s})[\frac{1}{2}(1.5\,\text{kg}) + 1\,\text{kg}]} = 0.676 \tag{s}$$

## Example 2.25

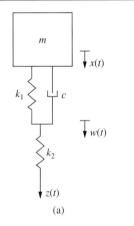

Figure 2.58(a) illustrates a simplified model of a vehicle suspension system. The vehicle of mass $m$ is attached to an element that is modeled as a spring of stiffness $k_1$ in parallel with a viscous damper of damping coefficient $c$. This element is in series with an element that is modeled as a spring of stiffness $k_2$. The other end of this spring is subject to a time-dependent displacement $z(t)$ caused by the terrain over which the vehicle is traveling. Determine a mathematical model for the motion of the vehicle.

### Solution

Let $x(t)$ represent the displacement of the vehicle's body from the system's equilibrium position. Let $w(t)$ represent the displacement of the junction between the two suspension elements. Free-body diagrams of the body and the junction at an arbitrary time are shown in Figures 2.58(b).

Applying Newton's second law to the vehicle body

$$\sum F = m\ddot{x} \tag{a}$$

$$k_1(w - x) + c(\dot{w} - \dot{x}) = m\ddot{x} \tag{b}$$

$$m\ddot{x} + c\dot{x} - c\dot{w} + k_1 x - k_1 w = 0 \tag{c}$$

Application of Newton's second law to the junction, which is massless, leads to

$$k_2(z - w) - k_1(w - x) - c(\dot{w} - \dot{x}) = 0 \tag{d}$$

$$-c\dot{x} + c\dot{w} - k_1 x + (k_1 + k_2)w = k_2 z \tag{e}$$

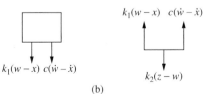

**FIG. 2.58** (a) The system of Example 2.25; (b) the free-body diagrams at an arbitrary instant.

Equations (c) and (e) provide a mathematical model for the system and can be summarized in matrix form as

$$\begin{bmatrix} m & 0 \\ 0 & 0 \end{bmatrix}\begin{bmatrix} \ddot{x} \\ \ddot{w} \end{bmatrix} + \begin{bmatrix} c & -c \\ -c & c \end{bmatrix}\begin{bmatrix} \dot{x} \\ \dot{w} \end{bmatrix} + \begin{bmatrix} k_1 & -k_1 \\ -k_1 & k_1 + k_2 \end{bmatrix}\begin{bmatrix} x \\ w \end{bmatrix} = \begin{bmatrix} 0 \\ k_2 z \end{bmatrix} \tag{f}$$

A single equation to solve for $x(t)$ can be obtained by eliminating $w(t)$ using Equations (b) and (d). The result is

$$\frac{cm}{k_2}\dddot{x} + m\left(1 + \frac{k_1}{k_2}\right)\ddot{x} + c\dot{x} + k_1 x = c\dot{z} + k_1 z \tag{g}$$

## Example 2.26

A centrifuge, illustrated schematically in Figure 2.59, of total mass $m$ operates at a constant rotational speed $\omega$ in rad/s. The centrifuge is mounted on a foundation that is modeled as a spring of stiffness $k$ in parallel with a viscous damper of damping coefficient $c$. The centrifuge has a rotational mass $m_0$ whose center of gravity is a distance $e$ from the axis of rotation. Determine a mathematical model for the motion of the centrifuge.

### Solution

Let $x(t)$ represent the displacement of the centrifuge from its equilibrium position. The matter in the centrifuge has a motion relative to the centrifuge. The acceleration of the matter is determined using the relative acceleration

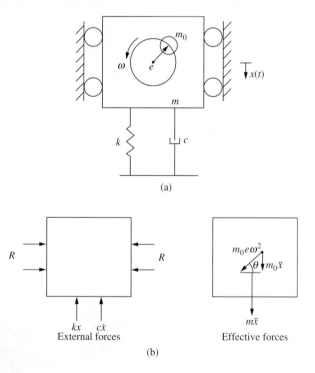

**FIG. 2.59** (a) The centrifuge with a rotating mass $m_0$ and a distance $e$ from the axis of rotation; (b) the free-body diagrams of the centrifuge, and the rotating mass at an arbitrary instant.

equation, Equation (2.16), as

$$\mathbf{a_m} = -e\omega^2 \cos\theta\mathbf{i} - (\ddot{y} + e\omega^2 \sin\theta)\mathbf{j} \tag{a}$$

where $\theta$ is the angle between a radius to the matter and the horizontal. By definition of the angular velocity

$$\omega = \frac{d\theta}{dt} \tag{b}$$

Since $\omega$ is constant and assuming $\theta = 0$ when $t = 0$, Equation (b) is integrated as

$$\theta = \omega t \tag{c}$$

Free-body diagrams of the centrifuge at an arbitrary instant are given in Figure 2.59(b). The external forces acting on the centrifuge are the spring force and viscous damping force. The effective forces include the inertia force of the machine and the inertia force of the matter. Application of D'Alembert's principle in the form of

$$\sum F_{ext} = \sum F_{eff} \tag{d}$$

in the vertical direction gives

$$-ky - c\dot{y} = m\ddot{y} + m_0\ddot{y} + m_0 e\omega^2 \sin(\omega t) \tag{e}$$

Equation (e) is rearranged as

$$(m + m_0)\ddot{y} + c\dot{y} + ky = m_0 e\omega^2 \sin(\omega t) \tag{f}$$

The unbalanced rotating mass in the centrifuge causes its motion induced by a single-frequency harmonic excitation.

## Example 2.27

Determine the natural frequency for the system of Figure 2.60. The identical thin disks of mass $m$ and radius $R$ are free to rotate and are constrained to move vertically.

### Solution

The mass moment of inertia of the disks is obtained from Table 2.1 as

$$\bar{I} = \tfrac{1}{2}mR^2 \tag{a}$$

Let $x$ be the displacement of the disk centered at $F$, $y$ the displacement of the disk centered at $E$, $z$ the displacement of the suspended mass, $\theta_1$ the counterclockwise angular rotation of the disk centered at $F$, and $\theta_2$ the counterclockwise angular rotation of the disk centered at $E$. All variables are measured from the system's equilibrium position. During the motion, the lengths of $AD$ and $EI$ remain constant. Segment $HI$ of cable $EI$ decreases in length by $y$, while segment $GE$ increases in length by $y - x$. Setting the total change in length of cable $EI$ to zero leads to

$$(y - x) - x = 0 \tag{b}$$
$$y = 2x \tag{c}$$

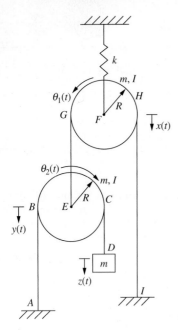

**FIG. 2.60** The system of Example 2.27.

Segment $AB$ of cable $AD$ decreases in length by $y$, while segment $CD$ increases in length by $z - y$. Setting the total change in length of cable $AD$ to zero leads to

$$(z - y) - y = 0 \tag{d}$$

$$z = 2y \tag{e}$$

Substitution of Equation (c) into Equation (e) leads to

$$z = 4x \tag{f}$$

Assuming no slip between the disks and the cables, the velocity of a point on the circumference of a disk must equal the velocity of the cable moving around the disk. This requirement leads to

$$R\dot{\theta}_1 = \dot{x} \tag{g}$$

$$R\dot{\theta}_2 = \dot{y} = 2\dot{x} \tag{h}$$

The differential equation is derived using the energy method. The kinetic energy of the system at an arbitrary instant is

$$T = \tfrac{1}{2}m\dot{x}^2 + \tfrac{1}{2}\bar{I}\dot{\theta}_1^2 + \tfrac{1}{2}m\dot{y}^2 + \tfrac{1}{2}\bar{I}\dot{\theta}_2^2 + \tfrac{1}{2}m\dot{z}^2 \tag{i}$$

Substituting Equations (a), (c), (f), (g), and (h) into Equation (i) leads to

$$T = \frac{1}{2}m\dot{x}^2 + \frac{1}{2}\left(\frac{1}{2}mR^2\right)\left(\frac{\dot{x}}{R}\right)^2 + \frac{1}{2}m(2\dot{x})^2 + \frac{1}{2}\left(\frac{1}{2}mR^2\right)\left(\frac{2\dot{x}}{R}\right)^2 + \frac{1}{2}m(4\dot{x})^2 \tag{j}$$

$$T = \frac{1}{2}\left(\frac{47}{2}m\right)\dot{x}^2 \tag{k}$$

Comparison of Equation (k) with Equation (2.118) leads to

$$m_{eq} = \frac{47}{2}m \tag{l}$$

The potential energy of the system at an arbitrary instant is

$$V = \frac{1}{2}kx^2 \tag{m}$$

Comparison of Equation (m) with Equation (2.119) leads to

$$k_{eq} = k \tag{n}$$

The differential equation modeling the system is obtained by substituting Equations (l) and (n) into Equation (2.130) leading to

$$\frac{47}{2}m\ddot{x} + kx = 0 \tag{o}$$

Equation (o) is put in the standard form of Equation (2.111) by dividing by $47m/2$, leading to

$$\ddot{x} + \frac{2k}{47m}x = 0 \tag{p}$$

The system's natural frequency is obtained from Equation (p) as

$$\omega_n = \sqrt{\frac{2k}{47m}} \tag{q}$$

## Example 2.28

A simplified two-degree-of-freedom model of a vehicle suspension system is shown in Figure 2.61. Let $x$ be the displacement of $G$, the mass center of the vehicle, and $\theta$, the clockwise angular displacement of the vehicle, both measured from the system's equilibrium position. The vehicle has a mass $m$ and centroidal moment of inertia $\bar{I}$. The front and rear wheels of the vehicle have vertical displacements of $y_1(t)$ and $y_2(t)$ respectively. Determine a mathematical model for the motion of the vehicle. Write the differential equations in matrix form.

### Solution

The free-body diagram of the system at an arbitrary instant is shown in Figure 2.61(b). Summation of forces gives

$$\sum F = m\ddot{x} \tag{a}$$

$$k_1[y_1 - (x - \ell_1\theta)] + c_1\left[\dot{y}_1 - (\dot{x} - \ell_1\dot{\theta})\right] + k_2[y_2 - (x + \ell_2\theta)]$$
$$+ c_2\left[\dot{y}_2 - (\dot{x} + \ell_2\dot{\theta})\right] = m\ddot{x} \tag{b}$$

$$m\ddot{x} + (c_1 + c_2)\dot{x} + (c_2\ell_2 - c_1\ell_1)\dot{\theta} + (k_1 + k_2)x + (k_2\ell_2 - k_1\ell_1)\theta =$$
$$k_1y_1 + k_2y_2 + c_1\dot{y}_1 + c_2\dot{y}_2 \tag{c}$$

Summing moments about the mass center gives

$$\sum M_G = \bar{I}\ddot{\theta} \tag{d}$$

$$-k_1[y_1 - (x - \ell_1\theta)]\ell_1 - c_1[\dot{y}_1 - (\dot{x} - \ell_1\dot{\theta})]\ell_1 + k_2[y_2 - (x + \ell_2\theta)]\ell_2$$
$$+ c_2[\dot{y}_2 - (\dot{x} + \ell_2\theta)]\ell_2 = \bar{I}\ddot{\theta} \tag{e}$$

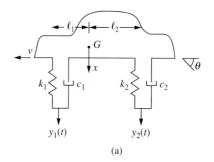

(a)

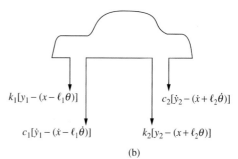

(b)

**FIG. 2.61** (a) The two-degree-of-freedom model of a vehicle suspension system; (b) the free-body diagrams at an arbitrary instant.

$$\bar{I}\ddot{\theta} + (c_2\ell_2 - c_1\ell_1)\dot{x} + (c_1\ell_1^2 + c_2\ell_2^2)\dot{\theta} + (k_2\ell_2 - k_1\ell_1)x + (k_2\ell_2^2 + k_1\ell_1^2)\theta$$
$$= -k_1\ell_1y_1 + k_2\ell_2y_2 - c_1\ell_1\dot{y}_1 + c_2\ell_2\dot{y}_2 \quad (f)$$

Equations (c) and (f) are summarized in matrix form as

$$\begin{bmatrix} m & 0 \\ 0 & \bar{I} \end{bmatrix}\begin{bmatrix} \ddot{x} \\ \ddot{\theta} \end{bmatrix} + \begin{bmatrix} c_1 + c_2 & c_2\ell_2 - c_1\ell_1 \\ c_2\ell_2 - c_1\ell_1 & c_1\ell_1^2 + c_2\ell_2^2 \end{bmatrix}\begin{bmatrix} \dot{x} \\ \dot{\theta} \end{bmatrix}$$
$$+ \begin{bmatrix} k_1 + k_2 & k_2\ell_2 - k_1\ell_1 \\ k_2\ell_2 - k_1\ell_1 & k_1\ell_1^2 + k_2\ell_2^2 \end{bmatrix}\begin{bmatrix} x \\ \theta \end{bmatrix} = \begin{bmatrix} k_1y_1 + k_2y_2 + c_1\dot{y}_1 + c_2\dot{y}_2 \\ -k_1\ell_1y_1 + k_2\ell_2y_2 - c_1\ell_1\dot{y}_1 + c_2\ell_2\dot{y}_2 \end{bmatrix} \quad (g)$$

## Example 2.29

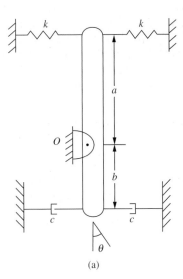

$ka\theta \longrightarrow \qquad \longrightarrow ka\theta$

$mg$

$R$

$cb\dot{\theta} \longleftarrow \qquad \longleftarrow cb\dot{\theta}$

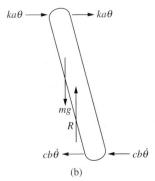

(b)

**FIG. 2.62** (a) The system of Example 2.29; (b) the free-body diagram at an arbitray instant.

Determine a mathematical model for the system of Figure 2.62. The springs are unstretched when the slender bar is in the vertical position. Assume $a > b$.

### Solution

A free-body diagram of the system at an arbitrary instant is shown in Figure 2.62(b). Consistent with the small angle assumption, the spring forces and viscous damping forces are assumed to remain horizontal. The applicable conservation law is Equation (2.103)

$$\sum M_O = I_O\ddot{\theta} \quad (a)$$

The centroidal moment of inertia of the slender bar is obtained from Table 2.1 as

$$\bar{I} = \frac{1}{12}m(a + b)^2 \quad (b)$$

The Parallel Axis Theorem is used to determine $I_O$, noting that the distance between the mass center and $O$ is $(a - b)/2$,

$$I_O = \bar{I} + m\left(\frac{a - b}{2}\right)^2 \quad (c)$$

$$I_O = \frac{1}{3}m(a^2 - ab + b^2) \quad (d)$$

The application of Equation (a) to the free-body diagram of Figure 2.62(b) using Equation (d) leads to

$$-2ka\theta(a) - 2cb\dot{\theta}(b) + mg\left(\frac{a - b}{2}\right)\sin\theta = \frac{1}{3}m(a^2 - ab + b^2)\ddot{\theta} \quad (e)$$

Using the small angle approximation, $\sin\theta \approx \theta$ in Equation (e) leads to

$$\frac{1}{3}m(a^2 - ab + b^2)\ddot{\theta} + 2cb^2\dot{\theta} + \left[2ka^2 - mg\left(\frac{a - b}{2}\right)\right]\theta = 0 \quad (f)$$

**Example 2.30**

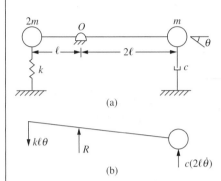

(a)

(b)

**FIG. 2.63** (a) The system of Example 2.30; (b) the free-body diagram of system at arbitrary instant.

Derive the mathematical model for the system of Figure 2.63(a) using $\theta$, the clockwise angular rotation of the bar as the dependent variable. The bar is massless with a particle attached at each end.

**Solution**

The Parallel Axis Theorem is used to determine the mass moment of inertia of the system about $O$. The moment of inertia of a particle about an axis through the particle is zero. Thus

$$I_O = (2m)(\ell)^2 + m(2\ell)^2 = 6m\ell^2 \tag{a}$$

A free-body diagram of the system, drawn at an arbitrary instant, is shown in Figure 2.63(b). The forces are labeled assuming small $\theta$. Since $O$ is a fixed axis of rotation, the appropriate conservation law is $\sum M_O = I_O\ddot{\theta}$. Assuming moments are taken as the positive clockwise application of the moment equation leads to

$$-(k\ell\theta)(\ell) - (2c\ell\dot{\theta})(2\ell) = 6m\ell^2\ddot{\theta}$$
$$6m\ddot{\theta} + 4c\dot{\theta} + k\theta = 0 \tag{b}$$

## 2.11 SUMMARY

### 2.11.1 Modeling Methods

This chapter focuses on the mathematical modeling of mechanical systems. Three methods have been developed.

- The direct application of Newton's second law to free-body diagrams of system components is the basic method. Appropriate forms of Newton's second law can be written for particles and rigid bodies, and the law can be specialized for rigid bodies undergoing planar motion. The method is fundamental and easy to remember, and it can always be applied. Its application requires drawing correct free-body diagrams of system components.
- D'Alembert's principle, developed from Newton's second law, states that for a rigid body in planar motion the system of external forces acting on a system component is equivalent to the system of effective forces for that component. Its application requires drawing two free-body diagrams at an arbitrary instant, one showing the external forces and the other showing the system's effective forces. There is no advantage to the use of D'Alembert's principle for the pure translation of a

single particle or rotation about a fixed axis. Use of D'Alembert's principle for planar motion of rigid bodies often eliminates the need for intermediate calculations, such as the determination of mass centers and the calculation of intermediate forces, necessary when applying Newton's second law.

- The energy method or the equivalent systems method is based on the model equation derived for a the motion of a linear one-degree-of-freedom system. The coefficients in the model equation are determined using the kinetic and potential energies of the system as well as the work done by viscous damping forces. An advantage of using the energy method is that its application does not require drawing free-body diagrams, and thus mistakes associated with drawing incorrect free-body diagrams (such as drawing forces in the wrong direction) are eliminated.

## 2.11.2 Chapter Highlights

Other important points developed in this chapter are as follows:

- A mechanical system is composed of inertia elements, stiffness elements, and damping elements. Forces and motion provide input to mechanical systems.
- A particle's mass is concentrated at a single point. The mass of a rigid body is distributed about its mass center. The distribution of mass is measured by the moment of inertia. Inertia elements have kinetic energy when in motion.
- Springs are flexible links in mechanical systems that store energy when stretched or compressed. A linear spring has a linear force-displacement relation.
- A combination of springs can be replaced by a single spring of an equivalent stiffness, which can be determined so that the equivalent spring has the same potential energy at any instant as the total potential energy of the combination.
- A viscous damper has a force proportional to the velocity of the particle to which it is attached. A viscous damper is useful because it leads to a linear term in the differential equation obtained through system modeling.
- A free-body diagram (FBD) is a diagram of a body at an arbitrary instant, abstracted from its surroundings and showing the effect of the surroundings in the form of forces. Both surface forces and body forces are illustrated and labeled consistently with chosen dependent variables.
- Newton's second law is the basic law of nature that forms the basis for the mathematical modeling of mechanical systems.

- A moment equation, as well as an appropriate form of Newton's second law, is necessary for the modeling of rigid bodies undergoing planar motion.
- D'Alembert's principle, an alternate formulation of Newton's second law, is useful for modeling systems with rigid bodies undergoing planar motion.
- An energy method, based on the Principle of Work and Energy, a spatially integrated form of Newton's second law, provides another alternate method for the mathematical modeling of mechanical systems.
- The differential equation governing the motion of a one-degree-of-freedom linear system can be written in a standard form in terms of two system parameters: the natural frequency and the damping ratio.
- States of a system correspond to independent energy storage components. The order of the system is equal to the number of states.

### 2.11.3 Important Equations

Chapter 2 contains the following important equations:

- Definition of velocity and acceleration

$$\mathbf{v} = \frac{d\mathbf{r}}{dt} = \frac{dx}{dt}\mathbf{i} + \frac{dy}{dt}\mathbf{j} + \frac{dz}{dt}\mathbf{k} \tag{2.2}$$

$$\mathbf{a} = \dot{\mathbf{v}} = \ddot{\mathbf{r}} \tag{2.4}$$

- Kinetic energy of particle and rigid body

$$\text{particle } T = \tfrac{1}{2}m\mathbf{v}\cdot\mathbf{v} \tag{2.6}$$

$$\text{rigid body } T = \tfrac{1}{2}m\bar{v}^2 + \tfrac{1}{2}\bar{I}\omega^2 \tag{2.27}$$

- Parallel Axis Theorem

$$I_{xx} = \bar{I}_{xx} + md_x^2 \tag{2.24}$$

- Linear force-displacement relation in a spring

$$F = kx \tag{2.40}$$

- Potential energy function for a spring

$$V = \tfrac{1}{2}kx^2 \tag{2.43}$$

- Equivalent stiffness of spring combinations

$$\text{parallel} \quad k_{eq} = \sum_{i=1}^{n} k_i \tag{2.47}$$

$$\text{series} \quad k_{eq} = \frac{1}{\sum\limits_{i=1}^{n} \frac{1}{k_i}} \tag{2.51}$$

- Force-velocity relation for a viscous damper

$$F = cv \tag{2.59}$$

- Newton's second law for a particle

$$\sum \mathbf{F} = m\mathbf{a} \tag{2.96}$$

- Newton's second law for a rigid body undergoing planar motion

$$\sum \mathbf{F} = m\mathbf{a} \tag{2.98}$$

$$\sum M_G = \bar{I}\alpha \tag{2.101}$$

- Moment equation for rotation about a fixed axis

$$\sum M_O = I_O\alpha \tag{2.102}$$

- D'Alembert's principle

$$\left(\sum \mathbf{F}\right)_{ext} = \left(\sum \mathbf{F}\right)_{eff} \tag{2.105}$$

$$\left(\sum M_A\right)_{ext} = \left(\sum M_A\right)_{eff} \tag{2.106}$$

- Standard form of differential equation for one-degree-of-freedom system

$$\ddot{x} + 2\zeta\omega_n\dot{x} + \omega_n^2 x = \frac{1}{\tilde{m}} F(t) \tag{2.111}$$

- Principle of Work and Energy

$$T_1 + V_1 + (W_{1\to 2})_{NC} = T_2 + V_2 \tag{2.116}$$

- Equivalent systems

$$T = \tfrac{1}{2} m_{eq} \dot{x}^2 \tag{2.118}$$

$$V = \tfrac{1}{2} k_{eq} x^2 \tag{2.119}$$

$$W_{1\to 2} = -\int_{x_1}^{x_2} c_{eq}\dot{x}\,dx \tag{2.120}$$

- Lagrange's equations

$$\frac{d}{dt}\left(\frac{\partial L}{\partial \dot{x}_i}\right) - \frac{\partial L}{\partial x_i} = 0 \tag{2.135}$$

# PROBLEMS

**2.1** The annular cylinder of Figure P2.1 is made of steel of mass density 7,600 kg/m³. Determine (a) $I_{xx}$ and (b) $I_{yy}$.

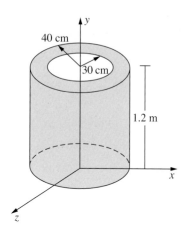

**FIG. P2.1**

**2.2** Determine the moment of inertia of the assembly of Figure P2.2 about (a) an axis perpendicular to the page through $A$ and (b) a horizontal axis at $A$.

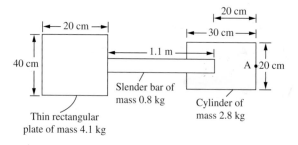

**FIG. P2.2**

**2.3** Determine the moment of inertia of the assembly of Figure P2.3 about an axis through $O$.

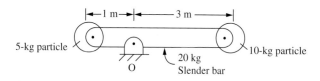

**FIG. P2.3**

**2.4** A 2-kg particle has a time-dependent velocity of

$$\mathbf{v}(t) = -2\sin(3t)\mathbf{i} + 4\cos(3t)\mathbf{j} + 2\mathbf{k}\ \text{m/s} \qquad (a)$$

The particle is at the origin of the coordinate system at $t = 0$. Determine (a) the position vector of the particle, (b) the acceleration of the particle, (c) the kinetic energy of the particle at $t = 1$ s, and (d) the linear momentum of the particle at $t = 1$ s.

**2.5** At the instant shown, the angular velocity of the 6-kg bar of Figure P2.5 is 10 r/s counterclockwise and its angular acceleration is 3 r/s² clockwise. Determine (a) the velocity of the center of the bar at this instant, (b) the acceleration of the center of the bar at this instant, (c) the kinetic energy of the bar at this instant, and (d) the angular momentum of the bar at this instant about an axis through the center of the bar.

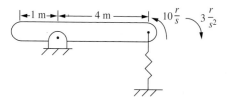

**FIG. P2.5**

**2.6** The fan shown in Figure P2.6 has four 1.2-kg blades of length 60 cm. The center of mass of each blade is 45 cm from the rotational axis. The centroidal moment of inertia of each blade is 0.4 kg·m². Determine the kinetic energy of the fan when it operates at 250 rpm.

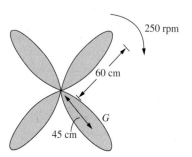

**FIG. P2.6**

**2.7** A schematic representation of a small single-cylinder engine is shown in Figure P2.7. The 0.5-kg crank is of length 10 cm and rotates about its mass center $O$ at a constant speed of 250 rpm clockwise. Its moment of inertia is $I_O = 0.4$ kg·m². The 0.4-kg connecting rod is a slender rod of length 30 cm. A 0.3-kg piston, constrained to move vertically, is attached to the connecting rod. Determine (a) the kinetic energy of the engine (crank, connecting rod, and piston) at the instant shown and (b) the acceleration of the piston at this instant.

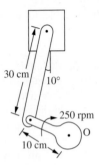

**FIG. P2.7**

**2.8** Determine the kinetic energy of the system of Figure P2.8 in terms of (a) $\dot{x}$, the velocity of the block, and (b) $\dot{\theta}$, the clockwise angular velocity of the disk.

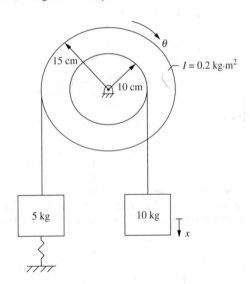

**FIG. P2.8**

**2.9** Determine the kinetic energy of the rotor and gear system of Figure P2.9 when shaft $AB$ rotates at 250 rpm.

**2.10** Determine the equivalent stiffness of the springs in the system of Figure P2.10.

**2.11** Determine the equivalent torsional stiffness of the springs in the system of Figure P2.11.

**2.12** Determine the equivalent stiffness of a spring placed at $A$ for the system of springs in the system of Figure P2.12.

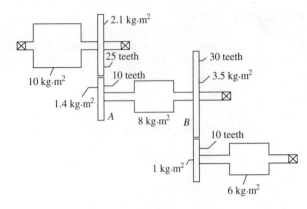

**FIG. P2.9**

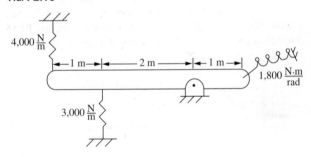

**FIG. P2.10**

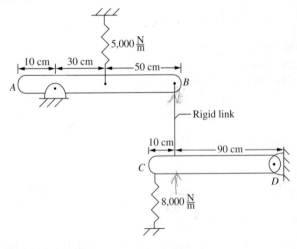

**FIG. P2.11**

**FIG. P2.12**

**2.13** Using a potential energy argument, show that the equivalent stiffness of two springs in series is

$$k_{eq} = \frac{k_1 k_2}{k_1 + k_2} \qquad (a)$$

**2.14** A model of a viscoelastic mechanical component is shown in Figure P2.14. Let $x(t)$ represent the displacement of the end of the component. Determine a relation between force developed in the component and $x$.

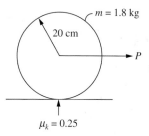

**FIG. P2.14**

**2.15** When a disk rolls on a surface without slip, the velocity of the point of contact of the disk with the surface is zero and the friction force is less than the maximum allowable of $\mu_k N$. The relative velocity equation is used to relate the velocity of the mass center to the angular velocity of the disk. (a) What is the maximum value of $P$ such that the disk of Figure P2.15 rolls without slip? (b) What is the angular acceleration of the disk for this value of $P$?

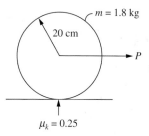

**FIG. P2.15**

**2.16** Determine the damping coefficient of an equivalent viscous damper that can be attached to the block of mass $M_A$ such that the work done by its damping force is equal to the work done by the viscous dampers shown in Figure P2.16.

**2.17** The coefficients of friction between a 45-kg block and the horizontal surface on which it rests are $\mu_s = 0.30$ and $\mu_k = 0.27$. (a) What force acting on the block, applied parallel to the surface, is required to initiate motion of the block? (b) What is the acceleration of the block when this force is applied?

**2.18** Repeat Example 2.6 if the gear reduction ratio is three to one.

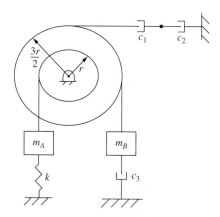

**FIG. P2.16**

**2.19** A 4-kg particle has a velocity of 2.5 m/s when the time-dependent force of Figure P2.19 is applied. (a) What is the velocity of the particle at $t = 0.5$ s? (b) What is the velocity of the particle at $t = 2.5$ s?

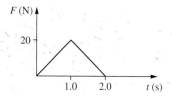

**FIG. P2.19**

**2.20** A cam and follower system is designed so that the motion of a particle in a mechanical system is as illustrated in Figure P2.20. (a) Write a mathematical representation for the displacement of the particle over the first period. (b) Determine the velocity of the particle over the first period. (c) Determine the acceleration of the particle over the first period.

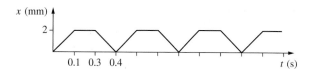

**FIG. P2.20**

**2.21** Derive a mathematical model for the system of Figure P2.21 using $x$ as the dependent variable.

**2.22** A 2,000-kg vehicle is traveling along a steep road at a speed of 60 m/s, as shown in Figure P2.22, when the driver spots a stalled car 300 m from the vehicle. The driver immediately applies the brakes resulting in a constant braking

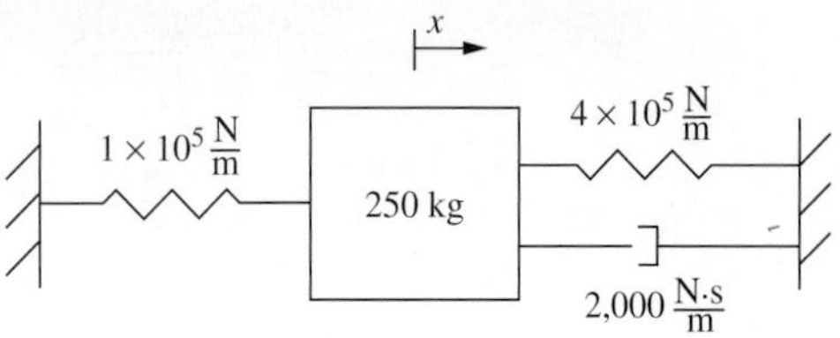

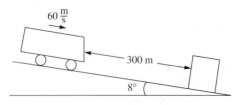

**FIG. P2.21**

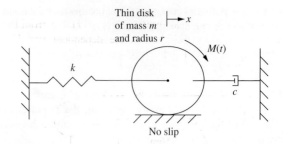

**FIG. P2.26**

**2.27** Derive a mathematical model for the system of Figure P2.27 where $y(t)$, the input to the system, is the prescribed displacement of the base of the system and $x(t)$ is the system output.

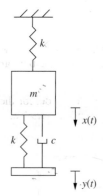

**FIG. P2.22**

force of $F_B$. Let $x(t)$ represent the vehicle's displacement measured from when the driver applies the brakes. (a) Derive a mathematical model for $x(t)$. (b) Use the model to determine the minimum value of $F_B$ such that the vehicle stops before it hits the stalled car.

**2.23** A worker is using a torque wrench to remove a bolt from the flange of a pressure vessel. The worker applies a torque of 500 N·m with the wrench. As it turns, a resisting moment of 480 N·m is applied to the bolt, which has a moment of inertia of 0.035 kg·m². It takes 50 revolutions to remove the bolt. How long will it take the worker to remove the bolt assuming the applied torque and the resisting torque are constant?

**2.24** A 10-kg projectile is fired at a speed of 800 m/s at an angle of 30° with the horizontal. The projectile will land 100 m below the elevation of firing. (a) How long will the projectile be in motion? (b) What is the horizontal range of the projectile? (c) What is the maximum altitude attained by the projectile?

**2.25** A 0.2-kg projectile is fired at a speed of 600 m/s at an angle of 40° with the horizontal. The projectile is fired into a wind, which is estimated to provide a constant horizontal force of 5 N opposing motion of the projectile. Let $x(t)$ represent the horizontal displacement of the projectile and $y(t)$ represent the vertical displacement of the projectile, both measured from the firing location. (a) Derive a mathematical model for $x(t)$ and $y(t)$. (b) What is the maximum altitude attained by the projectile? (c) What is the range of the projectile?

**2.26** Derive a mathematical model for the system of Figure P2.26 using $x$ as the dependent variable.

**FIG. P2.27**

**2.28** Derive a mathematical model for the system of Figure P2.28 using $x$ as the dependent variable.

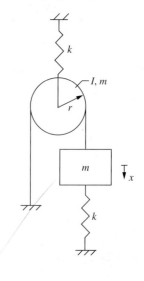

**FIG. P2.28**

**2.29** Derive a mathematical model for the system of Figure P2.29 using $\theta$, the angular displacement of the bar from the system's equilibrium position as the dependent variable. Assume the rod is rigid and massless and assume small $\theta$.

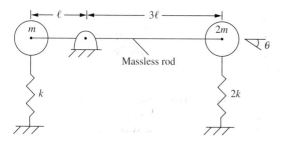

**FIG. P2.29**

**2.30** Derive a mathematical model for the system of Figure P2.29 and Problem 2.29 assuming the bar is a rigid slender bar of mass $m$.

**2.31** Derive a mathematical model for the system of Figure P2.31 using $x$ as the dependent variable.

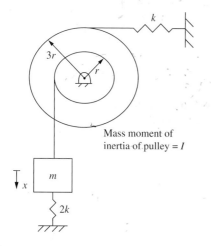

**FIG. P2.31**

**2.32** Derive a mathematical model for the system of Figure 2.32 using $\theta$ as the dependent and variable assuming small $\theta$. The bars are identical slender rods of mass $m$ and connected by a rigid massless rod.

**2.33** Derive a mathematical model for the compound pendulum of Figure P2.33. Use $\theta$, the counterclockwise angular displacement of the bar from the system's equilibrium position, as the dependent variable.

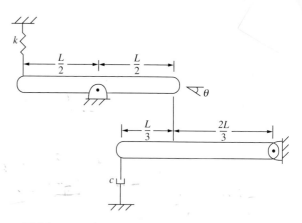

**FIG. P2.32**

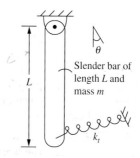

**FIG. P2.33**

**2.34** Derive a mathematical model for the system of Figure P2.34 using $x_1$ and $x_2$ as dependent variables.

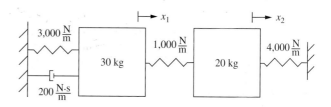

**FIG. P2.34**

**2.35** Derive a mathematical model for the system of Figure P2.35 using $\theta$ and $x$ as dependent variables.

**2.36** Derive a mathematical model for the system of Figure P2.36 using $x_1$, $x_2$, and $x_3$ as dependent variables.

**2.37** Derive a mathematical model for the system of Figure P2.37 using $x$ as the dependent variable.

**2.38** Derive a mathematical model for the system of Figure P2.38 using $\theta$ as the dependent variable noting that $y(t)$ is a prescribed displacement.

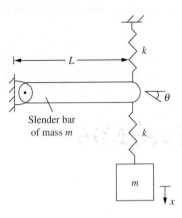

FIG. **P2.35**

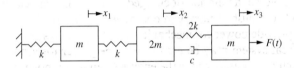

FIG. **P2.36**

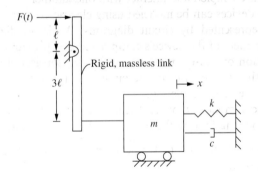

FIG. **P2.37**

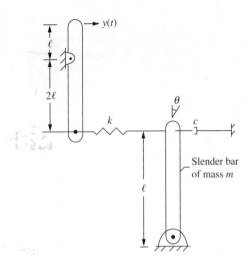

FIG. **P2.38**

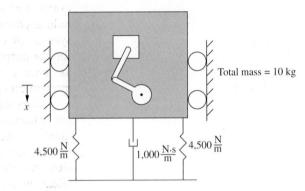

FIG. **P2.39**

**2.39** The single-cylinder engine of Problem 2.7 and Figure P2.7 is mounted in a housing, as illustrated in Figure P2.39, such that the total mass of the engine and the housing is $M$. The housing is mounted on an elastic foundation to allow vertical motion, but it is constrained from horizontal or rotational motion. The properties of the crank, connecting rod, and piston are those given in Problem 2.7. (a) Determine the displacement, velocity, and acceleration of the piston as a function of time if the crank rotates at a constant speed of 250 rpm. Assume that the crank, connecting rod, and piston are aligned at $t = 0$. (b) Derive a mathematical model for the motion of the engine in its housing and on the elastic foundation.

# 3

# Electrical Systems

Electrical systems involve the motion of charged particles through electric and magnetic fields established in system components. Electrical systems are often used to provide power for mechanical and thermal systems. For example an electrical system may be used to provide power for a household appliance such as a toaster or for the engine of an automobile. Such systems are often used in conjunction with mechanical or thermal systems for the purpose of power generation. In electromechanical systems such as motors, robots, transducers, and loudspeakers, the electrical and mechanical systems are coupled and interact with one another.

Many electrical devices can be modeled using electric circuits, which are schematically represented by circuit diagrams. A circuit diagram illustrates the arrangement of the device's components and provides a format for the application of conservation laws. A circuit consists of **active elements** (elements that add energy to the circuit) and **passive elements** (elements that dissipate or store energy).

The mathematical modeling of electrical systems is introduced in this chapter. The modeling follows the process outlined in Section 1.3.

1.  The system to be modeled is an electrical device that is represented by a circuit diagram.
2.  Implicit assumptions used in this study of modeling of electrical systems include the following:
    *   Circuit components are two-terminal elements.
    *   Circuit components obey linear laws.
    *   Circuit components are lumped and represented on a circuit diagram at specific locations the circuit.
    *   A circuit is not affected by external electromagnetic fields.
3.  The components of an electric circuit are discussed in Section 3.2.
4.  The dependent variables in circuit analysis are current and voltage, as introduced in Section 3.1. The parameters of circuit components are discussed in Section 3.2.
5.  The applicable physical laws, Kirchhoff's laws, are discussed in Section 3.3. Their application is illustrated in Sections 3.4–3.6.

6.  Appropriate initial conditions are specified.
7.  Mathematical analysis is discussed in Chapter 5.
8.  Analysis of system response is discussed in Chapters 6–8.

This chapter leads to an understanding of the fundamental principles of circuit analysis in order to understand the modeling of electromechanical systems and the fundamentals of electronic controllers.

Details of application of the steps 1–6 are presented in Sections 3.1–3.5. The process is applied to passive circuits used to model devices such as filters and power generation systems as well as active circuits containing operational amplifiers. Op-amp circuits that can be used as electronic controllers are disussed in Section 3.6. Analogies between electrical and mechanical systems are presented using energy methods is presented Electromechanical devices in Section 3.7 in which the motion of a mechanical system is coupled with an electromagnetic field is disussed in Section 3.8

## 3.1 CHARGE, CURRENT, VOLTAGE, AND POWER

Atoms consist of electrons orbiting a nucleus consisting of neutrons and protons. Forces between the particles in the nucleus and the electrons keep the electrons in orbit. These forces result due to the electric **charge** of the atomic particles. An electron is negatively charged, a proton is positively charged, and a neutron is neutrally charged. A repulsive force is developed between charged particles if the particles have the same charge; the force is attractive if the particles are oppositely charged. An object with an excess of electrons has a net negative charge, while an object with a deficiency of electrons has a net positive charge. The unit of charge is called a coulomb (C) and one electron has a charge of $-1.602 \times 10^{-19}$ C. The net charge of an object is the number of excess electrons times the charge of an electron.

The force between two particles of charge $q_1$ and $q_2$ is given by Coulomb's Law

$$F = \frac{q_1 q_2}{4\pi\varepsilon_0 r^2} \tag{3.1}$$

where $\varepsilon_0$ is the permittivity of free space, which has a numerical value of $8.85 \times 10^{-12} \mathrm{C^2/N \cdot m^2}$. The force is directed between the two particles if they are oppositely charged and away from the particles if they are like charged, as illustrated in Figure 3.1.

An **electric field** develops in the vicinity of a charged particle. The magnitude of the electric field at a given point due to the presence of a nearby charged particle is the force per unit charge of a force that would be developed at that point if a particle of unit charge was placed at that point. Electric field is a vector quantity. The total electric field at a point in space is the sum of the electric fields due to all charged particles.

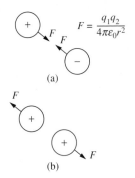

**FIG. 3.1** Forces between charged particles of (a) opposite charge and (b) like charge.

**Electric current** is the flow of charged particles. One ampere (A) of current results from 1 C of charge moving through a surface in a second

$$1 \text{ ampere (A)} = \frac{1 \text{ coulomb (C)}}{\text{second (s)}} \tag{3.2}$$

The algebraic symbol for electric current is $i$. Electric current is considered to be positive when it is in the direction of the motion of positively charged particles or opposite the direction of negatively charged particles.

**Electric potential** is the work required to move a 1-C charge against an electric field. The unit of electric potential, called the volt (V), is defined by

$$1 \text{ volt (V)} = \frac{1 \text{ N·m}}{1 \text{ C}} = 1 \frac{\text{joule (J)}}{\text{C}} \tag{3.3}$$

Electric potential is also called **voltage** and is represented by the algebraic symbol $v$.

The existence of an electric potential indicates the ability to do work to move particles against an electric field, thus inducing current. Electric potential is analogous to the potential energy due to gravity, which represents the ability to move a mass against a gravitational field, thus initiating kinetic energy.

**Electric power** is the rate at which energy is transferred from an electric source. The basic unit of electric power is the watt (W), defined by

$$1 \text{ watt (W)} = 1 \text{ J/s} \tag{3.4}$$

Equation (3.4) can be rewritten as

$$1 \text{ W} = 1 \text{ (J/C)(C/s)}$$
$$= (1 \text{ V})(1 \text{ A}) \tag{3.5}$$

Thus from Equation (3.4) electric power is the product of current and voltage or

$$P = vi \tag{3.6}$$

---

**Example 3.1**

---

When a 9-V battery is placed in a device such that the current is 3 A, it completely discharges in 20 hours (hr). What is the energy stored in the battery?

**Solution**

The power delivered during the discharge is calculated using Equation (3.6)

$$P = (9 \text{ V})(3 \text{ A}) = 27 \text{ W} \tag{a}$$

Thus the rate at which energy is transferred from the battery is 27 W. The energy stored in the battery is the total amount of energy transferred before it has completely discharged. Assuming the current is constant, the total energy transferred in 20 hr is

$$E = Pt$$
$$= (27 \text{ W})(20 \text{ hr})$$
$$= (27 \text{ J/s})(20 \text{ hr})(3600 \text{ s/hr})$$
$$= 1.94 \text{ MJ} \qquad \text{(b)}$$

## 3.2 CIRCUIT COMPONENTS

Active circuit components provide a source of energy to a circuit, while a passive circuit component stores or dissipates energy. Passive circuit components used in electrical and electromechanical systems modeled in this book and illustrated in Table 3.1 are resistors, capacitors, and inductors. Active circuit components, illustrated in Table 3.2, include voltage sources, current sources, and operational amplifiers. A circuit in which all sources provide constant input is called a direct current (dc) circuit. A circuit in which a source provides input that varies with time is called an alternating current (ac) circuit.

### 3.2.1 Resistors

A **resistor** is a circuit component that dissipates energy by converting electrical energy to heat. There is a decrease in voltage, a voltage drop, across the resistor in the direction of positive current. In general the voltage drop is a function of the current, $v = f(i)$. A linear resistor is one in which **Ohm's law**, which states that the voltage drop across a

**TABLE 3.1 PASSIVE CIRCUIT ELEMENTS**

| Element | Schematic | Voltage-Current Relations | Power | Energy |
|---|---|---|---|---|
| Resistor | $R$ | $v = iR$ $\quad$ $i = \dfrac{v}{R}$ | $P = i^2 R = \dfrac{v^2}{R}$ | Dissipated $E = \displaystyle\int_0^t i^2 R\,dt$ |
| Capacitor | $C$ | $i = C\dfrac{dv}{dt}$ $\quad$ $v = \dfrac{1}{C}\displaystyle\int_0^t i\,dt + v(0)$ | $P = Cv\dfrac{dv}{dt}$ | Stored in electric field $E = \tfrac{1}{2}Cv^2$ |
| Inductor | $L$ | $v = L\dfrac{di}{dt}$ $\quad$ $i = \dfrac{1}{L}\displaystyle\int_0^t v\,dt + i(0)$ | $P = Li\dfrac{di}{dt}$ | Stored in magnetic field $E = \tfrac{1}{2}Li^2$ |

**TABLE 3.2  ACTIVE CIRCUIT ELEMENTS**

| Element | Schematic | Comments |
|---|---|---|
| dc voltage source | | Voltage increases from − to +. |
| Time-varying voltage source | | Voltage increases from − to +. |
| Current source | | Current flow in direction of arrow. |
| Operational amplifier | | Voltages are referenced to ground, connected to external dc source. Output is proportional to difference of inputs. |

linear resistor is proportional to the current flowing through the resistor, is satisfied

$$v = iR \tag{3.7}$$

The constant of proportionality $R$ is called the resistance. The unit of resistance, the ohm ($\Omega$), is defined as

$$1 \text{ ohm } (\Omega) = \frac{1 \text{ V}}{1 \text{ A}} \tag{3.8}$$

The voltage drop in a resistor occurs in the direction of the current; the voltage is greater at the end of the resistor where the current enters.

Many components of electrical devices dissipate energy into heat. After use, the cord of an electric appliance is often warm as a result of energy dissipation within the wires. Batteries have internal resistances that dissipate energy when connected in a circuit. The resistance of an electrical component is dependent on the material of which the component is made. The resistance of a wire of length $L$ and cross-sectional area $A$ is calculated by

$$R = \frac{\rho L}{A} \tag{3.9}$$

where $\rho$ is the resistivity, with units of Ohms·m or Ohms·ft, of the material from which the wire is made. Conductivity $\sigma$ is the reciprocal of resistivity. Metals are good conductors of electricity and have high conductivity and low resistivity. Insulating materials have high resistivity and low conductivity.

The power dissipated by a resistor is the product of the current times the voltage drop across the resistor. Ohm's law, Equation (3.7), shows that the dissipated power can be written as either

$$P = i^2 R \tag{3.10}$$

or

$$P = \frac{v^2}{R} \tag{3.11}$$

**Example 3.2**

The voltage drop between two ends of a wire is measured as 3 V. The heat transferred from the wire in 1 s is measured as 90 J. Determine (a) the resistance of the wire and (b) the current in the wire.

**Solution**

The power dissipated in the wire is equal to the heat transferred from the wire in 1 s, $P = 90 \text{ J}/1 \text{ s} = 90 \text{ W}$.

(a) Solving Equation (3.11) for $R$ leads to

$$R = \frac{v^2}{P}$$

$$= \frac{(3 \text{ V})^2}{90 \text{ W}} = 0.1 \text{ } \Omega \qquad \text{(b)}$$

(b) The current in the wire is calculated using Ohm's law as

$$i = \frac{v}{R}$$

$$= \frac{3 \text{ V}}{0.1 \text{ } \Omega} = 30 \text{ A} \qquad \text{(c)}$$

### 3.2.2 Capacitors

A **capacitor** is a device that stores energy in the presence of an electric field. A capacitor develops a charge when it is subject to a potential difference. A **parallel plate capacitor**, illustrated in Figure 3.2, consists of two parallel plates made of a material that conducts electricity. The thin gap between the plates is of a nonconducting, or dielectric, material. When the plates of the capacitor are charged, one with a positive charge and one with a negative charge, an electric field develops in the dielectric material and stores energy. The stored energy is equal to the work required to charge the capacitor.

When a capacitor is connected to a battery, the battery does work to charge the plates of the capacitor until the potential difference across them equal to the battery voltage. When the voltage source is removed, the capacitor remains charged until discharged by equalizing the charges on the parallel plates.

The charge stored in the capacitor is proportional to the voltage

$$q = Cv \qquad (3.12)$$

The constant of proportionality in Equation (3.12) is called the **capacitance**, whose units are expressed in **farads** defined by

$$1 \text{ farad (F)} = 1 \text{ C/V} = 1 \text{ A·s/V} \qquad (3.13)$$

The capacitance of a parallel plate capacitor of plate area $A$ whose plates are separated by a distance $d$ is

$$C = K\varepsilon_0 \frac{A}{d} \qquad (3.14)$$

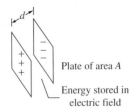

Plate of area $A$

Energy stored in electric field

**FIG. 3.2** Parallel plate capacitor, $C = K\frac{\varepsilon_0 A}{d}$.

where $K$ is the dielectric constant of the material in the gap separating the plates and $\varepsilon_0 = 8.85 \times 10^{-12} \ \mathrm{C^2/N \cdot m^2}$ is the permittivity of free space.

Differentiation of Equation (3.12) with respect to time leads to

$$\frac{dq}{dt} = C\frac{dv}{dt} \tag{3.15}$$

Current is the time rate of change of charge; thus using Equation (3.2) in Equation (3.15) leads to

$$i = C\frac{dv}{dt} \tag{3.16}$$

Integration of Equation (3.16) from $t = 0$ to an arbitrary time $t$ leads to

$$\int_0^t i\,dt = C\int_0^t \frac{dv}{dt}\,dt \tag{3.17}$$
$$= C[v(t) - v(0)]$$

where $i(0) = 0$ and $v(0)$ is the potential in the capacitor at $t = 0$. (Remember a capacitor may store energy even when current does not flow through the capacitor.) Rearranging Equation (3.17) to solve for for $v(t)$ leads to

$$v(t) = \frac{1}{C}\int_0^t i\,dt + v(0) \tag{3.18}$$

The power for the capacitor is obtained using Equations (3.6) and (3.16) as

$$P = vi$$
$$= v\left(C\frac{dv}{dt}\right) \tag{3.19}$$

The energy stored in a capacitor between $t = 0$ and an arbitrary time $t$ is

$$E = \int_0^t P\,dt$$
$$= \int_0^t Cv\frac{dv}{dt}\,dt$$
$$= \int_0^t C\frac{d}{dt}\left(\tfrac{1}{2}v^2\right)dt$$
$$= \tfrac{1}{2}C\left[v(t)^2 - v(0)^2\right] \tag{3.20}$$

Equation (3.19) gives the energy stored in the capacitor since $t = 0$. The energy stored in the capacitor at $t = 0$ is $\frac{1}{2}Cv(0)^2$ and the total energy stored in the capacitor is

$$E = \tfrac{1}{2}Cv^2 \tag{3.21}$$

---

**Example 3.3**

A capacitor of capacitance 40 µF initially has a charge of 200 µC when it is subject to an alternating current of

$$i(t) = 0.5\cos(100t) \text{ A} \tag{a}$$

Determine (a) $v(t)$, (b) $P(t)$, (c) $E(t)$, the energy stored since $t = 0$.

**Solution**

The initial voltage in the capacitor is

$$
\begin{aligned}
v(0) &= \frac{q(0)}{C} \\
&= \frac{200 \text{ µC}}{40 \text{ µF}} \\
&= 5 \text{ V}
\end{aligned}
\tag{b}
$$

(a) The voltage in the capacitor is obtained using Equation (3.18) as

$$
\begin{aligned}
v &= \frac{1}{40 \text{ µF}} \int_0^t [0.5\cos(100t) \text{ A}]dt + 5 \text{ V} \\
&= \frac{0.5}{(40 \times 10^{-6})(100)} \sin(100t)\big|_0^t + 5 \\
&= [125\sin(100t) + 5] \text{ V}
\end{aligned}
\tag{c}
$$

(b) The power in the capacitor is

$$
\begin{aligned}
P &= vi \\
&= \{[125\sin(100t) + 5] \text{ V}\}[0.5\cos(100t) \text{ A}] \\
&= [62.5\sin(100t)\cos(100t) + 2.5\cos(100t)] \text{ W} \\
&= [31.25\sin(200t) + 2.5\cos(100t)] \text{ W}
\end{aligned}
\tag{d}
$$

(c) The energy stored in the capacitor is obtained using Equation (3.20)

$$
\begin{aligned}
E &= \tfrac{1}{2}(40 \text{ µF})\left\{[125\sin(100t) + 5] \text{ V}^2 - (5 \text{ V})^2\right\} \\
&= (20 \times 10^{-6})\left[15625\sin^2(100t) + 1250\sin(100t)\right] \text{ J} \\
&= [0.3125\sin^2(100t) + 0.0250\sin(100t)] \text{ J} \\
&= \{(0.3125)\tfrac{1}{2}[1 - \cos(200t)] + 0.0250\sin(100t)\} \text{ J} \\
&= [156.25 - 156.25\cos(200t) + 25\sin(100t)] \text{ mJ}
\end{aligned}
\tag{e}
$$

Equation (e) shows that, for an alternating source, the energy stored in the capacitor increases while the voltage is increasing but is released when the voltage begins to decrease.

### 3.2.3 Inductors

When a coil of wire is placed in a moving magnetic field, current is induced in the wire. The moving magnetic field leads to a potential difference across the wire, which is referred to as an electromotive force (emf). The electromotive force induces current in the wire. Similarly when current moves through a coil of wire, as illustrated in Figure 3.3, a magnetic field is induced and energy is stored in the magnetic field. The process by which a current is developed due to a moving magnetic field or a magnetic field is developed in a coil of wire due to the motion of charged particles is called **induction**.

An **inductor** stores energy in the presence of a magnetic field. Most inductors are coils of wire. When current flows through the wire, work is done to establish a magnetic field and energy is stored in the magnetic field. When the current source is removed the magnetic field vanishes and the stored energy is released.

The magnetic flux $\phi$ developed in an inductor is proportional to the current such that $d\phi/di = \alpha$, a constant. **Faraday's law** states that the electric potential is proportional to the rate of change of the magnetic flux,

$$v = N\frac{d\phi}{dt} \tag{3.22}$$

where $N$ is the number of coils in the inductor. Using the chain rule for derivatives in Equation (3.22) leads to

$$\begin{aligned} v &= N\frac{d\phi}{di}\frac{di}{dt} \\ &= L\frac{di}{dt} \end{aligned} \tag{3.23}$$

where $L = N\frac{d\phi}{di}$ is a constant called the **inductance**, which is measured in henrys (H)

$$1 \text{ henry (H)} = 1\frac{\text{V}}{\text{A/s}} = 1\frac{\text{V·s}}{\text{A}} \tag{3.24}$$

The unit of V·s, is a called a weber (Wb), is a unit in which magnetic flux is measured. Thus Equation (3.24) can also be written as

$$1 \text{ H} = 1 \text{ Wb/A} \tag{3.25}$$

Integration of Equation (3.23) leads to

$$i(t) = \frac{1}{L}\int_0^t v(t)dt + i(0) \tag{3.26}$$

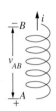

**FIG. 3.3** When current moves through a coil of wire, a magnetic field is induced in which energy is stored.

The power of the inductor is obtained using Equations (3.6) and (3.23) leading to

$$P = Li\frac{di}{dt} \tag{3.27}$$

The energy stored in the inductor is

$$E = \int_0^t P\,dt$$

$$= \tfrac{1}{2}L\Big\{[i(t)]^2 - [i(0)]^2\Big\} \tag{3.28}$$

When two inductors are near one another in the same circuit or in circuits of close proximity, the inductance of one inductor is affected by the magnetic field of the other. This induces a **mutual inductance** within each inductor. The mutual inductance between two inductors is denoted algebraically by $M$, measured in henrys. The sign on the mutual inductance is determined by the directions of the magnetic fluxes of the magnetic fields developed in the inductors.

Consider two inductors of inductances $L_1$ and $L_2$ with voltages $v_1$ and $v_2$ respectively. The currents flowing through the inductors are given by $i_1(t)$ and $i_2(t)$. The relations between the voltages and currents when mutual inductance is included are

$$v_1 = L_1\frac{di_1}{dt} + M\frac{di_2}{dt} \tag{3.29a}$$

$$v_2 = M\frac{di_1}{dt} + L_2\frac{di_2}{dt} \tag{3.29b}$$

---

**Example 3.4**

The current through a 4.2-mH inductor is

$$i(t) = 15(1 - e^{-1250t})\ \text{A} \tag{a}$$

Determine (a) the voltage in the inductor, (b) the power in the inductor, and (c) the energy stored in the inductor

**Solution**

(a) The voltage in the inductor is determined using Equation (3.23) as

$$v(t) = (4.2\ \text{mH})\frac{d}{dt}\big\{15(1 - e^{-1250t})\ \text{A}\big\}$$

$$= (4.2 \times 10^{-3})(15)(1250)e^{-1250t}\ \text{V}$$

$$= 78.75e^{-1250t}\ \text{V} \tag{b}$$

(b) The power in the inductor is obtained using Equation (3.6) as

$$P = (78.75e^{-1250t}\ \text{V})\big[15(1 - e^{-1250t})\ \text{A}\big]$$

$$= 1.18(e^{-1250t} - e^{-2500t})\ \text{kW} \tag{c}$$

(c) The energy stored in the inductor is calculated using Equation (3.28) as

$$E = \tfrac{1}{2}(4.2 \text{ mH})\left[15(1 - e^{-1250t}) \text{ A}\right]^2$$
$$= 472.5\left(1 - 2e^{-2500t} + e^{-1250t}\right) \text{ mJ} \tag{d}$$

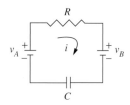

**FIG. 3.4** Circuit diagram for a device in which energy is provided by a direct current source $v_A$, is dissipated by a resistor, is absorbed by a source $v_B$, and is stored by a capacitor.

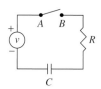

**FIG. 3.5** When the switch between $A$ and $B$ is open, no current flows through the circuit and the circuit is an open circuit. When the switch is closed, there is no voltage drop between $A$ and $B$ and the circuit is a short circuit.

### 3.2.4 Voltage and Current Sources

A **voltage source** is an external source of electric potential. The electric potential increases across a voltage source if the current is flowing from the negative ($-$) terminal of the source to the positive ($+$) terminal. If the current flows from the positive to the negative terminal, the voltage decreases across the source. This source is absorbing power. If, in the circuit of Figure 3.4, the current is flowing in the clockwise direction, $v_A$ is a time-varying voltage source, energy is absorbed and converted to heat by the resistor, a constant energy is absorbed by $v_B$, and energy is stored by the electric field developed by the capacitor.

A **current source** is an element that leads to a specific current flowing through a circuit. The direction of the current provided by a current source is indicated by an arrow on the source.

An **open circuit** is a circuit in which current is not flowing. A **short circuit** is a circuit component across which there is no voltage change. A **switch** is a circuit component that, when open, leads to an open circuit and, when closed, leads to a short circuit across the switch, as illustrated in Figure 3.5.

### 3.2.5 Operational Amplifiers

An **amplifier** is a device that magnifies a signal. An **operational amplifier** is a device with a positive terminal and a negative terminal and is connected to dc power sources. The output of the amplifier is proportional to the difference of the input voltages. Operational amplifiers are discussed in more detail in Section 3.7.

### 3.2.6 Electric Circuits and Mechanical Systems

Direct analogies between mechanical systems and electric circuits are developed in detail in Section 3.6. It is possible from the preceding discussion to draw similarities between the two.

The passive elements in the mechanical systems of Chapter 2 are inertia elements, stiffness elements, and damping elements. The active elements in mechanical systems are the force and motion input. The passive elements of a circuit are inductors, capacitors, and resistors. The active elements in a circuit are voltage and current sources.

In both types of systems the passive elements store or dissipate energy and the active elements provide energy to the system. In a mechanical system kinetic energy is stored by inertia elements, potential energy is stored by stiffness elements, and energy is dissipated by damping elements. In an electric circuit, energy is stored in magnetic fields by inductors, it is stored in electric fields by capacitors, and it is dissipated by resistors.

A state is defined in Section 2.9 as corresponding to independent energy storage components. This definition is generic, not specific for mechanical systems. Thus a state in an electrical system corresponds to independent inductors and capacitors.

## 3.3 KIRCHOFF'S LAWS

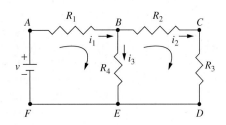

**FIG. 3.6** Circuit with nodes at $B$ and $E$ and loops $ABEF$ and $BCDE$.

A circuit diagram illustrates how components in an electrical circuit are connected to one another. An example circuit diagram is illustrated in Figure 3.6. The junction between three or more wires in a circuit is called a **node**. The nodes in the circuit of Figure 3.6 are at $B$ and $E$. A closed path in a circuit is called a **loop**. Independent loops in the circuit of Figure 3.6 are $ABEF$ and $BCDE$. The closed path $ACDF$ is also a loop, but is not independent of $ABEF$ and $BCDE$. The circuit in Figure 3.6 is an example of a two-loop circuit.

The motion of charged particles is governed by a set of equations called **Maxwell's equations**, which provide the basis for all electromagnetic theory. A disussion of the full set of Maxwell's equations is beyond the scope of this study; however, under certain assumptions, Maxwell's equations can be reduced to a simpler form that can be applied to model circuits. In a manner analogous to the development of the Principle of Work and Energy from Newton's second law, principles can be derived from Maxwell's equations, which can be applied to lumped parameter circuits in lieu of the full set of Maxwell's equations.

Circuits are idealized models of electrical devices consisting of discrete components with lumped properties. The application of Maxwell's equations is achieved by applying two principles:

1. *Conservation of Charge:* The charge does not accumulate at the junction between circuit components. Thus the rate of change of charge entering the junction is equal to the rate of change of charge leaving the junction.
2. *Conservation of Energy:* Energy is conserved around any loop in a circuit. The energy supplied by the sources is dissipated by resistors and stored by capacitors and inductors.

The laws of Conservation of Charge and Conservation of Energy are formulated into **Kirchoff's circuit laws**.

- **Kirchoff's current law (KCL)** states that the sum of currents at a node is zero. The sign of a current is determined by the direction of the current in the wire. A current flowing into the node is taken as positive, and a current flowing away from the node is taken to be negative.
- **Kirchoff's voltage law (KVL)** states that the total change in voltage around any loop in a circuit must be zero. A voltage change across a circuit component is positive if the current flow through the component leads to an increase in voltage in the circuit. A current flowing through a voltage source from the negative terminal to the positive terminal of the source leads to a voltage increase. A voltage change across a circuit component is negative if the component draws voltage from the circuit. Voltage changes across a resistor that dissipates energy into heat and across a capacitor that stores energy in an electric field are negative. KVL can be applied in either the clockwise or the counterclockwise direction around a loop. If the direction of application of KVL is opposite that of the assumed direction of a current, the current is taken to be negative.

The node law KCL, as stated, applies to the junction between three or more wires. However, Conservation of Charge also takes place at the junction of two wires. A trivial application of this principle is that the currents in two wires that meet at a simple junction (only two wires) have the same value and are in the same direction.

---

## Example 3.5

Apply Kirchoff's circuit laws to the two-loop circuit of Figure 3.6.

### Solution

Let $i_1$ be the current in wire $AB$ flowing from $A$ to $B$. Let $i_2$ be the current in wire $BC$ flowing from $B$ to $C$. Let $i_3$ be the current in wire $BE$ flowing from $B$ to $E$. Application of KCL at node $B$ yields

$$i_1 - i_2 - i_3 = 0 \tag{a}$$

Note that in Equation (a) the sign on $i_1$ is positive because it is flowing into the node, whereas the signs on $i_2$ and $i_3$ are negative because they are flowing away from $B$. Note that, because of the simple junctions at $A$ and $F$, the currents in wires $EF$ and $FA$ are $i_1$ and because of the simple junctions at $C$ and $D$, the currents in wires $CD$ and $DE$ are $i_2$. Thus application of KCL at node $E$ also leads to Equation (a).

Noting that Ohm's law, Equation (3.7), is used to calculate the voltage drop across a resistor, the application of KVL in the clockwise direction to the loop $ABEF$ leads to

$$v - R_1 i_1 - R_4 i_3 = 0 \tag{b}$$

The application of KVL in the clockwise direction to loop $BCDE$ leads to

$$-R_2 i_2 - R_3 i_2 + R_4 i_3 = 0 \qquad \text{(c)}$$

The application of KVL in the clockwise direction to loop $BCDE$ requires the determination of the voltage change across the resistor of resistance $R_4$ going from $E$ to $B$, which is opposite the assumed direction of $i_3$. Thus the current through the resistor is taken to be $-i_3$.

The currents in the circuit are obtained by the simultaneous solution of Equations (a)–(c). There are many methods for the solution of simultaneous equations. The application of Cramer's rule (see Appendix B) to a matrix formulation of the system of equations is used here. Equations (a)–(c) are summarized in matrix form as

$$\begin{bmatrix} 1 & -1 & -1 \\ R_1 & 0 & R_4 \\ 0 & -(R_2 + R_3) & R_4 \end{bmatrix} \begin{bmatrix} i_1 \\ i_2 \\ i_3 \end{bmatrix} = \begin{bmatrix} 0 \\ v \\ 0 \end{bmatrix} \qquad \text{(d)}$$

The application of Cramer's rule to Equation (d) to solve for $i_1$ leads to

$$i_1 = \frac{\begin{vmatrix} 0 & -1 & -1 \\ v & 0 & R_4 \\ 0 & -(R_2 + R_3) & R_4 \end{vmatrix}}{\begin{vmatrix} 1 & -1 & -1 \\ R_1 & 0 & R_4 \\ 0 & -(R_2 + R_3) & R_4 \end{vmatrix}}$$

$$= \frac{v(R_2 + R_3) + v(R_4)}{R_1(R_2 + R_3) + R_4(R_2 + R_3) + R_1 R_4}$$

$$= \frac{R_2 + R_3 + R_4}{(R_2 + R_3)(R_1 + R_4) + R_1 R_4} v \qquad \text{(e)}$$

Cramer's rule is used to solve for $i_2$ and $i_3$, leading to

$$i_2 = \frac{R_4}{(R_2 + R_3)(R_1 + R_4) + R_1 R_4} v \qquad \text{(f)}$$

$$i_3 = \frac{R_2 + R_3}{(R_2 + R_3)(R_1 + R_4) + R_1 R_4} v \qquad \text{(g)}$$

## 3.4 CIRCUIT REDUCTION

Components of electric circuits, like components of mechanical systems, can be placed in series or in parallel. As is the case for mechanical systems, series and parallel combinations of like components may be replaced by a

**FIG. 3.7** Circuit components in series: The same current passes through each component, and the voltage change from $A$ to $B$ is the sum of the voltage changes in the components.

single equivalent component. The criteria for the equivalence of two sets of electrical system components is that, when the same current is passed through each set, the same voltage change occurs.

### 3.4.1  Series and Parallel Components

Circuit components placed in series, as illustrated in Figure 3.7, have the same current flowing through each component and the total voltage change across the series combination is the sum of the voltage changes. Thus if there are $n$ series components in Figure 3.7

$$i_1 = i_2 = \cdots = i_n \tag{3.30}$$

$$v_{AB} = v_1 + v_2 + \cdots + v_n \tag{3.31}$$

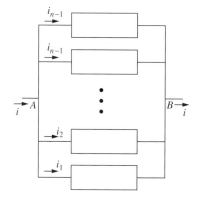

**FIG. 3.8** Circuit components in parallel: The voltage change is the same across each component, and the current is the sum of the currents in the components.

Circuit components placed in parallel are illustrated in Figure 3.8. Applying KCL to each of the nodes in the parallel combination, beginning with the node at $A$ leads to

$$i = i_1 + i_2 + \cdots + i_n \tag{3.32}$$

The voltage change from $A$ to $B$ is the same across each component

$$v_{AB} = v_1 = v_2 = \cdots = v_n \tag{3.33}$$

### 3.4.2  Series Combinations

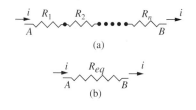

**FIG. 3.9** (a) $n$ resistors in series; (b) the equivalent resistance is the sum of resistances.

Figure 3.9(a) illustrates a series combination of $n$ resistors. It is desired to replace the combination by a single resistor of resistance $R_{eq}$, as illustrated in Figure 3.9(b). The voltage change across the resistor of Figure 3.9(b) when a current $i$ flows through the component is

$$v = R_{eq}i \tag{3.34}$$

The total voltage change across the series combination of Figure 3.9(a) is

$$v = R_1 i + R_2 i + \cdots + R_n i$$

$$= \left( \sum_{k=1}^{n} R_k \right) i \tag{3.35}$$

The comparison of Equations (3.34) and (3.35) leads to

$$R_{eq} = \sum_{k=1}^{n} R_k \qquad (3.36)$$

Equation (3.36) shows that the equivalent resistance of a series combination of resistors is the sum of the individual resistances. Equation (3.36) is similar to the equation for the equivalent stiffness of a parallel combination of springs.

A series combination of $n$ capacitors is illustrated in Figure 3.10(a). An equivalent system with a single capacitor of capacitance $C_{eq}$ is illustrated in Figure 3.10(b). The voltage change across the capacitor of Figure 3.10(b), assuming the capacitor is uncharged at $t = 0$, is

$$v_{AB} = \frac{1}{C_{eq}} \int_0^t i\, dt \qquad (3.37)$$

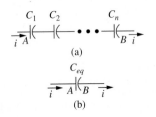

(a)

(b)

**FIG. 3.10** (a) $n$ capacitors in series; (b) the equivalent capacitance is given by Equation (3.39).

The voltage change across the series combination of Figure 3.10(a), assuming all capacitors are uncharged at $t = 0$, is

$$v_{AB} = \frac{1}{C_1} \int_0^t i\, dt + \frac{1}{C_2} \int_0^t i\, dt + \cdots + \frac{1}{C_n} \int_0^t i\, dt \qquad (3.38)$$

Comparison of Equations (3.37) and (3.38) leads to

$$C_{eq} = \frac{1}{\sum\limits_{k=1}^{n} \frac{1}{C_k}} \qquad (3.39)$$

Equation (3.39) shows that the equivalent capacitance of a series combination of capacitors is the reciprocal of the sum of the reciprocals of the capacitances of the members of the series combination. Equation (3.39) is similar to the equation for the equivalent stiffness of springs in parallel.

It is easy to show that the equivalent inductance of a combination of $n$ inductors placed in series with no mutual inductance is

$$L_{eq} = \sum_{k=1}^{n} L_k \qquad (3.40)$$

### 3.4.3 Parallel Combinations

The analysis of parallel combinations of circuit components uses Equations (3.32) and (3.33). Their applications to circuit components to determine equivalent values of the component parameters are left as exercises.

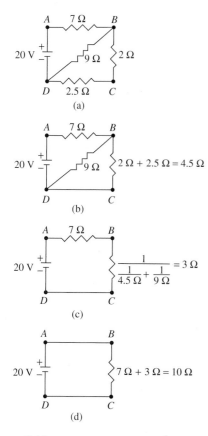

**FIG. 3.11** (a) The resistive circuit of Example 3.6; (b) resistors from $B$ to $C$ and $C$ to $D$ in series; (c) resistors from $B$ to $C$ and $B$ to $D$ in parallel; (d) resistors from $A$ to $B$ and $B$ to $C$ in series.

**TABLE 3.3 EQUIVALENCE EQUATIONS FOR CIRCUIT COMPONENTS**

| Component | Series Equivalence | Parallel Equivalence |
|---|---|---|
| Resistor | $R_{eq} = \sum_{k=1}^{n} R_k$ | $R_{eq} = \dfrac{1}{\sum_{k=1}^{n} \frac{1}{R_k}}$ |
| Capacitor | $C_{eq} = \dfrac{1}{\sum_{k=1}^{n} \frac{1}{C_k}}$ | $C_{eq} = \sum_{k=1}^{n} C_k$ |
| Inductor | $L_{eq} = \sum_{k=1}^{n} L_k$ | $L_{eq} = \dfrac{1}{\sum_{k=1}^{n} \frac{1}{L_k}}$ |

The equivalent resistance for $n$ resistors in parallel is

$$R_{eq} = \frac{1}{\displaystyle\sum_{k=1}^{n} \frac{1}{R_k}} \tag{3.41}$$

The equivalent capacitance for $n$ capacitors in parallel is

$$C_{eq} = \sum_{k=1}^{n} C_k \tag{3.42}$$

The equivalent inductance for $n$ inductors in parallel with no mutual inductance is

$$L_{eq} = \frac{1}{\displaystyle\sum_{k=1}^{n} \frac{1}{L_k}} \tag{3.43}$$

The equations for equivalent components for combinations of circuit elements are summarized in Table 3.3.

**Example 3.6**

Determine the power delivered by the 20-V source to the resistive circuit of Figure 3.11(a).

**Solution**

For purposes of analysis the combination of resistors can be replaced by a single resistor of resistance $R_{eq}$. The steps in the circuit reduction are shown in Figures 3.11(b)–3.11(d). The same current flows through the resistors between $B$ and $C$ and $C$ and $D$, and the total voltage drop between $B$ and $D$ is the sum of the voltage drops across the resistors. Thus these resistors are in series. Using KCL, the current entering node $B$ is divided into the current flowing from $B$ to $C$ and the current flowing from $B$ to $D$. Using KVL, the voltage drops across the resistors between $B$ and $D$ and across the equivalent resistor between $B$ and $C$ are equal. Thus these resistors are in parallel. Finally this equivalent resistor is in series with the resistor between $A$ and $B$. The equivalent resistance of the circuit is $R_{eq} = 10\,\Omega$. The power delivered to the circuit is obtained using Equation (3.11) as

$$
\begin{aligned}
P &= \frac{v^2}{R_{eq}} \\
&= \frac{(20\ \text{V})^2}{10\ \Omega} \\
&= 40\ \text{W}
\end{aligned}
$$
(a)

## Example 3.7

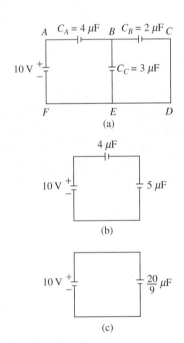

(a)

(b)

(c)

**FIG. 3.12** (a) The capacitive circuit of Example 3.7; (b) capacitors $B$ and $C$ are in parallel and replaced by an equivalent capacitor; (c) the equivalent capacitance of the circuits is $20/9\,\mu\text{F}$.

Consider the capacitive circuit of Figure 3.12. (a) Replace the capacitors with a single capacitor of equivalent capacitance. (b) What are the charges on each capacitor?

### Solution

(a) The 2-$\mu$F capacitor and the 3-$\mu$F capacitor have the same voltage change (from KVL) and thus are in parallel. Using Equation (3.42) they can be replaced by a single capacitor of $C_{eq1} = 5\mu$F, as illustrated in Figure 3.12(b). The capacitors in Figure 3.12(b) are in series (total voltage drop is same of voltage drops and current through each capacitor is the same) and using Equation (3.39) can be replaced by a single capacitor with

$$
C_{eq} = \frac{1}{\frac{1}{4\mu\text{F}} + \frac{1}{5\mu\text{F}}} = \frac{20}{9}\mu\text{F}
$$
(a)

The equivalent capacitance is illustrated in Figure 3.12(c).

(b) Let $v_A$, $v_B$, and $v_c$ represent the potential in the capacitors and let $q_A$, $q_B$, and $q_c$ represent the charges on each capacitor. From Conservation of Charge at node $B$

$$
q_A - q_B - q_C = 0
$$
(b)

Application of KVL around loop $ABEF$ gives

$$
10 - v_A - v_C = 0
$$
(c)

Since capacitors $B$ and $C$ are in parallel

$$
v_B - v_C = 0
$$
(d)

From Equation (3.12) the charge on each capacitor is $q_A = (4\mu\text{F})v_A$, $q_B = (2\mu\text{F})v_B$, and $q_C = (3\mu\text{F})v_C$. Substituting into Equation (b) and summarizing Equations (b)–(d) in matrix form leads to

$$
\begin{bmatrix} 4 & -2 & -3 \\ 1 & 0 & 1 \\ 0 & 1 & -1 \end{bmatrix} \begin{bmatrix} v_A \\ v_B \\ v_C \end{bmatrix} = \begin{bmatrix} 0 \\ 10 \\ 0 \end{bmatrix}
$$
(e)

Simultaneous solution of Equation (e) leads to $v_A = 50/9$ V, $v_B = v_C = 40/9$ V. The charges on the capacitors are

$$q_A = C_A v_A = (4\mu F)(50/9 \text{ V}) = 22.2 \ \mu C$$
$$q_B = C_B v_B = (2\mu F)(40/9 \text{ V}) = 8.89 \ \mu C$$
$$q_C = C_C v_C = (3\mu F)(40/9 \text{ V}) = 13.3 \ \mu C \qquad \text{(f)}$$

## 3.5 MODELING OF ELECTRIC CIRCUITS

Capacitors and inductors are circuit components that store energy. The voltage change across an inductor is proportional to the rate of change of current, while the current through a capacitor is proportional to the rate of change of voltage across the capacitor. Thus the presence of these components in an electric circuit leads to time-dependent current and voltage changes. Time-dependent sources also lead to time-dependent currents and voltage changes in a circuit.

The system response of an electric circuit is described by the time-dependent currents and voltage changes, which are determined through mathematical modeling of the circuit. Kirchoff's laws, both KCL and KVL, are applied to the circuit diagram, resulting in differential equations or integrodifferential equations whose solution provides the system response. The following examples illustrate the techniques for developing mathematical models for electric circuits.

### Example 3.8

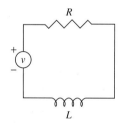

FIG. 3.13 The $LR$ circuit of Example 3.8.

Derive a mathematical model for the $LR$ circuit of Figure 3.13. The current is zero when the voltage source is connected at $t = 0$.

### Solution

Let $i(t)$ represent the current through the circuit. Application of KVL in a clockwise direction around the circuit leads to

$$v(t) - Ri - L\frac{di}{dt} = 0 \qquad \text{(a)}$$

$$L\frac{di}{dt} + Ri = v(t) \qquad \text{(b)}$$

Equation (b) is a first-order differential equation whose solution is the time-dependent current through the circuit. The determination of $i(t)$ requires the application of an appropriate initial condition. Since the source is connected at $t = 0$ is $i(0) = 0$.

## Example 3.9

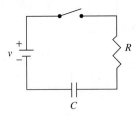

**FIG. 3.14** The $RC$ circuit of Example 3.9: Current flows through the circuit after the switch is closed.

Derive a mathematical model for the $RC$ circuit of Figure 3.14. The switch is open and then closed at $t = 0$. The capacitor is uncharged before the switch is closed. (a) Use the current through the circuit as the system output. (b) Use the voltage change across the capacitor as the system output.

### Solution

(a) Let $i(t)$ represent the current through the circuit. Since the circuit is open at $t = 0$, the source is not connected until $t = 0$. The appropriate mathematical representation of the voltage source is

$$v(t) = vu(t) \tag{a}$$

where $u(t)$ is the unit step function defined in Equation (1.33). The application of KVL in a clockwise fashion around the circuit leads to

$$vu(t) - Ri - \frac{1}{C}\int_0^t i\, dt = 0$$

$$Ri + \frac{1}{C}\int_0^t i\, dt = vu(t) \tag{b}$$

Equation (b) is an integrodifferential equation whose solution is the time-dependent current in the circuit.

Equation (b) can be converted into a first-order differential equation through differentiation with respect to time. Recalling that $du(t)/dt = \delta(t)$ where $\delta(t)$ is the unit impulse function, differentiation of Equation (b) leads to

$$R\frac{di}{dt} + \frac{1}{C}i = v\delta(t) \tag{c}$$

Solution of Equation (c) requires the application of the initial condition $i(0) = 0$.

The solution of Equation (b) does not require the application of an initial condition. Substitution of $t = 0$ into Equation (b) leads to $i(0) = 0$. However, the evaluation of Equation (b) at a time $t = 0^+$ shows that $i(0^+) = v/R$ and thus the solution of Equation (b) leads to a current that is discontinuous at $t = 0$. The discontinuous current is a result of the assumption that when the switch is closed the voltage source is instantaneously connected to the circuit with full potential being delivered. The solution of Equation (c), subject to the initial condition $i(0) = 0$, also leads to the prediction of a discontinuous current.

(b) Let $v_c(t)$ represent the voltage change across the capacitor. A trivial application of KCL shows that the current through the resistor is equal to the current through the capacitor

$$i_r = i_C \tag{d}$$

From Ohm's law the current through the resistor is

$$i_r = \frac{vu(t) - v_C}{R} \tag{e}$$

The current through the capacitor is obtained using Equation (3.16) as

$$i_C = C \frac{dv_C}{dt} \tag{f}$$

Substituting Equations (e) and (f) into Equation (d) gives

$$\frac{vu(t) - v_C}{R} = C \frac{dv_C}{dt}$$

$$C \frac{dv_C}{dt} + \frac{1}{R} v_C = \frac{v}{R} u(t) \tag{g}$$

---

**Example 3.10**

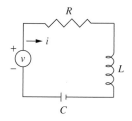

**FIG. 3.15** The series $LRC$ of Example 3.10.

Derive a mathematical model for the series $LRC$ circuit of Figure 3.15. There is no current flowing in the circuit at $t = 0$. The capacitor is uncharged at $t = 0$.

**Solution**

Application of KVL in the clockwise direction around the circuit leads to

$$v(t) - Ri - L \frac{di}{dt} - \frac{1}{C} \int_0^t i \, dt = 0 \tag{a}$$

$$L \frac{di}{dt} + Ri + \frac{1}{C} \int_0^t i \, dt = v(t) \tag{b}$$

Equation (b) is an integrodifferential equation whose solution, subject to $i(0) = 0$, is the time-dependent current through the circuit.

Equation (b) can be rewritten as a second-order differential equation through differentiation with respect to time

$$L \frac{d^2 i}{dt^2} + R \frac{di}{dt} + \frac{1}{C} i = \frac{dv}{dt} \tag{c}$$

If Equation (c) is used in lieu of Equation (b), a value of $di/dt(0)$ is necessary. Evaluation of Equation (b) at $t = 0$ leads to

$$L \frac{di}{dt}(0) = v(0) \tag{d}$$

---

**Example 3.11**

Obtain a mathematical model for the three-loop circuit of Figure 3.16.

**Solution**

The dependent variables for the model are $i_1(t)$, $i_2(t)$, and $i_3(t)$, as indicated on Figure 3.16. The application of KCL at nodes $B$ and $C$ leads to

$$i_1 - i_2 - i_4 = 0 \tag{a}$$

$$i_2 - i_3 - i_5 = 0 \tag{b}$$

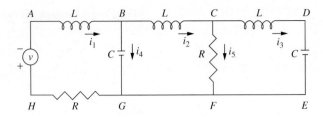

**FIG. 3.16** The three-loop circuit of Example 3.11.

The application of KVL in the clockwise direction around loop *ABGH* and using Equation (a) leads to

$$v(t) - L\frac{di_1}{dt} - \frac{1}{C}\int_0^t i_4 dt - Ri_1 = 0 \tag{c}$$

$$L\frac{di_1}{dt} + Ri_1 + \frac{1}{C}\int_0^t (i_1 - i_2) dt = v(t) \tag{d}$$

Application of KVL in the clockwise direction around loop *BCFG* and using Equations (a) and (b) leads to

$$-L\frac{di_2}{dt} - Ri_5 + \frac{1}{C}\int_0^t i_4 dt = 0 \tag{e}$$

$$L\frac{di_2}{dt} + R(i_2 - i_3) - \frac{1}{C}\int_0^t (i_1 - i_2) dt = 0 \tag{f}$$

Application of KVL in the clockwise direction around loop *CDEF* and using Equation (b) leads to

$$-L\frac{di_3}{dt} - \frac{1}{C}\int_0^t i_3 dt + Ri_5 = 0 \tag{g}$$

$$L\frac{di_3}{dt} - R(i_2 - i_3) + \frac{1}{C}\int_0^t i_3 dt = 0 \tag{h}$$

The circuit of Figure 3.16 is modeled by Equations (d), (f), and (h).

In each of the previous examples, nodal currents (currents entering or leaving nodes) were used as dependent variables. In some problems, especially those with current sources, it is more convenient to use voltages as dependent variables. The dependent variable chosen for Example 3.12 is the voltage drop across a circuit component, while the dependent variables in Example 3.13 are the voltages at nodes in the circuit.

## Example 3.12

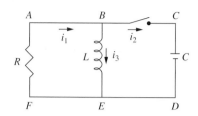

**FIG. 3.17** The parallel *LRC* circuit of Example 3.12.

Derive a mathematical model for the parallel *LRC* circuit of Figure 3.17. The capacitor has a charge $q$ at $t = 0$, when the switch is closed, allowing the capacitor to discharge.

### Solution

Define $v_R$ as the voltage change across the resistor, $v_C$ as the voltage change across the capacitor, and $v_L$ as the voltage change across the inductor. The application of KVL to loop *ABEF* in the clockwise direction leads to

$$v_R + v_L = 0 \tag{a}$$

The application of KVL to loop *BCDE* in the clockwise direction leads to

$$v_C - v_L = 0 \tag{b}$$

From Equations (a) and (b)

$$v_L = v_C = -v_R \tag{c}$$

The voltage change across the resistor is used as the dependent variable in the modeling of the parallel *LRC* circuit.

Application of KCL to the node at *B* using the definitions of currents shown in Figure 3.17 leads to

$$i_1 - i_2 - i_3 = 0 \tag{d}$$

The current-voltage laws for the circuit components [Equations (3.7), (3.16), and (3.26)] are used to give

$$i_1 = \frac{v_R}{R} \tag{e}$$

$$i_2 = C\frac{dv_C}{dt} = -C\frac{dv_R}{dt} \tag{f}$$

$$i_3 = \frac{1}{L}\int_0^t v_L dt = -\frac{1}{L}\int_0^t v_R dt \tag{g}$$

Substitution of Equations (e)–(g) into Equation (d) leads to

$$\frac{v_R}{R} + C\frac{dv_R}{dt} + \frac{1}{L}\int_0^t v_R dt = 0 \tag{h}$$

Equation (h) must be supplemented by an initial condition. The initial voltage in the capacitor due to the initial charge is obtained using Equation (3.10)

$$v_C(0) = \frac{q}{C} \tag{i}$$

Equations (c) and (i) are used to obtain

$$v_R(0) = -\frac{q}{C} \tag{j}$$

Equation (h) can be rewritten as a second-order differential equation through differentiation with respect to time

$$C\frac{d^2 v_R}{dt^2} + \frac{1}{R}\frac{dv_R}{dt} + \frac{1}{L}v_R = 0 \qquad \text{(k)}$$

Since Equation (k) is a second-order differential equation, two initial conditions must be specified to determine its response. One initial condition is provided by Equation (j). The second is obtained by evaluating Equation (h) at $t = 0$,

$$\frac{1}{R}v_R(0) + C\frac{dv_R}{dt}(0) = 0 \qquad \text{(l)}$$

$$\frac{dv_R}{dt}(0) = \frac{q}{RC^2} \qquad \text{(m)}$$

## Example 3.13

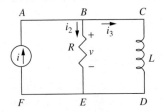

**FIG. 3.18** Circuit of Example 3.13.

Derive a mathematical model for the circuit of Figure 3.18.

### Solution

Application of KCL at node $B$ gives

$$i(t) - i_2 - i_3 = 0 \qquad \text{(a)}$$

Let $v$ be the voltage drop across the resistor. Then

$$i_2 = \frac{v}{R} \qquad \text{(b)}$$

KVL applied around loop $BCDE$ reveals that the voltage drop across the inductor is the same as that across the resistor. Thus

$$i_3 = \frac{1}{L}\int_0^t v(t)dt \qquad \text{(c)}$$

Substitution of Equations (b) and (c) into Equation (a) leads to

$$\frac{v}{R} + \frac{1}{L}\int_0^t v\,dt = i(t) \qquad \text{(d)}$$

## 3.6 MECHANICAL SYSTEMS ANALOGIES

### 3.6.1 Energy Principles

The second-order differential equation derived for the series $LRC$ circuit in Example 3.10, Equation (c), has the same form as the differential equation for a mass-spring-viscous damper system. Thus is reasonable to assume that the two systems are somehow analogous. The analogy between electrical and mechanical systems is best examined using energy methods.

It is shown in Section 2.9 that the kinetic and potential energies of a linear one-degree-of-freedom system have quadratic forms given by Equations (2.118) $\left(\frac{1}{2}m_{eq}\dot{x}^2\right)$ and (2.119) $\left(\frac{1}{2}k_{eq}x^2\right)$ respectively. The work done by the viscous damping forces has the form of Equation (2.120). The principle of work and energy is used to show that any linear one-degree-of-freedom mechanical system has a differential equation of the form

$$m_{eq}\ddot{x} + c_{eq}\dot{x} + k_{eq}x = F_{eq}(t) \tag{2.130}$$

Energies stored in electric circuit components also have quadratic forms. The energy stored in a capacitor in the presence of an electric field is $\frac{1}{2}Cv^2$. The energy stored in an inductor in the presence of a magnetic field is $\frac{1}{2}Li^2$. The energy lost across a resistor is of the form $\int Ri^2 dt$, which is similar to the expression for energy dissipated by a viscous damper when written as $\int c_{eq}\dot{x}^2 dt$. The similarities of the quadratic forms for stored energy and dissipated energy also suggest that making analogies between mechanical and electrical systems is valid.

## 3.6.2  Single-Loop Circuits with Voltage Sources

Consider a one-loop circuit with resistors, inductors, capacitors, and a voltage source. The analysis in Section 3.4 shows that the resistors can be replaced by a single resistor of equivalent resistance $R_{eq}$, the capacitors can be replaced by a single capacitor of equivalent capacitance $C_{eq}$, and the inductors can be replaced by a single inductor of equivalent inductance $L_{eq}$. Since energy is conserved in the circuit, the energy provided by the source $E_s$ is equal to the energy dissipated by the resistors $E_R$, plus the energy stored by the inductors $E_L$, plus the energy stored by the capacitors $E_C$

$$E_s = E_R + E_C + E_L \tag{3.44}$$

Differentiation of Equation (3.44) with respect to time leads to an equation for power in the circuit

$$P_s = P_R + P_C + P_L \tag{3.45}$$

The power from the source is $P_s = vi$ and the expressions for power from the circuit components are given in Equations (3.10), (3.19), and (3.27). Substitution into Equation (3.45) leads to

$$vi = i^2 R_{eq} + C_{eq}v\frac{dv}{dt} + L_{eq}i\frac{di}{dt} \tag{3.46}$$

Using Equation (3.17), the relation between current and voltage change across a capacitor, in Equation (3.46) leads to

$$vi = i^2 R_{eq} + \frac{1}{C_{eq}}i\int_0^t i\,dt + L_{eq}i\frac{di}{dt} \tag{3.47}$$

Factoring $i$ from both sides of Equation (3.47) and rejecting the possibility that $i(t) = 0$ for all $t$ leads to

$$L_{eq}\frac{di}{dt} + R_{eq}i + \frac{1}{C_{eq}}\int_0^t i\,dt = v(t) \tag{3.48}$$

Differentiation of Equation (3.48) with respect to time gives

$$L_{eq}\frac{d^2i}{dt^2} + R_{eq}\frac{di}{dt} + \frac{1}{C_{eq}}i = \frac{dv}{dt} \tag{3.49}$$

Equation (3.49) is of the same form as Equation (2.130), which governs the motion of a one-degree-of-freedom mechanical system. Both equations can be derived using energy principles, and their systems' behavior can be explained using energy principles.

When the mechanical system is subject to an external force resulting in motion, the spring stores potential energy and the inertia elements store kinetic energy. The viscous damper dissipates the energy. At any time the energy stored in the system is the sum of the kinetic energy and the potential energy.

When the series $LRC$ circuit is connected to a voltage source that results in current flow, the inductor stores energy in a magnetic field and the capacitor stores energy in an electric field. The resistor dissipates the energy. At any time the energy stored in the system is the sum of the energy stored in the inductor and the energy stored in the capacitor.

It is clear there is a direct analogy between the mechanical system and the electrical system. The resistor is analogous to the viscous damper, the capacitor is analogous to the spring, and the inductor is analogous to the mass. The preceding discussion shows the analogy between the parameters of a mechanical system and the parameters of a series $LRC$ circuit.

However, the analogy between the dependent variables is not quite as clear. An alternative formulation of Equation (3.48) is to use electric charge as a dependent variable rather than current. Recalling $q = di/dt$, where $q$ is electric charge, Equation (3.48) can be written as

$$L\frac{d^2q}{dt^2} + R\frac{dq}{dt} + \frac{1}{C}q = v(t) \tag{3.50}$$

Comparing Equation (3.50) with Equation (2.130), it is clear that there is a direct analogy between the external force in a mechanical system and the voltage source for a series $LRC$ circuit when electric charge is analogous to displacement. Note that the power delivered to a mechanical system by an external force is $F\dot{x}$ and the power delivered to an electric circuit by an external voltage source is $vi = v\dot{q}$. Thus the direct physical analogies are between displacement and electric charge and between force and voltage. However, since the dependent variable used in circuit analysis is usually current rather than electric charge, a mathematical analogy is used

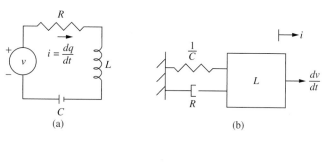

(a)                                                                (b)

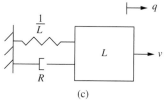

(c)

**FIG. 3.19**  (a) A series $LRC$ circuit; (b) an equivalent one-degree-of-freedom mechanical system using the analogy between current and displacement; (c) the equivalent mechanical system using the analogy between charge and displacement.

between current and displacement and force and rate of change of voltage. This analogy is illustrated in Figure 3.19.

### 3.6.3 Single-Loop Circuits with Current Sources

Now consider a circuit in which the external energy is supplied by a current source, such as the parallel $LRC$ circuit of Figure 3.20(a). The differential equation for this circuit in the absence of the current source, using KVL and KCL, is derived in Example 3.12. Using a similar analysis, note that if KVL is applied to each of the loops it is clear that the voltage change across each of the circuit components is equal. Application of KCL at the two nodes leads to

$$i_s = i_R + i_L + i_C \qquad (3.51)$$

Multiplying Equation (3.51) by $v$, the voltage across any of the system components, leads to

$$vi_S = vi_R + vi_L + vi_C \qquad (3.52)$$

Equation (3.52) represents a balance of power, or a time-differentiated form of Conservation of Energy. The left-hand side of Equation (3.52) is the power delivered to the circuit by the current source. The right-hand side is the sum of the power in the resistor, the capacitor, and the inductor. When these terms are replaced by Equations (3.10), (3.19), and (3.27) respectively, Equation (3.52) can be rewritten as

$$vi_s = \frac{v^2}{R} + Li_L \frac{di_L}{dt} + Cv \frac{dv}{dt} \qquad (3.53)$$

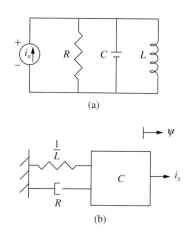

(a)

(b)

**FIG. 3.20**  (a) A parallel $LRC$ circuit; (b) the equivalent mechanical system using the analogy between magnetic flux and displacement.

Equation (3.26) is used in Equation (3.53) to substitute for the current in the inductor leading to

$$Cv\frac{dv}{dt} + \frac{v^2}{R} + \frac{1}{L}v\int_0^t v\,dt = vi_s \tag{3.54}$$

Factoring $v$ from both sides of Equation (3.54) and noting that $v(t) \neq 0$ for all $t$, leads to to

$$C\frac{dv}{dt} + \frac{1}{R}v + \frac{1}{L}\int_0^t v\,dt = i_s(t) \tag{3.55}$$

Differentiation of Equation (3.55) with respect to time gives

$$C\frac{d^2v}{dt^2} + \frac{1}{R}\frac{dv}{dt} + \frac{1}{L}v = \frac{di_s}{dt} \tag{3.56}$$

Equation (3.56) is also analogous to Equation (2.130). Energy principles are used to derive both equations. If a current source is applied to a parallel $LRC$ circuit, leading to the development of electric potential in the system, energy is stored in the capacitor and the inductor, and energy is dissipated in the resistor. The total stored energy is the sum of the energy in the capacitor and the energy in the inductor.

The mechanical-electrical analogy for a circuit with a current source is that the mass $m_{eq}$ corresponds to the capacitance $C$, the equivalent viscous damping coefficient corresponds to the reciprocal of the resistance $1/R$, and the equivalent stiffness corresponds to the reciprocal of the inductance $1/L$. However, as in the $LRC$ circuit with the voltage source, there is no direct physical correspondence between the dependent variables of displacement and electric potential. When Equation (3.55) is rewritten using the magnetic flux $\psi(x)$, which is related to the electric potential by $v = d\psi/dt$, it becomes

$$C\frac{d^2\psi}{dt^2} + \frac{1}{R}\frac{d\psi}{dt} + \frac{1}{L}\psi = i_s(t) \tag{3.57}$$

Equation (3.57) shows a direct correspondence between the magnetic flux and the displacement in a mechanical system and between the external force and the current source. The power delivered to the circuit by the external source is $vi_s = \dot\psi i_s$, which is analogous to the power $F\dot x$ delivered to the mechanical system by the external force. This analogy is illustrated in Figure 3.20(b).

Note that when the circuit has a voltage source, the inductance is analogous to the mass and the reciprocal of the capacitance is analogous to the stiffness; when the circuit has a current source, the capacitance is analogous to the mass and the reciprocal of the inductance is analogous to the stiffness. This switch in analogous components is explained by examining how the applied load leads to stored energy. In the mechanical system the application of an external force leads to a displacement and energy

**TABLE 3.4 SUMMARY OF MECHANICAL-ELECTRICAL ANALOGIES**

| Mechanical System | Circuit with Voltage Source | Circuit with Current Source |
|---|---|---|
| System parameters | | |
| Mass ($m$) | $L$ | $C$ |
| Stiffness ($k$) | $\dfrac{1}{C}$ | $\dfrac{1}{L}$ |
| Viscous damper ($c$) | $R$ | $\dfrac{1}{R}$ |
| Physical analogy | | |
| Displacement ($x$) | $q$ | $\psi$ |
| Force ($F$) | $v$ | $i$ |
| Velocity ($\dot{x}$) | $i$ | $v$ |
| Energy and power | | |
| Kinetic energy $\left(\frac{1}{2}m\dot{x}^2\right)$ | $\frac{1}{2}Li^2$ | $\frac{1}{2}Cv^2$ |
| Potential energy $\left(\frac{1}{2}kx^2\right)$ | $\frac{1}{2}\dfrac{q^2}{C}$ | $\frac{1}{2}\dfrac{\psi^2}{L}$ |
| Dissipated energy $\int c\dot{x}dx$ | $\int Ridq$ | $\displaystyle\int \frac{1}{R}vd\psi$ |
| Power ($F\dot{x}$) | $v\dot{q}=vi$ | $i\dot{\psi}=iv$ |

stored in the spring proportional to the square of the displacement and the kinetic energy proportional to the square of the time derivative of the displacement. A voltage source in the electric circuit leads to energy stored in the capacitor proportional to the square of the electric charge (the variable physically analogous to the displacement) and the energy stored in the inductor proportional to the square of the current (the square of the time derivative of charge). A current source in an electric circuit leads to energy stored in the inductor proportional to the magnetic flux (the variable physically analogous to the displacement) and the energy stored in the capacitor proportional to the square of the voltage (the square of the time derivative of the magnetic flux).

The mechanical-electrical analogies including stored energy for voltage sources and current sources are summarized in Table 3.4.

### 3.6.4 Multiple-Loop Circuits

Multiple-loop circuits have direct analogies to multidegree-of-freedom mechanical systems. When the circuit has a voltage source an inductor is analogous to a mass, a resistor is analogous to a viscous damper, and a capacitor is analogous to a spring. The circuit components common to multiple loops provide the coupling between the loops, and their analogous components provide the coupling between the masses.

## Example 3.14

(a) Draw the analogous mechanical system for the circuit of Figure 3.21(a).
(b) Write the differential equations for both the mechanical system and the electrical system.

### Solution

(a) The analogous mechanical system is shown in Figure 3.21(b). The differential equations for the circuit are derived by applying KVL and KCL are

$$L_1 \frac{di_1}{dt} + R_1 i_1 + R_2(i_1 - i_2) + \frac{1}{C} \int_0^t (i_1 - i_2) dt = v(t) \tag{a}$$

$$L_2 \frac{di_2}{dt} + R_3 i_2 + R_2(i_2 - i_1) + \frac{1}{C} \int_0^t (i_2 - i_1) dt = 0 \tag{b}$$

The differential equations for the analogous system are

$$m_1 \ddot{x}_1 + (c_1 + c_2)\dot{x}_1 - c_2 \dot{x}_2 + kx_1 - kx_2 = F(t) \tag{c}$$

$$m_2 \ddot{x}_2 - c_2 \dot{x}_1 + (c_2 + c_3)\dot{x}_2 - kx_1 + kx_2 = 0 \tag{d}$$

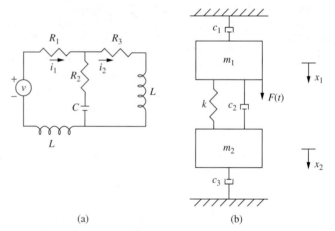

**FIG. 3.21** (a) The electric circuit of Example 3.14; (b) the analogous mechanical system.

(a)                    (b)

## Example 3.15

Draw the analogous electric circuit for the mechanical system of Figure 3.22(a). Write the differential equations for both the mechanical system and the electric circuit.

### Solution

The analogous electric circuit is shown in Figure 3.22(b). The differential equations for the mechanical system are derived by applying Newton's second law to free-body diagrams of the masses and the connection whose displacement

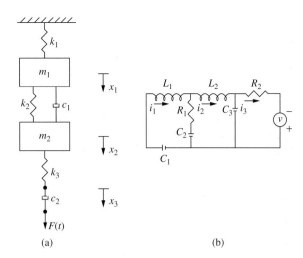

**FIG. 3.22** (a) The mechanical system of Example 3.15; (b) the analogous electric circuit.

(a)                    (b)

is $x_3$. The differential equations are

$$m_1\ddot{x}_1 + c_1\dot{x}_1 - c_1\dot{x}_2 + (k_1 + k_2)x_1 - k_2x_2 = 0 \tag{a}$$

$$m_2\ddot{x}_2 - c_1\dot{x}_1 + c_1\dot{x}_2 - k_2x_1 + (k_2 + k_3)x_2 - k_3x_3 = 0 \tag{b}$$

$$c_2\dot{x}_3 - k_3x_2 + k_3x_3 = F(t) \tag{c}$$

The differential equations of the analogous electrical system are

$$L_1\frac{di_1}{dt} + R_1(i_1 - i_2) + \frac{1}{C_1}\int_0^t i_1\,dt + \frac{1}{C_2}\int_0^t (i_1 - i_2)\,dt = 0 \tag{d}$$

$$L_2\frac{di_2}{dt} - R_1(i_1 - i_2) - \frac{1}{C_2}\int_0^t (i_1 - i_2)\,dt + \frac{1}{C_3}\int_0^t (i_2 - i_3)\,dt = 0 \tag{e}$$

$$R_2i_3 - \frac{1}{C_3}\int_0^t (i_2 - i_3)\,dt = v(t) \tag{f}$$

### 3.6.5 Mechanical Systems with Motion Input

Electric circuit analogies can also be made for mechanical systems with motion input. Consider the system of Figure 3.23 in which a mass is attached to a spring in parallel with a viscous damper attached to a base with a prescribed motion. The mathematical model for the displacement of the mass is

$$m\ddot{x} + c\dot{x} + kx = c\dot{y} + ky \tag{3.58}$$

Now consider the circuit of Figure 3.24. Application of KCL at node $B$ gives

$$i(t) - i_1 - i_2 = 0 \tag{3.59}$$

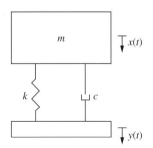

**FIG. 3.23** A mechanical system with motion input.

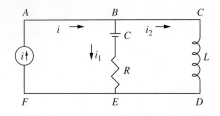

**FIG. 3.24** The electric circuit analogy for a mechanical system with motion input.

Application of KVL around loop $BCDE$ leads to

$$L\frac{di_2}{dt} + Ri_2 - \frac{1}{C}\int_0^t i_1\,dt = 0 \qquad (3.60)$$

Equation (3.59) is rearranged as $i_1 = i(t) - i_2$, which when substituted into Equation (3.60) leads to

$$L\frac{di_2}{dt} + Ri_2 + \frac{1}{C}\int_0^t i_2\,dt = Ri + \frac{1}{C}\int_0^t i\,dt \qquad (3.61)$$

Using the relationships between current and charge, $i_2 = dq_2/dt$ and $i = dq/dt$, in Equation (3.61) leads to

$$L\frac{d^2q_2}{dt^2} + R\frac{dq_2}{dt} + \frac{1}{C}q_2 = R\frac{dq}{dt} + \frac{1}{C}q \qquad (3.62)$$

Comparison of Equation (3.62) with Equation (3.58) illustrates the analogy. The displacement of the inertia element in the mechanical system is analogous to the charge flowing through the inductor and resistor, and the base displacement is analogous to the charge provided by the current source. The inertia element is analogous to the inductor, the viscous damper is analogous to the resistor, and the spring is analogous to the capacitor. These component analogies are the same as in the analogy between a mechanical system with force input and an electrical circuit with a voltage source.

In general the electric circuit analogy for a mechanical system with motion input is that of a current source in parallel with circuit components analogous to those of the mechanical system to which the motion input is prescribed. The analogy corresponding to the remainder of the mechanical system is placed in parallel with this combination. The component analogy is the same as that is used for the analogy between a mechanical system with a force input and an electric circuit with a voltage source.

---

**Example 3.16**

Draw the electric circuit analogy to the suspension system model of Figure 3.25(a).

**Solution**

The motion input for the suspension system is provided to the parallel-spring-viscous-damper combination. The analogous electric circuit has a current source in parallel with a capacitor and viscous damper. The analogy for the remainder of the mechanical system is that of a capacitor added in parallel and then an inductor added in parallel as illustrated in Figure 3.25(b).

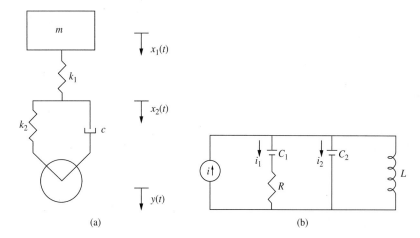

FIG. **3.25** (a) The suspension system mode of Example 3.16; (b) the analogous electric circuit.

The differential equations modeling the mechanical system are

$$m\ddot{x} + k_1(x_2 - x_1) = 0 \qquad\qquad \text{(a)}$$
$$k_1(x_2 - x_1) + c(\dot{y} - \dot{x}_1) + k_2(y - x_1) = 0 \qquad\qquad \text{(b)}$$

The differential equations modeling the electric circuit obtained through the application of KCL at the nodes and KVL around the loops are

$$L\frac{di_2}{dt} + \frac{1}{C_1}\int_0^t (i_2 - i_1)\,dt = 0 \qquad\qquad \text{(c)}$$

$$\frac{1}{C_1}\int_0^t (i_2 - i_1)\,dt + R(i - i_1) + \frac{1}{C_2}\int_0^t (i - i_1)\,dt = 0 \qquad\qquad \text{(d)}$$

### 3.6.6 States

As noted in Section 2.9 a state corresponds to an energy storage component. Possible choices for state variables in a mechanical system are velocity (kinetic energy is proportional to the square of velocity) and displacement (potential energy is proportional to the square of displacement). However, the choice of state variables is not unique.

States in an electric circuit also correspond to independent energy storage components. Each independent inductor and capacitor corresponds to a state. From the discussion in Section 3.6.2 it is clear that appropriate choices of state variables for a circuit with voltage sources are current through an inductor (stored energy is proportional to the square of current) and the change on a capacitor (stored energy is proportional to the square of charge). From the discussion of Section 3.6.3 it is clear that appropriate choices for state variables for a circuit with current sources are magnetic flux in an inductor and voltage change across a capacitor.

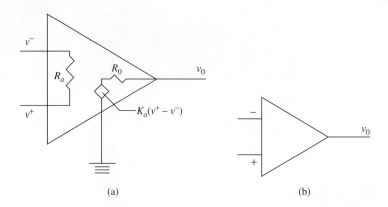

**FIG. 3.26** (a) The output from an operational amplifier is $K_a(v^+ - v^-)$; (b) an ideal operational amplifier has no current and $v^+ = v^-$.

(a)                                            (b)

---

## 3.7 OPERATIONAL AMPLIFIERS

An operational amplifier, schematically illustrated in Figure 3.26, is a device used in circuits for control systems and sensors. The operational amplifier has two input terminals, one positive and one negative. An input voltage passes through a resistor of very high resistance such that the current through the amplifier is

$$i_a = \frac{v^+ - v^-}{R_a} \tag{3.63}$$

Since the resistance is very high, the current is small. The operational amplifier contains a dependent source with a voltage

$$v_0 = K_a(v^+ - v^-) \tag{3.64}$$

where $K_a$ is the amplifier gain. If $|v^+ - v^-|$ exceeds that maximum voltage provided by an external power supply the amplifier is saturated and its voltage output is this maximum. The amplifier may have a second resistor with very low resistance such that unless saturated the output voltage from the amplifier is given by Equation 3.64.

    An ideal operational amplifier has $R_a = \infty$, $R_o = 0$, $K_a = \infty$, and $v^+ = v^-$. The mathematical modeling of circuits with operational amplifiers presented assumes ideal unsaturated amplifiers.

---

**Example 3.17**

Determine the output $v_2$ for each of the operational amplifier circuits of Figure 3.27.

**Solution**

    (a) Application of KCL at node $A$ gives

$$i_1 - i_2 - i_3 = 0 \tag{a}$$

Since the input to the positive terminal of the amplifier is zero and an ideal amplifier is assumed, the input to the negative terminal $v_A$ must also be zero and $i_3 = 0$. The voltage at $B$ is $v_2$; thus from Ohm's law $i_2 = \frac{v_A - v_2}{R_2} = \frac{-v_2}{R_2}$ and

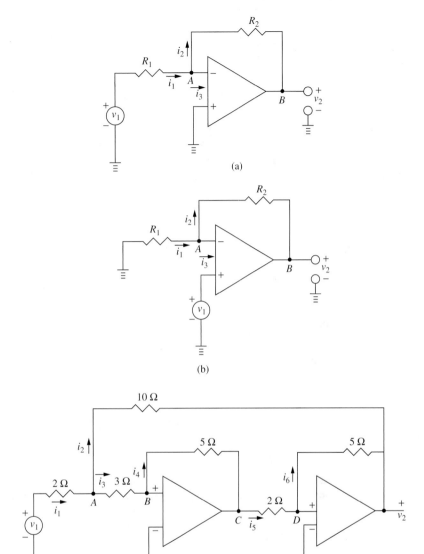

**FIG. 3.27** Circuits for Example 3.17.

$i_1 = \frac{v_1 - v_A}{R_1} = \frac{v_1}{R_1}$. Substituting into Equation (a) leads to

$$\frac{v_1}{R_1} + \frac{v_2}{R_2} = 0 \qquad (b)$$

Equation (b) is rearranged leading to

$$v_2 = -\frac{R_2}{R_1} v_1 \qquad (c)$$

The output from the circuit of Figure 3.27(a) is proportional to the input, but with the opposite sign. This circuit is called an inverting circuit.

(b) Application of KCL at node $A$ leads to Equation (a) of part a. The input to the positive terminal of the operational amplifier is $v_1$. Assuming an ideal amplifier, $i_3 = 0$ and $v_A = v_1$. The currents are then identified as

$$i_1 = \frac{-v_A}{R_1} = \frac{-v_1}{R_1} \tag{d}$$

$$i_2 = \frac{v_A - v_B}{R_2} = \frac{v_1 - v_2}{R_2} \tag{e}$$

Substitution of Equations (d) and (e) into Equation (a) leads to

$$\frac{-v_1}{R_1} - \frac{v_1 - v_2}{R_2} = 0 \tag{f}$$

Equation (f) is solved for $v_2$, yielding

$$v_2 = \left(1 + \frac{R_2}{R_1}\right)v_1 \tag{g}$$

Since the gain from the circuit of Figure 3.27(b) is positive it is called a noninverting circuit.

(c) Application of KCL at nodes $A$, $B$, and $D$, assuming ideal amplifiers, leads to

$$i_1 - i_2 - i_3 = 0 \tag{h}$$

$$i_3 - i_4 = 0 \tag{i}$$

$$i_5 - i_6 = 0 \tag{j}$$

The branch currents in terms of the voltages at the nodes are

$$i_1 = \frac{v_1 - v_A}{2} \tag{k}$$

$$i_2 = \frac{v_A - v_2}{10} \tag{l}$$

$$i_3 = \frac{v_A - v_B}{3} \tag{m}$$

$$i_4 = \frac{v_B - v_C}{5} \tag{n}$$

$$i_5 = \frac{v_C - v_D}{2} \tag{o}$$

$$i_6 = \frac{v_D - v_2}{5} \tag{p}$$

Since the input voltages to the positive terminals of each of the amplifiers is zero the voltages at the negative terminals are also zero, $v_B = 0$ and $v_D = 0$. Substituting Equations (k)–(p) into Equations (h)–(j) leads to

$$\frac{v_1 - v_A}{2} - \frac{v_A - v_2}{10} - \frac{v_A}{3} = 0 \tag{q}$$

$$\frac{v_A}{3} - \frac{-v_C}{5} = 0 \tag{r}$$

$$\frac{v_2}{2} - \frac{-v_2}{5} = 0 \tag{s}$$

Equations (r)–(s) are solved simultaneously leading to

$$v_2 = \frac{125}{11}v_1 \tag{t}$$

An alternate solution to part c is to note that both operational amplifiers in the system act as inverting amplifiers. Using the results of part a, $v_C = -\frac{5}{3}v_A$ and $v_2 = -\frac{3}{2}v_C$. These are used in conjunction with Equations (h), (k), (l), and (m) to arrive at the same solution.

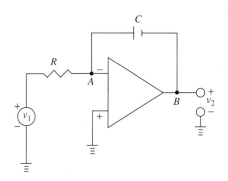

**FIG. 3.28** An operational amplifier circuit with a capacitor used to integrate input voltage.

Consider the operational amplifier circuit of Figure 3.28, in which a capacitor is used to bridge the amplifier. Application of KCL at node $A$ and recalling that the relation between current and voltage in a capacitor may be written as $i = C\frac{dv}{dt}$ leads to

$$\frac{v_1}{R} + C\frac{dv_2}{dt} = 0 \qquad (3.65)$$

Integrating both sides of Equation (3.65) with respect to time and solving for $v_2$ leads to

$$v_2 = -\frac{1}{RC}\int_0^t v_1 dt \qquad (3.66)$$

The output voltage from this circuit is proportional to the integral of the input. Thus the circuit of Figure 3.28 is an integrating circuit and in conjuction with an inverting circuit may be used as an electronic integral controller.

Consider the circuit of Figure 3.29 in which an inductor bridges an operational amplifier. Application of KCL at node $A$, recalling that the current-voltage relation in an inductor can be written as $i = \frac{1}{L}\int_0^t v\,dt$, leads to

$$\frac{v_1}{R} + \frac{1}{L}\int_0^t v_2 dt = 0 \qquad (3.67)$$

Differentiating Equation (3.67) with respect to time and solving for $v_2$ leads to

$$v_2 = -\frac{L}{R}\frac{dv_1}{dt} \qquad (3.68)$$

The output voltage from this circuit is the time derivative of the input. This circuit is a differentiating circuit.

Operational amplifier circuits with capacitors or inductors are used in the design of electronic control systems. Analog computers are designed

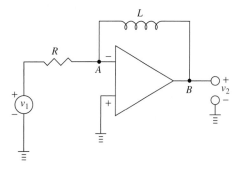

**FIG. 3.29** An operational amplifier circuit with an inductor used to differentiate input voltage.

using operational amplifiers, resistors, capacitors, and indicators to simulate the solution of differential equations.

## Example 3.18

Determine the relation between the input $v_1$ and output $v_2$ in the circuit of Figure 3.30.

### Solution

Application of KCL at node $A$ gives

$$i_1 - i_2 - i_3 = 0 \tag{a}$$

Since the positive terminal of the operational amplifier is connected to the ground, the voltage at $A$ is zero and since the amplifier is assumed to be ideal, $i_3 = 0$ and thus

$$i_1 = \frac{v_1}{R_1} \tag{b}$$

The current in the parallel combination of the inductor and the resistor is the sum of the currents in each component. The voltage change across each component is $v_A - v_2 = -v_2$. Thus

$$i_2 = -\frac{v_2}{R_2} - \frac{1}{L}\int_0^t v_2 dt \tag{c}$$

Substitution of Equations (b) and (c) into Equation (a) leads to

$$\frac{v_1}{R_1} + \frac{v_2}{R_2} + \frac{1}{L}\int_0^t v_2 dt = 0 \tag{d}$$

Equation (d) can be rewritten as

$$v_2 + \frac{R_2}{L}\int_0^t v_2 dt = -\frac{R_2}{R_1}v_1 \tag{e}$$

The output from the circuit is the solution to the integral equation, Equation (e).

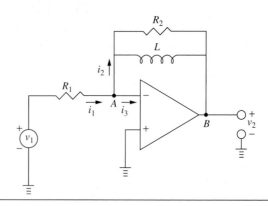

**FIG. 3.30** The operational amplifier circuit of Example 3.18 simulates the solution of the integral equation.

**Example 3.19**

Determine the relation between the input $v_1$ and output $v_2$ in the circuit of Figure 3.31.

**Solution**

Application of KCL at nodes $A$ and $C$ gives

$$i_1 - i_2 - i_3 = 0 \tag{f}$$
$$i_4 - i_5 - i_6 = 0 \tag{g}$$

The operational amplifiers are assumed to be ideal. Thus $i_3 = i_6 = 0$. Since the positive terminal of the first amplifier is connected to the ground $v_A = 0$ and since the positive terminal of the second amplifier is connected to the external source, $v_C = 100 \sin(80t)$. The other currents are

$$i_1 = \frac{100 \sin(100t)}{2{,}000} \tag{h}$$

$$i_2 = -\frac{v_B}{4{,}000} - 30 \times 10^{-6} \frac{dv_B}{dt} \tag{i}$$

$$i_4 = \frac{v_B - 100 \sin(80t)}{6{,}000} \tag{j}$$

$$i_5 = -30 \times 10^{-6} \frac{d}{dt}(v_2 - 100 \sin 80t) \tag{k}$$

Substitution of Equations (h)–(k) in Equations (f) and (g) leads to

$$\frac{100 \sin(100t)}{2{,}000} + \frac{v_B}{4{,}000} + 30 \times 10^{-6} \frac{dv_B}{dt} = 0 \tag{l}$$

$$\frac{v_B - 100 \sin(80t)}{6{,}000} + 30 \times 10^{-6} \frac{d}{dt}[v_2 - 100 \sin(80t)] = 0 \tag{m}$$

Equation (m) is used to solve for $v_B$, which is then substituted into Equation (l), leading to

$$\frac{100 \sin(100t)}{2{,}000} + \frac{100 \sin(80t)}{4{,}000} - \frac{6{,}000}{4{,}000}(30 \times 10^{-6})\frac{d}{dt}[v_2 - 100 \sin(80t)]$$

$$+ (30 \times 10^{-4})\frac{d}{dt}[\sin(80t)] - (30 \times 10^{-6})^2(6{,}000)\frac{d^2}{dt^2}[v_2 - 100 \sin(80t)] \tag{n}$$

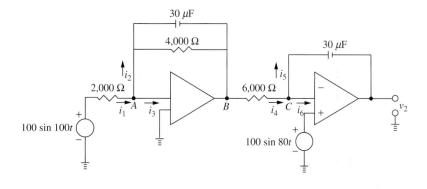

**FIG. 3.31** The operational amplifier circuit of Example 3.19.

Simplification of Equation (n) leads to

$$\frac{d^2v_2}{dt^2} + 8.33\frac{dv_2}{dt} = 1.11 \times 10^5\cos(80t) - 6.34 \times 10^5\sin(80t) + 9.26 \times 10^3\sin(100t)$$

(o)

The output from the circuit is the solution of the differential equation, Equation (o).

## 3.8 ELECTROMECHANICAL SYSTEMS

**FIG. 3.32** Schematic of an electromechanical system: Energy is transferred from the electrical and mechanical systems and stored in the coupling field.

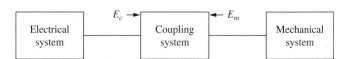

An **electromechanical system** is a system in which mechanical and electrical systems interact such that energy is transferred from one system to another. The energy transfer occurs in a **coupling field**, also called **a transducer**, as illustrated schematically in Figure 3.32. The coupling field may be an electrostatic or electromagnetic field. Whereas the general theory is the same regardless of which field is used, motors and many electromechanical energy conversion devices use electromagnetic fields. A brief review of the necessary concepts of magnetic fields is presented first, followed by an outline of the general theory of electromechanical systems. The theory is applied to loudspeakers, dc servomotors, and the developing field of microelectromechanical systems (MEMS).

### 3.8.1 Magnetic Fields

A magnetic field surrounds an electric charge in motion. A force acting perpendicular to the direction of motion acts on the charge. The force is given by Lorenz's law as

$$\mathbf{F} = q(\mathbf{V} \times \mathbf{B}) \tag{3.69}$$

where $q$ is the charge of the particle, $\mathbf{V}$ is its velocity, and $\mathbf{B}$ is the flux density of the magnetic field. When the charge is moving through a conductor perpendicular to the magnetic field, as shown in Figure 3.33, Lorenz's law shows that the magnitude of the force is $F = Bi\ell$ where $i$ is the current and $\ell$ is the length of the conductor. A voltage increase across the conductor of $v = B\ell V$ is induced.

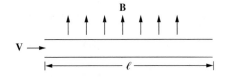

**FIG. 3.33** Current moving through a conductor of length $\ell$ with a velocity perpendicular to a magnetic field of flux density $\mathbf{B}$ has a magnetic force $F = Bi\ell$ and leads to an increase in voltage $v = B\ell V$ across the conductor.

This type of conductor may act as a generator or a motor. If it acts as a generator a mechanical system must deliver power equal to $P = FV = Bi\ell V$, and the power generated by the conductor is $P = vi = Bi\ell V$.

If the conductor acts as a motor it moves freely, being acted on and accelerated by the magnetic force. The conductor produces an increase in voltage, often called an electromotive force, to balance the magnetic force.

The simple descriptions of a generator and a motor can be generalized and made more complex by considering a time-varying magnetic field and a conductor consisting of a wire of $N$ coils. In this case the voltage is given by Faraday's law

$$v(t) = \frac{d\lambda}{dt} \tag{3.70}$$

where

$$\lambda = N\phi(t) \tag{3.71}$$

is called the flux linkage and $\phi(t)$ is the time-varying flux of the magnetic field (flux density equals flux/area).

Faraday's law is used to write the inductance of the conductor, defined as in Equation (3.23), in terms of the flux linkage as

$$L = \frac{\lambda}{i} \tag{3.72}$$

### 3.8.2  General Theory

A model of an electromechanical system is illustrated in Figure 3.32. The coupling field stores energy provided by the electrical system and the mechanical system. Assuming the coupling field is conservative, that is, no energy is dissipated in the coupling field, the application of Conservation of Energy to the coupling field leads to

$$E_c = E_e + E_m \tag{3.73}$$

where $E_c$ is the energy stored in the coupling field, $E_e$ is the energy transferred by the electric system, and $E_m$ is the energy transferred by the mechanical system. Differentiating Equation (3.73) with respect to time leads to a power form of the conservation law.

A simple electromechanical actuator is shown in Figure 3.34. The electrical system is a circuit with a voltage source, an inductor, and a resistor. The coupling field is a magnetic circuit connected to a mass-spring-viscous-damper system. The motion of the mechanical system induces a

**FIG. 3.34** The schematic of an electromechanical actuator. The electrical system, an *RL* circuit with a voltage source, transfers energy to the magnetic coupling field. The coupling field stores energy and transforms it into a coupling force acting on the mechanical system.

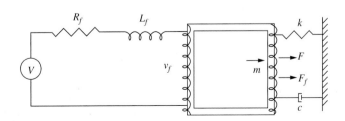

voltage drop $v_f$ in the coupling and the coupling field produces a force $F_f$, which acts on the mechanical system. Let $i(t)$ represent the current developed in the circuit and $x(t)$ represent the displacement of the mechanical system from its equilibrium position. The power developed in the transducer due to the induced voltage drop is $v_f i$, while the power used by the mechanical system is $F_f \dot{x}$. Thus the total power in the transducer is

$$\frac{dE_c}{dt} = v_f i - F_f \dot{x} \tag{3.74}$$

A differential form of Equation (3.74) is

$$dE_c = v_f i\, dt - F_f dx \tag{3.75}$$

Using the definition of flux linkage, Equation (3.70), Equation (3.75) becomes

$$dE_c = i\frac{d\lambda}{dt}dt - F_f dx$$
$$= i\, d\lambda - F_f dx \tag{3.76}$$

The energy stored in the transducer is a function of the flux linkage and the displacement, $E_c = E_c(\lambda, x)$. Since $E_c$ is a function of two variables, its total differential is

$$dE_c = \frac{\partial E_c}{\partial \lambda}d\lambda + \frac{\partial E_c}{\partial x}dx \tag{3.77}$$

Comparison of Equations (3.76) and (3.77) leads to

$$i = \frac{\partial E_c}{\partial \lambda} \tag{3.78}$$

$$F_f = -\frac{\partial E_c}{\partial x} \tag{3.79}$$

The force generated in the coupling field and acting on the mechanical system is determined using Equation (3.79).

---

**Example 3.20**

Consider the electromechanical actuator of Figure 3.34. The mechanical system is subject to an external force $F(t)$ and the electrical system has a voltage source $v(t)$. Let $x$ be the displacement of the mechanical system from where the spring is unstretched. It can be shown that the energy stored in the transducer is

$$E_c = \tfrac{1}{2}Li^2 \tag{a}$$

and that the inductance varies with $x$ according to

$$L = \frac{c_1}{c_2 + x} \tag{b}$$

Derive a mathematical model for the current in the circuit and the displacement of the mass from equilibrium.

### Solution

Combining Equations (a) and (b) gives

$$E_c = \tfrac{1}{2} \left( \frac{c_1}{c_2 + x} \right) i^2 \tag{c}$$

Equations (a) and (3.72) are combined to give

$$E_c = \tfrac{1}{2} \frac{\lambda^2}{L} \tag{d}$$

Using Equation (d), note that

$$\frac{\partial E_c}{\partial \lambda} = \frac{\lambda}{L} = i \tag{e}$$

Thus Equation (3.78) is satisfied.

The coupling force applied to the mechanical system is determined from Equations (3.72) and (c) as

$$F_c = -\frac{\partial E_c}{\partial x} = \frac{c_1 i^2}{2(c_2 + x)^2} \tag{f}$$

Application of Newton's law to the free-body diagram of Figure 3.35 gives

$$\sum F = ma$$
$$F(t) + F_c - kx + c\dot{x} = m\ddot{x}$$
$$m\ddot{x} + c\dot{x} + kx - \frac{c_1 i^2}{2(c_2 + x)^2} = F(t) \tag{g}$$

Application of Kirchoff's voltage law to the electrical system gives

$$v = Ri + L\frac{di}{dt} + v_c \tag{h}$$

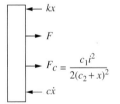

**FIG. 3.35** The free-body diagram of the mechanical system of Example 3.20.

---

## Example 3.21

Loudspeakers are electromechanical systems in which an electric circuit is used to produce a force that moves an object, usually a cone, in a direction to amplify sound waves. A voltage provided to the voice coil results in a coupling force that moves the cone, thus radiating acoustic waves. An electromechanical model for the loudspeaker of Figure 3.36(a) is illustrated in Figure 3.36(b). The coupling force acting on the cone is $K_1 i_2$ while the induced voltage in the voice coil (also called the back emf or back electromotive force) is $K_2 x$. Derive a mathematical model for the response of the loudspeaker.

### Solution

Let $x$ be the displacement of the cone from its equilibrium position. The application of Newton's law to a free-body diagram of the cone leads to

$$K_1 i_2 - kx - c\dot{x} = m\ddot{x}$$
$$m\ddot{x} + c\dot{x} + kx = K_1 i_2 \tag{a}$$

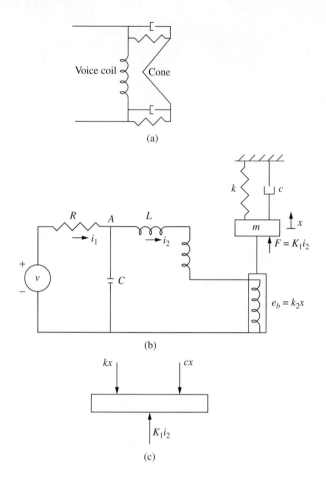

**FIG. 3.36** (a) A voltage is supplied to the voice coil of a loudspeaker, resulting in a coupling force that involves the loudspeaker cone; (b) the typical electric circuit of a loudspeaker; (c) the free-body diagram of the speaker cone.

The application of KCL to node $A$ shows that the current through the capacitor is $i_1 - i_2$. Subsequent application of KVL to each of the loops gives

$$Ri_1 + \frac{1}{C}\int_0^t (i_1 - i_2)\,dt = v \tag{b}$$

$$L\frac{di_2}{dt} - \frac{1}{C}\int_0^t (i_1 - i_2)\,dt = -K_2 x \tag{c}$$

## 3.8.3  Direct Current (dc) Servomotors

Direct current (dc) servomotors consist of a field circuit and an armature circuit. Figures 3.37 and 3.38 show separately controlled dc motors. Other possible configurations of the two circuits are shunt connection, series connection, and compound connection. For the separately connected

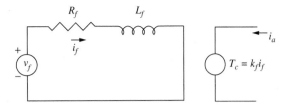

FIG. **3.37** A field-controlled DC motor with the armature current constant.

configurations let $i_a$ be the current in the armature circuit and let $i_f$ be the current in the field circuit. The armature circuit contains a motor of moment of inertia $J$, which rotates with an angular displacement $\theta(t)$. The technique for developing mathematical models of dc servomotors is as described in Section 3.8.2 except that the coupling field leads to a coupling torque applied to the armature and the torque is related to the coupling energy by

$$T_c = -\frac{\partial E_c}{\partial \theta} \tag{3.80}$$

Equation (3.80) is used to show that

$$T_c = M_{af} i_a i_f \tag{3.81}$$

where $M_{af}$ is the mutual inductance between the inductors in the armature circuit and the field circuit.

A field-controlled dc motor is one in which the armature current is constant and the field current varies with time. In this case Equation (3.81) is written as

$$\begin{aligned} T_c &= (M_{af} i_a) i_f \\ &= K_f i_f \end{aligned} \tag{3.82}$$

In an armature-controlled motor the field current is constant while the armature current varies with time. In this case Equation (3.81) is written as

$$\begin{aligned} T_c &= (M_{af} i_f) i_a \\ &= K_a i_a \end{aligned} \tag{3.83}$$

In an armature-controlled motor the rotation of the motor induces a voltage in the armature circuit, commonly called the back electromotive

FIG. **3.38** The armature-controlled dc motor field current is constant. Back emf is induced by the rotation of the motor.

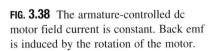
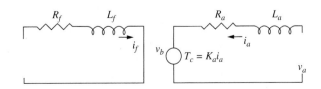

force $v_b$, which is proportional to the angular speed of the shaft

$$v_b = M_{af} i_f \omega \tag{3.84}$$

### Example 3.22

Derive a mathematical model for the field-controlled dc motor of Figure 3.39. The motor of moment of inertia $J$ is on a shaft with friction in its bearings with an equivalent torsional damping coefficient $c_t$. The shaft is subject to an external torque load $T$. The field circuit has a voltage source $v_f$, a resistance $R_f$, and an inductance $L_f$.

### Solution

The application of Kirchoff's loop law to the field circuit leads to

$$v_f = R_f i_f + L_f \frac{di_f}{dt} \tag{a}$$

The application of the moment equation to the motor leads to

$$\sum M = J\alpha$$

$$T_c + T - c_t\omega = J\frac{d\omega}{dt}$$

$$J\frac{d\omega}{dt} + c_t\omega - K_f i_f = T \tag{b}$$

Equations (a) and (b) constitute a mathematical model of the motor.

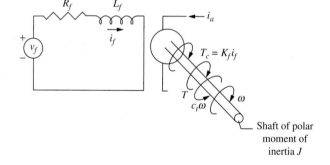

**FIG. 3.39** The field-controlled dc motor of Example 3.22. The shaft is mounted on bearings with torsional damping coefficient $c_t$ and subject to a load $T$.

### Example 3.23

Derive a mathematical model for the armature-controlled dc servomotor of Figure 3.40. The motor of moment of inertia $J$ is on a shaft with friction in its bearings with an equivalent torsional damping coefficient $c_t$. The armature circuit has a voltage source $v_a$, a resistance $R_a$, and an inductance $L_a$.

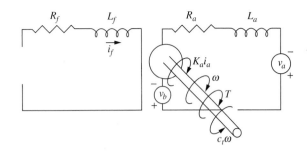

**FIG. 3.40** The dc servomotor of Example 3.23.

### Solution

The rotation of the motor induces a voltage in the armature circuit given by Equation (3.86). Application of Kirchoff's law to the armature circuit leads to

$$v_a = R_a i_a + L_a \frac{di_a}{dt} + v_b$$

$$L_a \frac{di_a}{dt} + R_a i_a + K_b \omega = v_a \tag{a}$$

Application of the moment equation to the motor gives

$$\sum M = J\alpha$$

$$T_c + T - c_t \omega = J \frac{d\omega}{dt}$$

$$J \frac{d\omega}{dt} + c_t \omega - K_a i_a = T \tag{b}$$

## 3.8.4 Microelectromechanical Systems (MEMSs) and Nanoelectromechanical Systems (NEMSs)

The terms "MEMS" and "NEMS" refer to electromechanical systems that are very small in physical size. MEMS refers to systems that operate on approximately the microscale (1 $\mu$m $= 10^{-6}$ m). An important example of a MEMS device is a chip on a silica substrate. Such chips have applications in computer technology. Other MEMS applications include micropumps driven by electrostatic or magnetic forces, micromirrors, and magnetic levitation devices. The principles used to model systems at the macroscale generally apply to MEMS systems. However, since displacements and motions are on a smaller scale, forces that are often ignored in macroscale modeling assume more importance in MEMS modeling. Various forms of friction, for example, are often ignored or assumed insignificant when modeling at the macroscale. However, the effect of energy dissipation caused by friction is often significant in MEMS systems. For this reason care is often taken to minimize the existence of unwanted nonconservative forces in MEMS systems.

NEMS operate on approximately the nanoscale (1 nm $= 10^{-9}$ m). The radius of one carbon atom is approximately 0.34 nm. Thus, modeling of NEMS is approaching modeling at the atomistic level at which the continuum assumption breaks down. Investigators have suggested that the limit at which the continuum model is valid is 50 nm. Modeling of devices with length scales below 50 nm requires applications of the principles of molecular dynamics. This study applies only to systems where continuum models are adequate. Assumptions used at the macroscale and even at the microscale break down at the nanoscale. For example, van der Waals forces, the repulsive and attractive forces between atoms, are often important in NEMS modeling.

The mathematical models derived for most MEMS and NEMS are nonlinear. The linearizing assumptions previously used and discussed in Chapter 1 are often not applicable. Small displacement assumptions are inherent in MEMS and NEMS modeling. Linear models obtained through perturbing about a steady state produces results of only limited applicability.

---

### Example 3.24

A schematic of a MEMS sensor, such as an accelerometer, is shown in Figure 3.41. A plate of mass $m$ and area $A$ is attached to a piezoelectric material whose effect is modeled as that of a spring of stiffness $k$ in parallel with a viscous damper of damping coefficient $c$. The bottom of the plate is subject to an applied voltage $v(t)$. The plate is parallel to a stationary, grounded plate. The total distance between the two fixed supports is $\ell$. Derive a mathematical model for $x(t)$, the distance from the charged plate to the upper fixed support.

#### Solution

The two plates, separated at time $t$ by a distance $d = \ell - x$, serve as a parallel plate capacitor. Assuming the gap between the plates has a dielectric constant of one, the capacitance is obtained using Equation (3.14) as

$$C = \varepsilon_0 \frac{A}{\ell - x} \tag{a}$$

The energy stored in the capacitor is obtained using Equation (3.21) as

$$E = \frac{1}{2} C v^2 = \frac{\varepsilon_0 A v^2}{2(\ell - x)} \tag{b}$$

The system is an example of an electromechanical system discussed earlier in this section with an electrical system supplying the voltage, the mass-spring-viscous-damper

**FIG. 3.41** The MEMS device of Example 3.24 consists of a mass suspended from a spring and viscous damper in parallel with an applied voltage to its bottom surface. As the plate moves, the capacitance between the plates varies and exerts an electrostatic coupling force on the plate.

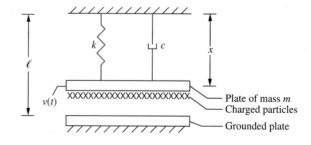

as the mechanical system, and the capacitor as the coupling field. The force acting on the mechanical system from the coupling is obtained by applying Equation (3.79) to Equation (a) resulting in

$$F_c = -\frac{\varepsilon_0 A v^2}{2} \frac{\partial}{\partial x}\left[\frac{1}{\ell - x}\right]$$

$$= -\frac{\varepsilon_0 A v^2}{2(\ell - x)^2} \tag{c}$$

Let $x_0$ be the value of $x$ when the system is in static equilibrium in the absence of an applied voltage and let $\ell_0$ be the unstretched length of the spring. A force balance on the free-body diagram of this equilibrium position leads to

$$k(x_0 - \ell_0) = mg \tag{d}$$

The application of Newton's second law to the free-body diagram of the charged plate at an arbitrary instant, shown in Figure 3.42, leads to

$$\sum F = ma$$

$$mg - k(x - \ell_0) - c\dot{x} + \frac{\varepsilon_0 A v^2}{2(\ell - x)^2} = m\ddot{x} \tag{e}$$

Equation (d) is used in Equation (e) leading to

$$m\ddot{x} + c\dot{x} + kx = kx_0 + \frac{\varepsilon_0 A v^2}{2(\ell - x)^2} \tag{f}$$

Equation (f) of Example 3.24 is a nonlinear differential equation. Since the dependent variable is not measured from the system's equilibrium position, the static spring force appears explicitly in the equation. However, if the dependent variable were measured from the static equilibrium position, the gravity force would explicitly appear in the differential equation. The statement put forth in Chapter 2 that the static spring force and gravity force cancel in the differential equations was under the caveats that the system is linear and that the dependent variable is measured from the system's equilibrium position; neither applies in this problem.

If the voltage source is a dc source, a steady state may occur after some time in which $\dot{x} = \ddot{x} = 0$. An equation for the steady-state length $x_s$ is determined using Equation (f)

$$kx_s = kx_0 + \frac{\varepsilon_0 A v^2}{2(\ell - x_s)^2} \tag{g}$$

The number of real roots with $x_s < \ell$ depends on the values of the parameters. Multiple, physically possible roots exist for certain values of the parameters. In such cases only one steady state is stable. If the electrostatic force is large compared to the spring force, Equation (g) may have no real solutions, indicating that no steady states exist. This is the origin of an instability problematic for MEMS called the "pull-in" instability in which the top plate is pulled into the lower plate.

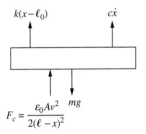

$k(x - \ell_0)$    $c\dot{x}$

$F_c = \frac{\varepsilon_0 A v^2}{2(\ell - x)^2}$    $mg$

**FIG. 3.42** The free-body diagram of the upper plate of Example 3.24, drawn at an arbitrary instant.

## 3.9 FURTHER EXAMPLES

The following assumptions are used in the succeeding examples

- All circuit components are linear and have lumped properties.
- Operational Amplifiers are ideal and unsaturated.
- Mutual inductances are negligible unless otherwise specified.
- KVL and KCL are applicable for all circuits with KVL always applied in a clockwise direction.

### Example 3.25

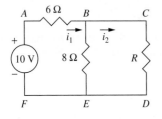

FIG. 3.43 The circuit for Example 3.25.

Determine the value of $R$ in the circuit of Figure 3.43 such that the power through the resistor is a maximum.

### Solution

Let $i_1$ be the current in the 6-$\Omega$ resistor, $i_2$ the current in the resistor of resistance $R$. The application of KCL at node $B$ shows that the current through the 8-$\Omega$ resistor is $i_1 - i_2$. Application of KVL to the two loops in the circuit leads to

$$10 - 6i_1 - 8(i_1 - i_2) = 0 \tag{a}$$

$$-Ri_2 + 8(i_1 - i_2) = 0 \tag{b}$$

Equations (a) and (b) are rearranged to

$$-14i_1 + 8i_2 = -10 \tag{c}$$

$$8i_1 - (8 + R)i_2 = 0 \tag{d}$$

Equations (c) and (d) are solved simultaneously leading to

$$i_1 = \frac{5(8 + R)}{24 + 7R} \tag{e}$$

$$i_2 = \frac{40}{24 + 7R} \tag{f}$$

The power through the resistor is

$$P = i_2^2 R$$

$$= \frac{1,600R}{(24 + 7R)^2} \tag{g}$$

The value of $R$ that leads to maximum power is obtained by setting $dP/dR = 0$ and solving for $R$. To this end, using the quotient rule for differentiation

$$\frac{dP}{dR} = \frac{1,600(24 + 7R)^2 - (1,600R)(2)(24 + 7R)(7)}{(24 + 7R)^4}$$

$$= \frac{1,600(24 + 7R)(24 - 7R)}{(24 + 7R)^4} \tag{h}$$

Setting $dP/dR = 0$ leads to $R = 24/7\ \Omega$. The maximum power is calculated as

$$P_{\max} = \frac{1,600(24/7)}{[24 + 7(24/7)]^2}\ \text{W}$$

(i)

$$= 2.38\ \text{W}$$

---

## Example 3.26

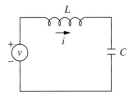

**FIG. 3.44** The $LC$ circuit of Example 3.26.

Derive a mathematical model for the $LC$ circuit of Figure 3.44.

### Solution

Define $i(t)$ as the current flowing through the circuit. Application of KVL in the clockwise direction leads to

$$v(t) - L\frac{di}{dt} - \frac{1}{C}\int_0^t idt = 0$$

(a)

Rearranging Equation (a) leads to

$$L\frac{di}{dt} + \frac{1}{C}\int_0^t idt = v(t)$$

(b)

---

## Example 3.27

Derive a mathematical model for the circuit of Example 3.11 and Figure 3.16, assuming mutual inductance of $M$ between each of the adjacent inductors and a mutual inductance of zero between the two nonadjacent inductors.

### Solution

Application of KVL for each loop assuming mutual inductance leads to

*Loop ABGH*

$$v(t) - \left(L\frac{di_1}{dt} + M\frac{di_2}{dt}\right) - \frac{1}{C}\int_0^t (i_1 - i_2)\, dt - Ri_1 = 0$$

(a)

$$L\frac{di_1}{dt} + M\frac{di_2}{dt} + Ri_1 + \frac{1}{C}\int_0^t i_1 dt - \frac{1}{C}\int_0^t i_2\, dt = v(t)$$

(b)

*Loop BCFG*

$$-\left(L\frac{di_2}{dt} + M\frac{di_1}{dt} + M\frac{di_3}{dt}\right) - R(i_2 - i_3) + \frac{1}{C}\int_0^t (i_1 - i_2)\, dt = 0$$

(c)

$$M\frac{di_1}{dt} + L\frac{di_2}{dt} + M\frac{di_3}{dt} + Ri_2 - Ri_3 - \frac{1}{C}\int_0^t i_1 dt + \frac{1}{C}\int_0^t i_2\, dt = 0$$

(d)

*Loop CDEF*

$$-\left(L\frac{di_3}{dt} + M\frac{di_2}{dt}\right) - \frac{1}{C}\int_0^t i_3 dt + R(i_2 - i_3) = 0 \tag{e}$$

$$M\frac{di_2}{dt} + L\frac{di_3}{dt} - Ri_2 + Ri_3 + \frac{1}{C}\int_0^t i_3 dt = 0 \tag{f}$$

The circuit is modeled using Equations (b), (d), and (f).

---

## Example 3.28

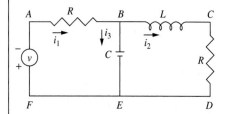

**FIG. 3.45**  The circuit for Example 3.28.

Derive a mathematical model for the circuit of Figure 3.45.

### Solution

The application of KCL at node *B* leads to

$$i_1 - i_2 - i_3 = 0 \tag{a}$$

Applying KVL around loop *ABEF* leads to

$$v - Ri_1 - \frac{1}{C}\int_0^t i_3\, dt = 0 \tag{b}$$

Applying KVL to loop *BCDE* leads to

$$-L\frac{di_2}{dt} - Ri_2 + \frac{1}{C}\int_0^t i_3\, dt = 0 \tag{c}$$

Substituting Equation (a) into Equations (b) and (c) and rearranging leads to

$$Ri_1 + \frac{1}{C}\int_0^t i_1\, dt - \frac{1}{C}\int_0^t i_2\, dt = v \tag{d}$$

$$-\frac{1}{C}\int_0^t i_1 dt + L\frac{di_2}{dt} + Ri_2 + \frac{1}{C}\int_0^t i_2\, dt = 0 \tag{e}$$

---

## Example 3.29

Determine the output voltage from the operational amplifier circuit of Figure 3.46. Assume both amplifiers are ideal.

### Solution

Since both amplifiers are ideal, the current through each amplifier is zero and the voltage at each terminal is the same. Note that both amplifiers are inverting amplifiers as in Example 3.17a. The output voltage from an inverting

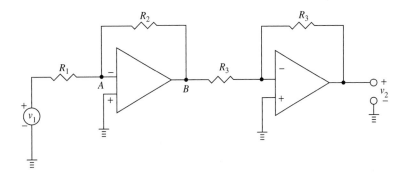

**FIG. 3.46** The two inverting operational amplifiers in series of Example 3.29.

amplifier is the negative of the ratio of the resistances times the input voltage. Thus

$$v_B = -\frac{R_2}{R_1} v_1 \qquad (a)$$

and

$$v_2 = -\frac{R_3}{R_3} v_B$$
$$= \frac{R_2}{R_1} v_1 \qquad (b)$$

As noted in Section 8.3 this circuit may be used as an electronic proportional controller.

## Example 3.30

Determine the relationship between the output voltage $v_o(t)$ and the input voltage $v_i(t)$ for the operational amplifier circuit of Figure 3.47.

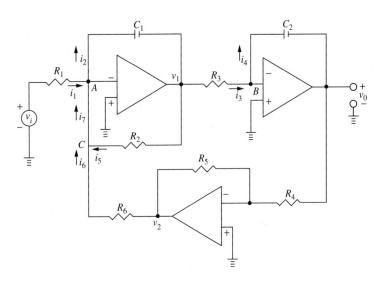

**FIG. 3.47** The operational amplifier circuit of Example 3.30.

## Solution

Defining currents as illustrated in Figure 3.47, the application of KCL at nodes $A$, $B$, and $C$ leads to

$$i_1 + i_7 - i_2 = 0 \tag{a}$$

$$i_3 - i_4 = 0 \tag{b}$$

$$i_5 + i_6 - i_7 = 0 \tag{c}$$

The currents are given by

$$i_1 = \frac{v_i}{R_1} \tag{d}$$

$$i_2 = -C_1 \frac{dv_1}{dt} \tag{e}$$

$$i_3 = \frac{v_1}{R_3} \tag{f}$$

$$i_4 = -C_2 \frac{dv_0}{dt} \tag{g}$$

$$i_5 = \frac{v_1}{R_2} \tag{h}$$

The operational amplifier spanned by the resistor $R_5$ is an inverting amplifier. Thus

$$v_2 = -\frac{R_5}{R_4} v_0 \tag{i}$$

$$i_6 = \frac{v_2}{R_6} = -\frac{R_5}{R_4 R_6} v_0 \tag{j}$$

Solving Equation (c) for $i_7$, substituting the result into Equation (a) and then substituting for the currents from Equations (d)–(i) leads to

$$\frac{v_i}{R_1} + \frac{v_1}{R_2} - \frac{R_5}{R_4 R_6} v_0 + C_1 \frac{dv_1}{dt} = 0 \tag{k}$$

Substitution of Equations (f) and (g) into Equation (b) leads to

$$\frac{v_1}{R_3} + C_2 \frac{dv_o}{dt} = 0 \tag{l}$$

Equation (l) is rearranged to give

$$v_1 = -R_3 C_2 \frac{dv_0}{dt} \tag{m}$$

Substitution of Equation (m) into Equation (k) gives

$$\frac{v_i}{R_1} - \frac{R_3 C_2}{R_2} \frac{dv_0}{dt} - \frac{R_5}{R_4 R_6} v_0 - C_1 C_2 R_3 \frac{d^2 v_0}{dt^2} = 0 \tag{n}$$

Rearrangement of Equation (m) leads to

$$C_1 C_2 R_3 \frac{d^2 v_0}{dt^2} + \frac{R_3 C_2}{R_2} \frac{dv_0}{dt} + \frac{R_5}{R_4 R_6} v_0 = \frac{v_i}{R_1} \tag{o}$$

The time-dependent output from the operational amplifier circuit is the solution to the differential equation, Equation (o).

## Example 3.31

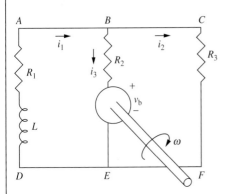

**FIG. 3.48** The generator circuit of Example 3.31.

Derive a mathematical model for the generator of Figure 3.48. The rotation of the shaft provides a voltage $v_b$ to the circuit.

**Solution**

Application of KCL at node $B$ leads to

$$i_1 - i_2 - i_3 = 0 \tag{a}$$

Application of KVL around loop $ABEF$ gives

$$-L\frac{di_1}{dt} - R_1 i_1 - R_2 i_3 - v_b = 0 \tag{b}$$

Application of KVL around loop $BCDE$ yields

$$-R_3 i_2 + v_b + R_2 i_3 = 0 \tag{c}$$

Substituting Equation (a) into Equations (b) and (c) and rearranging leads to

$$L\frac{di_1}{dt} + (R_1 + R_2)i_1 - R_2 i_2 + v_b = 0 \tag{d}$$

$$-R_1 i_1 + (R_2 + R_3)i_2 = v_b \tag{e}$$

## Example 3.32

Derive a mathematical model for the electromechanical system of Figure 3.49. The magnetic coupling field leads to a force $K_1 i_2$ applied to the end of the lever, and the motion of the plunger through the magnetic field leads to an induced electromotive force of $K_2 \dot{y}$, where $\dot{y}$ is the velocity of the plunger.

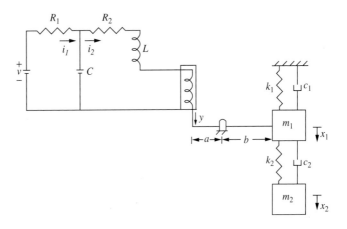

**FIG. 3.49** The system of Example 3.32.

### Solution

Application of KVL around the loops of the circuit noting that $y = \frac{b}{a}x_1$ leads to

$$R_1 i_1 + \frac{1}{C}\int_0^t i_1\, dt - \frac{1}{C}\int_0^t i_2\, dt = v \tag{a}$$

$$-\frac{1}{C}\int_0^t i_1\, dt + L\frac{di_2}{dt} + R_2 i_2 + \frac{1}{C}\int_0^t i_2\, dt - K_2\frac{b}{a}\dot{x}_1 = 0 \tag{b}$$

Summing moments on the free-body diagram of Figure 3.50(a) assuming small displacements gives

$$\sum M_O = I_O\alpha$$

$$-K_1 i_2 a - (k_1 x_1 + c_1\dot{x}_1)b + [k_2(x_2 - x_1) + c_2(\dot{x}_2 - \dot{x}_1)]b = m_1 b^2\left(\frac{\ddot{x}_1}{b}\right)$$

$$m_1\ddot{x}_1 + (c_1 + c_2)b^2\dot{x}_1 - c_2 b^2\dot{x}_2 + (k_1 + k_2)b^2 x_1 - k_2 b^2 x_2 + K_1 a b i_2 = 0 \tag{c}$$

Summing forces on the free-body diagram of Figure 3.50(b) gives

$$\sum F = m_2\ddot{x}_2$$

$$-k_2(x_2 - x_1) - c_2(\dot{x}_2 - \dot{x}_1) = m_2\ddot{x}_2$$

$$m_2\ddot{x}_2 - c_2\dot{x}_1 + c_2\dot{x}_2 - k_2 x_1 + k_2 x_2 = 0 \tag{d}$$

Equations (a)–(d) form the mathematical model for this system.

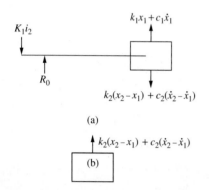

$K_1 i_2$

$k_1 x_1 + c_1\dot{x}_1$

$R_0$

$k_2(x_2 - x_1) + c_2(\dot{x}_2 - \dot{x}_1)$

(a)

$k_2(x_2 - x_1) + c_2(\dot{x}_2 - \dot{x}_1)$

(b)

**FIG. 3.50**  The free-body diagrams for system of Example 3.32.

### Example 3.33

Derive a mathematical model for the operational amplifier circuit of Figure 3.51.

### Solution

Application of KCL at node $A$ gives

$$i_1 - i_2 = 0 \tag{a}$$

The resistor $R_1$ and the capacitor $C_1$ are in parallel. Thus the voltage drop across the each element $v_1 - v_A$ is the same, and the current $i_1$ is the sum of the currents through each element. Assuming an ideal operational amplifier $v_A = 0$,

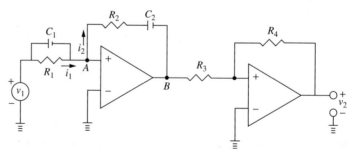

**FIG. 3.51**  The operational amplifier circuit of Example 3.33.

the total current through the parallel combination is

$$i_1 = \frac{v_1}{R_1} + C_1 \frac{dv_1}{dt} \tag{b}$$

The resistor $R_2$ and the capacitor $C_2$ are in series. Thus the current through each element $i_2$ is the same and the total voltage drop $-v_B$ is the sum of the voltage drops across each element. Thus

$$-v_B = R_2 i_2 + \frac{1}{C_2} \int_0^t i_2\, dt \tag{c}$$

Use of Equations (a) and (b) in Equation (c) leads to

$$-v_B = R_2 \left( \frac{v_1}{R_1} + C_1 \frac{dv_1}{dt} \right) + \frac{1}{C_2} \int_0^t \left( \frac{v_1}{R_1} + C_1 \frac{dv_1}{dt} \right) dt \tag{d}$$

$$= C_1 R_2 \frac{dv_1}{dt} + \left( \frac{R_2}{R_1} + \frac{C_1}{C_2} \right) v_1 + \frac{1}{R_1 C_2} \int_0^t v_1\, dt$$

The second operational amplifier acts as an inverter such that

$$v_2 = -\frac{R_4}{R_3} v_B \tag{e}$$

Substitution of Equation (d) into Equation (e) gives

$$v_2 = \frac{C_1 R_2 R_4}{R_3} \frac{dv_1}{dt} + \left( \frac{R_2 R_4}{R_1 R_3} + \frac{C_1 R_4}{C_2 R_3} \right) v_1 + \frac{R_4}{C_2 R_1 R_3} \int_0^t v_1\, dt \tag{f}$$

It is shown in Chapter 8 that this operational amplifier circuit functions as a proportional plus integral plus derivative controller.

# 3.10 SUMMARY

## 3.10.1 Mathematical Modeling of Electrical Systems

The principles used in the mathematical modeling of lumped parameter electric circuits are as follows:

- Kirchoff's current law (KCL) states that the sum of the currents entering a node in an electric circuit is zero.
- Kirchoff's voltage law (KVL) states that the sum of the voltage drops around any closed loop in a circuit is zero.
- Use of KVL and KCL to lumped parameter circuits leads to differential and integrodifferential equations whose independent variable is time and whose dependent variables are mesh currents of voltage drops across circuit components.

## 3.10.2 Other Chapter Highlights

- Electric current is a result of the motion of charged particles.
- Electric potential (voltage) is the work required to move a charged particle through an electric field.
- Passive circuit component are capacitors that store energy in an electric field, inductors that store energy in an electric field, and resistors that dissipate energy.
- Active circuit components are voltage sources, current sources, and operational amplifiers.
- Kirchoff's laws are statements of Conservation of Charge and Conservation of Energy.
- Circuit components are in series if they have the same current and the total voltage drop across the combination is the sum of the individual voltage drops.
- Circuit components are in parallel if each component has the same voltage drop and the total current through the combination is the sum of the currents through individual components.
- Components in series or parallel can be replaced by a single component with an equivalent parameter.
- The differential equations modeling circuits are analogous to the differential equations governing mechanical systems. The direct analogies for circuits with voltage sources are resistors are analogous to viscous dampers, capacitors are analogous to springs, inductors are analogous to inertia elements, charge is analogous to displacement, and current is analogous to velocity.
- Operational amplifiers have external voltage sources. An ideal amplifier has no current and the voltage at each terminal is the same.
- Electromechanical systems have an electrical system and a mechanical system coupled through a coupling field. The coupling field provides an electrostatic or electromagnetic force that acts on the mechanical system.
- Field-controlled dc servomotors have a coupling torque proportional to the current in the field circuit.
- Armature-controlled dc servomotors have a coupling torque proportional to the current in the armature circuit and an induced back emf.

## 3.10.3 Important Equations

- Electric power

$$P = vi \tag{3.6}$$

- Ohm's law

$$v = iR \tag{3.7}$$

- Voltage-current relation in a capacitor

$$i = C\frac{dv}{dt} \tag{3.16}$$

- Voltage-current relation in an inductor

$$v = L\frac{di}{dt} \qquad (3.23)$$

- Series LRC circuit

$$L_{eq}\frac{di}{dt} + R_{eq}i + \frac{1}{C_{eq}}\int_0^t i\,dt = v(t) \qquad (3.48)$$

- Electrostatic or electromagnetic force in coupling field

$$F_f = -\frac{\partial E_c}{\partial x} \qquad (3.79)$$

# PROBLEMS

**3.1** (a) What is the required separation distance for a parallel plate capacitor of plate area 0.5 m² such that its capacitance is 0.05 μF when glass $(K = 4.6)$ is used as a dielectric? (b) What is the charge on the capacitor when it is connected to a 12-V battery? (c) What is the energy stored in the capacitor when it is connected to a 12-V battery?

**3.2** When a 20-V battery is used to power a car, a current of 1.5 A runs through the car. The car runs for 3 hr before the battery is discharged. What is the energy stored in the battery?

**3.3** A 200-W CD player is powered from a household voltage source of 120 V. (a) What is the current in the CD player? (b) What is the resistance of the CD player?

**3.4** What is the required resistance of a heating coil that generates 1500 kJ of heat per hour when connected to a 120-V source?

**3.5** The time-dependent current at steady state in a series *RLC* circuit with a voltage source is

$$i(t) = 10\sin(100t + 0.4)\ \text{A}$$

The circuit parameters are $R = 100\ \Omega$, $C = 0.2\ \mu\text{F}$, and $L = 0.25\ \text{H}$. (a) What is the voltage change across the resistor? (b) What is the voltage change across the capacitor? (c) What is the voltage change across the inductor? (d) What is the time-dependent voltage, provided the source?

**3.6** For the circuit described in Problem 3.5 calculate and plot as a function of time (a) the energy stored in the capacitor, (b) the energy stored by the inductor, and (c) the energy dissipated by the resistor.

**3.7** (a) Determine the current through each resistor in the resistive circuit of Figure P3.7. (b) Determine the voltage drop across each resistor. (c) Determine the power in each resistor.

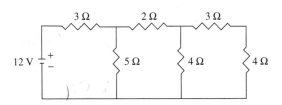

**FIG. P3.7**

**3.8** (a) Replace the resistors in the circuit of Figure P3.7 by a single resistor of an equivalent resistance. (b) Determine the power delivered to the circuit.

**3.9** Replace the resistors in the circuit of Figure P3.9 by a single resistor of an equivalent resistance.

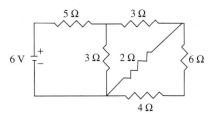

**FIG. P3.9**

**3.10** A 10-V battery serves as input to the circuit of Figure P3.10. The output from the circuit is to be the voltage across the 11-Ω resistor. What is the voltage measured across the resistor?

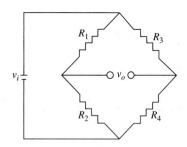

**FIG. P3.10**

**3.11** A Wheatstone bridge is a resistive circuit that is used in strain gauge transducers. A typical Wheatstone bridge circuit is shown in Figure P3.11 The resistor whose resistance is labeled as $R_4$ is the strain gauge. (a) Determine the output voltage $v_0$. (b) Derive a condition between the resistances such that the output voltage is zero. (c) Explain how this circuit serves as a strain gauge, noting that the strain in a wire is the ratio of its change in length to its original length.

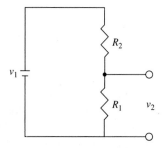

**FIG. P3.11**

**3.12** A circuit for a potentiometer is shown in Figure P3.12. The output is the voltage change measured across the resistor of resistance $R_2$. Determine this voltage in terms of $R_1$, $R_2$, and $v_1$.

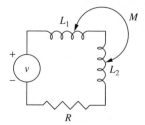

**FIG. P3.12**

**3.13** (a) Determine the voltage in each of the capacitors in the circuit of Figure P3.13. (b) Determine the charge on each of the capacitors.

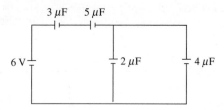

**FIG. P3.13**

**3.14** Replace the capacitors in the circuit of Figure P3.13 with a single capacitor of an equivalent capacitance.

**3.15** Derive a mathematical model for the $RC$ circuit of Figure P3.15. Use the voltage drop across the resistor as the dependent variable.

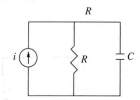

**FIG. P3.15**

**3.16** Derive a mathematical model for the circuit of Figure P3.16. Use the current in the circuit as the dependent variable.

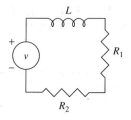

**FIG. P3.16**

**3.17** Derive a mathematical model for the circuit of Figure P3.17. The mutual inductance between the inductors is $M$. Use the current in the circuit as the dependent variable.

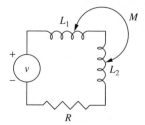

**FIG. P3.17**

**3.18** The switch in the circuit of Figure P3.18 is open and then closed at $t = 0$, connecting the capacitor to the circuit. (a) Determine the current through the $5\,\Omega$ resistor before the switch is closed. (b) Derive a mathematical model for the current in the capacitor after the switch is closed.

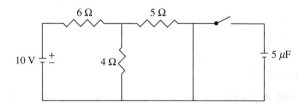

FIG. **P3.18**

**3.19** Derive a mathematical model for the circuit of Figure P3.19. Use the current through the inductor as the dependent variable.

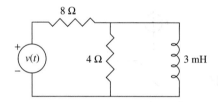

FIG. **P3.19**

**3.20** Derive a mathematical model for the circuit of Figure P3.20. Use the current through the inductor as the dependent variable.

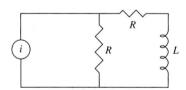

FIG. **P3.20**

**3.21** Derive a mathematical model for the circuit of Figure P3.21. Use $i_1, i_2$, and $i_3$ as the dependent variables.

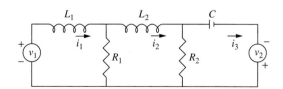

FIG. **P3.21**

**3.22** The capacitor $C_1$ in the circuit of Figure P3.22 has an initial charge $q$ while the capacitor $C_2$ is uncharged. Derive a mathematical model for the current in the circuit after the switch is closed at $t = 0$. Include the appropriate initial condition.

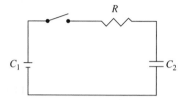

FIG. **P3.22**

**3.23** Derive a mathematical model for the circuit of Figure P3.23. Use the currents through the resistors as the dependent variables.

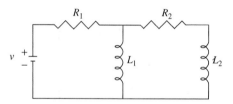

FIG. **P3.23**

**3.24** Derive a mathematical model for the circuit of Figure P3.24. Use the voltage across the 3-k$\Omega$ resistor and the current through the inductor as the dependent variables.

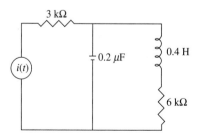

FIG. **P3.24**

**3.25** Derive a mathematical model for the circuit of Figure P3.25. Use $i_1$ and $i_2$ as the dependent variables.

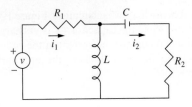

**FIG. P3.25**

**3.26** Derive a mathematical model for the circuit of Figure P3.26. Use the currents through the inductors as dependent variables.

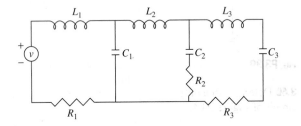

**FIG. P3.26**

**3.27** Derive a mathematical model for the circuit of Figure P3.27. Use the currents through the inductors as dependent variables.

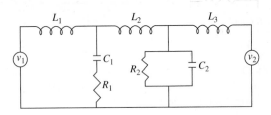

**FIG. P3.27**

**3.28** The mutual inductance between two adjacent inductors in the circuit of Figure P3.27 is $M$; the mutual inductance between non-adjacent inductors is zero. Derive a mathematical model for the circuit, including the effect of the mutual inductance.

**3.29** Determine a mechanical system analogous to the electrical system of Figure P3.21.

**3.30** Determine a mechanical system analogous to the electrical system of Figure P3.26.

**3.31** Determine a mechanical system analogous to the electrical system of Figure P3.27.

**3.32** Determine a mechanical system analogous to the electrical system of Figure P3.32

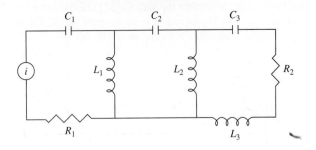

**FIG. P3.32**

**3.33** Determine an electrical circuit analogy to the mechanical system of Figure P3.33.

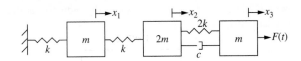

**FIG. P3.33**

**3.34** Determine an electrical circuit analogy to the mechanical system of Figure P3.34.

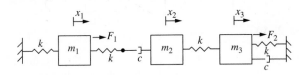

**FIG. P3.34**

**3.35** Determine the output $v_2$ from the operational amplifier circuit of Figure P3.35.

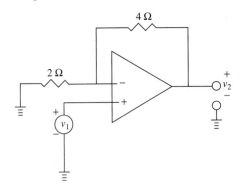

**FIG. P3.35**

**3.36** Determine the output $v_2$ from the operational amplifier circuit of Figure P3.36.

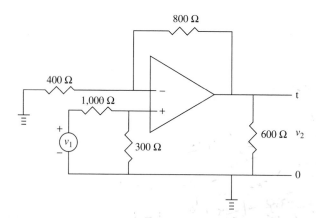

**FIG. P3.36**

**3.37** Determine the output $v_3$ from the operational amplifier circuit of Figure P3.37.

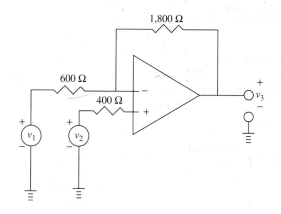

**FIG. P3.37**

**3.38** Determine the output $v_2$ from the operational amplifier circuit of Figure P3.38.

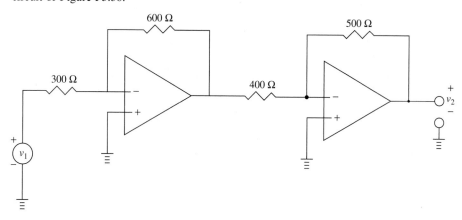

**FIG. P3.38**

**3.39** Determine the output $v_2$ from the operational amplifier circuit of Figure P3.39.

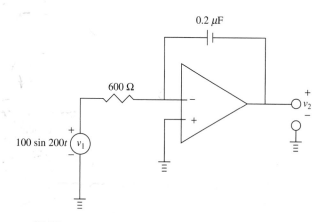

**FIG. P3.39**

**3.40** Determine the output $v_2$ from the operational amplifier circuit of Figure P3.40.

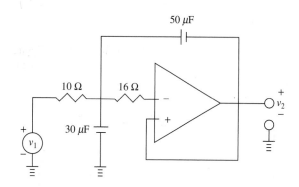

**FIG. P3.40**

**3.41** The circuit of Figure 3.41 is used in a capacitive proximity probe. Derive a mathematical model which relates the output $v_o$ to the input $v_i$.

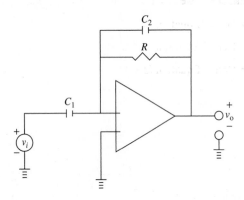

**FIG. P3.41**

**3.42** The circuit of Figure P3.42 can be used as a low-pass filter. Derive a mathematical model that relates the output $v_o$ to the input $v_i$.

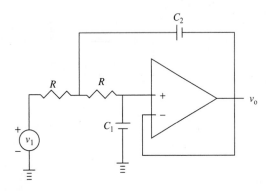

**FIG. P3.42**

**3.43.** Derive a mathematical model that relates the output $v_o$ to the input $v_i$ for the operational amplifier circuit of Figure P3.43.

**3.44** Design an operational amplifier circuit whose output $v_2$ simulates the response of the mechanical system of Figure P3.44.

**3.45** Design an operational amplifier circuit whose output $v_2$ is the solution of the differential equation $\dot{v}_2 + 10v_2 = 3v_1(t)$.

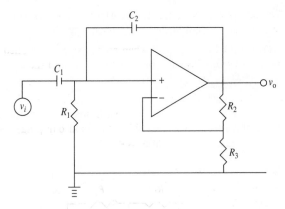

**FIG. P3.43**

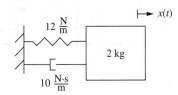

**FIG. P3.44**

**3.46** Design an operational amplifier circuit whose output $v_2$ is the solution of the differential equation $\ddot{v}_2 + 10\dot{v}_2 + 4v_1 = 3\dot{v}_1(t) + 2v_2(t)$.

**3.47** A field-controlled dc motor is used to power the gearing system of Figure P3.47. Derive a mathematical model

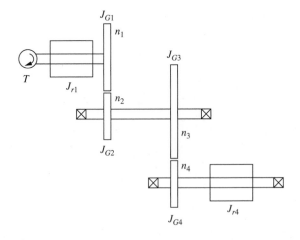

**FIG. P3.47**

for the system using the field current $i_f$ and the angular velocity of the shaft $\omega$ as the dependent variables.

**3.48** Repeat Problem 3.47 when an armature-controlled dc motor is used, with the armature current $i_a$ and the angular velocity of the shaft $\omega$ as dependent variables.

**3.49** Derive a mathematical model for the system of Problem 3.47 and Figure P3.47 when the shafts are supported by identical bearings, each with a torsional damping coefficient $c_t$ and a torsional stiffness $k_t$.

**3.50** Derive a mathematical model for the electromechanical system of Figure P3.50. Use $i_1, i_2, x_1$, and $x_2$ as dependent variables. The motion of the bar through the magnetic field of the inductor leads to a force $F_c$ acting on the horizontal bar as well as a back emf $e_b$ in the circuit.

**3.51** Derive Equation (3.41).

**3.52** Derive Equation (3.42).

**3.53** Derive Equation (3.43).

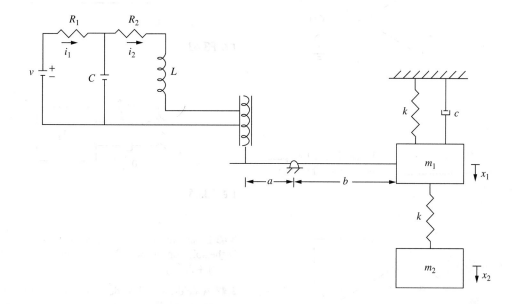

FIG. **P3.50**

# 4

# Fluid, Thermal, and Chemical Systems

Transport processes often occur in dynamic systems. The properties of fluid systems, such as density and temperature, change due to the transport of mass. Heat transfer processes, such as heating a tank of liquid, involve the transport of energy from a source to the liquid. Drugs are absorbed into biological tissue as the drug is transported throughout a body. This chapter is concerned with the modeling of dynamic systems in which transport processes occur.

## 4.1 INTRODUCTION

A fluid continuously deforms under the application of a force whereas a solid has a finite deformation. A microscopic fluid analysis involves the application of Newton's second law to a free-body diagram of a differential quantity of fluid. Analysis of the motion of a fluid often requires the application of other basic laws of nature in addition to or in lieu of Newton's second law. Conservation of Mass requires that the mass of a system of fluid particles is constant. Conservation of Energy, a statement of the First Law of Thermodynamics, when applied includes forms of energy other than kinetic and potential. Dependent variables in fluid flow problems may include velocity components and fluid pressure.

The motion of a fluid is very complex and depends on many parameters. Many simplifying assumptions are used to develop models of fluid systems. An incompressible fluid is one in which the density is approximately constant for a wide variety of fluid conditions. Most liquids, such as water and oils, are assumed to be incompressible fluids. The density of a compressible fluid changes as other properties, such as pressure and temperature, change. An ideal gas like air is an example of a compressible fluid. Pipe flows and liquid-level problems are studied in Sections 4.3 and 4.4. Hydraulic systems, which use liquids, and pneumatic systems, which use gases, often serve as actuators for mechanical systems and are studied in Section 4.5.

A thermal system is defined as a system in which heat transfer occurs. The heat transfer may occur through conduction in a solid or through convection in a fluid. Heat transfer leads to a change in temperature, a thermodynamic potential, which is a dependent variable in thermal systems. Thermal systems are studied in Section 4.6.

A chemical system is any system in which a chemical reaction occurs such that moles of one or more components are changed into moles of other components. A Conservation of Mass equation that includes the effect of the chemical reaction is necessary for every component in the system. Component concentrations are typically used as dependent variables in chemical systems. Chemical reactions often require heat to occur or give off heat while occurring. Thus many chemical systems are also thermal systems. Many chemical systems also have fluid flow. Chemical systems involving the continuous stirred tank reactor (CSTR) are studied in Section 4.7.

The principles of transport processes also apply to many other physical systems. A biological systems example is presented in Section 4.7.

The analyses of fluid, thermal, and chemical systems in this chapter use **lumped parameter assumptions**; that is, all dependent variables and all properties are constant across all spatial coordinates, and time is the only independent variable. Fluid properties (such as density and viscosity), thermal properties (such as thermal conductivity and specific heat), and chemical properties (such as rate of reaction) are all assumed to be constant in space but may vary with time. The lumped parameter assumption allows systems to be modeled using ordinary differential equations.

Consider a tank containing a liquid. The flow rate $Q$ into the tank is equal to the flow rate out. Thus the volume of liquid in the tank is constant. For a tank of uniform area height of the column of liquid in the tank, called the liquid level, is constant. The system is operating at a steady-state in which all properties are constant. Perturbations of properties of the fluid entering the tank leads to dynamic responses of the properties of the fluid in the tank. A change in flow rate affects the liquid level in the tank; a change in inlet stream temperature causes a time-dependent change in temperature in the tank; a change in the concentrations of the components of the incoming stream leads to changes in component concentrations in the tank. Whereas the properties of the incoming stream may be assumed to change instantaneously, the properties in the tank are continuous with time. Mathematical modeling of this system leads to ordinary differential equations whose solutions provide the time-dependent response of the dependent variables. Often perturbations in properties lead to a new steady state, but it is of interest to know how long it takes to reach the new steady state and the dynamic response until the steady state is reached.

In view of all this, most dependent variables are broken down into the original steady-state value plus a perturbation from steady state. For example if $H$ represents the level of liquid in a tank that is changing with time due to a time dependent in flow rate $Q$, then

$$H(t) = h_s + h(t) \tag{4.1a}$$

$$Q(t) = q_s + q(t) \tag{4.1b}$$

where $H(t)$ and $Q(t)$ are the total values of the quantities, $h_s$ and $q_s$ are the steady-state values, and $h(t)$ and $q(t)$ are the time-dependent perturbations from steady state. Often the differential equations for the total quantities are nonlinear. The equations are linearized, as shown in Section 1.5, by assuming that the perturbation quantities are small compared to the steady-state quantities and by using binominal or Taylor series expansions of the nonlinear terms about the steady state.

Lumped parameter modeling of fluid, thermal, and chemical systems requires a method of analysis different from that of mechanical and electrical systems. Instead of tracking the motion of a fixed set of particles, the changes occurring in a defined region in space, called a control volume, are tracked.

## 4.2 CONTROL VOLUME ANALYSIS

The term "system," as used so far in this book, refers to a set of integrated components designed to achieve a certain objective. A system converts an input into a desired output. An alternate usage of the word "system" is applied in for the mathematical modeling of fluid, thermal, and chemical problems. This alternate meaning is introduced to provide a framework for such analysis, but it retains its former definition throughout the remainder of the book.

A **system**, in the context just described, is defined as a fixed set of matter. A system does not gain or lose mass. Using this definition it is clear that the principles developed in Chapter 2 are applied to systems of particles including rigid bodies. The members of a system of particles are tracked in space with time. However, it is impractical to track every particle in a system of matter in a liquid or gas state. Consider, for example, the tank of Figure 4.1. Liquid enters the tank through the pipe at $A$ and exits through the pipe at $B$. Instead of tracking specific fluid particles after leaving the tank, it is of interest to analyze the properties of the liquid remaining in the tank. The properties of the liquid in the tank that may change with time include fluid level, the temperature of the fluid in the tank, and the concentrations of various species in the tank.

Consequently, a **control volume**, a region in space with defined boundaries, is defined and used for fluid, thermal, and chemical analysis in lieu of a system. An essential difference between a system and a control volume is that matter may cross the boundaries of a control volume whereas the matter of a system is fixed. The tank and the liquid in the tank of Figure 4.1 is an example of a control volume. Mass crosses the boundary of the control volume as fluid enters at $A$ and leaves at $B$.

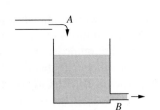

**FIG. 4.1** It is more practical to use a control volume analysis for the liquid in the tank rather than a system analysis.

A control volume is a region with defined boundaries; however, it is not necessary for it to be fixed in space or have fixed boundaries. A rocket is an example of an accelerating control volume where mass leaves the control volume as exhaust gases. A helium balloon, while it is being filled, is an example of a control volume whose boundaries are defined, but vary with time.

Systems that include fluid flow, heat transfer, or chemical reactions are usually best analyzed by using a control volume analysis. The basic laws of Conservation of Mass, Conservation of Energy, and Conservation of Momentum are modified from the conservation laws used for system analysis to account for mass and energy crossing the boundaries of a control volume.

## 4.2.1  Conservation of Mass

The control volume form of **Conservation of Mass** is summarized as follows

$$\begin{pmatrix} \text{Rate at which} \\ \text{mass enters a} \\ \text{control volume} \end{pmatrix} - \begin{pmatrix} \text{Rate at which} \\ \text{mass leaves a} \\ \text{control volume} \end{pmatrix} = \begin{pmatrix} \text{Rate at which mass} \\ \text{accumulates in a} \\ \text{control volume} \end{pmatrix}$$

(4.2)

Mass enters and leaves a control volume through its boundaries. The rate at which mass crosses the boundary of a control volume is called the **mass flow rate** $\dot{m}$, which has dimensions of $[MT^{-1}]$. Letting $m$ be the total mass of the control volume at any instant, Equation (4.2) is written as

$$\dot{m}_{in} - \dot{m}_{out} = \frac{dm}{dt} \tag{4.3}$$

The total mass in the control volume at any instant is

$$m = \int_V \rho \, dV \tag{4.4}$$

where $\rho$ is the mass density of the fluid in the control volume. The mass flow rate can be written as

$$\dot{m} = \rho Q \tag{4.5}$$

where $Q$ is the **volumetric flow rate**, which has dimensions of $[L^3 T^{-1}]$. Using Equations (4.4) and (4.5) in Equation (4.3) yields

$$\rho_{in} Q_{in} - \rho_{out} Q_{out} = \frac{d}{dt}\left( \int_V \rho dV \right) \tag{4.6}$$

If the fluid entering the control volume is of a constant species, the fluid in the control volume is homogeneous, and if the fluid is **incompressible**, the density is constant throughout the fluid and does not vary with time and $\rho = \rho_{in} = \rho_{out}$. Under these circumstances Equation (4.6) reduces to

$$Q_{in} - Q_{out} = \frac{dV}{dt} \qquad (4.7)$$

where $V$ is the total volume of fluid in the control volume.

Most liquids are incompressible and all gases are compressible. The Conservation of Mass equation is often referred to as the **continuity equation**. Thus Equation (4.7) is the control volume form of the continuity equation for an incompressible fluid. Equation (4.6) is the appropriate form of the continuity equation for a compressible fluid.

---

### Example 4.1

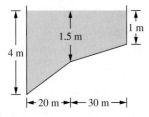

**FIG. 4.2** Swimming pool depth profile for Example 4.1.

A swimming pool of width 15 m has the depth profile shown in Figure 4.2. The pool, initially empty, is to be filled until the depth in the shallow end is 1 m. The pool is to be filled using a 2-cm diameter hose with a fluid velocity of 5 m/s. How long will it take to fill the pool?

**Solution**

Take the control volume as the pool and the water in the pool. The amount of water in the control volume increases as water enters through the hose. The total volume of water in the pool when filled is 1387.5 m³. The flow rate of the fluid entering through the hose is

$$Q = VA$$
$$= (5 \text{ m/s}) \left[\frac{\pi}{4}(0.02 \text{ m})^2\right] = 1.23 \times 10^{-3} \text{ m}^3/\text{s} \qquad (a)$$

Conservation of Mass in the form of Equation (4.7) is applied to give

$$1.23 \times 10^{-3} = \frac{dV}{dt} \qquad (b)$$

Equation (b) is integrated to determine the time to fill the pool as

$$\int_0^t 1.23 \times 10^{-3} dt = \int_0^{1387.5} dV$$
$$1.23 \times 10^{-3} t = 1387.5$$
$$t = 1.13 \times 10^6 \text{ s} = 313.3 \text{ hr} \qquad (c)$$

---

## 4.2.2  Energy Equation

The total energy in a control volume is

$$E = U + T + V + E_{other} \qquad (4.8)$$

where $U$ is the **internal energy** of all particles in the control volume, $T$ is the total kinetic energy, $V$ is the potential energy due to gravity, and $E_{other}$ represents other forms of energy such as energy stored in mechanical or electric components and energy associated with chemical or nuclear reactions. The total energy can be obtained by integrating the **specific energy**, the energy per unit mass $e$ over the control volume

$$E = \int_m e\, dm = \int_V e\rho\, dV \qquad (4.9)$$

The internal energy is the energy associated with the random motion of molecules and thus is a function of temperature. The specific internal energy is given by

$$u = c_p T \qquad (4.10)$$

where $c_p$ is the specific heat at constant pressure. The specific kinetic energy is $v^2/2$ and the specific potential energy is $gz$, where $z$ is the elevation above a defined datum. Using these definitions, Equation (4.9) becomes

$$E = \int_V \left( u + \frac{v^2}{2} + gz + e_{other} \right) \rho\, dV \qquad (4.11)$$

Energy is added to or subtracted from a control volume in four ways:

- Energy in the form of internal energy, kinetic energy, and potential energy is transported into or out of the control volume through its boundaries.
- Energy is transferred into the control volume in the form of heat.
- Work is done on the control volume by forces external to the control volume. These include pressure forces and viscous friction acting on the surface of the control volume.
- Energy is transferred out of the control volume due to work done by the components of the control volume on components external to the control volume. The work done during this energy transfer is referred to as **shaft work**. Examples of shaft work are energy required to drive a turbine or work done on the control volume by a pump.

The control volume form of the energy equation is summarized as

$$
\begin{pmatrix}
\text{Rate at which} \\
\text{energy is transported} \\
\text{into the control} \\
\text{volume through its} \\
\text{boundaries}
\end{pmatrix}
-
\begin{pmatrix}
\text{Rate at which} \\
\text{energy is transported} \\
\text{out of the control} \\
\text{volume through} \\
\text{its boundaries}
\end{pmatrix}
+
\begin{pmatrix}
\text{Rate at} \\
\text{which heat} \\
\text{is transferred} \\
\text{into the control} \\
\text{volume}
\end{pmatrix}
$$

$$
+
\begin{pmatrix}
\text{Rate at which} \\
\text{work is done} \\
\text{on the control} \\
\text{volume by pressure} \\
\text{and friction}
\end{pmatrix}
-
\begin{pmatrix}
\text{Rate at which} \\
\text{the control} \\
\text{volume does} \\
\text{work on its} \\
\text{surroundings}
\end{pmatrix}
=
\begin{pmatrix}
\text{Rate at} \\
\text{which energy} \\
\text{accumulates} \\
\text{within the} \\
\text{control volume}
\end{pmatrix}
$$

$$(4.12)$$

For a control volume with one inlet and one outlet, Equation (4.12) can be written as

$$\dot{m}_{in}\left(u + \frac{v^2}{2} + gz\right)_{in} - \dot{m}_{out}\left(u + \frac{v^2}{2} + gz\right)_{out} + \dot{Q} - \dot{W} + \dot{W}_p - \dot{W}_f$$

$$= \frac{d}{dt}\left[\int_V \left(u + \frac{v^2}{2} + gz + e_{other}\right)\rho dV\right] \tag{4.13}$$

The rate of heat transfer $\dot{Q}$ is positive when heat is transferred into the control volume and negative when heat is transferred from the control volume. The rate of shaft work $\dot{W}$ is positive when the control volume does work on its surroundings and negative when the surroundings do work on the control volume. The work done by friction forces $\dot{W}_f$ is always negative. The rate of work done by pressure forces is given by

$$\dot{W}_p = \left(\dot{m}\frac{p}{\rho}\right)_{in} - \left(\dot{m}\frac{p}{\rho}\right)_{out} \tag{4.14}$$

Using Equation (4.14) in Equation (4.13) leads to

$$\dot{m}_{in}\left(u + \frac{p}{\rho} + \frac{v^2}{2} + gz\right)_{in} - \dot{m}_{out}\left(u + \frac{p}{\rho} + \frac{v^2}{2} + gz\right)_{out} + \dot{Q} - \dot{W}_s - \dot{W}_f$$

$$= \frac{d}{dt}\left[\int_V \left(u + \frac{v^2}{2} + gz\right)\rho dV\right] \tag{4.15}$$

Equation (4.15) is the most general form of the energy equation for a control volume with one inlet and one outlet. Assumptions may be used to reduce the equation to simpler forms.

## 4.2.3 Bernoulli's Equation

The control volume form of the momentum equation is useful for problems involving accelerating control volumes such as rockets or for determining external forces acting on the control volume such as reactions at joints in pipes. However, the control volume form of the momentum equation is not useful for problems studied in this book and is thus not presented. Instead an equation that can be derived from the control volume momentum equation is presented: Bernoulli's equation.

Recall that for a mechanical system the Principle of Work and Energy is derived from Newton's second law by integrating the dot product of Newton's second law between two points in the path of motion. Bernoulli's equation is derived in a similar manner. A **streamline** in a flow is a line tangent to the velocity vector. For a steady flow the streamline is the same as a pathline, the path traveled by a fluid particle. The control volume form of the momentum equation can be applied to a control volume enclosing a streamline. For a frictionless flow of an incompressible fluid the following equation results

$$\frac{p_1}{\rho} + \frac{v_1^2}{2} + gz_1 = \frac{p_2}{\rho} + \frac{v_2^2}{2} + gz_2 \qquad (4.16)$$

where 1 and 2 refer to two points along a streamline in a flow. Equation (4.16) is referred to as **Bernoulli's equation**.

Under the assumptions of steady frictionless incompressible flow with no shaft work, the control volume energy equation, Equation (4.15) reduces to

$$\dot{m}_{in}\left(u + \frac{p}{\rho} + \frac{v^2}{2} + gz\right)_{in} - \dot{m}_{out}\left(u + \frac{p}{\rho} + \frac{v^2}{2} + gz\right)_{out} + \dot{Q} = 0 \qquad (4.17)$$

For steady flow there is no mass accumulation within the control volume and Equation (4.3) reduces to $\dot{m}_{in} = \dot{m}_{out} = \dot{m}$. Dividing Equation (4.17) by $\dot{m}$ and defining $q = \dot{Q}/\dot{m}$, Equation (4.17) reduces to

$$\left(u + \frac{p}{\rho} + \frac{v^2}{2} + gz\right)_{in} - \left(u + \frac{p}{\rho} + \frac{v^2}{2} + gz\right)_{out} + q = 0 \qquad (4.18)$$

Defining 1 as the inlet and 2 as the outlet, Equation (4.18) reduces to Bernoulli's equation [Equation (4.16)] if

$$q = u_2 - u_1 \qquad (4.19)$$

Equation (4.19) implies that heat transferred into the control volume leads to an increase of internal energy. Thus, Bernoulli's equation reconciles with the control volume form of the energy equation under the assumptions of steady incompressible flow with no friction and no shaft work. In such problems the energy equation and Bernoulli's equation, which is derived from Conservation of Momentum, are identical. The application of only Conservation of Mass and Bernoulli's equation is necessary when modeling such systems. Bernoulli's equation is analogous to the Principle of Work and Energy considered in Section 2.9 for mechanical systems.

Bernoulli's equation may be applied only along a streamline in the flow. The subscripts 1 and 2 in Equations (4.16) and (4.19) refer to two points on the same streamline. Since Bernoulli's equation involves the term $(p_2 - p_1)/\rho$, **gauge pressure**, the difference between absolute pressure and atmospheric pressure, may be used in expressing the pressure terms. Gauge pressure is zero when the flow is exposed to the open atmosphere.

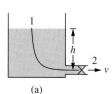

(a)

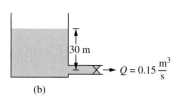

(b)

**FIG. 4.3** (a) The reservoir of Example 4.2. A streamline is taken from the free surface of the reservoir to the tank outlet for application of Bernoulli's equation. (b) The reservoir of Example 4.3. Head losses occur due to the valve and pipe.

## Example 4.2

The reservoir of Figure 4.3(a) has a valve that when open, allows water to flow out of the tank. Use Bernoulli's equation to determine the velocity of the fluid leaving the tank.

### Solution

The application of Bernoulli's equation requires the flow to be steady, which is a reasonable assumption provided the reservoir is large and the change in liquid level is negligible. Bernoulli's equation is applied along a streamline from 1, the free surface of the liquid to 2, the exit of the reservoir. Define $z = 0$ at the surface of the reservoir, thus $z_2 = -h$. The pressure at the free surface and at the exit and the pressure at the exit are both atmospheric. Since the reservoir is large, the velocity of the fluid on the free surface is negligible, $v_1 = 0$. Thus application of Bernoulli's equation from the free surface to the exit of the reservoir leads to

$$0 = \frac{v_2^2}{2} - gh \tag{a}$$

Equation (a) is solved for $v$ leading to

$$v_2 = \sqrt{2gh} \tag{b}$$

Dividing Equation (4.16) by $g$ leads to

$$h_1 = h_2 \tag{4.20}$$

where the **head** is defined by

$$h = \frac{p}{\rho g} + \frac{v^2}{2g} + z \tag{4.21}$$

The head, which has dimensions of $[L]$, is a property of the flow. Equation (4.21), the head form of Bernoulli's equation, shows that the head is a constant in a flow with no losses.

In a static case, $v = 0$ and Equation (4.21) reduces to

$$\frac{p_2 - p_1}{\rho g} = z_1 - z_2 \tag{4.22}$$

Equation (4.22) is called the **manometer equation** because it is the basis of the use of a manometer to measure static pressure.

A head in a flow is a measure of the total energy in the flow. A reduction in the head in a flow is called a **head loss**. If a head loss $h_\ell$ occurs between points 1 and 2 in a flow, the head equation is modified as

$$h_1 - h_\ell = h_2 \tag{4.23}$$

Equation (4.23) is called the **extended Bernoulli's equation**, that is, extended to take into account energy losses. The extended equation is really a head form of the energy equation. Head losses occur in a real flow due to friction, an obstacle in the flow, or a change in the flow. Head losses are often expressed in the form

$$h_\ell = K \frac{v^2}{2g} \tag{4.24}$$

where $K$ is a loss coefficient.

**Example 4.3**

Fluid exits the reservoir of Figure 4.3(b) when the valve is open. Energy is lost due to friction in the pipe and changes in the flow pattern in the valve. The fluid exits through a pipe of diameter 10 cm at a flow rate of 0.15 m³/s. Determine the loss coefficient for the pipe and valve system.

**Solution**

The application of the extended Bernoulli's equation between the free surface of the reservoir and the exit of the pipe using Equation (4.24) for the head loss gives

$$\frac{p_1}{\rho g} + \frac{v_1^2}{2g} + z_1 = K\frac{v_2^2}{2g} + \frac{p_2}{\rho g} + \frac{v_2^2}{2g} + z_2 \tag{a}$$

Both the free surface and the exit of the pipe are open to the atmosphere; thus $p_1 = p_2 = 0$. The reservoir is assumed to be large, thus $v_1 \approx 0$. Defining the datum at the level of the exit of the pipe leads to $z_1 = 30$ m and $z_2 = 0$ m. Using this information in Equation (a) leads to

$$z_1 = (K + 1)\frac{v_2^2}{2g} \tag{b}$$

The velocity of the fluid in the pipe is calculated as

$$v_2 = \frac{Q}{A}$$

$$= \frac{0.15 \text{ m}^3/\text{s}}{\frac{\pi}{4}(0.1 \text{ m})^2} = 19.1 \text{ m/s} \tag{c}$$

Solving Equation (b) for $K$ and substituting known values leads to

$$K = \frac{2gz_1}{v_2^2} - 1$$

$$= \frac{2(9.81 \text{ m/s}^2)(30 \text{ m})}{(19.1 \text{ m/s})^2} - 1 = 0.613 \tag{d}$$

## 4.3 PIPE FLOW

Most pipe flows may be analyzed using the head form of the extended Bernoulli's equation, Equation (4.23). Losses occur in a pipe due to friction and changes in flow pattern caused by elbows, bends, fittings, valves, and nozzles. Thus the head at the exit of the pipe is equal to the head at a the pipe inlet minus the head losses. If the pipe is of constant area and at a constant elevation, the head loss leads to a decrease in pressure along the pipe.

Streamlines for a fluid entering a sudden contraction are illustrated in Figure 4.4. The fluid near the contraction is entrained in a recirculating pattern, leading to a loss of kinetic energy.

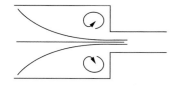

**FIG. 4.4** The energy is lost due to sudden contraction.

### 4.3.1 Losses

Head losses in pipes due to friction are called **major losses**, while losses, due to valves, expansions and contractions, bends, joints, and other changes in the flow are called **minor losses**. All losses are expressed in the form of Equation (4.24). Minor loss coefficients are highly dependent on geometry and vary greatly. Loss coefficients for pipe elbows range between 0.2 and 1.5, loss coefficients for contractions range up to 0.5, while loss coefficients for valves vary greatly depending on whether they are open or closed.

The total head loss in a piping system is

$$h_\ell = h_f + h_m \tag{4.25}$$

where $h_f$ is the head loss due to friction and $h_m$ is the sum of all minor head losses. The frictional head loss is dependent on the geometry of the pipe as well as on flow properties. The frictional head loss is written as

$$h_f = f \frac{L}{D} \frac{v^2}{2g} \tag{4.26}$$

where $L$ is the length of the pipe, $D$ is its diameter, and $f$ is called the friction factor. It has been shown that the friction factor is a function of two dimensionless parameters

$$f = f\left(\text{Re}, \frac{\varepsilon}{D}\right) \tag{4.27}$$

where Re is the **Reynolds number**

$$\text{Re} = \frac{\rho v D}{\mu} \tag{4.28}$$

and $\varepsilon/D$ is the ratio of the pipe's roughness to its diameter. In Equation (4.28) $\rho$ is the mass density of the fluid and $\mu$ is its dynamic viscosity. The Reynolds number represents the ratio of the flow's inertia forces to the friction forces.

A pipe flow is **laminar** if the flow moves along smooth parallel streamlines and the velocity of the fluid is constant at any point in the pipe. A pipe flow is **turbulent** if there are random fluctuations in the velocity profile and mixing of the fluid occurs at a macroscopic level. The Reynolds number provides a basis for predicting whether a pipe flow is laminar or turbulent. Laminar flow occurs in pipes for $\text{Re} < 2{,}300$ and turbulent flow occurs in pipes for $\text{Re} > 4{,}000$. For $2{,}300 < \text{Re} < 4{,}000$ the flow is transitional, fluctuating between laminar and turbulent.

For laminar flow, analytical methods can be used to show that the friction factor is

$$f = \frac{64}{\text{Re}} \tag{4.29}$$

Noting that the velocity of a fluid in a circular pipe is related to volumetric flow rate by

$$v = \frac{4Q}{\pi D^2} \tag{4.30}$$

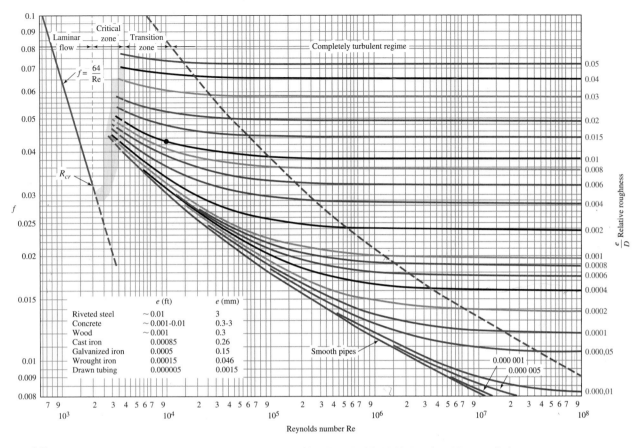

**FIG. 4.5** Moody diagram. *Source:* L. F. Moody from *Trans. ASME*, Vol. 66, 1944. Reprinted by permission.

the substitution of Equations (4.28) and (4.29) into Equation (4.26) leads to

$$h_f = \frac{128\mu L}{\rho g \pi D^4} Q \qquad (4.31)$$

Thus for laminar flow the frictional head loss in a pipe is proportional to the flow rate.

Empirical methods are used to quantify Equation (4.27) for turbulent flow. The **Colebrook equation** provides a quantitative fit to empirical data for a pressure drop in a pipe

$$\frac{1}{\sqrt{f}} = -2.0 \log \left( \frac{\varepsilon/D}{3.7} + \frac{2.51}{\mathrm{Re}\sqrt{f}} \right) \qquad (4.32)$$

It is not possible to solve Equation (4.32) in closed form to determine $f$ given values of Re and $\varepsilon/D$. The friction factor is obtained by solving the Colebrook equation using numerical methods or by referring to the **Moody diagram** (Figure 4.5), a chart that provides curves of $f$ vs. Re for different values of $\varepsilon/D$. When the fluid velocity is unknown, the Reynolds number is not known and the Moody diagram cannot be used to directly determine the friction factor. In such cases iterative methods are used to determine the velocity and the friction factor.

## Example 4.4

Water at 10°C ($\rho = 1,000$ kg/m$^3$, $\mu = 1.31 \times 10^{-3}$ N·s/m$^2$) enters the 600-m long pipe of Figure 4.6 at $A$, where the gauge pressure is 900 kPa. The fluid exits into the atmosphere at $B$, where the pressure is atmospheric. Pipe $AB$ is made of commercial steel with a roughness of $\varepsilon = 0.046$ mm and a diameter of 10 cm. The total of the minor loss coefficients is 12.4. Determine the flow rate through the pipe.

### Solution

The application of the extended Bernoulli's equation between $A$ and $B$ gives

$$\frac{p_A}{\rho g} + \frac{v_A^2}{2g} + z_A = h_\ell + \frac{p_B}{\rho g} + \frac{v_B^2}{2g} + z_B \tag{a}$$

The pipe is of constant diameter; thus Conservation of Mass implies $v_A = v_B$. The fluid exits into the atmosphere at $B$; thus the gauge pressure at $B$ is zero. The datum is taken at the level of the pipe at $A$: $z_A = 0$ m, $z_B = 40$ m. The total head loss is the sum of the minor losses and the major losses

$$h_\ell = h_f + h_m$$

$$= f \frac{L}{D} \frac{v^2}{2g} + \sum K \frac{v^2}{2g}$$

$$= \left( f \frac{600 \text{ m}}{0.1 \text{ m}} + 12.4 \right) \frac{v^2}{2g}$$

$$= (6,000f + 12.4) \frac{v^2}{2g} \tag{b}$$

The flow rate is related to the velocity by

$$v = \frac{Q}{A}$$

$$= \frac{4Q}{\pi D^2}$$

$$= \frac{4Q}{\pi (0.1 \text{ m})^2} = 1.27 \times 10^2 Q \tag{c}$$

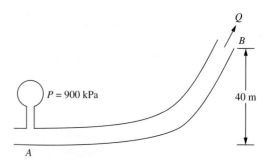

$P = 900$ kPa

$Q$

$B$

40 m

FIG. **4.6**  Pipe of Examples 4.4 and 4.5.

$A$

**TABLE 4.1  ITERATION TO DETERMINE FLOW RATE FOR EXAMPLE 4.4**

| $Q(\text{m}^3/\text{s})$ | Re [Equation (f)] | $f$ (Moody Diagram) | $Q(\text{m}^3/\text{s})$ [Equation (e)] |
|---|---|---|---|
| 1.0 | $9.69 \times 10^6$ | 0.0165 | $2.38 \times 10^{-2}$ |
| $2.38 \times 10^{-2}$ | $2.30 \times 10^5$ | 0.0185 | $2.26 \times 10^{-2}$ |
| $2.26 \times 10^{-2}$ | $2.19 \times 10^5$ | 0.019 | $2.23 \times 10^{-2}$ |

Substituting into Equation (a) leads to

$$\frac{900 \times 10^3 \text{ N/m}^2}{\left(1,000 \text{ kg/m}^3\right)\left(9.81 \text{ m/s}^2\right)} = (6,000f + 12.4)\frac{\left(1.27 \times 10^2 Q\right)^2}{2(9.81 \text{ m/s}^2)} + 40 \quad \text{(d)}$$

Equation (d) simplifies to

$$6.29 \times 10^{-2} = (6,000f + 12.4)Q^2 \quad \text{(e)}$$

The Reynolds number for the flow is calculated in terms of the flow rate as

$$\begin{aligned} \text{Re} &= \frac{\rho v D}{\mu} \\ &= \frac{(1,000 \text{ kg/m}^3)(1.27 \times 10^2 Q)(0.1 \text{ m})}{1.31 \times 10^{-3} \text{ N·s/m}^2} \\ &= 9.69 \times 10^6 Q \end{aligned} \quad \text{(f)}$$

The roughness ratio for the pipe is

$$\frac{\varepsilon}{D} = \frac{4.6 \times 10^{-5} \text{ m}}{0.1 \text{ m}} = 4.6 \times 10^{-4} \quad \text{(g)}$$

An iterative procedure is used to determine the flow rate using Equations (e) and (f) and the curve on the Moody diagram corresponding to the roughness ratio of Equation (g). A value of $Q$ is guessed. Equation (f) is used to determine the Reynolds number. The Moody diagram is used to determine the friction factor. Using this value of $f$, Equation (e) is used to determine a new value of $Q$. The iteration continues until convergence is achieved. The calculations for the iteration are given in Table 4.1. The process converges in three iterations to a flow rate of $2.23 \times 10^3 \text{ m}^3/\text{s}$.

The **resistance** of a pipe is defined as

$$R = \frac{dh_\ell}{dQ} \quad (4.33)$$

The resistance of a pipe with laminar flow and no minor losses is determined using Equation (4.31) as

$$R = \frac{128\mu L}{\rho g \pi D^4} \quad (4.34)$$

The resistance of a pipe with turbulent flow can be determined by developing a graph of $h_f$ as a function of $Q$. Note that for a pipe of constant diameter with minor losses, Equation (4.25) can be written in terms of flow rate as

$$h_\ell = \left( f\frac{L}{D} + \sum K \right) \frac{8}{\pi^2 D^4 g} Q^2 \tag{4.35}$$

Examination of the Moody diagram shows that for large Reynolds numbers the friction factor $f$ has little dependence on Re, varying mostly with $\varepsilon/D$. In this case Equation (4.35) can be written as

$$Q = B\sqrt{h_\ell} \tag{4.36}$$

where the constant of proportionality is

$$B = \left[ \left( f\frac{L}{D} + \sum K \right) \frac{8}{\pi^2 D^4 g} \right]^{-1/2} \tag{4.37}$$

Using Equation (4.37) the pipe's resistance is determined as

$$R = 2\frac{Q}{B^2} = 2\frac{h_\ell}{Q} \tag{4.38}$$

Equation (4.38) shows that the resistance of a pipe with turbulent flow is a function of flow rate. Given a curve for head loss vs. flow rate the resistance can be determined as the slope of the curve for a specified flow rate. A curve of frictional head loss vs. flow rate for a specified pipe could be developed using the Colebrook equation and the Moody diagram. However, in a practical situation where minor losses may be significant, the curve of head loss for the piping system vs. flow rate can be determined empirically.

## Example 4.5

Determine the resistance of the pipe of Example 4.4.

### Solution
Using Equations (b) and (c) of Example 4.4, the head loss is related to the flow rate as

$$h_\ell = (6{,}000f + 12.4)\frac{(1.27 \times 10^2 Q)^2}{2(9.81)}$$
$$= 8.22 \times 10^2 (6{,}000f + 12.4)Q^2 \tag{a}$$

Using the calculated values of $f = 0.019$ and $Q = 2.23 \times 10^{-2}$ m³/s in Equation (b) leads to $h_\ell = 51.7$ m. The pipe's resistance at this flow rate is obtained using Equation (4.38) as

$$R = 2\frac{h_\ell}{Q}$$

$$= 2\frac{51.7 \text{ m}}{2.23 \times 10^{-2} \text{ m}^3/\text{s}} = 4.64 \times 10^3 \text{ s/m}^2 \tag{b}$$

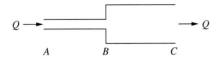

**FIG. 4.7** Pipes $AB$ and $BC$ are in series. The flow rate through each pipe is the same, and the head loss from $A$ to $C$ is the head loss from $A$ to $B$ plus the head loss from $B$ to $C$.

Pipes are similar to resistors in electric circuits in that they dissipate energy, which is converted into heat. Similar to resistors, pipe systems can be classified as being in series or parallel. Two pipes in series are illustrated in Figure 4.7. These pipes of different diameters are joined directly together such that use of the continuity equation implies that the flow rate in each pipe is the same. The total head loss across the system is the sum of the head losses in each pipe. Thus pipes in series are characterized as having the same flow rate but with the total head loss as the sum of the head losses.

An equivalent resistance for pipes with turbulent flow is defined by

$$R_{eq} = 2\frac{h_{\ell,tot}}{Q} \tag{4.39}$$

Referring to Figure 4.7 for the two pipes in series

$$h_{\ell,tot} = h_{\ell,AB} + h_{\ell,BC} \tag{4.40}$$

Using Equation (4.33) in Equation (4.40) leads to

$$h_{\ell,tot} = \tfrac{1}{2}R_{AB}Q + \tfrac{1}{2}R_{BC}Q \tag{4.41}$$

Substitution of Equation (4.41) in Equation (4.39) leads to

$$R_{eq} = R_{AB} + R_{BC} \tag{4.42}$$

Thus the equivalent resistance of pipes in series is the sum of the resistances of the pipes.

The application of the continuity equation to the parallel pipes of Figure 4.8 shows that

$$Q = Q_1 + Q_2 \tag{4.43}$$

The application of the extended Bernoulli's equation between $A$ and $B$ leads to

$$\frac{p_A}{\rho} + \frac{v_A^2}{2g} + z_A = h_\ell + \frac{p_B}{\rho} + \frac{v_B^2}{2g} + z_B \tag{4.44}$$

However, Equation (4.44) can be applied along a streamline through either pipe, which implies that the head loss in each pipe is the same. Thus parallel pipes are characterized as having the same head loss, and the total flow rate is the sum of the flow rates through the pipes.

Using a method similar to that to derive the equivalent resistance of resistors in parallel, it is determined that the equivalent resistance of $n$ pipes in parallel is

$$R_{eq} = \frac{1}{\displaystyle\sum_{i=1}^{n} \frac{1}{R_1}} \tag{4.45}$$

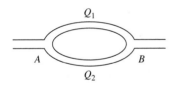

**FIG. 4.8** The two parallel pipes have the same head loss, and the total flow rate is the sum of the flow rates in the pipes.

### 4.3.2 Orifices

Consider the flow of an incompressible fluid in a pipe with an orifice, as illustrated in Figure 4.9. The application of Bernoulli's equation between

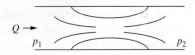

**FIG. 4.9** The relationship between flow rate and pressure drop for an incompressible flow through an orifice is given in Equation (4.49).

the upstream position 1 and the downstream position 2, assuming no change in elevation between positions, leads to

$$\frac{p_1}{\rho} + \frac{v_1^2}{2} = \frac{p_2}{\rho} + \frac{v_2^2}{2} \tag{4.46}$$

Since the flow is incompressible the density and thus the volumetric flow rate $Q$ are constant. Conservation of Mass is expressed as

$$Q = v_1 A_1 = v_2 A_2 \tag{4.47}$$

Equation (4.47) is used to express the velocities in terms of the flow rate and substituted into Equation (4.46), which can be rearranged as

$$Q = A_2 \sqrt{\frac{2(p_1 - p_2)}{\rho\left[1 - (A_2/A_1)^2\right]}} \tag{4.48}$$

Equation (4.48) is for an ideal flow without losses. Losses are taken into account by the introduction of a loss coefficient $C_d$ such that for a real flow

$$Q = \frac{C_d A_2}{\sqrt{1 - (A_2/A_1)^2}} \sqrt{\frac{2(p_1 - p_2)}{\rho}} \tag{4.49}$$

The flow coefficient $C_Q$ is defined as

$$C_Q = \frac{C_d}{\sqrt{1 - (A_2/A_1)^2}} \tag{4.50}$$

such that Equation (4.49) can be written as

$$Q = C_Q A_2 \sqrt{\frac{2(p_1 - p_2)}{\rho}} \tag{4.51}$$

### 4.3.3 Compressible Flows

The control volume form of the Conservation of Mass equation for a compressible flow in a pipe is

$$\rho_1 V_1 A_1 = \rho_2 V_2 A_2 \tag{4.52}$$

Since the density is not constant the volumetric flow rate is not constant. If the velocity of the fluid is greater than the speed of sound, the flow is said to be supersonic and a normal shock may form in a duct or nozzle. If the velocity of the fluid is less than the speed of sound the flow is subsonic. Only subsonic flows are considered here.

Consider the system of Figure 4.10. The system is at a steady state when the upstream pressure is increased, forcing air through the pipe into the reservoir. The mass of air in the reservoir increases, leading to an increase in pressure. Compressible flow, in a constant area pipe with

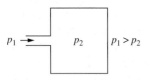

**FIG. 4.10** Compressible flow through a pipe or restriction is difficult to analyze. The resistance of the pipe is defined by Equation (4.53).

friction is called Fanno flow, which has complicated relationships between the pressure change across the length of the pipe and the mass flow rate. For a small increase in upstream pressure the flow is subsonic. The mass flow rate into the reservoir is roughly proportional to the square root of the difference between the upstream pressure $p_1$ and the pressure at the entrance to the reservoir $p_2$. This relationship is analogous to the relationship for volumetric flow rate and head loss for an incompressible flow, Equation (4.36). Thus the concept of resistance is defined for subsonic compressible flows in a fashion similar to that of liquid flow in pipes. Specifically the resistance is defined by

$$R = \frac{d}{d\dot{m}}(p_1 - p_2) \tag{4.53}$$

The resistance, defined by Equation (4.53), is not constant. For flow of a liquid in a pipe the resistance is calculated at the steady state, which linearizes the mathematical modeling of the system. However, for the system considered in Figure 4.10 the steady state corresponds to a mass flow rate of zero. Using Equation (4.51) at $\dot{m} = 0$ leads to significant error in the modeling. Thus an average value of $R$ is used. The linearizing approximation is

$$\dot{m} = \frac{p_1 - p_2}{R} \tag{4.54}$$

where $R$ is determined by developing a curve for $p_1 - p_2$ as a function of $\dot{m}$ (at the exit of the pipe) and using Equation (4.53) at an average value of $\dot{m}$.

## 4.4  MODELING OF LIQUID-LEVEL SYSTEMS

The level of the liquid in the tank of Figure 4.11 is constant when the flow rate of the inlet stream is equal to the flow rate through the pipe. When this steady state is disturbed, for example by a sudden increase or decrease in the inlet flow rate, the level of the liquid in the tank varies with time. Mathematical modeling of the change in liquid level requires application of the control volume form of conservation of mass as well as the pipe flow equations of Section 4.3. Such problems are referred to as liquid-level problems.

Liquid enters a tank of area $A$ at a volumetric flow rate $Q_i$. Liquid leaves the tank through a pipe of length $L$, diameter $D$, and roughness $\varepsilon$. The liquid is at a constant temperature and has a dynamic viscosity $\mu$ and a mass density $\rho$. Let $H(t)$ be the liquid level in the tank and $Q_o$ be the volumetric flow rate of the fluid leaving the pipe. A control volume consisting of the tank and the pipe is selected. Application of conservation of mass, in the form of Equation (4.7), to the control volume leads to

$$Q_i - Q_o = \frac{d}{dt}(AH) \tag{4.55}$$

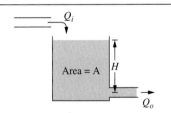

**FIG. 4.11**  The reservoir used as an example for modeling liguid-level systems. The system is operating at a steady state. When the inlet flow rate has a change, it leads to perturbations in $H$ and $Q_o$.

Consider a streamline from the entrance of the pipe to its exit. Since the area of the pipe is constant the velocity of the fluid at the inlet is the same as the velocity of the fluid at the outlet. The pressure head at the inlet of the pipe is $H$. The application of the extended Bernoulli's equation [Equation (4.23)] along this streamline leads to

$$H + \frac{v^2}{2g} - h_\ell = \frac{v^2}{2g}$$

$$h_\ell = H \qquad (4.56)$$

Thus the head loss in the pipe is equal to the level of the liquid in the tank.

Assuming turbulent flow the flow rate at the outlet is given by Equation (4.36), which, using Equation (4.56), leads to

$$Q_o = B\sqrt{H} \qquad (4.57)$$

Substitution of Equation (4.57) into Equation (4.55) leads to

$$A\frac{dH}{dt} + B\sqrt{H} = Q_i \qquad (4.58)$$

At $t = 0$ the system is at a steady state with an inlet flow rate of $q_{i,s}$ and a liquid level of $h_s$. The steady state implies $dh_s/dt = 0$ and thus Conservation of Mass requires that the outlet flow rate is equal to the inlet flow rate. Thus from Equation (4.57)

$$q_{o,s} = q_{i,s} = B\sqrt{h_s} \qquad (4.59)$$

Suppose the inlet flow rate changes at $t = 0$ and is given by $Q_i(t)$. The time-dependent level of the liquid in the tank is obtained by solving Equation (4.58) subject to the initial condition, $H(0) = h_s$. However, Equation (4.58) is a nonlinear differential equation for which a solution is difficult to obtain. If the expected change in liquid level is small compared to $h_s$ then Equation (4.58) may be linearized using the methods of Section 1.5. Indeed Equation (4.58) is the equation used in Section 1.5 to illustrate the method of linearization through the introduction of perturbation variables.

The steady state is disturbed due to a perturbation in the inlet flow rate defined such that

$$Q_i = q_{i,s} + q_i(t) \qquad (4.60)$$

The perturbation in flow rate induces a time-dependent perturbation in the liquid level $h(t)$, defined such that

$$H(t) = h_s + h(t) \qquad (4.61)$$

The outlet flow rate is obtained from Equation (4.57) as

$$Q_o(t) = B\sqrt{h_s + h(t)} \qquad (4.62)$$

Substituting Equations (4.60), (4.61), and (4.62) into Equation (4.58) leads to

$$q_{i,s} + q_i(t) - B\sqrt{h_s + h(t)} = A\frac{dh}{dt} \qquad (4.63)$$

Linearization of Equation (4.63) described in Example 1.5 leads to an approximate equation of

$$A\frac{dh}{dt} + \frac{B}{2\sqrt{h_s}}h = q_i(t) \tag{4.64}$$

The resistance of a pipe with turbulent flow, where the relationship between the flow rate and head loss is given by Equation (4.36), is developed in Equation (4.38), which leads to $R = 2\sqrt{h_s}/B$, where $R$ is the pipe resistance calculated for the original steady state. Using this result in Equation (4.64) gives

$$A\frac{dh}{dt} + \frac{1}{R}h = q_i(t) \tag{4.65}$$

Equation (4.65) is analogous to the differential equation for the voltage change across the capacitor of an $RC$ circuit, Equation (g) of Example 3.9. The head in the tank is analogous to the potential across the capacitor. The friction in the pipe dissipates energy into heat like a resistor. A loss of head occurs due to friction as flow passes through a pipe. The resistance of the pipe is the ratio of the change in liquid level to the change in flow rate. This is analogous to a resistor in an electrical circuit. Potential is lost as current passes through a resistor. A definition of electrical resistance for a linear resistor is the ratio of the change in current to the change in potential. Note, however, that the dimensions of flow resistance and electrical resistance are not the same.

The tank stores potential energy through gravity. The total potential energy of the liquid in the tank is the weight of the liquid in the tank multiplied by the distance from the mass center of the liquid to a reference plane or the total energy stored in the tank is

$$V = \tfrac{1}{2}\rho g A H^2 \tag{4.66}$$

The total energy stored in the electric field of a capacitor is $E = \tfrac{1}{2}Cv^2$, where the capacitance could be defined as the ratio of the change in the charge on the plates of a capacitor to the change in potential difference across the capacitor. The analogous definition of capacitance for the liquid-level system is the ratio of the change in volume of stored liquid to the head, leading to the capacitance of the tank being simply equal to its area.

The technique used to derive Equation (4.65) can be applied to any liquid-level problem, including problems involving multiple tanks. The control volume form of Conservation of Mass is applied to each tank. The extended Bernoulli's equation is used to relate the head loss in the pipes to the fluid levels. Perturbation variables, measured from the steady state, are introduced. Conservation of Mass equations are written using perturbation variables as dependent variables. The equations are linearized as in Section 1.5.

The resistance used in Equation (4.65) is evaluated at the original steady-state conditions. This is a result of the linearization process.

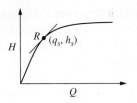

**FIG. 4.12** The resistance of a pipe is a function of flow rate through the pipe.

The resistance changes with flow rate as illustrated in Figure 4.12. At a given flow rate the resistance is the slope of the tangent to the $H$ vs. $Q$ curve. The resistance is a function of flow rate.

An alternate derivation of equations governing liquid-level problems involves the definition of resistance at the steady state, defined using Equation (4.33)

$$R = \frac{dH}{dQ_0}\bigg|_{Q_o = q_{o,s}} \tag{4.67}$$

This use of the resistance calculated for the original steady-state conditions is a result of the linearization process. The resistance changes with flow rate, as illustrated in Figure 4.12. At a given flow rate the resistance is the slope of the tangent to the $H$ vs. $Q$ curve. The a priori use of resistance at the steady state leads to an a priori alternate method for linearizing the differential equation; that is, the use of the linearizing assumption in deriving the differential equation leads to the formulation of a linear differential equation. Using the definition of the derivative, and since $h(t)$ and $q_o(t)$ are small perturbations from the steady state, the resistance is approximated as

$$
\begin{aligned}
R &\approx \frac{\Delta H}{\Delta Q_o} \\
&= \frac{H(q_{o,s} + q) - H(q_{o,s})}{q_{o,s} + q_o - q_{o,s}} \\
&= \frac{h_s + h(t) - h_s}{q_o} \\
&= \frac{h}{q_o}
\end{aligned} \tag{4.68}
$$

Equation (4.68) can be rewritten as

$$q_o = \frac{h}{R} \tag{4.69}$$

The output flow rate is then written as

$$Q_o = q_{o,s} + \frac{h}{R} \tag{4.70}$$

Using Equations (4.57) and (4.70) the Conservation of Mass equation, Equation (4.55) can be written as

$$q_{i,s} + q_i(t) - q_{o,s} - \frac{h}{R} = A\frac{dh}{dt} \tag{4.71}$$

Since $q_{i,s} = q_{o,s}$ is the Conservation of Mass equation for the steady state, Equation (4.71) reduces to Equation (4.65).

The linearization using this method occurs in the approximation of the resistance by Equation (4.68). In general the linear approximation for the resistance is the change in head loss due to the perturbations divided by the perturbation of the flow rate in the pipe. This, in turn, leads to a linear approximation for the perturbation in flow rate in a pipe as the change in head loss across the pipe divided by its resistance at steady state. Direct use of Equation (4.69) in the Conservation of Mass equation applied to a tank provides an alternate method for the mathematical modeling of liquid-level problems, which avoids the necessity for formal linearization of the resulting equations.

## Example 4.6

The tank of Figure 4.11 has an area of 900 m². The steady-state level is 50 m for an input flow rate of 2.5 m³/s. Derive a mathematical model for the perturbation in the level when the input flow rate is suddenly changed to 2.45 m³/s.

### Solution

The resistance of the pipe at steady state, assuming turbulent flow, is determined using Equation (4.39) as

$$R = 2\frac{h_s}{q_{o,s}}$$

$$= 2\frac{50 \text{ m}}{2.50 \text{ m}^3/\text{s}}$$

$$= 40 \text{ s/m}^2 \tag{a}$$

The perturbation in the input flow rate is written using the unit step function as

$$q_i(t) = -0.05u(t) \text{ m}^3/\text{s} \tag{b}$$

Using Equations (a) and (b) in Equation (4.65) yields

$$900\frac{dh}{dt} + \frac{1}{40}h = -0.05u(t) \tag{c}$$

$$3.6 \times 10^4 \frac{dh}{dt} + h = -2.0u(t) \tag{d}$$

## Example 4.7

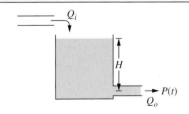

**FIG. 4.13** The system of Example 4.7. Pressure fluctuations at the tank's exit lead to perturbation in liquid level from the steady state.

The tank of Figure 4.13 is operating at a steady state when the pressure at the exit of the pipe begins to fluctuate according to $p = p_s + p(t)$. The input flow rate remains constant. Derive a mathematical model for the perturbation of the liquid level in the tank.

### Solution

The steady state is defined by $H = h_s$, $Q_i = q_{i,s}$, $Q_0 = q_{o,s}$, $dh_s/dt = 0$, and $Q_i = Q_0$. The perturbations in liquid-level and outlet flow rate are defined by

$$Q_o = q_{o,s} + q_o(t) \tag{a}$$
$$H = h_s + h(t) \tag{b}$$

The application of Conservation of Mass to the tank leads to Equation (4.7), which applied for this system becomes

$$q_{i,s} - Q_0 = A\frac{dh}{dt} \tag{c}$$

The application of the extended Bernoulli's equation along a streamline in the pipe gives

$$H = h_\ell + \frac{p(t)}{\rho g} \tag{d}$$

The outlet flow rate is related to head loss through Equation (4.57) leading to

$$H = \frac{Q_o^2}{B^2} + \frac{p}{\rho g} \tag{e}$$

Equation (e) is rearranged as

$$Q_o = B\sqrt{H - \frac{p}{\rho g}}$$
$$= B\sqrt{h_s + h - \frac{p}{\rho g}} \tag{f}$$

Substituting Equation (f) into Equation (c) leads to

$$q_{i,s} - B\sqrt{h_s + h - \frac{p}{\rho g}} = A\frac{dh}{dt} \tag{g}$$

Assuming $|h - \frac{p}{\rho g}| \ll h_s$ Equation (g) is linearized using the methods of Section 1.5 yielding

$$q_{i,s} - B\sqrt{h_s}\left(1 + \frac{1}{2}\frac{h - \frac{p}{\rho g}}{h_s}\right) = A\frac{dh}{dt} \tag{h}$$

Applying the steady-state condition [Equation (4.59)], to Equation (h) leads to

$$\frac{B}{2\sqrt{h_s}}\left(\frac{p}{\rho g} - h\right) = A\frac{dh}{dt} \tag{i}$$

The resistance of the pipe at the steady state is obtained using Equations (4.36) and (4.38) as

$$R = \frac{2\sqrt{h_s}}{B} \tag{j}$$

Using Equation (j) in Equation (i) leads to

$$A\frac{dh}{dt} + \frac{1}{R}h = \frac{p}{\rho g R} \tag{k}$$

**Example 4.8**

The two-tank system of Figure 4.14 is operating at a steady state when fluctuations in the inlet flow rates occur. The resistances of the pipes at steady state are $R_1$ and $R_2$. Derive a mathematical model for the perturbations in liquid level due to fluctuations in the inlet flow rates.

**Solution**

The following steady-state variables are defined:

| | |
|---|---|
| Liquid level in tank 1 | $h_{s1}$ |
| Liquid level in tank 2 | $h_{s2}$ |
| Inlet flow rate to tank 1 | $q_{si1}$ |
| Inlet flow rate to tank 2 | $q_{si2}$ |
| Oultet flow rate from pipe 2 | $q_{so} = q_{si_1} + q_{si_2}$ |

Perturbation variables are introduced

$$H_1(t) = h_{s1} + h_1(t) \qquad \text{(a)}$$
$$H_2(t) = h_{s2} + h_2(t) \qquad \text{(b)}$$
$$Q_{i1}(t) = q_{si1} + q_{i1}(t) \qquad \text{(c)}$$
$$Q_{i2}(t) = q_{si2} + q_{i2}(t) \qquad \text{(d)}$$
$$Q_o(t) = q_{so} + q_o(t) \qquad \text{(e)}$$

Define

$$Q_{p1} = q_{sp1} + q_{p1}(t) \qquad \text{(f)}$$

as the flow rate in pipe 1, connecting the two tanks.

The application of Conservation of Mass to each tank leads to

$$Q_{i1} - Q_{p1} = A_1 \frac{dH_1}{dt} \qquad \text{(g)}$$

$$Q_{i2} + Q_{p1} - Q_o = A_2 \frac{dH_2}{dt} \qquad \text{(h)}$$

The steady-state conditions are

$$q_{si1} - q_{sp1} = 0 \qquad \text{(i)}$$
$$q_{sp1} + q_{si2} - q_{so} = 0 \qquad \text{(j)}$$

The application of the extended Bernoulli's equation between the free surface of tank 1 and the exit of the pipe connecting the two tanks, noting that the pressure

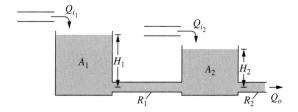

**FIG. 4.14** Two-tank system of Example 4.8.

head at the exit of the pipe is $P$ and assuming the kinetic energy of the flow is lost as the pipe exits into the second tank, leads to

$$H_1 = h_{\ell 1} + H_2 \tag{k}$$

The application of the extended Bernoulli's equation between the free surface of tank 2 and the outlet of the second pipe leads to

$$H_2 = h_{\ell 2} \tag{l}$$

Equations (k) and (l) can be rewritten as

$$h_1 = \Delta h_{\ell 1} + h_2 \tag{m}$$

$$h_2 = \Delta h_{\ell 2} \tag{n}$$

where $\Delta h_{\ell 1}$ and $\Delta h_{\ell 2}$ are the changes in head loss from the steady state. The linearized resistances, defined as in Equation (4.68), are related to the flow rate perturbations by

$$q_{p1} = \frac{\Delta h_{\ell 1}}{R_1} \tag{o}$$

$$q_o = \frac{\Delta h_{\ell 2}}{R_2} \tag{p}$$

Using Equations (o) and (p) in Equations (m) and (n) leads to

$$q_{p1} = \frac{1}{R_1}(h_1 - h_2) \tag{q}$$

$$q_o = \frac{1}{R_2}h_2 \tag{r}$$

Substitution of Equations (a)–(f) and (q) and (r) in Equations (g) and (h) leads to

$$q_{si1} + q_{i1} - q_{sp1} - \frac{1}{R_1}(h_1 - h_2) = A_1\frac{dh_1}{dt} \tag{s}$$

$$q_{si2} + q_{i2} + q_{sp1} + \frac{1}{R_1}(h_1 - h_2) - q_{so} - \frac{1}{R_2}h_2 = A_2\frac{dh_2}{dt} \tag{t}$$

Eliminating the steady-state terms from Equations (s) and (t) and rearranging leads to

$$A_1\frac{dh_1}{dt} + \frac{1}{R_1}h_1 - \frac{1}{R_1}h_2 = q_{i1}(t) \tag{u}$$

$$A_2\frac{dh_2}{dt} - \frac{1}{R_1}h_1 + \left(\frac{1}{R_1} + \frac{1}{R_2}\right)h_2 = q_{i2}(t) \tag{v}$$

## 4.5 PNEUMATIC AND HYDRAULIC SYSTEMS

Many dynamic systems have **pneumatic** components, which involve the flow of a compressed gas (usually air) thorough a passage or use air to supply power. Many dynamic systems have **hydraulic** components, which involve the flow of an incompressible liquid through a passage or use liquid to supply power. Pneumatic and hydraulic systems are used in motion control devices.

### 4.5.1 Pneumatic Systems

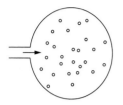

**FIG. 4.15** When mass flows into the pressure vessel, the density and pressure of the gas increase.

Consider the pressure vessel of Figure 4.15. When the valve between the pressure vessel and the supply pipe is open, gas flows into the pressure vessel at a flow rate $\dot{m}$. The application of the control volume form of Conservation of Mass to a control volume consisting of the pressure vessel leads to

$$\dot{m} = \frac{dm_v}{dt} \tag{4.72}$$

where $m_v$ is the mass of the gas in the vessel. As gas flows into the pressure vessel the total mass in the vessel increases. If the volume of the vessel is constant, then as the mass increases the density and pressure of the gas in the pressure vessel also increase. An **isothermal** process occurs at a constant temperature. Under certain conditions the temperature may also change, affecting the pressure.

The relationship between density, pressure, and temperature in a gas is called an **equation of state**. The equation of state for an ideal gas, such as air, helium, or nitrogen, is called the ideal gas law and is written as

$$p = \rho R_g T \tag{4.73}$$

where $\rho$ is the density of the gas, $p$ is the absolute pressure of the gas, and $T$ is the absolute temperature of the gas. The constant of proportionality $R_g$ is called the gas constant, which for an ideal gas is defined by

$$R_g = \frac{\bar{R}}{M} \tag{4.74}$$

where $M$ is the molecular mass of the gas and $\bar{R}$ is the universal gas constant,

$$\bar{R} = 8{,}314.3 \text{ J/kg·mol·K}$$
$$= 1{,}545.3 \text{ ft·lb/lbm·mol·R} \tag{4.75}$$

For air, often used in pneumatic systems, $M = 28.9/\text{mol}$ and the gas constant is

$$R = 287.0 \text{ J/kg·K} = 53.35 \text{ ft·lb/lbm·R} \tag{4.76}$$

A process is **isentropic** (constant entropy) if it is reversible. Throughout an isentropic process the pressure and density of an ideal gas are related by

$$\frac{p}{\rho^{\gamma}} = C \tag{4.77}$$

where $C$ is a constant for the process and $\gamma = c_p/c_v$, the specific heat ratio. For air, $\gamma = 1.4$.

Many processes are neither isothermal nor isentropic, but polytropic, in which the pressure and density are related throughout the process by

$$\frac{p}{\rho^n} = C \qquad (4.78)$$

where $C$ is a constant for the process and $n$ is the polytropic exponent. It is noted that isothermal processes ($n = 1$) and isentropic processes ($n = \gamma$) are special cases of polytropic processes.

## Example 4.9

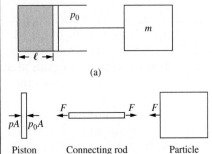

(a)

(b)

Piston        Connecting rod        Particle

**FIG. 4.16** (a) The system of Example 4.9, in which air expands and contracts in a sealed cylinder as a piston moves in the cylinder; (b) the free-body diagrams at an arbitrary instant.

The particle of mass $m$ of Figure 4.16(a) is connected to the end of a piston whose head of area $A$ is in a sealed cylinder of air. The system is in equilibrium when the pressure inside the cylinder is at the same pressure as the air surrounding the cylinder $p_0$. Let $x$ be the displacement of the particle from the system's equilibrium position. Derive a linear mathematical model for the system, assuming the expansion and contraction of the air in the cylinder is an isentropic process.

## Solution

The application of Newton's second law to the free-body diagram of the particle drawn at an arbitrary instant [Figure 4.16(b)], leads to

$$-F = m\ddot{x} \qquad (a)$$

where $F$ is the resultant force on the piston head. Assuming the piston head is massless

$$F = p_0 A - pA \qquad (b)$$

Since the cylinder is sealed, the mass of air contained in the cylinder is a constant

$$m_a = \rho_0 \ell A \qquad (c)$$

where $\ell$ is the length of the column of air when the system is in equilibrium. When the piston is displaced a distance $x$ from its equilibrium position, the length of the column of air is increased by $x$, leading to

$$m_a = \rho(\ell + x)A \qquad (d)$$

Equations (c) and (d) are used to determine the density as

$$\rho = \rho_0 \left(1 + \frac{x}{\ell}\right)^{-1} \qquad (e)$$

The process is assumed to be isentropic. Thus Equations (4.77) and (e) lead to

$$p = p_0 \left(1 + \frac{x}{\ell}\right)^{-\gamma} \qquad (f)$$

Substitution of Equations (b) and (f) into Equation (a) leads to

$$m\ddot{x} - p_0 A \left(1 + \frac{x}{\ell}\right)^{-\gamma} = -p_0 A \qquad (g)$$

Equation (g) is linearized by applying the binomial expansion

$$\left(1 + \frac{x}{\ell}\right)^{-\gamma} = 1 - \gamma\frac{x}{\ell} + \frac{\gamma(\gamma+1)}{2}\left(\frac{x}{\ell}\right)^2 + \cdots$$  (h)

Using Equation (h) in Equation (g) and truncating after the linear term leads to

$$m\ddot{x} + \frac{\gamma p_0 A}{\ell}x = 0$$  (i)

Equation (i) provides a linear mathematical model for the system in which the air in the piston and cylinder acts as a spring of stiffness $k = \gamma p_0 A/\ell$.

---

## Example 4.10

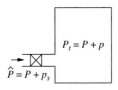

**FIG. 4.17** The system of Example 4.10: When the valve is opened and the supply pressure is greater than the tank pressure, gas flows into the pressure vessel.

The pressure vessel of Figure 4.17 is of volume $V$ and contains air at a pressure $P$ and temperature $T$ when a valve connecting the vessel to an external air supply is opened. The pressure of the supply remains constant at $\hat{P} = P + p_s$. Let $p$ be the perturbation in the pressure in the tank from its initial value defined such that

$$P_t = P + p$$  (a)

The flow into the tank is subsonic at a very low Mach number and the process can be assumed to be isothermal. An average resistance for the pipe and valve $R$, as in Equation (4.54), is defined such that

$$R = \frac{p_s - p}{\dot{m}}$$  (b)

Derive a mathematical model for $p(t)$.

### Solution

The application of the ideal gas law to the air in the tank before the valve is opened leads to

$$P = \rho_0 R_a T$$

$$T = \frac{P}{\rho_0 R_a}$$  (c)

The application of the control volume form of Conservation of Mass to the pressure vessel results in Equation (4.72) The mass of the air in the vessel at any time is

$$m_v = \rho V$$  (d)

The application of the ideal gas law at an arbitrary time during the process leads to

$$\rho = \frac{P_t}{R_a T}$$  (e)

Since the process is isothermal, Equation (c) is used in Equation (e) leading to

$$\rho = \rho_0\frac{P_t}{P} = \rho_0\left(1 + \frac{p}{P}\right)$$  (f)

Use of Equations (b), (d), and (f) in Equation (4.72) gives

$$\frac{d}{dt}\left[\rho_0 V\left(1+\frac{p}{P}\right)\right] = \frac{p_s - p}{R} \tag{g}$$

Noting from Equation (c) that $\rho_0/P = 1/R_a T$, Equation (g) is simplified to

$$\frac{V}{R_a T}\frac{dp}{dt} + \frac{p}{R} = \frac{p_s}{R} \tag{h}$$

Equation (b) is a first-order differential equation analogous to those for an $RC$ circuit or single-tank liquid-level system. The resistance of a pneumatic system is defined in general as the ratio of the change in pressure difference to the change in mass flow rate. The pressure difference is a nonlinear function of the mass flow rate. Thus the resistance, which is the slope of the line tangent to the $p - p_s$ vs. $\dot{m}$ curve, varies with mass flow rate. An average value of resistance is used to linearize the model.

Pursuing the $RC$ circuit analogy noted in Example 4.10 the capacitance of a pressure vessel is defined as the ratio of the change in the mass in the pressure vessel to the change in pressure

$$C = \frac{dm}{dp}$$

$$= \frac{d}{dp}(\rho V) \tag{4.79}$$

$$= V\frac{d\rho}{dp}$$

The application of Equation (4.79) to a pressure vessel being filled during an isothermal process at a temperature $T$ with an ideal gas of gas constant $R_g$ leads to

$$C = \frac{V}{R_g T} \tag{4.80}$$

## 4.5.2 Hydraulic Systems

Hydraulic power is used in machine tools, precision control systems, and a variety of applications in which high power and high precision are required. A unique application of hydraulic power was developed by Intamin AG of Switzerland. Intamin engineers designed the Top Thrill Dragster roller coaster located at Cedar Point Amusement Park in Sandusky, Ohio, with hydraulic power to accelerate a stationary roller coaster train to 120 mph in 4 s.

The pressure variation in an incompressible fluid varies linearly with depth into the liquid. If the pressure at a certain depth into a liquid of density $\rho$ is $p_0$ then the pressure at a depth $h$ further into the liquid is

$$p = p_0 + \rho g h \tag{4.81}$$

The pressure in a static liquid is constant at a given depth, not varying across the cross section of its container. These concepts are used to design hydraulic systems such as lifts that provide great mechanical advantage.

## Example 4.11

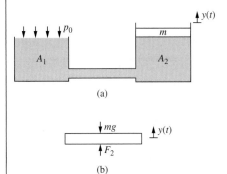

(a)

(b)

**FIG. 4.18** (a) The hydraulic lift of Example 4.11; (b) the free-body diagram of mass at an arbitrary instant.

A hydraulic lift is illustrated in Figure 4.18. At $t = 0$ a mass $m$ rests on the surface of the right tank while the left tank is subject to a free surface pressure such that the system is in equilibrium with both sides at the same level. Compressed air is then fed into the left tank such that its free surface pressure increases at a constant rate $\dot{p}_1$, which causes the mass to rise. Let $y(t)$ be the upward displacement of the mass from its initial position. Derive a mathematical model for $y(t)$, neglecting the inertia of the liquid.

### Solution

Since the liquid in both columns is at the same level at $t = 0$ the pressures on both surfaces are equal to

$$p_0 = \frac{mg}{A_2} \tag{a}$$

If the rate of change of pressure on the left surface is constant then

$$p_1 = p_0 + \dot{p}_1 t \tag{b}$$

The increase in pressure on the free surface of the left column leads to an increase in pressure on the right column, which in turn causes the mass to rise. Let $y$ be the upward displacement of the mass from its initial position and let $F_2$ be the force acting on the mass due to the pressure. The application of Newton's law to the free-body diagram of Figure 4.18(b) leads to

$$F_2 = m\ddot{y} + mg \tag{c}$$

The pressure on the surface of the right column is

$$p_2 = \frac{F_2}{A_2}$$

$$= \frac{m\ddot{y}}{A_2} + \frac{mg}{A_2}$$

$$= \frac{m\ddot{y}}{A_2} + p_0 \tag{d}$$

Let $x(t)$ be the downward displacement of the left column of liquid. Since the liquid is incompressible, the decrease in volume on the left side equals the increase in volume on the right side

$$A_1 x = A_2 y$$

$$x = \frac{A_1}{A_2} y \tag{e}$$

The difference in liquid level between both sides is $y + x$. The pressure varies linearly with depth into the liquid. Thus

$$p_1 = p_2 + \rho g(y + x) \tag{f}$$

Substitution of Equations (b), (d), and (e) into Equation (f) leads to

$$p_0 + \dot{p}_1 t = \frac{m\ddot{y}}{A_2} + p_0 + \rho g \left(1 + \frac{A_1}{A_2}\right) y$$

$$m\ddot{y} + \rho g (A_1 + A_2) y = \dot{p}_1 A_2 t \qquad (g)$$

Equation (g) of Example 4.11 is similar to the equation for a mechanical system composed of a mass $m$ attached to a spring of stiffness $\rho g (A_1 + A_2)$ acted on by an external force $\dot{p}_1 A_2 t$.

A basic hydraulic servomotor is illustrated in Figure 4.19. The motor is used to supply power to the mechanical system, often called the load. Initially the spool valve covers the opening to the cylinder containing the piston, admitting no flow. When liquid is supplied to the servomotor the change in pressure forces the spool valve to open, allowing flow into the cylinder. The pressure difference created across the head of the piston causes it to move and supply power to the load.

Liquid is supplied to the servomotor at a pressure $p_s$. The flow drains into a line of pressure $p_d$. The flow from the drain may be filtered and circulated to resupply the servomotor. The area of the face of the piston is $A_p$. The spool valve is of width $w$ and is assumed to be critically lapped such that any displacement of the spool valve allows flow into the cylinder. Let $x(t)$ represent the displacement of the spool valve and let $y(t)$ represent the displacement of the piston head.

Consider a control volume as defined in Figure 4.20, consisting of the fluid in the orifice flowing from the valve to the cylinder and the fluid to the right of the piston head. Let $Q_1$ be the flow rate through the spool valve into the cylinder. The rate at which mass accumulates in the control volume is $\rho A_p \dot{y}$. The application of Conservation of Mass to this control volume leads to

$$\rho Q_1 = \rho A_p \frac{dy}{dt} \qquad (4.82)$$

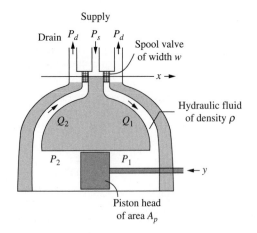

**FIG. 4.19**  A hydraulic servomotor.

$Q_1$

FIG. **4.20** A control volume used for the analysis of the hydraulic servometer.

The flow through the spool valve is that of flow through an orifice where the relation between the pressure drop across the orifice and the volumetric flow rate is given by Equation (4.52). The application of this equation to the spool valve leads to

$$Q_1 = C_Q A \sqrt{\frac{2(p_s - p_1)}{\rho}} \qquad (4.83)$$

where $C_Q$ is the flow coefficient, and $A$, the area of the valve opening is

$$A = wx \qquad (4.84)$$

Substitution of Equations (4.83) and (4.84) into Equation (4.82) leads to

$$C_Q wx \sqrt{\frac{2(p_s - p_1)}{\rho}} = A_p \frac{dy}{dt} \qquad (4.85)$$

The application of Conservation of Mass to a control volume encasing the entire servomotor and piston leads to $Q_1 = Q_2$. Assuming the discharge coefficients and areas of both spool valves are the same, the pressure drop across both valves must also be the same. Thus

$$p_2 - p_d = p_s - p_1 \qquad (4.86)$$

A steady-state condition occurs when the pressure on each side of the piston is equal and constant. In this case Equation (4.85) becomes

$$\hat{C}x = A_p \dot{y} \qquad (4.87)$$

where

$$\hat{C} = C_Q w \sqrt{\frac{2(p_s - p_1)}{\rho}} \qquad (4.88)$$

The flow rate through the orifice is a function of the length of the opening as well as the pressure difference, $\Delta p = p_s - p_1$

$$Q_1 = Q(x, \Delta p)$$
$$= C_d w \sqrt{\frac{2}{\rho}} x \sqrt{\Delta p} \qquad (4.89)$$

Equation (4.89) is linearized by expanding $Q_1$ using a two-variable Taylor series expansion about a steady-state operating point, defined by $x_s$, and $\Delta p_s$. To this end, neglecting nonlinear terms

$$Q_1 = Q_1(x_s, \Delta p_s) + x \frac{\partial Q_1}{\partial x}(x_s, \Delta p_s) + \Delta p \frac{\partial Q_1}{\partial \Delta p}(x_s, \Delta p_s)$$
$$= C_d w \sqrt{\frac{2}{\rho}} \left( x_s \sqrt{\Delta p_s} + \sqrt{\Delta p_s} x + \frac{1}{2} \frac{x_s}{\sqrt{\Delta p_s}} \Delta p \right) \qquad (4.90)$$

Equation (4.90) can be written as

$$Q_1 = Q_s + Q_x x + Q_p \Delta p \tag{4.91}$$

A linear model for the relation between the pressure difference across the orifice and the displacement of the piston is obtained by using Equation (4.91) in Equation (4.85), leading to

$$Q_s + Q_x x + Q_p \Delta p = A_p \dot{y} \tag{4.92}$$

---

## Example 4.12

The piston of the hydraulic servomotor of Figure 4.21 is connected to a mass-spring system. Derive a mathematical model for $y(t)$, the displacement of the mass from the system's equilibrium position.

### Solution

Since the piston is connected to the mass, its displacement is also $y(t)$. The force acting on the piston is

$$F_p = (p_2 - p_1)A_p \tag{a}$$

The application of Newton's second law to a free-body diagram of the load gives

$$F_p - ky = m\ddot{y} \tag{b}$$

The total pressure drop across the servomotor is $p_s - p_d$, which is the sum of the pressure drops across the orifices and across the head of the piston

$$p_s - p_d = p_s - p_1 + p_2 - p_1 + p_d - p_2 \tag{c}$$

Since the flow rates through the orifices must be equal $\Delta p = p_s - p_1 = p_d - p_2$, which when used in Equation (c) leads to

$$\Delta p = \tfrac{1}{2}[p_s - p_d - (p_2 - p_1)] \tag{d}$$

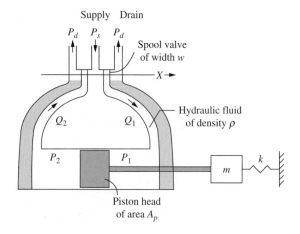

Supply   Drain

FIG. **4.21**  The hydraulic servometer of Example 4.12 is used to provide displacement to a mass-spring system.

Substitution of Equation (b) into Equation (d) gives

$$\Delta p = \tfrac{1}{2}\left[p_s - p_d - \frac{1}{A_p}(m\ddot{y} + ky)\right]$$ (e)

Using Equation (e) in Equation (4.92) leads to

$$Q_s + Q_x x + Q_p \tfrac{1}{2}\left[p_s - p_d - \frac{1}{A_p}(m\ddot{y} + ky)\right] = A_p \dot{y}$$ (f)

## Example 4.13

A walking beam is connected to the hydraulic servomotor, as shown in Figure 4.22. Let $z(t)$ be the input to the system and $y(t)$ be the output from the piston. Ignoring the load, derive a mathematical model for the system.

### Solution

Equation (4.87) defines the relation between the displacement of the spool valve and the displacement of the piston. Assume that the walking beam is vertical when the system is in equilibrium. The geometry of the walking beam is illustrated in Figure 4.22(b) for arbitrary values of $z$ and $y$. Applying the principles of similar triangles leads to

$$\frac{z + y}{a + b} = \frac{x + y}{b}$$ (a)

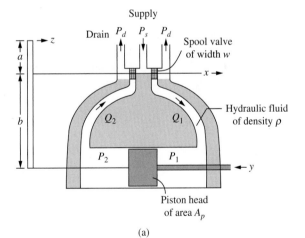

(a)

(b)

FIG. 4.22 (a) The hydraulic servometer with walking beam of Example 4.13; (b) the geometry of the walking beam of Example 4.13.

Equation (a) is solved for $x$ as

$$x = \frac{b}{a+b}z - \frac{a}{a+b}y \qquad (b)$$

Substitution of Equation (b) into Equation (4.87) leads to

$$\frac{\hat{C}}{a+b}(bz - ay) = A_p\dot{y}$$

$$A_p(a+b)\dot{y} + \hat{C}ay = \hat{C}bz \qquad (c)$$

Hydraulic servomotors, such as the one in Example 4.13, serve as actuators for control systems. The input to the servomotor is $z(t)$, its output is $y(t)$. A relation such as Equation (c) of Example 4.13 is derived between the input and the output. This relation leads to the development of a transfer function (Chapter 6) between the input and output. When $b \gg a$ the servomotor of Example 4.13 acts as an integral controller.

## 4.6 THERMAL SYSTEMS

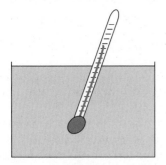

**FIG. 4.23** When a thermometer is placed in a bath of ambient temperature $T_\infty$, transient thermal behavior occurs as heat is transferred to the thermometer until steady state is reached.

A thermal system is a system in which heat transfer occurs. A thermometer is a classic example of a dynamic thermal system. When the thermometer of Figure 4.23, initially in a steady state at a uniform temperature, is suddenly placed in a medium at a higher temperature, heat is transferred to the thermometer. A transient response occurs as the temperature of the thermometer gradually approaches the temperature of the surrounding medium. The thermometer has a capacity for storing internal energy that accumulates due to the transfer of heat from the surrounding medium. The capacitance of the thermometer is defined as

$$C = \frac{dU}{dT} \qquad (4.93)$$

where $U$ is the total internal energy stored in the thermometer and $T$ is absolute temperature. The heat transfer is resisted from the medium with a thermal resistance defined as

$$R = \frac{\Delta T}{\dot{Q}} \qquad (4.94)$$

where $\Delta T$ is the temperature difference between the thermometer and the surrounding medium and $\dot{Q}$ is the rate of heat transfer between the two. When the temperatures of the thermometer and the surrounding medium are equal, thermal equilibrium occurs, the heat transfer ceases, and the thermometer stops storing energy.

The definitions of thermal capacitance [Equation (4.93)] and of thermal resistance [Equation (4.94)] are used for thermal systems in general. The total internal energy $U$ is the product of the mass of the body $m$ and the specific internal energy $u$. Recall from Equation (4.10) that the specific internal energy is equal to the specific heat $c_p$ times the temperature $T$. The application of Equation (4.93) leads to the thermal capacitance of a body as

$$C = mc_p \qquad (4.95)$$

The transient behavior of the thermometer of Figure 4.23 is modeled by applying Conservation of Energy to a control volume containing the thermometer. Energy, in the form of heat transfer, is transferred into the control volume through its boundary. Conservation of Energy reduces to a statement that the rate of change of internal energy is equal to the rate at which heat is transferred to the thermometer

$$\frac{dU}{dt} = \dot{Q} \qquad (4.96)$$

Assuming constant specific heat and using the definition of thermal resistance [Equation (4.94)], Equation (4.96) becomes

$$mc_p \frac{dT}{dt} = -\frac{1}{R}(T - T_\infty) \qquad (4.97)$$

where $T_\infty$ is the temperature of the surrounding medium. Equation (4.97) can be rearranged as

$$mc_p \frac{dT}{dt} + \frac{1}{R}T = \frac{1}{R}T_\infty \qquad (4.98)$$

Equation (4.98) is a first-order differential equation similar to that obtained for an $RC$ circuit, a single-tank liquid-level problem, and a linear pneumatic system.

Heat transfer occurs through three modes: conduction, convection, and radiation. Conduction heat transfer occurs when a body is not at a uniform temperature. Molecular motion results due to the difference in internal energy between higher temperature regions and lower temperature regions. Conduction heat transfer occurs between these molecules. The rate of conduction heat transfer per unit area is given by Fourier's conduction law as

$$\frac{\dot{Q}}{A} = -k\frac{\partial T}{\partial n} \qquad (4.99)$$

where $k$ is the thermal conductivity of the material and $n$ is the direction normal to the area through which heat is transferred. For one-dimensional heat transfer in the $x$ direction, Equation (4.99) becomes

$$\frac{\dot{Q}}{A} = -k\frac{dT}{dx} \qquad (4.100)$$

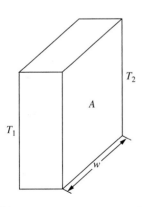

**FIG. 4.24** One-dimensional conduction heat transfer occurs through the wall of surface area $A$ and width $w$ when exterior surfaces are at different temperatures.

Consider a region of width $w$ and surface area $A$, as illustrated in Figure 4.24. The temperatures on the exterior surfaces of the region are $T_1$ and $T_2$. In the steady state the temperature gradient across the wall is a constant $\frac{dT}{dx} = \frac{T_2 - T_1}{w}$, in which case Equation (4.100) becomes

$$\dot{Q} = \frac{kA}{w}(T_1 - T_2) \tag{4.101}$$

Comparison of Equation (4.101) with Equation (4.94) shows that the resistance due to conduction is

$$R = \frac{w}{kA} \tag{4.102}$$

A similar analysis is used to show that the resistance due to conduction in an annular cylinder of length $L$, inner radius $r_i$ and outer radius $r_o$, which is made of a material of thermal conductivity $k$, is

$$R = \frac{1}{2\pi k L} \ln\left(\frac{r_o}{r_i}\right) \tag{4.103}$$

Conduction heat transfer occurs within the two adjacent regions illustrated in Figure 4.25. In the absence of other heat sources the rate of heat transfer across each region is the same and the total temperature change is the sum of the temperature changes across each region. Thus the temperature gradient across the entire region can be written as

$$\begin{aligned} \Delta T &= \Delta T_1 + \Delta T_2 \\ &= R_1 \dot{Q} + R_2 \dot{Q} \\ &= R_{eq} \dot{Q} \end{aligned} \tag{4.104}$$

Equation (4.104) shows that the equivalent resistance of the two regions is the sum of the individual resistances, which is the same relationship as the equivalent resistance of two resistors in series in an electrical circuit. Thus regions for which the rate of heat transfer is the same and the total temperature change is the sum of the temperature changes are said to be in series. Where the temperature change is the same in two regions but the total heat transfer rate across the regions is the sum of the heat transfer rates across each region, the regions are in parallel and have an equivalent resistance equal to the reciprocal of the sum of the reciprocals of the individual resistances.

Convection heat transfer occurs between a body and its surrounding medium when the body and the medium have different temperatures. While convection occurs between a moving fluid and a body, the following discussion is limited to bodies at rest. Fluid particles adjacent to a solid body must be at the same temperature as the body. Since the temperature of the surrounding medium is different from that of the body a thin region near the surface of the body must exist where a temperature gradient exists. This region is called a thermal boundary layer. The rate of heat transfer per unit area due to convection between a body whose surface is at a temperature $T$ and the surrounding medium at a temperature $T_\infty$ due to the existence of the thermal boundary layer is given by Newton's law of cooling as

$$\frac{\dot{Q}}{A} = h(T - T_\infty) \tag{4.105}$$

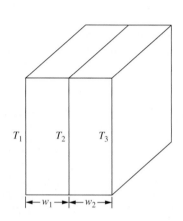

**FIG. 4.25** The equivalent thermal resistance of two regions in series is the sum of the resistances.

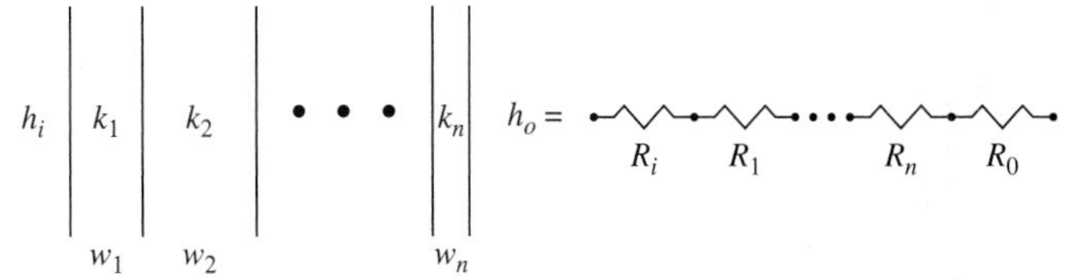

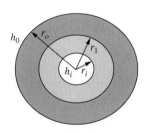

**FIG. 4.26** The equivalent thermal resistance of a region due to convection and conduction is modeled using resistors in series.

where $h$ is called the film coefficient. Equation (4.105) can be rewritten as

$$\dot{Q} = \frac{1}{R_c}(T - T_\infty) \tag{4.106}$$

where the convective resistance is

$$R_c = \frac{1}{hA} \tag{4.107}$$

The equivalent thermal resistance of the system of Figure 4.26 consisting of a multilayer wall with convection occurring at both the inner and outer surfaces of the wall is

$$R_{eq} = \frac{1}{h_i A} + \sum_j \frac{w_j}{k_j A} + \frac{1}{h_o A} \tag{4.108}$$

Equation (4.108) may be modified for the multilayer cylinder of Figure 4.27 using Equation (4.103) for the resistance for a layer of the cylinder and calculating the area over which convection occurs at the inner and outer radii of the cylinder

$$R = \frac{1}{h_i 2\pi r_i L} + \sum_j \frac{1}{2\pi k_j L} \ln\left(\frac{r_j}{r_{j-1}}\right) + \frac{1}{h_o 2\pi r_o L} \tag{4.109}$$

**FIG. 4.27** Resistance in an annular cylinder due to conduction and convection.

The **overall heat transfer coefficient** $U$ is defined between two media of temperatures $T_1$ and $T_2$ over an area $A$

$$\dot{Q} = UA(T_2 - T_1) \tag{4.110}$$

For a rectangular region, Equation (4.102) yields

$$U = \frac{1}{R_{eq} A} \tag{4.111}$$

When the external surface area of a cylinder, $A = 2\pi r_o L$, is used to define the overall heat transfer coefficient

$$U = \frac{1}{\dfrac{r_0}{r_i h_i} + r_o\left[\sum_j \dfrac{1}{k_j} \ln\left(\dfrac{r_j}{r_{j-1}}\right)\right] + \dfrac{1}{h_o}} \tag{4.112}$$

Radiation heat transfer occurs between two surfaces at different temperatures. The rate of radiation heat transfer per unit area between a surface at temperature $T_1$ and a surface at a lower temperature $T_2$ is

$$\frac{\dot{Q}}{A} = \sigma \Im (T_1^4 - T_2^4) \tag{4.113}$$

where $\sigma$ is the Stefan-Boltzmann constant and $\Im$ is a shape factor that depends on how one surface is viewed from the other. Radiation heat transfer is significant in the transient response of a thermal system only when one surface is at a much higher temperature. The modeling of a transient thermal system with radiation heat transfer is inherently nonlinear due to the presence of the fourth power of temperatures in Equation (4.113).

## Example 4.14

The exterior face of a building is made of 2-in thick brick with a thermal conductivity of 0.74 Btu/hr·ft·F. The wall, which, as illustrated in Figure 4.28, has a surface area of 450 ft$^2$, contains a cavity with insulation having an $R$ value of 12, and its interior surface is made of 0.5-in thick gypsum, which has a thermal conductivity of 0.24 Btu/hr·ft·F. The film coefficient between the ambient and the brick is 5.3 Btu/hr·ft$^2$·F. The film coefficient between the gypsum and the interior air is 2.8 Btu/hr·ft$^2$·F. (a) Determine the resistance and overall heat transfer coefficient between the ambient and the interior surfaces of the wall. (b) The ambient temperature in the summer varies over the course of a day as

$$T_\infty = 80 + 10 \sin\left(\frac{\pi t}{12}\right) \text{ F} \tag{a}$$

where $t$ is measured in hours. The interior room has a volume of 5,400 ft$^3$ and contains air with a density $2.38 \times 10^{-3}$ slug/ft$^3$ and has a specific heat of 7.72 Btu/slug·F. The room is not heated or cooled and is at a temperature of 74 F° at $t = 0$. Derive a mathematical model for the temperature in the interior room.

### Solution

(a) The $R$ value of the insulation as given is the resistance of 1 ft$^2$ of the material and has implied units of hr·ft$^2$·F/Btu. Thus the resistance of the insulation is

$$R_i = \frac{R}{A} = \frac{12 \text{ hr·ft}^2 \cdot \text{F/Btu}}{450 \text{ ft}^2} = 0.0267 \text{ hr·F/Btu} \tag{b}$$

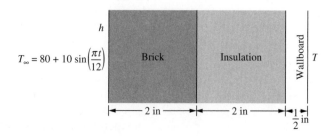

FIG. **4.28**  The system of Example 4.14: The exterior wall of a building is exposed to a periodic ambient temperature.

The resistances of the brick and gypsum are obtained using Equation (4.102),

$$R_b = \frac{(2 \text{ in})(1 \text{ ft}/12 \text{ in})}{(0.74 \text{ Btu/hr·ft·F})(450 \text{ ft}^2)} = 5.01 \times 10^{-4} \text{ hr·F/Btu} \qquad \text{(c)}$$

$$R_g = \frac{(0.5 \text{ in})(1 \text{ ft}/12 \text{ in})}{(0.24 \text{ Btu/hr·ft·F})(450 \text{ ft}^2)} = 3.86 \times 10^{-4} \text{ hr·F/Btu} \qquad \text{(d)}$$

The convective resistances are obtained using Equation (4.107),

$$R_{ci} = \frac{1}{(5.3 \text{ Btu/hr·ft}^2·\text{F})(450 \text{ ft}^2)} = 4.19 \times 10^{-4} \text{ hr·F/Btu} \qquad \text{(e)}$$

$$R_{co} = \frac{1}{(2.8 \text{ Btu/hr·ft}^2·\text{F})(450 \text{ ft}^2)} = 7.94 \times 10^{-4} \text{ hr·F/Btu} \qquad \text{(f)}$$

The equivalent resistance of the wall is

$$\begin{aligned}
R_{eq} &= R_{co} + R_b + R_i + R_g + R_{ci} \\
&= \big(4.19 \times 10^{-4} + 5.01 \times 10^{-4} + 2.67 \times 10^{-2} + 3.86 \times 10^{-4} \\
&\quad + 7.94 \times 10^{-4}\big) \text{ hr·F/Btu} \\
&= 2.88 \times 10^{-2} \text{ hr·F/Btu} \qquad \text{(g)}
\end{aligned}$$

The overall heat transfer coefficient is calculated from Equation (4.111) as

$$U = \frac{1}{R_{eq}A} = \frac{1}{(2.88 \times 10^{-2} \text{ hr·F/Btu})(450 \text{ ft}^2)} = 7.72 \times 10^{-2} \text{ Btu/hr·ft}^2·\text{F} \quad \text{(h)}$$

(b) The interior room is assumed to be at a uniform temperature at any time. The effect of temperature change on density is neglected and the density used is that at the initial temperature. The application of Conservation of Energy leads to Equation (4.98). To apply this equation it is noted that

$$mc_p = (2.38 \times 10^{-3} \text{ slug/ft}^3)(5,400 \text{ ft}^3)(7.72 \text{ Btu/slug·F}) = 99.2 \text{ Btu/F} \quad \text{(i)}$$

Thus Equation (4.91) becomes

$$99.2\frac{dT}{dt} + \frac{1}{2.88 \times 10^{-2}}T = \frac{1}{2.88 \times 10^{-2}}\left[80 + 10\sin\left(\frac{\pi t}{12}\right)\right] \qquad \text{(j)}$$

The problem formulation is completed by specifying the initial condition, $T(0) = 74$. It is noted that when Equation (j) is solved subject to the initial condition, the unit of time measurement is hr.

The modeling process used in Example 4.14 assumes lumped temperatures; that is, the transient conduction heat transfer through the brick, the insulation, and the wallboard is neglected. Equation (4.102) defining the resistance of a rectangular region is derived assuming a steady state and a constant temperature gradient. In a transient situation the temperature gradient is nonuniform through the region; that is, $\dot{Q}$ varies with $x$, a spatial coordinate measured into the wall from its exterior, as well as time. The modeling used in Example 4.14 neglects this variation and assumes an average value of the temperature gradient resulting in a lumped approximation for the temperature gradient. Such an assumption is reasonable for slowly varying changes in temperature and for thin regions.

Thermal energy can be added to a fluid in a control volume through an external heater, which transfers heat to the fluid at a constant rate. Application of the control volume form of the energy equation to a control volume with heat added and with one inlet and one outlet leads to

$$\frac{d}{dt}\int_V \rho u \, dV = \dot{Q} + \dot{m}_{in} u_{in} - \dot{m}_{out} u_{out} \tag{4.114}$$

where $u$ is the specific internal energy. Assuming the density and internal energy are constant throughout the control volume and the mass of the control volume does not change, Equation (4.114) is rewritten as

$$m\frac{du}{dt} = \dot{Q} + \dot{m}_{in} u_{in} - \dot{m}_{out} u_{out} \tag{4.115}$$

The internal energy is related to the temperature by Equation (4.10): $u = c_p T$. The specific heat for most liquids and gases varies with temperature. However, the assumption is often made that the specific heat is a constant or that, if the variation with temperature is included, when the equation is linearized the value of the specific heat at the steady-state condition is used. If the specific heat is assumed to be constant then Equation (4.115) becomes

$$mc_p\frac{dT}{dt} = \dot{Q} + \dot{m}_{in} c_p T_{in} - \dot{m}_{out} \, c_p T_{out} \tag{4.116}$$

---

### Example 4.15

Water enters a tank of volume 40 m³ at a flow rate of 0.5 kg/s at a temperature of 50°C. A heater delivers 4,000 J/s to the water in the tank. (a) Assuming a constant specific heat of $c_p = 4.19 \times 10^3$ J/kg·C, what is the temperature of the outlet flow? (b) The rate at which heat is added is suddenly changed to 4,500 J/kg·C. Derive a mathematical model for the temperature of the water in the tank as a function of time. The density of water is 1,000 kg/m³.

#### Solution

The application of Conservation of Mass to the tank, assuming no accumulation of water, shows that $\dot{m}_{in} = \dot{m}_{out}$. It is assumed that the heat is added such that the temperature in the tank is the same as the temperature of the water at the outlet, the specific heat is constant with temperature, the density of the water is constant, and no phase changes occur. Using these assumptions Equation (4.116) reduces to

$$\rho V c_p \frac{dT}{dt} = \dot{Q} + \dot{m} c_p (T_{in} - T) \tag{a}$$

(a) In the steady state the temperature is constant and Equation (a) reduces to

$$\dot{Q}_s + \dot{m}c_p(T_{in,s} - T_s) = 0 \tag{b}$$

Equation (b) is solved for the outlet temperature leading to

$$T_s = T_{in,s} + \frac{\dot{Q}_s}{\dot{m}c_v}$$

$$= 50°\text{C} + \frac{4{,}000 \text{ J/s}}{(0.5 \text{ kg/s})(4.19 \times 10^3 \text{ J/kg·C})} = 51.9°\text{C} \tag{c}$$

(b) The rate of heat transfer into the tank is written as the heat transfer at the initial steady state plus a perturbation

$$\dot{Q} = \dot{Q}_s + \dot{q} \tag{d}$$

where from the given information

$$\dot{q} = 500u(t) \tag{e}$$

The perturbation in the rate of heat transfer leads to a perturbation, $\theta(t)$, of temperature in the tank and the temperature in the outlet stream such that

$$T(t) = T_s + \theta(t) \tag{f}$$

Substitution of Equations (d) and (f) into Equation (a) leads to

$$\rho c_p V \frac{d}{dt}(T_s + \theta) = \dot{Q}_s + \dot{q} + \dot{m}c_p(T_{in} - T_s + \theta) \tag{g}$$

Noting that $dT_s/dt = 0$ and using Equation (b) in Equation (g) leads to

$$\rho V c_p \frac{d\theta}{dt} = \dot{q} + \dot{m}c_p\theta$$

$$\rho V c_p \frac{d\theta}{dt} + \dot{m}c_p\theta = \dot{q} \tag{h}$$

Substituting given values leads to

$$1.68 \times 10^8 \frac{d\theta}{dt} + 2.095 \times 10^3\theta = 500u(t) \tag{i}$$

The differential equation derived to model the change in temperature of the outlet stream in Example 4.15, Equation (i), is a first-order differential equation similar to that derived to describe the transient behavior of an $RC$ circuit and the dynamic response of the liquid level in a tank due to a perturbation in the inlet flow rate. Equation (i) is of the form

$$C\frac{d\theta}{dt} + \frac{1}{R}\theta = \dot{q}(t) \tag{4.117}$$

Analogous to the $RC$ circuit, the capacitance for the system is defined as $C = \rho A c_p$ and the resistance is defined as $R = 1/\dot{m}c_p$.

## 4.7 CHEMICAL AND BIOLOGICAL SYSTEMS

### 4.7.1 Continuous Stirred Tank Reactors (CSTRs)

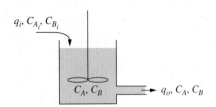

$q_i, C_{A_i}, C_{B_i}$

$C_A, C_B$          $q_o, C_A, C_B$

**FIG. 4.29** A chemical reaction occurs in a continous stirred tank reactor (CSTR), in which moles of component $A$ are converted to moles of component $B$.

A chemical system is a system in which a chemical reaction occurs by which moles of one component are converted into moles of another component. The specific chemical system modeled in this study is the continuous stirred tank reactor (CSTR), illustrated in Figure 4.29. The input stream to the CSTR shown has two components, which include the reactant $A$ and the product of the reaction $B$. The concentrations of the components in the input stream are known. The output stream includes the same components but at different concentrations than the input stream. The assumption that the reactor is a CSTR implies that the reactions occur continuously and that the composition of the mixture in the reactor is the same as the composition of the outlet stream.

The amount of a component in a mixture is quantified by its concentration, defined as the number of moles of the component per unit volume of the mixture. If $n_A$ is the number of moles of component $A$ in a mixture of volume $V$, the concentration of component $A$ is

$$C_A = \frac{n_A}{V} \tag{4.118}$$

The chemical reaction that converts component $A$ into component $B$ is represented as

$$A \xrightarrow{k} B \tag{4.119}$$

where $k$ is the rate at which the reaction occurs. The one-way arrow indicates that the reaction is irreversible. The rate at which the concentration of the reactant changes in a first-order reaction is

$$\frac{dC_A}{dt} = -kC_A \tag{4.120}$$

while for the product $B$

$$\frac{dC_B}{dt} = kC_A \tag{4.121}$$

Then negative sign in Equation (4.120) indicates that moles of component $A$ are being consumed by the reaction, while the positive sign in Equation (4.121) indicates that the reaction is producing moles of component $B$.

The rate of reaction has temperature dependence according to the Arrhenius law

$$k = \alpha e^{-E/(\bar{R}T)} \tag{4.122}$$

where $\alpha$ is a constant, $\bar{R}$ is the gas constant, and $E$ is the activation energy of the reaction. An isothermal reaction occurs at a constant temperature and its rate of reaction is constant. Endothermic reactions require heat to be added to effect the reaction, while exothermic reactions generate heat. Such reactions are not isothermal and their rate of reaction changes as the temperature changes according to Equation (4.122). The heat of reaction $\lambda$ is defined as the energy required per mole of reactant where $\lambda$ is positive for an endothermic reaction. The rate at which heat is generated in a CSTR of constant volume $V$ due to a reaction is obtained using Equations (4.118) and (4.120) as

$$\dot{Q} = \lambda \frac{dn_A}{dt}$$
$$= \lambda \frac{d}{dt}(VC_A)$$
$$= -\lambda V k C_A \tag{4.123}$$

Modeling of a CSTR requires the application of the control volume forms of Conservation of Mass and Conservation of Energy to the reactor. While the overall Conservation of Mass Equation is applicable, an equation balancing the number of moles of each component in the reactor can also be written. The number of moles of each component is not conserved because the chemical reaction leads to a decrease in the number of moles of the reactants and an increase in the number of moles of the products. The balance equation for the number of moles of component $A$ is

$$\begin{pmatrix} \text{Rate at which} \\ \text{moles of component } A \\ \text{enter the CSTR through} \\ \text{the inlet stream} \end{pmatrix} - \begin{pmatrix} \text{Rate at which the} \\ \text{moles of component } A \\ \text{exit the CSTR through} \\ \text{the outlet stream} \end{pmatrix}$$

$$+ \begin{pmatrix} \text{Rate at which the} \\ \text{moles of component } A \\ \text{are produced through} \\ \text{the chemical reaction} \end{pmatrix} = \begin{pmatrix} \text{Rate at which the} \\ \text{moles of component } A \\ \text{accumulate in} \\ \text{the CSTR} \end{pmatrix}$$

$$\tag{4.124}$$

An equation of the form of Equation (4.124) can be written for each component in the reactor. Thus for a reactor with $n$ components, $n$ equations of the form of Equation (4.124) may be written. When the $n$ equations are added together, they lead to the overall Conservation of Mass equation.

---

**Example 4.16**

An inlet stream for a CSTR of constant volume $V$ contains components $A$ and $B$ at concentrations $C_{Ai}$ and $C_{Bi}$ respectively. The volumetric flow rate into and out of the tank is $q$. An isothermal first-order irreversible reaction converts $A$ into $B$ within the tank, with a constant rate of reaction $k$. Derive a mathematical model for the concentrations of $A$ and $B$ in the outlet stream.

**Solution**

An assumption used in the modeling of a CSTR is that the concentrations of the components in the outlet stream are the same as those in the reactor. The rate at which moles of component $A$ enter the tank is $qC_{Ai}$ (volume/time times moles/volume) and the rate at which moles of component $A$ leave the tank is $qC_A$. The rate at which the chemical reaction changes the concentration of component $A$ is

$$\left(\frac{dC_A}{dt}\right)_{reaction} = -kC_A \tag{a}$$

Thus the rate of change in the number of moles of component $A$ due to the chemical reaction is

$$\left(\frac{dn_A}{dt}\right)_{reaction} = V\left(\frac{dC_A}{dt}\right)_{reaction} = -VkC_A \tag{b}$$

Thus the application of Equation (4.116) for component $A$ leads to

$$\frac{d}{dt}(VC_A) = qC_{Ai} - qC_A - VkC_A \tag{c}$$

Equation (4.116) is applied to component $B$ in a similar fashion with the exception that

$$\left(\frac{dn_B}{dt}\right)_{reaction} = VkC_A \tag{d}$$

Thus

$$\frac{d}{dt}(VC_B) = qC_{iB} - qC_B + VkC_A \tag{e}$$

Equations (d) and (e) constitute a mathematical model for this CSTR. If $V$ is constant these equations become

$$V\frac{dC_A}{dt} + (q + kV)C_A = qC_{Ai} \tag{f}$$

$$V\frac{dC_B}{dt} - kVC_A + qC_B = qC_{Bi} \tag{g}$$

## Example 4.17

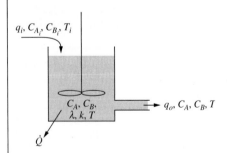

$q_i, C_{A_i}, C_{B_i}, T_i$

$C_A, C_B,$
$\lambda, k, T$

$q_o, C_A, C_B, T$

$\dot{Q}$

**FIG. 4.30** The CSTR of Example 4.17: The reaction is exothermic and a cooler is used to remove heat generated by the reaction.

An irreversible reaction of the form of Equation (4.119) occurs in the CSTR of Figure 4.30. The reaction is exothermic with a heat of reaction $\lambda$. A cooler removes heat from the tank at a constant rate $\dot{Q}$. (a) Derive a mathematical model for the system if the volume of the tank $V$ is constant, the input and output flow rates are $q$, the temperature of the inlet stream is $T_i$, and the concentrations of the inlet stream are $C_{Ai}$ and $C_{Bi}$. (b) Develop a set of equations defining the system's steady state. (c) The system is at steady state when the cooling rate decreases to $\dot{Q}_1$. Derive a linearized model for the perturbations in concentration and temperature that occur due to the change in the rate of cooling.

### Solution

(a) The equations for the rate of change of concentration for a constant volume CSTR are derived as Equations (f) and (g) in Example 4.16. Since the reaction gives off heat the system is not isothermal and the temperature in the reactor changes with time. Thus the Arrhenius law, Equation (4.122), is used in these equations for the rate of reaction resulting in

$$V\frac{dC_A}{dt} + \left(q + \alpha V e^{-E/(RT)}\right)C_A = qC_{Ai} \tag{a}$$

$$V\frac{dC_B}{dt} - \alpha V e^{-E/(RT)}C_A + qC_B = qC_{Bi} \tag{b}$$

The application of the energy equation, Equation (4.15), to a control volume enclosing the reactor leads to

$$\rho q c_p T_i - \rho q c_p T - \dot{Q} + \lambda V \alpha e^{-E/(RT)}C_A = \rho V c_p \frac{dT}{dt} \tag{c}$$

where Equation (4.123) is used for the rate at which heat is given off by the reaction. Equations (a), (b), and (c) constitute a mathematical model for the system with dependent variables $T$, $C_A$, and $C_B$.

(b) The steady state is defined by setting $dC_A/dt = 0$, $dC_B/dt = 0$, and $dT/dt = 0$. The nonlinear algebraic equations defining the steady-state variables $C_{As}, C_{Bs}$, and $T_s$ are obtained from Equations (a)–(c) as

$$\left(q + \alpha V e^{-E/(RT_s)}\right)C_{As} = qC_{Ai} \tag{d}$$

$$-qC_{as} + \alpha V e^{-E/(RT_s)}C_{Bs} = qC_{Bi} \tag{e}$$

$$\rho q c_p T_i - \rho q c_p T_s - \dot{Q} + \lambda \alpha V e^{-E/(RT_s)}C_{As} = 0 \tag{f}$$

(c) When the system is at the steady state described by Equations (d)–(f) and the rate of cooling decreases to $\dot{Q}_1$, transient behavior results. Perturbations of concentrations and temperature from the steady state are introduced by

$$C_A = C_{As} + C_{Ap} \tag{g}$$
$$C_B = C_{Bs} + C_{Bp} \tag{h}$$
$$T = T_s + T_p \tag{i}$$

Substitution of Equations (g)–(i) in Equations (a)–(c) leads to

$$V\frac{d}{dt}\left(C_{As} + C_{Ap}\right) + \left(q + \alpha V e^{-E/[r(T_s+T_p)]}\right)\left(C_{As} + C_{Ap}\right) = qC_{Ai} \tag{j}$$

$$V\frac{d}{dt}\left(C_{Bs} + C_{Bp}\right) - \alpha V e^{-E/[R(T_s+T_p)]}\left(C_{As} + C_{Ap}\right) + q\left(C_{Bs} + C_{Bp}\right) = qC_{Bi} \tag{k}$$

$$\rho V c_p \frac{d}{dt}\left(T_s + T_p\right) + \rho q c_p\left(T_s + T_p\right) - \lambda \alpha V e^{-E/[R(T_s+T_p)]}\left(C_{As} + C_{Ap}\right) = \rho q c_p T_i - \dot{Q}_1 \tag{l}$$

Setting the time derivatives of the steady-state terms to zero and using the linearization of the Arrhenius law obtained in Example 1.6 leads to

$$V\frac{dC_{Ap}}{dt} + qC_{As} + qC_{Ap} + \alpha V e^{-E/(RT_s)}\left(1 + \frac{E}{RT_s^2}T_p\right)\left(C_{As} + C_{Ap}\right) = qC_{Ai} \tag{m}$$

$$V\frac{dC_{Bp}}{dt} - \alpha V e^{-E/(RT_s)}\left(1 + \frac{E}{RT_s^2}T_p\right)\left(C_{As} + C_{Ap}\right) + qC_{Bs} + qC_{Bp} = qC_{Bi} \tag{n}$$

$$\rho V c_p \frac{dT_p}{dt} + \rho q c_p T_s + \rho q c_p T_p - \lambda \alpha V e^{-E/(RT_s)}\left(1 + \frac{E}{RT_s^2}T_p\right)\left(C_{As} + C_{Ap}\right)$$
$$= \rho c_p T_i - (\dot{Q}_1 - \dot{Q}) - \dot{Q} \tag{o}$$

Equations (d)–(f) are used in Equations (m)–(o) to subtract the steady state, leading to

$$V\frac{dC_{Ap}}{dt} + qC_{Ap} + \alpha Ve^{-E/(RT_s)}\left(C_{Ap} + \frac{E}{RT_s^2}T_pC_{As} + \frac{E}{RT_s^2}T_pC_{Ap}\right) = 0 \quad (p)$$

$$V\frac{dC_{Bp}}{dt} - \alpha Ve^{-E/(RT_s)}\left(C_{Ap} + \frac{E}{RT_s^2}T_pC_{As} + \frac{E}{RT_s^2}T_pC_{Ap}\right) + qC_{Bp} = 0 \quad (q)$$

$$\rho c_v V\frac{dT_p}{dt} + \rho qc_v T_p - \lambda \alpha Ve^{-E/(RT_s)}\left(C_{Ap} + \frac{E}{RT_s^2}T_pC_{As} + \frac{E}{RT_s^2}T_pC_{Ap}\right) = \dot{Q} - \dot{Q}_1$$

$$\text{(r)}$$

Equations (p)–(r) are nonlinear due to the presence of the product $T_pC_{Ap}$, which appears in each equation. The perturbation quantities are assumed to be small in comparison to the steady-state quantities. The product of two perturbation terms should be much smaller that one perturbation term. Equations (p) and (q) are linearized by neglecting the products of perturbations resulting in

$$V\frac{dC_{Ap}}{dt} + qC_{Ap} + \alpha Ve^{-E/(RT_s)}\left(C_{Ap} + \frac{E}{RT_s^2}T_pC_{As}\right) = 0 \quad (s)$$

$$V\frac{dC_{Bp}}{dt} - \alpha Ve^{-E/(RT_s)}\left(C_{Ap} + \frac{E}{RT_s^2}T_pC_{As}\right) + qC_{Bp} = 0 \quad (t)$$

$$\rho c_v V\frac{dT_p}{dt} + \rho qc_v T_p - \lambda \alpha Ve^{-E/(RT_s)}\left(C_{Ap} + \frac{E}{RT_s^2}T_pC_{As}\right) = \dot{Q} - \dot{Q}_1 \quad (u)$$

## 4.7.2 BIOLOGICAL SYSTEMS

Dynamic behavior occurs in many biological systems. Biological fluids are transported through the body, muscles and tissue move within the body, and body parts move relative to joints. The system dynamics of many biological processes are complex and beyond the scope of this study. However, the principles of chemical reaction kinetics may be applied to model processes involving the absorption and elimination of drugs in the body. The process of drug infusion is modeled in the following example.

**Example 4.18**

A drug is infused into the plasma through a controlled release mechanism such as through a patch. The concentration of the drug in the plasma is defined as

$$C = \frac{m_p}{V} \quad (a)$$

where $m_p$ is the mass of the drug in the plasma and $V$ is the volume of the plasma.

The drug is eliminated from the plasma due to normal biological processes at a rate $k_e$. The rate of infusion is $I$. Derive a mathematical model for the concentration of the drug in the plasma assuming that all the drug is infused to the plasma and none is infused to tissue.

### Solution

Consider a control volume consisting of the plasma. The appropriate form of Conservation of Mass applied to the drug in the control volume is

$$\text{Rate of accumulation} = \text{Rate of infusion} - \text{Rate of elimination} \qquad \text{(b)}$$

Using the defined parameters in Equation (b) leads to

$$\frac{d}{dt}(VC) = I - k_e(VC) \qquad \text{(c)}$$

$$V\frac{dC}{dt} + k_e VC = I \qquad \text{(d)}$$

## 4.8 FURTHER EXAMPLES

The following assumptions are used in the examples in this section:

- The applicable control volume forms of Conservation of Mass and Conservation of Energy are given by Equations (4.2) and (4.12) respectively.
- Liquids are incompressible.
- There is no shaft work acting on any control volume.
- The acceleration effects of liquids are negligible in large tanks.
- The equations developed using major and minor losses in pipes conveying liquids are linearized using Equation (4.68) and defining the resistance at the steady-state operating condition.
- The equations developed for the pressure drop in pipes conveying gases are linearized using Equation (4.54) and defining the resistance as determined for an average value of the mass flow rate.
- The variation of specific heat with temperature is negligible.
- The ideal gas law applies for all gases.
- Equation (4.91) provides a linearization for the pressure drop across an orifice in a hydraulic servomotor.
- Thermal properties are lumped spatially by defining the thermal resistance, as in Equations (4.102) and (4.103).
- Tanks in thermal and chemical systems are well stirred such that the properties of the exit stream are those in the tank.
- All chemical reactions are reversible and of the first order, and the rate of reaction is related to temperature by the Arrhenius law.
- Chemical reactions do not lead to phase changes.

**Example 4.19**

The two-tank system of Figure 4.31 is operating at a steady state as shown. (a) Derive a mathematical model for the changes in liquid levels that occur when the flow rates are suddenly changed to $Q_1 = 1.2$ m³/s and $Q_2 = 1.2$ m³/s. (b) Predict the liquid levels in the tank that result when the system reaches steady state at these flow rates.

**Solution**

(a) The notation used in the solution is the same as that in Example 4.8. The flow rate through the pipe at steady state is $q_{s1}$. The head loss through this pipe is $h_{s2} - h_{s1}$. The application of Conservation of Mass to the entire two-tank system reveals that the flow rate through the outlet pipe is $q_{s1} + q_{s2}$. The head loss through the exit pipe is $h_{s2}$. Assuming turbulent flow in the pipes their resistances at steady state are calculated using Equation (4.38)

$$R_1 = 2\frac{h_{s1} - h_{s2}}{q_{s1}} = 2\frac{(14 \text{ m}) - (12 \text{ m})}{1.0 \text{ m}^3/\text{s}} = 4.0 \text{ s/m}^2 \tag{a}$$

$$R_2 = 2\frac{h_{s2}}{q_{s1} + q_{s2}} = 2\frac{12 \text{ m}}{(1.0 \text{ m}^3/\text{s}) + (1.4 \text{ m}^3/\text{s})} = 10.0 \text{ s/m}^2 \tag{b}$$

The perturbations in the flow rates are

$$q_1(t) = 0.2u(t) \tag{c}$$
$$q_2(t) = -0.2u(t) \tag{d}$$

The modeling of this system is the same as that of the system of Example 4.8. Substitution of the given and calculated values into Equations (u) and (v) of Example 4.8 leads to

$$20\frac{dh_1}{dt} + 0.25h_1 - 0.25h_2 = 0.2u(t) \tag{e}$$

$$15\frac{dh_2}{dt} - 0.25h_1 + 0.35h_2 = -0.2u(t) \tag{f}$$

(b) Steady state occurs when $dh_1/dt = dh_2/dt = 0$, in which case Equations (e) and (f) become

$$0.25h_1 - 0.25h_2 = 0.2 \tag{g}$$
$$-0.25h_1 + 0.35h_2 = -0.2 \tag{h}$$

Simultaneous solution of Equations (g) and (h) yields $h_1 = 0.8$ m and $h_2 = 0$ m. Thus the new steady-state levels are $H_1 = h_{s1} + h_1 = 14.8$ m and $H_2 = h_{s2} + h_2 = 12$ m.

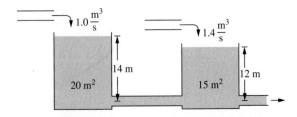

**FIG. 4.31** Two-tank liquid level system of Example 4.19.

Example 4.20

Figure 4.32 shows, at steady state, a series of $n$ cascading tanks. Derive a mathematical model to describe the change in liquid levels from steady state when the inlet flow rate into the first tank has a perturbation from steady state of $q_1(t)$. Define $R_i$ as the resistance in the pipe between tanks $i$ and $i+1$.

**Solution**

Let

$$H_i(t) = h_{is} + h_i(t) \tag{a}$$

represent the liquid level in tank $i, i = 1, 2 \ldots, n$. Let

$$Q_i(t) = q_{is} + q_i(t) \tag{b}$$

represent the flow rate into tank $i$, $i = 1, 2, \ldots, n$ and $Q_{n+1}(t)$ the flow rate out of the system. Using Equation (4.69) the perturbation flow rates and liquid levels are related through the resistances by

$$q_{i+1} = \frac{h_i}{R_i} \qquad i = 1, 2, \ldots, n \tag{c}$$

The application of the control volume form of Conservation of Mass applied to tank $i$ and linearization of the resulting equation leads to

$$A_i \frac{dh_i}{dt} = q_i - q_{i+1} \qquad i = 1, 2, \ldots, n \tag{d}$$

Using Equation (c) in Equation (d) for $i = 1$ leads to

$$A_1 \frac{dh_1}{dt} + \frac{1}{R_1} h_1 = q_1(t) \tag{e}$$

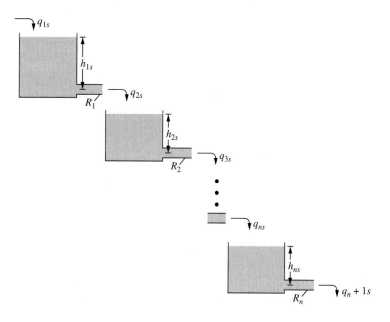

**FIG. 4.32** A series of $n$ tanks, shown at steady state, of Example 4.20.

Substitution of Equation (c) into Equation (d) for $i = 2, 3, \ldots, n$ yields

$$A_i \frac{dh_i}{dt} - \frac{1}{R_{i-1}} h_{i-1} + \frac{1}{R_i} h_i = 0 \tag{f}$$

Equations (e) and (f) provide the mathematical model for the system.

## Example 4.21

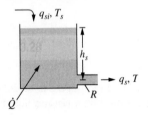

$q_{si}, T_s$

$h_s$

$q_s, T$

$R$

$\dot{Q}$

**FIG. 4.33** The steady-state operation of the heated tank of the Example 4.21.

Liquid enters the tank of Figure 4.33 at a flow rate $q_{si}$ and temperature $T_{si}$. The liquid is heated in the tank at a rate $\dot{Q}$ such that its temperature is $T_s$. The liquid is thoroughly heated such that the temperature of the outlet stream is also $T_s$. The steady state is disturbed when the flow rate and temperature of the inlet stream are suddenly changed such that

$$Q_i = q_{si} + q_i(t) \tag{a}$$
$$T_i = T_{si} + \theta_i(t) \tag{b}$$

Derive a linear mathematical model to describe the change in liquid level in the tank and the temperature of the outlet stream.

### Solution

Perturbations in the liquid level, the temperature of the liquid in the tank, and the outlet flow rate are introduced such that

$$H = h_s + h(t) \tag{c}$$
$$T(t) = T_s + \theta(t) \tag{d}$$
$$Q_o = q_s + q_o(t) \tag{e}$$

Assuming turbulent flow in the pipe the resistance at steady state is determined using Equation (4.38) as

$$R = 2 \frac{h_s}{q_{si}} \tag{f}$$

A linear model for the liquid level is obtained by using Equation (4.69) to relate the perturbation in the outlet flow to the perturbation in liquid level via the resistance determined at steady state

$$q_o(t) = \frac{h(t)}{R} \tag{g}$$

The application of Conservation of Mass to the tank, assuming the density of the fluid remains constant, and linearization of the resulting equation leads to

$$A \frac{dh}{dt} = q_i(t) - q_o(t) \tag{h}$$

Substitution of Equation (g) into Equation (h) leads to

$$A \frac{dh}{dt} + \frac{1}{R} h = q_i(t) \tag{i}$$

Application of the control volume form of Conservation of Energy at the initial steady state gives

$$\rho q_{is} c_p T_{is} - \rho c_p q_{is} T_s + \dot{Q} = 0 \tag{j}$$

Application of the control volume form of Conservation of Energy to the tank at an arbitrary time gives

$$\frac{d}{dt}(\rho A H c_p T) = \rho Q_i c_p T_i - \rho c_p Q_o T + \dot{Q} \tag{k}$$

Substitution of Equations (a)–(e) in Equation (k) leads to

$$\rho A c_p \frac{d}{dt}[(h_s + h)(T_s + \theta)] = \rho c_p(q_{is} + q_i)(T_{is} + \theta_i) - \rho c_p(q_{is} + q_o)(T_s + \theta) + \dot{Q} \tag{l}$$

The expansion of products in Equation (l), using the product rule for differentiation, setting derivatives of steady-state variables to zero, and using Equation (g) lead to

$$\rho A c_p \left[ h_s \frac{d\theta}{dt} + T_s \frac{dh}{dt} + \frac{d}{dt}(h\theta) \right] = \rho c_p(q_{is} T_{is} + q_i T_{is} + q_{is}\theta_i + q_i\theta_i)$$
$$- \rho c_p \left( q_{is} T_s + \frac{1}{R} h T_s + q_{is}\theta + \frac{1}{R} h\theta \right) + \dot{Q} \tag{m}$$

The nonlinear terms in Equation (m) arising from products of perturbation quantities are negligible in comparison to the perturbation quantities. When these adjustments are made and the steady-state condition [Equation (j)] is used, Equation (m) reduces to

$$A h_s \frac{d\theta}{dt} + A T_s \frac{dh}{dt} + \frac{T_s}{R} h + q_{is}\theta = T_{is} q_i + q_{is}\theta_i + q_i\theta_i \tag{n}$$

Equations (i) and (n) form the linear model for this system.

---

## Example 4.22

The system of Figure 4.34 contains two tanks with compressed air. The tanks of volume $V_1$ and $V_2$ are connected by a hose of average resistance $R_2$. The upstream tank is connected to a supply tank through a hose of resistance $R_1$. Initially the valve between the supply and the tanks is closed and both tanks are in equilibrium at equal pressure $P$. The valve is suddenly opened, allowing air to flow into the tanks. The upstream change in pressure is $p_s$. Derive a mathematical model for pressure in each of the tanks assuming the process is isothermal at a temperature $T$.

**FIG. 4.34** Two tanks of compressed air are at the same pressure when the upstream valve is opened.

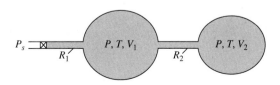

**Solution**

Perturbations in pressure in each of the tanks are defined such that

$$P_{t1} = P + p_1 \tag{a}$$

$$P_{t2} = P + p_2 \tag{b}$$

Let $\dot{m}_1$ be the mass flow rate into the first tank downstream of the supply and $\dot{m}_2$ the mass flow rate into the second tank. Equation (4.54) is used to relate the mass flow rates and pressures through the resistances by

$$R_1 = \frac{p_s - p_1}{\dot{m}_1} \tag{c}$$

$$R_2 = \frac{p_1 - p_2}{\dot{m}_2} \tag{d}$$

The application of the control volume form of Conservation of Mass to each tank leads to

$$\dot{m}_1 - \dot{m}_2 = \frac{dm_1}{dt} \tag{e}$$

$$\dot{m}_2 = \frac{dm_2}{dt} \tag{f}$$

The total mass of air in each tank is

$$m_1 = \rho_1 V_1 \tag{g}$$
$$m_2 = \rho_2 V_2 \tag{h}$$

The initial pressure and temperature in both tanks are the same. Thus the ideal gas law implies that the initial density in each tank is the same: $\rho_0 = P/R_a T$. As in Example 4.10 the process is isothermal, which implies that the densities are related to the pressures by

$$\rho_1 = \frac{P}{R_a T}\left(1 + \frac{p_1}{P}\right) \tag{i}$$

$$\rho_2 = \frac{P}{R_a T}\left(1 + \frac{p_2}{P}\right) \tag{j}$$

Use of Equations (c), (d), (g), (h), (i), and (j) in Equations (e) and (f) leads to

$$\frac{d}{dt}\left[\frac{PV_1}{R_a T}\left(1 + \frac{p_1}{P}\right)\right] = \frac{p_s - p_1}{R_1} - \frac{p_1 - p_2}{R_2} \tag{k}$$

$$\frac{d}{dt}\left[\frac{PV_2}{R_a T}\left(1 + \frac{p_2}{P}\right)\right] = \frac{p_1 - p_2}{R_2} \tag{l}$$

Simplification of Equations (k) and (l) leads to

$$\frac{V_1}{R_a T}\frac{dp_1}{dt} + \left(\frac{1}{R_1} + \frac{1}{R_2}\right)p_1 - \frac{1}{R_2}p_2 = \frac{1}{R_1}p_s \tag{m}$$

$$\frac{V_2}{R_a T}\frac{dp_2}{dt} - \frac{1}{R_2}p_1 + \frac{1}{R_2}p_2 = 0 \tag{n}$$

**Example 4.23**

A walking beam is attached to a hydraulic servomotor, as illustrated in Figure 4.35. One end of the walking beam is subject to the input $z(t)$, while the other end is attached to a viscous damper of damping coefficient $c$ and a linear spring of stiffness $k$. The spring is also attached to the piston, whose displacement is $y(t)$. Derive a mathematical model of the system relating $y(t)$ to $z(t)$.

### Solution

Let $w(t)$ be the displacement at the end of the walking beam that is attached to the spring and viscous damper. The application of similar triangles to relate the displacements of points on the walking beam, as in Example 4.13, leads to

$$\frac{z+w}{a+b} = \frac{x+w}{b}$$

$$w = \frac{b}{a}z - \frac{a+b}{a}x \tag{a}$$

For arbitrary values of $z$, $y$, and $w$ the force developed in the viscous damper is $c\dot{w}$, while the force developed in the spring is $k(w-y)$. Since the walking beam has no mass the forces must sum to zero

$$c\dot{w} + k(w-y) = 0 \tag{b}$$

The application of Conservation of Mass to the hydraulic servomotor leads to Equation (4.87)

$$\hat{C}x = A_p\dot{y} \tag{c}$$

Substitution of Equation (c) into Equation (a) leads to

$$w = \frac{b}{a}z - \frac{(a+b)A_p}{a\hat{C}}\dot{y} \tag{d}$$

Substitution of Equation (d) into Equation (b) leads to

$$\frac{(a+b)cA_p}{a\hat{C}}\ddot{y} + \frac{(a+b)k}{a\hat{C}}\dot{y} + ky = \frac{cb}{a}\dot{z} + \frac{kb}{a}z \tag{e}$$

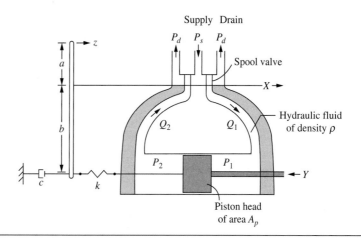

**FIG. 4.35** The hydraulic servometer and walking beam system of Example 4.23.

## Example 4.24

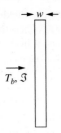

FIG. **4.36** The system of Example 4.24: The wall of thickness $w$ and surface area $A$ receives radiation from a body whose temperature is $T_b$ with a shape factor $\Im$.

The exterior surface of a wall of surface area $A$, illustrated in Figure 4.36, receives radiation heat transfer from a body whose temperature is $T_b$. The shape factor between the body and the wall is $\Im$. The wall has a small thickness $w$ and is made of a material of specific heat $c$. Assuming no other modes of heat transfer and a uniform wall temperature $T$, at any time (a) derive a mathematical model for the temperature in the wall and (b) derive a linear model for the temperature in the wall when the temperature of the radiating body is perturbed by $\theta_b(t)$ from steady state.

### Solution

(a) The total heat transfer to the wall due to radiation is given by the Stefan-Boltzmann equation as

$$\dot{Q}_w = \sigma \Im A \left(T_b^4 - T_w^4\right) \tag{a}$$

The application of Conservation of Energy to the wall leads to

$$\rho c A w \frac{dT_w}{dt} = \sigma \Im A \left(T_b^4 - T_w^4\right) \tag{b}$$

(b) When the system is in steady state, the temperature of the wall equals the temperature of the radiating body $T_w = T_b = T_s$. Let $\theta_o$ be perturbation in the wall temperature due to the perturbation in the temperature of the radiating body $\theta_b$

$$T_b = T_s + \theta_b \tag{c}$$

$$T_w = T_s + \theta_w \tag{d}$$

Substitution of Equations (c) and (d) in Equation (b) leads to

$$\rho c w \frac{d\theta_w}{dt} = \sigma \Im \left[ (T_s + \theta_b)^4 - (T_s + \theta_w)^4 \right] \tag{e}$$

Equation (e) is linearized using a binomial expansion truncated after the linear term as illustrated in Section 1.5

$$
\begin{aligned}
(T_s + \theta_w)^4 &= T_s^4 \left( 1 + \frac{\theta_w}{T_s} \right)^4 \\
&= T_s^4 \left( 1 + 4\frac{\theta_w}{T_s} + \cdots \right) \\
&= T_s^4 + 4T_s^3 \theta_w + \cdots
\end{aligned} \tag{f}
$$

Substitution of Equation (f) into Equation (e) and using a similar expansion for the temperature from the radiating body leads to

$$\rho c w \frac{d\theta_w}{dt} + 4\sigma \Im T_s^3 \theta_w = 4\sigma \Im T_s^3 \theta_b \tag{g}$$

## Example 4.25

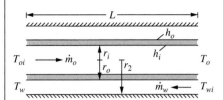

**FIG. 4.37** The double-pipe counter flow heat exchanger of Example 4.25 is used to cool oil using water as a cooling liquid.

A double-pipe counterflow heat exchanger is illustrated in Figure 4.37. The heat exchanger, which is used to cool oil, has a central pipe through which the oil of specific heat $c_o$ flows at a mass flow rate of $\dot{m}_o$. The temperature of the oil at the inlet is $T_{oi}$. The outer pipe has water at an inlet temperature $T_{wi}$ and mass flow rate $\dot{m}_w$ flowing in the opposite direction of the oil. The inner pipe has an inner radius $r_i$ and outer radius $r_o$ and is made of steel with a thermal conductivity $k$. The heat transfer coefficients on the interior and exterior of the interior pipe are $h_i$ and $h_o$ respectively. The outer pipe has an outer radius $r_2$ and is perfectly insulated, allowing no heat transfer from the water. The total length of the heat exchanger is $L$. (a) Assuming lumped parameters, derive a mathematical model for the temperatures of the oil and water in the heat exchanger and define the steady state. (b) Derive a model for the transient changes in temperatures if the temperature of the inlet stream of water is suddenly increased by $\theta_w$.

### Solution

(a) Heat transfer occurs though the inner pipe between the water and the oil due to both conduction and convection. The total resistance is obtained using Equation (4.109) as

$$R = \frac{1}{2\pi r_i L h_i} + \frac{1}{2\pi L k}\ln\left(\frac{r_o}{r_i}\right) + \frac{1}{2\pi r_o L h_o} \tag{a}$$

Thus the heat transfer between the oil and the water is

$$\dot{Q} = \frac{1}{R}(T_o - T_i) \tag{b}$$

An alternative to Equation (b) involving the overall heat transfer coefficient, as defined in Equation (4.112), is often used in heat exchanger analysis. Using this equation with

$$U = \frac{1}{\dfrac{r_0}{r_i h_i} + \dfrac{r_o}{k}\ln\left(\dfrac{r_o}{r_i}\right) + \dfrac{1}{h_o}} \tag{c}$$

leads to an alternative form of Equation (b) as

$$\dot{Q} = 2\pi r_o L U(T_o - T_w) \tag{d}$$

A lumped model is being used, which neglects the variation of temperature across the length of the heat exchanger. Consistent with this model is an assumption that the heat transfer occurs as soon as the fluid enters the heat exchanger and that the temperature of the fluid flowing through the heat exchanger is the same as the temperature of the fluid at the outlet.

The application of Conservation of Energy to a control volume consisting of the oil in the heat exchanger gives

$$\rho_o c_o \pi r_i^2 L \frac{dT_o}{dt} = c_0 \dot{m}_0(T_{oi} - T_o) - 2\pi r_o L U(T_o - T_w) \tag{e}$$

The application of Conservation of Energy to a control volume consisting of the water in the heat exchanger gives

$$\rho_w c_w \pi\left(r_2^2 - r_o^2\right) L \frac{dT_w}{dt} = c_w \dot{m}_w(T_{wi} - T_w) + 2\pi r_o L U(T_o - T_w) \tag{f}$$

The steady state is defined by $dT_o/dt = dT_w/dt = 0$. Adding Equations (e) and (f) together for the steady state leads to

$$c_o \dot{m}_o (T_{oi} - T_{os}) = c_w \dot{m}_w (T_{ws} - T_{wis}) \tag{g}$$

(b) Define perturbations in temperature from steady state by

$$T_w = T_{ws} + \theta_w \tag{h}$$

$$T_o = T_{ow} + \theta_o \tag{i}$$

Transient behavior is induced by a perturbation in the inlet temperature of the water such that

$$T_{wi} = T_{wis} + \theta_{wi} \tag{j}$$

Substitution of Equations (h)–(j) in Equations (e) and (f), using Equation (g), and rearranging lead to

$$\rho_o c_o \pi r_i^2 L \frac{d\theta_o}{dt} + (2\pi r_o L U + c_o \dot{m}_o)\theta_o - 2\pi r_o U L \theta_w = c_o \dot{m}_o \theta_{oi} \tag{k}$$

$$\rho_w c_w \pi (r_2^2 - r_o^2) L \frac{d\theta_w}{dt} - 2\pi r_o L U \theta_o + (2\pi r_o L U + c_w \dot{m}_w)\theta_w = c_w \dot{m}_w \theta_{wi} \tag{l}$$

## Example 4.26

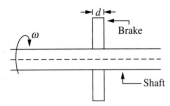

**FIG. 4.38** The system of Example 4.26: A brake is used to stop the rotation of the shaft of a machine tool.

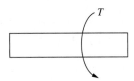

**FIG. 4.39** The free-body diagram of the shaft at an arbitrary instant.

A brake is used to stop the rotational motion of a shaft in a machine tool, as illustrated in Figure 4.38. The shaft of equivalent moment of inertia $I_s$ is initially rotating with an angular velocity $\omega_0$. The application of the brake leads to a constant braking torque $T$, applied to the shaft until its angular velocity is zero and the torque is removed. The kinetic energy of the shaft is dissipated by the frictional torque into heat, which is absorbed by the brake. The brake is a disk of diameter $D$ and thickness $\delta$. It is made from a material of mass density $\rho$ and of specific heat $c_p$. Heat is transferred between the brake and the ambient through a lateral surface area $A$ and with a heat transfer coefficient $h$. The disk is at an initial temperature $T_0$ and the ambient is at $T_\infty$. (a) Derive a mathematical model for the motion of the shaft as the brake is applied. (b) Determine the rate at which energy is dissipated from the shaft as a function of time. (c) Derive a mathematical model for the temperature distribution in the brake.

### Solution

(a) Equation (2.102) is applied to a free-body diagram of the shaft (Figure 4.39) at an arbitrary instant,

$$\sum M_o = I_o \alpha \tag{a}$$

where $\alpha = \frac{d\omega}{dt}$ is the angular acceleration of the shaft and the only external moment acting on the shaft is the constant braking torque. Thus Equation (a) becomes

$$I_{eq} \frac{d\omega}{dt} = -T \tag{b}$$

(b) The work done by the braking torque is

$$W = -\int T d\theta \tag{c}$$

The rate at which energy is dissipated by the torque is equal to the rate at which heat is transferred to the brake is

$$\dot{W} = \frac{d}{dt}\left(\int T d\theta\right) = T\frac{d\theta}{dt} = T\omega \tag{d}$$

Equation (b) is integrated to give

$$\omega(t) = -\frac{Tt}{I_{eq}} + \omega_0 \tag{e}$$

where the initial condition $\omega(0) = \omega_0$ has been applied. Thus the rate at which energy is dissipated is

$$\dot{W} = T\left(\omega_0 - \frac{Tt}{I_{eq}}\right) \tag{f}$$

The shaft stops and the torque is removed when $\omega = 0$. This time $t_1$ is obtained using Equation (e) as

$$t_1 = \frac{\omega_0 I_{eq}}{T} \tag{g}$$

The rate at which heat is transferred to the brake is written mathematically as

$$\dot{Q} = T\left(\omega_0 - \frac{Tt}{I_{eq}}\right)\left[1 - u\left(t - \frac{\omega_0 I_{eq}}{T}\right)\right] \tag{h}$$

(c) Application of the control volume form of Conservation of Energy to the disk leads to

$$mc_p\frac{dT}{dt} = \dot{Q} - hA(T - T_\infty) \tag{i}$$

Noting that the mass of the disk is $m = \frac{1}{4}\rho g\pi\delta D^2$, using Equation (h) in Equation (i) and simplifying leads to

$$\frac{1}{2}\rho g\pi\delta D^2 C_p\frac{dT}{dt} + hAT = hAT_\infty + T\left(\omega_0 - \frac{Tt}{I_{eq}}\right)\left[1 - u\left(t - \frac{\omega_0}{T}I_{eq}\right)\right] \tag{j}$$

## Example 4.27

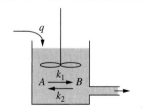

FIG. **4.40** The CSTR of Example 4.27.

The CSTR of Figure 4.40 has a two-way reaction represented by

$$A \underset{k_2}{\overset{k_1}{\rightleftharpoons}} B \tag{a}$$

Both reactions are first order. Derive a mathematical model for the concentrations of components $A$ and $B$ in the reactor assuming both reactions are isothermal.

### Solution

The two-way reaction implies that moles of $A$ are converted to moles of $B$ with a rate of reaction $k_1$ and that moles of $B$ are converted to moles of $A$ with a rate of reaction $k_2$. Thus the rate of change of the number of moles of $A$ in the reactor due to the reactions is

$$\left(\frac{dn_A}{dt}\right)_{reaction} = -k_1 V C_A + k_2 V C_B \tag{b}$$

Similarly the rate of change of the number of moles of $B$ in the reactor due to the reactions is

$$\left(\frac{dn_B}{dt}\right)_{reaction} = k_1 V C_A - k_2 V C_B \tag{c}$$

The application of the component material balances as in Example 4.16, but using Equations (b) and (c), leads to

$$V\frac{dC_A}{dt} + (q + k_1 V)C_A - k_2 V C_B = q C_{Ai} \tag{d}$$

$$V\frac{dC_B}{dt} - k_1 V C_A + (q + k_2 V)C_B = q C_{Bi} \tag{e}$$

## Example 4.28

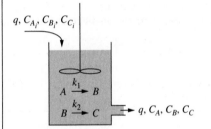

**FIG. 4.41** The CSTR of Example 4.28, in which two consecutive reactions occur.

The CSTR of Figure 4.41 is of constant volume $V$. Two consecutive reactions occur according to

$$A \xrightarrow{k_1} B \tag{a}$$

$$B \xrightarrow{k_2} C \tag{b}$$

where both reactions are first order. (a) Derive a mathematical model that describes the component concentrations assuming the reactions are isothermal. (b) Derive a mathematical model for the component concentrations assuming that the reactions are exothermic with heats of reaction $\lambda_1$ and $\lambda_2$ and that the tank is cooled with a rate of heat transfer $\dot{Q}$.

### Solution

(a) The rates of change of the number of moles of each component in the reactor due to the reactions are

$$\frac{d}{dt}(V C_A) = -k_1 V C_A \tag{c}$$

$$\frac{d}{dt}(V C_B) = k_1 V C_A - k_2 V C_B \tag{d}$$

$$\frac{d}{dt}(V C_C) = k_2 V C_B \tag{e}$$

The application of the equation to balance the number of moles of each component in the reactor [Equation (4.117)] gives

$$q C_{Ai} - q C_A - k_1 V C_A = \frac{d}{dt}(V C_A) \tag{f}$$

$$q C_{Bi} - q C_B + k_1 V C_A - k_2 V C_B = \frac{d}{dt}(V C_B) \tag{g}$$

$$q C_{Ci} - q C_C + k_2 V C_B = \frac{d}{dt}(V C_C) \tag{h}$$

Rearranging Equations (f)–(h), noting that $V$ is constant, leads to

$$V\frac{dC_A}{dt} + (q + k_1 V)C_A = qC_{Ai} \tag{i}$$

$$V\frac{dC_B}{dt} - k_1 VC_A + (q + k_2 V)C_B = qC_{Bi} \tag{j}$$

$$V\frac{dC_C}{dt} - k_2 VC_B + qC_C = qC_{Ci} \tag{k}$$

(b) The rates of reaction are assumed to follow the Arrhenius law such that

$$k_1 = \alpha e^{-E_1/(RT)} \tag{l}$$

$$k_2 = \alpha e^{-E_2/(RT)} \tag{m}$$

When the reactions are nonisothermal the temperature in the tank changes and becomes a fourth dependent variable. The component balance equations are the same as those for the nonisothermal case [Equations (f)–(h)], except that the rates of reaction are given by Equations (l) and (m). This substitution results in a nonlinear set of equations.

The fourth independent equation is obtained through the application of the control volume form of Conservation of Energy to the reactor. In the application of the energy equation it is assumed that the density and specific heat of the inlet stream are the same as those in the outlet stream. The resulting energy balance is

$$\rho q c_p T_i - \rho q c_p T - \dot{Q} + \lambda_1 V \alpha e^{-E_1/(RT)} C_A + \lambda_2 V \alpha e^{-E_2/(RT)} C_B = \rho V c_p \frac{dT}{dt} \tag{n}$$

Equations (f)–(h) and (n) are nonlinear and may be linearized by assuming perturbations from a steady state as in Example 4.17.

## Example 4.29

Consider the two-stage reactor of Figure 4.42. The first-order reaction

$$A \xrightarrow{k_1} B \tag{a}$$

occurs in the first reactor, which is of area $A_1$. In addition to Reaction (a), a second first-order reaction

$$B \xrightarrow{k_2} C \tag{b}$$

takes place in the second reactor, which is of area $A_2$. Assume the system is isothermal. (a) Derive a mathematical model for the concentrations of the components in each reactor assuming the volumes remain constant. (b) Derive a linear

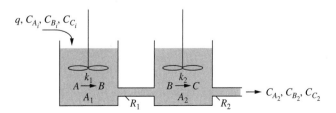

**FIG. 4.42**  The two-stage reactor of Example 4.29.

mathematical model for the system when the inlet flow rate is perturbed from steady state by $q_p$. The hydraulic resistances of the pipes at steady state are $R_1$ and $R_2$. (c) Note that it takes a finite amount of time for the mixture to travel along the pipe connecting the reactors. Thus the concentrations of the components in the inlet stream to the second reactor are those that were transported from the first reactor a finite time before. Modify the mathematical model of part (b) to account for this delay.

## Solution

(a) Since the volumes of both reactors remain constant, the flow rate through the system $q$ is a constant. However, for generality and for use of the modeling in part (b), let $q_1$, $q_2$, and $q_3$ be the volumetric flow rates into the first reactor, from the first reactor to the second reactor, and from the second reactor respectively. The application of the component balance equation, Equation (4.124) to each component in the first reactor gives

$$q_1 C_{Ai} - q_2 C_{A1} - k_1 V_1 C_{A1} = \frac{d}{dt}(V_1 C_{A1}) \tag{c}$$

$$q_1 C_{Bi} - q_2 C_{B1} + k_1 V_1 C_{A1} = \frac{d}{dt}(V_1 C_{B1}) \tag{d}$$

$$q_1 C_{Ci} - q_2 C_{C1} = \frac{d}{dt}(V_1 C_{C1}) \tag{e}$$

The concentrations of the components leaving the first reactor are the concentrations entering the second reactor. The application of the component balance equations to each component in the second reactor leads to

$$q_2 C_{A1} - q_3 C_{A2} - k_1 V_2 C_{A2} = \frac{d}{dt}(V_2 C_{A2}) \tag{f}$$

$$q_2 C_{B1} - q_3 C_{B2} + k_1 V_2 C_{A2} - k_2 V_2 C_{B2} = \frac{d}{dt}(V_2 C_{B2}) \tag{g}$$

$$q_2 C_{C1} - q_3 C_{C2} + k_2 V_2 C_{B2} = \frac{d}{dt}(V_2 C_{C2}) \tag{h}$$

(b) The system is operating at steady state when the inlet flow rate is perturbed by $q_p$. The steady state is defined by setting the time derivatives to zero in Equations (c)–(h) leading to

$$q_s C_{Ai} - q_s C_{A1s} - k_1 V_{1s} C_{A1s} = 0 \tag{i}$$
$$q_s C_{Bi} - q_s C_{B1s} + k_1 V_{1s} C_{A1s} = 0 \tag{j}$$
$$q_s C_{Ci} - q_s C_{C1s} = 0 \tag{k}$$

$$q_s C_{A1s} - q_s C_{A2s} - k_1 V_{2s} C_{A2s} = 0 \tag{l}$$
$$q_s C_{B1s} - q_s C_{B2s} + k_1 V_{2s} C_{A2s} - k_2 V_{2s} C_{B2s} = 0 \tag{m}$$
$$q_s C_{C1s} - q_s C_{C2s} + k_2 V_{2s} C_{C2s} = 0 \tag{n}$$

Perturbations from steady state are defined for the concentrations and volumes such that for a variable $z$ the perturbation $z_p$ is defined by $z = z_s + z_p$. Equations (c)–(h), as written, are still valid. However, when the volumes of the tanks are variable, they are nonlinear due to the products of concentration and volume. Consider a representative term

$$V_1 C_{A1} = (V_{1s} + V_{1p})(C_{A1s} + C_{A1p})$$
$$= V_{1s} C_{1s} + V_{1s} C_{A1p} + V_{1p} C_{A1s} + V_{1p} C_{A1p} \tag{o}$$

The first term in Equation (o) is the product of two steady-state terms and cancel when substituted into Equations (c)–(h) and Equations (i)–(n) are used. The last term in Equation (o) is the product of two perturbation terms. Such terms are small when compared to a perturbation term and are neglected, which is also used as a linearizing assumption.

The application of the control volume form of Conservation of Mass to each tank leads to

$$\frac{dV_{1p}}{dt} = q_{1p} - q_{2p} \tag{p}$$

$$\frac{dV_{2p}}{dt} = q_{2p} - q_{3p} \tag{q}$$

Noting that the liquid level in a tank is equal to the volume of the tank divided by the area, the flow rates are related to the hydraulic resistances and volumes by

$$q_{2p} = \frac{V_{2p}/A_2 - V_{1p}/A_1}{R_1} \tag{r}$$

$$q_{3p} = \frac{V_{2p}/A_2}{R_2} \tag{s}$$

Substitution of Equations (r) and (s) into Equations (p) and (q) leads to

$$\frac{dV_{1p}}{dt} = \frac{1}{R_1 A_2} V_{2p} - \frac{1}{R_1 A_1} V_{1p} + q_{1p} \tag{t}$$

$$\frac{dV_{2p}}{dt} = \frac{1}{R_1 A_1} V_{1p} - \left(\frac{1}{R_1} + \frac{1}{R_2}\right) \frac{V_{2p}}{A_2} \tag{u}$$

The use of approximations similar to that made with Equation (o) in Equations (c)–(h) and simplifying by using Equations (i)–(n) lead to

$$V_{1s}\frac{dC_{A1p}}{dt} + C_{A1s}\frac{dV_{1p}}{dt} + (q_s + k_1 V_{1s})C_{A1p} + C_{A1s}q_{2p} + k_1 C_{A1s}V_{1p} = q_p C_{Ai} \tag{v}$$

$$V_{1s}\frac{dC_{B1p}}{dt} + C_{B1s}\frac{dV_{1p}}{dt} - k_1 V_{1s}C_{A1p} - k_1 C_{A1s}V_{1p} + (q_s + k_1 V_{1s})C_{B1p}$$
$$+ C_{B1s}q_{2p} - k_1 C_{B1s}V_{1p} = q_p C_{Bi} \tag{w}$$

$$V_{1s}\frac{dC_{C1p}}{dt} + C_{C1s}\frac{dV_{1p}}{dt} + q_s C_{C1p} + C_{C1s}q_{2p} = q_p C_{Ci} \tag{x}$$

$$V_{2s}\frac{dC_{A2p}}{dt} + C_{A2s}\frac{dV_{2p}}{dt} + (q_s + k_1 V_{2s})C_{A2p} + C_{A2s}q_{3p} + k_1 C_{A2s}V_{2s}$$
$$- q_s C_{A1p} - C_{A1s}q_{2p} = 0 \tag{y}$$

$$V_{2s}\frac{dC_{B2}}{dt} + C_{B2s}\frac{dV_{2p}}{dt} + (q_s + k_2 V_{2s})C_{B2s} - k_1 V_{2s}C_{A2p} - k_1 C_{A2s}V_{2p}$$
$$- k_2 C_{B2s}V_{2p} - C_{B1s}q_{2p} - q_s C_{B1p} + C_{B2s}q_{3p} = 0 \tag{z}$$

$$V_{2s}\frac{dC_{C2p}}{dt} + C_{C2s}\frac{dV_{2p}}{dt} + q_s C_{C2p} + C_{C2s}q_{3p} - q_s C_{C1p} - C_{C1s}q_{2p}$$
$$- k_2 V_{2s}C_{B2p} - k_2 C_{B2s}V_{2p} = 0 \tag{aa}$$

Equations (v)–(aa) provide a mathematical model for the perturbed system. Note that Equations (p) and (q) may be used to further simplify the system.

(c) The perturbed variables are functions of time. Assuming no reactions occur in the pipe connecting the two reactors, the concentrations of the stream fed into the second reactor at time $t$ are not the concentrations in the stream leaving the reactor at time $t$, but rather the concentrations leaving the first reactor a time $t_d$ earlier. The delay time $t_d$ is calculated as

$$t_d = \frac{L}{q_2 A_p} \tag{bb}$$

where $A_p$ is the area of the pipe. The concentrations entering the second reactor at time $t$ are the concentrations leaving the first reactor at time $t - t_d$. For example $C_{A1}$ in Equation (f) should be replaced by

$$C_{A1}(t - t_d) = C_{A1s} + C_{A1p}(t - t_d) \tag{cc}$$

Linearization requires that $q_2$ in Equation (bb) be approximated by $q_{2s}$.

## Example 4.30

When a drug is infused into the plasma, as in Example 4.18, it is then transported by the plasma and distributed to the tissue in the body. This is a complex process, but a simple model uses only the concentration of the drug in the plasma $C_p$ and the concentration of the drug in the tissue $C_t$ as dependent variables. The concentrations are defined as

$$C_p = \frac{m_p}{V_p} \tag{a}$$

$$C_t = \frac{m_t}{V_t} \tag{b}$$

where $m_p$ is the mass of the drug in the plasma, $V_p$ is the volume of the plasma, $m_t$ is the mass of the drug in the tissue, and $V_t$ is the volume of the tissue. Through a transport process the drug is transported from the plasma to the tissue at a rate $k_1$. Simultaneously the drug is transported from the tissue to the plasma at a rate $k_2$. Normal biological processes cause the drug to be eliminated from the plasma at a rate of elimination $k_e$. The drug is infused into the plasma at a rate $I(t)$. Derive a mathematical model for the concentrations of the drug in the plasma and tissue. The mathematical model derived is called a two-compartment model for this pharmokinetic process.

### Solution

Conservation of Mass for the drug is applied to both the plasma and the tissue. The appropriate forms are

$$\begin{pmatrix} \text{Rate of} \\ \text{accumulation of the} \\ \text{drug in the plasma} \end{pmatrix} = \begin{pmatrix} \text{Rate of} \\ \text{infusion} \end{pmatrix} - \begin{pmatrix} \text{Rate of} \\ \text{elimination} \end{pmatrix} - \begin{pmatrix} \text{Rate of transport} \\ \text{from the plasma} \\ \text{to the tissue} \end{pmatrix}$$

$$+ \begin{pmatrix} \text{Rate of transport} \\ \text{from the tissue} \\ \text{to the plasma} \end{pmatrix} \tag{c}$$

$$\begin{pmatrix} \text{Rate of} \\ \text{accumulation of the} \\ \text{drug in the tissue} \end{pmatrix} = \begin{pmatrix} \text{Rate of transport} \\ \text{from the plasma} \\ \text{to the tissue} \end{pmatrix} - \begin{pmatrix} \text{Rate of transport} \\ \text{from the tissue} \\ \text{to the plasma} \end{pmatrix} \tag{d}$$

Use of the defined variables and parameters in Equations (c) and (d) lead to

$$V_p \frac{dC_p}{dt} = I(t) - k_e V_P C_p - k_1 V_p C_p + k_2 V_t C_t \tag{e}$$

$$V_t \frac{dC_t}{dt} = k_1 V_p C_p - k_2 V_t C_t \tag{f}$$

Equations (e) and (f) are rearranged to

$$V_p \frac{dC_P}{dt} + (k_e + k_1) V_p C_p - k_2 V_t C_t = I(t) \tag{g}$$

$$V_t \frac{dC_t}{dt} - k_1 V_p C_p + k_2 V_t C_t = 0 \tag{h}$$

## 4.9 SUMMARY

### 4.9.1 Mathematical Modeling of Transport Systems

The development of mathematical models for the dynamic behavior of fluid, thermal, chemical, and biological systems requires the modeling of transport processes. Macroscopic modeling of transport processes for lumped parameter systems requires the application of conservation laws to a distinct control volume. The general principles for the mathematical modeling of transport processes for lumped parameter systems are:

- A control volume, a region with defined boundaries through which mass can enter and leave, is identified.
- The control volume form of Conservation of Mass is applied, as stated in Equation (4.2).
- The control volume form of Conservation of Energy is applied, as stated in Equation (4.12).
- A mathematical model is obtained thorough the application of basic conservation laws to the control volume at an arbitrary instant.

### 4.9.2 Chapter Highlights

Some important concepts covered in Chapter 4 are:

- Bernoulli's equation is derived from the momentum equation and applies along a streamline in a flow. Under appropriate assumptions it can be reconciled with the control volume form of Conservation of Energy.
- Head is a measure of the total kinetic and potential energy and the potential of the pressure to do work.
- Minor head losses in piping systems occur due to changes in flow geometry at locations such as valves, elbows, and joints.
- Major head losses that occur due to friction are dependent on the Reynolds number and relative roughness ratio.

- The extended head form of Bernoulli's equation includes losses and is used to analyze flows in pipes.
- A pipe's resistance is the rate of change of head loss with flow rate.
- Liquid-level problems are the determination of transient perturbations of liquid level in a tank due to perturbation in a parameter such as flow rate.
- Liquid-level problems are linearized by using the system's resistances at steady state.
- Many pneumatic problems involve a polytropic process. Isothermal and isentropic processes are special cases of a polytropic process.
- A flow equation is developed for hydraulic servomotors that relates the rate of change of position of the piston to the opening of the spool valve.
- Heat transfer occurs in thermal systems by conduction, convection, or radiation.
- An equivalent thermal resistance can be determined for combined heat transfer by conduction and convection.
- A conservation of species equation, along the lines of Equation (4.124), can be developed for each species component in a CSTR.
- Nonisothermal reactions in CSTRs require the application of Conservation of Energy as well as Conservation of Species.
- Drug infusion is a biological transport process.

### 4.9.3 Important Equations

The following equations are intended to reinforce the important points in this chapter. When using any of these equations it is necessary to understand its inherent assumptions as well as the meaning of all parameters. This information is found in the text near the equation.

- Control volume form of Conservation of Mass

$$\dot{m}_{in} - \dot{m}_{out} = \frac{dm}{dt} \tag{4.3}$$

- Control volume form of Conservation of Energy

$$\dot{m}_{in}\left(u + \frac{p}{\rho} + \frac{v^2}{2} + gz\right)_{in} - \dot{m}_{out}\left(u + \frac{p}{\rho} + \frac{v^2}{2} + gz\right)_{out} + \dot{Q}$$

$$-\dot{W}_s - \dot{W}_f = \frac{d}{dt}\left[\int_V \left(u + \frac{v^2}{2} + gz\right)\rho dV\right] \tag{4.15}$$

- Bernoulli's equation

$$\frac{p_1}{\rho} + \frac{v_1^2}{2} + gz_1 = \frac{p_2}{\rho} + \frac{v_2^2}{2} + gz_2 \tag{4.16}$$

- Definition of head

$$h = \frac{p}{\rho g} + \frac{v^2}{2g} + z \tag{4.21}$$

- Extended Bernoulli's equation

$$h_1 - h_\ell = h_2 \tag{4.23}$$

- Friction head loss

$$h_f = f \frac{L}{D} \frac{v^2}{2g} \tag{4.26}$$

- Pipe resistance for turbulent flow

$$R = 2 \frac{h_\ell}{Q} \tag{4.38}$$

- Orifice equation

$$Q = C_Q A \sqrt{\frac{2(p_1 - p_2)}{\rho}} \tag{4.51}$$

- Resistance for compressible flows

$$\dot{m} = \frac{p_1 - p_2}{R} \tag{4.54}$$

- Mathematical model of liquid-level systems

$$A \frac{dh}{dt} + \frac{1}{R} h = q_i(t) \tag{4.65}$$

- Flow equation for hydraulic servomotors

$$\hat{C} x = A_p \dot{y} \tag{4.87}$$

- Thermal resistance for a rectangular wall

$$R_{eq} = \frac{1}{h_i A} + \sum_j \frac{w_j}{k_j A} + \frac{1}{h_o A} \tag{4.108}$$

- Thermal resistance for a cylinder

$$R_{eq} = \frac{1}{h_i 2\pi r_i L} + \sum_j \frac{1}{2\pi k_j L} \ln\left(\frac{r_j}{r_{j-1}}\right) + \frac{1}{h_o 2\pi r_o L} \tag{4.109}$$

- Rate of reaction

$$\frac{dC_A}{dt} = -k C_A \tag{4.120}$$

- Arrhenius law

$$k = \alpha e^{-E/(RT)} \tag{4.122}$$

## PROBLEMS

**4.1** The water clock of Figure P4.1 is of conical shape. The water exits through a 3-mm diameter hole. To what height should the clock be filled if the water is to drain in exactly 1 hr?

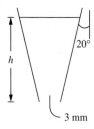

**FIG. P4.1**

**4.2** A spherical balloon is being filled with helium ($\rho = 0.169$ kg/m$^3$) at a constant mass flow rate of 0.1 kg/s. The balloon is considered to be filled when it reaches a diameter of 20 cm. How long does it take to fill the balloon?

**4.3** Water is contained in the large reservoir of Figure P4.3 under surface pressure of 2.0 kPa. Determine $\ell$, the range of the water jet.

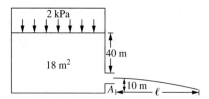

**FIG. P4.3**

**4.4** The reservoir of Figure P4.4 drains into a 20-cm pipe of length 26 m. What is the velocity of the water leaving the pipe if the pressure on the surface of the reservoir is atmospheric. Assume a friction factor of 0.003.

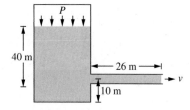

**FIG. P4.4**

**4.5** Repeat Problem 4.4 if the gauge pressure on the surface of the reservoir is 50 kPa, gauge.

**4.6** Repeat Problem 4.4 if the pipe has a relative roughness ratio of 0.002.

**4.7** What is the required diameter of a 10-m smooth pipe conveying water from a pressure head of 18 m that will deliver a flow rate of 0.2 m$^3$/s

**4.8** For purposes of modeling consider a toy water rocket to be a cylinder of diameter $D$ and length $L$, as illustrated in Figure P4.8. The rocket initially contains air at atmospheric pressure when it is filled to a height $h$ with water. (a) What is the pressure on the surface of the water after the rocket is filled and inverted? (b) After the rocket is released, water exits through a hole of diameter $d$. Let $z(t)$ be the instantaneous height of the water in the rocket. Determine the pressure on the surface of the water as a function of $z$. (c) Apply Bernoulli's equation between the surface of the water and the exit to approximate the velocity of the water leaving the rocket. (d) Apply the control volume form of Conservation of Mass to the rocket to derive a mathematical model for $z(t)$.

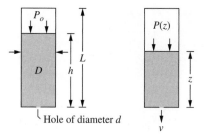

**FIG. P4.8**

**4.9** The pipe in the system of Figure P4.9 has a diameter of 30 cm and is made of galvanized iron ($\varepsilon = 0.15$ mm). Assuming no minor losses, determine (a) the flow rate through the pipe and (b) the resistance of the pipe at this flow rate.

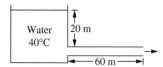

**FIG. P4.9**

**4.10** SAE 30 oil ($\rho = 917$ kg/m$^3$, $\mu = 0.290$ N·s/m$^2$) flows through a smooth 30-cm diameter pipe of length 30 m with a flow rate of 0.075 m³/s. Determine the resistance of the pipe in this flow situation. Assume no minor losses.

**4.11** The system of Figure P4.11 is operating at a steady state. Determine the resistances in pipes $A$ and $B$.

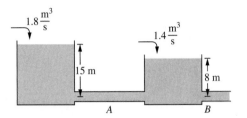

**FIG. P4.11**

**4.12** Determine the head loss between $A$ and $B$ and the flow rate in each of the pipes in the system of Figure P4.12.

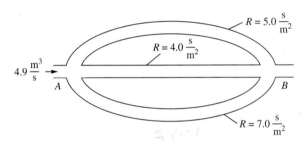

**FIG. P4.12**

**4.13** The system of Figure P4.13 is operating at a steady state. when the input flow rate is suddenly decreased to 2.2 m³/s. Derive a linearized mathematical model for the resulting perturbation of the liquid level in the tank.

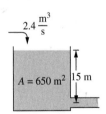

**FIG. P4.13**

**4.14** The two-tank system of Figure P4.14 is operating at a steady state when an inlet stream begins to flow into tank $B$ with a flow rate of 0.2 m³/s. Derive a mathematical model for the perturbations in the liquid levels in the tanks.

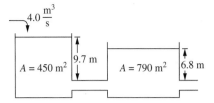

**FIG. P4.14**

**4.15** The two-tank system of Figure P4.15 is operating at a steady state when the inlet flow rate is increased to 4.6 m³/s. Derive a mathematical model for the perturbations in liquid levels in the tanks. Both tanks are circular in cross section.

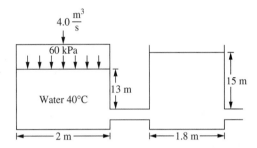

**FIG. P4.15**

**4.16** The system of Figure P4.16 is operating at a steady state when the inlet flow rate has a perturbation given by $q_p(t)$. Derive a linear mathematical model for the perturbations in liquid levels in the tanks. The tanks are connected by horizontal pipes.

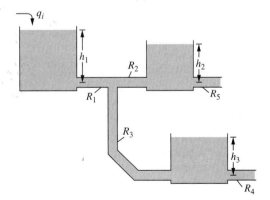

**FIG. P4.16**

**4.17** Derive a mathematical model for the system of Example 4.10 using temperature as the dependent variable assuming the process is isentropic.

**4.18** Derive a mathematical model for the system of Example 4.10 assuming the process is isothermal and using density as the dependent variable.

**4.19** The tank of Figure P4.19 is spherical with a diameter of 2 m. It contains helium $(R_{He} = 2,077 \text{ m}^2/\text{s}^2 \cdot K, \gamma = 1.66)$ at a pressure of 2.5 kPa and a temperature of 60°C. At $t = 0$ the valve is open, allowing helium to flow into the tank from an upstream source at a pressure of 10 kPa. Previous measurements have shown that the initial mass flow rate into the tank from the connecting pipe is 0.6 kg/s when the pressure in the tank is 2.5 kPa and the supply pressure is 8 kPa. Derive a linear mathematical model for the process assuming it is isothermal and using the pressure in the tank as the dependent variable.

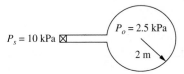

**FIG. P4.19**

**4.20** The tank of Figure P4.20 is of constant volume $V$ and contains an ideal gas of gas constant $R_g$ and viscosity $\mu$. The tank is being fed through a tube with a constant inlet flow rate $\dot{m}_i$. The tank is being bled through a pipe of length $L$ and diameter $D$ into a large reservoir at an exit pressure $p_e$. Assume the process is isothermal at a temperature $T$ and the flow through the pipe is laminar such that the friction factor is $f = 64/\text{Re}$. The gas in the tank is initially at a pressure $p_o$. Derive a mathematical model for the process using the pressure in the tank as the dependent variable.

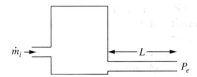

**FIG. P4.20**

**4.21** Derive a mathematical model for the system of Figure P4.21 using $x$, the displacement of the piston from the system's equilibrium position as the dependent variable and $F(t)$ as the input to the system. The hydraulic system lies in a horizontal plane.

**4.22** Derive a mathematical model for the system of Figure P4.22 using $x$, the displacement of the block of mass $m$ from the system's equilibrium position and $F(t)$, the force applied to the smaller piston as input to the system. Ignore the mass of the pistons. The hydraulic system lies in a horizontal plane.

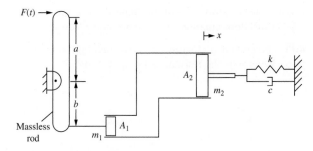

**FIG. P4.21**

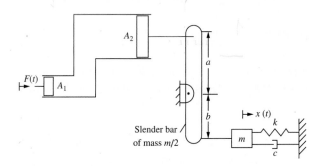

**FIG. P4.22**

**4.23** Use an energy method to derive a mathematical model for the system of Figure P4.22 using $x$ as the dependent variable and including the inertia of the pistons, but neglecting the inertia of the fluid. The pistons are of mass $m_1$ and $m_2$ respectively.

**4.24** The hydraulic servomotor of Figure P4.24 is connected to a walking beam as in Example 4.13. However, the end of the piston is connected to a mechanism as shown. Derive a mathematical model for the system using $x$, the displacement of the mass from the system's equilibrium position as the dependent variable, and $z$, the displacement of the end

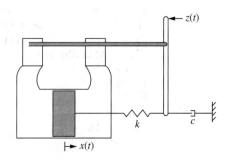

**FIG. P4.24**

of the walking beam as the input. Assume parameters of the hydraulic servomotor as in Examples 4.11–4.13.

**4.25** Determine the relationship between the input $y(t)$ and the output $x(t)$ for the hydraulic servomotor of Figure P4.25. Assume the appropriate flow equation is $\hat{C}x_A = A_P\dot{x}$ where $x_A$ is the displacement of particle $A$.

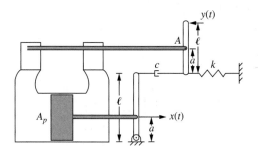

FIG. **P4.25**

**4.26** The thin plate of Figure P4.26 is at the ambient temperature $T_0$ when it begins to slide with constant velocity $V$ on a surface. The plate is made of a material of mass density $\rho$ and specific heat $c_p$. The plate is of uniform thickness $\ell$ and has a rectangular surface of width $w$ and depth $d$. The coefficient of kinetic friction between the plate and surface is $\mu$, and the heat transfer coefficient with the ambient is $h$. Derive a mathematical model for the temperature in the plate as it slides across the surface.

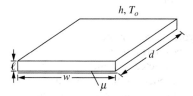

FIG. **P4.26**

**4.27** An aquarium houses a 600-m$^3$ tank of seawater. The aquatic life requires that the temperature be maintained near 30°C. A filtration and recirculation system maintains the tank at a constant volume. After filtration the entering stream has a flow rate of 1.8 m$^3$/s and a temperature of 20°C. Seawater has a density of $1.03 \times 10^3$ kg/m$^3$ and a specific heat of $4.0 \times 10^3$ J/kg·C. (a) What is the rate at which heat must be added to the tank to maintain the temperature of the tank at 30°C. (b) The system is operating at a steady state when the heater stops working. Derive a mathematical model for the change in temperature from the steady state, assuming constant specific heat.

**4.28** The aquarium in Problem 4.27 is operating at a steady state with heat added such that the temperature is maintained at 30°C when the filtration system stops working. The water is recirculated without cooling or filtration. Derive a mathematical model for the temperature in the tank after the filtration system stops, assuming that it takes 15 s to recirculate the flow through the system.

**4.29** The aquarium tank of Problem 4.27 is operating at steady state when the flow rate suddenly changes to 1.6 m$^3$/s. The heater continues to provide heat to the tank at the same rate. Derive a mathematical model for the change in temperature in the tank. Assume the volume of the tank and the temperature of the inlet stream are unchanged.

**4.30** The CSTR of Figure 4.30 has a first-order reaction represented by

$$2A \xrightarrow{k} B \qquad \text{(a)}$$

In the reaction described by Equation (a) two moles of component $A$ are used to form one mole of component $B$. Derive a mathematical model for the concentrations of components $A$ and $B$ in the reactor.

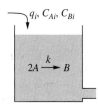

FIG. **P4.30**

**4.31** If a chemical reaction represented by $A \xrightarrow{k} B$ is of order $n$, the rate of consumption of reactant $A$ is given by $dC_A/dt = -kVC_A^n$. (a) Derive a mathematical model for the concentration of a reactant in a constant-volume CSTR with an isothermal reaction of order 1.5. (b) The reactor is operating at a steady state when a sudden perturbation in the concentration of the reactant in the input stream occurs such that $C_{Ai} = C_{Ai,s} + C_{Ai,p}$, where $C_{Ai,s}$ is the previous steady-state concentration of $A$ in the input stream and $C_{Ai,p}$ is its perturbation. Derive a linear model for the perturbation in the reactant concentration in the tank.

**4.32** The reaction in a CSTR is represented by

$$A + B \xrightarrow{k} C \qquad \text{(a)}$$

in which components $A$ and $B$ are the reactants and component $C$ is the product. The rate of change of the number of moles of reactant $A$ due to the reaction is

$$\left(\frac{dn_A}{dt}\right)_{reaction} = -kV^2 C_A C_B \qquad \text{(b)}$$

Develop a mathematical model for the concentrations of components $A$, $B$, and $C$ in the reactor assuming the reaction is isothermal.

**4.33** A first-order reaction given by $A \xrightarrow{k} B$ occurs in a CSTR. The reaction is exothermic with a heat of reaction of $\lambda$. The reactor is cooled by surrounding it with a jacket of cooling water, as illustrated in Figure P4.33. Water flows into the jacket at a rate $q_w$ at a temperature $T_{wi}$. A constant volume of water is maintained in the jacket. Let $R_w$ be the thermal resistance in the water and $R_1$ and $R_2$ be the thermal resistances from the interior and exterior walls of the jacket. Let $R_c$ be the convective resistance between the jacket and the component mixture in the reactor. Assume the exterior surface of the jacket is insulated. Derive a mathematical model for the process using the concentrations of the reactant and product $C_A$ and $C_B$, the temperature in the reactor $T$, and the temperature of the water in the jacket $T_w$ as dependent variables.

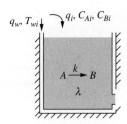

**FIG. P4.33**

**4.34** When a drug is taken orally or nasally (instead of being infused), it must first be absorbed and then diffused into the plasma. Let $m_a(t)$ represent the amount of the drug that is yet to be absorbed and let $k_a$ be the rate at which the drug is absorbed into the plasma. Define parameters as in Examples 4.18 and 4.30. (a) Derive a mathematical model for $m_a$ and $C_p$ assuming the one-compartment model of Example 4.18. (b) Derive a mathematical model for $m_a$, $C_p$, and $C_t$ assuming the two-compartment model of Example 4.30.

# 5

# Laplace Transforms

Methods of mathematical modeling of time-dependent physical systems are presented in Chapters 2–4. Mathematical modeling of a physical system whose dependent variables are time dependent leads to a differential equation, an integrodifferential equation, or a set of differential and integrodifferential equations. Methods of solution of the derived equations include exact analytical methods, approximate methods, and numerical methods. Exact analytical solutions, when available, are preferable. Approximate and numerical solutions are not exact and differ from the exact solution at each time. Errors in numerical solutions propagate with time. It is easier to study the effects of parameters on the response of a system when using an analytical solution.

The Laplace transform is the mathematical tool used in this book to solve linear differential equations and integrodifferential equations with constant coefficients. The Laplace transform method may be applied to determine the response of a linear dynamic system due to any physically possible system input. If an exact solution to the differential equation cannot be determined, the application of the Laplace transform method provides a solution in terms of an integral whose integrand is dependent on the system input. The integral could be evaluated numerically.

The Laplace transform method is not chosen arbitrarily. Subsequent chapters illustrate the value of the method in determining system response. The method is schematically illustrated in Figure 5.1. The mathematical models of Chapters 2–4 are derived in the "time domain," that is, time is the dependent variable. The Laplace transform is a mathematical procedure that transforms a function in the time domain into a function of a complex variable, usually designated as $s$. The Laplace transform exists in the $s$ domain, which is mathematically related to the physically defined frequency domain. The elegance of the method arises from its properties, which allow a differential equation for the system response in the time domain to be transformed into an algebraic equation for the transform of the system response in the $s$ domain. Algebraic methods are used to determine the transform of the system response.

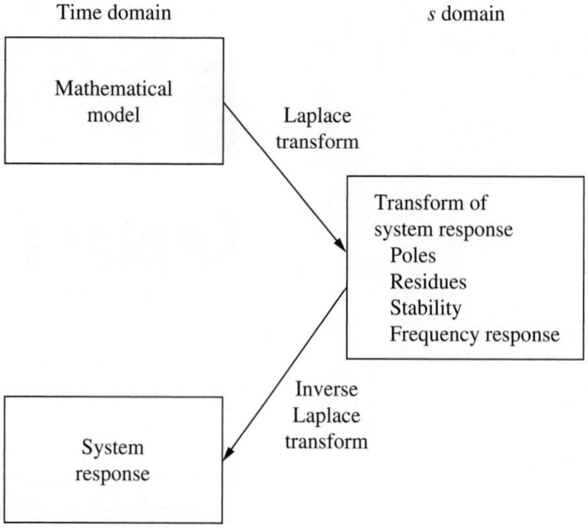

Time domain                                s domain

Mathematical
model
                    Laplace
                    transform

                              Transform of
                              system response
                                Poles
                                Residues
                                Stability
                                Frequency response

                    Inverse
                    Laplace
System          transform
response

**FIG. 5.1** The schematic of the Laplace transform method, applied to determine system response from a mathematical model.

Much can be learned about the system response from the $s$ domain, including end behavior of the response, system stability, and frequency response. The response in the time domain is obtained through the application of an inverse Laplace transform to the function in the $s$ domain.

The schematic application of this method to the model of an $RC$ circuit, after being connected to a constant source, is illustrated in Figure 5.2.

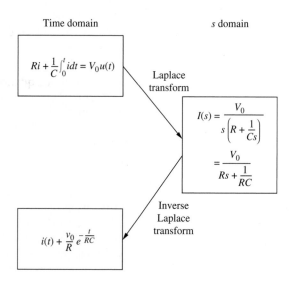

Time domain                                s domain

$$Ri + \frac{1}{C}\int_0^t i\,dt = V_0 u(t)$$

                    Laplace
                    transform

$$I(s) = \frac{V_0}{s\left(R + \dfrac{1}{Cs}\right)}$$

$$= \frac{V_0}{Rs + \dfrac{1}{RC}}$$

                    Inverse
                    Laplace
                    transform

$$i(t) + \frac{v_0}{R}\, e^{-\frac{t}{RC}}$$

**FIG. 5.2** Application of Laplace transform method to determine transient response of RC circuit.

## 5.1  DEFINITION AND EXISTENCE

The **Laplace transform** of a function $f(t)$ is defined by

$$\mathcal{L}\{f(t)\} = \int_0^\infty f(t)e^{-st}\,dt \tag{5.1}$$

where $\mathcal{L}$ is a symbolic operator representing the Laplace transform and $s$ is a complex variable. Since the operation defined by Equation (5.1) results in a function of the complex variable $s$, the following notation is used to denote the Laplace transform of $f(t)$

$$F(s) = \mathcal{L}\{f(t)\} \tag{5.2}$$

The Laplace transform of $f(t)$ exists if $f(t)$ is piecewise continuous and of **exponential order**. $f(t)$ is piecewise continuous if on the interval $a \le t \le b$, for any values of $a$ and $b$, it has a finite number of jump discontinuities (finite jumps). $f(t)$ is of exponential order if there exists an $\alpha > 0$ such that

$$\lim_{t\to\infty} e^{-\alpha t}|f(t)| = 0 \tag{5.3}$$

Clearly functions such as $\sin \omega t$, $\cos \omega t$, $e^{\lambda t}$, and $t^n$ $(n > 0)$ are of exponential order. Any function that can be physically generated is of exponential order.

The **inverse Laplace transform** is an operator denoted by $\mathcal{L}^{-1}$ that operates on a function of the complex variable $s$, whose result is a function of a real variable $t$. The operator is defined by

$$\mathcal{L}^{-1}\{G(s)\} = \frac{1}{2\pi i} \int_{\gamma-j\infty}^{\gamma+j\infty} G(s)e^{st}\,ds \tag{5.4}$$

The integral in Equation (5.4) is carried out in the complex plane. If

$$F(s) = \mathcal{L}\{f(t)\} \tag{5.5}$$

then

$$f(t) = \mathcal{L}^{-1}\{F(s)\} \tag{5.6}$$

where $f(t) = 0$ for all $t < 0$.

For every $f(t)$ there is one and only one $F(s)$ that satisfies Equations (5.5) and (5.6). That is, the Laplace transform of $f(t)$ is unique, as is the inverse Laplace transform of $F(s)$. If $f(t)$ and $F(s)$ satisfy Equations (5.5) and (5.6), they are called a **transform pair**.

## 5.2 DETERMINATION OF TRANSFORM PAIRS

### 5.2.1 Direct Integration

If $f(t)$ is a piecewise continuous function of exponential order, its Laplace transform can be determined by direct integration using Equation (5.1). This method works for the simple functions illustrated in this section. These results lead to the development of a **table of transform pairs** (Table 5.1). The properties of Laplace transforms derived in Section 5.3 and the use of tabulated transform pairs are used to add more entries to the table and can be used to determine transform pairs not listed in the table. From here on it is taken for granted that the Laplace transform exists for any function for which it is attempted.

The Laplace transform of $f(t)$ is a function of the complex variable $s$. The domain of $s$ is a subset of the complex plane. When a transform pair is specified, the domain of $s$ must also be specified. The integral in Equation (5.4) that defines the inverse Laplace transform is performed only over the domain of $s$.

The determination of the Laplace transforms for several of the examples in this section and the derivation of several properties in Section 5.3 uses the technique of integration by parts. The standard integration by parts formula is

$$\int u\,dv = uv - \int v\,du \tag{5.7}$$

Equation (5.7) can be applied to aid in the evaluation of both definite and indefinite integrals.

---

**Example 5.1**

---

Determine the Laplace transform of the exponential function, $f(t) = e^{at}$.

**Solution**

Application of Equation (5.1) for the given $f(t)$ leads to

$$\mathcal{L}\{e^{at}\} = \int_0^\infty e^{at} e^{-st} dt$$

$$= \int_0^\infty e^{-(s-a)t} dt$$

$$= -\frac{1}{s-a} e^{-(s-a)t}\Big|_0^\infty$$

$$= \frac{1}{s-a}\left(1 - \lim_{t\to\infty} e^{-(s-a)t}\right) \tag{a}$$

The limit in the last line of Equation (a) exists only if Re(s) >
limit is evaluated to zero leading to

$$\mathcal{L}\{e^{at}\} = \frac{1}{s-a} \qquad \text{Re}(s) > a$$

---

**TABLE 5.1  LAPLACE TRANSFORM PAIRS**

| $f(t)$ | $F(s) = \mathcal{L}\{f(t)\}$ |
|---|---|
| 1  $e^{at}$ | $\dfrac{1}{s-a}$ |
| 2  $\sin(\omega t)$ | $\dfrac{\omega}{s^2+\omega^2}$ |
| 3  $\cos(\omega t)$ | $\dfrac{s}{s^2+\omega^2}$ |
| 4  $t$ | $\dfrac{1}{s^2}$ |
| 5  $t^2$ | $\dfrac{2}{s^3}$ |
| 6  $t^n$ | $\dfrac{n!}{s^{n+1}}$ |
| 7  $\delta(t-a)$ | $e^{-as}$ |
| 8  $u(t-a)$ | $\dfrac{e^{-as}}{s}$ |
| 9  $\sinh(\omega t)$ | $\dfrac{\omega}{s^2-\omega^2}$ |
| 10  $\cosh(\omega t)$ | $\dfrac{s}{s^2-\omega^2}$ |
| 11  $te^{at}$ | $\dfrac{1}{(s-a)^2}$ |
| 12  $e^{-\zeta\omega_n t}\sin(\omega_n\sqrt{1-\zeta^2}t)$ | $\dfrac{\omega_n\sqrt{1-\zeta^2}}{s^2+2\zeta\omega_n s+\omega_n^2}$ |
| 13  $e^{-\zeta\omega_n t}\cos(\omega_n\sqrt{1-\zeta^2}t)$ | $\dfrac{s+\zeta\omega_n}{s^2+2\zeta\omega_n s+\omega_n^2}$ |
| 14  $t\sin(\omega t)$ | $\dfrac{2s\omega}{(s^2+\omega^2)^2}$ |
| 15  $t\cos(\omega t)$ | $\dfrac{s^2-\omega^2}{(s^2+\omega^2)^2}$ |
| 16  $e^{-at}-e^{-bt}$ | $\dfrac{(b-a)}{(s+a)(s+b)}$ |
| 17  $1-e^{-at}$ | $\dfrac{a}{s(s+a)}$ |
| 18  $ae^{-at}-be^{-bt}$ | $\dfrac{(a-b)s}{(s+a)(s+b)}$ |

Determine $\mathcal{L}\{t^n\}$ where $n$ is a non-negative integer.

### Solution

First consider the case when $n = 1$. Substituting $f(t) = t$ into Equation (5.1) leads to

$$\mathcal{L}\{t\} = \int_0^\infty t e^{-st}\, dt \tag{a}$$

The application of integration by parts, Equation (5.7), with $u = t$ and $dv = e^{-st}\, dt$ leads to Equation (a) being rewritten as

$$\mathcal{L}\{t\} = \left[ -\frac{1}{s} t e^{-st} \Big|_0^\infty + \frac{1}{s} \int_0^\infty e^{-st}\, dt \right]$$

$$= -\frac{1}{s} \lim_{t \to \infty} t e^{-st} + \frac{1}{s^2} - \frac{1}{s} \lim_{t \to \infty} e^{-st} \tag{b}$$

The limits in Equation (b) exist only for $\mathrm{Re}(s) > 0$ leading to

$$\mathcal{L}\{t\} = \frac{1}{s^2} \qquad \mathrm{Re}(s) > 0 \tag{c}$$

Now consider the general case

$$\mathcal{L}\{t^n\} = \int_0^\infty t^n e^{-st}\, dt \tag{d}$$

Using the integration by parts formula [Equation (5.7)] with $u = t^n$ and $dv = e^{-st} dt$ in evaluation of the integral in Equation (d) and assuming $\mathrm{Re}(s) > 0$ leads to

$$\mathcal{L}\{t^n\} = -\frac{1}{s} t^n e^{-st} \Big|_0^\infty + \frac{n}{s} \int_0^\infty t^{n-1} e^{-st}\, dt$$

$$= \frac{n}{s} \int_0^\infty t^{n-1} e^{-st}\, dt$$

$$= \frac{n}{s} \mathcal{L}\{t^{n-1}\} \tag{e}$$

Using Equation (e) with $n = 2$

$$\mathcal{L}\{t^2\} = \frac{2}{s} \mathcal{L}\{t\} = \left(\frac{2}{s}\right)\left(\frac{1}{s^2}\right) = \frac{2}{s^3} \tag{f}$$

then with $n = 3$

$$\mathcal{L}\{t^3\} = \frac{3}{s} \mathcal{L}\{t^2\} = \left(\frac{3}{s}\right)\left(\frac{2}{s^3}\right) = \frac{6}{s^4} \tag{g}$$

This process can be continued, leading to the following formula, which is verified by induction

$$\mathcal{L}\{t^n\} = \frac{n!}{s^{n+1}} \tag{5.9}$$

**Example 5.3**

Determine $\mathcal{L}\{\sin(\omega t)\}$.

**Solution**

Substituting $f(t) = \sin(\omega t)$ in Equation (5.1) leads to

$$\mathcal{L}\{\sin(\omega t)\} = \int_0^\infty \sin(\omega t)e^{-st}\,dt \tag{a}$$

The application of integration by parts [Equation (5.7)] to the integral in Equation (a) with $u = \sin(\omega t)$ and $dv = e^{-st}dt$ leads to

$$\mathcal{L}\{\sin(\omega t)\} = -\frac{1}{s}\sin(\omega t)e^{-st}\Big|_0^\infty + \frac{\omega}{s}\int_0^\infty \cos(\omega t)e^{-st}dt \tag{b}$$

The first term on the right-hand side of Equation (b) is zero if $\mathrm{Re}(s) > 0$. Assuming this is the case and using integration by parts on the integral in Equation (b) with $u = \cos(\omega t)$ and $dv = e^{-st}dt$ leads to

$$\mathcal{L}\{\sin(\omega t)\} = \frac{\omega}{s}\left[-\frac{1}{s}\cos(\omega t)e^{-st}\Big|_0^\infty - \frac{\omega}{s}\int_0^\infty \sin(\omega t)e^{-st}dt\right]$$

$$= \frac{\omega}{s^2} - \frac{\omega^2}{s^2}\mathcal{L}\{\sin(\omega t)\} \tag{c}$$

Rearranging Equation (c) and solving for $\{\sin(\omega t)\}$ leads to

$$\mathcal{L}\{\sin(\omega t)\} = \frac{\omega}{s^2 + \omega^2} \qquad \mathrm{Re}(s) > 0 \tag{5.10}$$

The Laplace transform of $\cos(\omega t)$ is obtained by a method similar to that used in Example 5.3. The result is

$$\mathcal{L}\{\cos(\omega t)\} = \frac{s}{s^2 + \omega^2} \tag{5.11}$$

**Example 5.4**

Determine $\mathcal{L}\{u(t - a)\}$ where $u(t)$ is the unit step function.

**Solution**

Substituting $f(t) = u(t - a)$ into Equation (5.1) and recalling that $u(t - a) = 0$ for $t < a$ and $u(t - a) = 1$ for $t > a$ leads to

$$\mathcal{L}\{u(t - a)\} = \int_0^\infty u(t - a)e^{-st}dt$$

$$= \int_a^\infty e^{-st}dt$$

$$= -\frac{1}{s}e^{-st}\Big|_a^\infty \tag{a}$$

Assuming $\text{Re}(s) > 0$, Equation (a) leads to

$$\mathcal{L}\{u(t - a)\} = \frac{1}{s}e^{-as} \tag{5.12}$$

---

**Example 5.5**

Determine $\mathcal{L}\{\delta(t - a)\}$, where $\delta(t)$ is the unit impulse function.

**Solution**

Substitution of $f(t) = \delta(t - a)$ in Equation (5.1) leads to

$$\mathcal{L}\{\delta(t - a)\} = \int_0^\infty \delta(t - a)e^{-st}dt \tag{a}$$

Using Equation (1.35) in Equation (a) leads to

$$\mathcal{L}\{\delta(t - a)\} = e^{-as} \tag{5.13}$$

---

Equations (5.8)–(5.13) provide Laplace transforms for basic functions and are summarized as transform pairs in Table 5.1. These equations and the properties derived in the next section are used to derive the remaining transform pairs in Table 5.1 as well as many others.

### 5.2.2  Use of MATLAB

MATLAB's symbolic toolbox has the capability to determine Laplace transforms of time-dependent functions. Since the Laplace transform is in the symbolic toolbox time and the transform variable must be declared as symbolic variables. The statement

syms s t

declares both $s$ and $t$ as symbolic variables. This allows MATLAB to perform algebraic manipulations on functions involving these variables.

The function to be transformed is defined as a function of $t$ using the standard hierarchy of operations. For example

$$F = \exp(-3*t)$$

defines the function $F = e^{-3t}$. The statement

$$H = \text{laplace}(F)$$

evaluates the Laplace transform of $F$ and assigns it to the function $H$, which will be displayed as an algebraic function of $s$.

## Example 5.6

Write a MATLAB M-file that determines the Laplace transform of each of the functions in Examples 5.1–5.5. Use $t^4$ as the function for Example 5.2 and $a = 3$ for the functions in Examples 5.4 and 5.5.

### Solution

The MATLAB script file Example5_6.m is illustrated in Figure 5.3, and output from its execution is illustrated in Figure 5.4. The results obtained are identical to those of Examples 5.1–5.5. The following is noted about the use of MATLAB for this example:

- The Heaviside function is another name for the unit step function, the name used by MATLAB: $\text{Heaviside}(t - a) = u(t - a)$.
- The Dirac function is another name for the unit impulse function, the name used by MATLAB: $\text{Dirac}(t - a) = \delta(t - a)$

```
% MATLAB program for Example 5.6
% Symbolic computation of Laplace transforms
%
% Declaration of symbolic variables
%   t=time
%   s=transform variable
%   w, a are parameters for symbolic evaluation of transforms
%
syms t s w a
% Example 5.1
F1=exp(a*t);
H1=laplace(F1)
% Example 5.2
F2=t^4;
H2=laplace(F2)
% Example 5.3
F3=sin(w*t);
H3=laplace(F3)
% Example 5.4
%   The Heaviside function is equivalent to the unit step function u(t-a)
%   See Example 5.6 notes for additional details
%
F4=sym('Heaviside(t-3)');
H4=laplace(F4)
% Example 5.5
%   The Dirac function is the Dirac delta function also known as the unit
%   impulse function
%   See Example 5.6 notes for additional details
%
F5=sym('Dirac(t-3)');
H5=laplace(F5)
%
% End of script file for Example 5.6
```

**FIG. 5.3** MATLAB script file for Example 5.6.

```
>> Example5_6

H1 =

1/(s-a)


H2 =

24/s^5


H3 =

w/(s^2+w^2)


H4 =

exp(-3*s)/s


H5 =

exp(-3*s)

>>
```

**FIG. 5.4** Output from execution of MATLAB script file Example5_6.m. The results are identical to those obtained in Examples 5.1–5.5.

- Heaviside and Dirac are special functions in MATLAB, obtained from the Maple library. They can be evaluated numerically at $t = 3$ using the commands mfun('Heaviside',3) or mfun('Dirac',3). 'mfun' is short for Maple function.
- The correct syntax for the use of the Heaviside or Dirac function symbolically is shown in Example 5.6. The statement F4=sym('Heaviside',3) allocates the variable F4 to u(t-3). However, symbolic evaluation of the Laplace transform of u(t-a) using MATALB is not successful if a is declared as a symbolic variable. It is successful only if a is defined by a numerical value.

## 5.3 LAPLACE TRANSFORM PROPERTIES

The properties developed in this section are useful in the derivation of transform pairs as well as in the application of Laplace transforms to the solution of differential and integrodifferential equations. Each property is stated, followed by a short proof of it. Relevant examples using the properties in determining transform pairs are presented. The application of the properties to the solution of differential equations is pursued in Section 5.5. The properties are summarized in Table 5.2.

The following properties are stated and derived using certain assumptions. When it is stated that $F(s) = \mathcal{L}\{f(t)\}$, it is implicitly

**TABLE 5.2 PROPERTIES OF LAPLACE TRANSFORMS**

| | Name | Equation |
|---|---|---|
| 1 | Definition of transform | $F(s) = \mathcal{L}\{f(t)\} = \int_0^\infty f(t)e^{-st}\,dt$ |
| 2 | Inverse transform | $f(t) = \mathcal{L}^{-1}\{F(s)\} = \dfrac{1}{2\pi i}\displaystyle\int_{\gamma-j\infty}^{\gamma+j\infty} F(s)e^{st}\,ds$ |
| 3 | Linearity of transform | $\mathcal{L}\{af(t) + bg(t)\} = aF(s) + bG(s)$ |
| 4 | Linearity of inverse transform | $\mathcal{L}^{-1}\{aF(s) + bG(s)\} = af(t) + bg(t)$ |
| 5 | First shifting theorem | $\mathcal{L}\{e^{at}f(t)\} = F(s - a)$ |
| 6 | Second shifting theorem | $\mathcal{L}\{f(t - a)u(t - a)\} = e^{-as}F(s)$ |
| 7 | Transform of first derivative | $\mathcal{L}\left\{\dfrac{df}{dt}\right\} = sF(s) - f(0)$ |
| 8 | Transform of second derivative | $\mathcal{L}\left\{\dfrac{d^2f}{dt^2}\right\} = s^2F(s) - sf(0) - \dot{f}(0)$ |
| 9 | Transform of $n$th derivative | $\mathcal{L}\left\{\dfrac{d^nf}{dt^n}\right\} = s^nF(s) - \displaystyle\sum_{k=1}^{n-1} s^{n-k}\dfrac{d^{k-1}f}{dt^{k-1}}(0)$ |
| 10 | Transform of integral | $\mathcal{L}\left\{\displaystyle\int_0^t f(t)\right\} = \dfrac{1}{s}F(s)$ |
| 11 | Derivatives of transform | $\mathcal{L}\{t^nf(t)\} = (-1)^n\dfrac{d^nF(s)}{ds^n}$ |
| 12 | Transform of periodic function | If $f(t)$ is periodic of period $T$, then $$\mathcal{L}\{f(t)\} = \frac{1}{1 - e^{-Ts}}\int_0^T f(t)e^{-st}\,dt$$ |
| 13 | Convolution | $\mathcal{L}\{f(t)*g(t)\} = F(s)G(s)$ |
| 14 | Final value theorem | $\displaystyle\lim_{t\to\infty} f(t) = \lim_{s\to 0} sF(s)$ |
| 15 | Initial value theorem | $\displaystyle\lim_{s\to\infty} sF(s) = f(0)$ |

assumed that $f(t)$ is of exponential order and its Laplace transform exists. When integration by parts is used and limits are taken to apply the upper limit of the transform integral, it is assumed these limits exist and are all zero.

**Linearity of the transform:** If $F(s) = \mathcal{L}\{f(t)\}$, $G(s) = \mathcal{L}\{g(t)\}$ and $a$ and $b$ are any scalars, then

$$\mathcal{L}\{af(t) + bg(t)\} = aF(s) + bG(s) \tag{5.14}$$

**Proof:** From the definition of the Laplace transform and using the basic properties of integrals,

$$\mathcal{L}\{af(t) + bg(t)\} = \int_0^\infty [af(t) + bg(t)]e^{-st}\,dt$$

$$= \int_0^\infty af(t)e^{-st}\,dt + \int_0^\infty bg(t)e^{-st}\,dt$$

$$= a\int_0^\infty f(t)e^{-st}\,dt + b\int_0^\infty g(t)e^{-st}\,dt$$

$$= aF(s) + bG(s) \tag{a}$$

### Example 5.7

Determine $\mathcal{L}\{\sinh(\omega t)\}$.

**Solution**

The hyperbolic sine function is written in terms of exponentials as

$$\sinh(\omega t) = \tfrac{1}{2}(e^{\omega t} - e^{-\omega t}) \tag{a}$$

Taking the Laplace transform of Equation (a) using the linearity property [Equation (5.14)] leads to

$$\mathcal{L}\{\sinh(\omega t)\} = \tfrac{1}{2}\mathcal{L}\{e^{\omega t}\} - \tfrac{1}{2}\mathcal{L}\{e^{-\omega t}\} \tag{b}$$

The transforms on the right-hand side of Equation (b) are evaluated using Equation (5.18). Substitution in Equation (b) leads to

$$\mathcal{L}\{\sinh(\omega t)\} = \frac{1}{2}\left(\frac{1}{s - \omega}\right) - \frac{1}{2}\left(\frac{1}{s + \omega}\right)$$

$$= \frac{1}{2}\left[\frac{s + \omega - (s - \omega)}{(s - \omega)(s + \omega)}\right]$$

$$= \frac{\omega}{s^2 - \omega^2} \tag{c}$$

**First shifting theorem:** If $F(s) = \mathcal{L}\{f(t)\}$ then

$$\mathcal{L}\{e^{at}f(t)\} = F(s - a) \tag{5.15}$$

**Proof:** By definition

$$F(s - a) = \int_0^\infty f(t)e^{-(s-a)t}\,dt$$

$$= \int_0^\infty [e^{at}f(t)]e^{-st}\,dt \tag{a}$$

The right-hand side of Equation (a) is by definition of the transform [Equation (5.1)] the Laplace transform of $e^{at}f(t)$. Hence Equation (5.15) is proved.

**Example 5.8**

Determine $\mathcal{L}\{e^{-2t}\cos(4t)\}$.

**Solution**

Equation (5.11) is used to determine

$$\mathcal{L}\{\cos(4t)\} = \frac{s}{s^2 + 16} \qquad \text{(a)}$$

The first shifting theorem indicates that the desired transform can be obtained by replacing $s$ by $s + 2$ in Equation (a) leading to

$$\mathcal{L}\{e^{-2t}\cos(4t)\} = \frac{s+2}{(s+2)^2 + 16}$$

$$= \frac{s+2}{s^2 + 4s + 20} \qquad \text{(b)}$$

**Second shifting theorem:** If $F(s) = \mathcal{L}\{f(t)\}$ then

$$\mathcal{L}\{f(t-a)u(t-a)\} = e^{-as}F(s) \qquad \text{(5.16)}$$

**Proof:** The definition of the Laplace transform, Equation (5.1) is used to obtain

$$\mathcal{L}\{f(t-a)u(t-a)\} = \int_0^\infty f(t-a)u(t-a)e^{-st}dt \qquad \text{(a)}$$

Since $u(t-a) = 0$ for $t < a$ and $u(t-a) = 1$ for $t > a$, Equation (a) can be written as

$$\mathcal{L}\{f(t-a)u(t-a)\} = \int_a^\infty f(t-a)e^{-st}dt \qquad \text{(b)}$$

The following change in variable

$$w = t - a \qquad \text{(c)}$$

is used to rewrite Equation (b) as

$$\mathcal{L}\{f(t-a)u(t-a)\} = \int_0^\infty f(w)e^{-s(w+a)}dw$$

$$= e^{-as}\int_0^\infty f(w)e^{-sw}dw \qquad \text{(d)}$$

The name of the variable of integration is irrelevant in the definition of the Laplace transform. Equation (5.1) could have been written with a $w$ replacing $t$ in the integral. Thus Equation (d) is the same as Equation (5.16).

## Example 5.9

Determine the Laplace transform of the triangular excitation of Example 1.7(c), which has the mathematical representation

$$f(t) = \frac{F_0}{t_0}[tu(t) - 2(t - t_0)u(t - t_0) + (t - 2t_0)u(t - 2t_0)] \tag{a}$$

### Solution

The application of linearity of transforms [Equation (5.14)] leads to

$$\mathcal{L}\{f(t)\} = \frac{F}{t_0}[\mathcal{L}\{tu(t)\} - 2\mathcal{L}\{(t - t_0)u(t - t_0)\} + \mathcal{L}\{(t - 2t_0)u(t - 2t_0)\}] \tag{b}$$

Each of the Laplace transforms in Equation (b) can be evaluated using Equation (5.9) with $n = 1$ and the second shifting theorem, Equation (5.16). For example using Equation (5.16) with $f(t) = t, F(s) = 1/s^2$, and $a = 2t_0$ leads to

$$\mathcal{L}\{(t - 2t_0)u(t - 2t_0)\} = \frac{e^{-2t_0s}}{s^2} \tag{c}$$

Using Equation (c) and similar evaluations for $\mathcal{L}\{tu(t)\}$ and $\mathcal{L}\{(t - t_0)u(t - t_0)\}$ in Equation (b) leads to

$$\begin{aligned}
\mathcal{L}\{f(t)\} &= \frac{F_0}{t_0}\left[\frac{1}{s^2} - 2\frac{e^{-t_0s}}{s^2} + \frac{e^{-2t_0s}}{s^2}\right] \\
&= \frac{F_0}{t_0}\frac{1 - 2e^{-t_0s} + e^{-2t_0s}}{s^2} \\
&= \frac{F_0}{t_0}\left(\frac{1 - e^{-t_0s}}{s}\right)^2
\end{aligned} \tag{d}$$

**Transform of derivatives:** If $F(s) = \mathcal{L}\{f(t)\}$ then

$$\mathcal{L}\left\{\frac{d^nf}{dt^n}\right\} = s^nF(s) - s^{n-1}f(0) - s^{n-2}\frac{df}{dt}(0)\cdots - s\frac{d^{n-2}f}{dt^{n-2}}(0) - \frac{d^{n-1}f}{dt^{n-1}}(0) \tag{5.17}$$

**Proof:** The proof of Equation (5.17) is by induction. First consider $n = 1$. Equation (5.1) is used to give

$$\mathcal{L}\left\{\frac{df}{dt}\right\} = \int_0^\infty \frac{df}{dt}e^{-st}dt \tag{a}$$

Integration by parts [Equation (5.7)] is used on Equation (a) with $u = e^{-st}$ and $dv = \frac{df}{dt}dt$ leading to

$$\begin{aligned}
\mathcal{L}\left\{\frac{df}{dt}\right\} &= f(t)e^{-st}\Big|_0^\infty + s\int_0^\infty f(t)e^{-st}dt \\
&= sF(s) - f(0)
\end{aligned} \tag{b}$$

Thus Equation (5.17) is proved for $n = 1$. Now assume Equation (5.17) is valid for $n = 1, 2, 3, \ldots k - 1$ for an arbitrary $k$ and consider

$$\mathcal{L}\left\{\frac{d^k f}{dt^k}\right\} = \int_0^\infty \frac{d^k f}{dt^k} e^{-st} dt \tag{c}$$

Integration by parts [Equation (5.7)] is used on Equation (c) with $u = e^{-st}$ and $dv = \frac{d^k f}{dt^k} dt$ leading to

$$\mathcal{L}\left\{\frac{d^k f}{dt^k}\right\} = e^{-st} \frac{d^{k-1} f}{dt^{k-1}} + s \int_0^\infty \frac{d^{k-1} f}{dt^{k-1}} e^{-st} dt$$

$$= -\frac{d^{k-1} f}{dt^{k-1}}(0) + s\mathcal{L}\left\{\frac{d^{k-1} f}{dt^{k-1}}\right\} \tag{d}$$

Use of the assumption that Equation (5.17) is true for $n = k - 1$ in Equation (d) leads to

$$\mathcal{L}\left\{\frac{d^k f}{dt^k}\right\} = -\frac{d^{k-1} f}{dt^{k-1}}(0) + s\left[s^{k-1} F(s) - s^{k-2} f(0) - \cdots - s\frac{d^{k-3} f}{dt^{k-3}}(0) - \frac{d^{k-2} f}{dt^{k-2}}(0)\right]$$

$$= s^k F(s) - s^{k-1} f(0) - s^{k-2} \frac{df}{dt}(0) - \cdots - s\frac{d^{k-2} f}{dt^{k-2}}(0) - \frac{d^{k-1} f}{dt^{k-1}}(0) \tag{e}$$

Equation (b) shows that Equation (5.17) is true for $n = 1$. Equation (e) shows that Equation (5.17) is true for any value $n = k$ for any value of $k$, assuming it is true for $n = k - 1$. Thus, by induction, Equation (5.17) is true for all $n$.

The property of the transform of the derivatives is useful in the solution of linear differential equations. It is convenient to write explicit expressions of Equation (5.17) for $n = 1$ and $n = 2$. If $\bar{F}(s) = \mathcal{L}\{F(t)\}$ then

$$\mathcal{L}\left\{\frac{df}{dt}\right\} = sF(s) - f(0) \tag{5.18}$$

$$\mathcal{L}\left\{\frac{d^2 f}{dt^2}\right\} = s^2 F(s) - sf(0) - \frac{df}{dt}(0) \tag{5.19}$$

## Example 5.10

Use Equation (5.11) for $\mathcal{L}\{\cos(\omega t)\}$ and Equation (5.18) to derive Equation (5.10).

### Solution

Applying Equation (5.18) with $f(t) = \cos(\omega t)$ and noting from Equation (5.11) that $F(s) = \dfrac{s}{s^2 + \omega^2}$ leads to

$$\mathcal{L}\{-\omega \sin(\omega t)\} = s\left(\frac{s}{s^2 + \omega^2}\right) - 1 \tag{a}$$

Rearranging Equation (a) leads to

$$\mathcal{L}\{\sin(\omega t)\} = \frac{1}{\omega}\left(1 - \frac{s^2}{s^2 + \omega^2}\right)$$

$$= \frac{1}{\omega}\left(\frac{s^2 + \omega^2 - s^2}{s^2 + \omega^2}\right)$$

$$= \frac{\omega}{s^2 + \omega^2} \tag{b}$$

**Transform of an integral:** If $F(s) = \mathcal{L}\{f(t)\}$ then

$$\mathcal{L}\left\{\int_0^t f(\tau)d\tau\right\} = \frac{1}{s}F(s) \tag{5.20}$$

**Proof:** The application of the definition of the Laplace transform to the left-hand side of Equation (5.20) leads to

$$\mathcal{L}\left\{\int_0^t f(\tau)d\tau\right\} = \int_0^\infty \left(\int_0^t f(\tau)d\tau\right)e^{-st}dt \tag{a}$$

The application of integration by parts [Equation (5.7)] to Equation (a) with $u = \int_0^t f(\tau)d\tau$ and $dv = e^{-st}dt$ leads to

$$\left\{\mathcal{L}\int_0^t f(\tau)d\tau\right\} = -\frac{1}{s}\left[\left(\int_0^t f(\tau)d\tau\right)e^{-st}\right]_{t=0}^{t=\infty} + \frac{1}{s}\int_0^\infty f(t)e^{-st}dt$$

$$= \frac{1}{s}F(s) \tag{b}$$

---

**Example 5.11**

It can be shown that

$$\mathcal{L}\left\{t^{-1/2}\right\} = \sqrt{\frac{\pi}{s}} \tag{a}$$

Use Equation (a) and Equation (5.20) to determine $\mathcal{L}\{t^{1/2}\}$.

**Solution**
It is noted that

$$\int_0^\tau \tau^{-1/2}d\tau = 2\tau^{1/2}\big|_0^t = 2t^{1/2} \tag{b}$$

Thus from Equation (b)

$$\mathcal{L}\left\{2t^{1/2}\right\} = \mathcal{L}\left\{\int_0^t \tau^{-1/2}d\tau\right\} \tag{c}$$

Application of [Equation (5.20)] to Equation (c) leads to

$$\mathcal{L}\{t^{1/2}\} = \frac{1}{2}\frac{1}{s}\sqrt{\frac{\pi}{s}}$$

$$= \frac{\sqrt{\pi}}{2s^{3/2}} \tag{d}$$

**Final value theorem:** If $F(s) = \mathcal{L}\{f(t)\}$ then if $\lim_{t \to \infty} f(t)$ exists,

$$\lim_{t \to \infty} f(t) = \lim_{s \to 0} sF(s) \tag{5.21}$$

**Proof:** Taking the limit of both sides of Equation (5.18) (transform of the first derivative) as $s$ approaches zero gives

$$\lim_{s \to 0} \int_0^\infty \frac{df}{dt} e^{-st} dt = \lim_{s \to 0} [sF(s) - f(0)] \tag{a}$$

Since the integral on the right-hand side of Equation (a) has a finite value, the order of the limit and the integration may be interchanged. This leads to

$$\lim_{s \to 0} sF(s) - f(0) = \int_0^\infty \frac{df}{dt} \left( \lim_{s \to 0} e^{-st} \right) dt$$

$$= \int_0^\infty \frac{df}{dt} dt$$

$$= f(t)|_0^\infty$$

$$= \lim_{t \to \infty} f(t) - f(0) \tag{b}$$

Equation (5.21) is obtained directly from Equation (b).

The final value theorem is useful in predicting the behavior of a system as time grows large using the Laplace transform of the system response. This is illustrated in the following example.

---

**Example 5.12**

The Laplace transform of the time-dependent response of a dynamic system is

$$X(s) = \frac{2(s^2 + 4s + 3)}{s(s^2 + 9)(s^2 + 4)(s + 5)} \tag{a}$$

Determine $\lim_{t \to \infty} x(t)$.

**Solution**

The application of the final value theorem to Equation (a) leads to

$$\lim_{t \to \infty} x(t) = \lim_{s \to 0} s \left[ \frac{2(s^2 + 4s + 3)}{s(s^2 + 9)(s^2 + 4)(s + 5)} \right]$$

$$= \frac{2(3)}{(9)(4)(5)} = \frac{1}{30} \tag{b}$$

The properties proved are the Laplace transform properties that are used most often in this book. Other useful properties, which follows, are stated without proof. The proofs are left as exercises.

**Initial value theorem:** If $F(s) = \mathcal{L}\{f(t)\}$ then

$$\lim_{s \to \infty} sF(s) = f(0) \tag{5.22}$$

**Change of scale:** If $F(s) = \mathcal{L}\{f(t)\}$ and $a \neq 0$ then

$$\mathcal{L}\left\{f\left(\frac{t}{a}\right)\right\} = aF(as) \tag{5.23}$$

**Differentiation of transform:** If $F(s) = \mathcal{L}\{f(t)\}$ then

$$\mathcal{L}\{tf(t)\} = -\frac{dF(s)}{ds} \tag{5.24}$$

$$\mathcal{L}\{t^2 f(t)\} = \frac{d^2 F(s)}{ds^2} \tag{5.25}$$

$$\mathcal{L}\{t^n f(t)\} = (-1)^n \frac{d^n F(s)}{ds^n} \tag{5.26}$$

**Transform of periodic function:** If $f(t)$ is a piecewise continuous function of period $T$ then

$$\mathcal{L}\{f(t)\} = \frac{1}{1 - e^{-Ts}} \int_0^T f(t) e^{-st} dt \tag{5.27}$$

**Convolution of two functions:** The convolution of two functions is defined as

$$f(t) * g(t) = \int_0^t f(t - \tau) g(\tau) d\tau \tag{5.28}$$

If $F(s) = \mathcal{L}\{f(t)\}$ and $G(s) = \mathcal{L}\{g(t)\}$ then

$$\mathcal{L}\{f(t) * g(t)\} = F(s)G(s) \tag{5.29}$$

The applications of these properties are illustrated in the following examples.

---

**Example 5.13**

Determine $F(s)$ if $f(t) = te^{-2t} \sin\left(2t + \frac{\pi}{4}\right)$.

**Solution**

The trigonometric identity for the sine of the sum of angles is used to write

$$\sin\left(2t + \frac{\pi}{4}\right) = \sin(2t)\cos\left(\frac{\pi}{4}\right) + \cos(2t)\sin\left(\frac{\pi}{4}\right)$$

$$= \frac{\sqrt{2}}{2}[\sin(2t) + \cos(2t)] \tag{a}$$

The application of the first shifting theorem [Equation (5.15)] gives

$$\mathcal{L}\{e^{-2t}\sin 2t\} = \frac{2}{(s+2)^2+4} = \frac{2}{s^2+4s+8} \tag{b}$$

$$\mathcal{L}\{e^{-2t}\cos(2t)\} = \frac{s+2}{(s+2)^2+4} = \frac{s+2}{s^2+4s+8} \tag{c}$$

Using Equation (a) and linearity of the transform

$$\mathcal{L}\left\{e^{-2t}\sin\left(2t+\frac{\pi}{4}\right)\right\} = \frac{\sqrt{2}}{2}\left[\frac{2}{s^2+4s+8}+\frac{s+2}{s^2+4s+8}\right]$$

$$= \frac{\sqrt{2}}{2}\frac{s+4}{s^2+4s+8} \tag{d}$$

The property of differentiation of the transform [Equation (5.24)] is used to yield

$$\mathcal{L}\{f(t)\} = -\frac{d}{ds}\left(\frac{\sqrt{2}}{2}\frac{s+4}{s^2+4s+8}\right)$$

$$= -\frac{\sqrt{2}}{2}\left[\frac{(1)(s^2+4s+8)-(s+4)(2s+4)}{(s^2+4s+8)^2}\right]$$

$$= \frac{\sqrt{2}}{2}\frac{s^2+8s+8}{(s^2+4s+8)^2} \tag{e}$$

## Example 5.14

Determine the Laplace transform of the periodic function of Figure 5.5.

### Solution

The function in Figure 5.5 is periodic of period $t_0$. Over one period it is defined as

$$f(t) = \begin{cases} F_0 & 0 \le t \le \dfrac{2t_0}{3} \\ 0 & \dfrac{2t_0}{3} < t < t_0 \end{cases} \tag{a}$$

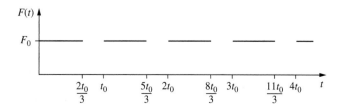

**FIG. 5.5** The periodic function for Example 5.14.

Equation (5.27) is used to determine the Laplace transform of $f(t)$ as

$$
\mathcal{L}\{f(t)\} = \frac{1}{1 - e^{-t_0 s}} \int_0^{t_0} f(t)e^{-st}\,dt
$$

$$
= \frac{1}{1 - e^{-t_0 s}} \int_0^{2t_0/3} F_0 e^{-st}\,dt
$$

$$
= \frac{F_0}{1 - e^{-t_0 s}} \left( -\frac{1}{s}e^{-st} \right)\Big|_0^{2t_0/3}
$$

$$
= \frac{F_0(1 - e^{-(2t_0/3)s})}{s(1 - e^{-t_0 s})} \tag{b}
$$

---

## Example 5.15

Use MATLAB to symbolically determine the Laplace transform of (a) the function of Example 5.9 with $F_0 = 1$ and $t_0 = 1$ and (b) the function of Example 5.13.

### Solution

   MATLAB M-files are convenient when a computation is to be performed many times. However, direct interactive work sessions using MATALB is convenient when the sequence of commands necessary to achieve an objective is not well defined. The Laplace transforms in this example are obtained through commands entered directly from the MATLAB prompt. The work sessions are shown in Figures 5.6(a) and (b). The following is noted about these work sessions:

```
>> % Solution to Example 5.15 (a)
>> syms s t
>> A=sym('Heaviside(t-1)');
>> B=sym('Heaviside(t-2)');
>> F=t-2*(t-1)*A+(t-2)*B;
>> G=laplace(F)

G =

1/s^2-2*exp(-s)/s^2+exp(-2*s)/s^2

>> G1=simplify(G)

G1 =

-(-1+2*exp(-s)-exp(-2*s))/s^2

>>
```

                    (a)

**FIG. 5.6**  MATLAB work sessions for the solution of Example 5.15, parts (a) and (b).

```
>> % Solution to Example 5.15 (b)
>> syms s t
>> F=t*exp(-2*t)*sin(2*t+pi/4);
>> G=laplace(F)

G =

1/8*2^(1/2)/(1/4*(s+2)^2+1)^2*(s+2)-
1/8*2^(1/2)/(1/4*(s+2)^2+1)+1/16*2^(1/2)*(s+2)^2/(1/4*(s+2)^2+1)^2

>> G1=simplify(G)

G1 =

1/2*2^(1/2)*(s^2+8*s+8)/(s^2+4*s+8)^2

>> G2=vpa(G1,3)

G2 =

.705*(s^2+8.*s+8.)/(s^2+4.*s+8.)^2

>>
```

(b)

**FIG. 5.6** *(Continued)*

- The function for part (a) is entered in several parts to reduce the possibility of error. If an error does occur in one of the entered functions, it is easier to correct than if the entire function were entered.
- MATLAB makes a number of symbolic commands available. When developing a MATLAB M-file it is difficult to know in advance which symbolic commands may be necessary to put the result in the desired form. During the work session for part (b) the command "simplify" resulted in a convenient algebraic form for the result. The result, when written in the form of G1, compares exactly with the result of Example 5.13.
- When MATLAB displays a symbolic result, it displays numerical terms as simple fractions or irrational terms as powers of rational numbers. It is often more convenient to write the result in a decimal form. The command "vpa(G1,3)" rewrites the symbolic expression in a decimal form with three significant digits. The term "vpa" stands for variable precision arithmetic.

## 5.4 INVERSION OF TRANSFORMS

The process of determining $f(t)$ from its Laplace transform $F(s)$ is called **inversion of the transform**. The relation between $f(t)$ and $F(s)$ is written symbolically in Equation (5.6), repeated here

$$f(t) = \mathcal{L}^{-1}\{F(s)\} \tag{5.2}$$

There are several methods of transform inversion. The direct method is to use the inversion integral, Equation (5.2). However, its use requires knowledge of integration in a complex plane, including residue analysis. Thus, it is not presented in this book.

The examples in this section are provided to illustrate the use of the table of transform pairs (Table 5.1) and the table of transform properties (Table 5.2). For the purposes of illustration the examples assume the availability of only transform pairs 1 through 10. The solution of some of the examples would be simpler if the remainder of the transform pairs in Table 5.1 were used.

The use of MATALB to determine the inverse transform is also presented.

### 5.4.1  Use of Tables and Properties

The most convenient method of transform inversion is to apply the properties of the transform in Table 5.2 and use the transform pairs in Table 5.1. This method requires logic that is the reverse of using the tables and properties to determine transform pairs. For example if $X(s)$ is of the form $e^{-as}F(s-a)$ where the transform pair for $f(t)$ and $F(s)$ can be located in Table 5.1, then the first shifting property from Table 5.2 is applied, leading to $x(t) = f(t)u(t-a)$.

The property of **linearity of the inverse transform** is that if $f(t) = \mathcal{L}^{-1}\{F(s)\}$ and $g(t) = \mathcal{L}^{-1}\{G(s)\}$ and $a$ and $b$ are scalars, then

$$\mathcal{L}^{-1}\{aF(s) + bG(s)\} = af(t) + bg(t) \qquad (5.30)$$

---

**Example 5.16**

---

Determine $x(t)$ if its Laplace transform is

$$X(s) = \frac{4s+6}{s^2+9} \qquad (a)$$

**Solution**

Equation (a) can be written as

$$\bar{x}(s) = 4\left(\frac{s}{s^2+9}\right) + 2\left(\frac{3}{s^2+9}\right) \qquad (b)$$

The application of linearity for inverse transforms [Equation (5.30)], in addition to transform pairs 3 and 2 from Table 5.1, leads to

$$x(t) = 4\cos(3t) + 2\sin(3t) \qquad (c)$$

**Example 5.17**

Determine $x(t)$ if its Laplace transform is

$$X(s) = \frac{2s}{s^2 + 6s + 13} \tag{a}$$

**Solution**

The quadratic polynomial in the denominator of Equation (a) does not have real roots. It is convenient to complete the square of the quadratic polynomial leading to

$$X(s) = \frac{2s}{(s+3)^2 + 4}$$

$$= 2\left[\frac{s+3}{(s+3)^2 + 4} - \frac{3}{(s+3)^2 + 4}\right] \tag{b}$$

Wherever $s$ appears on the right-hand side of the last step of Equation (b), it appears as $s + 3$. Thus if

$$F(s) = 2\left[\frac{s}{s^2 + 4} - \frac{3}{s^2 + 4}\right] \tag{c}$$

then $X(s) = F(s + 3)$ and the first shifting theorem is applied leading to

$$x(t) = e^{-3t} f(t) \tag{d}$$

where $f(t) = \mathcal{L}^{-1}\{F(s)\} = 2\cos(2t) - 3\sin(2t)$ is obtained using linearity, Equation (c), and transform pairs 3 and 2 from Table 5.1. Thus

$$x(t) = e^{-3t}[2\cos(2t) - 3\sin(2t)] \tag{e}$$

**Example 5.18**

Determine $x(t)$ if its Laplace transform is

$$X(s) = \frac{e^{-3s}}{(s+2)^2} \tag{a}$$

**Solution**

The presence of the exponential terms suggests use of the second shifting theorem. Its application leads to

$$x(t) = f(t - 3)u(t - 3) \tag{b}$$

where $f(t)$ is the inverse transform of

$$F(s) = \frac{1}{(s+2)^2} \tag{c}$$

The inverse transform of Equation (c) is obtained using transform pair 4 from Table 5.1 and the first shifting theorem as

$$f(t) = te^{-2t} \tag{d}$$

Combining Equations (b) and (d) gives

$$x(t) = (t-3)e^{-2(t-3)}u(t-3) \tag{e}$$

The preceding examples illustrate the basic procedure for using tables of transform pairs and properties to perform inversion of transforms. These examples also demonstrate some basic strategies for inverting transforms.

- If an exponential function of the transform variable appears in the numerator, use of the second shifting theorem is suggested.
- If the denominator contains a linear factor raised to an integer power, use transform pair 6 and the first shifting theorem.
- If the denominator contains a quadratic factor that does not have real roots, then complete the square and using the first shifting theorem in conjunction with transform pairs 2 and 3.
- If whenever $s$ appears in the transform it appears as $s - a$, apply the first shifting theorem.

## 5.4.2 Partial Fraction Decompositions

The inversion of the transform is more complicated when the denominator of the transform contains polynomials in $s$ higher than second order. Polynomials with real coefficients can always be factored into linear and quadratic factors. Consider as a general case a Laplace transform of the form

$$X(s) = \frac{Q(s)}{D(s)} \tag{5.31}$$

where $D(s)$ is a polynomial of order $n$ with a leading coefficient of one.

The function $Q(s)$ can be written as

$$Q(s) = R(s)N(s) \tag{5.32}$$

where $R(s)$ contains everything that is not part of a polynomial, such as exponential terms involving the transform variable. If $Q(s)$ is a polynomial, then $R(s) = 1$. $N(s)$ is a polynomial of order $p$. If $p \geq n$ then a long division or a synthetic division must be performed such that

$$\frac{N(s)}{D(s)} = W(s) + \frac{B(s)}{D(s)} \tag{5.33}$$

where $W(s)$ is a polynomial in $s$ of order $p - n$ (If $p = n$, $W(s)$ is a constant) and $B(s)$ is a polynomial of order $n - 1$ or less.

Let $s_1, s_2, \ldots, s_n$ be the roots of $D(s)$ such that it has a factorization of

$$D(s) = (s - s_1)(s - s_2) \cdots (s - s_n) \tag{5.34}$$

The roots of $D(s)$ are referred to as the **poles** of the transform.

A **partial fraction decomposition** of $B(s)/D(s)$ can be developed in the form

$$\frac{B(s)}{D(s)} = \sum_{k=1}^{n} \frac{A_k}{s - s_k} \tag{5.35}$$

where $A_k, k = 1, 2, \ldots, n$ are coefficients to be determined. These coefficients are called the **residues** of the transform.

Three cases are considered:

- All poles are real and distinct.
- Some of the poles are imaginary or complex.
- $D(s)$ has repeated roots. [In this case the form of $B(s)/D(s)$ is modified from that of Equation (5.35).]

---

**TABLE 5.3  PARTIAL FRACTION DECOMPOSITIONS**

| | $X(s)$ | Partial Fraction Decomposition for $X(s)$ |
|---|---|---|
| 1 | $\dfrac{1}{s(s+a)}$ | $\dfrac{1}{a}\left(\dfrac{1}{s} - \dfrac{1}{s+a}\right)$ |
| 2 | $\dfrac{1}{(s+a)(s+b)}$ | $\dfrac{1}{b-a}\left(\dfrac{1}{s+a} - \dfrac{1}{s+b}\right)$ |
| 3 | $\dfrac{s}{(s+a)(s+b)}$ | $\dfrac{1}{b-a}\left(\dfrac{b}{s+b} - \dfrac{a}{s+a}\right)$ |
| 4 | $\dfrac{1}{s(s^2+\omega^2)}$ | $\dfrac{1}{\omega^2}\left(\dfrac{1}{s} - \dfrac{s}{s^2+\omega^2}\right)$ |
| 5 | $\dfrac{1}{(s+a)(s^2+\omega^2)}$ | $\dfrac{1}{\omega^2+a^2}\left(\dfrac{1}{s+a} - \dfrac{s-a}{s^2+\omega^2}\right)$ |
| 6 | $\dfrac{s}{(s+a)(s^2+\omega^2)}$ | $\dfrac{1}{\omega^2+a^2}\left(\dfrac{as+\omega^2}{s^2+\omega^2} - \dfrac{a}{s+a}\right)$ |
| 7 | $\dfrac{1}{(s^2+\omega_1^2)(s^2+\omega_2^2)}$ | $\dfrac{1}{\omega_2^2-\omega_1^2}\left(\dfrac{1}{s^2+\omega_1^2} - \dfrac{1}{s^2+\omega_2^2}\right)$ |
| 8 | $\dfrac{1}{(s+a)(s^2+2\zeta\omega_n s+\omega_n^2)}$ | $\dfrac{1}{a^2+\omega_n^2-2\zeta a\omega_n}\left(\dfrac{1}{s+a} - \dfrac{s+2\zeta\omega_n-a}{s^2+2\zeta\omega_n s+\omega_n^2}\right)$ |
| 9 | $\dfrac{s}{(s+a)(s^2+2\zeta\omega_n s+\omega_n^2)}$ | $\dfrac{1}{a^2+\omega_n^2-2\zeta a\omega_n+\omega_n^2}\left(\dfrac{as+\omega_n^2}{s^2+2\zeta\omega_n s+\omega_n^2} - \dfrac{a}{s+a}\right)$ |
| 10 | $\dfrac{1}{(s^2+\omega^2)(s^2+2\zeta\omega_n s+\omega_n^2)}$ | $\dfrac{1}{(\omega_n^2-\omega^2)^2+(2\zeta\omega\omega_n)^2}\left[\dfrac{-2\zeta\omega_n s+\omega_n^2-\omega^2}{s^2+\omega^2} + \dfrac{2\zeta\omega_n s-(2\zeta\omega_n)^2+(\omega_n^2-\omega^2)}{s^2+2\zeta\omega_n s+\omega_n^2}\right]$ |
| 11 | $\dfrac{s}{(s^2+\omega^2)(s^2+2\zeta\omega_n s+\omega_n^2)}$ | $\dfrac{1}{(\omega_n^2-\omega^2)^2+(2\zeta\omega\omega_n)^2}\left[\dfrac{(\omega_n^2-\omega^2)s-2\zeta\omega_n+\omega^2}{s^2+\omega^2} - \dfrac{(\omega_n^2-\omega^2)s-2\zeta\omega_n^3}{s^2+2\zeta\omega_n s+\omega_n^2}\right]$ |

**5.4.2.1 Real Distinct Poles**   Multiplying Equation (5.35) by $D(s)$ and using the form of Equation (5.34) for $D(s)$ leads to

$$B(s) = \sum_{i=1}^{n} \left[ A_i \prod_{\substack{j=1 \\ j \neq i}}^{n} (s - s_j) \right] \tag{5.36}$$

Equation (5.36) is valid for all $s$. Consider its evaluation for $s = s_k$ for any $k, 1 \leq k \leq n$. The only nonzero term in the resulting summation corresponds to $i = k$, leading to

$$A_k = \frac{B(s_k)}{\prod_{\substack{j=1 \\ j \neq k}}^{n} (s_k - s_j)} \tag{5.37}$$

An alternate representation of Equation (5.37) is

$$A_k = \lim_{s \to s_k} \frac{(s - s_k) B(s)}{D(s)} \tag{5.38}$$

---

### Example 5.19

Determine $x(t)$ if

$$X(s) = \frac{3s^2 + 2s + 1}{s^3 + 7s^2 + 14s + 8} \tag{a}$$

**Solution**

Equation (a) is of the form of Equation (5.31) with

$$Q(s) = B(s) = 3s^2 + 2s + 1 \tag{b}$$

$$D(s) = s^3 + 7s^2 + 14s + 8 \tag{c}$$

The roots of $D(s)$ are determined as $s_1 = -4, s_2 = -2,$ and $s_3 = -1$ such that

$$D(s) = (s + 4)(s + 2)(s + 1) \tag{d}$$

The coefficients in the partial fraction decomposition of $X(s)$ are determined using Equation (5.37) as

$$A_1 = \frac{3s_1^2 + 2s_1 + 1}{(s_1 + 2)(s_1 + 1)}$$

$$= \frac{3(-4)^2 + 2(-4) + 1}{(-4 + 2)(-4 + 1)}$$

$$= \frac{41}{6} \tag{e}$$

$$A_2 = \frac{3s_2^2 + 2s_2 + 1}{(s_2 + 4)(s_2 + 1)}$$

$$= \frac{3(-2)^2 + 2(-2) + 1}{(-2 + 4)(-2 + 1)}$$

$$= -\frac{9}{2} \tag{f}$$

$$A_3 = \frac{3s_3^2 + 2s_3 + 1}{(s_3 + 4)(s_3 + 2)}$$

$$= \frac{3(-1)^2 + 2(-1) + 1}{(-1 + 4)(-1 + 2)}$$

$$= \frac{2}{3} \tag{g}$$

The partial fraction decomposition of $X(s)$ is

$$X(s) = \frac{41}{6}\frac{1}{s+4} - \frac{9}{2}\frac{1}{s+2} + \frac{2}{3}\frac{1}{s+1} \tag{h}$$

Transform pair 1 of Table 5.1 is used to invert Equation (h) resulting in

$$x(t) = \frac{41}{6}e^{-4t} - \frac{9}{2}e^{-2t} + \frac{2}{3}e^{-t} \tag{i}$$

**5.4.2.2 Complex Poles**   If $D(s)$ has complex roots, they occur in complex conjugate pairs; that is, if $D(s_k) = 0$ then $D(\bar{s}_k) = 0$ where $\bar{s}_k$ is the complex conjugate of $s_k$. Equations (5.35) and (5.37) may still be used to determine the partial fraction decomposition of $B(s)/D(s)$, but evaluation of Equation (5.37) leads to complex coefficients when $s_k$ is complex. The use of complex algebra can be avoided by noting that

$$\frac{A_k}{s - s_k} + \frac{\bar{A}_k}{s - \bar{s}_k} = \frac{Cs + D}{(s - s_k)(s - \bar{s}_k)} \tag{5.39}$$

where $C$ and $D$ are real coefficients and $(s - s_k)(s - \bar{s}_k)$ is a quadratic factor with real coefficients

$$(s - s_k)(s - \bar{s}_k) = s^2 - 2\text{Re}(s_k)s + [\text{Re}(s_k)]^2 + [\text{Im}(s_k)]^2$$

$$= s^2 + 2as + b \tag{5.40}$$

Any polynomial with real coefficients can be written as a product of linear factors with real roots and quadratic factors. The partial fraction decomposition of Equation (5.35) can be modified using Equation (5.39) to handle the quadratic factors. The appropriate term in the partial fraction decomposition corresponding to a quadratic factor is $(Cs + D)/(s^2 + 2as + b)$. This term is easily inverted by completing the square in the denominator, as in Example 5.17, and using the first shifting theorem and transform pairs for $\sin(\omega t)$ and $\cos(\omega t)$.

The coefficients corresponding to the real factors can be determined using Equation (5.37). The coefficients, such as those of Equation (5.39), corresponding to the quadratic factors can be determined by multiplying Equation (5.35) by $N(s)$ and then evaluating the resulting expression at independent values of $s$ that do not coincide with the real roots.

## Example 5.20

Determine $x(t)$ if

$$X(s) = \frac{4s + 2}{s^3 + 11s^2 + 44s + 60} \tag{a}$$

### Solution

The denominator of Equation (a) can be factored as

$$s^3 + 11s^2 + 44s + 60 = (s + 3)(s^2 + 8s + 20) \tag{b}$$

The quadratic factor of Equation (b) does not have real roots. Thus the appropriate partial fraction decomposition for Equation (a) is

$$\frac{4s + 2}{s^3 + 11s^2 + 44s + 60} = \frac{A}{s + 3} + \frac{Cs + D}{s^2 + 8s + 20} \tag{c}$$

The coefficient of the linear factor is determined as

$$A = \frac{4s + 2}{s^2 + 8s + 20}\bigg|_{s = -3}$$

$$= \frac{4(-3) + 2}{(-3)^2 + 8(-3) + 20}$$

$$= -2 \tag{d}$$

Substituting Equation (d) into Equation (c) and multiplying by the denominator of the left-hand side leads to

$$4s + 2 = -2(s^2 + 8s + 20) + (Cs + D)(s + 3) \tag{e}$$

Evaluation of Equation (e) for $s = 0$ leads to

$$2 = -2(20) + D(3) \Rightarrow D = 14 \tag{f}$$

Evaluation of Equation (e) for $s = 1$ leads to

$$4(1) + 2 = -2[(1)^2 + 8(1) + 20] + [C(1) + 14](1 + 3) \Rightarrow C = 2 \tag{g}$$

Substituting Equations (d), (f), and (g) into Equation (c) and completing the square of the quadratic factor in the denominator leads to

$$X(s) = -\frac{2}{s + 3} + \frac{2s + 14}{s^2 + 8s + 20}$$

$$= -\frac{2}{s + 3} + \frac{2(s + 4)}{(s + 4)^2 + 4} + \frac{6}{(s + 4)^2 + 4} \tag{h}$$

The first shifting theorem and transform pairs 1, 3, and 2 respectively from Table 5.1 are used to invert Equation (h) leading to

$$x(t) = -2e^{-3t} + 2e^{-4t}\cos(2t) + 3e^{-4t}\sin(2t) \tag{i}$$

**5.4.2.3 Repeated Poles**   Consider the case where $s_k$ is a root of $D(s)$ of multiplicity $p > 1$. Equation (5.35) is modified to account for repeated roots by replacing the terms in the equation corresponding to the repeated roots by

$$\frac{C_1}{s - s_k} + \frac{C_2}{(s - s_k)^2} + \cdots + \frac{C_p}{(s - s_k)^p} \tag{5.41}$$

Suppose the roots of $D(s)$ are numbered such that $s_1, s_2, \ldots, s_{n-p}$ are distinct roots and $s_k = s_{n-p+1}$ is a repeated root of multiplicity $p$. Using Equation (5.41) in Equation (5.35) leads to

$$\frac{B(s)}{D(s)} = \sum_{i=1}^{n-p} \frac{A_i}{s - s_i} + \sum_{j=1}^{p} \frac{C_j}{(s - s_k)^j} \tag{5.42}$$

Multiplying Equation (5.42) by $(s - s_k)^p$ leads to

$$(s - s_k)^p \frac{B(s)}{D(s)} = \sum_{i=1}^{n-p} \frac{A_i (s - s_k)^p}{s - s_i} + \sum_{j=1}^{p} C_j (s - s_k)^{p-j} \tag{5.43}$$

Evaluation of Equation (5.43) for $s = s_k$ leads to

$$C_p = \left[ (s - s_k)^p \frac{B(s)}{D(s)} \right]_{s = s_k} \tag{5.44}$$

Differentiation of Equation (5.43) with respect to $s$ and then evaluating for $s = s_k$ leads to

$$C_{p-1} = \left\{ \frac{d}{ds} \left[ (s - s_k)^p \frac{B(s)}{D(s)} \right] \right\}_{s = s_k} \tag{5.45}$$

Continued differentiation and evaluation for $s = s_k$ leads to

$$C_{p-\ell} = \frac{1}{\ell!} \left\{ \frac{d^\ell}{ds^\ell} \left[ (s - s_k)^p \frac{B(s)}{D(s)} \right] \right\}_{s = s_k} \qquad \ell = 0, 1, 2, \ldots, p \tag{5.46}$$

The coefficients in the partial fraction decomposition corresponding to repeated roots are obtained using Equation (5.46).

**Example 5.21**

Determine $x(t)$ if

$$X(s) = \frac{5s^3 + 2s^2 + 3s - 1}{(s + 2)(s + 1)^3} \tag{a}$$

**Solution**

The appropriate partial fraction decomposition for $X(s)$ is

$$\frac{5s^3 + 2s^2 + 3s - 1}{(s + 2)(s + 1)^3} = \frac{A}{s + 2} + \frac{C_1}{s + 1} + \frac{C_2}{(s + 1)^2} + \frac{C_3}{(s + 1)^3} \tag{b}$$

The coefficient $A$ is determined using Equation (5.37) as

$$A = \left[\frac{5s^3 + 2s^2 + 3s - 1}{(s+1)^3}\right]_{s=-2}$$

$$= \frac{5(-2)^3 + 2(-2)^2 + 3(-2) - 1}{(-2+1)^3} = 39 \qquad \text{(c)}$$

The coefficients for the repeated root $s = -1$ are determined using Equation (5.46)

$$C_3 = \left[\frac{5s^3 + 2s^2 + 3s - 1}{s+2}\right]_{s=-1}$$

$$= \frac{5(-1)^3 + 2(-1)^2 + 3(-1) - 1}{-1+2} = -7 \qquad \text{(d)}$$

$$C_2 = \left\{\frac{d}{ds}\left[\frac{5s^3 + 2s^2 + 3s - 1}{s+2}\right]\right\}_{s=-1}$$

$$= \left[\frac{(s+2)(15s^2 + 4s + 3) - (5s^3 + 2s^2 + 3s - 1)(1)}{(s+2)^2}\right]_{s=-1}$$

$$= \left[\frac{10s^3 + 32s^2 + 8s + 7}{(s+2)^2}\right]_{s=-1}$$

$$= \frac{10(-1)^3 + 32(-1)^2 + 8(-1) + 7}{(-1+2)^2} = 21 \qquad \text{(e)}$$

$$C_1 = \frac{1}{2}\left\{\frac{d^2}{ds^2}\left[\frac{5s^3 + 2s^2 + 3s - 1}{s+2}\right]\right\}_{s=-1}$$

$$= \frac{1}{2}\left\{\frac{d}{ds}\left[\frac{10s^3 + 32s^2 + 8s + 7}{(s+2)^2}\right]\right\}_{s=-1}$$

$$= \frac{1}{2}\left[\frac{(s+2)^2(30s^2 + 64s + 8) - 2(s+2)(10s^3 + 32s^2 + 8s + 7)}{(s+2)^4}\right]_{s=-1}$$

$$= \frac{1}{2}\left[\frac{(-1+2)^2(30(-1)^2 + 64(-1) + 8) - 2(-1+2)(10(-1)^3 + 32(-1)^2 + 8(-1) + 7)}{(-1+2)^4}\right]$$

$$= -34 \qquad \text{(f)}$$

Using Equations (b)–(f) in Equation (a) leads to

$$X(s) = \frac{39}{s+2} - \frac{34}{s+1} + \frac{21}{(s+1)^2} - \frac{7}{(s+1)^3} \qquad \text{(g)}$$

The transform of Equation (g) is inverted using the first shifting theorem and transform pairs 1 and 4 from Table 5.1. The result is

$$x(t) = 39e^{-2t} - 34e^{-t} + 21te^{-t} - \tfrac{7}{2}t^2 e^{-t} \qquad \text{(h)}$$

**5.4.2.4 Brute Force Methods**   The methods just described are elegant methods for determining residues for partial fraction decomposition. The formulas are easy to forget and can be tedious to apply. The brute force methods described in this section are also tedious, but they do not require special knowledge.

Once the poles are determined the appropriate partial fraction decomposition is assumed as a sum of fractions of linear factors of $s$, nonreducible quadratic factors of $s$, and powers of repeated factors. The numerator for each of the linear factors contains one unknown coefficient, and the numerator for each quadratic factor contains two unknown coefficients. There are $n$ unknown coefficients in the partial fraction decomposition when $D(s)$ is a polynomial of order $n$.

The assumed partial fraction decomposition is multiplied by $D(s)$. Both sides of the resulting equation are polynomials in $s$ of order $n - 1$. Since $s$ is a variable that can change independently, the equations must be satisfied for all possible values of $s$. Two methods can be used to determine a set of $n$ simultaneous linear equations to solve for the unknown coefficients.

- Since powers of $s$ are linearly independent, coefficients of like powers of $s$ on each side must be equal. Equating coefficients of $s^k$ for $k = 0, 1, \ldots, n - 1$ leads to $n$ equations.
- The equations may be evaluated for $n$ values of $s$, which are not poles. Numerical evaluation for each value of $s$ leads to one equation.

## Example 5.22

Determine $x(t)$ if

$$X(s) = \frac{s^4 + 8s^3 + 29s^2 + 40s + 36}{(s+2)^2(s+3)(s+4)(s^2 + 4s + 9)} \tag{a}$$

**Solution**

The appropriate partial fraction decomposition for Equation (a) is

$$\frac{s^4 + 8s^3 + 29s^2 + 40s + 36}{(s+2)^2(s+3)(s+4)(s^2 + 4s + 9)} = \frac{A}{s+3} + \frac{B}{s+4} + \frac{C}{s+2} + \frac{D}{(s+2)^2} + \frac{Es + F}{s^2 + 4s + 9} \tag{b}$$

Multiplication of Equation (b) by $D(s)$ leads to

$$
\begin{aligned}
s^4 + 8s^3 + 29s^2 + 40s + 36 &= A(s+4)(s+2)^2(s^2 + 4s + 9) \\
&\quad + B(s+3)(s+2)^2(s^2 + 4s + 9) \\
&\quad + C(s+3)(s+4)(s+2)(s^2 + 4s + 9) \\
&\quad + D(s+3)(s+4)(s^2 + 4s + 9) \\
&\quad + (Es + F)(s+3)(s+4)(s+2)^2 \tag{c}
\end{aligned}
$$

Evaluation of Equation (c) for $s = -1, 0, 1, 2, 3$, and 4 respectively leads to the following set of equations

$$
\begin{bmatrix}
18 & 12 & 36 & 36 & -6 & 6 \\
144 & 108 & 216 & 108 & 0 & 48 \\
630 & 504 & 840 & 280 & 180 & 180 \\
2{,}016 & 1{,}680 & 2{,}520 & 630 & 960 & 480 \\
5{,}250 & 4{,}500 & 6{,}300 & 1{,}260 & 3{,}150 & 1{,}050 \\
11{,}808 & 10{,}332 & 13{,}776 & 2{,}296 & 8{,}064 & 2{,}016
\end{bmatrix}
\begin{bmatrix}
A \\ B \\ C \\ D \\ E \\ F
\end{bmatrix}
=
\begin{bmatrix}
18 \\ 36 \\ 114 \\ 312 \\ 714 \\ 1{,}428
\end{bmatrix}
\tag{d}
$$

The solution of Equation (d) is $A = -9.80$, $B = 5.13$, $C = 3.70$, $D = -0.480$, $E = 1.01$, $F = 3.05$. Substituting these values into Equation (b) leads to

$$
X(s) = -\frac{9.80}{s+3} + \frac{5.13}{s+4} + \frac{3.70}{s+2} - \frac{0.480}{(s+2)^2} + \frac{1.01s + 3.05}{s^2 + 4s + 9}
\tag{e}
$$

Inversion of Equation (e) yields

$$
x(t) = -9.80e^{-3t} + 5.13e^{-4t} + 3.70e^{-2t} - 0.480te^{-2t}
$$
$$
+ 1.01e^{-2t}\cos\left(\sqrt{5}t\right) + 0.461e^{-2t}\sin\left(\sqrt{5}t\right)
\tag{f}
$$

## 5.4.3 Inversion of Transforms of Periodic Functions

The Laplace transform of a periodic function is obtained using property 12 of Table 5.2. From property 12 it is clear that the transform of a periodic function has a term of the form of $1 - e^{-Ts}$ in its denominator. Thus if a transform has this type of term in its denominator, its inverse is a periodic function.

Consider a transform of the form

$$
X(s) = \frac{F(s)}{1 - e^{-Ts}}
\tag{5.47}
$$

The binomial expansion of the denominator is

$$
(1 - e^{-Ts})^{-1} = 1 + e^{-Ts} + e^{-2Ts} + e^{-3Ts} + \cdots
$$
$$
= \sum_{k=0}^{\infty} e^{-kTs}
\tag{5.48}
$$

Use of Equation (5.48) in Equation (5.47) leads to

$$
X(s) = \sum_{k=0}^{\infty} \left[ e^{-kTs} F(s) \right]
\tag{5.49}
$$

The second shifting theorem and linearity of the inverse transform are applied to Equation (5.49) leading to

$$
x(t) = \sum_{k=0}^{\infty} f(t - kT)u(t - kT)
\tag{5.50}
$$

where $f(t) = \mathcal{L}^{-1}\{F(s)\}$.

**Example 5.23**

Determine $x(t)$ if

$$X(s) = \frac{1}{1 - e^{-3s}} \left( \frac{4}{s(s^2 + 4)} \right) \tag{a}$$

**Solution**

A partial fraction decomposition of Equation (a) leads to

$$X(s) = \frac{1}{1 - e^{-3s}} \left[ \frac{1}{s} - \frac{s}{s^2 + 4} \right] \tag{b}$$

Equation (b) is in the form of Equation (5.48) with $T = 3$ and

$$F(s) = \frac{1}{s} - \frac{s}{s^2 + 4} \tag{c}$$

Equation (c) is inverted using the transform pairs 8 and 3 of Table 5.1 leading to

$$f(t) = 1 - \cos 2t \tag{d}$$

Using Equation (d) in Equation (5.50) leads to

$$x(t) = \sum_{k=0}^{\infty} \{[1 - \cos(2t - 6k)]u(t - 3k)\} \tag{e}$$

## 5.4.4  Use of MATLAB

Just as MATLAB is capable of symbolically determining the Laplace transform of a function of a real variable $t$, it is capable of determining the inverse transform of a function of a complex variable $s$. Assuming $s$ is declared as a symbolic variable and $G$ is a defined function of $s$, the appropriate MATLAB command to determine the inverse transform of $G(s)$ and assign it to the variable $x$ is

$$x = \text{ilaplace}(G)$$

**Example 5.24**

Use MATLAB to determine and plot the inverse transforms of the functions in (a) Example 5.18 and (b) Example 5.21.

**Solution**

A MATLAB script file, Example5_24.m, which provides the inverse transform and plots the results as a function of time, is illustrated in Figure 5.7. The output from the execution of the M-file and the resulting plots are shown in Figure 5.8. Note the following regarding Example5_24.m:

```
% Example 5.24
% Determination and plotting of inverse transforms
%
% Declaration of symbolic varaibles
%    s=transform variable
%    t=time
syms s t
%
% Example 5.24(a)
%    F1 is function in transform domain
%    G1 is inverse transform of F1
%    G1A is decimal form of G1 with 4 significant figures
%
F1=exp(-3*s)/(s+2)^2;
G1=ilaplace(F)
G1A=vpa(G,4)
%
% Example 5.24 (b)
%
F2=(5*s^3+2*s^2+3*s-1)/(s+2)/(s+1)^3;
G2=ilaplace(F2)
G2A=vpa(G2,4)
%
% Defining t to range between 0 and 10 in increments of 0.1
%
t=0:.01:10;
%
% Defining vectors of numerical values for purposes of plotting
% subs(G1A,t) numerically evaluates the symbolic function G1A at the
%    current value of t
%
x1=subs(G1A,t);
x2=subs(G2A,t);
% Since more than one figure is to be plotted the figure statement is
%    necessary such that both figures are retained
%
figure
plot(t,x1)
xlabel('t')
ylabel('x')
title('Inverse transform of e^-^2^s/(s+2)^2')
figure
plot(t,x2)
xlabel('t')
ylabel('x')
title('Inverse transform of (5s^3+2s^2+3s-1)/[(s+2)(s+1)^3]')
%
% End of Example5_24.m
```

**FIG. 5.7** Script for Example5_24.m.

```
>> Example5_24

G1 =

(exp(-2*t+6)*t-3*exp(-2*t+6))*Heaviside(t-3)


G1A =

(exp(-2.*t+6.)*t-3.*exp(-2.*t+6.))*Heaviside(t-3.)


G2 =

39*exp(-2*t)+(-7/2*t^2-34+21*t)*exp(-t)


G2A =

39.*exp(-2.*t)+(-3.500*t^2-34.+21.*t)*exp(-1.*t)

>>
```

(a)

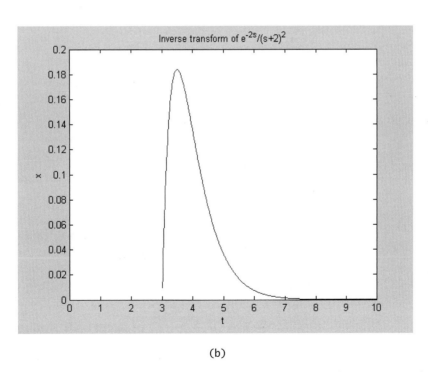

(b)

**FIG. 5.8** Output from the execution of Example5_24.m: (a) the screen output, (b) the plot of the inverse transform of part (a), and (c) the plot of the inverse transform of part (b). (*Continued*)

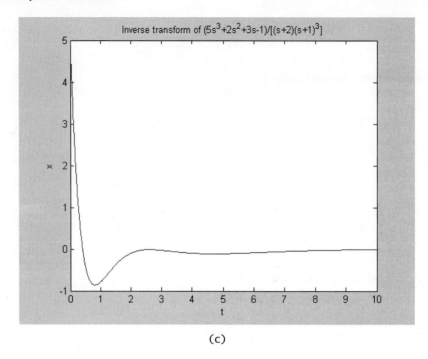

Inverse transform of $(5s^3+2s^2+3s-1)/[(s+2)(s+1)^3]$

(c)

**FIG. 5.8** (*Continued*)

- Since G1A and G2A are symbolic functions it is necessary to numerically evaluate them at discrete values of $t$ for purposes of plotting. The command "x1=subs(G1A,t)" numerically evaluates the symbolic function G1A at the current value of $t$ and assigns the value to $x1$.
- The "figure" statements are necessary so that both figures are retained as output. Without the second "figure" statement the first plot would be erased.
- In text in annotating graphs the caret (^) means to use the next character as a superscript and the underscore (_) means to use the next character as a subscript.

MATLAB is also useful in performing partial fraction decompositions on ratios of polynomials of the form of Equation (5.31). The polynomials in the numerator and denominator are defined by their coefficients. The coefficients of a polynomial are stored in a row vector. If $N(s) = a_3 s^3 + a_2 s^2 + a_1 s + a_0$ then $N(s)$ is defined in MATLAB by defining a row vector with four columns according to

$$N = [a_3 \ a_2 \ a_1 \ a_0]$$

Once $N(s)$ and $D(s)$ are defined, MATLAB calculates the poles and residues using the statement

$$[r,p,k] = \text{residue}(N,D)$$

where $r$ is a column vector of the residues of $N(s)/D(s)$ corresponding to the elements in the vector of poles [**p**]. The vector **k** is returned as the null vector if the order of $N(s)$ is less than the order of $D(s)$. If the order of

$N(s)$ is equal to or greater than the order of $D(s)$ then **k** is the vector of coefficients in the function $W(s)$ defined in Equation (5.33). If the poles are complex MATLAB provides complex residues. The partial fraction decomposition is built using the elements of $r$, $p$, and $k$. The order of the elements in $r$ correspond to the order of the elements in $p$.

## Example 5.25

Use the MATLAB "residue" command to help determine the partial fraction decompositions for:

(a)  $X(s) = \dfrac{3s^2 + 2s + 1}{s^3 + 7s^2 + 14s + 8}$ of Example 5.19

(b)  $X(s) = \dfrac{4s + 2}{s^3 + 11s^2 + 44s + 60}$ of Example (5.20)

(c)  $X(s) = \dfrac{3s^2 + 3s + 4}{s^2 + 5s + 6}$

### Solution

The work sessions for the use of the "residue" command to determine the poles and residues of the transforms are shown in Figure 5.9. The partial fraction decompositions are determined from the output as

(a)
$$X(s) = \frac{6.8333}{s+4} - \frac{4.5000}{s+2} + \frac{0.6667}{s+1}$$
(a)

```
>> N=[3 2 1];
>> D=[1 7 14 8];
>> [r,p,k]=residue(N,D)

r =

    6.8333
   -4.5000
    0.6667


p =

   -4.0000
   -2.0000
   -1.0000


k =

    []

         (a)
```

**FIG. 5.9** The solution of Example 5.25, parts (a), (b), and (c): The MATLAB "residue" command aids in partial fraction decompositions. (*Continued*)

```
>> N=[4 2];
>> D=[1 11 44 60];
>> [r,p,k]=residue(N,D)

r =

   1.0000 - 1.5000i
   1.0000 + 1.5000i
  -2.0000


p =

  -4.0000 + 2.0000i
  -4.0000 - 2.0000i
  -3.0000


k =

   []

>>
```

(b)

```
>> N=[3 3 4];
>> D=[1 5 6];
>> [r,p,k]=residue(N,D)

r =

  -22.0000
   10.0000


p =

  -3.0000
  -2.0000


k =

   3

>>
```

**FIG. 5.9**  (*Continued*)                                (c)

**(b)**
$$X(s) = \frac{1 - 1.5j}{s - (-4 + 2j)} + \frac{1 + 1.5j}{s - (-4 - 2j)} - \frac{2}{s + 3}$$

$$= \frac{2s + 14}{s^2 + 8s + 20} - \frac{2}{s + 3} \tag{b}$$

**(c)**
$$X(s) = 3 - \frac{22}{s + 3} + \frac{10}{s + 2} \tag{c}$$

## 5.5 LAPLACE TRANSFORM SOLUTION OF DIFFERENTIAL EQUATIONS

Linear differential equations with constant coefficients and integrodifferential equations obtained through the mathematical modeling of dynamic systems can be solved using the **Laplace transform method**. When the Laplace transform of the differential equation is taken, linearity of the transform allows the transform of each term to be taken individually. The property of transform of derivatives property 9 of Table 5.2 allows a differential equation in the time domain to be transformed into an algebraic equation in the $s$ domain. The property of transform of integrals property 10 of Table 5.2, allows integrodifferential equations to be transformed into algebraic equations in the $s$ domain. Once the transform is obtained in the $s$ domain the methods of Section 5.4 are employed to determine the response in the time domain.

### 5.5.1 Systems with One Dependent Variable

Given the differential equation and appropriate initial conditions, the following steps summarize the application of the Laplace transform method to a linear differential equation with constant coefficients where $x(t)$ is the dependent variable and defining $X(s) = \mathcal{L}\{x(t)\}$.

1. The Laplace transform is applied to both sides of the differential equation.
2. The property of linearity of the transform is applied.
3. If nonzero, the Laplace transform of the nonhomogeneous term is obtained.
4. The property of transform of derivatives is applied using appropriate initial conditions.
5. Algebraic methods are used to determine $X(s)$.
6. $x(t)$ is determined using the inversion methods of Section 5.4.

This procedure is illustrated in the following examples.

**Example 5.26**

A one-degree-of-freedom mechanical system has a natural frequency of 20 r/s and a damping ratio of 0.1. The system has initial conditions $x(0) = 0.01$ m and $\dot{x}(0) = 2.0$ m/s. Determine the free response of the system.

**Solution**

The differential equation governing the free response of a linear one-degree-of-freedom system is Equation (2.111), with $F(t) = 0$,

$$\ddot{x} + 2\zeta\omega_n\dot{x} + \omega_n^2 x = 0 \qquad (a)$$

Substituting given values for the natural frequency and damping ratio leads to

$$\ddot{x} + 4\dot{x} + 400x = 0 \qquad (b)$$

Applying the Laplace transform to Equation (b) leads to

$$\mathcal{L}\{\ddot{x} + 4\dot{x} + 400x\} = \mathcal{L}\{0\} \qquad (c)$$

Linearity of the transform is used in Equation (c) to give

$$\mathcal{L}\{\ddot{x}\} + 4\mathcal{L}\{\dot{x}\} + 400\mathcal{L}\{x\} = 0 \qquad (d)$$

Defining $X(s) = \mathcal{L}\{x(t)\}$ and using the property of transforms of derivatives in equation (d) yields

$$s^2 X(s) - sx(0) - \dot{x}(0) + 4[sX(s) - x(0)] + 400X(s) = 0 \qquad (e)$$

Substituting the given initial conditions into Equation (e) and rearranging leads to

$$(s^2 + 4s + 400)X(s) = 0.01s + 2.04 \qquad (f)$$

Solving for $X(s)$ from Equation (f) yields

$$X(s) = \frac{0.01s + 2.04}{s^2 + 4s + 400} \qquad (g)$$

The quadratic polynomial in the denominator does not have real roots. Completing the square in the denominator of Equation (g) gives

$$X(s) = \frac{0.01s + 2.04}{(s + 2)^2 + 396} \qquad (h)$$

Equation (h) can be rewritten as

$$X(s) = \frac{0.01(s + 2)}{(s + 2)^2 + 396} + \frac{2.02}{(s + 2)^2 + 396} \qquad (i)$$

The inverse transform of Equation (i) is obtained using linearity and transform pairs 12 and 13 yielding

$$x(t) = 0.01e^{-2t}\cos\left(\sqrt{396}t\right) + \frac{2.02}{\sqrt{396}}e^{-2t}\sin(\sqrt{396}t)$$

$$= 0.01e^{-2t}\cos\left(19.90t\right) + 0.102e^{-2t}\sin(19.90t) \qquad (j)$$

**Example 5.27**

The differential equation governing the current in a series $RL$ circuit with a direct current source connected at $t = 0$ is

$$L\frac{di}{dt} + Ri = V_0 u(t) \tag{a}$$

Where $i(t)$ is the current in the circuit, $L$ is the inductance, $R$ is the resistance, and $V_0$ is the potential in the source. If there is no current in the circuit when the source is connected at $t = 0$, $i(0) = 0$, determine the time-dependent current in the circuit.

**Solution**

Taking the Laplace transform of both sides of Equation (a) leads to

$$\mathcal{L}\left\{L\frac{di}{dt} + Ri\right\} = \mathcal{L}\{V_0 u(t)\} \tag{b}$$

Using linearity of the transform, Equation (b) becomes

$$L\mathcal{L}\left\{\frac{di}{dt}\right\} + R\mathcal{L}\{i(t)\} = \mathcal{L}\{V_0 u(t)\} \tag{c}$$

Defining $I(s) = \mathcal{L}\{i(t)\}$ and using the transform of derivative property and transform pair 8 of Table 5.1 in Equation (c)

$$L[sI(s) - i(0)] + RI(s) = \frac{V_0}{s} \tag{d}$$

Applying the initial condition and solving for $I(s)$ yields

$$I(s) = \frac{V_0}{(Ls + R)s}$$

$$= \frac{V_0}{L}\frac{1}{\left(s + \dfrac{R}{L}\right)s} \tag{e}$$

A partial fraction decomposition of the right-hand side of Equation (e) leads to

$$I(s) = \frac{V_0}{L}\left[\frac{\dfrac{L}{R}}{s} - \frac{\dfrac{L}{R}}{s + \dfrac{R}{L}}\right]$$

$$= \frac{V_0}{R}\left[\frac{1}{s} - \frac{1}{s + \dfrac{R}{L}}\right] \tag{f}$$

The transform in Equation (f) is inverted using transform pairs 8 and 1 of Table 5.1 resulting in

$$i(t) = \frac{V_0}{R}\left[u(t) - e^{-(R/L)t}\right] \tag{g}$$

### Example 5.28

Use the Laplace transform method to determine the solution of the differential equation

$$\ddot{x} + 25x = 0.9 \sin 4t \tag{a}$$

subject to the initial conditions $x(0) = 0$ and $\dot{x}(0) = 0$.

**Solution**

Taking the Laplace transform of Equation (a) and using the linearity of the transform gives

$$\mathcal{L}\{\ddot{x}\} + 25\mathcal{L}\{x\} = 0.9\mathcal{L}\{\sin (4t)\} \tag{b}$$

Defining $X(s) = \mathcal{L}\{x(t)\}$, using transform pair 2 from Table 5.1 to determine the Laplace transform of the nonhomogeneous term, and applying the property of transform of derivatives to Equation (b) leads to

$$s^2 X(s) - sx(0) - \dot{x}(0) + 25X(s) = \frac{3.6}{s^2 + 16} \tag{c}$$

Applying the given initial conditions and solving Equation (c) for $X(s)$ yields

$$X(s) = \frac{3.6}{(s^2 + 16)(s^2 + 25)} \tag{d}$$

A partial fraction decomposition of the right-hand side of Equation (d) results in

$$X(s) = \frac{0.4}{s^2 + 16} - \frac{0.4}{s^2 + 25} \tag{e}$$

The inversion of Equation (e) leads to

$$x(t) = 0.1 \sin(4t) - 0.08 \sin(5t) \tag{f}$$

## 5.5.2 Systems of Differential Equations

The Laplace transform method can be applied to determine the solution of a system of linear differential equations with constant coefficients. Let $x_1(t)$, $x_2(t)$, ..., $x_n(t)$ be the dependent variables whose Laplace transforms are $X_1(s), X_2(s), \ldots, X_n(s)$ respectively. A system of $n$ differential equations with appropriate initial conditions has been formulated. Use the following steps to apply the Laplace transform method to solve for the dependent variables.

1. Apply the Laplace transform to both sides of all differential equations in the system.
2. The property of linearity of the transforms is applied to each differential equation.

3. The Laplace transforms of all nonhomogeneous terms are determined.
4. The property of derivatives of transforms is applied to all equations using the initial conditions.
5. A system of simultaneous linear algebraic equations is developed. The unknowns are the transforms of the dependent variables.
6. The simultaneous solution of the linear equations using algebraic methods leads to $X_1(s), X_2(s), \ldots, X_n(s)$.
7. The inversion methods of Section 5.4 are used to determine $x_1(t)$, $x_2(t), \ldots, x_n(t)$.

The procedure is illustrated in the following example.

**Example 5.29**

Determine the solution to the differential equations

$$\ddot{x}_1 + 5x_1 - 2x_2 = 2e^{-t} \tag{a}$$

$$\ddot{x}_2 - 2x_1 + 2x_2 = 0 \tag{b}$$

subject to the initial conditions $x_1(0) = 0$, $x_2(0) = 0$, $\dot{x}_1(0) = 0$, $\dot{x}_2(0) = 0$.

**Solution**

Taking the Laplace transform of Equations (a) and (b) and using linearity of the transform leads to

$$\mathcal{L}\{\ddot{x}_1\} + 5\mathcal{L}\{x_1\} - 2\mathcal{L}\{x_2\} = 2\mathcal{L}\{e^{-t}\} \tag{c}$$

$$\mathcal{L}\{\ddot{x}_2\} - 2\mathcal{L}\{x_1\} + 2\mathcal{L}\{x_2\} = 0 \tag{d}$$

Using transform pair 1 from Table 5.1 in Equation (c) and the property of transform of derivatives in Equations (c) and (d) leads to

$$s^2 X_1(s) - sx_1(0) - \dot{x}_1(0) + 5X_1(s) - 2X_2(s) = \frac{2}{s+1} \tag{e}$$

$$s^2 X_2(s) - sx_2(0) - \dot{x}_2(0) - 2X_1(s) + 2X_2(s) = 0 \tag{f}$$

The application of the initial conditions to Equations (e) and (f) and rearranging leads to

$$(s^2 + 5)X_1(s) - 2X_2(s) = \frac{2}{s+1} \tag{g}$$

$$-2X_1(s) + (s^2 + 2)X_2(s) = 0 \tag{h}$$

The matrix form of Equations (g) and (h) is

$$\begin{bmatrix} s^2 + 5 & -2 \\ -2 & s^2 + 2 \end{bmatrix} \begin{bmatrix} X_1(s) \\ X_2(s) \end{bmatrix} = \begin{bmatrix} \dfrac{2}{s+1} \\ 0 \end{bmatrix} \tag{i}$$

Since the solution of the preceding equations leads to functions of the variable $s$, the algebra is minimized by solving the equations using Cramer's rule.

Its application leads to

$$X_1(s) = \frac{\begin{vmatrix} \dfrac{2}{s+1} & -2 \\ 0 & s^2+2 \end{vmatrix}}{\begin{vmatrix} s^2+5 & -2 \\ -2 & s^2+2 \end{vmatrix}}$$ (j)

$$X_2(s) = \frac{\begin{vmatrix} s^2+5 & \dfrac{2}{s+1} \\ -2 & 0 \end{vmatrix}}{\begin{vmatrix} s^2+5 & -2 \\ -2 & s^2+2 \end{vmatrix}}$$ (k)

The determinants in Equations (j) and (k) are expanded leading to

$$X_1(s) = \frac{\left(\dfrac{2}{s^2+1}\right)(s^2+2) - (-2)(0)}{(s^2+5)(s^2+2) - (-2)(-2)}$$

$$= \frac{2(s^2+2)}{(s+1)(s^4+7s^2+6)}$$

$$= \frac{2(s^2+2)}{(s+1)(s^2+1)(s^2+6)}$$ (l)

$$X_2(s) = \frac{(s^2+5)(0) - \left(\dfrac{2}{s+1}\right)(-2)}{(s^2+5)(s^2+2) - (-2)(-2)}$$

$$= \frac{4}{(s+1)(s^2+1)(s^2+6)}$$ (m)

The appropriate partial fraction decomposition for Equation (l) is

$$\frac{2(s^2+2)}{(s+1)(s^2+1)(s^2+6)} = \frac{A}{s+1} + \frac{Bs+C}{s^2+1} + \frac{Ds+E}{s^2+6}$$ (n)

The residue for the linear factor is determined using Equation (5.37)

$$A = \frac{2(s^2+2)}{(s^2+1)(s^2+6)}\bigg|_{s=-1}$$

$$= \frac{2\left[(-1)^2+2\right]}{\left[(-1)^2+1\right]\left[(-1)^2+6\right]}$$

$$= \tfrac{3}{7} = 0.429$$ (o)

Substituting for $A$ in Equation (n) and multiplying both sides by the denominator of the left-hand side leads to

$$2(s^2+2) = \tfrac{3}{7}(s^2+1)(s^2+6) + (Bs+C)(s+1)(s^2+6)$$
$$+ (Ds+E)(s+1)(s^2+1)$$ (p)

Since Equation (p) is valid for all $s$, four independent equations are developed by giving $s$ four different numerical values. This leads to a set of four simultaneous equations to solve for $B$, $C$, $D$, and $E$. To this end setting $s = 0$, 1, 2, and $-2$ leads to the following equations respectively

$$6C + E = \frac{10}{7} \tag{q}$$

$$14B + 14C + 4D + 4E = 0 \tag{r}$$

$$60B + 30C + 30D + 15E = -\frac{66}{7} \tag{s}$$

$$20B - 10C + 10D - 5E = -\frac{66}{7}. \tag{t}$$

Simultaneous solution leads to $B = -0.200$, $C = 0.200$, $D = -0.2286$, and $E = 0.2286$. Thus Equation (l) can be written as

$$X_1(s) = \frac{0.429}{s+1} + \frac{-0.200s + 0.200}{s^2 + 1} + \frac{-0.229s + 0.229}{s^2 + 6} \tag{u}$$

A similar partial fraction decomposition leads to

$$X_2(s) = \frac{0.286}{s+1} + \frac{-0.400s + 0.400}{s^2 + 1} + \frac{0.114s - 0.114}{s^2 + 6} \tag{v}$$

Inversion of the transforms in Equations (u) and (v) yields

$$x_1(t) = 0.429e^{-t} - 0.200 \cos t + 0.200 \sin t - 0.229 \cos(2.45t)$$
$$+ 0.0935 \sin(2.45t) \tag{w}$$

$$x_2(t) = 0.286e^{-t} - 0.400 \cos t + 0.400 \sin t + 0.114 \cos(2.45t)$$
$$- 0.0466 \sin(2.45t) \tag{x}$$

## 5.5.3 Integrodifferential Equations

An integrodifferential equation contains derivatives of the dependent variable as well as an indefinite integral whose integrand contains the dependent variable. Integrodifferential equations arise in the study of *LRC* circuits. The Laplace transform method may be applied to determine the solution of a linear integrodifferential equation with constant coefficients. Given the integrodifferential equation and appropriate initial conditions, the following steps are used in the application of the Laplace transform method to a linear integrodifferential equation with constant coefficients, where $x(t)$ is the dependent variable and defining $X(s) = \mathcal{L}\{x(t)\}$:

1. The Laplace transform is applied to both sides of the integrodifferential equation.
2. Linearity of the transform is applied.
3. The Laplace transform of the nonhomogeneous terms are obtained.
4. The properties of transform of derivatives and transform of indefinite integrals are applied, including application of initial conditions.
5. Algebraic methods are applied to determine $X(s)$.
6. $x(t)$ is determined using the inversion methods of Section 5.4.

This procedure is illustrated in the following example.

## Example 5.30

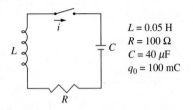

$L = 0.05$ H
$R = 100\ \Omega$
$C = 40\ \mu F$
$q_0 = 100$ mC

**FIG. 5.10**  The $RLC$ circuit of
Example 5.30: The capacitor has a
charge $q_0$ before the switch is closed.

The integrodifferential equation governing the current in the series $RLC$ circuit
of Figure 5.10 is

$$L\frac{di}{dt} + Ri + \frac{1}{C}\int_0^t i(t)dt + v(0) = 0 \tag{a}$$

where $v(0)$ is the voltage on the capacitor at $t = 0$. At $t = 0$, when the switch is
closed, the capacitor has a charge of $q_0 = 4\,\text{mC}$. Determine $i(t)$.

**Solution**

The voltage on the capacitor at $t = 0$ is obtained using Equation (3.12) as

$$v(0) = \frac{q_0}{C}$$

$$= \frac{4\ \text{mC}}{40\ \mu\text{F}}$$

$$= 100\ \text{V} \tag{b}$$

Taking the Laplace transform of Equation (a) and using linearity of the transform leads to

$$L\mathcal{L}\left\{\frac{di}{dt}\right\} + R\mathcal{L}\{i\} + \frac{1}{C}\mathcal{L}\left\{\int_0^t i(t)\,dt\right\} + \mathcal{L}\{v(0)\} = 0 \tag{c}$$

Using the properties of transform of the derivative and transform of the integral
in Equation (c) leads to

$$L[sI(s) - i(0)] + RI(s) + \frac{1}{Cs}I(s) + \frac{v(0)}{s} = 0 \tag{d}$$

The circuit is an open circuit with no current before the switch is closed. Thus,
$i(0) = 0$. Solving Equation (d) for $I(s)$ leads to

$$I(s) = -\frac{v(0)}{\left(Ls^2 + Rs + \dfrac{1}{Cs}\right)} \tag{e}$$

Substituting known values into Equation (e) leads to

$$I(s) = -\frac{100}{\left(0.05s^2 + 100s + \dfrac{1}{40 \times 10^{-6}s}\right)}$$

$$= -\frac{100}{0.05(s^2 + 2{,}000s + 500{,}000)}$$

$$= -\frac{2{,}000}{s^2 + 2{,}000s + 500{,}000} \tag{f}$$

The roots of the polynomial in the denominator are $s_1 = -1.71 \times 10^3$ and
$s_2 = -292.9$. Thus Equation (f) is rewritten as

$$I(s) = -\frac{2{,}000}{(s + 1.71 \times 10^3)(s + 292.9)} \tag{g}$$

A partial fraction decomposition of the right-hand side of Equation (g) leads to

$$I(s) = \frac{-1.414}{s + 292.9} + \frac{1.414}{s + 1.71 \times 10^3} \tag{h}$$

The inversion of Equation (g) is performed using transform pair 1 from Table 5.1 leading to

$$i(t) = 1.414e^{-1.71 \times 10^3 t} - 1.414e^{-242.9t} \text{ A} \tag{i}$$

## 5.5.4  Use of MATLAB

MATLAB is useful in reducing the amount of tedious algebra required in applying the Laplace transform method as well as for the graphical illustration of the solution. MATALB can be applied as described in Section 5.4. Once $X(s)$ has been determined, the symbolic capabilities of MATALB may be used to invert the transform. MATALB is particularly useful for systems with more than one dependent variable that require symbolic solution of a set of algebraic equations.

---

**Example 5.31**

Use MATALB to aid in the solution of Example 5.29. Write an M-file that, beginning with Equation (i), solves for the transforms, inverts the transforms, and plots the response.

**Solution**

The MATLAB file Example5_31.m is shown in Figure 5.11. The output from the execution and the plot are shown in Figure 5.12. Note the following regarding this program:

- The elements of the matrix **A** and the column vector **b** are functions of the symbolic variable $s$.
- The vector xbar is the symbolic solution of the set of algebraic equations.

```
% Example5.31
% Use of MATLAB to aid in the Laplace transform solution of a coupled set
%      of differential equations
%
% Defining symbolic variables
%   s=transform variable
%   t=time
syms s t
%
% Defining coefficient matrix
%
A=[s^2+5 -2;-2 s^2+2];
%
% Defining right-hand side vector
```

**FIG. 5.11**  The file Example5_31.m, which determines inverse transforms for the system of Example 5.29 and plots the responses. (*Continued*)

```
%
b=[2/(s+1);0];
%
% Solving for transforms
%
xbar=A^-1*b
%
% Inverse transform of vector of solutions
%
x=ilaplace(xbar)
x1=vpa(x,3)
%
% Plotting system response
%
t=0:0.05:10;
%
% Determining numerical values of response at discrete times
y1=subs(x1(1),t);
y2=subs(x1(2),t);
plot(t,y1,'-',t,y2,'.')
xlabel('t')
ylabel('x')
title('Solution of Example 5.31')
legend('x_1(t)','x_2(t)')
%
% End of Example5_31.m
```

**FIG. 5.11** (*Continued*)

```
>> Example5_31

xbar =

[ 2*(s^2+2)/(s^4+7*s^2+6)/(s+1)]
[     4/(s^4+7*s^2+6)/(s+1)]


x =

[ -1/5*cos(t)+1/5*sin(t)-8/35*cos(6^(1/2)*t)+4/105*6^(1/2)*sin(6^(1/2)*t)+3/7*exp(-t)]
[ -2/5*cos(t)+2/5*sin(t)+4/35*cos(6^(1/2)*t)-2/105*6^(1/2)*sin(6^(1/2)*t)+2/7*exp(-t)]


x1 =

[ -.200*cos(t)+.200*sin(t)-.229*cos(2.45*t)+.933e-1*sin(2.45*t)+.429*exp(-1.*t)]
[ -.400*cos(t)+.400*sin(t)+.114*cos(2.45*t)-.466e-1*sin(2.45*t)+.286*exp(-1.*t)]

>>
```

(a)

**FIG. 5.12** (a) The screen output from the execution of Example5_31.m (the results are identical to those obtained in Example 5.29); (b) the plots of system response.

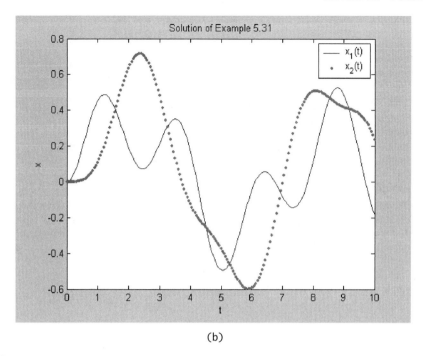

(b)

**FIG. 5.12** (*Continued*)

## 5.6 FURTHER EXAMPLES

The examples in this section illustrate the use of the Laplace transform method in the solution of differential equations or integrodifferential equations obtained through the mathematical modeling of physical systems. The equations presented have all been derived in examples in Chapters 3, 4, and 5. The previous examples in this chapter have paid attention to detail. Partial fraction decompositions have been developed in detail. The use of Table 5.1 has been limited to transform pairs 1–10.

The steps followed in these examples have been outlined in Section 5.5. The procedure is repetitive. The examples in the preceding sections provided familiarity and practice with the procedure so that now several steps may be performed simultaneously. Partial fraction decompositions are presented without the detail of previous sections, sometimes using Table 5.3. All entries in Table 5.1 are available for use in determining transforms and inverting transforms.

MATLAB is used to perform computations, numerically and symbolically, and to plot system response.

**Example 5.32**

The differential equation governing the response of a centrifuge due to the unbalanced rotating mass in the centrifuge is derived as Equation (f) of Example 2.26 as

$$(m + m_0)\ddot{y} + c\dot{y} + ky = m_0 e\omega^2 \sin(\omega t) \tag{a}$$

Use the Laplace transform method to determine the response of the centrifuge using the following values

| | |
|---|---|
| Mass of centrifuge | $m = 48$ kg |
| Rotating mass | $m_0 = 2$ kg |
| Eccentricity | $e = 0.1$ m |
| Damping coefficient | $c = 4000$ N·s/m |
| Stiffness | $k = 2 \times 10^6$ N/m |
| Angular velocity | $\omega = 100$ r/s |
| Initial displacement | $y(0) = 0$ m |
| Initial velocity | $\dot{y}(0) = 0$ m/s |

**Solution**

Substitution of the given values into Equation (a) leads to

$$50\ddot{y} + 4{,}000\dot{y} + 2 \times 10^6 y = 2{,}000 \sin(100t) \tag{b}$$

Equation (a) is put in the standard form of Equation (2.111) by dividing by 50

$$\ddot{y} + 80\dot{y} + 40{,}000y = 40 \sin(100t) \tag{c}$$

Define $Y(s) = \mathcal{L}\{y(t)\}$. Taking the Laplace transform of both sides of Equation (c) and using linearity of the transform leads to

$$\mathcal{L}\{\ddot{y}\} + 80\mathcal{L}\{\dot{y}\} + 40{,}000\mathcal{L}\{y\} = 40\mathcal{L}\{\sin(100t)\} \tag{d}$$

Applying the property of transform of derivatives and using transform pair 2 from Table 5.1 in Equation (d) leads to

$$[s^2 Y(s) - sy(0) - \dot{y}(0)] + 80[s Y(s) - y(0)] + 40{,}000 Y(s) = \frac{4{,}000}{s^2 + 10{,}000} \tag{e}$$

Using the initial conditions In Equation (e) and solving for $Y(s)$ yields

$$Y(s) = \frac{4{,}000}{(s^2 + 80s + 40{,}000)(s^2 + 10{,}000)} \tag{f}$$

The quadratic polynomials in the denominator do not have real roots. The partial fraction decomposition of the right-hand side of Equation (f) leads to

$$Y(s) = \frac{3.32 \times 10^{-4}s - 9.79 \times 10^{-2}}{s^2 + 80s + 40{,}000} + \frac{-3.32 \times 10^{-4}s + 1.2 \times 10^{-1}}{s^2 + 10{,}000} \tag{g}$$

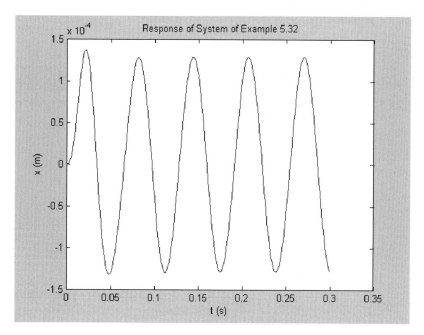

**FIG. 5.13** The MATLAB-generated plot of response of the system of Example 5.32. The initial transient response decays quickly, leaving only a steady-state response.

Equation (g) can be rewritten as

$$Y(s) = \frac{3.32 \times 10^{-4}(s+40)}{s^2 + 80s + 40,000} - \frac{1.11 \times 10^{-1}}{s^2 + 80s + 40,000} - \frac{3.32 \times 10^{-4}s}{s^2 + 10,000} + \frac{1.2 \times 10^{-1}}{s^2 + 10,000}$$

(h)

Transform pairs 12, 11, 3, and 2 from Table 5.1 are used to invert Equation (h) resulting in

$$y(t) = 3.32 \times 10^{-4} e^{-40t} \cos\left(1.96 \times 10^3 t\right) - 5.66 \times 10^{-4} e^{-40t} \sin\left(1.96 \times 10^3 t\right)$$
$$- 3.32 \times 10^{-4} \cos\left(100t\right) + 1.2 \times 10^{-3} \sin\left(100t\right)$$

(i)

The centrifuge is at rest in its equilibrium position when its drum starts rotating. The first two terms in Equation (i) represent the initial transient response of the centrifuge that occurs because of the movement away from equilibrium. These terms die out quickly, leaving only the last two terms as a steady-state response. The response is illustrated in Figure 5.13.

**Example 5.33**

The differential equations governing the motion of the simplified vehicle suspension system derived Equation (f) of Example 2.25 are

$$m\ddot{x} + c\dot{x} - c\dot{w} + k_1 x - k_1 w = 0$$

(a)

$$-c\dot{x} + c\dot{w} - k_1 x + (k_1 + k_2)w = k_2 z$$

(b)

The system parameters have the following numerical values

| | |
|---|---|
| Vehicle mass | $m = 500$ kg |
| Damping coefficient | $c = 17500$ N·s/m |
| Upper stiffness | $k_1 = 2.45 \times 10^6$ N/m |
| Lower stiffness | $k_2 = 1.47 \times 10^6$ N/m |

Determine $x(i)$ under the following conditions:

(a) $z(t) = 0$, $x(0) = 1.6$ mm, $\dot{x}(0) = 0$ m/s, $w(0) = 1$ mm. The solution of this problem is the free response of the system if the vehicle has a sudden displacement of 1 mm at $t = 0$. Note that at $t = 0$ there is no force in the viscous damper and the springs act as two springs in series.

(b) $z(t) = 0.001[u(t) - u(t - 0.5)]$ m and $x(0) = 0$ m, $\dot{x}(0) = 0$ m/s, $w(0) = 0$ m. The solution of this problem is the system response if the wheel has a 1-mm displacement for 0.5 s.

## Solution

Substituting the values of the parameters into Equations (a) and (b) leads to

$$500\ddot{x} + 17{,}500\dot{x} - 17{,}500\dot{w} + 2.45 \times 10^6 x - 2.45 \times 10^6 w = 0 \tag{c}$$

$$-17{,}500\dot{x} + 17{,}500\dot{w} - 2.45 \times 10^6 x + 3.92 \times 10^6 w = 1.47 \times 10^6 z \tag{d}$$

Define $X(s) = \mathcal{L}\{x(t)\}$, $W(s) = \mathcal{L}\{w(t)\}$, and $Z(s) = \mathcal{L}\{z(t)\}$. Taking the Laplace transform of Equations (c) and (d) and using the property of linearity of the transform leads to

$$500\mathcal{L}\{\ddot{x}\} + 17{,}500\mathcal{L}\{\dot{x}\} - 17{,}500\mathcal{L}\{\dot{w}\} + 2.45 \times 10^6 \mathcal{L}\{x\} - 2.45 \times 10^6 \mathcal{L}\{w\} = 0 \tag{e}$$

$$-17{,}500\mathcal{L}\{\dot{x}\} + 17{,}500\,\mathcal{L}\{\dot{w}\} - 2.45 \times 10^6 \mathcal{L}\{x\} + 3.92 \times 10^6 \mathcal{L}\{w\} = 1.47 \times 10^6 \mathcal{L}\{z\} \tag{f}$$

Applying the property of transform of derivatives to Equations (e) and (f) leads to

$$500\left[s^2 X(s) - sx(0) - \dot{x}(0)\right] + 17{,}500[sX(s) - x(0)] - 17{,}500[sW(s) - w(0)]$$
$$+ 2.45 \times 10^6 X(s) - 2.45 \times 10^6 W(s) = 0 \tag{g}$$

$$-17{,}500[sX(s) - x(0)] + 17{,}500[sW(s) - w(0)] - 2.45 \times 10^6 X(s)$$
$$+ 3.92 \times 10^6 W(s) = 1.47 \times 10^6 Z(s) \tag{h}$$

(a) Setting $z(t) = 0$ and applying the initial conditions to Equations (g) and (h) leads to

$$500\left[s^2 X(s) - 0.0016s\right] + 17{,}500[sX(s) - 0.0016] - 17{,}500[sW(s) - 0.001]$$
$$+ 2.45 \times 10^6 X(s) - 2.45 \times 10^6 W(s) = 0 \tag{i}$$

$$-17{,}500[sX(s) - 0.0016] + 17{,}500[sW(s) - 0.001] - 2.45 \times 10^6 X(s)$$
$$+ 3.92 \times 10^6 W(s) = 0 \tag{j}$$

Equations (i) and (j) are written in matrix form as

$$\begin{bmatrix} 500s^2 + 17{,}500s + 2.45 \times 10^6 & -17{,}500s - 2.45 \times 10^6 \\ -17{,}500s - 2.45 \times 10^6 & 17{,}500s + 3.92 \times 10^6 \end{bmatrix} \begin{bmatrix} X(s) \\ W(s) \end{bmatrix} = \begin{bmatrix} 0.8s + 10.5 \\ -10.5 \end{bmatrix}$$

(k)

The MATLAB file Example5._33.m used to finish this problem is illustrated in Figure 5.14. The output from the execution of Example5_33.m is given in Figure 5.15.

(b) Transform pair 8 of Table 5.1 is used to obtain

$$Z(s) = 0.001 \left[ \frac{1 - e^{-0.5s}}{s} \right]$$

(l)

```
% Example 5.33
%
% Defining symbolic variables
syms s t
% Defining coefficient matrix
A=[500*s^2+17500*s+2.45E6 -17500*s-2.45E6;-17500*s-2.457E6 17500*s+3.92E6];
% Part (a)
ba=[0.8*s+10.5;-10.5];
xbar=A^-1*ba
xa=ilaplace(xbar(1));
xa1=vpa(simplify(xa),3);
disp('The response of the system of Example 5.33(a) is')
xa2=vpa(simplify(xa1),3);
xa2
% Part(b)
bb=[0; 1.47E3/s*(1-exp(-0.5*s))]
xbarb=A^-1*bb
xb=ilaplace(xbarb(1));
xb1=vpa(simplify(xb),3);
disp('The response of the system of Example 5.33(b) is')
xb2=vpa(simplify(xb1),3);
xb2
% Plotting resposnes
t=0:.01:1;
ya=subs(vpa(xa,3),t);
yb=subs(vpa(xb,3),t);
plot(t,ya,'-',t,yb,'--')
xlabel('t (s)')
ylabel('x (m)')
legend('x_a(t)','x_b(t)')
%
% End of Example 5.33
```

**FIG. 5.14** The script of Example5_33.m.

```
>> Example5_33
xbar =

[          1/500* (s+224)/ (s^3+224*s^2+2926*s+409640)*(4/5*s+21/2)-21/1000*(s+140)/(s^3+2
24*s^2+2926*s+409640)]
[ 1/2500* (5*s+702)/(s^3+224*s^2+2926*s+409640)*(4/5*s+21/2)-3/5000*(s^2+35*s+4900)/(s^3+2
24*s^2+2926*s+409640)]
```

The response of the system of Example 5.33 (a) is

```
xa2 =

.150e-5*exp(-219.*t)+.160e-2*exp(-2.41*t)*cos(43.2*t)+.970e-4*exp(-2.41*t)*sin(43.2*t)

bb =

[               0]
[ 1470/s*(1-exp(-1/2*s))]

xbarb =
[       147/50*(s+140)/(s^3+224*s^2+2926*s+409640)/s*(1-exp(-1/2*s))]
[ 21/250*(s^2+35*s+4900)/(s^3+224*s^2+2926*s+40964 0)/s*(1-exp(-1/2*s))]
```

The response of the system of Example 5.33(b) is
```
xb2 =

-.218e-4*Heaviside(t-.500)*exp(-219.*t+110.)+.102e-2*Heaviside(t-.500)*exp(-2.41*t+1.21)*
cos(43.2*t-21.6)-.532e-4*Heaviside(t-.500)*exp(-2.41*t+1.21)*sin(43.2*t-21.6)-.100e-2*Hea
viside(t-.500)+.218e-4*exp(-219.*t)-.102e-2*exp(-2.41*t)*cos(43.2*t)+.530e-4*exp(-2.41*t)
*sin(43.2*t)+.100e-2
>>
```

(a)

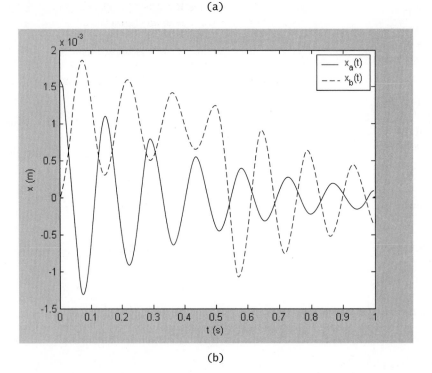

(b)

**FIG. 5.15** (a) Output from the execution of Example5_33.m; (b) the system response for parts (a) and (b).

Substituting Equation (l) into Equations (g) and (h), noting that all initial conditions are zero and writing the resulting equations in matrix form, leads to

$$
\begin{bmatrix} 500s^2 + 17{,}500s + 3.92 \times 10^6 & -17{,}500s - 1.47 \times 10^6 \\ -17{,}500s - 1.47 \times 10^6 & 17{,}500s + 1.47 \times 10^6 \end{bmatrix} \begin{bmatrix} X(s) \\ W(s) \end{bmatrix}
$$

$$
= \begin{bmatrix} 0 \\ 1.47 \times 10^3 \left( \dfrac{1 - e^{-0.5s}}{s} \right) \end{bmatrix} \tag{m}
$$

The remainder of the solution is determined using MATLAB and illustrated in Figure 5.15.

---

## Example 5.34

The differential equation governing the current through a series $RL$ circuit is derived in Example 3.8 as

$$
L\frac{di}{dt} + Ri = v(t) \tag{a}
$$

The parameters for a specific circuit have the following values

$$
\begin{aligned}
\text{Inductance} &\qquad L = 0.2\ \text{H} \\
\text{Resistance} &\qquad R = 100\ \Omega
\end{aligned}
$$

No current is in the circuit at $t = 0$. Determine the current through the circuit and the voltage across the conductor if
(a) $v(t) = W\delta(t)$, where $W = 200$ V·s. The solution to this problem provides the response of the circuit due to an impulsive voltage source.
(b) $v(t) = V_0 u(t)$, where $V_0 = 10$ V. The solution of this problem provides the response of the circuit to a 10-V dc source connected at $t = 0$.
(c) $v(t) = V_0 \sin \omega t$, where $V_0 = 10$ V and $\omega = 100$ rad/s. The solution to this problem provides the response of the $RL$ circuit to an alternating voltage source.

### Solution

Taking the Laplace transform of Equation (a), using linearity and the application of the initial condition $i(0) = 0$, leads to

$$
LsI(s) + RI(s) = V(s) \tag{b}
$$

Equation (b) is solved for $\bar{I}(s)$ leading to

$$
I(s) = \frac{V(s)}{L\left(s + \dfrac{R}{L}\right)} \tag{c}
$$

(a) For $v(t) = W\delta(t)$, $V(s) = W$. Thus Equation (a) becomes

$$
I(s) = \frac{W}{L\left(s + \dfrac{R}{L}\right)}
$$

$$
= \frac{200}{0.2\left(s + \dfrac{100}{0.2}\right)}
$$

$$
= \frac{1{,}000}{s + 500} \tag{d}
$$

The Laplace transform of Equation (d) is inverted leading to

$$i(t) = 1,000e^{-500t} \text{ A} \tag{e}$$

(b) For $v(t) = V_0u(t)$, $V(s) = V_0/s$. Thus Equation (a) becomes

$$I(s) = \frac{V_0}{Ls\left(s + \dfrac{R}{L}\right)}$$

$$= \frac{10}{0.2s\left(s + \dfrac{R}{L}\right)}$$

$$= \frac{50}{s(s + 500)} \tag{f}$$

A partial fraction decomposition of Equation (f) leads to

$$I(s) = \frac{0.1}{s} - \frac{0.1}{s + 500} \tag{g}$$

Inversion of the Laplace transform of Equation (g) leads to

$$i(t) = 0.1u(t) - 0.1e^{-500t} \tag{h}$$

(c) For $v(t) = V_0 \sin(\omega t)$, $V(s) = \frac{V_0\omega}{s^2 + \omega^2}$. Thus Equation (a) becomes

$$I(s) = \frac{V_0\omega}{L(s^2 + \omega^2)\left(s + \dfrac{R}{L}\right)}$$

$$= \frac{10(100)}{0.2(s^2 + 10,000)(s + 500)}$$

$$= \frac{5,000}{(s^2 + 10,000)(s + 500)} \tag{i}$$

A partial fraction decomposition of Equation (i) using entry (5) of Table 5.3 leads to

$$I(s) = \frac{1}{7}\left(\frac{1}{s + 500} - \frac{s - 500}{s^2 + 10,000}\right) \tag{j}$$

Inversion of the transform in Equation (j) leads to

$$i(t) = \tfrac{1}{7}\left[e^{-500t} - \cos(100t) + 5\sin(100t)\right] \text{ A} \tag{k}$$

## Example 5.35

The intergo differential equations governing the current flow through the three-loop circuit of Example 3.11 are

$$L\frac{di_1}{dt} + Ri_1 + \frac{1}{C}\int_0^t (i_1(t) - i_2(t))dt = v(t) \tag{a}$$

$$L\frac{di_2}{dt} + R(i_2 - i_3) - \frac{1}{C}\int_0^t (i_1 - i_2)dt = 0 \tag{b}$$

$$L\frac{di_3}{dt} - R(i_2 - i_3) + \frac{1}{C}\int_0^t i_3 dt = 0 \tag{c}$$

A specific circuit has components with the following values

| | |
|---|---|
| Inductance | $L = 0.5$ H |
| Resistance | $R = 3$ k$\Omega$ |
| Capacitatnce | $C = 2.5$ µF |

Determine $I_1(s), I_2(s)$, and $I_3(s)$ if $v(t) = 100(1 - e^{-.05t})$ V and there is no current in the circuit at $t = 0$. Determine $i_3(t)$.

## Solution

Taking the Laplace transform of Equations (a)–(c), using linearity of the transform and the properties of transform of derivatives and transform of integrals, and setting the initial conditions to zero leads to

$$LsI_1(s) + RI_1(s) + \frac{1}{Cs}[I_1(s) - I_2(s)] = V(s) \tag{d}$$

$$LsI_2(s) + R[I_2(s) - I_3(s)] - \frac{1}{Cs}[I_1(s) - I_2(s)] = 0 \tag{e}$$

$$LsI_3(s) - R[I_2(s) - I_3(s)] + \frac{1}{Cs}I_3(s) = 0 \tag{f}$$

Substitution of parameter values and $V(s)$ obtained using transform pairs 1 and 8 from Table 5.1 into Equations (d)–(f), multiplying by $s$, and collecting terms leads to

$$(0.5s^2 + 3,000s + 400,000)I_1(s) - 400,000I_2(s) = 100s\left(\frac{1}{s} - \frac{1}{s+0.05}\right) \tag{g}$$

$$-400,000I_1(s) + (0.5s^2 + 3,000s + 400,000)I_2(s) - 3,000sI_3(s) = 0 \tag{h}$$

$$-3,000sI_2(s) + (0.5s^2 + 3,000s + 400,000)I_3(s) = 0 \tag{i}$$

Equations (g)–(i) are summarized in matrix form as

$$\begin{bmatrix} 0.5s^2 + 3,000s + 400,000 & -400,000 & 0 \\ -400,000 & 0.5s^2 + 3,000s + 400,000 & -3,000s \\ 0 & -3,000s & 0.5s^2 + 3,000s + 400,000 \end{bmatrix}$$

$$\times \begin{bmatrix} I_1(s) \\ I_2(s) \\ I_3(s) \end{bmatrix} = \begin{bmatrix} \dfrac{5}{s(s+0.05)} \\ 0 \\ 0 \end{bmatrix} \tag{j}$$

Cramer's rule is used to solve for $I_1(s)$ as

$$I_1(s) = \frac{\begin{vmatrix} \dfrac{5}{s(s+0.05)} & -400,000 & 0 \\ 0 & 0.5s^2 + 3,000s + 400,000 & -3,000s \\ 0 & -3,000s & 0.5s^2 + 3,000s + 400,000 \end{vmatrix}}{\begin{vmatrix} 0.5s^2 + 3,000s + 400,000 & -400,000 & 0 \\ -400,000 & 0.5s^2 + 3,000s + 400,000 & -3,000s \\ 0 & -3,000s & 0.5s^2 + 3,000s + 400,000 \end{vmatrix}}$$

$$= \frac{5(s^4 + 1.2 \times 10^4 s^3 + 1.6 \times 10^6 s^2 + 9.6 \times 10^9 s + 6.4 \times 10^{11})}{s^2(s + 0.05)(s^3 + 1.2 \times 10^4 s^2 + 1.6 \times 10^6 s + 9.6 \times 10^9)} \tag{k}$$

$I_1(s)$ and $I_2(s)$ are obtained in a similar fashion as

$$I_2(s) = \frac{4 \times 10^6 (s^2 + 6 \times 10^3 s + 8 \times 10^5)}{s^2 (s + 0.05)(s^3 + 1.2 \times 10^4 s^2 + 1.6 \times 10^6 s + 9.6 \times 10^9)} \tag{1}$$

$$I_3(s) = \frac{2.4 \times 10^{10}}{s(s + 0.05)(s^3 + 1.2 \times 10^4 s^2 + 1.6 \times 10^6 s + 9.6 \times 10^9)} \tag{m}$$

The cubic term in the denominator of Equations (k)–(l) can be factored as $(s = 1.19 \times 10^4)(s^2 + 7.08s + 868.6)$. Thus the partial fraction decomposition for $I_3(s)$ is

$$I_3(s) = \frac{2.5}{s} - \frac{50.0}{s + 0.05} + \frac{169.5}{s + 1.19 \times 10^4} + \frac{149.6s + 5.36 \times 10^3}{s^2 + 7.08s + 868.6} \tag{n}$$

The inversion of Equation (n) leads to

$$i_3(t) = 2.5u(t) - 50.0e^{-.05t} + 169.5e^{-1.19 \times 10^4 t} + 149.6e^{-3.54t} \cos(29.3t)$$
$$+ 164.8e^{-3.54t} \sin(29.3t) \tag{o}$$

## Example 5.36

The voltage source in an $LC$ circuit is the square waveform of Figure 5.16. The specific circuit has components with the following parameters

| | |
|---|---|
| Inductance | $L = 0.64$ H |
| Capacitance | $C = 100$ µF |
| Maximum voltage | $V_0 = 6.4$ V |
| Period | $T = 0.06$ s |
| Initial current | $i(0) = 0$ A |

The equation governing the current in the circuit is

$$L\frac{di}{dt} + \frac{1}{C}\int_0^t i\,dt = v(t) \tag{a}$$

Determine and plot $i(t)$.

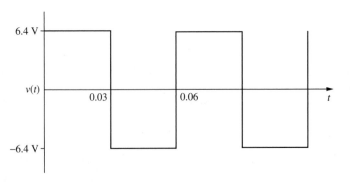

**FIG. 5.16** The square wave form of Example 5.36.

## Solution

Define $I(s) = \{i(t)\}$ and $V(s) = \{v(t)\}$. Taking the Laplace transform of Equation (a) and using linearity of the transform leads to

$$L\left\{\frac{di}{dt}\right\} + \frac{1}{C}\left\{\int_0^t i\,dt\right\} = V(s) \tag{b}$$

$V(s)$ is determined using the property of transform of periodic functions

$$V(s) = \frac{1}{1 - e^{-Ts}} \int_0^T V(t)^{e^{-st}}dt$$

$$= \frac{1}{1 - e^{-Ts}}\left[\int_0^{T/2} V_0 e^{-st}dt + \int_{T/2}^T (-V_0)e^{-st}dt\right]$$

$$= \frac{1}{1 - e^{-Ts}}\left[-\frac{V_0}{s}e^{-st}\Big|_0^{T/2} + \frac{V_0}{s}e^{-st}\Big|_{T/2}^T\right]$$

$$= \frac{V_0\left(1 - 2e^{-(T/2)s} + e^{-Ts}\right)}{s(1 - e^{-Ts})}$$

$$= \frac{V_0\left(1 - e^{-(T/2)s}\right)^2}{s\left[1 - \left(e^{-(T/2)s}\right)^2\right]}$$

$$= \frac{V_0\left(1 - e^{-(T/2)s}\right)}{s(1 + e^{-(T/2)s})} \tag{c}$$

Using Equation (c) and the properties of transform of derivatives and transform of integrals in Equation (b) and applying the initial condition leads to

$$LsI(s) + \frac{1}{Cs}I(s) = \frac{V_0\left(1 - e^{-(T/2)s}\right)}{s(1 + e^{-(T/2)s})} \tag{d}$$

Solving Equation (d) for $I(s)$ leads to

$$I(s) = \frac{V_0\left(1 - e^{-(T/2)s}\right)}{L\left(s^2 + \dfrac{1}{LC}\right)(1 + e^{-(T/2)s})} \tag{e}$$

Application of the binomial expansion

$$\left(1 + e^{-(T/2)s}\right)^{-1} = \sum_{k=0}^{\infty}(-1)^k e^{-k(T/2)s} \tag{f}$$

in Equation (e) leads to

$$I(s) = \frac{V_0}{L\left(s^2 + \dfrac{1}{LC}\right)}\left[1 - 2\sum_{k=1}^{\infty}(-1)^k e^{-k(T/2)s}\right] \tag{g}$$

The inversion of Equation (g) using the second shifting theorem and transform pair 2 leads to

$$i(t) = V_0\sqrt{\frac{C}{L}}\left[\sin\left(\sqrt{\frac{1}{LC}}t\right) - 2\sum_{k=1}^{\infty}(-1)^k\sin\left(\sqrt{\frac{1}{LC}}\frac{t-kT}{2}\right)u\frac{t-kT}{2}\right] \quad (h)$$

Substitution of given values leads to

$$i(t) = 0.08\left[\sin(80t) - 2\sum_{k=1}^{\infty}(-1)^k\sin(80(t-0.03k))u(t-0.03k)\right] \quad (i)$$

The MATLAB file Example5_36.m, written to develop a plot of Equation (i), is shown in Figure 5.17. The plot generated from its execution is shown in Figure 5.18. Since the Heaviside function is defined for symbolic computation only, it is difficult to use in developing a plot. However, for any $t$, the upper limit on the summation for which terms are nonzero is finite. The floor function, which finds the largest integer smaller than the argument, is used to determine this upper limit for each $t$, thus avoiding the need for evaluation of the Heaviside function.

```
% Example 5.36
%
% Set up counting loop for time
i=0
for i=1:700
  i=i+1;
  t(i)=.001*i;
% Evaluation of Equation (i) of Example 5.36
  Cur(i)=sin(80*t(i));
% The use of the floor function avoids use of unit step functions
  j=floor(t(i)/.03)
  if j>0
  for k=1:j
    Cur(i)=Cur(i)-2*(-1)^k*sin(80*(t(i)-0.03*k));
  end
end
  Cur(i)=Cur(i)*0.08;
end
plot(t,Cur)
xlabel('t (s)')
ylabel('i (A)')
title('Response of LC circuit due to periodic square wave')
%
% End of Example535.m
```

**FIG. 5.17** The script for Example5_36.m.

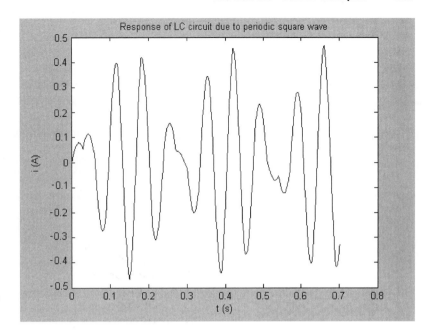

**FIG. 5.18** The MATLAB-generated plot for the response of system of Example 5.36.

## Example 5.37

The differential equations governing the perturbations in liquid levels in the two-tank system of Example 4.8 are

$$A_1 \frac{dh_1}{dt} + \frac{1}{R_1} h_1 - \frac{1}{R_1} h_2 = q_1(t) \tag{a}$$

$$A_2 \frac{dh_2}{dt} - \frac{1}{R_1} h_1 + \left( \frac{1}{R_1} + \frac{1}{R_2} \right) h_2 = q_2(t) \tag{b}$$

Determine the response for a system with $A_1 = 900$ m$^2$, $A_2 = 1{,}500$ m$^2$, $R_1 = 4.0$ s/m$^2$, $R_2 = 5.0$ s/m$^2$, when the flow rate perturbations are

$$q_1(t) = -1.0u(t) \text{ m}^3/\text{s} \tag{c}$$
$$q_2(t) = 1.0u(t - 5) \text{ m}^3/\text{s} \tag{d}$$

### Solution

Substitution of the given values and of Equations (c) and (d) into Equations (a) and (b) leads to

$$18{,}000 \frac{dh_1}{dt} + 5h_1 - 5h_2 = -20u(t) \tag{e}$$

$$30{,}000 \frac{dh_2}{dt} - 5h_1 + 9h_2 = 20u(t - 5) \tag{f}$$

Taking the Laplace transform of Equations (e) and (f), noting that $dh_1/dt(0) = dh_2/dt(0) = 0$ m/s, and defining $H_1(s) = \mathcal{L}\{h_1(t)\}$ and $H_2(s) = \mathcal{L}\{h_2(t)\}$ leads to

$$18{,}000sH_1(s) + 5H_1(s) - 5H_2(s) = -\frac{20}{s} \tag{g}$$

$$30{,}000sH_2(s) - 5H_1(s) + 9H_2(s) = \frac{20e^{-5s}}{s} \tag{h}$$

Equations (g) and (h) are summarized in matrix form as

$$\begin{bmatrix} 18{,}000s + 5 & -5 \\ -5 & 30{,}000s + 9 \end{bmatrix} \begin{bmatrix} H_1(s) \\ H_2(s) \end{bmatrix} = \frac{20}{s} \begin{bmatrix} -1 \\ e^{-5s} \end{bmatrix} \tag{i}$$

Cramer's rule is applied to Equation (i) leading to

$$H_1(s) = \frac{\begin{vmatrix} \dfrac{20}{s} & -1 & -5 \\ & e^{-5s} & 30{,}000s + 9 \end{vmatrix}}{\begin{vmatrix} 18{,}000s + 5 & -5 \\ -5 & 30{,}000s + 9 \end{vmatrix}}$$

$$= \frac{\dfrac{20}{s}\left(5e^{-5s} - 30{,}000s - 9\right)}{5.4 \times 10^8 s^2 + 3.12 \times 10^5 s + 20} \tag{j}$$

and

$$H_2(s) = \frac{\dfrac{20}{s}\left[(18{,}000s + 5)e^{-5s} - 5\right]}{5.4 \times 10^8 s^2 + 3.12 \times 10^5 s + 20} \tag{k}$$

Equations (j) and (k) are rewritten as

$$H_1(s) = 3.70 \times 10^{-8} \frac{5e^{-5s} - 30{,}000s - 9}{s(s + 5.04 \times 10^{-4})(s + 7.34 \times 10^{-5})} \tag{l}$$

$$H_2(s) = 3.70 \times 10^{-8} \frac{(18{,}000s + 5)e^{-5s} - 5}{s(s + 5.04 \times 10^{-4})(s + 7.34 \times 10^{-5})} \tag{m}$$

Partial fraction decompositions lead to

$$H_1(s) = \frac{-9.0}{s} + \frac{1.04}{s + 5.04 \times 10^{-4}} + \frac{7.96}{s + 7.34 \times 10^{-5}}$$

$$+ e^{-5s}\left(\frac{5.0}{s} + \frac{0.852}{s + 5.04 \times 10^{-4}} - \frac{5.85}{s + 7.34 \times 10^{-5}}\right) \tag{n}$$

$$H_2(s) = \frac{-5.0}{s} - \frac{0.852}{s + 5.04 \times 10^{-4}} + \frac{5.85}{s + 7.34 \times 10^{-5}}$$

$$+ e^{-5s}\left(\frac{5.0}{s} - \frac{0.694}{s + 5.04 \times 10^{-4}} - \frac{4.31}{s + 7.34 \times 10^{-5}}\right) \tag{o}$$

Inversion of the transforms leads to

$$h_1(t) = \left(-9.0 + 1.04e^{-5.04 \times 10^{-4}t} + 7.96e^{-7.34 \times 10^{-5}t}\right)u(t)$$

$$+ \left(5 + 0.852e^{-5.04 \times 10^{-4}(t-5)} - 5.85e^{-7.34 \times 10^{-5}(t-5)}\right)u(t - 5) \tag{p}$$

$$h_2(t) = \left(-5 - 0.852e^{-5.04 \times 10^{-4}t} + 5.85e^{-7.34 \times 10^{-5}t}\right)u(t)$$
$$+ \left(5 - 0.694e^{5.04 \times 10^{-4}(t-5)} - 4.31e^{-7.34 \times 10^{-5}(t-5)}\right)u(t-5) \qquad \text{(q)}$$

## 5.7 SUMMARY

### 5.7.1 Mathematical Solutions for Response of Dynamic Systems

The Laplace transform method is applied to solve the differential and integrodifferential equations obtained in Chapters 2–4 through the mathematical modeling of dynamic systems. Important points regarding Laplace transforms and their application to the solution of differential equations are as follows:

- The Laplace transform of a function of time $t$ is a function of a complex variable $s$. It is defined in terms of an integral over time, but with the integrand dependent on $s$. The Laplace transform is unique and exists for any function that can be physically generated.
- The inverse transform of a function of the complex variable $s$ is a function of time. The inverse transform is defined by an integral in the complex plane with $t$ as a parameter. The inverse transform is unique.
- If $F(s) = \mathcal{L}\{f(t)\}$ then $f(t) = \mathcal{L}^{-1}\{F(s)\}$ and $f(t)$ and $F(s)$ form a transform pair. Table 5.1 presents a useful list of transform pairs.
- Laplace transforms may be determined directly from its definition.
- The properties of Laplace transforms are derived that aid in the determination of transform pairs. Table 5.2 presents a list of useful properties.
- Inverse transforms are determined using Tables 5.1 and 5.2.
- Partial fraction decompositions are often required to write the inverse transform in a form in which the tables can be used.
- The properties of linearity of the transform and transform of derivatives allow the Laplace transform method to be successful in obtaining solutions of linear differential equations with constant coefficients.
- When taking the Laplace transform of both sides of a differential equation, the application of the transform of derivatives property leads to an algebraic equation that is solved for the transform of the dependent variable. The transform is inverted to obtain the solution of the differential equation.
- The initial conditions are applied during the application of the transform of derivatives property.
- When applying the Laplace transform method to integrodifferential equations, the property of transform of integrals is used.
- Taking the Laplace transform of a coupled set of differential equations results in a set of simultaneous algebraic equations to solve for the transforms of the dependent variables.

- The Symbolic Toolbox of MATLAB can be used to determine symbolically Laplace transforms and inverse Laplace transforms. It is also useful for the symbolic solutions of sets of simultaneous algebraic equations that arise when the method is applied to a coupled set of differential equations. The symbolic equations may be evaluated at discrete values of time to allow plotting the response.

### 5.7.2 Important Equations

The important equations for Chapter 5 are given in Table 5.1, a table of transform pairs, and in Table 5.2, a table of transform properties.

## PROBLEMS

**5.1** Use the definition of the transform to determine $\mathcal{L}\{\sinh(\omega t)\}$.

**5.2** Use the definition of the transform to determine $\mathcal{L}\{t\sin(\omega t)\}$.

**5.3** Evaluate $\mathcal{L}\{t^2 e^{3t}\}$.

**5.4** Evaluate $\mathcal{L}\{\cos(2t)u(t-1)\}$.

**5.5** Evaluate $\mathcal{L}\{te^{-3t}[u(t) - u(t-2)]\}$.

**5.6** Evaluate $\mathcal{L}\{2t\cos(4t)\}$.

**5.7** Determine $F(s)$ for $f(t)$, given in Figure P5.7.

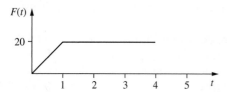

**FIG. P5.7**

**5.8** Determine $F(s)$ for $f(t)$ given in Figure P5.8.

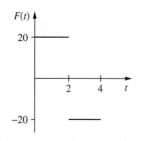

**FIG. P5.8**

**5.9** Determine $F(s)$ if $f(t) = 2e^{-2t}[u(t) - u(t-3)]$.

**5.10** Determine $F(s)$ if $f(t) = \int_0^t \tau e^{-(t-\tau)} \sin[2(t-\tau)]d\tau$.

**5.11** Noting that $\mathcal{L}\{t^{-1/2}\} = \sqrt{\pi/s}$, evaluate (a) $\mathcal{L}\{t^{1/2}\}$ using the property of differentiation of the transform, (b) $\{t^{3/2}\}$, and (c) $\mathcal{L}\{t^{1/2}e^{-3t}\}$.

**5.12** The Laplace transform of the response of a dynamic system is

$$X(s) = \frac{s^2 + 5}{s(s+3)(s+4)(s^2+15)}$$

Determine $\lim_{t\to\infty} x(t)$.

**5.13** The Laplace transform of the response of a dynamic system is

$$X(s) = \frac{s+3}{4s^2 + 6s + 17}$$

What is $x(0)$?

**5.14** The Bessel function of order 0 of $t$, denoted by $J_0(t)$, is a special function that has many engineering applications. It can be shown that $\mathcal{L}\{J_0(t)\} = 1/\sqrt{s^2+1}$. (a) Evaluate $\mathcal{L}\{tJ_0(t)\}$. (b) Evaluate $\mathcal{L}\{J_0(2t)\}$. (c) It can be shown that the Bessel function of order 1 is related to the Bessel function of order 1 by $J_1(t) = -\frac{d}{dt}(J_0(t))$. Given that $J_0(0) = 1$, evaluate $\mathcal{L}\{J_1(t)\}$.

**5.15** Show that

$$\mathcal{L}\{\sin(\omega t)\cosh(\omega t) - \cos(\omega t)\sinh(\omega t)\} = \frac{4\omega^3}{s^2 + 4\omega^2}$$

**5.16** Show that

$$\mathcal{L}\left\{\frac{2}{t}[1 - \cos(\omega t)]\right\} = \ln\left(\frac{s^2 + \omega^2}{s^2}\right)$$

**5.17** Show that

$$\mathcal{L}\left\{\frac{1}{t}\sin{(\omega t)}\right\} = \tan^{-1}\left(\frac{\omega}{s}\right)$$

**5.18** Prove the change of scale property of Laplace transforms.

**5.19** Prove the initial value theorem.

**5.20** Prove the property of differentiation of the transform.

**5.21** Determine the transform of the periodic function of Figure P5.21.

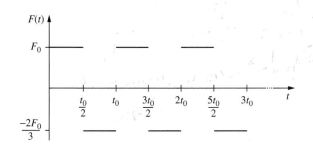

**FIG. P5.21**

**5.22** Determine $x(t)$ if

$$X(s) = \frac{3s+1}{s^2+9}$$

**5.23** Determine $x(t)$ if

$$X(s) = \frac{3s+1}{s(s^2+9)}$$

**5.24** Determine $x(t)$ if

$$X(s) = \frac{3s+1}{s^2-9}$$

**5.25** Determine $x(t)$ if

$$X(s) = \frac{3s+1}{s(s^2-9)}$$

**5.26** Determine $x(t)$ if

$$X(s) = \frac{3s+1}{s(s^2+9)}\left(1-e^{-2s}\right)$$

**5.27** Determine $x(t)$ if

$$X(s) = \frac{s^2+2}{(s+3)(s+1)(s^2+16)}$$

**5.28** Determine $x(t)$ if

$$X(s) = \frac{2s+3}{s^2+6s+13}$$

**5.29** Determine $x(t)$ if

$$X(s) = \frac{5s^2+4}{s(s^2+2s+26)}$$

**5.30** Determine $x(t)$ if

$$X(s) = \frac{2s+7}{(s+2)(s+4)^2}$$

**5.31** Determine $x(t)$ if

$$X(s) = \frac{1}{(s^2+2s+4)(s+1)^2}$$

**5.32** Use the MATLAB "residue" command to construct partial fraction decompositions for $X(s)$, as given in (a) Problem 5.23, (b) Problem 5.25, (c) Problem 5.29, and (d) Problem 5.30.

**5.33** Use the Laplace transform method to solve the differential equation

$$\dot{x} + 3x = 2e^{-3t}$$

subject to $x(0) = 0$.

**5.34** Use the Laplace transform method to solve the differential equation

$$\ddot{x} + 2\dot{x} + 17x = 0$$

subject to the initial conditions $x(0) = 0$ and $\dot{x}(0) = 1$.

**5.35** Use the Laplace transform method to solve the differential equation

$$\ddot{x} + 25x = 3\sin{(5t)}$$

subject to the initial conditions $x(0) = 0$ and $\dot{x}(0) = 0$.

**5.36** Use the Laplace transform method to solve the differential equation

$$\ddot{x} + 49x = 7e^{-2t}$$

subject to the initial conditions $x(0) = 0$ and $\dot{x}(0) = 0$.

**5.37** Use the Laplace transform method to solve the differential equation

$$\dddot{x} + 2\ddot{x} + 25\dot{x} + 50x = 0$$

subject to the initial conditions $x(0) = 1$, $\dot{x}(0) = 0$, and $\ddot{x}(0) = 0.5$.

**5.38** Use the Laplace transform method to solve the differential equations

$$\dot{x}_1 + 2x_1 - x_2 = 2e^{-t}$$
$$\dot{x}_2 - x_1 + 3x_2 = 0$$

subject to the initial conditions $x_1(0) = 0$ and $x_2(0) = 0$.

Problems 5.39–5.41 refer to the second-order mechanical system of Figure P5.39 whose mathematical model is

$$10\ddot{x} + 20\dot{x} + 250x = F(t) \qquad (a)$$

**5.39** Determine the free response of the system $(F = 0)$ when $x(0) = 1$ mm and $\dot{x}(0) = 0$.

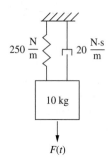

**FIG. P5.39**

**5.40** Determine the response of the system when $F(t) = 200 \sin(4t)$ N with $x(0) = 0$ and $\dot{x}(0) = 0$.

**5.41** Determine the response of the system when $F(t)$ is as given in Figure P5.41 with $x(0) = 0$ and $\dot{x}(0) = 0$.

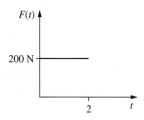

**FIG. P5.41**

Problems 5.42–5.44 refer to the series $LRC$ circuit of Figure P5.42, whose mathematical model is

$$0.1\frac{di}{dt} + 1,000i + 4 \times 10^5 \int_0^t i\,dt = v(t) \qquad (a)$$

**5.42** Determine $i(t)$ when $v(t) = 1,000 \sin(500t)$ V with $i(0) = 0$.

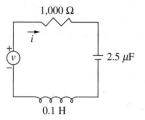

**FIG. P5.42**

**5.43** Determine $i(t)$ when $v(t) = 1,000e^{-.02t}$ V with $i(0) = 0$.

**5.44** Determine $i(t)$ when $v(t)$ is as given in Figure P5.44 with $i(0) = 0$.

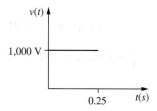

**FIG. P5.44**

For Problems 5.45–5.49 use MATLAB as needed.

**5.45** The integrodifferential equations obtained for the currents in the two-loop circuit of Example 3.28 are

$$Ri_1 + \frac{1}{C}\int_0^t i_1\,dt - \frac{1}{C}\int_0^t i_2\,dt = v(t) \qquad (a)$$

$$-\frac{1}{C}\int_0^t i_1\,dt + L\frac{di_2}{dt} + Ri_2 + \frac{1}{C}\int_0^t i_2\,dt = 0 \qquad (b)$$

Solve Equations (a) and (b) when $v(t) = 500\,u(t)$ V with $i_1(0) = 0$ and $i_2(0) = 0$. The circuit parameters are $R = 500$ Ω, $C = 20$ μF, and $L = 0.2$ H. Plot $i_1(t)$ and $i_2(t)$ on the same axes.

**5.46** The differential equations obtained for the displacements of the blocks in Example 2.22 are

$$m_1\ddot{x}_1 + (k_1 + k_2)x_1 - k_2x_2 = 0 \qquad (a)$$
$$m_2\ddot{x}_2 - k_2x_1 + (k_2 + k_3)x_2 = 0 \qquad (b)$$

Use the Laplace transform method to solve Equations (a) and (b) for $x_1(t)$ and $x_2(t)$ with the initial conditions $x_1(0) = 1$ mm, $\dot{x}_1(0) = 0$, $x_2(0) = 0$, and $\dot{x}_2(0) = 0$ when

$m_1 = 10$ kg, $m_2 = 20$ kg, $k_1 = 10{,}000$ N/m, $k_2 = 15{,}000$ N/m, and $k_3 = 10{,}000$ N/m. Plot $x_1(t)$ and $x_2(t)$ on the same axes.

**5.47** The differential equations derived in Example 4.8 for the perturbations in liquid level in a two-tank system due to perturbations in inlet flow rates are

$$A_1 \frac{dh_1}{dt} + \frac{1}{R_1} h_1 - \frac{1}{R_1} h_2 = q_{i1}(t) \qquad \text{(a)}$$

$$A_2 \frac{dh_2}{dt} - \frac{1}{R_1} h_1 + \left( \frac{1}{R_1} + \frac{1}{R_2} \right) h_2 = q_{i2}(t) \qquad \text{(b)}$$

Determine the liquid-level perturbation $h_1(t)$ when $q_{i1}(t) = 0$, $q_{i2}(t) = -0.5\, u(t)$ m³/s when $A_1 = 100$ m², $A_2 = 150$ m², $R_1 = 10$ s/m², and $R_2 = 8$ s/m².

**5.48** The equations obtained for the perturbation in temperatures of oil and water in the double pipe heat exchanger of Example 4.25 are

$$\rho_o c_o \pi r_i^2 L \frac{d\theta_o}{dt} + (2\pi r_o LU + c_o \dot{m}_o)\theta_o - 2\pi r_o UL\theta_w = c_o \dot{m}_o \theta_{oi}$$

$$\rho_w c_w \pi (r_2^2 - r_o^2) L \frac{d\theta_w}{dt} - 2\pi r_o LU\theta_o + (2\pi r_o LU + c_w \dot{m}_w)\theta_w$$
$$= c_w \dot{m}_w \theta_{wi}$$

Determine the perturbations in temperatures of the oil and water when $\theta_{oi}(t) = 0$, $\theta_{wi}(t) = 3u(t)$, and

$\rho_w = 9.60 \times 10^2$ kg/m³, $\rho_o = 7.50 \times 10^2$ kg/m³,

$c_w = 4.19 \times 10^3$ J/kg·C, $c_o = 2.4 \times 10^3$ J/kg·C,

$U = 8.65 \times 10^2$ W/m²·C, $\dot{m}_w = 2.51$ kg/s, $\dot{m}_o = 4.85$ kg/s,

$r_i = 2$ cm, $r_o = 2.2$ cm, $r_2 = 4.2$ cm, $L = 2.5$ m

Assume $\theta_o(0) = \theta_w(0) = 0$.

**5.49** The differential equations obtained for the concentrations of the reactant and product in the constant volume CSTR of Example 4.16 are

$$V \frac{dC_A}{dt} + (q + kV)C_A = qC_{Ai}$$

$$V \frac{dC_B}{dt} - kVC_A + qC_B = qC_{Bi}$$

Determine the concentrations of the reactant $A$ and the product $B$ when $V = 1.00 \times 10^{-3}$ m³, $q = 2.0 \times 10^{-6}$ m³/s, $k = 2.0 \times 10^{-3}$s⁻¹, $C_{Ai}(t) = 0.2u(t)$ mol/L, and $C_{Bi}(t) = 0$. Plot $C_{Ai}(t)$ and $C_{Bi}(t)$.

# 6

# Transient Analysis and Time Domain Response

A system's **transient response** is its response due to changes in input. A transient response may occur when a nonzero initial condition is imposed on a system. A transient response may also occur due to a perturbation from a steady-state operating condition.

## 6.1 TRANSFER FUNCTIONS

### 6.1.1 Definition and Determination

The **transfer function** for a system is defined as the ratio of the Laplace transform of the output of the system to the Laplace transform of the input of the system for an arbitrary input. If $x(t)$ represents the output of the system and $f(t)$ represents the input to the system then the transfer function $G(s)$ is defined as

$$G(s) = \frac{\mathcal{L}\{x(t)\}}{\mathcal{L}\{f(t)\}} = \frac{X(s)}{F(s)} \tag{6.1}$$

The transfer function is defined assuming all initial conditions are zero. If the system has more than one dependent variable representing its output, then a transfer function may be defined for each output variable. In addition if a system has more than one independent input, then a transfer function may be defined for each output variable corresponding to each input.

The Laplace transform of the output of a linear system is directly proportional to the Laplace transform of the input. Thus in determining the transfer function using Equation (6.1), $F(s)$ appears in the numerator, and the denominator is canceled from the expression. The transfer function is dependent on system parameters and independent of the input. The transfer function contains information about the system dynamics.

**Example 6.1**

The differential equation governing the motion of a single-degree-of-freedom mass-spring-viscous damper system subject to an external force is derived in Example 2.10 as

$$m\ddot{x} + c\dot{x} + kx = f(t) \tag{a}$$

Determine the transfer function of the system defined by

$$G(s) = \frac{X(s)}{F(s)} \tag{b}$$

**Solution**

Taking the Laplace transform of Equation (a) using linearity of the transform and the property of derivative of the transform while taking all initial conditions as zero leads to

$$ms^2 X(s) + csX(s) + kX(s) = F(s) \tag{c}$$

Solving Equation (c) for $X(s)$ leads to

$$X(s) = \frac{F(s)}{ms^2 + cs + k} \tag{d}$$

The transfer function is obtained using Equations (b) and (d) as

$$G(s) = \frac{1}{ms^2 + cs + k} \tag{e}$$

**Example 6.2**

The integrodifferential equation for a series $LRC$ circuit with a voltage source is derived in Example 3.9 as

$$L\frac{di}{dt} + Ri + \frac{1}{C}\int_0^t i\, dt = v(t) \tag{a}$$

Determine the transfer function for the circuit defined by

$$G(s) = \frac{I(s)}{V(s)} \tag{b}$$

**Solution**

Taking the Laplace transform of Equation (a) using linearity of the transform and the properties of transform of derivatives and transform of integrals leads to

$$sLI(s) + RI(s) + \frac{1}{Cs}I(s) = V(s) \tag{c}$$

Solving Equation (c) for $I(s)$ leads to

$$I(s) = \frac{V(s)}{sL + R + \dfrac{1}{Cs}}$$

$$= \frac{sV(s)}{Ls^2 + Rs + \dfrac{1}{C}} \tag{d}$$

The transfer function is obtained from Equation (d) as

$$G(s) = \frac{s}{Ls^2 + Rs + \dfrac{1}{C}} \tag{e}$$

---

## Example 6.3

The differential equation governing the motion of a vehicle with a suspension system modeled as a parallel spring-viscous damper combination in series with another spring is derived in Example 2.25 as

$$\frac{cm}{k_2}\dddot{x} + m\left(1 + \frac{k_1}{k_2}\right)\ddot{x} + c\dot{x} + k_1 x = c\dot{z} + k_1 z \tag{a}$$

where $z(t)$ is the input to the system that specifies the road contour. Determine the transfer function $G(s) = X(s)/Z(s)$ for this system.

### Solution

Taking the Laplace transform of Equation (a) using linearity of the transform and the property of transform of derivatives, assuming all initial conditions have values of zero, leads to

$$\frac{cm}{k_2}s^3 X(s) + m\left(1 + \frac{k_1}{k_2}\right)s^2 X(s) + csX(s) + k_1 X(s) = csZ(s) + k_1 Z(s) \tag{b}$$

Solving Equation (b) for $X(s)$ yields

$$X(s) = \frac{(cs + k_1)Z(s)}{\dfrac{cm}{k_2}s^3 + m\left(1 + \dfrac{k_1}{k_2}\right)s^2 + cs + k_1} \tag{c}$$

The transfer function is determined using Equation (c) as

$$G(s) = \frac{cs + k_1}{\dfrac{cm}{k_2}s^3 + m\left(1 + \dfrac{k_1}{k_2}\right)s^2 + cs + k_1} \tag{d}$$

---

## Example 6.4

The hydraulic servomotor of Figure 6.1, under certain conditions, acts as a proportional plus differential controller. Derive the transfer function $G(s) = X(s)/Y(s)$.

### Solution

For convenience introduce the intermediate variables $z(t)$ as the displacement of the point on the walking beam where the spring and viscous damper

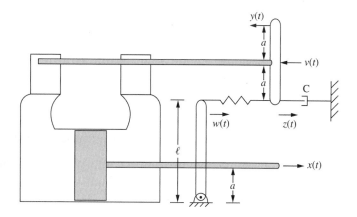

FIG. **6.1**

are attached, let $w(t)$ represent the point on the lever of length $\ell$ where the spring is attached, and let $v(t)$ represent the displacement of the rod connected to the valve. From geometry, using similar triangles.

$$w = \frac{\ell}{a}x \qquad (a)$$

Geometry is also used to obtain

$$v = \frac{y - z}{2} \qquad (b)$$

The flow equation for the servomotor is

$$\hat{C}v = A\dot{x} \qquad (c)$$

where $\hat{C}$ is the flow coefficient of the motor and $A$ is the area of the piston. A force balance on the walking beam leads to

$$k(w - z) = c\dot{z} \qquad (d)$$

Taking the Laplace transform of Equations (a)–(d) leads to

$$W(s) = \frac{\ell}{a}X(s) \qquad (e)$$

$$V(s) = \tfrac{1}{2}Y(s) - \tfrac{1}{2}Z(s) \qquad (f)$$

$$\hat{C}V(s) = AsX(s) \qquad (g)$$

$$k[W(s) - Z(s)] = csZ(s) \qquad (h)$$

Equations (e)–(h) are rearranged to yield

$$G(s) = \frac{1 + \dfrac{cs}{k}}{\dfrac{\ell}{a} + \dfrac{2A}{\hat{C}}s + \dfrac{2cA}{\hat{C}k}s^2} \qquad (i)$$

### 6.1.2 Multiple Inputs and Multiple Outputs

If a system has one input $f(t)$ and multiple outputs $x_1(t), x_2(t), \ldots, x_n(t)$, then a transfer function can be defined for each output

$$G_1(s) = \frac{X_1(s)}{F(s)}, G_2(s) = \frac{X_2(s)}{F(s)}, \ldots, G_n(s) = \frac{X_n(s)}{F(s)} \qquad (6.2)$$

If a system has multiple inputs and multiple outputs then for each output an independent transfer function may be defined for each input. For a system with $m$ inputs $f_1(t), f_2(t), \ldots, f_m(t)$ and $n$ outputs $x_1(t), x_2(t), \ldots, x_n(t)$, an $m \times n$ matrix of transfer functions $\mathbf{K}(s)$ is defined as

$$\mathbf{K}(s) = \begin{bmatrix} \dfrac{X_1(s)}{F_1(s)} & \dfrac{X_1(s)}{F_2(s)} & \cdots & \dfrac{X_1(s)}{F_m(s)} \\ \dfrac{X_2(s)}{F_1(s)} & \dfrac{X_2(s)}{F_2(s)} & \cdots & \dfrac{X_2(s)}{F_m(s)} \\ \dfrac{X_3(s)}{F_1(s)} & \dfrac{X_3(s)}{F_2(s)} & \cdots & \dfrac{X_3(s)}{F_m(s)} \\ \vdots & \vdots & \ddots & \vdots \\ \dfrac{X_n(s)}{F_1(s)} & \dfrac{X_n(s)}{F_2(s)} & \cdots & \dfrac{X_n(s)}{F_m(s)} \end{bmatrix} \qquad (6.3)$$

Equation (6.3) can be summarized by defining the element in the $i$th row and $j$th column of $\mathbf{K}(s)$ as $G_{i,j}(s)$ where

$$G_{i,j}(s) = \frac{X_i(s)}{F_j(s)} \qquad (6.4)$$

In determining a transfer function of the form of Equation (6.4), each input except $f_j(t)$ is taken to be zero.

If the transfer function matrix is determined and the inputs known then the Laplace transforms of the responses are determined using linear superposition as

$$X_i(s) = \sum_{j=1}^{m} G_{i,j}(s) F_j(s) \qquad i = 1, 2, \ldots, n \qquad (6.5)$$

The equations in Equation (6.5) are summarized in matrix form as

$$\mathbf{X}(s) = \mathbf{K}(s)\mathbf{F}(s) \qquad (6.6)$$

where the $i$th element of the $n \times 1$ column vector $\mathbf{X}(s)$ is $X_i(s)$ and the $j$th element of the $m \times 1$ column vector $\mathbf{F}(s)$ is $F_j(s)$.

---

**Example 6.5**

The two-tank liquid-level problem of Examples 4.8 and 4.19 has two independent inputs $q_{i1}(t)$ and $q_{i2}(t)$ and two outputs $h_1(t)$ and $h_2(t)$. The mathematical model for the system is derived as Equations (u) and (v) of Example 4.8, which are rewritten using the numerical values of Example 4.19 as

$$20\frac{dh_1}{dt} + 0.25h_1 - 0.25h_2 = q_{i1}(t) \qquad (a)$$

$$15\frac{dh_2}{dt} - 0.25h_1 + 0.35h_2 = q_{i2}(t) \tag{b}$$

Determine the matrix of transfer functions $\mathbf{K}(s)$ for this system.

## Solution

Taking the Laplace transform of Equations (a) and (b) assuming all initial conditions are zero leads to

$$\begin{bmatrix} 20s + 0.25 & -0.25 \\ -0.25 & 15s + 0.35 \end{bmatrix} \begin{bmatrix} H_1(s) \\ H_2(s) \end{bmatrix} = \begin{bmatrix} Q_{i1}(s) \\ Q_{i2}(s) \end{bmatrix} \tag{c}$$

Cramer's rule is used leading to

$$H_1(s) = \frac{\begin{bmatrix} Q_{i1}(s) & -0.25 \\ Q_{i2}(s) & 15s + 0.35 \end{bmatrix}}{\begin{vmatrix} 20s + 0.25 & -0.25 \\ -0.25 & 15s + 0.35 \end{vmatrix}}$$

$$= \frac{(15s + 0.35)Q_{i1}(s) + 0.25Q_{i2}(s)}{(20s + 0.25)(15s + 0.35) - (-0.25)(-0.25)}$$

$$= \frac{(15s + 0.35)Q_{i1}(s) + 0.25Q_{i2}(s)}{300s^2 + 10.75s + 0.025} \tag{d}$$

and

$$H_2(s) = \frac{\begin{vmatrix} 20s + 0.25 & Q_{i1}(s) \\ -0.25 & Q_{i2}(s) \end{vmatrix}}{300s^2 + 14.25s + 0.025}$$

$$= \frac{0.25Q_{i1}(s) + (20s + 0.25)Q_{i2}(s)}{300s^2 + 10.75s + 0.025} \tag{e}$$

The first column of the matrix of transfer functions is obtained by setting $Q_{i2}(s) = 0$, which leads to

$$G_{1,1}(s) = \frac{H_1(s)}{Q_{i1}(s)} = \frac{15s + 0.35}{300s^2 + 10.75s + 0.025} \tag{f}$$

$$G_{2,1}(s) = \frac{H_2(s)}{Q_{i1}(s)} = \frac{0.25}{300s^2 + 10.75s + 0.025} \tag{g}$$

The second column of the matrix of transfer functions is obtained by setting $Q_{i1}(s) = 0$, which leads to

$$G_{1,2}(s) = \frac{H_1(s)}{Q_{i2}(s)} = \frac{0.25}{300s^2 + 10.75s + 0.025} \tag{h}$$

$$G_{2,2}(s) = \frac{H_2(s)}{Q_{i2}(s)} = \frac{20s + 0.25}{300s^2 + 10.75s + 0.025} \tag{i}$$

The transfer function matrix is

$$\mathbf{K}(s) = \frac{1}{300s^2 + 10.75s + 0.025} \begin{bmatrix} 15s + 0.35 & 0.25 \\ 0.25 & 20s + 0.25 \end{bmatrix} \tag{j}$$

Note that $\mathbf{K}(s)$ is the inverse of the coefficient matrix in Equation (c).

### 6.1.3 System Order

The transfer functions of the systems of Examples 6.1–6.5 are the ratio of two polynomials, as are most transfer functions. Thus $G(s) = N(s)/D(s)$, where $N(s)$ and $D(s)$ are polynomials in $s$. Exceptions are the time delay systems discussed in Section 6.8, where the transfer function includes exponential terms.

The **order** of a system is the order of $D(s)$, the polynomial in the denominator of the transfer function. The systems of Examples 6.1, 6.2, 6.4, and 6.5 are second-order systems while the system of Example 6.3 is a third-order system. In previous chapters the order of a system has been related to the number of independent energy storage components in the system. The determination of order using the denominator of the transfer functions is consistent with the energy storage definition. The mechanical system of Example 6.1 has two energy storage components: the mass that stores kinetic energy and the spring that stores potential energy. The $LRC$ circuit of Example 6.2 has two energy storage components: the inductor that stores energy in a magnetic field and the capacitor that stores energy in an electric field. The two springs in the system of Example 6.3 are independent energy storage components in that they cannot be replaced by a single spring. Thus the system of Example 6.3 has three energy storage components. The hydraulic servomotor of Example 6.4 is an energy storage component in that it stores energy in the fluid through the pressure difference. Since the spring also stores potential energy; the system of Example 6.4 has two energy storage components. Each of the tanks in the liquid-level system of Example 6.5 stores potential energy; thus the system has two energy storage components.

The **poles** of the system are the roots of $D(s)$, the values of $s$ such that $D(s) = 0$. The poles of a dynamic system may be real, purely imaginary, complex, or some combination of these possibilities. Since the coefficients of $D(s)$ are all real, complex poles must occur in complex conjugate pairs.

The **zeroes** of a transfer function are the roots of $N(s)$. Not all transfer functions have zeroes, as illustrated by Example 6.1 and Example 6.5.

## 6.2 TRANSIENT RESPONSE SPECIFICATION

The transient response of a dynamic system depends on the input. The types of transient response and the parameters defining transient response are considered in this section.

When the transfer function for a system is known, the transient response for a variety of inputs can be determined using the transfer function. Equation (6.1) can be rearranged as

$$X(s) = G(s)F(s) \tag{6.7}$$

The system response is determined by taking the inverse transform of Equation (6.7)

$$x(t) = \mathcal{L}^{-1}\{G(s)F(s)\} \tag{6.8}$$

### 6.2.1 Free Response

The free response occurs due to nonzero initial conditions and in the absence of any additional input. The free response of a system is determined by applying the Laplace transform method, as discussed in Chapter 5.

### 6.2.2 Impulsive Response

Noting that $\{\mathcal{L}\{\delta(t)\}\} = 1$, it is clear from Equation (6.8) that the transfer function is equal to the Laplace transform of the system response when the input is the unit impulse function. Thus the system response due to an input of a unit impulse, called the **impulsive response**, is

$$x_i(t) = \mathcal{L}^{-1}\{G(s)\} \tag{6.9}$$

The impulsive response of a system may be discontinuous. The principle of impulse and momentum shows that the application of a unit impulse to a mechanical system leads to a discontinuity in velocity. However, displacement, usually used as the dependent variable, is continuous. Recall, in the analogy between electrical systems powered by voltage sources and mechanical systems, that the velocity in a mechanical system is analogous to the current and that displacement is analogous to electric charge. Continuing with this analogy an impulsive voltage leads to a discontinuity in current, but a continuous charge. In the analogy between a mechanical system and a circuit with a current source, velocity is analogous to voltage and displacement is analogous to magnetic flux. Thus if a circuit is subject to an impulsive current, the voltage is discontinuous but the magnetic flux is continuous.

The value of the impulsive response immediately after the application of the impulse is obtained using the initial value theorem

$$x_i(0) = \lim_{s \to \infty} sG(s) \tag{6.10}$$

Suppose $G(s) = N(s)/D(s)$ is the ratio of two polynomials. The following can be deduced from Equation (6.10):

- The response is continuous at $t = 0$, $x_i(0) = 0$, if the order of $D(s)$ is at least two greater than the order of $N(s)$.
- The response is discontinuous, but finite at $t = 0$, $x_i(0) = C$, where $C$, is a finite numerical value, if the order of $D(s)$ is one greater than the order of $N(s)$.
- The response is unbounded at $t = 0$, $x_i(0) = \infty$ if the order of $D(s)$ is less than or equal to the order of $N(s)$.

This shows that the response of a first-order system due to a unit impulse input is at best discontinuous at $t = 0$. Since $D(s)$ is a first-order polynomial, $N(s)$ cannot be more than one order less than $D(s)$.

### 6.2.3 Step Response

The **step response** of a system is its response when the input is the unit step function $u(t)$. Noting that $\mathcal{L}\{u(t)\} = 1/s$, Equation (6.7) shows that the transform of the step response is

$$X_s(s) = \frac{1}{s} G(s) \tag{6.11}$$

The step responses of many systems have a definite limit for large $t$; that is, they approach a final value. The application of the final value theorem to Equation (6.11) shows that the final value of the step response is

$$
\begin{aligned}
x_{s,f} &= \lim_{t \to \infty} x_s(t) \\
&= \lim_{s \to 0} s X_s(s) \\
&= \lim_{s \to 0} s\left(\frac{1}{s} G(s)\right) \\
&= G(0) \tag{6.12}
\end{aligned}
$$

Parameters defining the transient response of a dynamic system are often defined in terms of its step response. Since, for a stable system (see Section 6.3), the step response has a defined final value and certain benchmark times, and parameters are defined.

The **2 percent settling time** is the time $t_s$ such that the step response is permanently within 2 percent of its final value, $G(0)$. The 2 percent settling time is used as a benchmark in the transient response of dynamic systems, but certainly settling times can be defined using different percentages.

The **10–90 percent rise time** is the time $t_r$ it takes for the response to initially grow from 10 percent of its final value to 90 percent of its final value. The selection of 10 percent and 90 percent to define the settling time provides benchmark values of the rise time, but certainly a rise time can be defined using other percentages.

If, during the step response, $x(t)$ exceeds $G(0)$, then the system is said to have **overshoot**. In such cases the **percentage overshoot** is defined by

$$\eta_o = 100\left(\frac{x_p - G(0)}{G(0)}\right) \tag{6.13}$$

where $x_p$ is the maximum, or peak, value of $x(t)$. The time at which $x$ attains $x_p$ is called the **peak time** $t_p$.

Equations (6.9) and (6.11) show that

$$X_i(s) = s X_s(s) \tag{6.14}$$

The application of the property of transform of derivative leads to

$$\mathcal{L}\left\{\frac{dx_s}{dt}\right\} = s X_s(s) - x_s(0) \tag{6.15}$$

Substitution of Equation (6.14) into Equation (6.15) leads to

$$\mathcal{L}\left\{\frac{dx_s}{dt}\right\} = X_i(s) - x_s(0) \tag{6.16}$$

Inversion of both sides of Equation (6.16) leads to

$$x_i(t) = \frac{dx_r}{dt} + x_s(0)\delta(t) \tag{6.17}$$

Thus $x_i(t) = dx_s/dt$ except possibly at $t = 0$.

### 6.2.4 Ramp Response

The ramp response of a system, useful in control system design, is the response due to a unit ramp input, $r(t) = t$. Since $\mathcal{L}\{t\} = 1/s^2$ the transform of the ramp response of a dynamic system is

$$X_r(s) = \frac{1}{s^2} G(s) \tag{6.18}$$

Use of Equation (6.11) in Equation (6.18) leads to

$$X_r(s) = \frac{1}{s} X_s(s) \tag{6.19}$$

Equation (6.19) implies that ramp response of a system is the step response of a system whose transfer function is $X_s(s)$, the transform of the step response of the system. A procedure similar to the procedure used to derive Equation (6.17) shows that

$$\frac{dx_r}{dt} = x_s(t) \tag{6.20}$$

In deriving Equation (6.20) it is assumed that $x_r(0) = 0$, which is expected since the ramp input is continuous at $t = 0$.

### 6.2.5 Convolution Integral

The transient response to an arbitrary input $F(t)$ is obtained through rewriting Equation (6.1) in the form of Equation (6.7), $X(s) = G(s)F(s)$. Property 13 of Table 5.2 implies that the inverse transform of a product of transforms is the convolution of the inverse transforms. Noting that the transfer function is the transform of the system's impulsive response implied by the convolution property

$$x(t) = x_i(t) * F(t) \tag{6.21}$$

or

$$x(t) = \int_0^t F(\tau)x_i(t - \tau)d\tau \tag{6.22}$$

Equation (6.22) is called the **convolution integral** solution for the response of a system. In the application of the integral it is noted that $F(t)$ is the system input and $x_i(t)$ is the response due to a unit impulse. Equation (6.22) can be used for numerical integration if the Laplace transform $X(s)$ is not invertible.

## 6.2.6 Transient System Response Using MATLAB

There are two ways to define the transfer function in MATLAB. One is simply to use the Symbolic Toolbox and enter the transfer function as a function of a symbolic variable $s$. The second is to define the transfer function from its numerator and denominator when both are polynomials in $s$, $G(s) = N(s)/D(s)$. The coefficients of the polynomials in $N(s)$ and $D(s)$ are entered as vectors as in Example 5.25. Then the transfer function is defined using the 'tf' command from the Control System Toolbox

$$G = \text{tf}(N,D)$$

where $N$ and $D$ are the vectors defining $N(s)$ and $D(s)$ respectively.

Given the transfer function, MATLAB is able to determine and plot the impulsive response and the step response using the following sets of commands

$$\text{impulse}(N,D)$$
$$\text{step}(N,D)$$

or

$$\text{impulse}(\text{tf})$$
$$\text{step}(\text{tf})$$

---

**Example 6.6**

A system has a transfer function of

$$G(s) = \frac{s+3}{s^2 + 6s + 20} \qquad (a)$$

(a) Determine the initial value of the response when the system is subjected to a unit impulse input. (b) Determine the final value of the response when the system is subjected to a unit step input. (c) Define the transfer function using MATALB, and use the transfer function to determine the system's impulsive response, its step response, and its ramp response. (d) Use the MATLAB-generated plot of the step response to approximate the 2 percent settling time, the 10–90 percent rise time, the percent overshoot, and the peak time.

**Solution**

(a) The initial value of the response when the system is subject to a unit impulse is obtained using Equation (6.3)

$$x_i(0) = \lim_{s \to \infty} s \frac{s+3}{s^2 + 6s + 20}$$

$$= \lim_{s \to \infty} \frac{s^2 + 3s}{s^2 + 6s + 20}$$

$$= \lim_{s \to \infty} \frac{1 + \dfrac{3}{s}}{1 + \dfrac{6}{s} + \dfrac{20}{s^2}} = 1 \qquad (b)$$

(b) The final value of the response when the system is subject to a unit step input is obtained using Equation (6.5) as

$$x_f = G(0) = \frac{3}{20} = 0.15 \qquad \text{(c)}$$

(c) The MATLAB work session is shown in Figure 6.2. The resulting plots are shown in Figure 6.3. Note that to determine the ramp response a transfer function G1(s) is defined, which has the same numerator as G(s) but its denominator is the denominator of G(s) multiplied by s. The ramp response of the system

```
>> % Defining numerator and denominator of transfer function
>> N=[1 3]

N =

   1   3

>> D=[1 6 20]

D =

   1   6   20

>> % Defining transfer function
>> G=tf(N,D)

Transfer function:
  s + 3
--------------
s^2 + 6 s + 20

>> % Impulsive response
>> impulse(G)
>> %Step response
>> step(G)
>> % Defining transfer function for ramp response
>> % Numerator is the same, but denominator is multiplied by s
>> D1=[1 6 20 0]

D1 =

   1   6   20   0
>> G1=tf(N,D1)

Transfer function:
    s + 3
------------------
s^3 + 6 s^2 + 20 s

>> step(G1)
>>
```

**FIG. 6.2** The MATLAB work session for the solution of Example 6.6.

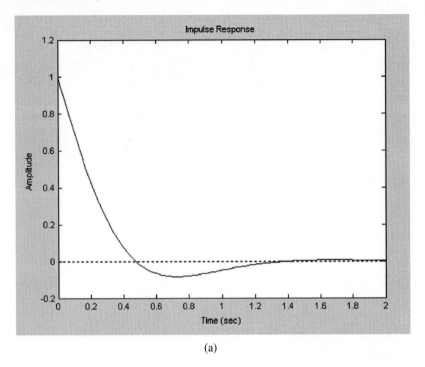

(a)

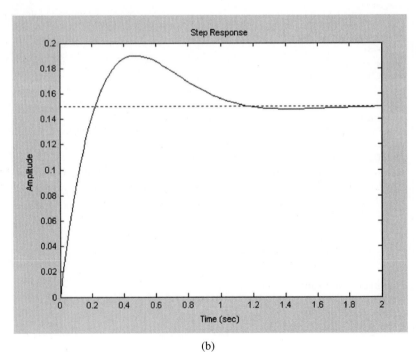

(b)

**FIG. 6.3** The MATLAB output for Example 6.6: (a) the impulsive response; (b) the step response; (c) the ramp response.

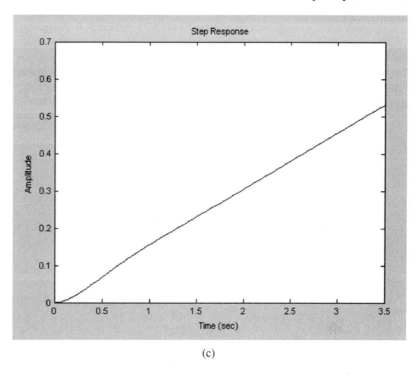

Step Response

(c)

**FIG. 6.3** (*Continued*)

whose transfer function is G(s) is the step response of the system whose transfer function is G1(s).

(d) It is, of course, difficult to determine the values of the response specifications exactly from Figure 6.3(b). Since $x_f = 0.15$ the 2 percent settling time is the time that the step response is permanently within 0.003 of 0.15. From Figure 6.3(b) it appears that $t_s = 1.4$ s. The 10–90 percent rise time is the time that it takes for the response to rise from 0.015 to 0.135. From Figure 6.3(b) the rise time is approximately $t_r = 0.2$ s. The maximum displacement is obtained from Figure 6.3(b) as $x_{max} = 0.19$. The percentage offshoot is obtained using Equation (6.6) as

$$\eta_o = 100\left(\frac{0.19 - 0.15}{0.15}\right) = 26.7\% \tag{d}$$

The peak time is $t_p = 0.5$ s.

## 6.3 STABILITY ANALYSIS

### 6.3.1 General Theory

The denominator of the transfer function for an $n$th order system is an $n$th order polynomial in $s$, $D(s)$, where without loss of generality, the coefficient of $s^n$ is assumed to be one. The polynomial has $n$ linear factors of the form $s - s_k$ where $s_1, s_2, \ldots, s_n$ are the roots of $D(s)$, also called the **poles of the transfer function**. Thus

$$D(s) = (s - s_1)(s - s_2)(s - s_3) \cdots (s - s_n) \tag{6.23}$$

The poles may be real or complex, but if $D(s)$ has complex roots they occur in complex conjugate pairs. If $s_k = s_{kr} + js_{kj}$ is a pole then its complex conjugate $\bar{s}_k = s_{kr} - js_{kj}$ is also a pole. The product of the two linear factors corresponding to these poles is

$$(s - s_k)(s - \bar{s}_k) = (s - s_{kr})^2 + s_{kj}^2 \qquad (6.24)$$

Without loss of generality the discussion may proceed by considering the transfer function for a third-order system with one real root and a pair of complex conjugate roots such that $D(s) = (s - s_1)\left[(s - s_{2r})^2 + s_{2j}^2\right]$. A partial fraction decomposition of the transfer function, as described in Section 5.4, leads to

$$G(s) = \frac{A}{s - s_1} + \frac{Bs + C}{(s - s_{2r})^2 + s_{2j}^2} \qquad (6.25)$$

It is shown in Section 6.2 that $x_i(t) = \mathcal{L}^{-1}\{G(s)\}$ is the response of the system due to the input of a unit impulse. $x_i(t)$ is obtained for this system by inverting the transforms in Equation (6.25) using transform pairs 1, 12, and 13 from Table 5.1 resulting in

$$G(t) = Ae^{s_1 t} + e^{s_{2r}t}\left[B\cos\left(s_{2j}t\right) + \frac{C + Bs_{2r}}{s_{2j}} \sin\left(s_{2j}t\right)\right] \qquad (6.26)$$

The inherent stability of a system is determined by $x_{i,f} = \lim_{t \to \infty} x_i(t)$, the final value of its impulsive response. If $x_{i,f} = 0$ the system is stable; if $x_{i,f} = \infty$ the system is unstable; and if $x_{i,f} = C$ where $C$ is a nonzero constant the system is neutrally stable.

If either $s_1 > 0$ or $s_{2r} > 0$ then $x_i(t)$ is unbounded as $t$ increases. In this case the system is unstable. If $s_1 < 0$ and $s_{2r} < 0$, then $\lim_{t \to \infty} x_i(t) = 0$ and the system is stable. If either $s_1 = 0$ or $s_{2r} = 0$ and neither is positive, then $x_i(t)$ is bounded but does not approach 0 as $t$ increases. In this case the system reaches a nonzero steady state and the system is neutrally stable.

The preceding result can be generalized for any system with the following statements. A system is stable if and only if all poles of its transfer function have negative real parts. If any pole has a positive real part, then the system is unstable. If any pole has a real part of zero and all other poles have nonpositive real parts, then the system is neutrally stable.

**Example 6.7**

The differential equation derived in Example 2.29, which governs the motion of the system of Figure 6.4, an inverted compound pendulum with springs and viscous dampers, is

$$\tfrac{1}{3}m(a^2 - ab + b^2)\ddot{\theta} + 2cb^2\dot{\theta} + \left[2ka^2 - mg\left(\frac{a - b}{2}\right)\right]\theta = 0 \qquad (a)$$

Determine a stability criterion for this system involving $k$, $a$, $b$, $m$, and $c$.

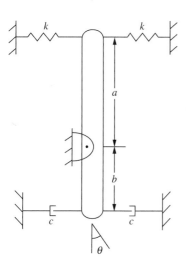

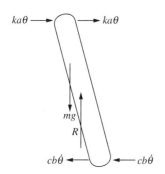

**FIG. 6.4** System is unstable if $ka < mg\frac{(b-a)}{2}$ and neutrally stable if $c = 0$.

## Solution

The differential equation for the pendulum when it is subject to a moment $M(t)$ is

$$\tfrac{1}{3}m(a^2 - ab + b^2)\ddot{\theta} + 2cb^2\dot{\theta} + \left[2ka^2 - mg\left(\frac{a-b}{2}\right)\right]\theta = M(t) \tag{b}$$

Taking the Laplace transform of Equation (b) leads to

$$\Theta(s) = \frac{M(s)}{\tfrac{1}{3}m(a^2 - ab + b^2)s^2 + 2cbs + 2ka^2 - mg\left(\frac{a-b}{2}\right)} \tag{c}$$

The transfer function $G(s) = \Theta(s)/M(s)$ is determined using Equation (c) as

$$G(s) = \frac{1}{\tfrac{1}{3}m(a^2 - ab + b^2)s^2 + 2cbs + 2ka^2 - mg\left(\frac{a-b}{2}\right)} \tag{d}$$

The poles are the values of $s$ such that

$$\tfrac{1}{3}m(a^2 - ab + b^2)s^2 + 2cbs + 2ka^2 - mg\left(\frac{a-b}{2}\right) = 0 \tag{e}$$

The quadratic formula is used to determine these values as

$$s_1 = \frac{3}{m(a^2 - ab + b^2)}\left(-cb - \sqrt{D}\right) \tag{f}$$

$$s_2 = \frac{3}{m(a^2 - ab + b^2)}\left(-cb + \sqrt{D}\right) \tag{g}$$

where

$$D = c^2b^2 - \tfrac{1}{3}(a^2 - ab + b^2)\left[2ka^2 - mg\left(\frac{a-b}{2}\right)\right] \tag{h}$$

If $D < 0$ then both poles are complex with negative real parts and thus the system is stable. If $D > 0$ then both poles are real, $s_1$ is negative while the sign of $s_2$ depends on the value of $D$. The system is stable if $s_2 < 0$ which requires

$$-cb + \sqrt{c^2b^2 - \tfrac{1}{3}(a^2 - ab + b^2)\left[2ka^2 - mg\left(\frac{a-b}{2}\right)\right]} < 0 \tag{i}$$

Algebraic manipulation of Equation (i) leads to a stability criterion of

$$2ka^2 > mg\left(\frac{a-b}{2}\right) \tag{j}$$

If the system is undamped, $c = 0$, then both poles are purely imaginary and the system is neutrally stable.

The stability criterion developed in Example 6.7 [Equation (j)] is a statement that the moment of the spring forces about the pin support is greater than the moment of the gravity force. The potential energy of the system at an arbitrary instant, assuming small displacements, is $V = \tfrac{1}{2}\{2ka^2 - mg[(a - b)/2]\}\theta^2$. Thus the stability criterion is also the condition under which the potential energy is positive.

**Example 6.8**

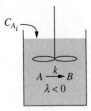

$C_{A_i}$

$A \xrightarrow{k} B$
$\lambda < 0$

**FIG. 6.5** The CSTR with an exothermic reaction is unstable when the heat of the reaction is greater than a critical value.

Consider the CSTR of Example 4.17 and Figure 6.5, in which the reaction is exothermic. The linearized equations describing the perturbations in the concentration of reactant $A$ $C_{Ap}$ and reactor temperature $T_p$ are given in Equations (s) and (u) of Example 4.18 as

$$V\frac{dC_{Ap}}{dt} + qC_{Ap} + \alpha Ve^{-E/(RT_s)}\left(C_{Ap} + \frac{E}{RT_s^2}T_pC_{As}\right) = 0 \qquad (a)$$

$$\rho Vc_p\frac{dT_p}{dt} + \rho qc_p T_p - \lambda\alpha Ve^{-E/(RT_s)}\left(C_{Ap} + \frac{E}{RT_s^2}T_pC_{As}\right) = \Delta\dot{Q} \qquad (b)$$

where the parameters are as described in Example 4.17 and $\Delta\dot{Q}$ is the perturbation of the rate at which heat is removed from the system. Describe the stability of the system in terms of system parameters.

**Solution**

The transfer function $G(s) = \mathcal{L}\{T_p\}/\mathcal{L}(\Delta\dot{Q})$ is determined as

$$G(s) = \frac{Vs + q + b}{\rho c_p V^2 s^2 + \left[(2q + b)\rho c_p V - \lambda\mu b V\right]s + \rho c_p q^2 + b\rho c_p q - \lambda b\mu q} \qquad (c)$$

where

$$b = \alpha Ve^{-E/(RT_s)} \qquad (d)$$

$$\mu = \frac{EC_{As}}{RT_s^2} \qquad (e)$$

The quadratic formula is applied to determine the poles of the transfer function as

$$s_1 = -\frac{q + b}{V} \qquad (f)$$

$$s_2 = \frac{\lambda\mu b - q\rho c_p}{\rho c_p V} \qquad (g)$$

The pole $s_2$ is positive and the system unstable when

$$\lambda > \frac{q\rho c_p}{\mu b} \qquad (h)$$

The parameter $\lambda$ is the rate at which heat is released from the reaction. If $\lambda$ exceeds the critical value of Equation (h) the temperature increases without bound. However, Equations (a) and (b) are obtained from the linearization of nonlinear differential equations. In the process the temperature will increase until the linearizing assumptions are no longer valid and the system's inherent nonlinearity takes over. The nonlinearity will lead to a saturation of the temperature and the concentration. While the temperature will remain bounded it will be at a high value and the reactor will not function as desired.

### 6.3.2 Routh's Method

Consider an $n$th-order system with $j$ pairs of complex conjugate poles and $n - 2j$ real poles. In this case, Equation (6.24) can be written as

$$D(s) = \left\{ \prod_{k=1}^{j} \left[ (s - s_{kr})^2 + s_{kj}^2 \right] \right\} \left[ \prod_{k=1}^{n-2j} (s - s_k) \right] \qquad (6.27)$$

When the products in Equation (6.27) are expanded the result is an $n$th order polynomial of the form

$$D(s) = s^n + a_1 s^{n-1} + a_2 s^{n-2} + \cdots + a_k s^{n-k} + \cdots a_{n-1} s + a_n \qquad (6.28)$$

The coefficients $a_1, a_2, \ldots, a_n$ are sums of products of the negatives of the real parts of the poles and the square of the imaginary parts. If the system is stable then the real parts of all poles are negative. Thus all coefficients are sums of positive numbers and are thus all positive.

This shows that a necessary condition for stability is that all the coefficients of $D(s)$, the polynomial in the denominator of the transfer function, are positive. Thus a necessary condition for the stability of a second-order system governed by the equation $\ddot{x} + 2\zeta\omega_n \dot{x} + \omega_n^2 x = 0$ is that both $\zeta$ and $\omega_n^2$ be positive numbers. If $\omega_n^2 < 0$ then the potential energy of a mechanical system, as measured from the system's equilibrium position, could be negative and the system would be unstable, as is possible in Example 6.7. If the damping ratio is negative (negative damping), then instead of dissipating energy the damping mechanism is actually adding energy to the system. If the damping ratio of a second-order system is zero (undamped) then $D(s) = s^2 + \omega_n^2$; thus $a_1 = 0$ and the system is neutrally stable.

While the requirement that all coefficients in the polynomial of the denominator of the transfer function be positive is a necessary condition for stability, it is not a sufficient condition. If the necessary condition for stability is met, then the application of Routh's method, presented without derivation, determines whether the system is stable without solving for the poles. Routh's method is an algebraic method that determines the number of poles with positive real parts.

To apply Routh's method develop a matrix of numbers formed from the coefficients of $D(s)$. The matrix has $n + 1$ rows. If $n$ is even, the matrix has $k = (n/2) + 1$ columns. If $n$ is odd the matrix has $k = (n + 1)/2$ columns. The matrix for an even value of $n$ is shown in Equation (6.29).

$$\mathbf{B} = \begin{bmatrix} 1 & a_2 & a_4 & \cdots & a_{n-2} & a_n \\ a_1 & a_3 & a_5 & \cdots & a_{n-1} & 0 \\ b_{3,1} & b_{3,2} & b_{3,3} & \cdots & b_{3,n-1} & 0 \\ b_{4,1} & b_{4,2} & b_{4,3} & \cdots & b_{4,n-1} & 0 \\ \vdots & \vdots & \vdots & \cdots & \vdots & \vdots \\ b_{n+1,1} & b_{n+1,2} & b_{n+1,3} & \cdots & b_{n+1,n-1} & 0 \end{bmatrix} \qquad (6.29)$$

where

$$b_{i,j} = \frac{b_{i-1,1}b_{i-2,j+1} - b_{i-2,1}b_{i-1,j+1}}{b_{i-1,1}} \qquad i = 3, ..., n+1; \; j = 1, ..., k-1$$

(6.30)

The matrix for an odd value of $n$ is

$$\mathbf{B} = \begin{bmatrix} 1 & a_2 & a_4 & \cdots & a_{n-3} & a_{n-1} \\ a_1 & a_3 & a_5 & \cdots & a_{n-2} & a_n \\ b_{3,1} & b_{3,2} & b_{3,3} & \cdots & b_{3,n-1} & 0 \\ b_{4,1} & b_{4,2} & b_{4,3} & \cdots & b_{4,n-1} & 0 \\ \vdots & \vdots & \vdots & \cdots & \vdots & \vdots \\ b_{n+1,1} & b_{n+1,2} & b_{n+1,3} & \cdots & b_{n+1,n-1} & 0 \end{bmatrix}$$

(6.31)

where Equation (6.30) is used to determine the entries after the second row. Routh's method specifies that the number of poles with positive real parts is equal to the number of sign changes in the elements of the first column of **B**. Thus if all elements of the first column of **B** are positive, then there are no sign changes in the column, the system has no poles with positive real parts, and it is stable. If any element of the first column of **B** is negative, then there is at least one pole with a positive real part and the system is unstable.

A special case occurs when an element of the first row column of **B** is zero. This occurs when the system has either a pair of purely imaginary roots or a repeated root. To determine which case occurs, replace the zero in the first column by an arbitrarily small positive value, say $\delta$ and continue to calculate the remainder of the elements of **B** according to Equation (6.30). If the next element in the first column is zero, the system has a pair of purely imaginary roots and is neutrally stable. If the next element in the first column is negative, the system has a repeated positive root and is unstable.

## Example 6.9

Use Routh's method to determine a stability criterion for a third-order system with $D(s) = s^3 + a_1 s^2 + a_2 s + a_3$.

### Solution

For $n = 3$, **B** is a $4 \times 2$ matrix, initially set up as

$$\mathbf{B} = \begin{bmatrix} 1 & a_2 \\ a_1 & a_3 \\ b_{3,1} & 0 \\ b_{4,1} & 0 \end{bmatrix}$$

(a)

where from Equation (6.30)

$$b_{3,1} = \frac{b_{2,1}b_{1,2} - b_{1,1}b_{2,2}}{b_{2,1}}$$

$$= \frac{a_1 a_2 - (1)a_3}{a_1} \tag{b}$$

$$b_{4,1} = \frac{b_{3,1}b_{2,2} - b_{2,1}b_{3,2}}{b_{3,1}}$$

$$= \frac{\dfrac{a_1 a_2 - a_3}{a_1}a_3 - a_1(0)}{\dfrac{a_1 a_2 - a_3}{a_1}}$$

$$= a_3 \tag{c}$$

Thus the first column of **B** is

$$\begin{bmatrix} 1 \\ a_1 \\ \dfrac{a_1 a_2 - a_3}{a_1} \\ a_3 \end{bmatrix} \tag{d}$$

All elements of the column are positive and the system stable if

$$a_1 a_2 > a_3 \tag{e}$$

---

## Example 6.10

Use Routh's method to determine the stability of a system with the transfer function $G(s) = (s + 2)/(s^4 + 5.5s^3 + 8s^2 + 0.5s + 1)$.

### Solution

The matrix **B** for this transfer function is determined as

$$\mathbf{B} = \begin{bmatrix} 1 & 8 & 1 \\ 5.5 & 0.5 & 0 \\ 7.910 & 1 & 0 \\ -0.1953 & 0 & 0 \\ 1 & 0 & 0 \end{bmatrix} \tag{a}$$

The first column of **B** has two sign changes. Thus $G(s)$ has two poles with positive real parts and the system is unstable.

## 6.3.3  Relative Stability

Routh's method, as described, determines the absolute stability of a system. In some cases it is also necessary to determine the relative stability of a system, that is, how close the system is to being unstable. A system is stable when all its poles have negative real parts. For a stable system a generalization of Equation (6.26) is used to determine the system's impulsive response. The dominant pole in a stable system is the pole whose real part has the smallest absolute value. The term in the impulsive response corresponding to this pole has the slowest rate of decay and is dominant in the solution for large $t$. In subsequent sections it is shown how properties of transient response such as settling time and peak time depend on the dominant pole. A design consideration in a control system may be whether the settling time is less than a given value. If the dominant pole is too close to the imaginary axis the system may not satisfy a design constraint.

To address these concerns the concept of relative stability is used. A system is stable relative to the line $s = -p$ if all poles lie to the left of $s = -p$. To determine the relative stability consider $D(s - p)$. Equation (6.23) is used to show that if $s_k$ is a root of $D(s)$ then $s_k + p$ is a root of $D(s - p)$. Thus the stability of the system relative to $s = -p$ may be determined by applying Routh's method to $D(s - p)$.

## Example 6.11

A control system for a guidance system is being designed such that its transfer function is

$$G(s) = \frac{1}{s^3 + 10s^2 + (K + 3)s + 50} \tag{a}$$

where $K$ is a design parameter in the control system.

(a) Determine the values of $K$ for which the system is absolutely stable. (b) Determine the values of $K$ for which the system is stable relative to $s = -5$. (c) Determine the values of $K$ for which the system is stable relative to $s = -1$.

### Solution

(a) Equation (e) of Example 6.9 is a stability criterion, derived using Routh's method, for third-order systems. The application of Equation (e) to the transfer function of Equation (a) leads to

$$10(K + 3) > 50$$
$$K > 2 \tag{b}$$

Thus the system is absolutely stable if $K > 2$.

(b) The stability of the system relative to $s = -5$ is determined by applying Routh's method to $D(s - 5)$. To this end

$$D(s - 5) = (s - 5)^3 + 10(s - 5)^2 + (K + 3)(s - 5) + 50$$
$$= s^3 - 5s^2 + (K - 22)s + 160 - 5K \tag{c}$$

A necessary requirement for stability is that all coefficients in the polynomial are positive. Since $D(s - 5)$ has a negative coefficient, independent of the value of $K$, there are no values of $K$ such that the system is stable relative to $s = -5$.

(c) The stability of the system relative to $s = -1$ is determined by applying Routh's method to $D(s - 1)$. To this end

$$D(s - 1) = (s - 1)^3 + 10(s - 1)^2 + (K + 3)(s - 1) + 50$$
$$= s^3 + 7s^2 + (K - 14)s + 56 - K \tag{d}$$

Application of Equation (e) of Example 6.9 to Equation (d) shows that the system is stable relative to $s = -1$ when

$$7(K - 14) > 56 - K$$

$$K > 19.25 \tag{e}$$

### 6.3.4  An Introduction to Root-Locus Analysis

Transfer functions determined during control system design often contain parameters, similar to $K$ in Example 6.11, which are to be determined such that the transient response meets certain specifications. Major concerns in control system design include stability and relative stability. Stability and transient response are dependent on the poles of the transfer function. If $D(s)$ includes a parameter then it is convenient to have knowledge of how the locations of the poles change as the parameter changes.

A root-locus diagram is a plot of the location of the poles of the transfer function as a parameter varies. The diagram gives a locus of all possible pole locations for all possible values of the parameter.

The basics of root-locus analysis are presented in this section. The method is visited again in Chapter 8 where it is used in the analysis and design of control systems. Root-locus diagrams for second- and third-order systems are presented as well as the application of MATLAB to the development of root-locus diagrams. The general method of construction of root-locus diagrams is presented in Appendix D.

Let $K$ be a parameter that appears linearly in the denominator of a transfer function $D(s)$. The location of the system poles depend on the value of $K$ and are obtained by setting $D(s) = 0$. Polynomials $Q(s)$ and $R(s)$ can be defined such that the polynomial equation defining the poles can be written as

$$Q(s) + KR(s) = 0 \tag{6.32}$$

The root-locus diagram for an $n$th order system has $n$ branches, one corresponding to each of the poles. In most systems $Q(s)$ is a polynomial of order $n$ while $R(s)$ is a polynomial of order $m < n$.

It is determined from Equation (6.32) that for $K = 0$ the poles are the roots of $Q(s)$ and for $K = \infty$ the poles are the roots of $R(s)$. Thus the loci begin at the roots of $Q(s)$ and end at either the roots of $R(s)$ or at infinity.

An alternate form of Equation (6.32) that is used in root-locus diagram construction is

$$K \frac{R(s)}{Q(s)} = -1 \tag{6.33}$$

**Example 6.12**

Develop the root-locus diagram for the transfer function

$$G(s) = \frac{K}{s^2 + 5s + 6 + K} \tag{a}$$

**Solution**

$D(s)$ is written in the form of Equation (6.32) with $Q(s) = s^2 + 5s + 6$ and $R(s) = 1$. However, the quadratic equation may be used to determine the roots of $D(s)$ as

$$s_1 = \tfrac{1}{2}\left(-5 - \sqrt{1 - 4K}\right) \tag{b}$$

$$s_2 = \tfrac{1}{2}\left(-5 + \sqrt{1 - 4K}\right) \tag{c}$$

When $K = 0$, $s_1 = -3$, $s_2 = -2$. As $K$ increases $s_1$ increases and $s_2$ decreases and both are real until $K = \tfrac{1}{4}$. For $K = \tfrac{1}{4}$, $s_1 = s_2 = -2.5$. This is the breakaway point, at which the root-locus curve leaves the real axis. For $K > \tfrac{1}{4}$ both $s_1$ and $s_2$ are imaginary with a negative real part and they are complex conjugates. The root-locus becomes the line in the complex plane $s = -2.5$. As $K$ increases the imaginary part of $s_1$ approaches $-\infty$ while the imaginary part of $s_2$ approaches $\infty$. The root-locus diagram is illustrated in Figure 6.6.

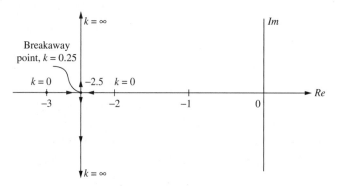

**FIG. 6.6** The Root-Locus diagram for the quadratic transfer function.

**Example 6.13**

Draw the root-locus diagram for the system of Example 6.11.

**Solution**

Writing the denominator of the transfer function of Example 6.11 in the form of Equation (6.32) leads to

$$Q(s) = s^3 + 10s^2 + 3s + 50 \tag{a}$$

$$R(s) = s \tag{b}$$

**TABLE 6.1 POLES OF TRANSFER FUNCTION OF EXAMPLE 6.13**

| $K$ | $S_1$ | $S_2$ | $S_3$ |
|---|---|---|---|
| 0 | −10.187 | $0.0936 − 2.1234j$ | $0.0936 + 2.1234j$ |
| 1 | −10.0944 | $0.0472 − 2.2251j$ | $0.0472 + 2.2251j$ |
| 2 | −10.000 | $−2.236j$ | $2.236j$ |
| 5 | −9.706 | $−0.147 − 2.265j$ | $−0.147 + 2.265j$ |
| 10 | −9.177 | $−0.4144 − 2.297j$ | $−0.4144 + 2.297j$ |
| 20 | −7.8877 | $−1.0561 − 2.2855j$ | $−1.0561 + 2.2855j$ |
| 50 | −1.1723 | $−4.4137 − 4.813j$ | $−4.4137 + 4.813j$ |
| 100 | −0.5093 | $−4.7453 − 8.697j$ | $−4.7453 + 8.697j$ |

The roots of $Q(s)$ are obtained using MATLAB's 'roots' command as $−10.187$ and $0.0936 \pm 2.123j$. The root of $R(s)$ is $s = 0$.

A detailed procedure for the construction of a root-locus diagram, along with several examples, is presented in Appendix D. The method used here is to vary the value of $K$, use MATLAB's 'roots' command to determine the roots for each $K$, and then plot the root-locus diagram from the results. The roots for several values of $K$ are shown in Table 6.1 and the resulting root-locus diagram is given in Figure 6.7.

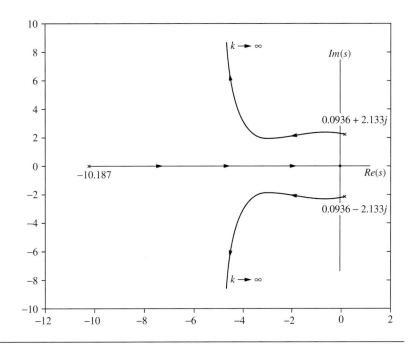

**FIG. 6.7** The root-locus diagram for the transfer function of Example 6.12.

MATLAB has the capability of drawing the root-locus diagram given the vector forms of $Q(s)$ and $R(s)$. The appropriate command is

$$\text{rlocus(R,Q)}$$

## Example 6.14

```
>> Q=[1 3 2 0]

Q =

   1    3    2    0

>> R=[2 3]

R =

   2    3

>> rlocus(R,Q)
>>
```

**FIG. 6.8** The MATLAB work session to generate the root-locus diagram for the transfer function of Example 6.14.

Use MATLAB to draw the root-locus diagram for the transfer function

$$G(s) = \frac{2s + 3}{s^3 + 3s^2 + (2 + K)s + 3K} \tag{a}$$

### Solution

Writing the denominator of the transfer function in the form of Equation (6.32) gives

$$Q(s) = s^3 + 3s^2 + 2s \tag{b}$$

$$R(s) = s + 3 \tag{c}$$

The MATLAB work session to create the root-locus diagram is reproduced in Figure 6.8, and the resulting root-locus diagram is given in Figure 6.9. Note that x marks the point on each branch corresponding to $K = 0$, the roots of $Q(s)$ are 0, −1, and −2. An o marks the point on a branch corresponding to $K = \infty$, which for the branch along the real axis is the root of $R(s)$, which is −3. Since all roots are in the left half of the complex plane for all values of $K$, there is no value of $K$ for which this system is unstable.

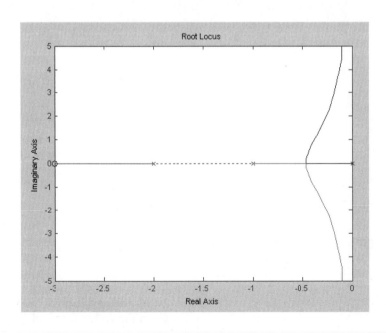

**FIG. 6.9** The MATLAB-generated root-locus diagram for the transfer function of Example 6.14.

## 6.4 FIRST-ORDER SYSTEMS

The standard form of the differential equation governing the behavior of a linear first-order system is

$$T\frac{dx}{dt} + x = f(t) \tag{6.34}$$

The parameter $T$ is called the system's **time constant**. First-order systems studied in Chapters 2–4 are summarized in Table 6.2. Specification of an initial condition of the form

$$x(0) = x_0 \tag{6.35}$$

is necessary to determine a unique solution to Equation (6.34).

**TABLE 6.2 FIRST-ORDER SYSTEMS**

| Name | Schematic | Differential Equation |
|------|-----------|----------------------|
| Spring-viscous damper | | $c\dot{x} + kx = F(t)$ |
| LR Circuit | | $L\dfrac{di}{dt} + Ri = v(t)$ |
| RC Circuit | | $C\dfrac{dv}{dt} + \dfrac{1}{R}v = i(t)$ |
| Liquid-level problem | | $A\dfrac{dh}{dt} + \dfrac{1}{R}h = q$ |
| Lumped thermal system | | $\rho c V\dfrac{d\theta}{dt} + hA\theta = hA\theta_\infty$ |

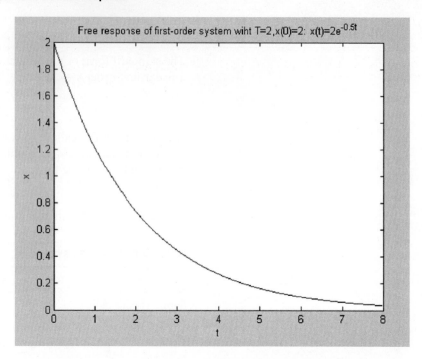

**FIG. 6.10** The free response of a first-order system.

Taking the Laplace transform of Equation (6.34) using Equation (6.35) leads to

$$X(s) = \frac{F(s) + x_0 T}{T\left(s + \frac{1}{T}\right)} \qquad (6.36)$$

The transfer function for a first-order system is determined from Equation (6.36) as

$$G(s) = \frac{1}{T\left(s + \frac{1}{T}\right)} \qquad (6.37)$$

A first-order system has one pole

$$s_1 = -\frac{1}{T} \qquad (6.38)$$

Thus a first-order system is stable if $T > 0$.

### 6.4.1 Free Response

The free response of a first-order system is obtained by inverting Equation (6.36) with $F(s) = 0$, leading to

$$x(t) = x_0 e^{-\frac{t}{T}} \qquad (6.39)$$

The free response of a first-order system, illustrated in Figure 6.10, decays exponentially and approaches zero asymptotically. The free-response of a linear first-order system depends on one parameter, the time constant. Table 6.3 illustrates the number of time constants it takes the response to decay to a percentage of its initial value.

**TABLE 6.3 TRANSIENT RESPONSE OF FIRST-ORDER SYSTEM, $x(t) = x_0 e^{-t/T}$**

| $x(t)/x_0$ | $t/T$ |
|---|---|
| 0.1 | 2.30 |
| 0.2 | 1.61 |
| 0.3 | 1.20 |
| 0.4 | 0.916 |
| 0.5 | 0.693 |
| 0.6 | 0.511 |
| 0.7 | 0.357 |
| 0.8 | 0.223 |
| 0.9 | 0.105 |

## Example 6.15

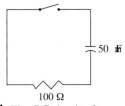

100 Ω

FIG. **6.11** The $RC$ circuit of Example 6.15.

The capacitor in the $RC$ circuit of Figure 6.11 has an initial charge of 2 mC. The switch is initially open then closed at $t = 0$, allowing the capacitor to discharge. (a) Determine the power delivered by the capacitor as a function of time. (b) How long does it take for the charge to drop to 0.5 mC?

### Solution

(a) The differential equation for the voltage in the circuit is

$$C\frac{dv}{dt} + \frac{1}{R}v = 0 \qquad (a)$$

$$RC\frac{dv}{dt} + v = 0 \qquad (b)$$

From Equation (b) the time constant for this first-order system is determined as

$$T = RC = (100 \ \Omega)(50 \times 10^{-6} \ F) = 5 \times 10^{-3} \ s \qquad (c)$$

The initial voltage in the capacitor is

$$v(0) = \frac{q(0)}{C} = \frac{2 \times 10^{-3}C}{50 \times 10^{-6}F} = 40 \ V \qquad (d)$$

Equations (c) and (d) are used in Equation (6.39) leading to

$$v(t) = 40e^{-200t} \ V \qquad (e)$$

The power in the capacitor is determined using Equation (3.19) as

$$P = Cv\frac{dv}{dt}$$
$$= (50 \times 10^{-6} \ F)(40e^{-200t} \ V)(-8{,}000e^{-200t} \ V/s)$$
$$= -16e^{-400t} \ W \qquad (f)$$

(b) The charge in the capacitor is

$$q = Cv$$
$$= (50 \times 10^{-6} \ F)(40e^{-200t} \ V)$$
$$= 2e^{-200t} \ mC \qquad (f)$$

Setting the charge to 0.5 mC leads to

$$0.5 \ mC = 2e^{-200t} \ mC$$
$$0.25 = e^{-200t} \qquad (g)$$
$$t = -\frac{\ln(0.25)}{200} = 6.93 \times 10^{-3} \ s$$

The free-response of a first-order system is a model for the dynamic response of any system in which the rate of change of the dependent variable is proportional to the dependent variable. Examples include continuously compounded interest in which the rate of change of an investment $m(t)$ is proportional to the current value of the investment with the interest rate $i$ at the constant of proportionality

$$\frac{dm}{dt} = im \qquad (6.40)$$

The solution of Equation (6.40) is $m(t) = m_0 e^{it}$ where $m_0$ is the amount initially invested. This is an example of an unstable first-order system. Theoretically the investment grows without bound over time. One would not put money in a bank if the amount deposited decayed with time.

---

## Example 6.16

The rate of decay of a radioactive isotope is proportional to the amount of the instantaneous amount of the isotope. The half-life of an isotope is defined as the time it takes for a quantity of the isotope to decay to half of its initial mass. The isotope $^{14}C$, used in dating artifacts, has a half-life of 5,700 years. How long will it take 20 g of the isotope to decay to 18 g?

### Solution

Since the rate of change of the mass of the isotope is proportional to the amount of the isotope $m(t)$

$$\frac{dm}{dt} = -rm \qquad (a)$$

where $r$ is the rate of decay. The rate of change is negative because the isotope is losing mass. Equation (a) can be rewritten in the form of Equation (6.34) as

$$\frac{1}{r}\frac{dm}{dt} + m = 0 \qquad (b)$$

Comparing Equation (b) with Equation (6.34) it is clear that the time constant for the system is equal to the reciprocal of the rate of decay. The solution of Equation (b) is obtained using Equation (6.39) as

$$m(t) = m_0 e^{-rt} \qquad (c)$$

Let $t_h$ be the half-life of an isotope. Evaluation of Equation (c) at $t = t_h$ leads to

$$m(t_h) = \tfrac{1}{2} m_0 = m_0 e^{-rt_h} \qquad (d)$$

Equation (d) is used to solve for $r$ as

$$r = \frac{1}{t_h} \ln(2) \qquad (e)$$

Substitution of Equation (e) into Equation (c) leads to

$$m(t) = m_0 e^{-\ln(2)\frac{t}{t_h}} \qquad (f)$$

Application of Equation (f) to the data given for the $^{14}C$ isotope leads to

$$18\text{ g} = (20\text{ g}) e^{-\ln(2)\frac{t}{5700\,\text{yr}}} \qquad (g)$$

The solution of Equation (g) leads to $t = 866$ yr.

### 6.4.2 Impulsive Response

The response of a first-order system when $f(t) = I\delta(t)$ is obtained by inverting $IG(s)$, which is obtained using Equation (6.37) as

$$x_i(t) = \frac{I}{T} e^{-\frac{t}{T}} u(t) \tag{6.41}$$

Evaluation of Equation (6.41) at $t = 0$ leads to $x(0) = I/T$, which is in conflict with the applied initial condition $x(0) = 0$. This illustrates the necessity of using the unit step function in the response given by Equation (6.41). It is clear that in this context $u(0) = 0$ and $u(t) = 1$ for $t > 0$. Differentiating Equation (6.41) with respect to time leads to

$$\dot{x}_i(t) = -\frac{I}{T^2} e^{-\frac{t}{T}} u(t) + \frac{I}{T} \delta(t) \tag{6.42}$$

Equation (6.42) shows that the derivative of the dependent variable is infinite at $t = 0$, leading to a discontinuity in the dependent variable at $t = 0$. It is shown in Section 6.2 that the response of a first-order system with an impulsive input is discontinuous at $t = 0$.

The discontinuity in the dependent variable can be understood by considering the first-order mechanical system in Table 6.2. The application of the principle of impulse and momentum to a mechanical system reveals that application of an impulse leads to a finite change in linear momentum. The linear momentum of a system is mass times velocity. Since a first-order system has no inertia element and hence no mass, a finite change in linear momentum occurs only if the velocity is infinite. A mechanical system without an inertia element is an idealized system in which the inertia of the spring and viscous damper are neglected. In a real system the mass of these elements absorbs the impulse, leading to a finite change in velocity and a continuous displacement.

---

**Example 6.17**

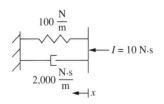

**FIG. 6.12** The bumper, modeled by a spring in parallel with a viscous damper, is subject to an impulse when struck by a vehicle.

The bumper of Figure 6.12 is modeled as a spring in parallel with a viscous damper. Determine the response of the system predicted by this model when it is struck by a body that imparts an impulse of 10 N·s to the bumper.

**Solution**

The differential equation governing the displacement of the bumper is

$$2000\frac{dx}{dt} + 100x = 10\delta(t)$$
$$20\frac{dx}{dt} + x = 0.1\delta(t) \tag{a}$$

The time constant for the system is

$$T = \frac{2{,}000 \text{ N·s/m}}{100 \text{N/m}} = 20 \text{ s} \tag{b}$$

The system response is obtained using Equation (6.41) as

$$x(t) = \frac{0.1\,\text{m·s}}{20\,\text{s}} e^{-.05t}$$
$$= 0.005 e^{-0.05t} \text{ m} \tag{c}$$

### 6.4.3 Step Response

If $f(t) = Zu(t)$ and $x_0 = 0$, Equation (6.36) becomes

$$X_s(s) = \frac{Z}{Ts\left(s + \dfrac{1}{T}\right)} \tag{6.43}$$

Equation (6.43) is inverted using transform pair 11 of Table 5.1 leading to

$$x_s(t) = Z\left(1 - e^{-\frac{t}{T}}\right)u(t) \tag{6.44}$$

The response given by Equation (6.44) and illustrated in Figure 6.13 shows that a first-order system subject to a step input approaches a new steady state. It is shown in Section 6.2, through application of the final value theorem, that $\lim_{t\to\infty} X_s(t) = G(0)$. For a first-order system $X_{s,t} = Z$.

Equation (6.42) is used to determine that the 2 percent settling time for a first-order system is

$$t_s = 3.92T \tag{6.45}$$

and that the 10–90 percent rise time is

$$t_r = [\ln(0.9) - \ln(0.1)]T = 2.20T \tag{6.46}$$

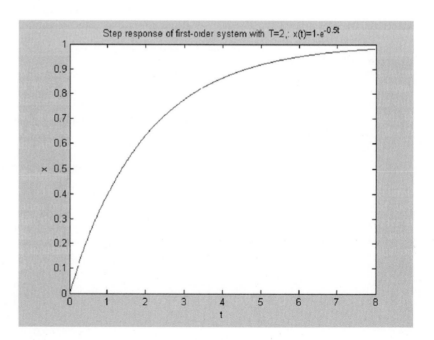

**FIG. 6.13** The step response of a first-order system.

**Example 6.18**

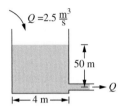

$Q = 2.5 \frac{m^3}{s}$

50 m

$\leftarrow$ 4 m $\rightarrow$

$\rightarrow Q$

**FIG. 6.14** The steady-state operation of the tank of Example 6.18.

The circular tank of Figure 6.14 has a steady-state level of 50 m for a flow rate of 2.5 m³/s. The flow rate is suddenly increased to 2.52 m³/s. Determine (a) the level of the liquid when the system reaches steady state, (b) the time-dependent response of the system, and (c) the 2 percent settling time for the system.

**Solution**

Assuming turbulent flow of the liquid in the pipe, its resistance at the initial steady state is determined using Equation (4.38) as

$$R = 2\frac{H}{Q}$$

$$= 2\frac{50 \text{ m}}{2.5 \text{ m}^3/\text{s}} = 40 \text{ s/m}^2 \qquad \text{(a)}$$

The linearized differential equation for the perturbation in liquid level is

$$A\frac{dh}{dt} + \frac{1}{R}h = qu(t) \qquad \text{(b)}$$

where $q = 0.02 \text{ m}^3/\text{s}$ is the perturbation of the flow rate and $A = \pi/4\,(4 \text{ m})^2 = 12.6 \text{ m}^2$ is the cross-sectional area of the tank. Using these values Equation (b) becomes

$$12.6\frac{dh}{dt} + 0.025h = 0.02u(t)$$

$$5.04 \times 10^2\frac{dh}{dt} + h = 0.8u(t) \qquad \text{(c)}$$

(a) The value of Z used in Equation (6.43) for this liquid level system is obtained from Equation (c) as 0.8. Equation (b) is used to show that, in general, the final value of a single tank liquid level system is $Rq$.

(b) The time constant for the system is $T = 5.04 \times 10^2$ s. The system response is obtained using (6.44) as

$$h(t) = 0.8\left(1 - e^{-1.99 \times 10^{-3}t}\right)u(t) \qquad \text{(d)}$$

The 2 precent settling time is calculated using Equation (6.45) as

$$t_s = 3.92T = 3.92(5.04 \times 10^2 \text{ s}) = 1.98 \times 10^3 \text{ s} \qquad \text{(e)}$$

## 6.4.4 Ramp Response

The ramp response is obtained by inverting

$$X_r(s) = \frac{1}{Ts^2\left(s + \dfrac{1}{T}\right)} \qquad (6.47)$$

Inversion of Equation (6.47) leads to

$$x_r(t) = t - T + Te^{-\frac{t}{T}} \qquad (6.48)$$

## 6.5  SECOND-ORDER SYSTEMS

The standard form of the differential equation for the response of a second-order mechanical system determined in Equation (2.110) is

$$\ddot{x} + 2\zeta\omega_n\dot{x} + \omega_n^2 x = \frac{1}{m} f(t) \tag{6.49}$$

The response of a first-order system is dependent on one system parameter: the time constant. The response of a second-order system is dependent on two system parameters: the damping ratio $\zeta$ and the natural frequency $\omega_n$. Table 6.4 illustrates second-order systems encountered in Chapters 2–4 and the algebraic forms of the system parameters.

Initial conditions required to determine the response of a second-order system are

$$x(0) = x_0 \tag{6.50a}$$
$$\dot{x}(0) = \dot{x}_0 \tag{6.50b}$$

Taking the Laplace transform of Equation (6.49) using appropriate transform properties and applying the initial conditions of Equations (6.50) leads to

$$X(s) = \frac{\dfrac{F(s)}{m} + (s + 2\zeta\omega_n)x_0 + \dot{x}_0}{s^2 + 2\zeta\omega_n s + \omega_n^2} \tag{6.51}$$

**TABLE 6.4 SECOND-ORDER SYSTEMS**

| Name | Schematic | Differential Equation | Natural Frequency | Damped Ratio |
|---|---|---|---|---|
| Mass-spring-viscous damper | | $m\ddot{x} + c\dot{x} + kx = F(t)$ | $\sqrt{\dfrac{k}{m}}$ | $\dfrac{c}{2\sqrt{mk}}$ |
| Series $LRC$ circuit | | $L\dfrac{di}{dt} + Ri + \dfrac{1}{C}\displaystyle\int_0^t i\,dt = v(t)$ | $\sqrt{\dfrac{1}{LC}}$ | $\dfrac{R}{2}\sqrt{\dfrac{C}{L}}$ |
| Parallel $LRC$ circuit | | $C\dfrac{dv}{dt} + \dfrac{v}{R} + \dfrac{1}{L}\displaystyle\int_0^t v\,dt = i(t)$ | $\dfrac{1}{\sqrt{LC}}$ | $\dfrac{1}{2R}\sqrt{\dfrac{L}{C}}$ |

The transfer function of a second-order system governed by Equation (6.49) is

$$G(s) = \frac{1}{m(s^2 + 2\zeta\omega_n s + \omega_n^2)} \tag{6.52}$$

A second-order system is defined as any system having a transfer function with a quadratic denominator that can be written in the form

$$D(s) = s^2 + 2\zeta\omega_n s + \omega_n^2 \tag{6.53}$$

The standard form of the transfer function for a second-order system is

$$G(s) = \frac{As + B}{s^2 + 2\zeta\omega_n s + \omega_n^2} \tag{6.54}$$

For a mechanical system with force input $A = 0$ and $B = 1/m$. For a series $LRC$ circuit $A = 1/L$ and $B = 0$. For a mechanical system with motion input, $A = 2\zeta\omega_n$ and $B = \omega_n^2$. Second-order systems include systems whose mathematical modeling results in two first-order differential equations. In such cases it is usually more convenient to develop transfer functions from the two first-order equations rather than algebraically combine the equations into a second-order differential equation of the form of Equation (6.49). The system's natural frequency and damping ratio are then determined from the transfer function.

## Example 6.19

Consider the two-tank liquid level system of Examples 4.8 and 5.3 and Figure 4.14. The differential equations derived for the perturbations in liquid level are

$$A_1 \frac{dh_1}{dt} + \frac{1}{R_1} h_1 - \frac{1}{R_1} h_2 = q_1(t) \tag{a}$$

$$A_2 \frac{dh_2}{dt} - \frac{1}{R_1} h_1 + \left(\frac{1}{R_1} + \frac{1}{R_2}\right) h_2 = q_2(t) \tag{b}$$

Identify the natural frequency and damping ratio for the system.

### Solution

The transfer functions, defined as when $q_2(t) = 0$

$$G_1(s) = \frac{H_1(s)}{Q_1(s)} \qquad G_2(s) = \frac{H_2(s)}{Q_1(s)} \tag{c}$$

are obtained by taking the Laplace transforms of Equations (a) and (b) and solving the resulting system simultaneously for $H_1(s)$ and $H_2(s)$. The results are

$$G_1(s) = \frac{A_2 s + \dfrac{1}{R_1} + \dfrac{1}{R_2}}{\left(A_1 s + \dfrac{1}{R_1}\right)\left(A_2 s + \dfrac{1}{R_1} + \dfrac{1}{R_2}\right) - \dfrac{1}{R_1^2}}$$

$$= \frac{A_2 s + \dfrac{1}{R_1} + \dfrac{1}{R_2}}{A_1 A_2 \left[s^2 + \left(\dfrac{1}{A_2 R_1} + \dfrac{1}{A_2 R_2} + \dfrac{1}{A_1 R_1}\right) s + \dfrac{1}{A_1 A_2 R_1 R_2}\right]} \tag{d}$$

$$G_2(s) = \cfrac{1}{R_1 A_1 A_2 \left[ s^2 + \left( \cfrac{1}{A_2 R_1} + \cfrac{1}{A_2 R_2} + \cfrac{1}{A_1 R_1} \right) s + \cfrac{1}{A_1 A_2 R_1 R_2} \right]} \tag{e}$$

The two-tank liquid level system is a second-order system with the natural frequency

$$\omega_n = \sqrt{\frac{1}{A_1 A_2 R_1 R_2}} \tag{f}$$

and damping ratio

$$\zeta = \frac{1}{2\omega_n} \left( \frac{1}{A_2 R_1} + \frac{1}{A_2 R_2} + \frac{1}{A_1 R_1} \right)$$

$$= \frac{1}{2} \sqrt{\frac{A_1 R_2}{A_2 R_1} + \frac{A_1 R_1}{A_2 R_2} + \frac{A_2 R_1}{A_1 R_1}} \tag{g}$$

For $G_1(s)$, $A = 1/A_1$, $B = \frac{1}{A_1 A_2} \left( \frac{1}{R_1} + \frac{1}{R_2} \right)$ while for $G_2(s)$, $A = 0$, $B = 1/R_1 A_1 A_2$.

## 6.5.1 Free Response

The free response of a system is a transient response that occurs due to nonzero initial conditions. Nonzero initial conditions are the result of an initial energy. When the mechanical system in Table 6.4 is displaced from its equilibrium position, potential energy is stored in the spring. When the inertia element has a nonzero initial velocity the system has an initial kinetic energy. The total energy in the system at $t = 0$ is

$$E = \tfrac{1}{2} k x_0^2 + \tfrac{1}{2} m \dot{x}_0^2 \tag{6.55}$$

If the capacitor in the series *LRC* circuit is initially charged then energy is stored in an electric field and the system has a nonzero value of $q(0)$. If current is flowing in the circuit at $t = 0$ then energy is stored in the inductor, resulting in a nonzero value of $dq/dt(0)$.

$$E = \tfrac{1}{2} L \left[ \frac{dq}{dt}(0) \right]^2 + \tfrac{1}{2} \frac{1}{C} [q(0)]^2 \tag{6.56}$$

The poles of the transfer function for a second-order system are the roots of $D(s)$,

$$s_1 = -\zeta \omega_n - \omega_n \sqrt{\zeta^2 - 1} \tag{6.57a}$$

$$s_2 = -\zeta \omega_n + \omega_n \sqrt{\zeta^2 - 1} \tag{6.57b}$$

The mathematical form of the free response depends on the value of the damping ratio $\zeta$. Four cases are considered.

**Case 1: $\zeta = 0$ (Undamped Response)** For an undamped system, $\zeta = 0$, the poles of the transfer function are purely imaginary, $s_{1,2} = \pm j\omega_n$,

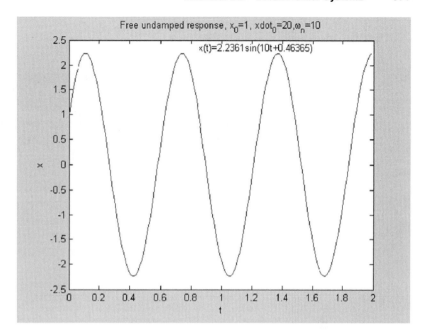

**FIG. 6.15** The free undamped response of a second-order system.

and Equation (6.51) becomes

$$X(s) = \frac{x_0 s + \dot{x}_0}{s^2 + \omega_n^2} \tag{6.58}$$

Equation (6.58) is inverted using transform pairs (2) and (3) from Table 5.1 yielding

$$x(t) = x_0 \cos(\omega_n t) + \frac{\dot{x}_0}{\omega_n} \sin(\omega_n t) \tag{6.59}$$

An alternate representation of the undamped response is

$$x(t) = A \sin(\omega_n t + \phi) \tag{6.60}$$

where the **amplitude** $A$ and the **phase** $\phi$ are given by

$$A = \sqrt{x_0^2 + (\dot{x}_0/\omega_n)^2} \tag{6.61}$$

$$\phi = \tan^{-1}(\omega_n x_0/\dot{x}_0) \tag{6.62}$$

The free response of a second-order undamped system, illustrated in Figure 6.15, is periodic of period

$$T = \frac{2\pi}{\omega_n} \tag{6.63}$$

One period has occurred when the system returns to the state defined by the initial conditions. This is equivalent to the system executing one **cycle**. Thus the period represents the time it takes the system to execute one cycle, and the reciprocal of the period, called the **frequency**, is the number of cycles executed per unit of time. Using seconds as the unit of time the frequency is measured in cycles per second [Hertz (Hz)]. The frequency in

Hertz is determined as

$$f = \frac{\omega_n}{2\pi} \tag{6.64}$$

After execution of one cycle the argument of the sine function increases by $2\pi$ radians. Thus 1 cycle $= 2\pi$ radians and from Equation (6.64) it is clear that $\omega_n$ is the frequency in radians per second (r/s). The system parameter $\omega_n$ is called the system's **natural frequency** because it is the frequency of the system response in the absence of viscous damping and external excitation. The natural frequency is the frequency of the response due to nonzero initial conditions. The natural frequency is dependent on system parameters and has no dependence on the initial conditions.

During the free response of an undamped system, energy is neither added to the system by an external force nor dissipated from the system. Thus energy is conserved. The amplitude of the system response being constant is a statement of Conservation of Energy. The amplitude of an undamped system is proportional to the energy stored in the system. Equations (6.55) and (6.60) are used to show that for a mechanical system

$$A = \sqrt{\frac{2E}{k}} \tag{6.65}$$

while for a series $LRC$ circuit

$$A = \sqrt{2CE} \tag{6.66}$$

The phase represents the difference between the system response and a purely sinusoidal response. If $x_0 = 0$ then $\phi = 0$. If $\dot{x}_0 = 0$ and $x_0 > 0$, then $\phi = \pi/2$. If $\dot{x}_0 = 0$ and $x_0 < 0$ then $\phi = -\pi/2$. If both initial conditions are nonzero and of the same sign, then the phase is positive, $0 < \phi < \pi/2$. If both initial conditions are nonzero and of opposite signs, then the phase is negative, $-\pi/2 < \phi < 0$. If the phase is positive the response is said to lead a sinusoidal response. If the phase is negative the response is said to lag a sinusoidal response.

**Case 2: $0 < \zeta < 1$ (Underdamped Response)**   For $0 < \zeta < 1$, the poles of the transfer function are complex conjugates with negative real parts. The inversion of Equation (6.51) is achieved by completing the square in $D(s)$

$$X(s) = \frac{(s + \zeta\omega_n)x_0 + \zeta\omega_n x_0 + \dot{x}_0}{(s + \zeta\omega_n)^2 + \omega_d^2} \tag{6.67}$$

where

$$\omega_d = \omega_n\sqrt{1 - \zeta^2} \tag{6.68}$$

Equation (6.67) is inverted using Table 5.1 leading to

$$x(t) = A_d e^{-\zeta\omega_n t}\sin(\omega_d t + \phi_d) \tag{6.69}$$

where the amplitude and phase for the damped system are

$$A_d = \sqrt{x_0^2 + \left(\frac{\dot{x}_0 + \zeta\omega_n x_0}{\omega_d}\right)^2} \tag{6.70}$$

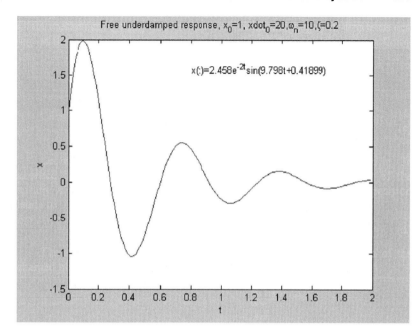

Free underdamped response, $x_0=1$, xdot$_0=20$, $\omega_n=10$, $\zeta=0.2$

$x(:)=2.458e^{-2t}\sin(9.798t+0.41899)$

**FIG. 6.16** The free underdamped response of a second-order system.

$$\phi_d = \tan^{-1}\left(\frac{x_0\omega_d}{\dot{x}_0 + \zeta\omega_n x_0}\right) \tag{6.71}$$

The response of an underdamped system is illustrated in Figure 6.16. The response is not periodic, but cyclic with an exponentially decaying amplitude. The time between peaks on the response curve constitutes one cycle. The time to execute one cycle is the **damped period**

$$T_d = \frac{2\pi}{\omega_d} \tag{6.72}$$

and $\omega_d$ is called the **damped natural frequency**.

An underdamped system has a component that dissipates energy, such as a viscous damper in a mechanical system or a resistor in an electric circuit. The resistance provided by the dissipative mechanism slows the system down. The larger the resistance is, the longer it takes the system to execute one cycle and the smaller the damped frequency. For an underdamped system the energy dissipated during a cycle is less than the energy stored in the system. The energy dissipated over one cycle is a constant fraction of the energy stored in the system at the beginning of the cycle.

**Example 6.20**

A simplified model of a vehicle suspension system is illustrated in Figure 6.17. For parameter values of $m = 200$ kg, $k = 5 \times 10^5$ N·m and $c = 1.5 \times 10^4$ N·s/m, determine (a) the damping ratio for the system, (b) the system's damped period, (c) the system response when the vehicle experiences a sudden change in road contour that leads to the initial conditions $x(0) = 0.001$ m, $\dot{x}(0) = 0$.

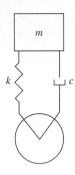

**FIG. 6.17** The simplified one-degree-of-freedom model for the vehicle suspension system in Example 6.20.

## Solution

The differential equation governing the free response of the vehicle after it encounters a sudden change in road contour is

$$m\ddot{x} + c\dot{x} + kx = 0 \tag{a}$$

which is put in the standard form of Equation (6.49) by dividing by $m$ leading to

$$\ddot{x} + \frac{c}{m}\dot{x} + \frac{k}{m}x = 0 \tag{b}$$

The natural frequency of the system is determined from Equation (b) as

$$\omega_n = \sqrt{\frac{k}{m}}$$

$$= \sqrt{\frac{5 \times 10^5 \text{ N/m}}{200 \text{ kg}}} = 50 \text{ r/s} \tag{c}$$

(a) The damping ratio is obtained from Equations (b) and (c) as

$$\zeta = \frac{c}{2m\omega_n}$$

$$= \frac{1.5 \times 10^4 \text{ N·s/m}}{2(200 \text{ kg})(50 \text{ r/s})} = 0.75 \tag{d}$$

(b) The damped natural frequency is calculated using Equation (6.68) as

$$\omega_d = \omega_n\sqrt{1 - \zeta^2}$$

$$= 50 \text{ r/s}\sqrt{1 - (0.75)^2} = 33.1 \text{ r/s} \tag{e}$$

The damped period is calculated from Equation (6.72) as

$$T_d = \frac{2\pi}{\omega_d}$$

$$= \frac{2\pi}{33.1 \text{ r/s}} = 0.190 \text{ s} \tag{f}$$

(c) For the given initial conditions the amplitude and phase are calculated using Equations (6.70) and (6.71) as

$$A_d = \sqrt{(0.001)^2 + \left(\frac{0.75(50)(0.001)}{33.1}\right)^2} = 1.51 \times 10^{-3} \text{ m} \tag{g}$$

$$\phi_d = \tan^{-1}\left[\frac{(0.001)(33.1)}{(0.75)(50)(0.001)}\right] = 0.723 \text{ r} \tag{h}$$

The system response is calculated using Equation (6.69) as

$$x(t) = 1.51 \times 10^{-3}e^{-37.5t}\sin(33.1t + 0.723) \text{ m} \tag{i}$$

The **logarithmic decrement** $\delta$ is defined as the natural logarithm of the ratio of amplitudes on successive cycles and can be evaluated by

$$\delta = \ln\left(\frac{x(t)}{x(t + T_d)}\right) \tag{6.73}$$

Substitution of Equation (6.69) into Equation (6.73) leads to

$$\delta = \ln\left(\frac{A_d e^{-\zeta \omega_n t} \sin(\omega_d t)}{A_d e^{-\zeta \omega_n (t + T_d)} \sin(\omega_d t + \omega_d T_d)}\right)$$

$$= \zeta \omega_n T_d$$

$$= \frac{2\pi\zeta}{\sqrt{1 - \zeta^2}} \tag{6.74}$$

The logarithmic decrement provides a means to measure the damping ratio, when it is not known, for an underdamped system.

The repeated application of Equation (6.73) over $n$ full cycles of motion leads to an alternate formulation for the logarithmic decrement as

$$\delta = \frac{1}{n}\ln\left(\frac{x(t)}{x(t + nT_d)}\right) \tag{6.75}$$

---

### Example 6.21

A 150-kg machine is mounted on an elastic foundation. A test is performed to determine the stiffness and damping properties of the foundation. The machine is given a 100-N·s impulse. The resulting response is oscillatory with exponentially decreasing amplitude. It is observed that the amplitude decreases by 85 percent over 5 cycles, which take 1.2 s to execute. Determine the stiffness and damping coefficient of the foundation.

#### Solution

The system is modeled as a one-degree-of-freedom mass-spring-viscous damper system. Since the motion is oscillatory with exponentially decaying amplitude the system is underdamped. The damped period is calculated as

$$T_d = \frac{1.2\,\text{s}}{5\,\text{cycles}} = 0.24\,\text{s} \tag{a}$$

The damped natural frequency is then computed as

$$\omega_d = \frac{2\pi}{T_d} = 26.2\,\text{r/s} \tag{b}$$

The logarithmic decrement is calculated using Equation (6.75) as

$$\delta = \tfrac{1}{5}\ln\left(\frac{A}{0.15A}\right) = 0.379 \tag{c}$$

where $A$ is the amplitude at the beginning of the first cycle. Equation (6.72) is rearranged to calculate the damping ratio

$$\zeta = \frac{\delta}{\sqrt{\delta^2 + 4\pi^2}} = 0.0603 \qquad (d)$$

The natural frequency is calculated using Equation (6.68)

$$\omega_n = \frac{\omega_d}{\sqrt{1 - \zeta^2}} = 26.2 \text{ r/s} \qquad (e)$$

The stiffness and damping coefficient of the foundation are then determined by

$$k = m\omega_n^2 = (150 \text{ kg})(26.2 \text{ r/s})^2 = 1.03 \times 10^5 \text{ N/m} \qquad (f)$$

$$c = 2\zeta m\omega_n = 2(0.0603)(150 \text{ kg})(26.2 \text{ r/s}) = 4.74 \times 10^2 \text{ N·s/m} \qquad (g)$$

**Case 3: $\zeta = 1$ (Critically Damped System)**   For $\zeta = 1$ the poles of the transfer function for a second-order system are equal, $s_{1,2} = -\omega_n$. In this case Equation (6.41) becomes

$$X(s) = \frac{x_0}{s + \omega_n} + \frac{\dot{x}_0 + \omega_n x_0}{(s + \omega_n)^2} \qquad (6.75)$$

The transform in Equation (6.75) is inverted leading to

$$x(t) = x_0 e^{-\omega_n t} + (\dot{x}_0 + \omega_n x_0) t e^{-\omega_n t} \qquad (6.76)$$

Equation (6.76) is plotted in Figure 6.18 for several sets of initial conditions. The response of a critically damped system decays exponentially. The system does not execute one full cycle of motion. If neither initial condition is zero and the signs of the initial conditions are opposite, then

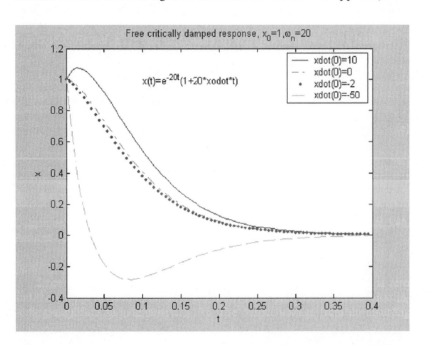

**FIG. 6.18** The free response of a critically damped system.

the system overshoots its equilibrium position if

$$\frac{x_0}{\dot{x}_0 + \omega_n x_0} < 0 \tag{6.77}$$

**Case 4: $\zeta > 1$ (Overdamped System)** For $\zeta > 1$ the poles of the transfer function are real and positive.

$$s_1 = -\omega_n\left(\zeta + \sqrt{\zeta^2 - 1}\right) \tag{6.78a}$$

$$s_2 = -\omega_n\left(\zeta - \sqrt{\zeta^2 - 1}\right) \tag{6.78b}$$

Equation (6.51) is written as

$$X(s) = \frac{(s + \zeta\omega_n)x_0 + \dot{x}_0}{(s - s_1)(s - s_2)} \tag{6.79}$$

Use of transform pairs 16 and 18 from Table 5.1 in the inversion of the transform of Equation (6.79) yields

$$x(t) = \frac{1}{2\sqrt{\zeta^2 - 1}}\left\{\left[x_0\left(-\zeta + \sqrt{\zeta^2 - 1}\right) - \frac{\dot{x}_0}{\omega_n}\right]e^{s_1 t}\right.$$
$$\left. + \left[x_0\left(\zeta + \sqrt{\zeta^2 - 1}\right) + \frac{\dot{x}_0}{\omega_n}\right]e^{s_2 t}\right\} \tag{6.80}$$

The free response of an overdamped system, as illustrated in Figure 6.19 for several sets of initial conditions, decays exponentially. An overdamped system takes longer to approach equilibrium than a critically damped system.

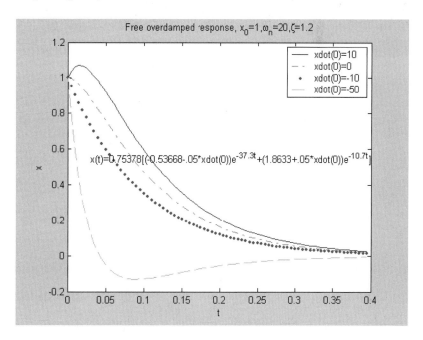

**FIG. 6.19** The free response of an overdamped system.

**Example 6.22**

The parallel $LRC$ circuit of Example 3.12 and Figure 3.17 has a resistance $R = 500\ \Omega$ and inductance $L = 0.25$ H. The capacitor has an initial charge of $q = 12.5\ \mu$C. Determine the voltage after the switch is closed at $t = 0$. if (a) $C = 25\ \mu$F, (b) $C = 0.25\ \mu$F, and (c) $C = 0.16\ \mu$F.

**Solution**

It is shown in Example 3.12 that the voltage change across each of the circuit elements is the same for the parallel $LRC$ circuit. The differential equation governing the voltage change across the resistor is obtained in Equation (k) of Example 3.12 as

$$C\frac{d^2v}{dt^2} + \frac{1}{R}\frac{dv}{dt} + \frac{1}{L}v = 0 \tag{a}$$

The initial conditions corresponding to an initial charge of $q$ on the capacitor are given as Equations (j) and (m) of Example 3.10 as

$$v(0) = -\frac{q}{C} \tag{b}$$

$$\frac{dv}{dt}(0) = \frac{q}{RC^2} \tag{c}$$

The natural frequency and damping ratio for this second-order system are determined from Equation (a) as

$$\omega_n = \sqrt{\frac{1}{LC}} \tag{d}$$

$$\zeta = \frac{1}{2RC\omega_n}$$

$$= \frac{1}{2R}\sqrt{\frac{L}{C}} \tag{e}$$

(a) For $C = 25\ \mu$F the natural frequency, damping ratio, and initial conditions are calculated as

$$\omega_n = \sqrt{\frac{1}{(0.25\ \text{H})(25 \times 10^{-6}\ \text{F})}} = 400\ \text{r/s} \tag{f}$$

$$\zeta = \frac{1}{2(500\ \Omega)}\sqrt{\frac{0.25\ \text{H}}{25 \times 10^{-6}\ \text{F}}} = 0.1 \tag{g}$$

$$v(0) = -\frac{12.5 \times 10^{-6}\ \text{C}}{25 \times 10^{-6}\ \text{F}} = -0.5\ \text{V} \tag{h}$$

$$\frac{dv}{dt}(0) = \frac{12.5 \times 10^{-6}\ \text{C}}{(500\ \Omega)(25 \times 10^{-6}\ \text{F})^2} = 40\ \text{V/s} \tag{i}$$

Since $\zeta < 1$ the system is underdamped. The damped natural frequency is

$$\omega_d = 400\ \text{r/s}\sqrt{1 - (0.1)^2} = 398.0\ \text{r/s} \tag{j}$$

The response is determined from Equation (6.69) as

$$v(t) = A_d e^{-40t} \sin(398t + \phi_d) \qquad \text{(k)}$$

where the initial conditions are used in Equations (6.70) and (6.71) to obtain

$$A_d = \sqrt{(-0.5 \text{ V})^2 + \left[\frac{(40 \text{ V/s}) + (0.1)(400 \text{ r/s})(-0.5 \text{ V})}{398.0 \text{ r/s}}\right]^2} = 0.503 \text{ V} \qquad \text{(l)}$$

$$\phi_d = \tan^{-1}\left[\frac{(-0.5 \text{ V})(398.0 \text{ r/s})}{(40 \text{ V/s}) + (0.1)(400 \text{ r/s})(-0.5 \text{ V})}\right] = -1.47 \text{ r} \qquad \text{(m)}$$

Substituting Equations (l) and (m) into Equation (k) leads to

$$v(t) = 0.503 e^{-40t} \sin(398.0t - 1.47) \text{ V} \qquad \text{(n)}$$

(b) For $C = 0.25 \text{ μF}$ the natural frequency, damping ratio, and initial conditions are calculated using Equations (d), (e), (b), and (c) as $\omega_n = 4{,}000 \text{ r/s}$, $\zeta = 1$, $v(0) = -50 \text{ V}$, and $dv/dt(0) = 4 \times 10^5 \text{ V/s}$. The damping ratio is one; thus the system is critically damped. The response is determined using Equation (6.76) as

$$v(t) = (-50 \text{ V})e^{-4000t} + \left[4 \times 10^5 \text{ V/s} + (4{,}000 \text{ r/s})(-50 \text{ V})\right]t e^{-4000t}$$

$$= \left(2 \times 10^5 t - 50\right)e^{-4000t} \text{ V} \qquad \text{(o)}$$

(c) For $C = 0.16 \text{ μF}$ the natural frequency, damping ratio, and initial conditions are determined using Equations (d), (e), (b), and (c) as $\omega_n = 5{,}000 \text{ r/s}$, $\zeta = 1.25$, $v(0) = -78.1 \text{ V}$, and $dv/dt(0) = 9.77 \times 10^5 \text{ V/s}$. Since the damping ratio is greater than one, the system is overdamped. The response is given by Equation (6.80) with

$$s_1 = -5{,}000\left[1.25 + \sqrt{(1.25)^2 - 1}\right] = -10{,}000 \qquad \text{(p)}$$

$$s_2 = -5{,}000\left[1.25 - \sqrt{(1.25)^2 - 1}\right] = -2{,}500 \qquad \text{(q)}$$

Equation (6.80) is evaluated as

$$v(t) = \frac{1}{2(0.75)}\left\{\left[(-78.1 \text{ V})(-0.5) - \frac{9.77 \times 10^5 \text{ V/s}}{5{,}000 \text{ r/s}}\right]e^{-10{,}000t}\right.$$

$$\left. + \left[(-78.1 \text{ V})(2) + \frac{9.77 \times 10^5 \text{ V/s}}{5{,}000 \text{ r/s}}\right]e^{-2{,}500t}\right\}$$

$$= \left(26.16 e^{-2500t} - 104.2 e^{-10000t}\right) \text{ V} \qquad \text{(r)}$$

## 6.5.2 Impulsive Response

An impulsive force applied to a mechanical system leads to an instantaneous discrete change in velocity. An impulsive voltage applied to a parallel *LRC* circuit leads to an instantaneous discrete change in the current through the inductor. An impulsive current applied to a series $LRC$ circuit leads to an instantaneous discrete change in voltage in the capacitor.

These changes can be observed by examining the differential equations with an impulsive input. Consider for example the integrodifferential equation for the parallel $LRC$ circuit when subject to an impulsive voltage

$$L\frac{di}{dt} + Ri + \frac{1}{C}\int_0^t idt = I_V\delta(t) \tag{6.81}$$

If there is no current in the circuit when the impulse is applied at $t = 0$, then evaluation of Equation (6.81) at $t = 0$, leads to

$$L\frac{di}{dt}(0) = I_V\delta(0) \tag{6.82}$$

Thus $di/dt(0)$ is infinite and the current is discontinuous at $t = 0$. The current at $t = 0^+$ is $I_V/L$. The impulsive response of this second-order system is similar to that of a circuit with $i(0) = I_V/L$. The mathematical forms of the responses are identical for $t > 0$.

**TABLE 6.5 IMPULSIVE RESPONSE OF SECOND-ORDER SYSTEMS**

| Transfer Function | Condition | $x(t)$ |
|---|---|---|
| $\dfrac{1}{s^2 + 2\zeta\omega_n s + \omega_n^2}$ | Underdamped $\omega_d = \omega_n\sqrt{1-\zeta^2}$ | $\dfrac{1}{\omega_d}e^{-\zeta\omega_n t}\sin(\omega_d t)$ |
| $\dfrac{1}{s^2 + 2\zeta\omega_n s + \omega_n^2}$ | Critically damped | $te^{-\omega_n t}$ |
| $\dfrac{1}{s^2 + 2\zeta\omega_n s + \omega_n^2}$ | Overdamped $s_1 = -\omega_n\left(\zeta + \sqrt{\zeta^2 - 1}\right)$ $s_2 = -\omega_n\left(\zeta - \sqrt{\zeta^2 - 1}\right)$ | $\dfrac{1}{s_2 - s_1}\left(e^{s_2 t} - e^{s_1 t}\right)$ |
| $\dfrac{s}{s^2 + 2\zeta\omega_n s + \omega_n^2}$ | Underdamped $\omega_d = \omega_n\sqrt{1-\zeta^2}$ | $e^{-\zeta\omega t}\left[\cos(\omega_d t) + \dfrac{\zeta\omega_n}{\omega_d}\sin(\omega_d t)\right]$ |
| $\dfrac{s}{s^2 + 2\zeta\omega_n s + \omega_n^2}$ | Critically damped | $e^{-\omega_n t}(1 + \omega_n t)$ |
| $\dfrac{s}{s^2 + 2\zeta\omega_n s + \omega_n^2}$ | Overdamped $s_1 = -\omega_n\left(\zeta + \sqrt{\zeta^2 - 1}\right)$ $s_2 = -\omega_n\left(\zeta - \sqrt{\zeta^2 - 1}\right)$ | $\dfrac{1}{s_2 - s_1}\left(s_2 e^{s_1 t} - s_1 e^{s_2 t}\right)$ |

In a similar fashion it is reasoned that the impulsive response of a second-order mechanical system is equal to the free response due to a non-zero initial velocity equal to $1/m$ and multiplied by $u(t)$. Impulsive responses of second-order systems are catalogued in Table 6.5 for systems with $A = 1$ and $B = 0$ and for systems with $A = 0$ and $B = 1$. The impulsive response for nonzero values of both $A$ and $B$ is obtained by forming a linear combination of the appropriate responses.

## Example 6.23

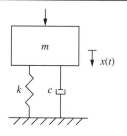

**FIG. 6.20** The schematic representation of the stamping machine of Example 6.23.

A stamping machine, illustrated schematically in Figure 6.20, is of mass 200 kg and rests on an isolator that is modeled as a spring of stiffness $3 \times 10^5$ N/m in parallel with a viscous damper of damping coefficient 5,000 N·s/m. During its operation the machine is subject to an impulse of magnitude 20 N·s. (a) Determine the response of the machine after the application of the impulse. (b) Determine the maximum displacement of the machine after the application of the impulse and the time at which the maximum displacement occurs. (c) Determine the maximum value of the force developed in the isolator, which is calculated by

$$F = c\dot{x} + kx \tag{a}$$

(d) Determine the isolator efficiency, which is defined by

$$E = \frac{I^2}{2mF_{\max}x_{\max}} \tag{b}$$

### Solution

The mathematical model for the system is

$$m\ddot{x} + c\dot{x} + kx = I\delta(t) \tag{c}$$

which is written in the standard form of Equation (6.49) as

$$\ddot{x} + 2\zeta\omega_n\dot{x} + \omega_n^2 x = \frac{I}{m}\delta(t) \tag{d}$$

The natural frequency and damping ratio of the system are

$$\omega_n = \sqrt{k/m}$$

$$= \sqrt{\frac{3 \times 10^5 \text{ N/m}}{200 \text{ kg}}} = 38.7 \text{ r/s} \tag{e}$$

$$\zeta = \frac{c}{2m\omega_n}$$

$$= \frac{5,000 \text{ N·s/m}}{2(200 \text{ kg})(38.7 \text{ r/s})} = 0.323 \tag{f}$$

The damped natural frequency for this underdamped system is

$$\omega_d = \omega_n \sqrt{1 - \zeta^2}$$

$$= 38.7 \text{ r/s} \sqrt{1 - (0.323)^2} = 36.6 \text{ r/s} \tag{g}$$

(a) The impulsive response of an underdamped system is obtained using Table 6.5 as

$$x(t) = \frac{I}{m\omega_d} e^{-\zeta\omega_n t} \sin(\omega_d t) \tag{h}$$

Substitution of the given and calculated values in Equation (h) leads to

$$x(t) = \frac{20 \text{ N·s}}{(200 \text{ kg})(36.6 \text{ r/s})} e^{-(0.323)(38.7)t} \sin(36.6t)$$

$$= 2.73 e^{-12.5t} \sin(36.6t) \text{ mm} \tag{i}$$

(b) The maximum displacement occurs the first time the velocity is zero. From Equation (i) the velocity is determined as

$$\dot{x}(t) = e^{-12.5t}[-34.1 \sin(36.6t) + 99.9 \cos(36.6t)] \text{ mm/s} \tag{j}$$

The time $t_p$ at which the maximum displacement occurs is obtained by setting $\dot{x}$ to zero, which leads to

$$-34.1 \sin(36.6t_p) + 99.9 \cos(36.6t_p) = 0$$

$$\tan(36.6t_p) = 99.9/34.1 = 2.91$$

$$t_p = \frac{1}{36.6} \tan^{-1}(2.91) = 3.39 \times 10^{-2} \text{ s} \tag{k}$$

The maximum displacement is

$$x_{\max} = x(t_p)$$

$$= 2.73 e^{-12.5(3.39 \times 10^{-2})} \sin[36.6(3.39 \times 10^{-2})] = 1.69 \text{ mm} \tag{l}$$

(c) The force developed in the isolator is

$$F = c\dot{x} + kx$$

$$= 5,000 \text{ N·s/m}\{e^{-12.5t}[-34.1 \sin(36.6t) + 99.9 \cos(36.6t)] \text{ mm/s}\}$$

$$+ 3 \times 10^5 \text{ N/m}[2.73 e^{-12.5t} \sin(36.6t) \text{ mm}]$$

$$= e^{-12.5t}[648.5 \sin(36.6t) + 499.5 \cos(36.6t)] \text{ N} \tag{m}$$

The time $t_f$ at which the maximum force occurs is obtained by setting $\dot{F}$ to zero leading to

$$0 = -12.5 e^{-12.5t_f}\left[648.5 \sin(36.6t_f) + 499.5 \cos(36.6t_f)\right]$$

$$+ e^{-12.5t_f}\left[(648.5)(36.6) \cos(36.6t_f) - (499.5)(36.6) \sin(36.6t_f)\right]$$

$$= e^{-12.5t_f}\left[-2.64 \times 10^4 \sin(36.6t_f) + 1.75 \times 10^4 \cos(36.6t_f)\right]$$

$$t_f = \frac{1}{36.6} \tan^{-1}\left(\frac{1.75 \times 10^4}{2.64 \times 10^4}\right) = 1.60 \times 10^{-2} \text{ s} \tag{n}$$

The maximum force is

$$F_{\max} = F(t_f) = e^{-12.5(1.60 \times 10^{-2})}\{648.5 \sin\left[(36.6)(1.60 \times 10^{-2})\right]$$
$$+ 499.5 \cos\left(36.6\right)(1.60 \times 10^{-2})\}$$
$$= 634.2 \text{ N} \tag{o}$$

(d) The isolator efficiency is calculated as

$$E = \frac{I^2}{2mF_{\max}x_{\max}}$$

$$= \frac{(20 \text{ N/s})^2}{2(200 \text{ kg})(634.2 \text{ N})(1.69 \times 10^{-3} \text{ m})} = 0.933 \tag{p}$$

### 6.5.3  Step Response

The response of a second-order mechanical system due to a step input of the form $F(t) = F_0 u(t)$ is obtained by inverting

$$X_s(s) = \frac{F_0}{ms(s^2 + 2\zeta\omega_n s + \omega_n^2)} \tag{6.83}$$

Recalling that the final value of a stable system after application of a step input is $G(0)$ leads to $x_{s,f} = F_0/m\omega_n^2$. If the system is undamped ($\zeta = 0$), the response reaches a steady state, in which periodic oscillations occur with a mean value of $F_0/m\omega_n^2$. If $\zeta > 0$ the response reaches a steady state, in which the response asymptotically approaches $F_0/m\omega_n^2$.

The step response of a series $LRC$ circuit is obtained by inverting

$$I(s) = \frac{V_0 s}{L(s^2 + 2\zeta\omega_n s + \omega_n^2)} \tag{6.84}$$

Application of the final value theorem to Equation (6.84) shows that $\lim_{t \to \infty} i(t) = 0$. Thus the current in a circuit with a step increase in voltage eventually decays to zero.

The responses of undamped, underdamped, critically damped, and overdamped systems due to a step input are catalogued in Table 6.6 for the cases $A = 0$ and $B = 1$ and $A = 1$ and $B = 0$. The step response for nonzero values of both $A$ and $B$ is obtained by forming a linear combination of the appropriate response. The step responses are illustrated in Figure 6.21.

The response specifications of settling time, rise time, overshoot, and peak time are determined from the system response and depend on the damping ratio and natural frequency. Since the final value of a system with $B = 0$ is zero, the definitions of rise time percent overshoot do not apply to the response.

**TABLE 6.6 STEP RESPONSE OF SECOND-ORDER SYSTEMS**

| Transfer Function | Condition | $x(t)$ |
|---|---|---|
| $\dfrac{1}{s^2 + 2\zeta\omega_n s + \omega_n^2}$ | Underdamped<br><br>$\omega_d = \omega_n\sqrt{1-\zeta^2}$ | $\dfrac{1}{\omega_n^2}\left[1 - e^{-\zeta\omega_n t}\cos(\omega_d t) + \dfrac{\zeta\omega_n}{\omega_d}e^{-\zeta\omega_n t}\sin(\omega_d t)\right]$ |
| $\dfrac{1}{s^2 + 2\zeta\omega_n s + \omega_n^2}$ | Critically damped | $\dfrac{1}{\omega_n^2}\left[1 - e^{-\omega_n t} - \omega_n t e^{-\omega_n t}\right]$ |
| $\dfrac{1}{s^2 + 2\zeta\omega_n s + \omega_n^2}$ | Overdamped<br><br>$s_1 = -\omega_n\left(\zeta + \sqrt{\zeta^2 - 1}\right)$<br><br>$s_2 = -\omega_n\left(\zeta - \sqrt{\zeta^2 - 1}\right)$ | $\dfrac{1}{s_1 s_2(s_1 - s_2)}\left(s_1 - s_2 + s_2 e^{s_1 t} - s_1 e^{s_2 t}\right)$ |
| $\dfrac{s}{s^2 + 2\zeta\omega_n s + \omega_n^2}$ | Underdamped<br><br>$\omega_d = \omega_n\sqrt{1-\zeta^2}$ | $e^{-\zeta\omega_n t}\sin(\omega_d t)$ |
| $\dfrac{s}{s^2 + 2\zeta\omega_n s + \omega_n^2}$ | Critically damped | $te^{-\omega_n t}$ |
| $\dfrac{s}{s^2 + 2\zeta\omega_n s + \omega_n^2}$ | Overdamped<br><br>$s_1 = -\omega_n\left(\zeta + \sqrt{\zeta^2 - 1}\right)$<br><br>$s_2 = -\omega_n\left(\zeta - \sqrt{\zeta^2 - 1}\right)$ | $\dfrac{1}{s_2 - s_1}\left(e^{s_2 t} - e^{s_1 t}\right)$ |

**FIG. 6.21** The responses of second-order systems due to unit step input: (a) The underdamped mechanical system with $\omega_n = 10$ r/s and $\zeta = 0.1$; (b) the critically damped mechanical system with $\omega_n = 10$ r/s; (c) the overdamped mechanical system with $\omega_n = 10$ r/s and $\zeta = 1.3$; (d) the undamped series $LRC$ circuit with $\omega_n = 10$ r/s and $\zeta = 0.1$; (e) the critically damped series $LRC$ circuit with $\omega_n = 10$ r/s; (f) the overdamped series $LRC$ circuit with $\omega_n = 10$ r/s and $\zeta = 1.3$.

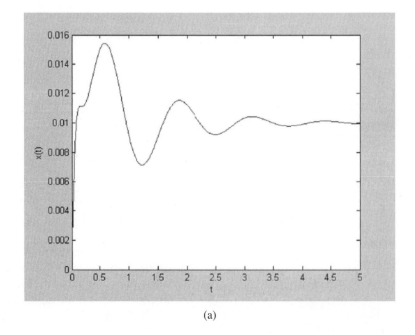

(a)

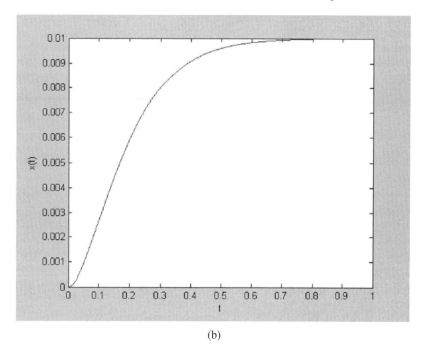

(b)

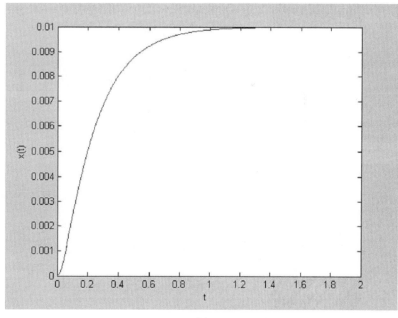

(c)

**FIG. 6.21** (*Continued*)

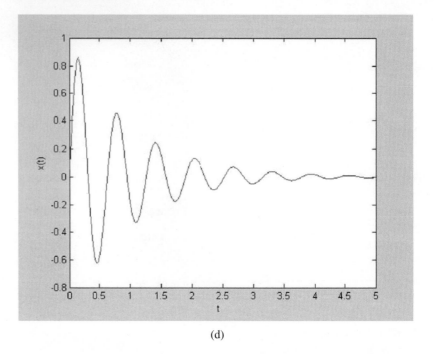

(d)

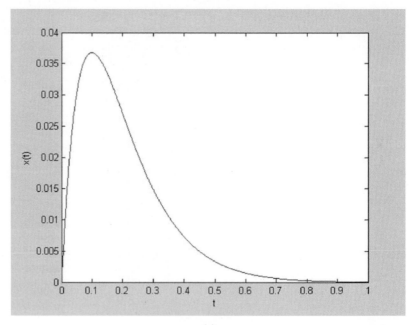

**FIG. 6.21** (*Continued*)

(e)

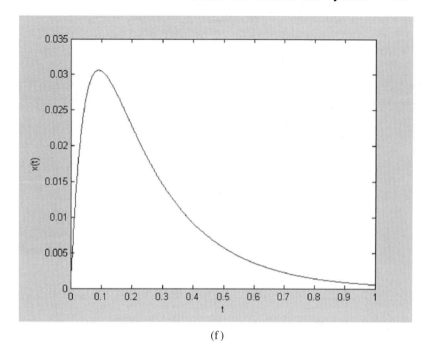

**FIG. 6.21** (*Continued*)                                                        (f)

## Example 6.24

The steady-state operation of a two-tank system is illustrated in Figure 6.22. Both tanks are circular in cross section. The input flow rate has a step increase of $0.2$ m$^3$/s. Determine (a) the steady-state liquid levels in each tank after the increase in flow rate, (b) the time-dependent response of each tank, (c) the 2 percent settling time, and (d) the 10–90 percent rise time. (e) Use MATLAB to plot the system response.

### Solution

The application of Conservation of Mass to the tanks in the steady state shows that the flow rates through each of the pipes is equal to the input flow rate. Thus the head losses in the pipes at the original steady state are

$$h_{\ell 1} = H_2 - H_1 = 5.1 \text{ m} \tag{a}$$

$$h_{\ell 2} = H_2 = 8.5 \text{ m} \tag{b}$$

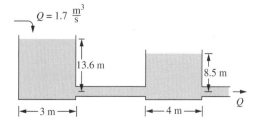

**FIG. 6.22** The steady-state operation of the two-tank liquid-level system of Example 6.24.

The resistances of the pipes at steady state are calculated using Equation (4.41) as

$$R_1 = 2\frac{h_{\ell 1}}{Q} = 2\frac{5.1 \text{ m}}{1.7 \text{ m}^3/\text{s}} = 6.0 \text{ s/m}^2 \tag{c}$$

$$R_2 = 2\frac{h_{\ell 2}}{Q} = 2\frac{8.5 \text{ m}}{1.7 \text{ m}^3/\text{s}} = 10.0 \text{ s/m}^2 \tag{d}$$

The areas of the tanks are calculated as

$$A_1 = \frac{\pi}{4}(3 \text{ m})^2 = 7.07 \text{ m}^2 \tag{e}$$

$$A_2 = \frac{\pi}{4}(4 \text{ m})^2 = 12.6 \text{ m}^2 \tag{f}$$

It is shown in Example 6.19 that a two-tank liquid-level system is a second-order system. The natural frequency and damping ratio are determined using Equations (f) and (g) from Example 6.19

$$
\begin{aligned}
\omega_n &= \sqrt{\frac{1}{A_1 A_2 R_1 R_2}} \\
&= \sqrt{\frac{1}{(7.07 \text{ m}^2)(12.6 \text{ m}^2)(6.0 \text{ s/m}^2)(10 \text{ s/m}^2)}} \\
&= 1.37 \times 10^{-2} \text{ r/s}
\end{aligned}
\tag{g}
$$

$$
\begin{aligned}
\zeta &= \frac{1}{2\omega_n}\sqrt{\frac{1}{A_2 R_1} + \frac{1}{A_2 R_2} + \frac{1}{A_1 R_1}} \\
&= \frac{1}{2(1.37 \times 10^{-2} \text{ r/s})} \\
&\quad \times \sqrt{\frac{1}{12.6 \text{ m}^2(6.0 \text{ s/m}^2)} + \frac{1}{12.6 \text{ m}^2(10 \text{ s/m}^2)} + \frac{1}{7.07 \text{ m}^2(6.0 \text{ s/m}^2)}} \\
&= 7.75
\end{aligned}
\tag{h}
$$

The input to the system is

$$q_1(t) = 0.2u(t) \tag{i}$$

which has a Laplace transform of

$$Q_1(s) = \frac{0.2}{s} \tag{j}$$

The transforms of the liquid levels are determined using the transfer functions, Equations (d) and (e), obtained in Example 6.19

$$H_1(s) = G_1(s)Q(s)$$

$$= \frac{0.2\left(A_1 s + \frac{1}{R_1} + \frac{1}{R_2}\right)}{A_1 A_2 s(s^2 + 2\zeta\omega_n s + \omega_n^2)} \tag{k}$$

$$H_2(s) = G_2(s)Q(s)$$

$$= \frac{0.2}{R_1 A_1 A_2 s(s^2 + 2\zeta\omega_n s + \omega_n^2)} \tag{1}$$

(a) The final value theorem is applied to Equations (k) and (l), leading to the steady-state values of $h_1(t)$ and $h_2(t)$ as

$$\lim_{t\to\infty} h_1(t) = \lim_{s\to 0} sH_1(s)$$

$$= \lim_{s\to 0} \frac{0.2\left(A_1 s + \dfrac{1}{R_1} + \dfrac{1}{R_2}\right)}{A_1 A_2\left(s^2 + 2\zeta\omega_n s + \omega_n^2\right)}$$

$$= \frac{0.2\left(\dfrac{1}{R_1} + \dfrac{1}{R_2}\right)}{A_1 A_2 \omega_n^2}$$

$$= \frac{0.2 \text{ m}^3/\text{s}\left(\dfrac{1}{6 \text{ s/m}^2} + \dfrac{1}{10 \text{ s/m}^2}\right)}{(7.07 \text{ m}^2)(12.6 \text{ m}^2)(1.37 \times 10^{-2} \text{ r/s})^2} = 3.19 \text{ m} \tag{m}$$

$$\lim_{t\to\infty} h_2(t) = \lim_{s\to 0} sH_2(s)$$

$$= \frac{0.2}{R_1 A_1 A_2 \omega_n^2}$$

$$= \frac{0.2 \text{ m}^3/\text{s}}{(6 \text{ s/m}^2)(7.07 \text{ m}^2)(12.6 \text{ m}^2)(1.37 \times 10^{-2} \text{ r/s})^2}$$

$$= 1.99 \text{ m} \tag{n}$$

The new steady-state liquid levels are

$$H_1 = 13.6 \text{ m} + 3.19 \text{ m} = 16.79 \text{ m} \tag{o}$$

$$H_2 = 8.5 \text{ m} + 1.99 \text{ m} = 10.49 \text{ m} \tag{p}$$

(b) Since the system is overdamped the poles of the transfer function are

$$s_1 = -\omega_n\left(\zeta + \sqrt{\zeta^2 - 1}\right)$$

$$= -1.37 \times 10^{-2}\left(7.75 + \sqrt{(7.75)^2 - 1}\right) = -0.211 \tag{q}$$

$$s_2 = -\omega_n\left(\zeta - \sqrt{\zeta^2 - 1}\right)$$

$$= -1.37 \times 10^{-2}\left(7.75 - \sqrt{(7.75)^2 - 1}\right) = -8.88 \times 10^{-4} \tag{r}$$

Partial fraction decompositions of $H_1(s)$ and $H_2(s)$ are obtained using entry 12 of Table 5.3 as

$$H_1(s) = \frac{0.2}{A_1 A_2}\left(\left(\frac{1}{R_1}+\frac{1}{R_2}\right)\frac{1}{s_1 s_2 s} + \frac{1}{s_1 - s_2}\left\{\left[A_1 + \left(\frac{1}{R_1}+\frac{1}{R_2}\right)\frac{1}{s_1}\right]\frac{1}{s - s_1}\right.\right.$$

$$\left.\left. - \left[A_1 + \left(\frac{1}{R_1}+\frac{1}{R_2}\right)\frac{1}{s_2}\right]\frac{1}{s - s_2}\right\}\right)$$

$$H_1(s) = \frac{3.18}{s} - \frac{0.0622}{s+0.211} - \frac{3.13}{s+8.88 \times 10^{-4}} \tag{s}$$

$$H_2(s) = \frac{0.2}{R_1 A_1 A_2}\left[\frac{1}{s_1 s_2 s} + \frac{1}{s_1 - s_2}\left(\frac{1}{s_1(s - s_1)} - \frac{1}{s_2(s - s_2)}\right)\right]$$

$$= \frac{1.99}{s} + \frac{0.00844}{s+0.211} - \frac{2.00}{s+8.88 \times 10^{-4}} \tag{t}$$

The transforms in Equations (s) and (t) are inverted leading to

$$h_1(t) = 3.18u(t) - 0.0622e^{-0.211t} - 3.13e^{-8.91 \times 10^{-4}t} \tag{u}$$

$$h_2(t) = 1.99u(t) + 0.00844e^{-0.211t} - 2.00e^{-8.91 \times 10^{-4}t} \tag{v}$$

(c) The 2 percent settling time is the time that it takes for both tanks to reach 98 percent of their new steady-state values

$$0.98(3.18) = 3.18 - 0.0622e^{-0.211t_s} - 3.13e^{-8.91 \times 10^{-4}t_s} \tag{w}$$

The solution of Equation (w) is $t_s = 4.37 \times 10^3$ s.

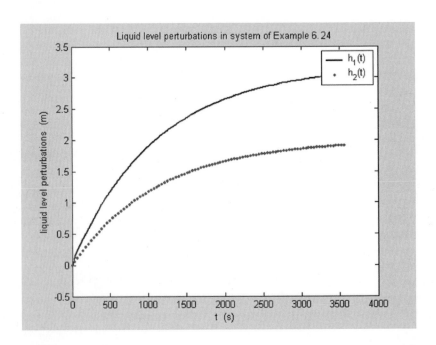

Liquid level perturbations in system of Example 6.24

**FIG. 6.23**

(d) In a manner similar to the determination of the 2 percent settling time, the time required to rise to 10 percent of the system's final value is 100 s and the time required to rise to 90 percent of the system's final value is 2,570 s. Thus the rise time is $t_r = 2{,}570 \text{ s} - 100 \text{ s} = 2{,}470 \text{ s}$.

(e) MATLAB is used to develop plots for the time-dependent responses of the liquid level perturbations as shown in Figure 6.23.

## 6.5.4 General Transient Response

The transient response of a second-order system due to any input can be determined using either the Laplace transform method or the convolution integral solution, which is derived using the Laplace transform method.

### Example 6.25

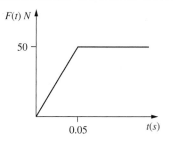

**FIG. 6.24** The input force for Example 6.25.

The time-dependent force applied to a one-degree-of-freedom mass-spring-viscous damper system is illustrated in Figure 6.24. The system parameters are $m = 1.0$ kg, $k = 1 \times 10^4$ N/m, $c = 600$ N·s/m. Determine (a) the steady-state displacement of the system from its equilibrium position and (b) the time-dependent response of the system.

**Solution**

The transfer function for the system is

$$G(s) = \frac{1}{m(s^2 + 2\zeta\omega_n s + \omega_n^2)} \tag{a}$$

where the natural frequency and damping ratio are

$$\omega_n = \sqrt{\frac{k}{m}}$$

$$= \sqrt{\frac{1 \times 10^4 \text{ N/m}}{1 \text{ kg}}} = 100 \text{ r/s} \tag{b}$$

$$\zeta = \frac{c}{2m\omega_n}$$

$$= \frac{600 \text{ N·s/m}}{2(10 \text{ kg})(100 \text{ r/s})} = 0.3 \tag{c}$$

The damped natural frequency for the system is calculated as

$$\omega_d = (100 \text{ r/s})\sqrt{1 - (0.3)^2} = 95.4 \text{ r/s} \tag{d}$$

The input force is written using unit step functions as

$$\begin{aligned} F(t) &= 1{,}000t[u(t) - u(t - 0.05)] + 50u(t - 0.05) \\ &= 1{,}000tu(t) - 1{,}000(t - 0.05)u(t - 0.05) \end{aligned} \tag{e}$$

The Laplace transform of the input force is

$$F(s) = \frac{1,000}{s^2}(1 - e^{-0.05s}) \tag{f}$$

The transform of the displacement is obtained using Equations (a) and (d) as

$$X(s) = G(s)F(s)$$

$$= \frac{1,000(1 - e^{-0.05s})}{s^2(s^2 + 2\zeta\omega_n s + \omega_n^2)} \tag{g}$$

(a) The displacement from equilibrium of the mass in the steady state is obtained by the application of the final value theorem

$$\lim_{t \to \infty} x(t) = \lim_{s \to 0} sX(s)$$

$$= \lim_{s \to 0} \frac{1,000(1 - e^{-0.05s})}{s(s^2 + 2\zeta\omega_n s + \omega_n^2)}$$

$$= \lim_{s \to 0} \frac{1,000(1 - e^{-0.05s})}{s\omega_n^2} \tag{h}$$

The limit in Equation (h) is indeterminate but can be evaluated using L'Hospital's rule

$$\lim_{t \to \infty} x(t) = \frac{1,000(0.05)}{\omega_n^2} = 5 \text{ mm} \tag{i}$$

(b) A partial fraction decomposition of Equation (g) leads to

$$X(s) = \frac{1,000}{\omega_n^4}\left(\frac{\omega_n^2}{s^2} - \frac{2\zeta\omega_n}{s} + \frac{2\zeta\omega_n s}{s^2 + 2\zeta\omega_n s + \omega_n^2}\right)(1 - e^{-0.05s}) \tag{j}$$

Inversion of the transform in Equation (j) leads to

$$x(t) = \frac{1,000}{\omega_n^4}\Bigg\{\left[\omega_n^2 t - 2\zeta\omega_n + 2\zeta\omega_n e^{-\zeta\omega_n t}\cos(\omega_d t)\right.$$

$$\left. - 2\frac{\zeta^2\omega_n^2}{\omega_d}e^{-\zeta\omega_n t}\sin(\omega_d t)\right]u(t) - \left[\omega_n^2(t - 0.05) - 2\zeta\omega_n\right.$$

$$+ 2\zeta\omega_n e^{-\zeta\omega_n(t-0.05)}\cos(\omega_d(t - 0.05))$$

$$\left. - 2\frac{\zeta^2\omega_n^2}{\omega_d}e^{-\zeta\omega_n(t-0.05)}\sin(\omega_d(t - 0.05))\right]u(t - 0.05)\Bigg\} \tag{k}$$

Substitution of numerical values leads to

$$x(t) = \left\{\left[0.1t - 6 \times 10^{-4} + 6 \times 10^{-4}e^{-30t}\cos(95.4t) - 1.89 \times 10^{-4}e^{-30t}\sin(95.4t)\right]\right\}$$

$$- \left[0.1(t - 0.05) - 6 \times 10^{-4} + 6 \times 10^{-4}e^{-30(t-0.05)}\cos[95.4(t - 0.05)]\right.$$

$$\left. - 1.89 \times 10^{-4}e^{-30(t-0.05)}\sin[95.4(t - 0.05)]\right]u(t - 0.05) \tag{l}$$

The script of a MALAB program, Example6_25.m, which symbolically determines the system response from Equation (g) and then plots the response, is shown in Figure 6.25(a). The symbolic form of the response obtained from the execution of Example6_25.m, shown in Figure 6.25(b), agrees with Equation (l). The response is plotted in Figure 6.25(c).

```
% Example 6.25
%
% Declaring symbolic variables
syms s,t
% Defining parameters
zeta=0.3;
wn=100;
% Defining transform, Equation (g) of Example 6.25
xbar=1000*(1-exp(-0.05*s))/(s^2*(s^2+2*zeta*wn*s+wn^2))
% Taking inverse transform
x=ilaplace(xbar)
% Simplifying symbolic form of response
x2=simplify(x1)
x3=vpa(x1,3)
% Plotting response
t=0:.0001:.2;
x4=subs(x3,t);
plot(t,x4)
xlabel('t (s)')
ylabel('x (m)')
title('Response of system of Example 6.25')
%
% End of Example6_25.m
```

(a)

x3=

.560e-2*Heaviside(t-.500e-1)-.600e-3*Heaviside(t-.500e-1)*exp(-
30.*t+1.50)*cos(95.4*t-4.77)+ .860e-3*Heaviside(t-.500e-1)*exp(-
30.*t+1.50)*sin(95.4*t-4.77)-.100*Heaviside(t-.500e-1)*t-.600e-3+.600e-3*exp(-
30.*t)*cos(95.4*t)-.860e-3*exp(-30.*t)*sin(95.4*t)+.100*t

(b)

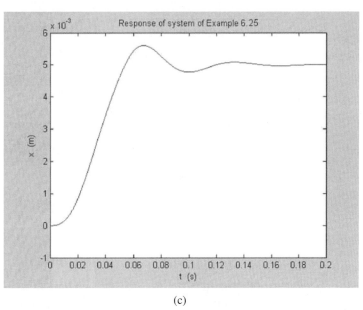

(c)

**FIG. 6.25** (a) Script of the file Example6_25.m; (b) partial output from the execution of the file; (c) the plot of system response.

## 6.6 HIGHER-ORDER SYSTEMS

The transfer function, $G(s)$, of an $n$th-order system has $n$ poles: $s_i$, $i = 1, 2, \ldots, n$. If $k$ poles $s_1, s_2, \ldots, s_k$ are real, then $n - k$ poles are complex and occur in conjugate pairs. It is assumed that the system is stable or neutrally stable and thus all poles have nonpositive real parts.

### 6.6.1 General Case

The time constant of a first-order system is the negative of the reciprocal of the pole of its transfer function. By analogy the time constants for a higher-order system can be defined as the negative reciprocals of its real poles,

$$T_i = -\frac{1}{s_i} \qquad i = 1, 2, \ldots, k \qquad (6.85)$$

The complex conjugate poles of a second-order system are represented in terms of a natural frequency $\omega_n$ and a damping ratio $\zeta$. By analogy, natural frequencies and damping ratios for higher-order systems are defined according to

$$s_\ell, \bar{s}_\ell = \omega_\ell \left( -\zeta_\ell \pm j\sqrt{1 - \zeta_\ell^2} \right) \qquad \ell = 1, 2, \ldots, \tfrac{1}{2}(n - k) \qquad (6.86)$$

It is noted that $(s - s_\ell)(s - \bar{s}_\ell) = s^2 + 2\zeta_\ell \omega_\ell s + \omega_\ell^2 (1 - \zeta_\ell^2)$.

The general form of a transfer function for an $n$th order system is

$$G(s) = \frac{N(s)}{\left(s + \dfrac{1}{T_1}\right) \cdots \left(s + \dfrac{1}{T_k}\right) \left[s^2 + 2\zeta_1 \omega_1 s + \omega_1^2(1 - \zeta_1^2)\right] \cdots \left[s^2 + 2\zeta_p \omega_p s + \omega_p^2(1 - \zeta_p^2)\right]}$$

$$(6.87)$$

where $p = \tfrac{1}{2}(n - k)$. A partial fraction decomposition of Equation (6.87) is of the form

$$G(s) = \sum_{i=1}^{k} \frac{A_i}{s - \frac{1}{T_i}} + \sum_{\ell=1}^{p} \frac{B_\ell s + C_\ell}{s^2 + 2\zeta_\ell \omega_\ell s + \omega_\ell^2(1 - \zeta_\ell^2)} \qquad (6.88)$$

For a system input $F(s)$ the system output is

$$X(s) = \sum_{i=1}^{k} \frac{A_i F(s)}{s - \frac{1}{T_i}} + \sum_{\ell=1}^{p} \frac{(B_\ell s + C_\ell)F(s)}{s^2 + 2\zeta_\ell \omega_\ell s + \omega_\ell^2(1 - \zeta_\ell^2)} \qquad (6.89)$$

Equation (6.89) shows that the response of a higher-order system whose initial conditions are all zero is the sum of responses of first- and second-order systems.

**Example 6.26**

Determine the step response of the system of Figure 6.26(a).

**Solution**

The application of Newton's law to the free-body diagrams of Figure 6.26(b) leads to

$$m\ddot{x} + (k_1 + k_2)x - k_2 y = F(t) \tag{a}$$
$$-k_2 x + k_2 y + c\dot{y} = 0 \tag{b}$$

Tanking the Laplace transforms of Equations (a) and (b) leads to

$$(ms^2 + k_1 + k_2)X(s) - k_2 Y(s) = F(s) \tag{c}$$
$$-k_2 X(s) + (cs + k_2)Y(s) = 0 \tag{d}$$

Simultaneous solution of Equations (c) and (d) leads to

$$X(s) = \frac{(cs + k_2)F(s)}{mc\left(s^3 + \dfrac{k_2}{c}s^2 + \dfrac{k_1 + k_2}{m}s + \dfrac{k_1 k_2}{mc}\right)} \tag{e}$$

Substitution of given values leads to the transfer function

$$G(s) = \frac{X(s)}{F(s)} = \frac{0.01s + 1,500}{s^3 + 100s^2 + 2,500s + 100,000} \tag{f}$$

Using the MATLAB program of Figure 6.27(a), the poles of the transfer function are calculated as $-84.82$, $-7.79 + j33.5$, and $-7.79 - j33.5$. The step response of the system is presented in Figure 6.27(c).

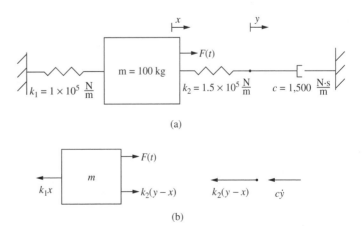

(a)

(b)

**FIG. 6.26** (a) The system of Example 6.26; (b) the free-body diagrams at an arbitrary instant.

```
% Example6_26.m
% System parameters
m=100;             % mass in kg
k1=1.E5;            % stiffness in N/m
k2=1.5E5;          % stiffness in N/m
c=1500;            % damping coefficient in N-s/m
% Numerator of transfer function
N1=1/m;
N0=k2/m;
% Denominator of transfer function
D3=1;
D2=k2/c;
D1=(k1+k2)/m;
D0=k1*k2/(m*c);
% Transfer function
N=[N1 N0];
D=[D3 D2 D1 D0];
% Poles of transfer fucntion
[r,p,k]=residue(N,D)
% Step response
step(N,D)
title('Step response for system of Example 6.26')
ylabel('x (m)')
xlabel('t')
%
% End of Example6_26.m
```

(a)

```
>> Example_626

r =


  0.2143
 -0.1072 - 0.2451i
 -0.1072 + 0.2451i


p =

 -84.4178
  -7.7911 +33.5244i
  -7.7911 -33.5244i

k =
   []

>>
```

(b)

**FIG. 6.27** (a) Script of Example6_26.m, which determines the poles of the transfer function for Example 6.26 and plots the step response; (b) the output from the execution of Example6_26.m; (c) the step response plotted using Example6_26.m.

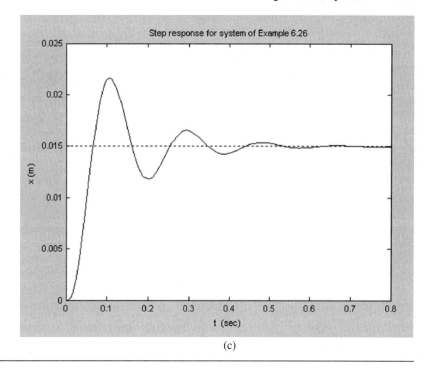

**FIG. 6.27** (*Continued*)                                                                (c)

## Example 6.27

Consider a cascade of *n* CSTRs, illustrated in Figure 6.28, in which the reactant *A* is fed into the first tank. Reactant *A* is consumed in the *i*th reactor, of volume $V_i$, at a rate $k_i$. The tanks all have constant volume and all reactions are isothermal. (a) Determine the transient behavior of the reactant in each reactor when the concentration in the feed stream suddenly changes by $\Delta C_{Ai}$. (b) Determine the final concentration of reactant *A* in each tank.

### Solution

The molecular balance equation for component *A* in the first reactor is

$$V_1 \frac{dC_{A1}}{dt} + (q + k_1 V_1) = q\Delta C_{ai} \tag{a}$$

The molecular balance equation for component *A* in reactor *p* for $p = 2, 3, \ldots$ is

$$V_p \frac{dC_{Ap}}{dt} - qC_{Ap-1} + (q + k_p V_p)C_{Ap} = 0 \tag{b}$$

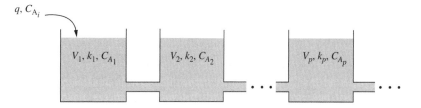

**FIG. 6.28** The cascade of CSTRs for Example 6.27.

where $C_{Ap}$ is the perturbation in the concentration of component $A$ from the steady state due to the perturbation in the inlet stream.

Taking the Laplace transform of Equations (a) and (b) leads to

$$(V_1 s + q + k_1 V_1)\bar{C}_{A1} = q\Delta\bar{C}_{Ai} \tag{c}$$

$$(V_p s + q + k_p V_p)\bar{C}_{Ap} - q\bar{C}_{Ap-1} = 0 \qquad p = 2, 3, \ldots \tag{d}$$

The first transfer function is determined using Equation (c) as

$$G_1(s) = \frac{q}{V_1\left(s + \dfrac{q}{V_1} + k_1\right)} \tag{e}$$

Equation (d) is rearranged to

$$G_p(s) = \frac{q}{V_p\left(s + \dfrac{q}{V_p} + k_p\right)} G_{p-1}(s) \tag{f}$$

Mathematical induction, using Equations (e) and (f), is used to show

$$G_p(s) = \frac{q^p}{V_1 V_2 \cdots V_p\left(s + \dfrac{q}{V_1} + k_1\right)\left(s + \dfrac{q}{V_2} + k_2\right)\cdots\left(s + \dfrac{q}{V_p} + k_p\right)} \tag{g}$$

Thus the transfer function $G_p(s)$ is $p$th order and has $p$ real poles

$$s_k = -\left(\frac{q}{V_k} + k_k\right) \qquad k = 1, 2, \ldots, p \tag{h}$$

The time constants for the $p$th reactor are

$$T_k = \frac{V_k}{q + k_k V_k} \qquad k = 1, 2, \ldots, p \tag{i}$$

A partial fraction decomposition of Equation (g) leads to

$$G_p(s) = \sum_{k=1}^{p} \frac{A_{k,p}}{s + \dfrac{1}{T_k}} \tag{j}$$

where

$$A_{k,p} = \frac{q^p}{\left(\displaystyle\prod_{i=1}^{p} V_i\right)\left[\displaystyle\prod_{\substack{i=1 \\ i\neq k}}^{p}\left(\dfrac{q}{V_p} + k_p - \dfrac{q}{V_k} - k_k\right)\right]} \tag{k}$$

The response due to a step change in inlet concentration, $F(s) = \Delta C_{Ai}/s$ is

$$C_p(t) = \Delta C_{Ai}\sum_{k=1}^{p} A_{k,p}\left(1 - e^{-\frac{t}{T_k}}\right) \tag{l}$$

(b) The final value theorem can be applied to Equation (g) to determine the final value for the perturbation in the concentration of component $A$ in each reactor

$$\Delta C_{ps} = \Delta C_{Ai}\lim_{s\to 0} G(s)$$

$$= \Delta C_{Ai}\frac{q^p}{\displaystyle\prod_{k=1}^{p}(q + k_k V_k)} \tag{m}$$

## 6.6.2 Multidegree-of-Freedom Mechanical Systems

The general form of the differential equations governing the response of a linear $n$-degree-of-freedom mechanical system is

$$\mathbf{M\ddot{x} + C\dot{x} + Kx = F} \tag{6.90}$$

where $\mathbf{M}$ is an $n \times n$ mass matrix, $\mathbf{C}$ is an $n \times n$ damping matrix, $\mathbf{K}$ is an $n \times n$ stiffness matrix, $\mathbf{F}$ is an $n \times 1$ vector of forces, and $\mathbf{x}$ is an $n \times 1$ vector of dependent variables

$$\mathbf{x} = \begin{bmatrix} x_1(t) \\ x_2(t) \\ \vdots \\ x_n(t) \end{bmatrix} \tag{6.91}$$

The vector of inputs $\mathbf{F}$ is of the form

$$\mathbf{F} = \begin{bmatrix} F_1(t) \\ F_2(t) \\ \vdots \\ F_n(t) \end{bmatrix} \tag{6.92}$$

**6.6.2.1 Transfer Functions** Taking the Laplace transform of Equation (6.90), assuming all initial conditions are zero leads to

$$\left(s^2\mathbf{M} + s\mathbf{C} + \mathbf{K}\right)\mathbf{X}(s) = \mathbf{F}(s) \tag{6.93}$$

where

$$\mathbf{X}(s) = \begin{bmatrix} X_1(s) \\ X_2(s) \\ \vdots \\ X_n(s) \end{bmatrix} \qquad \mathbf{F}(s) = \begin{bmatrix} F_1(s) \\ F_2(s) \\ \vdots \\ F_n(s) \end{bmatrix} \tag{6.94}$$

The **impedance matrix** is defined as

$$\mathbf{Z}(s) = s^2\mathbf{M} + s\mathbf{C} + \mathbf{K} \tag{6.95}$$

Equation (6.93) is rewritten using of the impedance matrix as

$$\mathbf{Z}(s)\mathbf{X}(s) = \mathbf{F}(s) \tag{6.96}$$

Cramer's rule may be used to determine each of the components of $\mathbf{X}(s)$

$$X_i(s) = \frac{|\mathbf{W_i}(s)|}{D(s)} \tag{6.97}$$

where $\mathbf{W_i}(s)$ is the matrix obtained by replacing the $i$th column of $\mathbf{Z}(s)$ by $\mathbf{F}(s)$ and

$$D(s) = |\mathbf{Z}(s)| \tag{6.98}$$

The determinant of $\mathbf{W_i}(s)$ can be evaluated by column expansion using the $i$th column leading to

$$|\mathbf{W_i}(s)| = \sum_{j=1}^{n} F_j(s) U_{i,j}(s) \tag{6.99}$$

Equation (6.97) is written as

$$X_i(s) = \sum_{j=1}^{n} G_{i,j}(s) F_j(s) \tag{6.100}$$

where $G_{i,j}(s)$ is the transfer function for $x_i$ corresponding to an input $F_j$ and is given by

$$G_{i,j}(s) = \frac{U_{i,j}(s)}{D(s)} \tag{6.101}$$

Each transfer function has the same denominator and the same poles. For an $n$-degree-of-freedom system $D(s)$ is a polynomial in $s$ of order $2n$ and thus has $2n$ roots.

If $F_k(t) = \delta(t)$ and $F_j(t) = 0$, $j = 1, 2, \ldots, n$ except $j = k$. Then $F_j(s) = 1$ and all other $F_j(s) = 0$. The impulsive responses are obtained as

$$X_{i,k}(s) = G_{i,k}(s) \tag{6.102}$$

as the response for $x_i(t)$ due to a unit impulse input for $F_k(t)$.

**6.6.2.2 Undamped Systems** For an undamped system, $D(s)$ contains only even powers of $s$ and the transfer functions have poles of the form

$$\begin{aligned} s_{k,1} &= -j\omega_k \\ s_{k,2} &= j\omega_k \end{aligned} \qquad k = 1, 2, \ldots, n \tag{6.103}$$

In this case $D(s)$ can be factored as

$$D(s) = v \sum_{i=1}^{n} \left(s^2 + \omega_i^2\right) \tag{6.104}$$

where $v$ is a constant.

The transfer function $G_{i,j}(s)$ can be written as

$$G_{i,j}(s) = \sum_{\ell=1}^{n} \frac{A_{i,l}s + B_{i,l}}{\left(s^2 + \omega_\ell^2\right)} \tag{6.105}$$

Thus the response of $x_i(t)$ due to a unit impulse input for $F_j(t)$ is

$$x_i(t) = \sum_{\ell=1}^{n} \left[A_{i,l} \cos(\omega_\ell t) + \frac{B_{i,l}}{\omega_\ell} \sin(\omega_\ell t)\right] \tag{6.106}$$

Use of the trigonometric identity for the sine of a sum leads to an alternate representation of Equation (6.106) as

$$x_i(t) = \sum_{\ell=1}^{n} X_{i,l} \sin(\omega_\ell t + \phi_{\ell,j}) \tag{6.107}$$

An applied impulse imparts an initial energy to a system, initiating motion of the system. The resulting response is analogous to the free response of the system due to an initial velocity. Equation (6.106) illustrates that this response is the linear combination of trigonometric terms at different frequencies. The frequencies in Equations (6.106) and (6.107) are the natural frequencies of the multidegree-of-freedom system.

The poles of a stable undamped multidegree-of-freedom system are of the form of Equations (6.103) where the $\omega_i$ are the natural frequencies of the system. As with a one-degree-of-freedom system, when an undamped multidegree-of-freedom system has an initial kinetic or potential energy, the free response of the system is sustained at the natural frequencies without additional energy input. An $n$-degree-of-freedom has up to $n$ distinct natural frequencies. The free response is a linear combination of trigonometric terms over all natural frequencies.

---

## Example 6.28

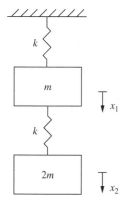

**FIG. 6.29** The two-degree-of-freedom system of Example 6.28.

Consider the two-degree-of-freedom system of Figure 6.29. (a) Determine the system's natural frequencies. (b) Determine the response of the system when a unit impulse is applied to the particle of mass $m$.

### Solution

The differential equations governing the motion of the two-degree-of-freedom system are

$$\begin{bmatrix} m & 0 \\ 0 & 2m \end{bmatrix} \begin{bmatrix} \ddot{x}_1 \\ \ddot{x}_2 \end{bmatrix} + \begin{bmatrix} 2k & -k \\ -k & k \end{bmatrix} \begin{bmatrix} x_1 \\ x_2 \end{bmatrix} = \begin{bmatrix} F_1 \\ F_2 \end{bmatrix} \tag{a}$$

The impedance matrix for the system is

$$\mathbf{Z}(s) = s^2 \mathbf{M} + \mathbf{K}$$
$$= \begin{bmatrix} ms^2 + 2k & -k \\ -k & 2ms^2 + k \end{bmatrix} \tag{b}$$

The determinant of the impedance matrix is

$$D(s) = (ms^2 + 2k)(2ms^2 + k) - (-k)(-k)$$
$$= 2m^2 s^4 + 3kms^2 + k^2$$
$$= 2m^2 \left( s^4 + 3\frac{k}{m}s^2 + \frac{k^2}{m^2} \right) \tag{c}$$

The natural frequencies are determined from the roots of $D(s)$. To this end the quadratic formula is applied yielding

$$s^2 = \frac{1}{2}\left[ -3\frac{k}{m} \pm \sqrt{\left(3\frac{k}{m}\right)^2 - 4\frac{k^2}{m^2}} \right]$$
$$= \frac{k}{2m}\left(-3 \pm \sqrt{5}\right) \tag{d}$$

Equation (d) is used to determine

$$s_1 = \sqrt{\frac{k}{2m}\left(-3 + \sqrt{5}\right)} = j0.618\sqrt{\frac{k}{m}} \tag{e}$$

$$s_2 = \sqrt{\frac{k}{2m}\left(-3 - \sqrt{5}\right)} = j1.618\sqrt{\frac{k}{m}} \tag{f}$$

Thus the natural frequencies are

$$\omega_1 = 0.618\sqrt{\frac{k}{m}} \tag{g}$$

$$\omega_2 = 1.618\sqrt{\frac{k}{m}} \tag{h}$$

(b) If $F_1(t) = \delta(t)$ and $F_2(t) = 0$ then

$$\mathbf{F}(s) = \begin{bmatrix} 1 \\ 0 \end{bmatrix} \tag{i}$$

Cramer's rule is used to determine

$$X_1(s) = \frac{\begin{vmatrix} 1 & -k \\ 0 & 2ms^2 + k \end{vmatrix}}{2m^2(s^2 + \omega_1^2)(s^2 + \omega_2^2)}$$

$$= \frac{2ms^2 + k}{2m^2(s^2 + \omega_1^2)(s^2 + \omega_2^2)} \tag{j}$$

$$X_2(s) = \frac{\begin{vmatrix} ms^2 + k & 1 \\ -k & 0 \end{vmatrix}}{2m^2(s^2 + \omega_1^2)(s^2 + \omega_2^2)}$$

$$= \frac{k}{2m^2(s^2 + \omega_1^2)(s^2 + \omega_2^2)} \tag{k}$$

Partial fraction decomposition of the transforms in Equations (j) and (k) using Table 5.3 leads to

$$X_1(s) = \frac{1}{\omega_2^2 - \omega_1^2}\left[\frac{k - 2m\omega_1^2}{s^2 + \omega_1^2} - \frac{k - 2m\omega_2^2}{s^2 + \omega_2^2}\right]$$

$$= \frac{0.106m}{s^2 + 0.382\dfrac{k}{m}} + \frac{1.89m}{s^2 + 2.618\dfrac{k}{m}} \tag{l}$$

$$X_2(s) = \frac{k}{\omega_2^2 - \omega_1^2}\left[\frac{1}{s^2 + \omega_1^2} - \frac{1}{s^2 + \omega_2^2}\right]$$

$$= \frac{0.447m}{s^2 + 0.382\dfrac{k}{m}} - \frac{0.447m}{s^2 + 2.618\dfrac{k}{m}} \tag{m}$$

Inversion of Equations (l) and (m) gives

$$x_1(t) = \left[ 0.171 \sin\left( 0.618 \sqrt{\frac{k}{m}}\, t \right) + 0.723 \sin\left( 1.618 \sqrt{\frac{k}{m}}\, t \right) \right] \qquad \text{(n)}$$

$$x_2(t) = \left[ 0.723 \sin\left( 0.618 \sqrt{\frac{k}{m}}\, t \right) - 0.171 \sin\left( 1.618 \sqrt{\frac{k}{m}}\, t \right) \right] \qquad \text{(o)}$$

The transfer functions given by Equations (l) and (m) in Example 6.28 are the same as $G_{1,1}(s)$ and $G_{2,1}(s)$ as defined in Equation (6.105).

If the transfer functions $G_{i,j}(s)$, defined in Equation (6.105), are obtained and inverted leading to impulsive responses of the form of Equation (6.107), then the transient response of the system due to a general forcing can be obtained. From Equation (6.107)

$$x_i(t) = \mathcal{L}^{-1}\left\{ \sum_{j=1}^{n} G_{i,j}(s) F_j(s) \right\}$$

$$= \sum_{j=1}^{n} \mathcal{L}^{-1}\left\{ G_{i,j}(s) F_j(s) \right\} \qquad (6.108)$$

Recalling that the Laplace transform of the convolution of two functions is the product of their Laplace transforms leads to the inversion of Equation (6.108) in the form

$$x_i(t) = \sum_{j=1}^{n} G_{i,j}(t) * F_j(t)$$

$$= \sum_{j=1}^{n} \int_{0}^{t} G_{i,j}(t - \tau) F_j(\tau) d\tau \qquad (6.109)$$

## 6.7  SYSTEMS WITH TIME DELAY

A dynamic system experiences a **time delay** when its response at time $t$ is explicitly affected by the system's response at a previous time, $t - \tau$ for a fixed value of $\tau$. If $x(t)$ is a dependent variable in a system with a time delay of $\tau$ then the mathematical model for the system may include a term of the form $x(t - \tau)$. It is necessary to employ a modified version of the second shifting theorem to determine the transfer function for the system; if $x(t) = 0$ for $t < 0$ then

$$\mathcal{L}\{x(t - \tau)\} = e^{-\tau s} X(s) \qquad (6.110)$$

The pipe connecting the two CSTRs of Figure 6.30 is of finite length $L$. If the average velocity of the flow through the pipe is $v$, then the time

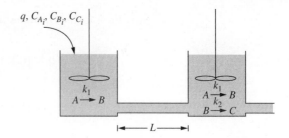

**FIG. 6.30** Two CSTRs are connected by a pipe of finite length. The transport of the mixture from reactor 1 through pipe leads to a delay in the mixture reaching reactor 2. The concentration entering reactor 2 at time $t$ is that leaving reactor 1 at time $t - L/v$.

required for fluid particles to travel between the two reactors is $\tau = L/v$. The properties of the mixture entering the second reactor at time $t$ are the properties of the mixture leaving the first reactor at time $t - \tau$.

**Example 6.29**

The double CSTR system of Figure 6.30 is operating at steady state when the concentration of component $A$ in the inlet stream is suddenly changed by $C_{Aip}$. The flow rate and concentrations of all other components in the inlet stream are unchanged. All reactions are first order, the system is isothermal, and the volume of the mixtures in each reactor is constant. The time required for fluid particles to travel between the reactors is $\tau$. (a) Determine the transfer functions for the perturbations in concentrations of $A$, $B$, and $C$ in each reactor. (b) Determine the response of the perturbation of concentration for component $A$ in each reactor if $V_1 = 1.25 \times 10^{-3}$ m$^3$, $V_2 = 2.08 \times 10^{-3}$ m$^3$, $q = 2.4 \times 10^{-6}$ m$^3$/s, $k_1 = 2.05 \times 10^{-3}$ s$^{-1}$, $\tau = 10.0$ s, and the inlet concentration of component $A$ has a step increase of 0.2 mol/L at $t = 0$.

**Solution**

(a) Conservation equations are applied to each component in each tank. Each concentration is written as the sum of the steady-state concentration and a perturbation in concentration. The steady state is subtracted leading to the mathematical model for the perturbations in concentration,

$$V_1 \frac{dC_{A1p}}{dt} + (q + k_1 V_1) C_{A1p}(t) = q C_{Aip} \quad \text{(a)}$$

$$V_1 \frac{dC_{B1p}}{dt} - k_1 V_1 C_{A1p}(t) + q C_{B1p}(t) = 0 \quad \text{(b)}$$

$$V_1 \frac{dC_{C1p}}{dt} + q C_{C1p}(t) = 0 \quad \text{(c)}$$

$$V_2 \frac{dC_{A2p}}{dt} - q C_{A1p}(t - \tau) + (q + k_1 V_2) C_{A2p}(t) = 0 \quad \text{(d)}$$

$$V_2 \frac{dC_{B2p}}{dt} - q C_{B1p}(t - \tau) - k_1 V_2 C_{A2p}(t) + (q + k_2 V_2) C_{B2p}(t) = 0 \quad \text{(e)}$$

$$V_2 \frac{dC_{C2p}}{dt} - k_2 V_2 C_{Bp2}(t) + q C_{C2p}(t) = 0 \quad \text{(f)}$$

Taking the Laplace transforms of Equations (a)–(f), noting that the appropriate initial conditions are

$$C_{A1}(0) = C_{B1p}(0) = C_{C1p}(0) = C_{A2p}(0) = C_{B2p}(0) = C_{C2p}(0) = 0 \qquad \text{(g)}$$

and using Equation (6.110) in transforming Equations (d) and (e) leads to

$$\bar{C}_{A1p}(s) = \frac{q\bar{C}_{Aip}(s)}{V_1\left(s + k_1 + \dfrac{q}{V_1}\right)} \qquad \text{(h)}$$

$$\bar{C}_{B1p}(s) = \frac{k_1\bar{C}_{A1p}(s)}{s + \dfrac{q}{V_1}}$$

$$= \frac{k_1 q\bar{C}_{Aip}(s)}{V_1\left(s + k_1 + \dfrac{q}{V_1}\right)\left(s + \dfrac{q}{V_1}\right)} \qquad \text{(i)}$$

$$\bar{C}_{C1p}(s) = 0 \qquad \text{(j)}$$

$$\bar{C}_{A2p}(s) = \frac{q\bar{C}_{A1p}(s)e^{-\tau s}}{V_2\left(s + k_1 + \dfrac{q}{V_2}\right)}$$

$$= \frac{q^2 e^{-\tau s}\bar{C}_{Aip}(s)}{V_1 V_2\left(s + k_1 + \dfrac{q}{V_1}\right)\left(s + k_1 + \dfrac{q}{V_2}\right)} \qquad \text{(k)}$$

$$\bar{C}_{B2p}(s) = \frac{q\bar{C}_{B1p}(s)e^{-\tau s} + k_1 V_2 \bar{C}_{A2p}(s)}{V_2\left(s + k_2 + \dfrac{q}{V_2}\right)}$$

$$= \frac{q^2 k_1}{V_1 V_2} \frac{\left(2s + 2k_1 + \dfrac{q}{V_2}\right)\bar{C}_{Aip}(s)e^{-\tau s}}{\left(s + k_2 + \dfrac{q}{V_1}\right)\left(s + \dfrac{q}{V_1}\right)\left(s + k_1 + \dfrac{q}{V_1}\right)\left(s + k_1 + \dfrac{q}{V_2}\right)} \qquad \text{(l)}$$

$$\bar{C}_{C2}(s) = \frac{k_2\bar{C}_{B2p}(s)}{s + \dfrac{q}{V_2}}$$

$$= \frac{q^2 k_1 k_2}{V_1 V_2} \frac{\left(2s + 2k_1 + \dfrac{q}{V_2}\right)\bar{C}_{Aip}(s)e^{-\tau s}}{\left(s + k_2 + \dfrac{q}{V_1}\right)\left(s + \dfrac{q}{V_1}\right)\left(s + k_1 + \dfrac{q}{V_1}\right)\left(s + k_1 + \dfrac{q}{V_2}\right)\left(s + \dfrac{q}{V_2}\right)} \qquad \text{(m)}$$

(b) Noting that $C_{Aip}(s) = 0.2/s$, the substitution of numerical values into Equations (h) and (k) leads to

$$\bar{C}_{A1p}(s) = \frac{3.84 \times 10^{-4}}{s(s + 3.97 \times 10^{-3})} \qquad \text{(n)}$$

$$\bar{C}_{A2p}(s) = \frac{8.51 \times 10^{-10} e^{-10.0s}}{s(s + 3.97 \times 10^{-3})(s + 3.20 \times 10^{-3})} \qquad \text{(o)}$$

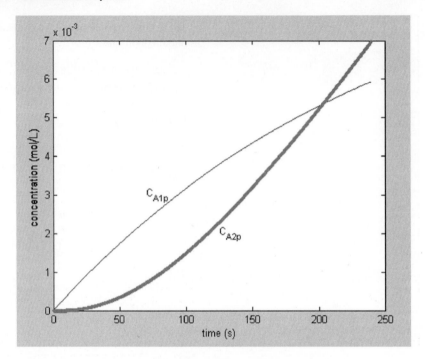

**FIG. 6.31** The perturbation concentrations of component $A$ in each reactor of Example 6.29. The perturbation concentration in reactor $B$ is zero until $t = 10$ s due to the dead time, during which the mixture from reactor $A$ is transported to reactor $B$.

Partial fraction decomposition and inversion of the transforms leads to

$$C_{A1p}(t) = 9.67 \times 10^{-2}\left(1 - e^{-3.97 \times 10^{-3}t}\right)u(t) \tag{p}$$

$$C_{A2p}(t) = \left(6.67 \times 10^{-7} + 2.79 \times 10^{-4}e^{3.97 \times 10^{-3}(t-10.0)}\right.$$
$$\left. -5.32 \times 10^{-3}e^{3.20 \times 10^{-3}(t-10.0)}\right)u(t - 10.0) \tag{q}$$

The concentration perturbations are plotted in Figure 6.31. The perturbation in the second reactor is zero until $t = 0.5$ s. The time between the perturbation in the inlet stream and the response of the second reactor is called dead time.

Mathematical modeling of vibrations of machine tools often leads to delay differential equations. Consider, for example, a cutting operation on a lathe, illustrated in Figure 6.32. The workpiece rotates at a constant speed $\omega$. During the cutting operation the machine tool provides a cutting force to the workpiece. The machine tool is modeled using a one-degree-of-freedom mass-spring-viscous damper model. The dependent variable $x(t)$ represents the penetration of the cutting tool into the workpiece. The workpiece resists the penetration and develops a cutting force that acts on the tool. The cutting force is modeled by the force of a spring in parallel with a viscous damper, $F_c = F_s + F_d$. The spring force $F_s$ is proportional to the instantaneous change in thickness of the workpiece, which is the difference between the current penetration of the tool

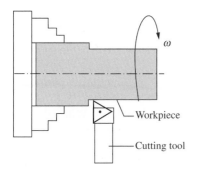

**FIG. 6.32** As the cutting tool penetrates the workpiece, which is rotating in the lathe at constant speed $\omega$, a cutting force is imparted to the tool from the workpiece.

and its penetration one revolution earlier

$$F_s = k_w[x(t) - x(t - \tau)] \qquad (6.111)$$

where $k_w$ is the cutting stiffness and $\tau = 2\pi/\omega$ is the period of revolution of the workpiece. The damping force is proportional to the period of revolution of the workpiece and the rate of penetration

$$F_d = c_w\tau\dot{x}(t) \qquad (6.112)$$

where $c_w$ is the penetration damping coefficient.

The mathematical model for the system using the physical model of Figure 6.33 is

$$m\ddot{x}(t) + (c + c_w\tau)\dot{x}(t) + (k + k_w)x(t) - k_wx(t - \tau) = F(t) \qquad (6.113)$$

where $F(t)$ is an externally applied force. Taking the Laplace transform of Equation (6.113) using Equation (6.110) leads to the system transfer function

$$G(s) = \frac{X(s)}{F(s)} = \frac{1}{ms^2 + (c + c_w\tau)s + (k + k_w) - k_we^{-\tau s}} \qquad (6.114)$$

The denominator of the transfer function for the penetration of the machine tool into the workpiece is not a polynomial in $s$, as in all previous examples, but contains an exponential function. The poles of the transfer function are still determined as the values of $s$ for which the denominator of the transfer function is zero. To this end the poles are determined by solving

$$ms^2 + (c + c_w\tau)s + (k + k_w) - k_we^{-\tau s} = 0 \qquad (6.115)$$

Equation (6.115) has real and complex solutions. Let $s = a + jb$ be a solution of Equation (6.115). Substitution into Equation (6.115), using Euler's identity to replace $e^{-jsb}$ and setting the real and imaginary parts of the resulting equation to zero independently leads to

$$m(a^2 - b^2) + (c + c_w\tau)a + k + k_w - k_we^{a\tau}\cos(b\tau) = 0 \qquad (6.116)$$

$$2mab + (c + c_w\tau)b + k_we^{a\tau}\sin(b\tau) = 0 \qquad (6.117)$$

Equations (6.116) and (6.117) have an infinity of solutions. Recalling from Section 6.3 that a system is stable only if all its poles have negative real parts leads to the conclusion that vibrations of a machine tool are stable only if all solutions of Equations (6.116) and (6.117) have a value of $a < 0$. The nature of the stability of the system depends on the numerical

**FIG. 6.33** A cutting tool is modeled as a one-degree-of-freedom system with $x(t)$ as penetration into the workpiece, which provides cutting force $F_c$ to the tool.

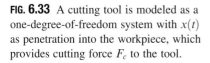

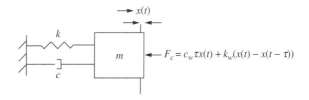

values of the system parameters $m, c, k, c_w, k_w$, and $\tau$. Instability of machine tool vibrations, when parameters are such that $G(s)$ has poles with positive real parts, is called machine tool chatter and leads to errors in the cutting process.

Inversion of Equation (6.114) to determine the impulsive response of the system requires the formal application of the inversion integral, Equation (5.4), and is beyond the scope of this study. When the inversion is performed the impulsive response is of the form

$$x_i(t) = \sum_i C_i e^{a_i t} \sin(b_i t + \phi_i) \qquad (6.118)$$

The summation in Equation (6.118) is carried out over all sets of solutions $(a_i, b_i)$ of Equations (6.116) and (6.117). The values of the constants $C_i$ and $\phi_i$ are determined from the inversion process.

Many time delay systems, such as the machine tool cutting process, have transfer functions with exponential terms in the denominator. Such transfer functions have an infinite number of poles. This is in contrast to the transfer functions for most lumped parameter or discrete systems, which have a finite number of poles, but is similar to those for distributed or continuous systems. Equation (6.118) illustrates that the impulsive response of a stable system is a linear combination of damped modes. The frequencies of the modes correspond to the imaginary parts of the poles while the real parts provide information regarding damping.

## 6.8 FURTHER EXAMPLES

The examples in this section develop the transient response for systems whose models are developed in examples in Chapters 2–4.

### Example 6.30

The differential equation derived in Example 4.15 for the perturbation in the temperature of water being heated in a tank when the rate at which heat is added to the tank suddenly changes is

$$1.68 \times 10^8 \frac{d\theta}{dt} + 2.10 \times 10^3 \theta = 500 u(t) \qquad \text{(a)}$$

(a) Determine the time constant for the system. (b) Determine the response of the system. (c) Determine the increase in temperature when the system reaches its new steady state. (d) Determine the 8 percent settling time for the system.

**Solution**

(a) Equation (a) is rewritten in the form of Equation (6.34) as

$$8 \times 10^4 \frac{d\theta}{dt} + \theta = 0.238 u(t) \qquad \text{(b)}$$

from which the time constant is determined as $T = 80,000$ s.

(b) Application of Equation (6.44), the response of a first-order system due to a unit step input, to this system leads to

$$\theta(t) = 0.238\left(1 - e^{-t/80,000}\right) \tag{c}$$

(c) From Equation (c) it is clear that $\lim_{t\to\infty} \theta(t) = 0.238$. Thus the increase in the rate at which heat is transferred to the tank leads to an increase in the steady-state temperature of the water of $\theta_s = 0.238$ C.

(d) The 8 percent settling time is the time $t_s$ such that

$$\theta(t_s) = 0.92\theta_s \tag{d}$$

Substitution of Equation (c) into Equation (d) leads to

$$0.238\left(1 - e^{-t_s/80,000}\right) = (0.92)(0.238) \tag{e}$$

The solution of Equation (e) is $t_s = 2.01 \times 10^5$ s

---

## Example 6.31

The CSTR of Example 4.17 has a constant volume of $1.32 \times 10^{-3}$ m$^3$ with an inlet flow rate $q = 1.5 \times 10^{-6}$ m$^3$/s. Initially, neither the tank nor the inlet stream contains any reactant $A$. The initial concentration of the product $B$ in the inlet stream and tank is 0.12 mol/L. At $t = 0$ a step change occurs such that the concentrations of the reactant $A$ and the product $B$ in the inlet stream are 0.25 mol/L and 0.12 mol/L respectively. The rate of the first-order reaction that occurs in the CSTR is $1.8 \times 10^{-3}$ s$^{-1}$. Assume the process is isothermal. (a) Determine the time constant for the CSTR for the reactant $A$. (b) Determine $C_A(t)$. (c) Determine $\bar{C}_B(s)$. (d) Determine $C_B(t)$.

### Solution

The differential equations derived in Example 4.17 are

$$V\frac{dC_A}{dt} + (q + kV)C_A = qC_{Ai} \tag{a}$$

$$V\frac{dC_B}{dt} - kVC_A + qC_B = qC_{Bi} \tag{b}$$

The initial conditions given in the problem statement are $C_A(0) = 0$, $C_B(0) = 0.12$ mol/L.

(a) Equation (a) is rewritten in the standard from of the equation for a first-order system as

$$\frac{V}{q + kV}\frac{dC_A}{dt} + C_A = \frac{q}{q + kV}C_{Ai} \tag{c}$$

The time constant is obtained from Equation (c) as

$$T = \frac{V}{q + kV}$$

$$= \frac{1.32 \times 10^{-3}\ \text{m}^3}{(1.5 \times 10^{-6}\ \text{m}^3/\text{s}) + (1.8 \times 10^{-3}\ \text{s}^{-1})(1.32 \times 10^{-3}\ \text{m}^3)} = 3.41 \times 10^2\ \text{s} \tag{d}$$

(b) The input to the system is $C_{Ai}(t) = 0.25u(t)$ mol/L. Substituting numerical values into Equation (c) leads to

$$3.41 \times 10^2 \frac{dC_A}{dt} + C_A = 0.387u(t) \tag{e}$$

The response of a first-order system due to a unit step input is given by Equation (6.44), whose application to this system leads to

$$C_A(t) = 0.387\left(1 - e^{-\frac{t}{3.41 \times 10^2}}\right)u(t) \text{ mol/L} \tag{f}$$

(c) Taking the Laplace transform of Equation (a) with $C_{Ai}(t) = C_{Ai}u(t)$ leads to

$$\bar{C}_A(s) = \frac{qC_{Ai}}{s(Vs + q + kV)} \tag{g}$$

Taking the Laplace transform of Equation (b) applying the initial condition $C_B(0) = C_{Bi}$ and using $C_{Bi}(t) = C_{Bi}u(t)$ leads to

$$Vs\bar{C}_B(s) - VC_{Bi} - kV\bar{C}_A + q\bar{C}_B = \frac{qC_{Bi}}{s} \tag{h}$$

Substituting Equation (g) into Equation (h) and rearranging leads to

$$\bar{C}_B(s) = \frac{qC_{Bi}}{s(Vs + q)} + \frac{VC_{Bi}}{Vs + q} + \frac{kVqC_{Ai}}{s(Vs + q)(Vs + q + kV)} \tag{i}$$

(d) Partial fraction decomposition of Equation (i) leads to

$$\bar{C}_B(s) = \frac{C_{Bi}}{s} + C_{Ai}\left[\frac{1}{\left(1 + \frac{q}{kV}\right)s} - \frac{1}{s + \frac{q}{V}} + \frac{1}{\left(1 + \frac{kV}{q}\right)\left(s + k + \frac{q}{V}\right)}\right] \tag{j}$$

Inversion of Equation (j) leads to

$$C_B(t) = C_{Bi} + C_{Ai}\left[\frac{1}{1 + \frac{q}{kV}} - e^{-\frac{q}{V}t} + \frac{1}{1 + \frac{kV}{q}}e^{-\left(k + \frac{q}{V}\right)t}\right] \tag{k}$$

Substitution of numerical values into Equation (k) leads to

$$C_B(t) = 0.12 + 0.25\left(0.613 - e^{-1.14 \times 10^{-3}t} + 0.387e^{-2.94 \times 10^{-3}t}\right)\frac{\text{mol}}{\text{L}} \tag{l}$$

---

## Example 6.32

The hydraulic servomotor of Example 4.13 contains oil of density $\rho = 900$ kg/m$^3$. The discharge coefficient through the orifice is $C_Q = 0.65$, the width of the spool valve is $w = 30$ mm, the area of the piston is $A_p = 14$ cm$^2$, the dimensions of the segments of the walking beam are $a = 2$ cm and $b = 12$ cm. The pressure difference is $p_s - p_1 = 400$ kPa. (a) Determine the time constant for the system. (b) Determine $y(t)$ when the end of the beam is given a sudden displacement $z(t) = 0.003u(t)$ m.

### Solution

The differential equation derived in Example 4.13 is

$$A_p(a + b)\dot{y} + \hat{C}ay = \hat{C}bz \tag{a}$$

where

$$\hat{C} = C_Q w \sqrt{\frac{2(p_s - p_1)}{\rho}}$$

$$= (0.65)(0.030 \text{ m})\sqrt{\frac{2(4.0 \times 10^5 \text{ N/m}^2)}{900 \text{ kg/m}^3}} = 0.581 \text{ m}^2/\text{s} \quad \text{(b)}$$

Equation (a) is rewritten in the standard form of a first-order equation as

$$\frac{A_p}{\hat{C}}\left(1 + \frac{b}{a}\right)\dot{y} + y = \frac{b}{a}z(t) \quad \text{(c)}$$

(a) The system's time constant is obtained from Equation (c) as

$$T = \frac{A_p}{\hat{C}}\left(1 + \frac{b}{a}\right)$$

$$\frac{(14 \text{ cm}^2)(1 \text{ m}/100 \text{ cm})^2}{0.581 \text{ m}^2/\text{s}}\left(1 + \frac{12 \text{ cm}}{2 \text{ cm}}\right) = 1.69 \times 10^{-2} \text{ s} \quad \text{(d)}$$

(b) Substitution of numerical values into Equation (c) leads to

$$1.69 \times 10^{-2}\dot{y} + y = 0.018u(t) \quad \text{(e)}$$

The solution of Equation (e) is obtained using Equation (6.44) as

$$y(t) = 0.018\left(1 - e^{-\frac{t}{1.69 \times 10^{-2}}}\right)\text{m} \quad \text{(f)}$$

---

## Example 6.33

The pressure vessel of Example 4.10 and Figure 4.17 has a volume $V = 1.2 \text{ m}^3$ and contains air at a pressure $P = 100$ kPa and temperature $T_a = 25$ C. An upstream valve is open from a supply of pressure $\hat{P} = 500$ kPa. The pipe between the supply and the pipe has an average resistance $R = 1 \times 10^4 \text{ m}^{-1}\text{s}^{-1}$. How long will it take the pressure in the pressure vessel to reach 450 kPa? Assume the flow through the connector is subsonic and the process is isothermal.

### Solution

The differential equation for the perturbation pressure in the pressure vessel is derived in Example 4.10 as

$$\frac{V}{R_a T_a}\frac{dp}{dt} + \frac{p}{R} = \frac{p_s}{R} \quad \text{(a)}$$

where $p_s = 400$ kPa is the difference from the initial pressure in the tank. Equation (a) is rewritten in the standard form of the differential equation for a first-order system as

$$\frac{VR}{R_a T_a}\frac{dp}{dt} + p = p_s \quad \text{(b)}$$

The time constant is determined from Equation (b) as

$$T = \frac{VR}{R_a T_a}$$

$$= \frac{(1.2 \text{ m}^3)(1 \times 10^4 \text{ m}^{-1}\text{s}^{-1})}{(287 \text{ J·kg/K})(298 \text{ K})} = 0.140 \text{ s} \qquad (c)$$

Substitution of numerical values into Equation (b) leads to

$$0.140 \frac{dp}{dt} + p = 400u(t) \qquad (d)$$

The solution of Equation (d) subject to $p(0) = 0$ is obtained using Equation (6.44) as

$$p(t) = 400\left(1 - e^{-t/0.140}\right) \text{kPa} \qquad (e)$$

The pressure in the pressure vessel reaches 450 kPA when the perturbation pressure is 350 kPa. Thus the time required to reach this pressure is obtained by solving

$$350 = 400\left(1 - e^{-t/0.140}\right) \qquad (f)$$

The solution of Equation (f) is $t = 0.291$ s.

---

## Example 6.34

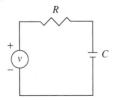

**FIG. 6.34** The $RC$ circuit of Example 6.34.

Consider the $RC$ circuit of Figure 6.34. Determine the current in the circuit and the voltage drop across the capacitor due to an impulsive source.

### Solution

Application of KVL around the loop leads to

$$Ri + \frac{1}{C}\int_0^t i\,dt = v(t) \qquad (a)$$

The transfer function for the circuit is

$$G(s) = \frac{I(s)}{V(s)} = \frac{\frac{1}{R}s}{s + \frac{1}{RC}}$$

$$= \frac{1}{R}\left(1 - \frac{1}{RC}\frac{1}{s + \frac{1}{RC}}\right) \qquad (b)$$

The impulsive response is determined by inverting the transfer function resulting in

$$i(t) = \frac{1}{R}\left[\delta(t) - \frac{1}{RC}e^{-t/(RC)}u(t)\right] \qquad (c)$$

The voltage drop across the capacitor is

$$v(t) = \frac{1}{C} \int_0^t i\, dt$$

$$= \frac{1}{RC} \int_0^t \left[ \delta(t) - \frac{1}{RC} e^{-t/(RC)} u(t) \right] dt$$

$$= \frac{1}{RC} \left\{ u(t) - \left[ e^{-t/RC} u(t) \right]_{t=0}^{t=t} \right\}$$

$$= \frac{1}{RC} e^{-t/(RC)} u(t) \tag{d}$$

Equation (c) shows that an impulsive voltage applied to an $RC$ circuit leads to an impulsive current, causing an instantaneous change in current from 0 before the impulse to $-1/RC^2$ after application of the impulse. The voltage across the capacitor has a discrete jump at $t = 0$ to $1/RC$.

---

## Example 6.35

Consider the pharmokinetic problem of Examples 4.18 and 4.30. A drug is infused into the plasma at a constant rate of 0.302 mg/hr until it is expended after 12 hr. The rate of elimination is 0.105/hr. Assume the available volume of plasma is 40 L. (a) Using the one-component model of Example 4.18 determine the concentration of the drug in the plasma as a function of time. (b) Use the two-component model of Example 4.31 to determine the concentration of the drug in the plasma and in the tissue, assuming the available volume of tissue is 4 L with $k_1 = 0.202$/hr and $k_2 = 0.133$/hr. (c) Plot and compare the concentration of the drug in the plasma using the two models.

### Solution

(a) The rate of infusion is $I(t) = 0.302[1 - u(t - 12)]$ mg/hr. (a) Equation (d) of Example 4.18 is rearranged as

$$\frac{1}{k_e} \frac{dC}{dt} + C = \frac{I(t)}{k_e V} \tag{a}$$

The time constant for the first-order system is $T = 1/k_e = 9.52$ hr. Substitution of values into Equation (a) leads to

$$9.52 \frac{dC}{dt} + C = 0.0719[1 - u(t - 12)] \tag{b}$$

where $t$ is measured in hours and the concentration is in milligrams per liter. The response of the system during the time that the drug is being infused is that of a first-order system with a step input. After the drug is expended the response of the system is that of a first-order system with a nonzero initial condition and decays, and it asymptotically approaches zero for a long time. Taking the Laplace transform of Equation (b) leads to

$$C(s) = \frac{0.0719\left(1 - e^{-12s}\right)}{s(9.52s + 1)} = \frac{7.553 \times 10^{-3}\left(1 - e^{-12s}\right)}{(s + 0.105)s} \tag{c}$$

Inversion of Equation (c) leads to

$$C(t) = 7.19 \times 10^{-2} \left[ \left( 1 - e^{-0.105t} \right) u(t) - \left( 1 - e^{-0.105(t-12)} \right) u(t - 12) \right] \text{ mg/L} \quad \text{(d)}$$

(b) The transfer functions for the system are determined using Equations (g) and (h) of Example 4.30 as

$$G_1(s) = \frac{C_p(s)}{I(s)} = \frac{s + k_2}{V_p[s^2 + (k_e + k_1 + k_2)s + k_e k_2]} \quad \text{(e)}$$

$$G_2(s) = \frac{C_t(s)}{I(s)} = \frac{k_1}{V_t[s^2 + (k_e + k_1 + k_2)s + k_e k_2]} \quad \text{(f)}$$

The two-compartment model is a second-order system of natural frequency

$$\omega_n = \sqrt{k_e k_2} = \sqrt{(0.105/\text{hr})(0.133/\text{hr})} = 0.118 \text{ r/hr} \quad \text{(g)}$$

and damping ratio

$$\zeta = \frac{k_e + k_1 + k_2}{2\omega_n} = \frac{(0.105/\text{hr}) + (0.202/\text{hr}) + (0.133/\text{hr})}{2(0.118/\text{hr})} = 1.86 \quad \text{(h)}$$

The poles of the transfer function are

$$s_{1,2} = \omega_n \left( -\zeta \pm \sqrt{\zeta^2 - 1} \right) = -0.405, -0.0344 \quad \text{(i)}$$

Substitution of the given and calculated values into Equations (e) and (f) leads to

$$G_1(s) = \frac{0.025(s + 0.133)}{(s + 0.405)(s + 0.0344)} \quad \text{(j)}$$

$$G_2(s) = \frac{0.0333}{(s + 0.405)(s + 0.133)} \quad \text{(k)}$$

Noting that $I(s) = 0.302(1 - e^{-12s})/s$ and performing partial fraction decompositions on Equations (j) and (k) lead to

$$\bar{C}_p(s) = \left( \frac{0.0721}{s} - \frac{0.0137}{s + 0.405} - \frac{0.0589}{s + 0.0344} \right) \left( 1 - e^{-12s} \right) \quad \text{(l)}$$

$$\bar{C}_t(s) = \left( \frac{0.722}{s} + \frac{0.0670}{s + 0.405} - \frac{0.788}{s + 0.0344} \right) \left( 1 - e^{-12s} \right) \quad \text{(m)}$$

Equations (l) and (m) are inverted leading to

$$C_p(t) = \left( 0.0721 - 0.0137 e^{-0.405t} - 0.0589 e^{-0.0344t} \right) u(t)$$
$$- \left( 0.0721 - 0.0137 e^{-0.405(t-12)} - 0.0589 e^{-0.0344(t-12)} \right) u(t - 12) \quad \text{(n)}$$

$$C_t(t) = \left( 0.722 + 0.0670 e^{-0.405t} - 0.788 e^{-0.0344t} \right) u(t)$$
$$- \left( 0.722 + 0.0670 e^{-0.405(t-12)} - 0.788 e^{-0.0344(t-12)} \right) u(t - 12) \quad \text{(o)}$$

(c) Equations (d) and (n) are plotted using MATLAB on the same graph in Figure 6.35.

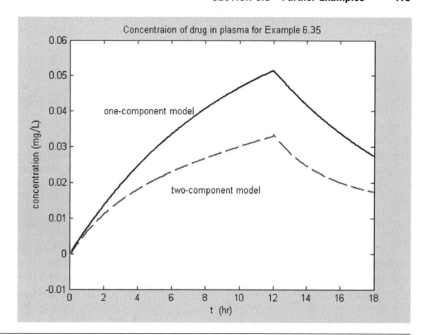

Concentraion of drug in plasma for Example 6.35

one-component model

two-component model

**FIG. 6.35** The time-dependent response for the concentration of a drug in plasma during and after infusion.

## Example 6.36

Consider the second-order system of Example 2.24 when the linear spring has a stiffness of $2.30 \times 10^2$ N/m and the viscous damping coefficient is $1.40 \times 10^2 \ \frac{\text{N·s}}{\text{m}}$. All other parameters are those specified in Figure 2.57. The natural frequency of this system is 94.9 r/s and it's damping ratio of 0.421. Determine the response of the system (a) when the disk has an initial angular displacement of 0.05 rad and an initial angular velocity of zero, (b) when the disk is at rest in equilibrium and is subject to an angular impulse of magnitude 2.8 N·m·s, and (c) when the disk is in equilibrium and a moment of magnitude 14 N·m is applied.

### Solution

(a) Since the damping ratio of the system is less than one, the free response is underdamped and the response is given by Equation (6.69). The system's damped natural frequency is

$$\omega_d = \omega_n \sqrt{1 - \zeta^2} = (94.9 \text{ r/s})\sqrt{1 - (0.421)^2} = 86.1 \text{ r/s} \qquad (a)$$

For the initial conditions, $\theta(0) = \theta_0$ and $\dot{\theta}(0) = 0$, Equations (6.70) and (6.71) lead to

$$A_d = \theta_0 \sqrt{1 + (\zeta\omega_n/\omega_d)^2}$$

$$= (0.05 \text{ r})\sqrt{1 + \left[\frac{(0.421)(94.9 \text{ r/s})}{86.1 \text{ r/s}}\right]^2} = 5.51 \times 10^{-2} \text{ r} \qquad (b)$$

$$\phi_d = \tan^{-1}\left(\frac{\omega_d}{\zeta\omega_n}\right)$$

$$= \tan^{-1}\left(\frac{86.1 \text{ r/s}}{(0.421)(94.9 \text{ r/s})}\right) = 1.14 \text{ rad} \qquad (c)$$

Substitution of numerical values into Equation (6.69) leads to the free response of the disk as

$$\theta(t) = 5.51 \times 10^{-2} e^{-40.0t} \sin(86.1t + 1.14) \text{ rad} \tag{d}$$

(b) The differential equation governing the angular displacement of the disk when it is subject to a moment $M(t)$ is

$$\ddot{\theta} + 2\zeta\omega_n\dot{\theta} + \omega_n^2\theta = \frac{M(t)}{I_{eq}} \tag{e}$$

Numerical values given in Example 2.24 and Equation (l) of the same example are used to calculate the equivalent mass moment of inertia of the system as

$$I_{eq} = (\tfrac{1}{2}m_1 + m_2)r^2 = [\tfrac{1}{2}(1.5 \text{ kg}) + (1 \text{ kg})](0.2 \text{ m})^2 = 0.07 \text{ kg·m}^2 \tag{f}$$

If $M(t) = 2.8\delta(t)$ then Equation (e) becomes

$$\ddot{\theta} + 80\dot{\theta} + 9{,}000\theta = 40\delta(t) \tag{g}$$

The system's transfer function corresponds to that of the first three entries in Table 6.5, which is used to obtain the response of the system as

$$\theta(t) = \frac{40}{86.1} e^{-(0.421)(94.9)} \sin(86.1t)$$

$$= 0.465 e^{-40t} \sin(86.1t) \text{ r} \tag{h}$$

(c) If $M(t) = 14u(t)$ then Equation (e) becomes

$$\ddot{\theta} + 80\dot{\theta} + 9{,}000\theta = 200u(t) \tag{i}$$

The first entry of Table 6.6 is used to determine the response of the system as

$$\theta(t) = \frac{200}{(94.9)^2}\left[1 - e^{-40t}\left(\cos 86.1t + \frac{(0.421)(94.9)}{(86.1)}\sin 86.1t\right)\right]$$

$$= 0.0222\left[1 - e^{-40t}(\cos 86.1t + 0.471 \sin 86.1t)\right] \text{ r} \tag{j}$$

## Example 6.37

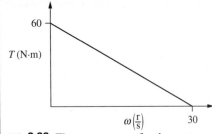

**FIG. 6.36** The torque curve for the motor of Example 6.37.

A dynamometer test is run on an armature-controlled dc servomotor, as modeled in Example 3.23. During the test the dynamometer applies a constant armature voltage, and the output torque and shaft speed are measured. For a certain motor when a dynamometer test is run with $v_a = 50$ V, the torque curve, the plot of torque vs. speed as shown in Figure 6.36, is approximately linear. The armature resistance is 100 $\Omega$ and the armature inductance is 0.15 H. (a) Use the torque curve to determine the values of $K_a$ and $K_b$. (b) The motor is used to run a shaft with $J = 1.8$ kgm² and $c_t = 2.15$ Nms/r. Determine the transient response of the armature current and the shaft speed when a 100-V source is connected to the circuit. (c) How long will it take for the speed to become within 2 percent of its eventual steady-state value?

## Solution

(a) The system is operating at steady state for the dynamometer test. Equation (a) of Example 3.23, evaluated at steady state, becomes

$$R_a i_a + K_b \omega = v_a \tag{a}$$

Equation (3.83) is used to relate the armature current to the torque $i_a = T/K_a$, which when used in Equation (a) leads to

$$\frac{R_a}{K_a} T + K_b \omega = v_a \tag{b}$$

Equation (b) is rearranged in the form of the equation of a line as

$$T = \frac{K_a K_b}{R_a} \omega + \frac{K_a}{R_a} v_a \tag{c}$$

The slope of the torque curve, determined from Figure 6.36, is $-2.0$ N·m·s/r and the intercept with the torque axis is 60 N·m. Thus Equation (c) implies

$$\frac{K_a}{R_a} v_a = 60 \text{ N·m}$$

$$K_a = \frac{(60 \text{ N·m})(100 \text{ } \Omega)}{50 \text{ V}} = 120 \text{ N·m/A} \tag{d}$$

and

$$-\frac{K_a K_b}{R_a} = -2 \text{ N·m·s/r}$$

$$K_b = \frac{(2 \text{ N·m·s/r})(100 \text{ } \Omega)}{120 \text{ N·m/A}} = 1.67 \text{ V·s/r} \tag{e}$$

(b) Substitution of given and calculated values into Equations (a) and (b) of Example 3.23 with $v_a = 100u(t)$ leads to

$$0.15 \frac{di_a}{dt} + 100 i_a + 1.67 \omega = 100u(t) \tag{f}$$

$$1.8 \frac{d\omega}{dt} + 2.15 \omega - 120 i_a = 0 \tag{g}$$

Taking the Laplace transform of Equations (f) and (g) assuming all initial conditions are zero leads to

$$\begin{bmatrix} 0.15s + 100 & 1.67 \\ -120 & 1.8s + 2.15 \end{bmatrix} \begin{bmatrix} I_a(s) \\ \Omega(s) \end{bmatrix} = \begin{bmatrix} \dfrac{100}{s} \\ 0 \end{bmatrix} \tag{h}$$

The system of Equation (h) is solved simultaneously yielding

$$I_a(s) = \frac{666.7s + 796.2}{s(s^2 + 666.7s + 1.54 \times 10^3)} \tag{i}$$

$$\Omega(s) = \frac{4.44 \times 10^4}{s(s^2 + 666.7s + 1.54 \times 10^3)} \tag{j}$$

The response is that of a second-order system of natural frequency $\omega_n = \sqrt{1.54 \times 10^3} = 34.2$ r/s and damping ratio $\zeta = 666.7/2(34.2) = 9.75$. Partial fraction decomposition of Equations (i) and (j) leads to

$$I_a(s) = \frac{0.517}{s} - \frac{1.01}{s + 6.64 \times 10^2} + \frac{0.489}{s + 2.32} \tag{k}$$

$$\Omega(s) = \frac{28.83}{s} + \frac{0.101}{s + 6.64 \times 10^2} - \frac{28.93}{s + 2.32} \tag{l}$$

Equations (k) and (l) are inverted leading to

$$i_a(t) = \left( 0.517 - 1.01 e^{-6.64 \times 10^2 t} + 0.489 e^{-2.32 t} \right) u(t) \tag{m}$$

$$\omega(t) = \left( 28.83 + 0.101 e^{-6.64 \times 10^2 t} - 28.93 e^{-2.32 t} \right) u(t) \tag{n}$$

(c) The eventual steady-state speed is 28.83 r/s. The time required for the speed to reach 98 percent of this value is obtained by solving

$$(0.98)(28.83) = 28.83 + 0.101 e^{-6.64 \times 10^2 t} - 28.83 e^{-2.32 t}$$

$$-0.577 = 0.101 e^{-6.64 \times 10^2 t} - 28.83 e^{-2.32 t} \tag{o}$$

The solution to Equation (o) is $t = 1.69$ s.

---

## Example 6.38

Design the operational amplifier circuit of Example 3.30 such that it simulates a critically damped second-order system with natural frequency of 250 r/s.

### Solution

Equation (o) of Example 3.30 is the differential equation simulated by the operational amplifier circuit,

$$C_1 C_2 R_3 \frac{d^{2} v_0}{dt^2} + \frac{R_3 C_2}{R_2} \frac{dv_0}{dt} + \frac{R_5}{R_4 R_6} v_0 = \frac{v_i}{R_1} \tag{a}$$

Dividing Equation (a) by $C_1 C_2 R_3$ leads to

$$\frac{d^2 v_0}{dt^2} + \frac{1}{C_1 R_2} \frac{dv_0}{dt} + \frac{R_5}{R_3 R_4 R_6 C_1 C_2} v_0 = \frac{1}{R_1 R_3 C_1 C_2} v_i \tag{b}$$

The transfer function for the circuit is obtained from Equation (b) as

$$G(s) = \frac{V_0(s)}{V_i(s)} = \frac{\dfrac{1}{R_1 R_3 C_1 C_2}}{s^2 + \dfrac{1}{C_1 R_2} s + \dfrac{R_5}{R_3 R_4 R_6 C_1 C_2}} \tag{c}$$

Equation (c) is the standard form of the transfer function for a second-order system. System parameters are identified by comparing Equation (c) with the standard form of the denominator of the transfer function for a second-order system [Equation 6.54]. Through this comparison the natural frequency and damping ratio are

$$\omega_n = \sqrt{\frac{R_5}{R_3 R_4 R_6 C_1 C_2}} \tag{d}$$

$$\zeta = \frac{1}{2 \omega_n C_1 R_2} \tag{e}$$

Since there are seven circuit parameters in Equations (d) and (e) and only two conditions to satisfy the design is not unique. Arbitrarily choose $R_4 = R_5$, $C_1 = C_2 = 1$ μF. Substituting into Equation (d) and setting the natural frequency to 250 r/s leads to

$$\sqrt{\frac{1 \times 10^{12}}{R_3 R_6}} = 250 \qquad (f)$$

Choosing $R_3 = R_6$ in Equation (f) gives $R_3 = R_6 = 4$ kΩ.

A second-order system is critically damped when its damping ratio is one. Thus from Equation (e)

$$\frac{1}{2(250 \text{ r/s})(1 \times 10^{-6} \text{ F})R_2} = 1 \qquad (g)$$

$$R_2 = 2 \text{ kΩ}$$

## Example 6.39

(a) Determine the natural frequencies of the three-degree-of-freedom system of Example 2.19 when $m_1 = m_2 = m_3 = 1$ kg and $k_1 = k_2 = k_3 = 10000$ N/m. (b) Use MATLAB to determine the step response for $x_1(t)$, for the undamped system.

**Solution**

Substitution of given values into Equation (j) of Example 2.19 with $c_1 = c_2 = c_3 = 0$ leads to

$$\begin{bmatrix} 1 & 0 & 0 \\ 0 & 1 & 0 \\ 0 & 0 & 1 \end{bmatrix} \begin{bmatrix} \ddot{x}_1 \\ \ddot{x}_2 \\ \ddot{x}_3 \end{bmatrix} + \begin{bmatrix} 2000 & -1000 & 0 \\ -1000 & 2000 & -1000 \\ 0 & -1000 & 1000 \end{bmatrix} \begin{bmatrix} x_1 \\ x_2 \\ x_3 \end{bmatrix} = \begin{bmatrix} 0 \\ 0 \\ F(t) \end{bmatrix} \qquad (a)$$

Taking the Laplace transform of each of the equations of Equation (a) with all initial conditions set equal to zero gives>

$$\begin{bmatrix} s^2 + 2000 & -1000 & 0 \\ -1000 & s^2 + 2000 & -1000 \\ 0 & -1000 & s^2 + 1000 \end{bmatrix} \begin{bmatrix} \bar{x}_1(s) \\ \bar{x}_2(s) \\ \bar{x}_3(s) \end{bmatrix} = \begin{bmatrix} 0 \\ 0 \\ F_1(s) \end{bmatrix} \qquad (b)$$

Cramer's rule is used to solve for $X_1(s)$

$$X_1(s) = \frac{\begin{vmatrix} 0 & -1,000 & 0 \\ 0 & 2,000 & -1,000 \\ F(s) & -1,000 & 1,000 \end{vmatrix}}{\begin{vmatrix} s^2 + 2,000 & -1,000 & 0 \\ -1,000 & s^2 + 2,000 & -1,000 \\ 0 & -1,000 & s^2 + 1,000 \end{vmatrix}} \qquad (c)$$

from which a transfer function is obtained as

$$G_1(s) = \frac{X_1(s)}{F(s)} = \frac{1 \times 10^6 F(s)}{s^6 + 5 \times 10^3 s^4 + 6 \times 10^6 s^2 + 1 \times 10^9} \qquad (d)$$

```
>> N=[1E6];
>> D=[1 0 5E3 0 6E6 0 1E9];
>> [r,p,k]=residue(N,D)

r =

        0 - 0.0017i
        0 + 0.0017i
        0 + 0.0055i
        0 - 0.0055i
        0 - 0.0086i
        0 + 0.0086i


p =

        0 +56.9823i
        0 -56.9823i
        0 +39.4330i
        0 -39.4330i
        0 +14.0735i
        0 -14.0735i


k =

    []

>> step(N,D)
>>
```

(a)

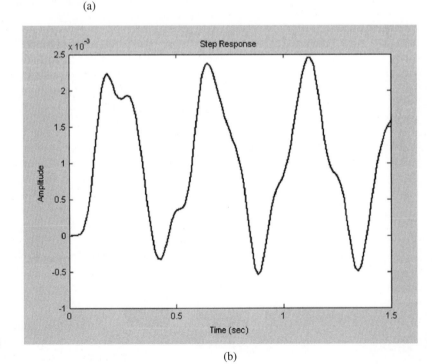

(b)

**FIG. 6.37** (a) The MATLAB work session used to determine the natural frequencies of the system of Example 6.39 and (b) the step response for $x_1(t)$.

The poles of the transfer function are obtained using MATLAB as $s = \pm j57.0$, $\pm j39.4$, $\pm j14.1$. Thus the natural frequencies are $\omega_1 = 14.1$ r/s, $\omega_2 = 39.4$ r/s, $\omega_3 = 57.0$r/s.

(b) The MATLAB work session to determine the poles of the transfer function and the step response for $x_1(t)$, as well as the step response, are shown in Figure 6.37.

## Example 6.40

Determine the step responses for the currents if the circuit of Example 3.11 when $R = 2,000\ \Omega$, $C = 0.2\ \mu F$, and $L = 0.5$ H.

### Solution

The substitution of given parameters into Equations (d), (f), and (h) of Example 3.11 leads to

$$0.5\frac{di_1}{dt} + 2,000i_1 + 5 \times 10^6 \int_0^t (i_1 - i_2)dt = v(t) \tag{a}$$

$$0.5\frac{di_2}{dt} + 2,000(i_2 - i_3) - 5 \times 10^6 \int_0^t (i_1 - i_2)dt = 0 \tag{b}$$

$$0.5\frac{di_3}{dt} - 2,000(i_2 - i_3) + 5 \times 10^6 \int_0^t i_3 dt = 0 \tag{c}$$

Taking the Laplace transforms of Equations (a)–(c) and writing the resulting set of simultaneous algebraic equations for the transforms in matrix form leads to

$$\begin{bmatrix} 0.5s + 2,000 + \dfrac{5 \times 10^6}{s} & -\dfrac{5 \times 10^6}{s} & 0 \\[2mm] -\dfrac{5 \times 10^6}{s} & 0.5s + 2,000 + \dfrac{5 \times 10^6}{s} & -2,000 \\[2mm] 0 & -2,000 & 0.5s + 2,000 + \dfrac{5 \times 10^6}{s} \end{bmatrix}$$

$$\times \begin{bmatrix} I_1(s) \\ I_2(s) \\ I_3(s) \end{bmatrix} = \begin{bmatrix} V(s) \\ 0 \\ 0 \end{bmatrix} \tag{d}$$

Since the inverses of the transfer functions are the impulsive responses, the transfer functions are obtained by simultaneous solution of the preceding equations with $V(s) = 1$. MATLAB program Example6_40a, whose script is listed in Figure 6.38, is used to symbolically determine the transfer functions resulting in

$$G_1(s) = \frac{I_1(s)}{V(s)}$$

$$= \frac{2(s^4 + 8,000s^3 + 2 \times 10^7 s^2 + 8 \times 10^{10}s + 1 \times 10^{14})}{s^5 + 1.2 \times 10^4 s^4 + 6.2 \times 10^7 s^3 + 2.4 \times 10^{11} s^2 + 5.2 \times 10^{14} + 8 \times 10^{17}} \tag{e}$$

```
% Example 6.40a
% Transfer functions for currents in three mesh circuit
%
% Defining impedance matrix
%
F1=0.5*s+2000+5E6/s;
F2=-5E6/s;
F3=-2000;
A=[F1 F2 0;F2 F1 F3;0 F3 F1]
%
% Calculating transfer functions
%
b=[1; 0; 0];
G=A^-1*b
G1=simplify(G)
G2=vpa(G1,4)
%
% End of Exmaple6_40a.m
```

**FIG. 6.38** Script of Example6_40a.m used to symbolically determine the transfer functions for a three-loop circuit.

$$G_2(s) = \frac{I_2(s)}{V(s)} = \frac{2 \times 10^7}{s^3 + 8{,}000s^2 + 2 \times 10^7 s + 8 \times 10^{10}} \tag{f}$$

$$G_3(s) = \frac{I_3(s)}{V(s)} = \frac{8 \times 10^{10} s}{s^5 + 1.2 \times 10^4 s^4 + 6.2 \times 10^7 s^3 + 2.4 \times 10^{11} s^2 + 5.2 \times 10^{14} s + 8 \times 10^{17}} \tag{g}$$

```
% Example6.40b
% Program to plot step responses on same graph
% Defining numerators and denominators of transfer functions
N1=2*[1 8000 0.2E8 0.8E11 0.1E15];
N2=[0.2E8];
N3=[0.8E11];
D1=[1 1.2E4 6.2E7 2.4E11 5.2E14 8E17];
D2=[1 8000 .2E8 .8E11];
t=0:.00005:.01;
% Defining transfer functions
G1=tf(N1,D1)
G2=tf(N2,D2)
G3=tf(N3,D1)
% Calculating step responses
i1=step(G1,t);
i2=step(G2,t);
i3=step(G3,t);
% Plotting ster responses
plot(t,i1,'-',t,i2,'.',t,i3,'--')
xlabel('t (s)')
ylabel('current (A)')
title('Step responses for circuit of Example 6.40')
% Since the magnitude of the step response for i3 is much less than those
% for i1 and i2 the step response for i3 is plotted separately
legend('i_1(t)','i_2(t)','i_3(t)')
figure
plot(t,i3)
xlabel('t (s)')
ylabel('current (A)')
title('Current in third loop, i_3 (t), for circuit of Example 6.40')
%
% End of Example6_40.m
```

**FIG. 6.39** Script of Example6_40b.m used to plot the step responses for currents in the three-mesh circuit of Example 6.40.

The script of Example6_40b.m is shown in Figure 6.39. This script is used to determine and plot the step responses, which are illustrated in Figure 6.40. The following is noted about the script and the responses:

- Figure 6.40(a) shows that the magnitude of the step response for $i_3(t)$ is much less than the magnitudes of the step responses for the other currents. Thus a separate plot is generated for its step response.

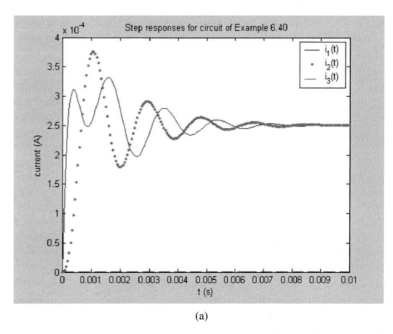

(a)

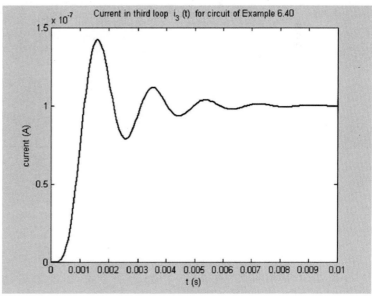

**FIG. 6.40** Step responses for the currents in the three-loop circuit of Example 6.40.

(b)

- The numerator and denominator forms for the transfer functions are converted into defined transfer functions to determine each individually and then are plotted on the same set of axes.
- The command 'i1=step(G1,t)' sets up a vector of values of the step response for $i_1(t)$ corresponding to the defined values of $t$.

## Example 6.41

The heat exchanger of Example 4.25 uses water at 20 C to cool oil at 300 C. The properties of the heat exchanger and the flow are $\rho_w = 9.60 \times 10^2$ kg/m³, $\rho_o = 7.50 \times 10^2$ kg/m³, $c_w = 4.19 \times 10^3$ J/kg·C, $c_o = 2.4 \times 10^3$ J/kg·C, $U = 8.65 \times 10^2$ W/m²·C, $\dot{m}_w = 2.51$ kg/s, $\dot{m}_o = 4.85$ kg/s, $r_i = 2$ cm, $r_o = 2.2$ cm, $r_2 = 4.2$ cm, $L = 2.5$ m. (a) Determine the steady-state outlet temperatures of the oil and water. (b) Determine the transient perturbations in outlet temperatures when the temperature of the water is suddenly increased to 30 C. Assume the properties of both fluids remain constant.

### Solution

(a) Substituting the given values into Equations (e) and (f) of Example 4.25 with the time derivatives set to zero leads to the steady-state equations

$$(2.43 \times 10^3 \text{ J/kg·C})(4.85 \text{ kg/s})(300\,C - T_{os}) = 2\pi(0.022 \text{ m})$$
$$\times (2.5 \text{ m})(8.65 \times 10^2 \text{ W/m}^2\text{·C})(T_{os} - T_{ws}) \qquad \text{(a)}$$

$$(4.19 \times 10^3 \text{ J/kg·C})(2.51 \text{ kg/s})(20\,C - T_{ws}) = -2\pi(0.022 \text{ m})$$
$$\times (2.5 \text{ m})(8.65 \times 10^2 \text{ W/m}^2\text{·C})(T_{os} - T_{ws}) \qquad \text{(b)}$$

Equations (a) and (b) are solved yielding $T_{os} = 293.3$ C, $T_{ws} = 27.5$ C.
(b) Substitution of given values into Equations (k) and (l) of Example 4.25 gives

$$(7.50 \times 10^2 \text{ kg/m}^3)(2.43 \times 10^3 \text{ J/kg·C})\pi(0.02 \text{ m})^2(2.5 \text{ m})\frac{d\theta_o}{dt}$$
$$+ \left[2\pi(0.022 \text{ m})(2.5 \text{ m})(8.65 \times 10^3 \text{ W/m}^2\text{·C})\right.$$
$$+ (2.43 \times 10^3 \text{ J/kg·C})(4.85 \text{ kg/s})\left.\right]\theta_o - 2\pi(0.022 \text{ m})$$
$$\times (8.65 \times 10^3 \text{ W/m}^2\text{·C})(2.5 \text{ m})\theta_w = 0 \qquad \text{(c)}$$

$$(9.60 \times 10^2 \text{ kg/m}^3)(4.19 \times 10^3 \text{ J/kg·C})\pi\left[(0.042 \text{ m})^2 - (0.022 \text{ m})^2\right](2.5 \text{ m})\frac{d\theta_w}{dt}$$
$$- 2\pi(0.022 \text{ m})(8.65 \times 10^3 \text{ W/m}^2\text{·C})(2.5 \text{ m})\theta_0 + \left[2\pi(0.022 \text{ m})(2.5 \text{ m})\right.$$
$$\times (8.65 \times 10^3 \text{ W/m}^2\text{·C}) + (4.19 \times 10^3 \text{ J/kg·C})(2.51 \text{ kg/s})\left.\right]\theta_w$$
$$= (4.19 \times 10^3 \text{ J/kg·C})(2.51 \text{ kg/s})(10\,C)u(t)$$
$$\text{(d)}$$

Simplification of Equations (c) and (d) leads to

$$5.73 \times 10^3 \frac{d\theta_o}{dt} + 1.48 \times 10^4 \theta_o - 2.99 \times 10^3 \theta_w = 0 \tag{e}$$

$$4.04 \times 10^4 \frac{d\theta_w}{dt} - 2.99 \times 10^3 \theta_o + 1.35 \times 10^4 \theta_w = 1.05 \times 10^5 u(t) \tag{f}$$

Taking the Laplace transforms of Equations (e) and (f) leads to

$$\begin{bmatrix} 5.73 \times 10^3 s + 1.48 \times 10^4 & -2.99 \times 10^3 \\ -2.99 \times 10^3 & 4.04 \times 10^4 s + 1.35 \times 10^4 \end{bmatrix} \begin{bmatrix} \Theta_o(s) \\ \Theta_w(s) \end{bmatrix} = \begin{bmatrix} 0 \\ \dfrac{1.05 \times 10^5}{s} \end{bmatrix} \tag{g}$$

The system of Equation (g) is solved simultaneously leading to

$$\Theta_o(s) = \frac{1.36}{s(s^2 + 2.92s + 0.827)} \tag{h}$$

$$\Theta_w(s) = \frac{2.61s + 6.73}{s(s^2 + 2.92s + 0.827)} \tag{i}$$

Use of MATLAB to invert the transforms leads to

$$\theta_o(t) = \left(1.64 + 0.222e^{-2.63t} - 1.87e^{-0.314t}\right) u(t) \tag{j}$$

$$\theta_w(t) = \left(8.14 - 0.246e^{-0.263t} - 8.11e^{-0.314t}\right) u(t) \tag{k}$$

The eventual steady-state temperatures are $T_o = 293.3 \text{ C} + 1.64 \text{ C} = 295.0 \text{ C}$ and $T_w = 27.5 \text{ C} + 8.14 \text{ C} = 35.6 \text{ C}$.

---

## Example 6.42

After the water leaves the heat exchanger of Example 4.25 it is cooled to the inlet temperature and recirculated through the heat exchanger and used for the inlet stream. The heat exchanger system of Example 4.25 is operating at steady state under the conditions described in Example 6.41 when the system cooling the water breaks. The water is recirculated through the system at the temperature it leaves the heat exchanger but takes 5 s to again reach the heat exchanger inlet. Determine the transforms of the perturbation in temperatures due to the breakdown of the cooling system. Determine the eventual steady-state temperatures.

### Solution
After the cooling system stops functioning, the inlet water temperature is that of the exit water temperature, delayed by $\tau = 5$ s, $T_{wi} = T_w(t - \tau)$. From the definitions of the temperature perturbations of Equations (h)–(j) of Example 4.25,

$$\theta_{wi} = T_{wi} - T_{wis}$$

$$= T_w(t - \tau) - T_{wis}$$

$$= T_{ws} + \theta_w(t - \tau)T_{wis}$$

$$= (T_{ws} - T_{wis}) + \theta_w(t - \tau) \tag{a}$$

Substituting Equation (a) into Equation (l) of Example 4.25 leads to

$$\rho_w c_w \pi (r_2^2 - r_o^2) L \frac{d\theta_w}{dt} - 2\pi r_o L U \theta_o + (2\pi r_o L U + c_w \dot{m}_w)\theta_w$$

$$= c_w \dot{m}_w [T_{ws} - T_{wis} + \theta_w(t - \tau)] \qquad \text{(b)}$$

The substitution of the given values of Example 6.41 into Equation (k) of Example 4.25 and Equation (a) of this example leads to

$$5.73 \times 10^3 \frac{d\theta_o}{dt} + 1.48 \times 10^4 \theta_o - 2.99 \times 10^3 \theta_w = 0 \qquad \text{(c)}$$

$$4.04 \times 10^4 \frac{d\theta_w}{dt} - 2.99 \times 10^3 \theta_o + 1.35 \times 10^4 \theta_w = 1.05 \times 10^4 [7.5 + \theta_w(t - 5)] \qquad \text{(d)}$$

Taking the Laplace transform of Equations (c) and (d) using Equation (6.110) leads to

$$\begin{bmatrix} 5.73 \times 10^3 s + 1.48 \times 10^4 & -2.99 \times 10^3 \\ -2.99 \times 10^3 & 4.04 \times 10^4 s + 1.35 \times 10^4 - 1.05 \times 10^4 e^{-5s} \end{bmatrix}$$

$$\times \begin{bmatrix} \Theta_o(s) \\ \Theta_w(s) \end{bmatrix} = \begin{bmatrix} 0 \\ \dfrac{7.35 \times 10^4}{s} \end{bmatrix} \qquad \text{(e)}$$

Simultaneous solution of the equations represented by Equation (e) leads to

$$\Theta_o(s) = \frac{0.949}{s(s^2 + 2.92s + 0.827 - 0.260se^{-5s} - 0.671e^{-5s})} \qquad \text{(f)}$$

$$\Theta_w(s) = \frac{1.82s + 4.70}{s(s^2 + 2.92s + 0.827 - 0.260se^{-5s} - 0.671e^{-5s})} \qquad \text{(g)}$$

The inversion of Equations (f) and (g) is difficult and will not be attempted. However, stability can be determined by finding the poles of the transforms. It can be reasoned physically that stability must be achieved. Assuming no losses to the surroundings, only a finite amount of energy in the form of heat is transferred from the oil to the water. When the cooling system is not functioning this heat is returned to the system and the heat transfer from the oil is reduced. It is not possible to increase the temperature of the oil above its initial temperature of 300 C. The final value theorem is applied to determine the eventual steady-state temperatures of the oil and water $\theta_{fo} = \lim_{s \to 0} s\Theta_o(s) = 6.09$, $\theta_{fw} = \lim_{s \to 0} s\Theta_w(s) = 30.1$. Hence, using the results of Example 6.35(a) the eventual temperatures are $T_o = 293.3 \text{ C} + 6.09 \text{ C} = 299.4 \text{ C}$ and $T_w = 27.5 \text{ C} + 30.1 \text{ C} = 57.6 \text{ C}$.

## 6.9 SUMMARY

### 6.9.1 Chapter Highlights

- Chapter 6 is the first part of the study of system response: the transient response of dynamic systems.

- The chapter uses the Laplace transform method to derive and study transient response.

- The system transfer function is defined as the ratio of the transform of the system output to the transform of the system input.
- For most cases the transfer function is a ratio of two polynomials. $N(s)$ is the polynomial in the numerator, and $D(s)$ is the polynomial in the denominator.
- The order of the system is defined as the order $D(s)$.
- The poles of the system are the roots of $D(s)$.
- A matrix of transfer functions may be defined for multiple input, multiple output systems.
- The free response is the system response due to nonzero initial conditions and in the absence of additional system input.
- The impulsive response is the system response due to the input of a unit impulse.
- The impulsive response is obtained as the inverse of the transfer function.
- The impulsive response is continuous if the order of $N(s)$ is at least two less than the order of $D(s)$. Impulsive responses of first-order systems are always discontinuous.
- The step response is the system response due to the input of a unit step function.
- Transient response specifications, defined for the step response, include settling time, rise time, percent overshoot, and peak time.
- The ramp response is the system response due to the input of a unit ramp function.
- A system is stable if and only if all its poles have negative real parts.
- A necessary, but not sufficient, condition for stability is if all coefficients of $D(s)$ are positive.
- Routh's criterion is used to determine stability from $D(s)$.
- Routh's criterion may be used to determine relative stability with respect to $s = -p$ by examining absolute the stability of $D(s - p)$.
- Root-locus diagrams are diagrams of poles of transfer functions that are dependent on a parameter $K$.
- Root-locus diagrams can be used to determine the values of a parameter for which a system is stable.
- The transient response of a first-order system is dependent on one parameter, the time constant.
- The transient response of a stable first-order system exponentially approaches its final value.
- The transient response of a second-order system is dependent on two system parameters: the natural frequency $\omega_n$ and the damping ratio $\zeta$.
- The free response of an undamped system ($\zeta = 0$) is periodic.
- The free response of an underdamped system ($0 < \zeta < 1$) is cyclic, but not periodic and decays exponentially.
- The free-response of a critically damped system ($\zeta = 1$) decays exponentially.
- The free response of an overdamped system ($\zeta > 1$) decays exponentially, but at a slower rate than the response of a critically damped system.

- The impulsive and step responses of second-order systems are given in Tables 6.4 and 6.5 respectively.
- The transient response of a higher-order system is a linear combination of first-order and second-order system responses.
- The transfer function for systems with time delay includes exponential terms.
- Transfer functions can be defined in MATLAB from vectors of coefficients of $N(s)$ and $D(s)$.
- MATLAB can be used to determine and plot the impulsive, step, and ramp responses, given the system's transfer function.
- MATLAB's symbolic capabilities can be used to determine the transfer functions for multidegree-of-freedom systems from the impedance matrix.

## 6.9.2 Important Equations

- Definition of transfer function

$$G(s) = \frac{X(s)}{F(s)} \tag{6.1}$$

- Transfer functions for MIMO systems

$$G_{i,j}(s) = \frac{X_i(s)}{G_j(s)} \tag{6.4}$$

- Impulsive response

$$x_i(t) = \mathcal{L}^{-1}\{G(s)\} \tag{6.9}$$

- Initial value of the impulsive response

$$x_i(0) = \lim_{s \to \infty} sG(s) \tag{6.10}$$

- Final value after step input

$$x_f = G(0) \tag{6.12}$$

- Convolution integral

$$x(t) = \int_0^t F(\tau)x_i(t - \tau)d\tau \tag{6.22}$$

- Equation for the development of the root-locus diagram

$$Q(s) + KR(s) = 0 \tag{6.32}$$

- General form of the differential equation for a first-order system

$$T\frac{dx}{dt} + x = f(t) \tag{6.34}$$

- Transfer function for a first-order system

$$G(s) = \frac{1}{T\left(s + \dfrac{1}{T}\right)} \tag{6.37}$$

- Free response of a first-order system

$$x(t) = x_0 e^{-\frac{t}{T}} \tag{6.39}$$

- Impulsive response of a first-order system

$$x(t) = \frac{I}{T} e^{-\frac{t}{T}} u(t) \tag{6.41}$$

- Step response of a first-order system

$$x(t) = Z\left(1 - e^{-\frac{t}{T}}\right) u(t) \tag{6.44}$$

- Standard form of a differential equation for a second-order system

$$\ddot{x} + 2\zeta\omega_n\dot{x} + \omega_n^2 x = \frac{1}{m} f(t) \tag{6.49}$$

- Standard form of a transfer function for a second-order system

$$G(s) = \frac{As + B}{s^2 + 2\zeta\omega_n s + \omega_n^2} \tag{6.54}$$

- Poles of a transfer function of a second-order system

$$s_1 = -\zeta\omega_n - \omega_n\sqrt{\zeta^2 - 1} \tag{6.57a}$$

$$s_2 = -\zeta\omega_n + \omega_n\sqrt{\zeta^2 - 1} \tag{6.57b}$$

- Undamped free response of a second-order system

$$x(t) = A \sin(\omega_n t + \phi) \tag{6.60}$$

- Underdamped free response of a second-order system

$$x(t) = A_d e^{-\zeta\omega_n t} \sin(\omega_d t + \phi_d) \tag{6.69}$$

- Critically damped free response of a second-order system

$$x(t) = e^{-\omega_n t} + (\dot{x}_0 + \omega_n x_0) e^{-\omega_n t} \tag{6.76}$$

- Overdamped free response of a second-order system

$$x(t) = \frac{1}{2\sqrt{\zeta^2 - 1}} \left\{ \left[ x_0\left(-\zeta + \sqrt{\zeta^2 - 1}\right) - \frac{\dot{x}_0}{\omega_n} \right] e^{s_1 t} \right.$$

$$\left. + \left[ x_0\left(\zeta + \sqrt{\zeta^2 - 1}\right) + \frac{\dot{x}_0}{\omega_n} \right] e^{s_2 t} \right\} \tag{6.80}$$

- Impedance matrix

$$\mathbf{Z}(s) = s^2\mathbf{M} + s\mathbf{C} + \mathbf{K} \tag{6.95}$$

- Time delay systems

$$\mathcal{L}\{x(t - \tau)\} = e^{-\tau s} X(s) \tag{6.110}$$

## PROBLEMS

**6.1** The suspension system of Example 2.26 has the following parameters: $m = 1{,}500$ kg, $I = 250$ kg·m², $\ell_1 = 0.8$ m, $\ell_2 = 1.3$ m, $k_1 = 1.3 \times 10^5$ N/m, $k_2 = 1.5 \times 10^5$ N/m, $c_1 = 900$ N·s/m, $c_2 = 800$ N·s/m. Determine the transfer functions $G_1(s) = X(s)/Y(s)$ and $G_2(s) = \Theta(s)/Y(s)$.

**6.2** Determine the transfer functions $G_1(s) = I_1(s)/V(s)$ and $G_2(s) = I_2(s)/V(s)$ for the circuit of Example 3.28. ·

**6.3** Determine the transfer functions $G_1(s) = I_f(s)/V_f(s)$ and $G_2(s) = \Omega(s)/V_f(s)$ for the field-controlled dc motor of Example 3.22.

**6.4** Determine the transfer functions $G_1(s) = T_p(s)/Q(s)$, $G_2(s) = C_{Ap}(s)/Q(s)$, and $G_3(s) = C_{Bp}(s)/Q(s)$ from Equations (s), (t), and (u) from Example 4.17, where $Q(s) = \mathcal{L}\{\dot{Q} - \dot{Q}_1\}$.

**6.5** Determine the transfer functions $G_1(s) = \Theta_o(s)/\Theta_{wi}(s)$ and $G_2(s) = \Theta_w(s)/\Theta_{wi}(s)$ for the counterflow heat exchanger of Example 4.25.

**6.6** Use Routh's stability criterion to show that the suspension system of Problem 6.1 is stable.

**6.7** The transfer function of a third-order system is

$$G(s) = \frac{s + 4}{s^3 + 8s^2 + 15s + 20} \tag{a}$$

Use Routh's stability criterion to determine if the system is stable.

**6.8** The transfer function of a second-order system in a feedback loop with an integral controller is

$$G(s) = \frac{K_i(4s + 25)}{s^3 + 5s^2 + (10 + 4K_i)s + 25K_i} \tag{a}$$

(a) Determine a criteria for $K_i$, the integral gain such that the system is stable. (b) Use MATLAB to draw impulsive responses of the system for several values of $K_i$ to confirm the result of part (a). (c) Use MATLAB to draw the root-locus diagram for the system to also confirm the result of part (a).

**6.9** The transfer function of a third-order system is

$$G(s) = \frac{2s + 4}{s^3 + 4s^2 + 5s + 3} \tag{a}$$

Use Routh's criteria to determine (a) the absolute stability of the system and (b) the stability of the system relative to $s = -1$.

**6.10** Use MATLAB to draw the root-locus diagram for the transfer function of the system of Problem 6.8. Use the root-locus diagram to discuss the stability of the system as $K$ varies.

**6.11** The closed-loop transfer function for a plant placed in a feedback control loop with a proportional controller is

$$H(s) = \frac{KG(s)}{1 + KG(s)} \tag{a}$$

where $K$ is the proportional gain and $G(s)$ is the transfer function of the plant. The transfer function of a fourth order plant is

$$G(s) = \frac{s + 2}{s(s - 2)(s^2 + 6s + 5)} \tag{b}$$

For what values of $K$ is the closed-loop system stable?

**6.12** The two-tank liquid-level system of Example 5.37 is placed in a feedback control loop with a proportional controller as discussed in Problem 6.11. For what values of the proportional gain $K$ is the closed loop system stable?

**6.13** Consider the circuit of Figure P6.13. (a) Determine the transfer function $G(s) = I_1(s)/V(s)$. (b) Determine the time constant for the system. (c) Determine the response of the system when $v(t) = 10u(t)$ V.

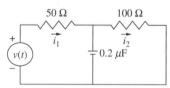

**FIG. P6.13**

**6.14** The capacitor in the system of Figure P6.14 has an initial charge of $1.3 \times 10^{-3}$ C. The switch is closed at $t = 0$, allowing the capacitor to discharge. How long after the switch is closed does it take for the charge on the capacitor to drop to $5 \times 10^{-6}$ C?

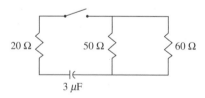

**FIG. P6.14**

**6.15** Consider the $RL$ circuit of Figure P6.15 when the input current is a unit impulse. Determine (a) the voltage drop across the inductor and (b) the current through the inductor.

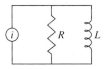

**FIG. P6.15**

**6.16** A first-order system has the transfer function

$$G(s) = \frac{2}{4s + 15}$$

Determine (a) the system's step response; (b) the system's impulsive response; (c) the system's ramp response; (d) the system's 2 percent settling time; (e) the system's 10–90 percent rise time.

**6.17** Reconsider the pharmokinetic system of Example 6.35. (a) Determine the concentration in the plasma as a function of time using the first-order model when the drug is infused again 2 hr after being expended. (b) Set up a general model for the plasma concentration when this process of infusion is continually repeated; that is, the drug is infused at a constant rate for 12 hr, no drug is infused for 2 hr, then the drug is infused at a constant rate for the next 12 hr and so on.

**6.18** The system of Figure P6.18 is at steady state when the exit pressure is suddenly changed from atmospheric pressure to a gauge pressure of 2.5 kPa. The differential equation obtained in Equation (k) of Example 4.8 for the perturbation in liquid level due to a perturbation in pressure is

$$A\frac{dh}{dt} + \frac{1}{R}h = \frac{p}{\rho g R} \qquad (a)$$

(a) Determine the level of the liquid when a new steady state is reached; (b) the time-dependent response of the system; and (c) the 4 percent settling time.

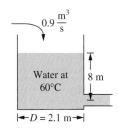

**FIG. P6.18**

**6.19** A 20-ft steel 32-Btu/hr·ft·F pipe has an inner diameter of 12 in and an outer diameter of 14 in. A 0.5-in layer of insulation of R value 15 covers the pipe. The film coefficients at the inner and outer surfaces of the system are

2.1 Btu/hr·ft$^2$·F and 3.2 Btu/hr·ft$^2$·F respectively. The pipe is at a uniform temperature of 70 F when the external temperature is suddenly changed to 68 F. (a) What is the time constant for the system? (b) Determine the time-dependent temperature in the interior of the pipe. (c) How long does it take for the temperature in the interior of the pipe to reach 68.3 F?

**6.20** The disk brake of Example 4.26 is used to stop a shaft rotating at 350 rpm. The brake has a diameter of 10 in and thickness 0.5 in, is made of a steel of density 0.283 lbm/in$^3$, and has a specific heat of 0.12 Btu/hr·ft·F. The heat transfer coefficient is 4.9 Btu/hr·ft$^2$·F. The shaft has an equivalent moment of inertia of 1.25 slugs·ft$^2$ and the ambient temperature is 25 C. (a) Using the model of Equation (j) of Example 4.26 determine the temperature in the brake if the shaft is to be brought to rest in 1 s. (b) Determine the minimum time the shaft can be brought to rest if the maximum temperature in the brake is to be 110° F.

**6.21** The brake of Problem 6.20(a) is to be applied every 3 min. Determine the time-dependent temperature in the brake over 1 hr of operation.

**6.22** A tank of volume 20 m$^3$ initially contains air at a temperature of 40 C and a pressure of 12 kPa. An upstream valve is opened, allowing air from a supply of pressure 20 kPa to flow into the tank. The resistance between the supply and the tank is 16.0 1/m·s. The process is isothermal. (a) What is the time constant for the system? (b) Determine the pressure in the tank as a function of time. (c) How long does it take for the pressure in the tank to reach 18 kPa?

**6.23** The transfer function for a second-order system is

$$G(s) = \frac{2s + 3}{4s^2 + 10s + 45}$$

Determine the natural frequency and damping ratio for the system.

**6.24** The transfer function for a second-order system is

$$G(s) = \frac{3s + 5}{s^2 + 20s + 500}$$

(a) Determine the impulsive response of the system. (b) Determine the step response of the system. (c) Determine the 2 percent settling time. (d) Determine the 10–90 percent rise time. (e) Determine the percent overshoot of the step response. (f) Determine the peak time of the step response.

**6.25** For the system for Figure P6.25, determine (a) its free response when the block is displaced 2 mm from equilibrium and then released; (b) its impulsive response; (c) its step response; and (d) its ramp response.

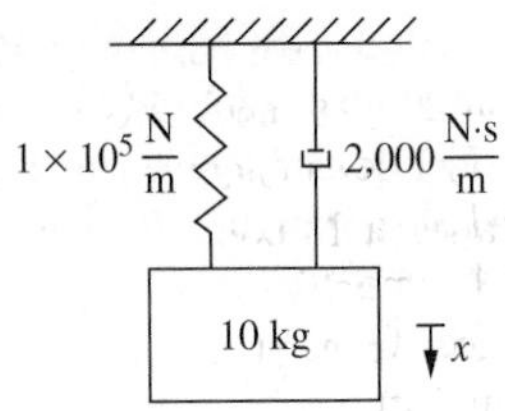

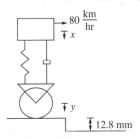

**FIG. P6.25**

**6.26** The bar of Example 2.14 has a length of 60 cm and a mass of 1.2 kg. The spring stiffness is $2.3 \times 10^4$ N/m. For what values of $c$, the viscous damping coefficient, is its free response overdamped?

**6.27** Determine the free response of the system of Example 2.17 if the disk is rotated 1° clockwise from equilibrium and then released. Use $r = 10$ cm, $m_1 = 1$ kg, $m_2 = 0.2$ kg, $I = 0.05$ kg·m², $k_1 = k_2 = 1 \times 10^5$ N/m, and $c_1 = c_2 = 225$ N·s/m.

**6.28** The door of Figure P6.28 is free to rotate about an axis through its hinges. The door has a moment of inertia about an axis through its hinges of 25.2 kg·m². Each hinge has a torsional stiffness of 15 N·m/r. A door damper provides torsional damping such that the system is critically damped. (a) If the door is held open at an angle of 50° and then released, how long does it take for the door to close to an angle of 5° where it automatically latches shut. (b) If the door is initially closed, what initial angular velocity must be imparted to it to cause it to open to an angle of 70°?

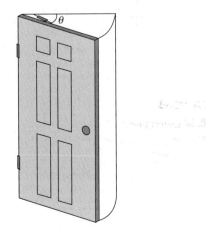

**FIG. P6.28**

**6.29** Repeat Problem 6.28 if the damper provides torsional damping such that the damping ratio of the system is 1.15.

**6.30** Figure P6.30 illustrates a simplified model of a vehicle suspension system. Let $y(t)$ is the displacement of the wheel as it traverse the road contour, and $x(t)$ is the displacement of the vehicle. (a) Show that the mathematical model for $x(t)$ is

$$m\ddot{x} + c\dot{x} + kx = c\dot{y} + ky$$

(b) Consider a vehicle with $m = 500$ kg, $k = 3.2 \times 10^5$ N/m, and $c = 10,000$ N·s/m. The vehicle is traveling with a constant horizontal velocity of 80 km/hr. Determine the response of the vehicle after it encounters a sudden dip of 12.8 mm in the road.

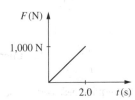

**FIG. P6.30**

**6.31** A manufacturing machine of mass 500 kg is mounted on an isolator, which is modeled by a spring of stiffness $1.2 \times 10^6$ N/m in parallel with a viscous damper such that the damping ratio is 0.28. During operation the machine is subject to the ramp excitation of Figure P6.31. The force is removed after 2 s. (a) Determine and plot $x(t)$, the displacement of the machine for $0 < t < 4$ s. (b) Determine and plot the force transmitted to the foundation through the isolator $kx + c\dot{x}$ for $0 < t < 4$ s.

**FIG. P6.31**

**6.32** The capacitor in the circuit of Figure P6.32 has an initial charge $q_0$. The switch is open and the closed at $t = 0$. Determine the resulting current through the inductor if $L = 0.2$ H, $C = 0.4$ μF, $q_0 = 3.1$ μC, and (a) $R = 1,000$ Ω, (b) $R = 2,000$ Ω, and (c) $R = 3,000$ Ω.

**6.33** The switch in the circuit is of Figure P6.33 is open and then closed at $t = 0$. (a) Determine the transfer function $G(s) = V_2(s)/V_1(s)$. (b) Determine the response $v_2(t)$ if $L = 0.25$ H, $R = 3.6$ kΩ, $C = 0.15$ μH, and $V = 12$ V.

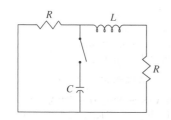

**FIG. P6.32**

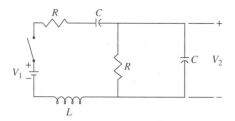

**FIG. P6.33**

**6.34** Determine $v_2(t)$ when $v_1(t) = 20u(t)$ V for the operational amplifier circuit of Figure P6.34.

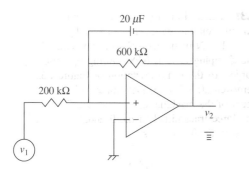

**FIG. P6.34**

**6.35** Determine the liquid-level perturbations $h_1(t)$ and $h_2(t)$ for the system of Problem 4.14.

**6.36** Determine the liquid-level perturbations in each of the tanks of Problem 4.16 if the steady-state levels are $h_{1s} = 12$ m, $h_{2s} = 15$ m, and $h_{3s} = 12$ m. The steady-state inlet flow rate is 2.25 m³/s and the flow rate suddenly increases to 2.31 m³/s. The tanks are all circular with a diameter of 6.5 m.

**6.37** Determine the natural frequencies of the two-degree-of-freedom mechanical system of Figure P6.37.

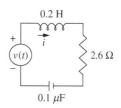

**FIG. P6.37**

**6.38** The machine of Figure P6.38 is mounted on a visco-elastic foundation that is modeled as a spring in parallel with a spring and viscous damper in series. This model of the isolator is called a Maxwell model. Let $x(t)$ represent the response of the machine when subject to a force $F(t)$. (a) Determine the transfer function $G_1(s) = X(s)/F(s)$. (b) The force transmitted to the foundation through the isolator is $F_T = k_1 x + c\dot{z}$, where $z(t)$ is the displacement of the joint between the spring and the viscous damper. Determine the transfer function $G_2(s) = F_T(s)/F(s)$. (c) Determine the impulsive response for $x(t)$ and $F_T(t)$.

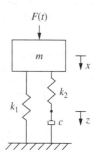

**FIG. P6.38**

**6.39** Determine and plot, for the system of Figure P6.39, its response $\theta(t)$ (a) when $F(t) = 10\delta(t)$ and (b) when $F(t) = 100u(t)$.

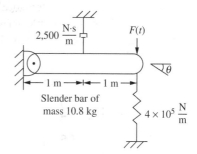

**FIG. P6.39**

**6.40** Determine and plot, for the system of Figure P6.40, its response $i(t)$ (a) when $v(t) = 10\delta(t)$, (b) when $v(t) = 10u(t)$, and (c) when $v(t) = 0.5t$.

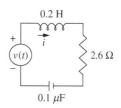

**FIG. P6.40**

**6.41** Determine the step response for $v_2$ in the circuit of Figure P6.41.

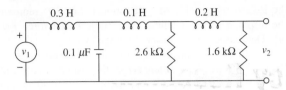

**6.42** Determine the step response of the operational amplifier circuit of Figure P6.42.

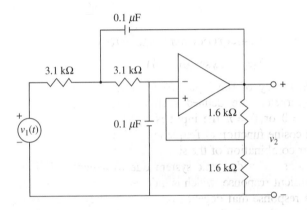

**FIG. P6.42**

**6.43** Determine the step response of the two-CSTR system of Figure P6.43 when the concentration of the reactant $A$ in the inlet stream suddenly changes. The mixture takes 2.5 s to travel between the reactors.

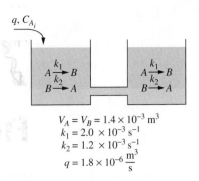

$$V_A = V_B = 1.4 \times 10^{-3} \text{ m}^3$$
$$k_1 = 2.0 \times 10^{-3} \text{ s}^{-1}$$
$$k_2 = 1.2 \times 10^{-3} \text{ s}^{-1}$$
$$q = 1.8 \times 10^{-6} \frac{\text{m}^3}{\text{s}}$$

**FIG. P6.43**

**6.44** Consider the system of Problem 6.33. Write a MATLAB program that does all the following: (a) inputs values of all parameters; (b) specifies the transfer function from the numerator and denominator; (c) determines the poles of the transfer function; (d) from the poles decides on the appropriate form of the free response; (e) plots the free response, which is dependent on the poles of the transfer function.

**6.45** Write a MATLAB program that symbolically determines the matrix of transfer functions defined by $G_{i,j}(s) \equiv X_i(s)/F_j(s)$ for the three-degree-of freedom system of Figure P6.45.

**6.46** Write a MATLAB program that determines and plots the impulsive response of the door of Problem 6.29.

**6.47** Write a MATLAB program that computes and plots the step responses, on the same set of axes, for the oil and water temperature perturbations of the counterflow heat exchanger of Example 6.41.

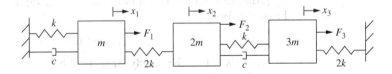

**FIG. P6.45**

# 7

# Frequency Response

The general form of a single-frequency harmonic input is

$$F(t) = F_0 \sin(\omega t + \psi) \tag{7.1}$$

where $\omega$ is the frequency of the input, $F_0$ is its amplitude, and $\psi$ is its phase. The parameters are defined such that $\omega > 0$, $F_0 > 0$, and $-\pi < \psi \le \pi$. If $\psi = 0$ or $\psi = \pi$ the input is purely sinusoidal. If $\psi = \pi/2$, the input is a cosine function of frequency $\omega$. For other values of $\psi$ the input is a linear combination of the sine and cosine functions.

The total response of a dynamic system due to a harmonic input is the sum of a transient response, which is independent of the frequency of the input and a response that depends on the frequency of the input. For a stable system, as shown in Chapter 6, the transient response decays exponentially and eventually becomes insignificant compared to the response that is specific to the input. When this occurs the system has reached steady state and the remaining response is called the **steady-state response**.

The steady-state response of a linear system with dependent variable $x(t)$, due to a sinusoidal input of the form of Equation (7.1), is shown to be of the form

$$x(t) = X \sin(\omega t + \psi + \phi) \tag{7.2}$$

where $X$ is called the **steady-state amplitude** and $\phi$ is called the **steady-state phase**. The steady-state amplitude and steady-state phase are functions of the input frequency, the input amplitude, and system parameters, but are independent of initial conditions and input phase. Thus, without loss of generality, initial conditions and input phase are taken to be zero when studying steady-state response.

The **frequency response** is the study of the dependence of the steady-state response on the input frequency. The frequency response is studied in the **time domain** by determining the steady-state response of the system as a function of time, as well as the **frequency domain** (also called the **s domain**) by studying the Laplace transform of the response. The frequency domain response is often studied using graphical tools.

## 7.1 UNDAMPED SECOND-ORDER SYSTEMS

The response of an undamped second-order or higher system due to a harmonic input is a special case in studying frequency response and is treated separately. The general form of the differential equation for a second-order system with sinusoidal input is

$$\ddot{x} + \omega_n^2 x = \frac{1}{m} F_0 \sin(\omega t) \tag{7.3}$$

Taking the Laplace transform of Equation (7.3) assuming both initial conditions are zero leads to

$$X(s) = \frac{F_0 \omega}{m(s^2 + \omega^2)(s^2 + \omega_n^2)} \tag{7.4}$$

A partial fraction decomposition of Equation (7.4) using entry 7 of Table 5.3 is obtained as

$$X(s) = \frac{F_0 \omega}{m(\omega_n^2 - \omega^2)} \left[ \frac{1}{s^2 + \omega^2} - \frac{1}{s^2 + \omega_n^2} \right] \tag{7.5}$$

For $\omega \neq \omega_n$ inversion of Equation (7.5) leads to

$$x(t) = \frac{F_0}{m(\omega_n^2 - \omega^2)} \left[ \sin(\omega t) - \frac{\omega}{\omega_n} \sin(\omega_n t) \right] \tag{7.6}$$

The $\sin(\omega_n t)$ term in Equation (7.6) represents the transient response, whereas the $\sin(\omega t)$ term is the part of the response specific to the input. An undamped system model is often used to simplify the system analysis, but all real systems have some form of energy dissipation. Damping leads to the decay of the transient solution, leaving only the steady-state response after a long time. This response is

$$x(t) = \frac{F_0}{m(\omega_n^2 - \omega^2)} \sin(\omega t) \tag{7.7}$$

Equation (7.7) is of the form of Equation (7.2) with

$$X = \left| \frac{F_0}{m(\omega_n^2 - \omega^2)} \right| \tag{7.8}$$

$$\phi = \begin{cases} 0 & \omega < \omega_n \\ \pi & \omega > \omega_n \end{cases} \tag{7.9}$$

The steady-state amplitude is defined as positive. Then the response is in phase with the input when the input frequency is less than the natural frequency and 180° out of phase with the input when the input frequency is greater than the natural frequency.

A special case occurs when the input frequency coincides with the natural frequency. In this case, Equation (7.4) is written as

$$X(s) = \frac{F_0 \omega_n}{m(s^2 + \omega_n^2)^2} \tag{7.10}$$

Equation (7.10) can be rewritten as

$$X(s) = \frac{F_0}{2m\omega_n}\left[\frac{1}{s^2 + \omega_n^2} - \frac{s^2 - \omega_n^2}{(s^2 + \omega_n^2)^2}\right] \tag{7.11}$$

Inversion of Equation (7.11) is performed using transform pairs 2 and 15 of Table 5.1 leading to

$$x(t) = \frac{F_0}{2m\omega_n^2}[\sin(\omega_n t) - \omega_n t \cos(\omega_n t)] \tag{7.12}$$

Equation (7.12) illustrates that when the input frequency for an undamped system coincides with its natural frequency, the system response grows without bound. This condition is called **resonance**. When resonance occurs the system response grows without bound until either the system fails or the response becomes so large that basic assumptions made to linearize the system are no longer valid and nonlinear behavior occurs.

When an undamped second-order system has an initial energy input, its resulting free response is periodic with a frequency equal to the natural frequency. Since an undamped system has no dissipative mechanism, the response sustains itself without additional energy input. When the system is subject to a sinusoidal input the response is a linear combination of a sinusoidal term at the natural frequency and a sinusoidal term at the input frequency. Some of the energy initiates the free response while much of the input energy leads to a response at the input frequency. However, when the input frequency coincides with the natural frequency the energy input is not necessary to sustain the response. Thus energy continues to be stored in the system leading to a system response that grows without bound.

When the input frequency is close to, but not exactly equal to, the natural frequency the frequency ratio $r = \omega/\omega_n$ can be written as $r = 1 + \varepsilon$ where $\varepsilon$ is a small number. In this case a trigonometric identity can be used in Equation (7.6) as

$$x(t) = \frac{F_0}{m(\omega_n^2 - \omega^2)}[\sin(\omega t) - (1 + \varepsilon)\sin(\omega_n t)]$$

$$= \frac{F_0}{m(\omega_n^2 - \omega^2)}\left\{\sin\left[\left(\frac{\omega - \omega_n}{2}\right)t\right]\cos\left[\left(\frac{\omega + \omega_n}{2}\right)t\right] - \varepsilon \sin(\omega_n t)\right\}$$

$$\approx \frac{F_0}{m(\omega_n^2 - \omega^2)}\sin\left[\left(\frac{\omega - \omega_n}{2}\right)t\right]\cos\left[\left(\frac{\omega + \omega_n}{2}\right)t\right] \tag{7.13}$$

The response given by Equation (7.13) is characterized by a slow buildup and decrease in amplitude. The response is periodic of period $T = 4\pi/(\omega + \omega_n)$. but is cyclic of period $T_b = 2\pi/|\omega - \omega_n|$. The type of response is called **beating** and $T_b$ is the **period of beating**.

## Example 7.1

A 40-kg machine is mounted on springs of equivalent stiffness $1 \times 10^5$ N/m. During operation it is subject to a harmonic force $F(t) = 100 \sin(\omega t)$ N. Determine and plot the system response for (a) $\omega = 25$ r/s, (b) $\omega = 50$ r/s, (c) $\omega = 52$ r/s, and (d) $\omega = 100$ r/s.

### Solution

The system is modeled by a differential equation of the form of Equation (7.3) with $F_0 = 100$ N and

$$\omega_n = \sqrt{\frac{k}{m}}$$

$$= \sqrt{\frac{1 \times 10^5 \text{ N/m}}{40 \text{ kg}}} = 50 \text{ r/s} \qquad (a)$$

(a) For $\omega = 25$ r/s the response is given by Equation (7.6). The substitution of given values into Equation (7.5) leads to

$$x(t) = \frac{100 \text{ N}}{40 \text{ kg}\left[(50 \text{ r/s})^2 - (25 \text{ r/s})^2\right]}\left[\sin(25t) - \frac{25 \text{ r/s}}{50 \text{ r/s}}\sin(50t)\right]$$

$$= 1.33[\sin(25t) - 0.5\sin(50t)] \text{ mm} \qquad (b)$$

The effects of both the input frequency and the natural frequency are evident in the response, which is plotted in Figure 7.1(a). The response is periodic of period $T = 2\pi/25$ s.

(b) The input frequency $\omega = 50$ r/s coincides with the system's natural frequency leading to resonance. The substitution of the given values into Equation (7.12) gives

$$x(t) = \frac{100 \text{ N}}{2(40 \text{ kg})(50 \text{ r/s})^2}[\sin(50t) - 50t\cos(50t)]$$

$$= 0.5[\sin(50t) - 50t\cos(50t)] \text{ mm} \qquad (c)$$

The response when the input frequency coincides with the natural frequency is illustrated in Figure 7.1(b). The response is cyclic but has a continual increase in amplitude illustrating the phenomenon of resonance.

(c) The input frequency $\omega = 52$ r/s is close, but not exactly equal to, the natural frequency. Thus the response is closely approximated by Equation (7.13). Substituting the given values into Equation (7.13) leads to

$$x(t) = \frac{100 \text{ N}}{(40 \text{ kg})\left[(50 \text{ r/s})^2 - (52 \text{ r/s})^2\right]}\left[\sin\left(\frac{52-50}{2}t\right)\cos\left(\frac{52+50}{2}t\right)\right]$$

$$= -12.3\sin(t)\cos(51t) \text{ mm} \qquad (d)$$

Equation (d) is plotted in Figure 7.1(c), which illustrates the phenomenon of beating where there is a continual buildup and decay of amplitude. The period of the response is $T = (4\pi)/(\omega + \omega_n) = 0.123$ s. The period of beating is $T_b = (2\pi)/|\omega - \omega_n| = \pi$ s.

(d) The input frequency $\omega = 100$ r/s is greater than the natural frequency. The substitution of values into Equation (7.6) leads to

$$x(t) = \frac{100 \text{ N}}{40 \text{ kg}\left[(50 \text{ r/s})^2 - (100 \text{ r/s})^2\right]}\left[\sin(100t) - \frac{100 \text{ r/s}}{50 \text{ r/s}}\sin(50t)\right]$$

$$= -0.333[\sin(100t) - 2\sin(50t)] \text{ mm} \tag{e}$$

Equation (e) is plotted in Figure 7.1(d).

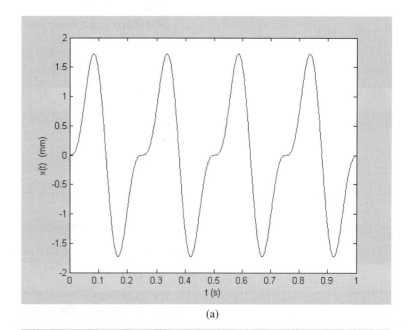

(a)

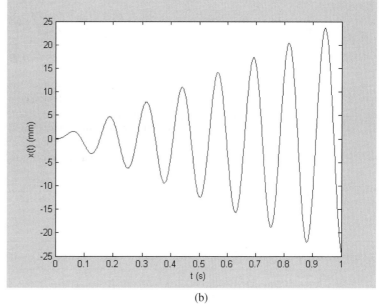

(b)

**FIG. 7.1** (a) The response of the undamped system of Example 7.1 with $\omega = 25$ r/s. The response is periodic of period $T = 2\pi/25$ s. (b) The response of the undamped system of Example 7.1 when the input frequency coincides with natural frequency. The response illustrates the phenomenon of resonance characterized by unbounded amplitude growth. (c) The response of the undamped system of Example 7.1 when the input frequency is near, but not exactly equal to, the natural frequency. The beating phenomenon illustrated features a continual buildup and decrease in amplitude. (d) The response of the undamped system of Example 7.1 when the input frequency is twice the natural frequency. (*Continued*)

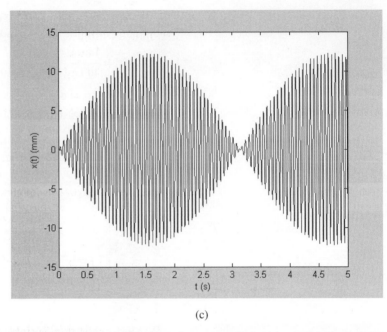

(c)

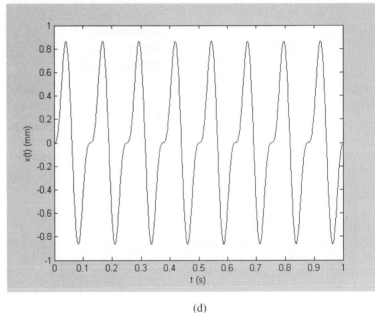

**FIG. 7.1**  (*Continued* )

(d)

## 7.2 SINUSOIDAL TRANSFER FUNCTION

Consider a stable $n$th-order system with a transfer function $G(s)$ defined for a dependent variable $x(t)$ such that $X(s) = F(s)G(s)$ for any system input $F(t)$. Let $s_1, s_2, \ldots, s_n$ be the poles of $G(s)$. Since the system is stable all poles have negative real parts. If the system is subject to a sinusoidal input of the form of Equation (7.1) with $\psi = 0$, then $F(s) = F_0/(s^2 + \omega^2)$ and

$$X(s) = \frac{F_0 \omega}{s^2 + \omega^2} G(s) \tag{7.14}$$

The poles of $X(s)$ are those of $G(s)$ as well as $j\omega$ and $-j\omega$. Thus Equation (7.14) can be written as

$$X(s) = \frac{F_0 \omega}{(s - j\omega)(s + j\omega)} G(s) \tag{7.15}$$

which has a partial fraction decomposition of the form

$$X(s) = \frac{B_1}{s - j\omega} + \frac{B_2}{s + j\omega} + \sum_{k=1}^{n} \frac{A_k}{s - s_k} \tag{7.16}$$

Inversion of Equation (7.16) leads to

$$x(t) = B_1 e^{j\omega t} + B_2 e^{-j\omega t} + \sum_{k=1}^{n} A_k e^{s_k t} \tag{7.17}$$

Since the system is stable and all poles of the transfer function have negative real parts, all terms in the summation in Equation (7.17) decay exponentially. After some period of time these terms will be insignificant compared to the first two terms, which constitute the steady-state response

$$x_s(t) = B_1 e^{j\omega t} + B_2 e^{-j\omega t} \tag{7.18}$$

The coefficients $B_1$ and $B_2$ are the residues of linear factors and are determined using the method developed in Section 5.4 as

$$B_1 = (s - j\omega)X(s)|_{s=j\omega} \tag{7.19}$$

$$B_2 = (s + j\omega)X(s)|_{s=-j\omega} \tag{7.20}$$

Use of Equation (7.15) in Equations (7.19) and (7.20) leads to

$$B_1 = \frac{F_0 \omega}{2j\omega} G(j\omega)$$

$$= -j\frac{F_0}{2} G(j\omega) \tag{7.21}$$

$$B_2 = \frac{F_0 \omega}{-2j\omega} G(-j\omega)$$

$$= j\frac{F_0}{2} G(-j\omega) \tag{7.22}$$

Substitution of Equations (7.21) and (7.22) in Equation (7.18) leads to

$$x_s(t) = j\frac{F_0}{2} \left[ G(-j\omega)e^{-j\omega t} - G(j\omega)e^{j\omega t} \right] \tag{7.23}$$

$G(j\omega)$ can be written as

$$G(j\omega) = \text{Re}[G(j\omega)] + j\,\text{Im}[G(j\omega)] \qquad (7.24)$$

or in a polar form

$$G(j\omega) = |G(j\omega)|e^{j\phi} \qquad (7.25)$$

where the magnitude of $G(j\omega)$ is

$$|G(j\omega)| = \sqrt{\{\text{Re}[G(j\omega)]\}^2 + \{\text{Im}[G(j\omega)]\}^2} \qquad (7.26)$$

and its phase is

$$\phi = \tan^{-1}\left\{\frac{\text{Im}[G(j\omega)]}{\text{Re}[G(j\omega)]}\right\} \qquad (7.27)$$

$G(-j\omega)$ is the complex conjugate of $G(j\omega)$. Thus from Equations (7.24) and (7.25)

$$G(-j\omega) = \text{Re}[G(j\omega)] - j\text{Im}[G(j\omega)] \qquad (7.28)$$

and

$$G(-j\omega) = |G(j\omega)|e^{-j\phi} \qquad (7.29)$$

Using Equations (7.25) and (7.27) in Equation (7.23) leads to

$$x_s(t) = j\frac{F_0}{2}|G(j\omega)|\left[e^{-j(\omega t + \phi)} - e^{j(\omega t + \phi)}\right] \qquad (7.30)$$

Equation (7.30) is equivalent to

$$x_s(t) = F_0|G(j\omega)|\sin(\omega t + \phi) \qquad (7.31)$$

Equation (7.31) is of the form of Equation (7.2) where the steady-state amplitude is $F_0|G(j\omega)|$ and the phase is the same as the phase of $G(j\omega)$.

Thus the steady-state response of a stable system due to a sinusoidal input is obtained from the system's transfer function. The steady-state amplitude of the system is the amplitude of the input times the magnitude of the transfer function evaluated for $s = j\omega$ and the phase in the steady state is the phase of the same transfer function.

$G(j\omega)$ is referred to as the **sinusoidal transfer function** because it is used to determine the steady-state response of a system due to a sinusoidal input. It is also used to determine the frequency response of a system, which is the variation of the steady-state amplitude and phase with $\omega$.

---

**Example 7.2**

The transfer function for the displacement of the vehicle of Examples 2.25 and 5.33 is derived in Example 6.3 as

$$G(s) = \frac{cs + k_1}{\dfrac{cm}{k_2}s^3 + m\left(1 + \dfrac{k_1}{k_2}\right)s^2 + cs + k_1} \qquad (a)$$

Using the numerical values of Example 5.33, Equation (a) becomes

$$G(s) = \frac{1.75 \times 10^4 s + 2.45 \times 10^6}{5.95 s^3 + 1.33 \times 10^3 s^2 + 1.75 \times 10^4 s + 2.45 \times 10^6} \tag{b}$$

Determine the steady-state response of the system when

$$z(t) = 0.01 \sin(100t) \tag{c}$$

## Solution

The steady-state response is of the form of Equation (7.31) where $\phi$ is determined using Equation (7.27), $F_0 = 0.01$, and $\omega = 100$. Using Equation (b)

$$G(j100) = \frac{1.75 \times 10^4 (100j) + 2.45 \times 10^6}{5.95(100j)^3 + 1.33 \times 10^3 (100j)^2 + 1.75 \times 10^4 (100j) + 2.45 \times 10^6}$$

$$= \frac{2.45 \times 10^6 + 1.75 \times 10^6 j}{-5.95 \times 10^6 j - 1.33 \times 10^7 + 1.75 \times 10^6 j + 2.45 \times 10^6}$$

$$= \frac{2.45 \times 10^6 + 1.75 \times 10^6 j}{-1.09 \times 10^7 - 4.20 \times 10^6 j} \tag{d}$$

Equation (d) is put in the form of Equation (7.24) by multiplying numerator and denominator by the complex conjugate of the denominator. To this end

$G(j100)$

$$= -\frac{2.45 \times 10^6 + 1.75 \times 10^6 j}{1.09 \times 10^7 + 4.20 \times 10^6 j} \cdot \frac{1.09 \times 10^7 - 4.20 \times 10^6 j}{1.09 \times 10^7 - 4.20 \times 10^6 j}$$

$$= -\frac{[(2.45 \times 10^6)(1.09 \times 10^7) - (1.75 \times 10^6)(-4.20 \times 10^6)] + j[(1.75 \times 10^6)(1.09 \times 10^7) + (2.455 \times 10^6)(-4.20 \times 10^6)]}{(1.09 \times 10^7)^2 + (4.20 \times 10^6)^2}$$

$$= -0.251 - 0.064j \tag{e}$$

The magnitude and phase of $G(j100)$ are calculated as

$$|G(j100)| = \sqrt{(-0.251)^2 + (-0.064)^2} = 0.259 \tag{f}$$

$$\phi = \tan^{-1}\left(\frac{-0.064}{-0.251}\right) = 194.4° \tag{g}$$

Since both the numerator and denominator of the argument of the inverse tangent in Equation (g) are negative, the appropriate phase is in the third quadrant. The steady-state response of the vehicle is

$$x_s(t) = (0.01)(0.259) \sin(100t + 194.4°)$$

$$= 2.59 \times 10^{-3} \sin(100t + 194.4°) \tag{h}$$

MATLAB is useful for performing the tedious computations required in numerical evaluation of the sinusoidal transfer function. If $z$ is a complex number then the commands

$$A = \text{abs}(z)$$

$$B = \text{angle}(z)$$

respectively evaluate the magnitude and phase of $z$.

## Example 7.3

Write a MATLAB program that determines and plots on the same graph the sinusoidal responses of the system of Example 7.2 for three different input frequencies.

### Solution

Example 7_3.m, whose script is shown in Figure 7.2(a), does the following:

- Defines the transfer function of Equation (b) of Example 7.2 using the symbolic variable s.
- Inputs the amplitude of the input displacement and three values for input frequency.

```
% Example 7.3
% Numerical evaluation of sinusoidal transfer function
%
% Defining symbolic variable
syms s
% Transfer fucntion of Example 7.2 in terms of s
G=(1.75E4*s+2.45E6)/(5.95*s^3+1.33E3*s^2+1.75E4*s+2.45E6);
%
% Input amplitude and input frequencies
%
Y=input('Input amplitude of displacement in m ');
disp('Input three frequencies in r/s for which steady-state response is to be evaluated')
for k=1:3
  str=['omega(',num2str(k),')=  '];
  w(k)=input(str);
% Evaluate sinusoidal transfer function
  A(k)=subs(G,j*w(k));
% Determine amplitude and phase
  phi(k)=angle(A(k));
  Q(k)=Y*abs(A(k));
% Print amplitude and phase
  str1=['Steady state amplitude=',num2str(Q(k)),' m'];
  str2=['Steady-state phase=',num2str(phi(k)),' rad'];
  disp(str1)
  disp(str2)
end
% Determine time scale for plotting
% The value of dt is chosen such that the graph will illustrate the
%     response over at least three periods for each value of omega
C=min(w);
Tmax=2*pi/C;
dt=3*Tmax/100;
for i=1:101
  t(i)=(i-1)*dt;
  for k=1:3
    x(k,i)=Q(k)*sin(w(k)*t(i)+phi(k));
  end
end
```

(a) (*continued*)

**FIG. 7.2** (a) MATLAB script for Example7_3.m; (b) the output from the execution of the script; (c) the steady-state responses for the three values of input frequency

```
%
% Plot steady-state responses
%
plot(t,x(1,:),'-',t,x(2,:),'-.',t,x(3,:),'--')
xlabel('t (s)')
ylabel('x (m)')
title('Steady-state response for Example 7.3')
str1=['\omega=',num2str(w(1)),' rad/s'];
str2=['\omega=',num2str(w(2)),' rad/s'];
str3=['\omega=',num2str(w(3)),' rad/s'];
legend(str1,str2,str3)
%
% End of Example7_3.m
```

**FIG. 7.2**  (Part a *Concluded* )

```
>>Example 7_3
Input amplitude of displacement in m  .001
Input three frequencies in r/s for which steady-state response is to be evaluated
omega(1)=  100
Steady state amplitude=0.00025878 m
Steady-state phase=-2.8907 rad
omega(2)=  120
Steady state amplitude=0.0001735 m
Steady-state phase=-2.8885 rad
omega(3)=  150
Steady state amplitude=0.00011031 m
Steady-state phase=-2.8877 rad
>>
```

(b)

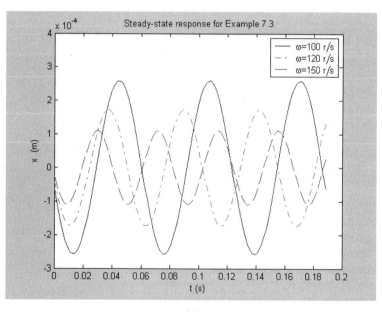

(c)

**FIG. 7.2**  (*Continued* )

- Evaluates the sinusoidal transfer function for each input frequency.
- Determines the steady-state amplitude and steady-state phase for each input frequency.
- Calculates and plots the steady-state response for each input frequency.

The output from the execution of Example7_3.m is shown in Figure 7.2(b) and the steady-state responses for the input frequency is illustrated in Figure 7.2(c). The following is noted regarding the program:

- MATLAB allows use of either i or j to represent the complex number $\sqrt{-1}$. Note, however, that the output presents complex numbers using i. It is possible, as is done in this program to redefine either i or j to represent another variable. In Example 7_3.m, i is used as a counter.
- The command 'subs(G,j*w(k))' numerically evaluates the symbolic function G at a value of s=j*w(k).
- The command 'C=min(w)' determines the minimum value of the elements of w. This command is used to determine the minimum value of the input frequency, which corresponds to the maximum value of the period of the input. The time scale for the plots is adjusted such that the steady-state response is plotted for at least three periods for each input frequency.

## Example 7.4

Determine the frequency response for a series $LRC$ circuit.

### Solution

The integrodifferential equation for the series $LRC$ circuit with a sinusoidal voltage source is

$$L\frac{di}{dt} + Ri + \frac{1}{C}\int_0^t idt = V_0 \sin(\omega t) \tag{a}$$

The transfer function is obtained from the differential equation as

$$G(s) = \frac{I(s)}{V(s)}$$

$$= \frac{1}{Ls + R + \dfrac{1}{Cs}}$$

$$= \frac{s}{L\left(s^2 + \dfrac{R}{L}s + \dfrac{1}{CL}\right)} \tag{b}$$

Equation (b) can be rewritten as

$$G(s) = \frac{s}{L\left(s^2 + 2\zeta\omega_n + \omega_n^2\right)} \tag{c}$$

where the natural frequency for the circuit is

$$\omega_n = \sqrt{\frac{1}{CL}} \tag{d}$$

and its damping ratio is

$$\zeta = \frac{R}{2}\sqrt{\frac{C}{L}} \tag{e}$$

$G(j\omega)$ is obtained from Equation (c) as

$$G(j\omega) = \frac{j\omega}{L\left[(j\omega)^2 + 2\zeta\omega_n(j\omega) + \omega_n^2\right]}$$

$$= \frac{j\omega}{L\left[(\omega_n^2 - \omega^2) + j2\zeta\omega\omega_n\right]}$$

$$= \frac{\omega\left[2\zeta\omega\omega_n + j(\omega_n^2 - \omega^2)\right]}{L\left[(\omega_n^2 - \omega^2)^2 + (2\zeta\omega\omega_n)^2\right]} \tag{f}$$

Equation (f) leads to

$$\text{Re}[G(j\omega)] = \frac{2\zeta\omega^2\omega_n}{L\left[(\omega_n^2 - \omega^2)^2 + (2\zeta\omega\omega_n)^2\right]} \tag{g}$$

$$\text{Im}[G(j\omega)] = \frac{\omega(\omega_n^2 - \omega^2)}{L\left[(\omega_n^2 - \omega^2)^2 + (2\zeta\omega\omega_n)^2\right]} \tag{h}$$

The steady-state amplitude and phase are determined using Equations (g) and (h) as

$$I = V_0|G(j\omega)|$$

$$= V_0\sqrt{\left\{\frac{2\zeta\omega^2\omega_n}{L\left[(\omega_n^2 - \omega^2)^2 + (2\zeta\omega\omega_n)^2\right]}\right\}^2 + \left\{\frac{\omega(\omega_n^2 - \omega^2)}{L\left[(\omega_n^2 - \omega^2)^2 + (2\zeta\omega\omega_n)^2\right]}\right\}^2}$$

$$= \frac{V_0\omega}{L}\sqrt{\frac{1}{(\omega_n^2 - \omega^2)^2 + (2\zeta\omega\omega_n)^2}}$$

$$\phi = \tan^{-1}\left\{\frac{\text{Im}[G(j\omega)]}{\text{Re}[G(j\omega)]}\right\} \tag{i}$$

$$= \tan^{-1}\left\{\frac{\dfrac{\omega(\omega_n^2 - \omega^2)}{L\left[(\omega_n^2 - \omega^2)^2 + (2\zeta\omega\omega_n)^2\right]}}{\dfrac{2\zeta\omega^2\omega_n}{L\left[(\omega_n^2 - \omega^2)^2 + (2\zeta\omega\omega_n)^2\right]}}\right\}$$

$$= \tan^{-1}\left(\frac{\omega_n^2 - \omega^2}{2\zeta\omega\omega_n}\right) \tag{j}$$

## 7.3 GRAPHICAL REPRESENTATION OF THE FREQUENCY RESPONSE

Graphical presentation of the frequency response is valuable for understanding the steady-state response of a dynamic system. Various forms of graphical presentation are useful in different situations. Three methods of graphical presentation are presented: frequency response curves, Bode diagrams, and Nyquist plots. Bode diagrams and Nyquist plots are useful in control system analysis and design. These methods of representation are explained in this section and applied to the frequency response for the series $LRC$ circuit obtained in Example 7.4 of Section 7.2. They are applied to other systems in later sections. All graphs are developed using MATLAB.

### 7.3.1 Frequency Response Curves

**Frequency response curves** are plots of the steady-state amplitude vs. input frequency and phase angle vs. input frequency. Frequency response curves are often presented as a family of curves on a common set of coordinate axes, each curve representing the frequency response for a specific value of a system parameter such as the damping ratio. It is usually desirable to nondimensionalize the relationship between the steady-state amplitude and the input frequency. The vertical axis used for the amplitude curves is this nondimensional parameter. The horizontal axis used for both the amplitude and phase curves is a nondimensional parameter that, for second-order systems, is the ratio of the input frequency to the natural frequency, called the **frequency ratio**

$$r = \frac{\omega}{\omega_n} \tag{7.32}$$

The relationship between the steady-state amplitude of the current in a series $LRC$ circuit and the frequency of a sinusoidal voltage source is determined as Equation (i) in Example 7.3

$$I = \frac{V_0 \omega}{L} \sqrt{\frac{1}{(\omega_n^2 - \omega^2)^2 + (2\zeta\omega\omega_n)^2}} \tag{7.33}$$

Equation (7.33) is rearranged by multiplying both sides by $L\omega_n/V_0$ leading to

$$\frac{L\omega_n I}{V_0} = \omega\omega_n \sqrt{\frac{1}{(\omega_n^2 - \omega^2)^2 + (2\zeta\omega\omega_n)^2}}$$

$$= \frac{\omega}{\omega_n} \sqrt{\frac{1}{\left[1 - \left(\frac{\omega}{\omega_n}\right)^2\right]^2 + \left(2\zeta\frac{\omega}{\omega_n}\right)^2}} \tag{7.34}$$

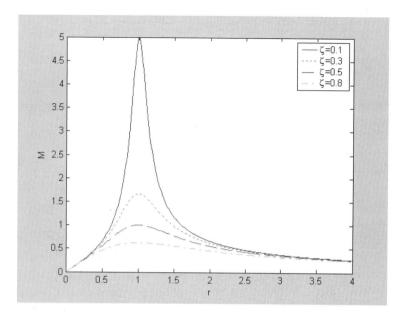

**FIG. 7.3** The frequency response curves for a series *LRC* circuit.

Both sides of Equation (7.34) are nondimensional and the equation is written using the frequency ratio as

$$M = \frac{L\omega_n I}{V_0} = \frac{r}{\sqrt{(1 - r^2)^2 + (2\zeta r)^2}} \tag{7.35}$$

A family of curves illustrating the behavior of $L\omega_n I / V_0$ vs. $r$ for several values of $\zeta$ is presented in Figure 7.3.

From Figure 7.3 and Equation (7.35) it is ascertained that

- $M(0) = 0$ for all $\zeta$;
- $\lim_{r \to \infty} M = 0$ and behaves as $1/r$ for large $r$;
- $M$ is a maximum near $r = 1$ and decreases as $r$ increases from there. Calculus is used to determine the value of $r$ where the steady-state amplitude is a maximum and the corresponding maximum amplitude.

The phase angle for the series *LRC* circuit is determined as Equation (j) in Example 7.3 as

$$\phi = \tan^{-1}\left(\frac{\omega_n^2 - \omega^2}{2\zeta\omega\omega_n}\right) \tag{7.36}$$

Equation (7.36) is rewritten using the frequency ratio as

$$\phi = \tan^{-1}\left(\frac{1 - r^2}{2\zeta r}\right) \tag{7.37}$$

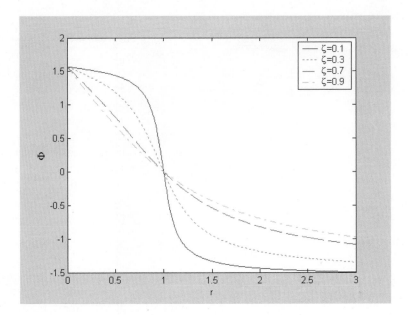

**FIG. 7.4** Phase angle vs. frequency ratio for an *LRC* circuit.

Equation (7.37) is illustrated in Figure 7.4, which shows $\phi$ vs. $r$ for several values of $\zeta$. For all values of $\zeta$ the curves of Figure 7.4 show

- $\phi = \pi/2$ for $r = 0$;
- $\phi = 0$ for $r = 1$;
- $\phi \to -\pi/2$ as $r$ grows large.

### 7.3.2 Bode Diagrams

#### 7.3.2.1 Construction and Asymptotes
Bode diagrams are similar to frequency response curves, but they are drawn on logarithmic scales and use the magnitude of the sinusoidal transfer function instead of the amplitude. A Bode diagram consists of two plots. One is a diagram plotting

$$L(\omega) = 20 \log |G(j\omega)| \qquad (7.38)$$

vs. $\log(\omega)$. The other is a diagram of $\phi$ vs. $\log(\omega)$. The units of $L(\omega)$ are decibels (dB). The vertical scale of the amplitude Bode diagram is in decibels, while a nondimensional parameter such as the frequency ratio is often used on the horizontal axis.

The conversion of the sinusoidal transfer function into decibels using Equation (7.38) allows an easy comparison of frequency responses for different systems. A numerical value of the sinusoidal transfer function of one translates into zero decibels. A change in the order of magnitude in the sinusoidal transfer function (factor of 10) translates into a change of 20 dB. A tenfold increase in the sinusoidal transfer function leads to a 20-dB increase in $L(\omega)$ while a reduction in the sinusoidal transfer function by a factor of 10 leads to a 20-dB decrease in $L(\omega)$. A range of a factor of 10 in the frequency is called a decade, which is a unit change on the horizontal logarithmic scale. Since the horizontal scale is logarithmic a Bode diagram can illustrate the frequency response over a wide range of frequencies. However, unlike frequency response curves, the Bode diagram cannot show the response for $\omega = 0$.

Bode diagrams for the series $LRC$ circuit, determined using Equations (i) and (j) of Example 7.4, are illustrated in Figure 7.5 for a natural frequency of $\omega_n = 1$.

Low-frequency asymptotes and high-frequency asymptotes of the Bode diagram are important in control system analysis. As their name implies the low- and high-frequency asymptotes are the lines to which Equation (7.38) are asymptotic to as $\omega \to 0$ and $\omega \to \infty$ respectively.

Using Equation (i) of Example 7.4 for the series $LRC$ circuit

$$L(\omega) = 20 \log \left[ \frac{\omega}{\sqrt{(\omega_n^2 - \omega^2)^2 + (2\zeta\omega\omega_n)^2}} \right] \tag{7.39}$$

Properties of logarithms are used to rewrite Equation (7.40) as

$$L(\omega) = 20 \log(\omega) - 10 \log \left[ (\omega_n^2 - \omega^2)^2 + (2\zeta\omega\omega_n)^2 \right] \tag{7.40}$$

The low-frequency asymptote of the series $LRC$ circuit is obtained from Equation (7.40) as

$$L(\omega)|_{\omega \to 0} = 20 \log(\omega) - 40 \log(\omega_n) \tag{7.41}$$

Equation (7.41), when plotted on a logarithmic scale is a line of slope of 20 dB per decade of frequency change. That is $L(\omega)$ increases by 20 dB as $\omega$ increases by a factor of 10. The high-frequency asymptote is obtained using Equation (7.40) as

$$L(\omega)|_{\omega \to \infty} = -20 \log(\omega) \tag{7.42}$$

which on the Bode diagram is a line of slope $-20$ dB per decade of frequency change.

### 7.3.2.2  Products of Transfer Functions

Consider a system in which the transfer function $G(s)$ can be written as

$$G(s) = G_1(s)G_2(s) \tag{7.43}$$

The sinusoidal transfer function for the system is

$$\begin{aligned}
G(j\omega) &= G_1(j\omega)G_2(j\omega) \\
&= |G_1(j\omega)|e^{j\phi_1}|G_2(j\omega)|e^{j\phi_2} \\
&= |G_1(j\omega)||G_2(j\omega)|e^{j(\phi_1 + \phi_2)}
\end{aligned} \tag{7.44}$$

Thus

$$|G(j\omega)| = |G_1(j\omega)||G_2(j\omega)| \tag{7.45}$$

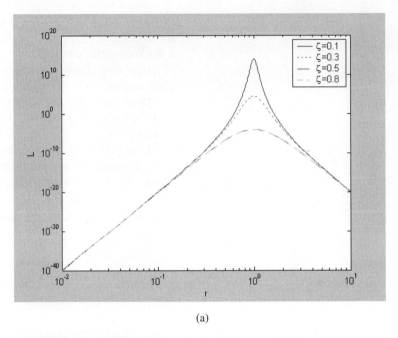

(a)

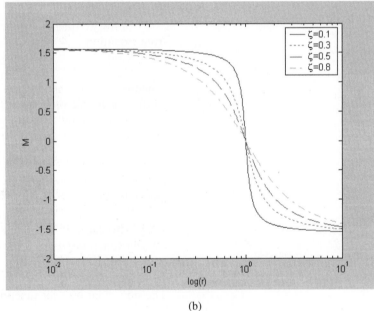

(b)

**FIG. 7.5** (a) The Bode plot of $|G(j\omega)|$ for the series $LRC$ circuit: (b) The phase part of the Bode diagram for the series $LRC$ circuit.

and defining $\phi$ as the phase of the response

$$\phi = \phi_1 + \phi_2 \qquad (7.46)$$

From Equation (7.46) it is clear that the phase angle component of the Bode plot of a transfer function given by Equation (7.43) can be obtained by summing the phases for the individual transfer functions.

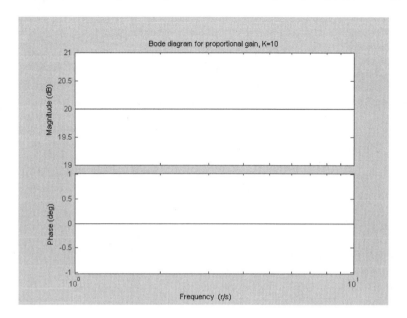

**FIG. 7.6** The Bode diagram for the transfer
function corresponding to the proportional
control with $K = 10$.

Using Equation (7.38)

$$L(\omega) = 20\log\left[|G(j\omega)|\right]$$
$$= 20\log\left[|G_1(j\omega)|G_2(j\omega)|\right]$$
$$= 20\log\left[|G_1(j\omega)|\right] + 20\log\left[|G_2(j\omega)|\right]$$
$$= L_1(\omega) + L_2(\omega) \tag{7.47}$$

Thus the amplitude component of the Bode plot can be obtained by simply adding the amplitude components of the Bode plots for the individual transfer functions.

### 7.3.2.3 Bode Diagrams for Common Transfer Functions

Bode diagrams for some basic transfer functions can be drawn and catalogued. The property of products of transfer functions can be used to draw Bode diagrams for more complicated transfer functions. Some of the transfer functions considered are those corresponding to controllers considered in Chapter 8.

The transfer function of an actuator that provides proportional control is of the form $G(s) = K$, where $K$ is a constant called the gain. For this transfer function $|G(j\omega)| = K$ and $\phi = 0$. The Bode diagram for proportional gain is constructed from

$$L(\omega) = 20\log(K) \tag{7.48a}$$
$$\phi = 0° \tag{7.48b}$$

The Bode diagram for proportional gain is a horizontal line, as illustrated in Figure 7.6. The title was inserted form the workspace using the 'title' command. The phase portion of the Bode diagram is a horizontal line of $\phi = 0$.

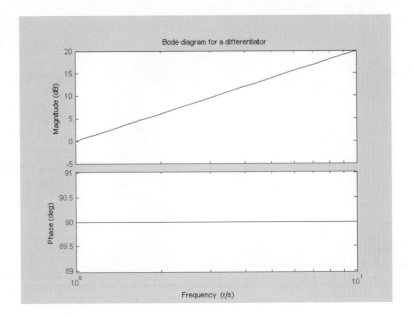

**FIG. 7.7** The Bode diagram for a differentiator. The amplitude part is line with a slope of 20 dB/decade. The phase portion is the horizontal line $\phi = \pi/2$.

The transfer function of a differential controller (a differentiator) is $G(s) = s$. It can be shown that $G(j\omega) = j\omega, |G(j\omega)| = \omega$, and $\phi = 90°$. The Bode diagram for a differentiator is constructed from

$$L(\omega) = 20 \log(\omega) \tag{7.49a}$$

$$\phi = 90° \tag{7.49b}$$

The Bode diagram for a differentiator is a line of slope 20 dB/decade, which passes through $L = 0$ when $\omega = 1$, as illustrated in Figure 7.7. The phase portion of the Bode diagram is a horizontal line of $\phi = 90°$.

The transfer function $G(s) = 1/s$ is the transfer function provided by an integral controller (an integrator). It can be shown that $G(j\omega) = -j(1/\omega)$, $|G(j\omega)| = 1/\omega$, and $\phi = -90°$. The Bode diagram for an integrator is constructed from

$$L(\omega) = -20 \log(\omega) \tag{7.50a}$$

$$\phi = -90° \tag{7.50b}$$

The Bode diagram for an integrator is a line of slope $-20$ dB/decade that passes through $L = 0$ when $\omega = 1$, as illustrated in Figure 7.8. The phase portion of the Bode diagram is a horizontal line $\phi = -90°$.

The numerator of transfer functions often includes a term of the form $G(s) = 1 + \tau s$, which is often referred to a first-order lead. The sinusoidal transfer functions for first-order lead is $G(j\omega) = 1 + j\tau\omega$, which gives $|G(j\omega)| = \sqrt{1 + \tau^2\omega^2}$, $\phi = \tan^{-1}(\omega\tau)$. The Bode diagram for first-order lead is constructed from

$$L(\omega) = 10 \log(1 + \omega^2\tau^2) \tag{7.51a}$$

$$\phi = \tan^{-1}(\omega\tau) \tag{7.51b}$$

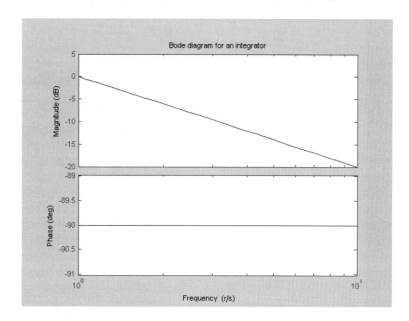

**FIG. 7.8** The amplitude part of a Bode diagram for an integrator is a line of slope $-20$ dB/decade. The phase part of a Bode diagram is a horizontal line $\phi = -\pi/2$.

The low-frequency asymptote for first-order lead is the horizontal line $L = 0$, while the high-frequency asymptote is $L = 20\log(\tau) + 20\log(\omega)$, which is a line of slope of 20 dB/decade and $L = 20\log(\tau)$ when $\omega = 1$. The Bode diagram for first-order lead is illustrated in Figure 7.9.

The transfer function corresponding to the denominator of a first-order system, called first-order lag is $G(s) = 1/(1 + \tau s)$. The corresponding sinusoidal transfer function is $G(j\omega) = (1 - j\omega\tau)/(1 + \omega^2\tau^2)$,

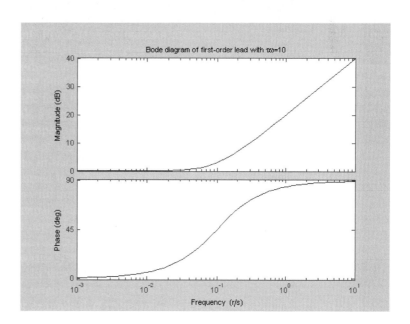

**FIG. 7.9** The Bode diagram of a first-order lead. The low-frequency asymptote is $L = 0$, the while high-frequency asymptote is a line of slope 20 dB/decade.

**FIG. 7.10** the Bode diagram for a first-order lag with $\tau\omega = 10$ has a low-frequency asymptote of $L(\omega) = 0$ and a high-frequency asymptote of slope $-20$ dB/dec.

which leads to $|G(j\omega)| = 1/\sqrt{\omega^2 + \tau^2}$ and $\phi = \tan^{-1}(-\omega\tau)$. The Bode diagram for first-order lag is constructed from

$$L(\omega) = -10\log\left(1 + \omega^2\tau^2\right) \tag{7.52a}$$

$$\phi = \tan^{-1}(-\omega\tau) \tag{7.52b}$$

The low-frequency asymptote is the line $L = 0$ while the high-frequency asymptote is $L = -20\log(\tau) - 20\log(\omega)$, which on the Bode diagram is a line of slope $-20$ dB/decade and passes through $L = -20\log(\tau)$ when $\omega = 1$. It is noted that the phase for a first-order lag is always negative while the phase for a first-order lead is always positive; thus these transfer functions are aptly named. The Bode diagram for a first-order lag is shown in Figure 7.10.

The transfer function corresponding to a second-order numerator with a complex zeros is of the form $G(s) = s^2 + 2\alpha\beta s + \beta^2$. This transfer function corresponds to a second-order lead. The corresponding sinusoidal transfer function is $G(j\omega) = \beta^2 - \omega^2 + j2\alpha\beta\omega$, which leads to $|G(j\omega)| = \sqrt{(\beta^2 - \omega^2)^2 + (2\alpha\beta\omega)^2}$ and $\phi = \tan^{-1}[2\alpha\beta\omega/(\beta^2 - \omega^2)]$. The Bode diagram for a second-order lead is constructed from

$$L(\omega) = 10\log\left[\left(\beta^2 - \omega^2\right)^2 + (2\alpha\beta\omega)^2\right] \tag{7.53a}$$

$$\phi = \tan^{-1}\left(\frac{2\alpha\beta\omega}{\beta^2 - \omega^2}\right) \tag{7.53b}$$

The Bode diagram for a second-order lead is illustrated in Figure 7.11. The low-frequency asymptote is $L = 40\log(\beta)$ while the high-frequency asymptote is $L = 40\log(w)$.

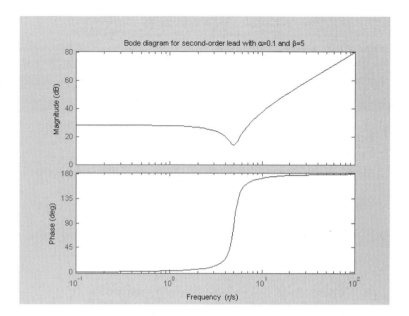

**FIG. 7.11** The Bode diagram for a second-order lead has a low-frequency asymptote of $40 \log (5) = 27.99$. Its high-frequency asymptote is a line of slope 40 dB/decade.

The transfer function corresponding to the denominator of a second-order system with complex poles (an underdamped system) is of the general form $G(s) = 1/(s^2 + 2\zeta\omega_n s + \omega_n^2)$. This system is also called second-order lag. Its sinusoidal transfer function is $G(j\omega) = (\omega_n^2 - \omega^2 - j2\zeta\omega\omega_n)/[(\omega_n^2 - \omega^2)^2 + (2\zeta\omega\omega_n)^2]$, which leads to $|G(j\omega)| = 1/\sqrt{(\omega_n^2 - \omega^2)^2 + (2\zeta\omega\omega_n)^2}$ and $\phi = \tan^{-1}[-2\zeta\omega\omega_n/(\omega_n^2 - \omega^2)]$. The Bode diagram for a second-order underdamped system is constructed from

$$L(\omega) = -10 \log \left[ \left( \omega_n^2 - \omega^2 \right)^2 + (2\zeta\omega\omega_n)^2 \right] \tag{7.54a}$$

$$\phi = \tan^{-1} \left( -\frac{2\zeta\omega\omega_n}{\omega_n^2 - \omega^2} \right) \tag{7.54b}$$

The low-frequency asymptote for an underdamped second-order system is $L = -40 \log (\omega_n)$ while its high-frequency asymptote is $L = -40 \log (\omega)$. The Bode diagram for an underdamped second-order system is shown in Figure 7.12.

The transfer function for a time delay is $G(s) = e^{-\tau s}$. The corresponding sinusoidal transfer function is $G(j\omega) = e^{-j\tau\omega} = \cos (\omega\tau) - j\sin (\omega\tau)$, which leads to $|G(j\omega)| = 1$ and $\phi = -\omega\tau$. The Bode diagram for a time delay is constructed from

$$L(\omega) = 0 \tag{7.55a}$$

$$\phi = -\omega\tau \tag{7.55b}$$

The amplitude portion of the Bode diagram for a time delay, as illustrated in Figure 7.13, is a horizontal line.

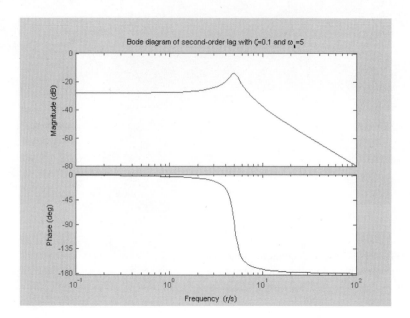

**FIG. 7.12** The Bode diagram of a second-order lag.

Bode diagrams for many transfer functions may be constructed using the superposition formulas of Equations (7.46) and (7.47) and the forms of $L(\omega)$ and $\phi$ for the transfer functions of Equations (7.48)–(7.55). It is noted that if $G(s) = N(s)/D(s)$, where $N(s)$ and $D(s)$ are polynomials with real coefficients, then $N(s)$ and $D(s)$ can be factored using only

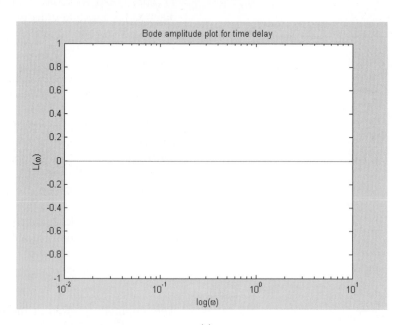

**FIG. 7.13** The Bode plot for the time delay transfer function.

(a)

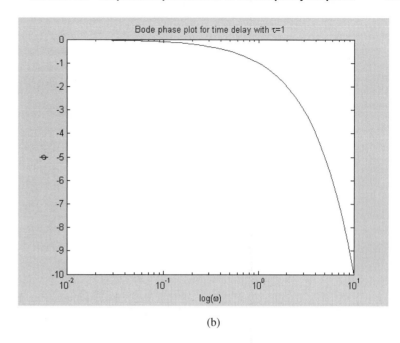

FIG. 7.13  (*Continued*)

(b)

linear and quadratic factors. Then $G(s)$ can be written as the product of first-order and second-order lead and lag transfer functions.

## Example 7.5

Consider the series *LRC* circuit when the parameters are prescribed such that $\zeta = 1.2$ and $\omega_n = 100$ r/s. Develop the Bode plot for this system and discuss its low- and high-frequency asymptotes.

### Solution

Since $\zeta > 1$, the system is overdamped and the real poles of the transfer function are calculated using Equation (6.78) as

$$s_1 = -\omega_n \left( \zeta + \sqrt{\zeta^2 - 1} \right) = -186.3 \tag{a}$$

$$s_2 = -\omega_n \left( \zeta - \sqrt{\zeta^2 - 1} \right) = -53.7 \tag{b}$$

For the overdamped system the transfer function can be written as

$$G(s) = \frac{s}{(s - s_1)(s - s_2)} \tag{c}$$

Equation (c) can be rewritten as the product of transfer functions just defined by dividing the numerator and denominator by $s_1 s_2$ leading to

$$G(s) = \frac{\dfrac{1}{s_1 s_2} s}{(1 + \tau_1 s)(1 + \tau_2 s)} \tag{d}$$

where

$$\tau_1 = -\frac{1}{s_1} = 5.37 \times 10^{-3} \tag{e}$$

$$\tau_2 = -\frac{1}{s_2} = 1.86 \times 10^{-2} \tag{f}$$

Equation (d) can be written as

$$G(s) = G_1(s)G_2(s)G_3(s)G_4(s) \tag{g}$$

where

$$G_1(s) = \frac{1}{s_1 s_2} = 1 \times 10^{-4} \tag{h}$$

$$G_2(s) = s \tag{i}$$

$$G_3(s) = \frac{1}{1 + \tau_1 s} \tag{j}$$

$$G_4(s) = \frac{1}{1 + \tau_2 s} \tag{k}$$

Thus the transfer function for the system is composed of the transfer functions for a proportional gain, a differentiator, and two first-order lags. Its Bode diagram is constructed using the superposition of Equations (7.46) and (7.47) as well as Equations (7.48), (7.49), and (7.52).

$$L(\omega) = 20\log\left(1 \times 10^{-4}\right) + 20\log\left(\omega\right) - 10\log\left(1 + 2.88 \times 10^{-5}\omega^2\right)$$
$$- 10\log\left(1 + 3.47 \times 10^{-4}\omega^2\right) \tag{l}$$

$$\phi = 0° + \tan^{-1}(-5.37 \times 10^{-3}\omega) + \tan^{-1}(1.86 \times 10^{-2}\omega) + 90° \tag{m}$$

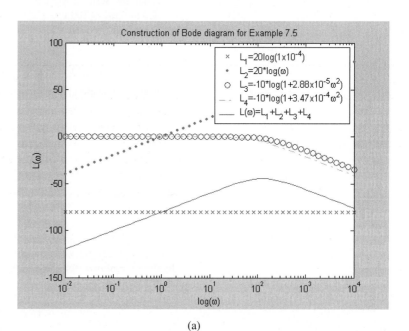

**FIG. 7.14** The construction of the Bode diagram using the properties for Bode diagrams from products of transfer functions: (a) The amplitude plot; (b) the phase plot.

(a)

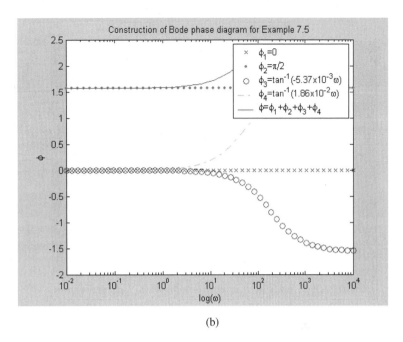

**FIG. 7.14** *(Continued)*

(b)

The development of the Bode diagram for this overdamped system is illustrated in Figure 7.14. The low-frequency asymptote is obtained using Equation (l) as

$$L = -80 + 20 \log (\omega) \qquad (n)$$

The high-frequency asymptote is obtained using Equation (l) as

$$L = -80 + 20 \log (\omega) - 10 \log (2.88 \times 10^{-5} \omega^2) - 10 \log (3.47 \times 10^{-4} \omega^2)$$
$$= -80 + 20 \log (\omega) - 20 \log (\omega) + 45.41 - 20 \log (\omega) + 34.59$$
$$= -20 \log (\omega) \qquad (o)$$

#### 7.3.2.4 Bode Diagram Parameters    Several parameters are defined to describe the frequency response as presented on the Bode diagram.

The **corner frequency**, also called the breakpoint frequency, is the frequency at which the low-frequency asymptote intersects the high-frequency asymptote. For example the corner frequency for a first-order lag is obtained by requiring $0 = -20 \log (\tau) - 20 \log (\omega)$, which leads to a corner frequency of $\omega = 1/\tau$.

The **cutoff frequency** $\omega_c$ is the frequency at which

$$L(0) - L(\omega_c) = 3 \text{ dB} \qquad (7.56)$$

For a first order-lag, $L(0) = 0$ and the corner frequency is obtained from $10 \log (1 + \omega_c^2 \tau^2) = 3$, which is solved leading to $\omega_c = 0.9976/\tau$.

The **bandwidth** is the range of $\omega$ over which $L(\omega)$ is greater than 3 dB below $L(0)$. The bandwidth for a first-order lag is from $\omega = 0$ to $\omega = \omega_c = 0.9976/\tau$.

If all the poles and zeroes of a transfer function have negative real parts the transfer function is a **minimum phase transfer function**. If any of the poles or zeroes of a transfer function has a non-negative real part the transfer function is a **nonminimum phase transfer function**. As its name implies the phase angle for a minimum phase transfer function has a defined minimum.

---

**Example 7.6**

Determine the corner frequency for the system of Example 7.5.

**Solution**

The low-frequency asymptote and the high-frequency asymptote for $L(\omega)$ are given by Equations (n) and (o) of Example 7.5 respectively. The corner frequency is the frequency at which they intersect. Thus the corner frequency is obtained by solving

$$-20\log(\omega) = -80 + 20\log(\omega)$$
$$80 = 40\log(\omega) \tag{a}$$

Equation (a) is solved leading to a corner frequency of $\omega = 100\,\text{r/s}$.

---

### 7.3.3 Nyquist Diagrams

A Nyquist diagram is a plot of $\text{Re}[G(j\omega)]$ vs. $\text{Im}[G(j\omega)]$ with $\text{Re}[G(j\omega)]$ as the abscissa and $\text{Im}[G(j\omega]$ as the ordinate. A Nyquist diagram is equivalent to a plot of $|G(j\omega)|$ vs. $\phi$ in polar coordinates. Since $\text{Re}[G(j\omega)]$ and $\text{Im}[G(j\omega)]$ have parametric representations in terms of the frequency $\omega$, a Nyquist diagram is developed by varying $\omega$ from $-\infty < \omega < \infty$. Key points on a Nyquist diagram are the points corresponding to $\omega = 0$ and $\omega \to \pm\infty$, as well as points where the diagram intercepts the axes.

The real and imaginary parts of the transfer function for the series *LRC* circuit were developed in Equations (g) and (h) of Example 7.3 and are repeated below

$$\text{Re}[G(j\omega)] = \frac{2\zeta\omega^2\omega_n}{\left[(\omega_n^2 - \omega^2)^2 + (2\zeta\omega\omega_n)^2\right]} \tag{7.57a}$$

$$\text{Im}[G(j\omega)] = \frac{\omega(\omega_n^2 - \omega^2)}{\left[(\omega_n^2 - \omega^2)^2 + (2\zeta\omega\omega_n)^2\right]} \tag{7.57b}$$

Equations (7.57) are used to determine the following regarding the Nyquist diagram:

- When $\omega = 0$, $\text{Re}[G(0)] = 0$ and $\text{Im}[G(0)] = 0$.
- These equations also show that as $\omega \to \pm\infty$, $\text{Re}[G(j\omega)] \to 0$ and $\text{Im}[G(j\omega)] \to 0$.

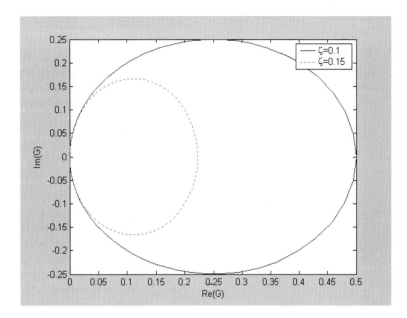

**FIG. 7.15** The Nyquist diagram for the series $LRC$ circuit with $\omega_n = 10$.

- $\text{Im}[G(j\omega)] = 0$ only for $\omega = 0$ and $\omega = \omega_n$. When $\omega = \omega_n$, $\text{Re}[G(j\omega)] = 1/2\zeta\omega_n$.
- $\text{Re}[G(j\omega)] = 0$ only for $\omega = 0$.
- $\text{Re}[G(j\omega)] > 0$ for all $\omega$, while $\text{Im}[G(j\omega)] < 0$ for $\omega < \omega_n$ and $\text{Im}[G(j\omega)] < 0$ for $\omega > \omega_n$.

These statements suggest that the Nyquist plot for the transfer function for the series $LRC$ circuit should be a closed curve (since it approaches the origin for both large and small $\omega$) existing in the first and fourth quadrants of the $G$ plane. It crosses the Re axis only at 0 and $1/(2\zeta\omega_n)$. The Nyquist plot is tangent to the Im axis at the origin. These observations are confirmed in Figure 7.15, which shows the Nyquist plot for the transfer function for a series $LRC$ circuit for $\omega_n = 10$ r/s and for several values of $\zeta$.

### 7.3.4 Use of MATLAB to Develop Bode Plots and Nyquist Diagrams

MATLAB's Controls Toolbox has direct commands for developing Bode plots and Nyquist diagrams from definition of the transfer function. If $N$ is a vector containing the coefficients in polynomial in the numerator of the transfer function, and $D$ is a vector containing the coefficients in the polynomial in the denominator of the transfer function the commands

$$\text{bode(N,D)}$$

$$\text{nyquist(N,D)}$$

plot on the screen system's Bode plot and Nyquist diagram respectively.

The default range for the Bode diagram is $-1 < \log(\omega) < 2$ while the Nyquist diagram is drawn for $-\infty < \omega < \infty$. The user may define a specific points to use in developing either diagram. This allows, for example, the Bode diagram to be focused on a smaller range of frequencies or the Nyquist diagram to be drawn using only positive values of $\omega$. The user supplies a vector specifying the values of $\omega$ to be used, say

$$\text{omega} = 0.001.0.001.10$$

The commands

$$\text{bode(N,D,omega)}$$

$$\text{nyquist(N,D,omega)}$$

draw the Bode plot and Nyquist diagram using points calculated corresponding to only the values specified by the vector omega.

The preceding commands draw plots on the screen without saving any data used to generate the plots. It is often convenient, for comparison purposes, to draw several plots on the same set of axes. The command

$$[\text{G,phi}] = \text{bode(N,D,omega)}$$

calculates the magnitude and phase of the sinusoidal transfer function for the defined values in omega and stores them in the vectors $G$ and phi respectively. If the argument corresponding to omega is not specified the values are calculated for the default values used by MATLAB in developing a Bode plot. The vector of magnitudes $G$ can be converted to decibels using Equation (7.39).

The command

$$[\text{R,I}] = \text{nyquist(N,D,omega)}$$

calculates the real and imaginary parts of the sinusoidal transfer function for the specified values of omega and stores them in the vectors R and I respectively.

---

**Example 7.7**

A proportional controller of proportional gain $A$ is being designed for use in a feedback control system. When implemented the transfer function for the system is

$$G(s) = \frac{Ks}{s^2 + (K+2)s + 25} \tag{a}$$

Bode diagrams and Nyquist plots both play an important role in control system design. Thus it is desired to draw these plots for different values of $K$. Write a MATALB program that (a) inputs three values of $K$ for which the system is to be analyzed, (b) draws the Bode plots for each value of $K$ on the same axes, and (c) draws the Nyquist diagram for each value of $K$ on the same axes.

```
% Example 7.7
% Bode plots and Nyqusit diagrams
%
% Input values of K
disp('Input three values of proportional gain')
for i=1:3
  str=['K(',num2str(i),')=  '];
  K(i)=input(str);
end
%
% Set range of omegas for Bode plot
omegab=logspace(-3,3,200);
% Set range of omegas for Nyquist diagram
omegan=linspace(0,200,2000);
for i=1:3
% Define transfer function
  N=[K(i) 0];
  D=[1 K(i)+2 25];
% Determine vectors for Bode plots
  [G(:,i),phi(:,i)]=bode(N,D,omegab);
%   L(:,i)=20*log10(G(:,i));
% Determine vectors for Nyquist plots
  [R(:,i),I(:,i)]=nyquist(N,D,omegan);
end
% Plot Nyquist diagrams
plot(R(:,1),I(:,1),'-',R(:,2),I(:,2),'-.',R(:,3),I(:,3),'--')
xlabel('Re')
ylabel('Im')
title('Nyquist plots for Example 7.7')
str1=['K=',num2str(K(1))];
str2=['K=',num2str(K(2))];
str3=['K=',num2str(K(3))];
legend(str1,str2,str3)
figure
% Plot amplitude portion of Bode plots
semilogx(omegab,20*log(G(:,1)),'-',omegab,20*log(G(:,2)),'-.',omegab,20*log(G(:,3)),'--')
xlabel('log(\omega)')
ylabel('L(\omega) (dB)')
title('Amplitude Bode plots for Example 7.7')
legend(str1,str2,str3)
figure
% Plot phase portion of Bode plots
semilogx(omegab,phi(:,1),'-',omegab,phi(:,2),'-.',omegab,phi(:,3),'--')
xlabel('log(\omega)')
ylabel('phi')
title('Phase Bode plots for Example 7.7')
legend(str1,str2,str3)
%
% End of Example7_7.m
```

(a)

**FIG. 7.16** (a) The script of Example7_7.m; (b) the Nyquist diagrams generated from the execution of program; (c) the amplitude portion of Bode diagram generated from the execution of program; (d) the phase portion of the Bode diagram generated from the execution of the program. (*Continued*)

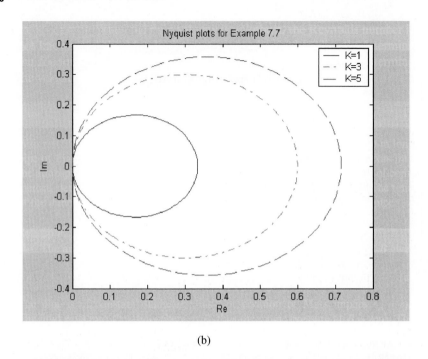

(b)

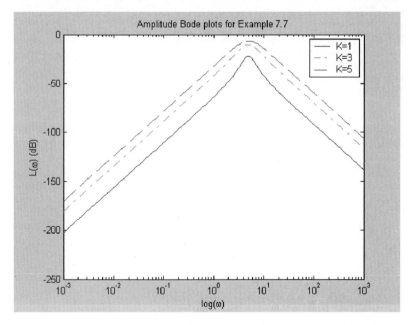

**FIG. 7.16** (*Continued*)                                                        (c)

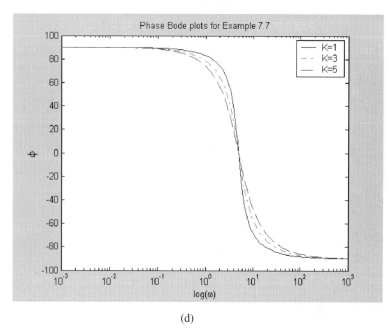

FIG. **7.16** (*Continued*)

(d)

### Solution

The script for the MATLAB file Example7_7.m is given in Figure 7.16(a). The Nyquist diagrams and Bode plots generated from the execution of this script are given in Figures 7.16(b)–(d). The following is noted about the script:

- The command 'logspace(-3,3,200)' is used to generate the vector of values for which the calculations to generate the Bode diagrams are made. This command generates a vector of 200 equally spaced values on a logarithmic scale between $10^{-3}$ and $10^3$.
- The command 'linspace(0,200,2000)' is used to generate the vector of values for which the calculations to generate the Nyquist diagrams are made. This command generates a vector of 2,000 equally spaced values between 0 and 200.
- The amplitude portions of the Bode diagrams are developed using a semi-logarithmic plot. The abscissa is a logarithmic scale whereas the ordinate is a linear scale since it measures $L(\omega)$, which has already been converted to a logarithmic scale.
- The phase portion of the Bode diagrams are developed using a semilogarithmic scale.
- A variable represented by R(:,2) refers to the second column of a double subscripted variable R.

## 7.4 FIRST-ORDER SYSTEMS

The transfer function for a first-order system, such as an *LR* circuit, is of the form

$$G(s) = \frac{1}{RT\left(s + \dfrac{1}{T}\right)} \tag{7.58}$$

where $T$ is the system's time constant. Evaluation of $G(j\omega)$ leads to

$$G(j\omega) = \frac{1}{RT\left(j\omega + \dfrac{1}{T}\right)}$$

$$= \frac{\dfrac{1}{T} - j\omega}{RT\left(\omega^2 + \dfrac{1}{T^2}\right)} \tag{7.59}$$

Thus

$$\mathrm{Re}[G(j\omega)] = \frac{1}{RT^2\left(\omega^2 + \dfrac{1}{T^2}\right)} \tag{7.60a}$$

$$\mathrm{Im}[G(j\omega)] = \frac{-\omega}{RT\left(\omega^2 + \dfrac{1}{T^2}\right)} \tag{7.60b}$$

The steady-state amplitude of a first-order $LR$ circuit is

$$I = V_0|G(j\omega)|$$

$$= \frac{V_0}{RT}\sqrt{\frac{1}{\omega^2 + \dfrac{1}{T^2}}} \tag{7.61}$$

The phase of the first-order system is

$$\phi = -\tan^{-1}(\omega T) \tag{7.62}$$

Defining the nondimensional frequency $\beta = \omega T$, Equations (7.61) and (7.62) can be written in nondimensional form as

$$M = \frac{RI}{V_0} = \sqrt{\frac{1}{\beta^2 + 1}} \tag{7.63}$$

$$\phi = -\tan^{-1}(\beta) \tag{7.64}$$

Frequency response curves for the first-order system are illustrated in Figure 7.17, which represents the frequency response curves for all first-order systems. The following is noted from these curves:

- As $\omega$ increases the steady-state amplitude decreases.
- $M = 1/\sqrt{2}$ for $\beta = 1$.
- $\lim_{\beta \to \infty} M = 0$, but $M$ behaves as $1/r$ for large $r$.
- $\phi$ is small and negative for small $\beta$.
- $\phi = -\pi/4$ for $\beta = 1$.
- $\lim_{\beta \to \infty} \phi = -\pi/2$.

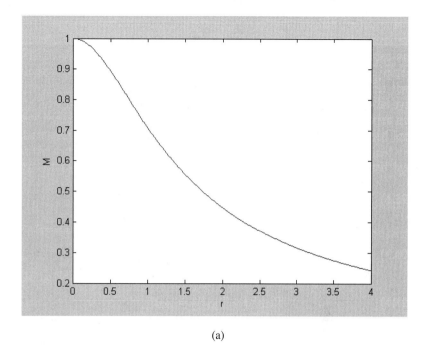

(a)

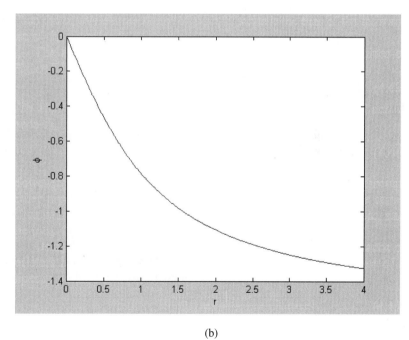

**FIG. 7.17** The frequency response curves for a first-order system: (a) The steady-state amplitude; (b) the phase angle.

(b)

```
>> N=[1];
>> D=[1 1];
>> bode(N,D)
>> grid
>>
```

(a)

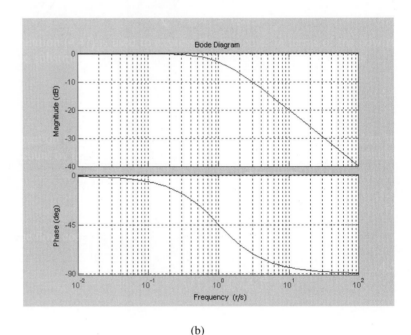

(b)

**FIG. 7.18**  (a) The MATLAB work session used to generate the Bode diagram of a first-order system; (b) the Bode diagram.

Bode plots for a first-order system are shown in Figure 7.18(b). The script of the MATLAB work session used to generate these diagrams is given in Figure 7.18(a). The low-frequency asymptote is horizontal at $\log(M) = 0$. The high-frequency asymptote is a line with a slope of $-20$ dB. The two asymptotes intersect at $\beta = 1$. The Bode diagrams are consistent with those generated for a first-order lag in Section 7.3.

The imaginary part of $G(j\omega)$ is negative for all positive values of $\beta$ and positive for all negative values of $\beta$. When $\beta = 0$, $\mathrm{Re}[G(j\omega)] = 1$ and $\mathrm{Im}[G(j\omega)] = 0$. As $\beta$ gets large $\mathrm{Re}[G(j\omega)] \to 0$ and $\mathrm{Im}[G(j\omega)] \to 0$. It can be shown algebraically that

$$\left\{\mathrm{Re}[G(j\omega)] - \tfrac{1}{2}\right\}^2 + \left\{\mathrm{Im}[G(j\omega)]\right\}^2 = \tfrac{1}{4} \qquad (7.65)$$

Thus it is deducted that the Nyquist plot for a first-order system is a semicircle of radius 1/2 existing in the first and fourth quadrants of the $G$ plane with its center on the Re axis at Re = 1/2. The Nyquist plot of a first-order system drawn using MATLAB is illustrated in Figure 7.19.

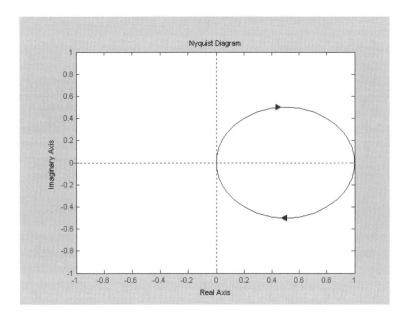

**FIG. 7.19** The Nyquist diagram for a first-order system.

---

## Example 7.8

The differential equation for the perturbation in liquid level $h(t)$ in a tank of area $A$ due to a perturbation in exit pressure $p(t)$ is derived in Example 4.7 as

$$A\frac{dh}{dt} + \frac{1}{R}h = \frac{p(t)}{\rho g R} \tag{a}$$

where $R$ is the resistance in the exit pipe and $\rho$ is the mass density of the liquid. Determine the steady-state response for $h(t)$ for a circular tank of diameter 4 m filled with water ($\rho = 1{,}000$ kg/m$^3$) when the pipe resistance is 4.0 s/m$^2$ and the perturbation in exit pressure is

$$p(t) = 250\sin(0.05t) \text{ Pa} \tag{b}$$

where $t$ is measured in seconds.

### Solution

Equation (a) is rewritten as

$$RA\frac{dh}{dt} + h = \frac{p(t)}{\rho g} \tag{c}$$

The time constant for the system is determined from Equation (c) as

$$T = RA$$

$$= \left(4.0\,\frac{\text{s}}{\text{m}^2}\right)\left[\frac{\pi}{4}(4\text{ m})^2\right] = 50.3 \text{ s} \tag{d}$$

The transfer function $G(s) = H(s)/P(s)$ is determined from Equation (c) as

$$G(s) = \frac{1}{\rho g T \left( s + \dfrac{1}{T} \right)} \tag{e}$$

The transfer function of Equation (e) is the same as the transfer function in Equation (7.58), but with $\rho g$ replacing $R$. The analogous form of Equation (7.63) is

$$\frac{\rho g H}{p_0} = \sqrt{\frac{1}{1 + \beta^2}} \tag{f}$$

where $p_0$ is the amplitude of the pressure perturbation and $H$ is the steady-state amplitude of the perturbation in the liquid level. The frequency parameter for this problem is

$$\begin{aligned} \beta &= \omega T \\ &= (0.05 \text{ r/s})(50.3 \text{ s}) = 2.52 \end{aligned} \tag{g}$$

Solving Equation (f) for $H$ leads to

$$\begin{aligned} H &= \frac{p_0}{\rho g} \sqrt{\frac{1}{1 + \beta^2}} \\ &= \frac{250 \text{ N/m}^2}{(1{,}000 \text{ kg/m}^3)(9.81 \text{ m/s}^2)} \sqrt{\frac{1}{1 + (2.2)^2}} = 10.6 \text{ mm} \end{aligned} \tag{h}$$

The phase angle is obtained from Equation (7.64) as

$$\begin{aligned} \phi &= \tan^{-1}(-\beta) \\ &= \tan^{-1}(-2.52) = -1.19 \text{ rad} \end{aligned} \tag{i}$$

The steady-state perturbation in the liquid level is

$$\begin{aligned} h(t) &= H \sin(\omega t + \phi) \\ &= 10.6 \sin(0.05t - 1.19) \text{ mm} \end{aligned} \tag{j}$$

## 7.5 SECOND-ORDER SYSTEMS

The frequency response for the series $LRC$ circuit was used to illustrate the development of frequency response diagrams, Bode diagrams, and Nyquist diagrams. The frequency response for other important second-order systems is examined in this section. All plots are generated using MATLAB.

### 7.5.1 One-Degree-of-Freedom Mechanical System

The standard form of the differential equation for the displacement of a linear one-degree-of-freedom mechanical system is Equation (2.111) which is repeated below

$$\ddot{x} + 2\zeta\omega_n\dot{x} + \omega_n^2 x = \frac{1}{m}F(t) \tag{2.111}$$

The transfer function $G(s) = X(s)/F(s)$ determined for Equation (2.111) is

$$G(s) = \frac{1}{m(s^2 + 2\zeta\omega_n s + \omega_n^2)} \tag{7.66}$$

The steady-state response of the system due to a sinusoidal input of the form $F(t) = F_0 \sin(\omega t)$ is obtained through the evaluation of $G(j\omega)$, which leads to

$$G(j\omega) = \frac{(\omega_n^2 - \omega^2) - j(2\zeta\omega\omega_n)}{m\left[(\omega_n^2 - \omega^2)^2 + (2\zeta\omega\omega_n)^2\right]} \tag{7.67}$$

The steady-state response of the second-order system is of the form of Equation (7.2) where

$$X = F_0|G(j\omega)|$$

$$= \frac{F_0}{m}\sqrt{\frac{1}{(\omega_n^2 - \omega^2)^2 + (2\zeta\omega\omega_n)^2}} \tag{7.68}$$

and

$$\phi = \tan^{-1}\left(\frac{2\zeta\omega\omega_n}{\omega - \omega_n^2}\right) \tag{7.69}$$

Nondimensional forms of Equations (7.68) and (7.69) are obtained in terms of the frequency ratio $r = \omega/\omega_n$ as

$$\frac{m\omega_n^2 X}{F_0} = \frac{1}{\sqrt{(1 - r^2)^2 + (2\zeta r)^2}} \tag{7.70}$$

$$\phi = \tan^{-1}\left(\frac{2\zeta r}{r^2 - 1}\right) \tag{7.71}$$

Frequency response curves for the second-order system are shown in Figure 7.20. Analysis of these curves, as well as Equations (7.70) and (7.71), shows the following:

- For $\zeta < 1/\sqrt{2}$ the steady-state amplitude decreases with increasing frequency, approaching zero for large $r$.
- For $\zeta < 1/\sqrt{2}$ the steady-state amplitude increases with increasing $r$ until it reaches a maximum and then decreases approaching zero for large $r$.
- For $\zeta > 1/\sqrt{2}$ the steady-state amplitude has its maximum of $F_0/(m\omega_n^2)$ at $r = 0$ and decreases with increasing $r$, approaching zero for large $r$.

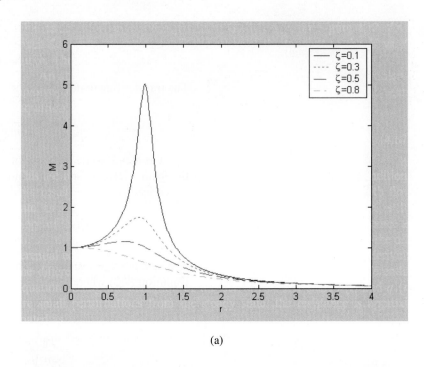

(a)

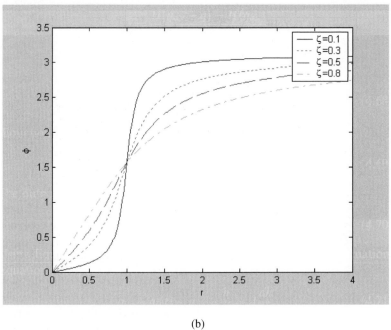

**FIG. 7.20** The frequency response curves
for mass-spring-viscous damper system:
(a) The steady-state amplitude vs.
frequency ratio; (b) the phase angle vs.
frequency ratio.

(b)

- The maximum value of $X$ over all values of $r$ occurs for the value of $r$ for which $dX/dr = 0$. The maximum value occurs for $r = \sqrt{1 - 2\zeta^2}$ and is equal to

$$X_{\max} = \frac{F_0}{m\omega_n^2(2\zeta)\sqrt{1 - \zeta^2}} \tag{7.72}$$

- For a fixed $r$, the larger the value of $\zeta$, the smaller the value of $X$.
- The phase is near zero for $r$ near zero and is near $\pi$ for large $r$.

## Example 7.9

Determine the steady-state response of the machine of Example 7.1 if the 40-kg machine is mounted on a spring of stiffness $1 \times 10^5$ N/m in parallel with a viscous damper of damping coefficient 400 N·s/m when it is subject to an input force of $F(t) = 100 \sin(\omega t)$ N for (a) $\omega = 25$ r/s, (b) $\omega = 50$ r/s, and (c) $\omega = 100$ r/s. In addition, (d) draw its Bode diagram, specifying its low-frequency and high-frequency asymptotes, corner frequency, and cutoff frequency.

**Solution**

The steady-state response of the system for an input frequency $\omega$ is

$$x(t) = X \sin(\omega t + \phi) \tag{a}$$

where from Equations (7.70) and (7.71)

$$X = \frac{F_0}{m\omega_n^2} \sqrt{\frac{1}{(1 - r^2)^2 + (2\zeta r)^2}} \tag{b}$$

$$\phi = \tan^{-1}\left(\frac{2\zeta r}{r^2 - 1}\right) \tag{c}$$

The natural frequency is determined in Example 7.1 as 50 r/s. The damping ratio is

$$\zeta = \frac{c}{2m\omega_n}$$

$$= \frac{400 \text{ N·s/m}}{2(40 \text{ kg})(50 \text{ r/s})} = 0.1 \tag{d}$$

It is noted that

$$\frac{F_0}{m\omega_n^2} = \frac{100 \text{ N}}{(40 \text{ kg})(100 \text{ r/s})^2} = 0.25 \text{ mm} \tag{e}$$

(a) For $\omega = 25$ r/s the frequency ratio is $r = (25 \text{ r/s})/(50 \text{ r/s}) = 0.5$. The evaluation of Equations (b) and (c) lead to

$$X = (0.25 \text{ mm})\sqrt{\frac{1}{[1 - (0.5)^2]^2 + [2(0.1)(0.5)]^2}} = 0.330 \text{ mm} \tag{f}$$

$$\phi = \tan^{-1}\left(\frac{2(0.1)(0.5)}{(0.5)^2 - 1}\right) = -0.133 \text{ rad} \tag{g}$$

The steady-state response for an input frequency of 25 r/s is

$$x(t) = 0.330 \sin (25t - 0.133) \text{ mm} \tag{h}$$

(b) For $\omega = 50$ r/s the frequency ratio is $r = (50 \text{ r/s})/(50 \text{ r/s}) = 1$. The evaluation of Equations (b) and (e) lead to

$$X = (0.25 \text{ mm})\sqrt{\frac{1}{[1 - (1)^2]^2 + [2(0.1)(1)]^2}} = 1.25 \text{ mm} \tag{i}$$

$$\phi = \tan^{-1}\left(\frac{2(0.1)(1)}{1 - (1)^2}\right) = -\frac{\pi}{2} \text{ rad} \tag{j}$$

The steady-state response for an input frequency of 50 r/s is

$$x(t) = 1.25 \sin\left(50t - \frac{\pi}{2}\right) \text{ mm} \tag{k}$$

(c) For $\omega = 100$ r/s the frequency ratio is $r = (100 \text{ r/s})/(50 \text{ r/s}) = 2$. The evaluation of Equations (b) and (e) lead to

$$X = (0.25 \text{ mm})\sqrt{\frac{1}{[1 - (2)^2]^2 + [2(0.1)(2)]^2}} = 0.083 \text{ mm} \tag{l}$$

$$\phi = \tan^{-1}\left(\frac{2(0.1)(2)}{(2)^2 - 1}\right) = 0.133 \text{ rad} \tag{m}$$

The steady-state response for an input frequency of 100 r/s is

$$x(t) = 0.083 \sin (100t + 0.133) \text{ mm} \tag{n}$$

Bode diagrams corresponding to the second-order system of Example 7.9 are shown in Figure 7.21. The Bode diagram is that of the second-order underdamped systems given in Equations (7.54). Its low-frequency asymptote is

$$L = -20 \log (\omega_n^2) = -68.0 \text{ dB} \tag{o}$$

while its high-frequency asymptote is

$$L = -40 \log (\omega) \tag{p}$$

The corner frequency is obtained by equating Equations (o) and (p) leading to

$$-40 \log (\omega) = -68.0$$
$$\omega = 50 \text{ r/s} \tag{q}$$

It is noted that the system's corner frequency is equal to its natural frequency, which is true for all transfer functions for a second-order lag system.

The system's cutoff frequency is the frequency when $L(\omega)$ is equal to 3 dB less than $L(0)$. To this end

$$-68.0 + 10 \log \left[(50^2 - \omega_c^2)^2 + [2(0.1)(50)\omega_c]^2\right] = 3$$
$$10 \log \left[(2{,}500 - \omega_c^2)^2 + 100\omega_c^2\right] = 71.0$$
$$\omega_c^4 - 4{,}900\omega_c^2 + 6.25 \times 10^6 = 1.25 \times 10^7 \tag{r}$$

Equation (r) is solved leading to its only real and positive solution of $\omega_c = 77.1$ r/s.

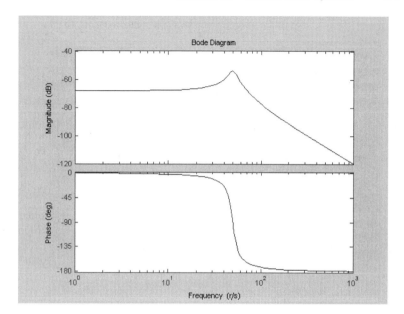

**FIG. 7.21** The Bode diagram for second-order mechanical system of Example 7.9.

For $\omega = 0$, $\mathrm{Re}[G(j\omega)] = 1/m$ and $\mathrm{Im}[G(j\omega)] = 0$. As $\omega$ grows large $\mathrm{Re}[G(j\omega)] \to 0$ and $\mathrm{Im}[G(j\omega)] \to 0$. For all values of $\omega$ $\mathrm{Im}[G(j\omega)] \leq 0$. For $\omega < \omega_n$, $\mathrm{Re}[G(j\omega)] > 0$ and for $\omega > \omega_n$, $\mathrm{Re}[G(j\omega)] < 0$. Thus the Nyquist diagram for the second-order system exists in the third and fourth quadrants of the $G$ plane. The curve touches the Re axis at $1/m$ and approaches the origin. It is noted that for $\omega = \omega_n$, $\mathrm{Re}[G(j\omega)] = 0$ and $\mathrm{Im}[G(j\omega)] = -1/\lfloor 2\zeta m\omega_n^2 \rfloor$ and thus intersects the Im axis at this point. Nyquist plots for this second-order system for several values of $\zeta$ are shown in Figure 7.22.

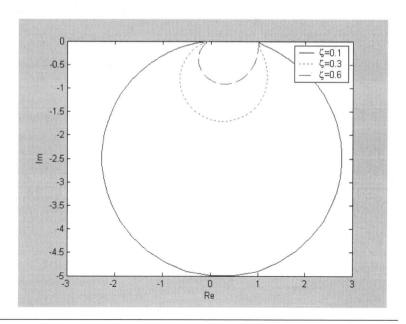

**FIG. 7.22** The Nyquist diagrams for a second-order mechanical system, for $m = 1$.

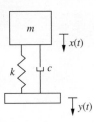

**FIG. 7.23** The mass-spring-viscous damper system with a prescribed motion of support.

## 7.5.2 Motion Input

The differential equation governing the response of a mass-spring-viscous damper system due to a prescribed motion of its base, as illustrated in Figure 7.23, is

$$\ddot{x} + 2\zeta\omega_n\dot{x} + \omega_n^2 x = 2\zeta\omega_n\dot{y} + \omega_n^2 y \qquad (7.73)$$

It is often of interest to determine the acceleration of the mass

$$a(t) = \ddot{x}(t) \qquad (7.74)$$

The displacement of the mass relative to its base is defined as

$$z(t) = x(t) - y(t) \qquad (7.75)$$

Taking the Laplace transform of Equations (7.73)–(7.75) leads to the transfer functions

$$G_1(s) = \frac{X(s)}{Y(s)}$$

$$= \frac{2\zeta\omega_n s + \omega_n^2}{s^2 + 2\zeta\omega_n s + \omega_n^2} \qquad (7.76)$$

$$G_2(s) = \frac{A(s)}{Y(s)}$$

$$= \frac{s^2(2\zeta\omega_n s + \omega_n^2)}{s^2 + 2\zeta\omega_n s + \omega_n^2} \qquad (7.77)$$

$$G_3(s) = \frac{Z(s)}{Y(s)}$$

$$= \frac{s^2}{s^2 + 2\zeta\omega_n s + \omega_n^2} \qquad (7.78)$$

The evaluation of the transfer functions at $s = j\omega$ leads to

$$|G_1(j\omega)| = \sqrt{\frac{\omega_n^4 + (2\zeta\omega\omega_n)^2}{(\omega_n^2 - \omega^2)^2 + (2\zeta\omega\omega_n)^2}} \qquad (7.79)$$

$$|G_2(j\omega)| = \omega^2\sqrt{\frac{\omega_n^4 + (2\zeta\omega\omega_n)^2}{(\omega_n^2 - \omega^2)^2 + (2\zeta\omega\omega_n)^2}} \qquad (7.80)$$

$$|G_3(j\omega)| = \frac{\omega^2}{\sqrt{(\omega_n^2 - \omega^2)^2 + (2\zeta\omega\omega_n)^2}} \qquad (7.81)$$

Equations (7.79)–(7.81) are written in terms of the frequency ratio $r = \omega/\omega_n$ as

$$T = |G_1(j\omega)| = \sqrt{\frac{1 + (2\zeta r)^2}{(1 - r^2)^2 + (2\zeta r)^2}} \qquad (7.82)$$

$$\omega_n^2 \Gamma = |G_2(j\omega)| = \omega_n^2 r^2 \sqrt{\frac{1 + (2\zeta r)^2}{(1 - r^2)^2 + (2\zeta r)^2}} \tag{7.83}$$

$$\Lambda = |G_3(j\omega)| = \frac{r^2}{\sqrt{(1 - r^2)^2 + (2\zeta r)^2}} \tag{7.84}$$

The steady-state amplitudes due to a sinusoidal motion input of the form $y(t) = Y \sin(\omega t)$ are

$$X = TY \tag{7.85}$$

$$A = \omega_n^2 \Gamma Y \tag{7.86}$$

$$Z = \Lambda Y \tag{7.87}$$

The ratio $T = X/Y$ and is called the transmissibility because it represents the ratio of the amplitude of the output to the amplitude of the input. When T < 1 the amplitude of the steady-state displacement of the mass is less than the amplitude of the support. In this case, the spring and viscous damper system acts to **isolate** the mass from the motion of its base. When $T > 1$ the amplitude of the steady-state displacement of the mass is larger than the amplitude of the base. When this occurs the spring and viscous damper act to **amplify** the motion transmitted from the base to the mass. The transmissibility is illustrated in Figure 7.24 as a function of the frequency ratio for several values of the damping ratio. From Equation (7.82) and Figure 7.24 it is noted that

- $T = 0$ when $r = 0$ for all values of $\zeta$;
- $T = 1$ when $r = \sqrt{2}$ for all values of $\zeta$;

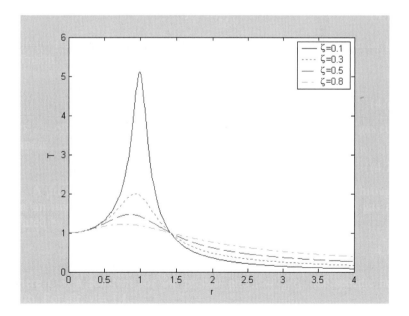

**FIG. 7.24** The transmissibility ratio as function of frequency ratio for several values of the damping ratio.

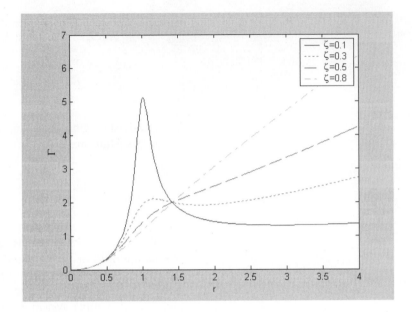

**FIG. 7.25** $\Gamma$ as a function of the frequency ratio for several values of the damping ratio.

- $T > 1$ for all $r < \sqrt{2}$ for all values of $\zeta$;
- $T < 1$ for all $r > \sqrt{2}$ for all values of $\zeta$;
- $T$ approaches zero for large $r$ for all values of $\zeta$;
- For $r > \sqrt{2}$, $T$ is smaller for smaller values of $\zeta$.

Thus it is clear $r > \sqrt{2}$ is the **range of isolation** and $r < \sqrt{2}$ is the **range of amplification**.

For a specific system Equation (7.86) can be used along with Equation (7.83) to determine the steady-state amplitude of the acceleration of the mass. Figure 7.25 illustrates the variation of $\Gamma$ with the frequency ratio for several values of the damping ratio. The following is obtained from Figure 7.25 and Equation (7.83):

- $\Gamma = 0$ for $r = 0$ for all values of $\zeta$.
- As $r \to \infty$, $\Gamma \to 2\zeta r$.
- $\Gamma = 2$ for $r = \sqrt{2}$ for all values of $\zeta$.
- For $\zeta < \sqrt{2}/4$ the function $\Gamma$ has a relative maximum for some value of $r < \sqrt{2}$ and a relative minimum for some value of $r > \sqrt{2}$.
- For $\zeta > \sqrt{2}/4$ the value of $\Gamma$ increases with increasing $r$, achieving no relative maximum.

The steady-state amplitude of the displacement of the mass relative to the support is given by Equation (7.87), where $\Lambda$ is defined in Equation (7.84). The variation of $\Lambda$ with $r$ is illustrated in Figure 7.26, which shows that:

- $\Lambda = 0$ when $r = 0$ for all values of $\zeta$;
- $\Lambda \to 1$ as $r \to \infty$ for all values of $\zeta$;

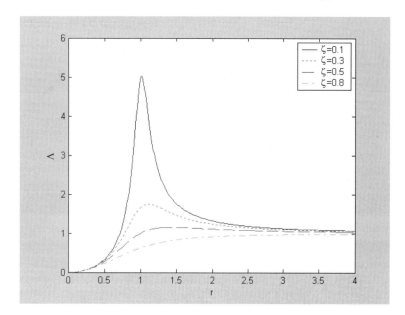

**FIG. 7.26** $\Lambda$ as a function of the frequency ratio for several values of the damping ratio.

- For $\zeta < 1/\sqrt{2}$, $\Lambda$ has a maximum of $1/2\zeta\sqrt{1-\zeta^2}$ corresponding to a value of $r = 1/\sqrt{1-2\zeta^2}$.
- For $\zeta > 1/\sqrt{2}$, $\Lambda$ increases with increasing $r$, never attaining its maximum.

## Example 7.10

A 50-kg machine is mounted on an isolator of stiffness $1.8 \times 10^5$ N/m and damping coefficient 360 N·s/m. The floor on which the isolator is mounted has a sinusoidal displacement of amplitude 0.2 mm and frequency 150 r/s. (a) Determine the steady-state amplitude of the displacement of the machine. (b) Determine the steady-state amplitude of the machine's acceleration. (c) Determine the steady-state amplitude of the displacement of the machine relative to the floor.

### Solution

The system is modeled as a mass-spring-viscous damper system subject to a sinusoidal motion input. The system's natural frequency and damping ratio are

$$\omega_n = \sqrt{\frac{k}{m}}$$

$$= \sqrt{\frac{1.8 \times 10^5 \text{ N/m}}{50 \text{ kg}}} = 60 \text{ r/s} \tag{a}$$

$$\zeta = \frac{c}{2m\omega_n}$$

$$= \frac{360 \text{ N·s/m}}{2(50 \text{ kg})(60 \text{ r/s})} = 0.06 \tag{b}$$

The frequency ratio for the input is

$$r = \frac{\omega}{\omega_n}$$

$$= \frac{150 \text{ r/s}}{60 \text{ r/s}} = 2.5 \tag{c}$$

(a) The steady-state amplitude of the machine's displacement is obtained using Equations (7.82) and (7.85) as

$$X = (Y)(T)$$

$$= (0.2 \text{ mm})\sqrt{\frac{1 + [2(0.06)(2.5)]^2}{[1 - (2.5)^2]^2 + [2(0.06)(2.5)]^2}} = 0.0397 \text{ mm} \tag{d}$$

(b) The steady-state amplitude of the machine's acceleration is

$$A = \omega^2 X$$

$$= (150 \text{ r/s})^2 (0.0397 \text{ mm}) = 0.893 \text{ m/s}^2 \tag{e}$$

(c) The steady-state amplitude of the displacement of the machine relative to the floor is obtained using Equations (7.84) and (7.87) as

$$Z = (Y)(\Lambda)$$

$$= (0.2 \text{ mm})\frac{(2.5)^2}{\sqrt{[1 - (2.5)^2]^2 + [2(0.06)(2.5)]^2}} = 0.238 \text{ mm} \tag{f}$$

## 7.5.3  Filters

Electric circuits can be designed to act as filters. The input to such a circuit is a sinusoidal voltage source of constant amplitude. The output from the filter is a sinusoidal voltage at some amplitude. A filter amplifies the voltage from sources of some frequencies and greatly reduces the amplitude from sources at other frequencies. A **low-pass filter** amplifies low frequencies and filters higher frequencies. A **high-pass filter** amplifies higher frequencies and filters lower frequencies. A **band-pass filter** amplifies frequencies in a certain frequency range and filters frequencies outside the designated range. A **band-reject filter** filters frequencies in a certain range and amplifies frequencies outside the range. The range of frequencies amplified by a circuit and the range of frequencies filtered by a circuit are determined from a frequency response analysis of the circuit.

**Example 7.11**

The output from the circuit of Figure 7.27 is $v_2(t)$, the voltage across terminals at the ends of the resistor. (a) Derive the transfer function $G(s) = V_2(s)/V_1(s)$. (b) Determine the steady-state response for $v_2(t)$ when $R = \sqrt{L/C}$. (c) Discuss why this circuit is called a second-order low-pass filter (called a Butterworth filter).

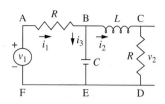

**FIG. 7.27** A second-order Butterworth filter.

## Solution

(a) The application of KCL at node B leads to

$$i_1 - i_2 - i_3 = 0$$
$$i_3 = i_1 - i_2 \qquad \text{(a)}$$

The application of KVL around loop ABEF gives

$$v_1(t) - Ri_1 - \frac{1}{C}\int_0^t i_3\,dt = 0 \qquad \text{(b)}$$

The application of KVL around loop BCDE leads to

$$-L\frac{di_2}{dt} - Ri_2 + \frac{1}{C}\int_0^t i_3\,dt = 0 \qquad \text{(c)}$$

The substitution of Equation (a) into Equations (b) and (c) leads to

$$R_1 i_1 + \frac{1}{C}\int_0^t (i_1 - i_2)\,dt = v_1(t) \qquad \text{(d)}$$

$$L\frac{di_2}{dt} + Ri_2 - \frac{1}{C}\int_0^t (i_1 - i_2)\,dt = 0 \qquad \text{(e)}$$

Taking the Laplace transforms of Equations (d) and (e) assuming $i_1(0) = i_2(0) = 0$ leads to

$$\left(R + \frac{1}{Cs}\right)I_1 - \frac{1}{Cs}I_2(s) = V_1(s) \qquad \text{(f)}$$

$$-\frac{1}{Cs}I_1(s) + \left(Ls + R + \frac{1}{Cs}\right)I_2(s) = 0 \qquad \text{(g)}$$

Solving Equations (f) and (g) for $I_2(s)$ leads to

$$I_2(s) = \frac{V_1(s)}{LRC\left[s^2 + \left(\dfrac{R}{L} + \dfrac{1}{RC}\right)s + \dfrac{2}{LC}\right]} \qquad \text{(h)}$$

The voltage from $C$ to $D$ is

$$v_2(t) = Ri_2 \qquad \text{(i)}$$

Taking the Laplace transform of Equation (i) leads to

$$V_2(s) = RI_2(s) \qquad \text{(j)}$$

Equations (h) and (j) are combined to determine the desired transfer function

$$G(s) = \frac{1}{LC\left[s^2 + \left(\dfrac{R}{L} + \dfrac{1}{RC}\right)s + \dfrac{2}{LC}\right]} \qquad \text{(k)}$$

Equation (k) is the transfer function of a second-order system of natural frequency

$$\omega_n = \sqrt{\frac{2}{LC}} \tag{l}$$

and damping ratio

$$\zeta = \frac{1}{2\omega_n}\left(\frac{R}{L} + \frac{1}{RC}\right)$$

$$= \frac{1}{2\sqrt{2}}\left(R\sqrt{\frac{C}{L}} + \frac{1}{R}\sqrt{\frac{L}{C}}\right) \tag{m}$$

(b) For $R = \sqrt{L/C}$ Equation (m) evaluates to a damping ratio of $\zeta = 1/\sqrt{2}$ and the transfer function given by Equation (k) becomes

$$G(s) = \frac{\omega_n^2}{2\left(s^2 + \sqrt{2}\omega_n s + \omega_n^2\right)} \tag{n}$$

where the natural frequency is given by Equation (l).

The steady-state response of the output voltage due to a sinusoidal input voltage of the form $v_1(t) = V_1 \sin(\omega t)$ is

$$v_2(t) = V_2 \sin(\omega t + \phi) \tag{o}$$

where

$$V_2 = V_1|G(j\omega)| \tag{p}$$

and $\phi$ is the phase of $G(j\omega)$. Substituting $s = j\omega$ in Equation (n) leads to

$$G(j\omega) = \frac{\omega_n^2}{2}\left[\frac{\left(\omega_n^2 - \omega^2\right) - j\sqrt{2}\omega\omega_n}{\left(\omega_n^2 - \omega^2\right)^2 + 2\omega_n^2\omega^2}\right] \tag{q}$$

Equation (q) is used to obtain

$$V_2 = \frac{V_1}{2\sqrt{1 + \left(\dfrac{\omega}{\omega_n}\right)^4}} \tag{r}$$

$$\phi = \tan^{-1}\left(\frac{-\sqrt{2}\dfrac{\omega}{\omega_n}}{1 - \left(\dfrac{\omega}{\omega_n^2}\right)}\right) \tag{s}$$

(c) The Bode diagram for the transfer function of Equation (n) is given in Figure 7.28. Equation (r) and Figure 7.28 both show that the steady-state amplitude of the output for small frequencies is near the amplitude of the steady-state input, while the steady-state amplitude of the response corresponding to high-frequency input is very small. Thus this circuit filters out high-frequency input and passes on low-frequency input; hence the name low-pass filter. The circuit is a second-order filter because its transfer function is that of a second-order system.

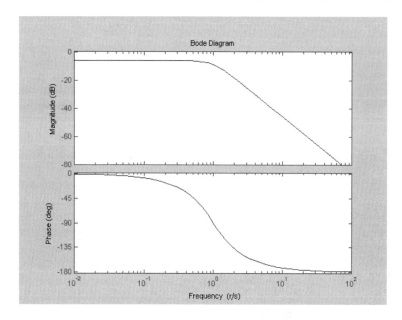

**FIG. 7.28** The Bode diagram for a second-order Butterworth filter.

## 7.6 HIGHER-ORDER SYSTEMS

The same tools are used to develop the steady-state response of higher-order systems (systems greater than second order) as used for first- and second-order systems. The sinusoidal transfer function is used to determine the steady-state amplitudes and phase angles. Frequency response curves, Bode plots, and Nyquist plots illustrate qualitative and quantitative features of the frequency response.

A higher-order system is usually modeled using more than one dependent variable. Examples of higher-order systems include multidegree-of-freedom mechanical systems and multiloop electric circuits with capacitors or inductors.

As shown in Section 6.6, an undamped $n$-degree-of-freedom system has $n$ natural frequencies. Resonance occurs when the frequency of a sinusoidal input coincides with any of the system's natural frequencies. As for a one-degree-of-freedom system, resonance is characterized by unbounded growth in system response. When the input frequency does not coincide with a natural frequency, a steady-state response exists for each dependent variable and is of the form

$$x_i(t) = X_i \sin(\omega t + \phi_i) \tag{7.88}$$

where for an undamped system $\phi_i$ may either be 0 or $\pi$.

A proportionally damped $n$-degree-of-freedom system is characterized by the $n$ natural frequencies of the undamped system and $n$ modal damping ratios. The steady-state response of a proportionally damped system is of the form of Equation (7.88) where the steady-state amplitude and steady-state phase are dependent on the natural frequencies and modal damping ratios. The response of a $n$-degree-of-freedom with a general form of viscous damping is not as easy to characterize. Frequencies may be identified for underdamped modes, but not for overdamped modes. The steady-state response is of the form of Equation (7.88).

The algebra required to determine system response for a higher-order system is complex and tedious. Higher-order systems are now often modeled using the state-space methods of Chapter 9. Further discussion of the frequency response of higher-order systems with a general viscous damping is delayed until Chapter 9.

Section 6.6 presents a general discussion of the free response of a higher-order system. The discussion of the frequency response of higher-order systems is limited to the specific applications of dynamic vibration absorbers for mechanical systems and higher-order filters.

## 7.6.1 Dynamic Vibration Absorbers

Consider the two-degree-of-freedom system of Figure 7.29. The matrix form of the differential equations governing the response of the system is

$$\begin{bmatrix} m_1 & 0 \\ 0 & m_2 \end{bmatrix} \begin{bmatrix} \ddot{x}_1 \\ \ddot{x}_2 \end{bmatrix} + \begin{bmatrix} k_1 + k_2 & -k_2 \\ -k_2 & k_2 \end{bmatrix} \begin{bmatrix} x_1 \\ x_2 \end{bmatrix} = \begin{bmatrix} F_0 \sin{(\omega t)} \\ 0 \end{bmatrix} \quad (7.89)$$

The transfer functions for the system are obtained as

$$G_1(s) = \frac{X_1(s)}{F(s)} = \frac{m_2 s^2 + k_2}{(m_1 s^2 + k_1 + k_2)(m_2 s^2 + k_2) - (k_2)^2} \quad (7.90)$$

$$G_2(s) = \frac{X_2(s)}{F(s)} = \frac{-k_2}{(m_1 s^2 + k_1 + k_2)(m_2 s^2 + k_2) - (k_2)^2} \quad (7.91)$$

The steady-state amplitudes are obtained as the real parts of the sinusoidal transfer functions. After some algebra

$$X_1 = F_0 |G_1(j\omega)|$$

$$= \frac{F_0 (k_2 - m_2 \omega^2)}{m_1 m_2 \omega^4 - [(k_1 + k_2)m_2 + k_2 m_1]\omega^2 + k_1 k_2} \quad (7.92)$$

$$X_2 = F_0 |G_2(j\omega)|$$

$$= \frac{F_0 k_2}{m_1 m_2 \omega^4 - [(k_1 + k_2)m_2 + k_2 m_1]\omega^2 + k_1 k_2} \quad (7.93)$$

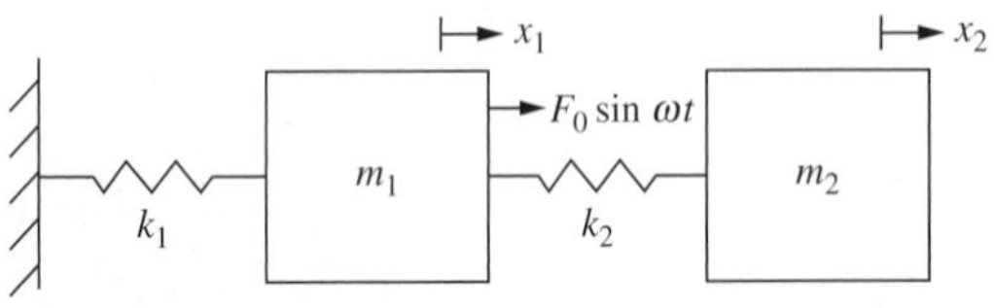

**FIG. 7.29** A two-degree-of-freedom mechanical system. If $k_2/m_2 = \omega^2$ then the steady-state amplitude $X_1 = 0$ and the mass $m_2$ acts as a vibration absorber.

Equation (7.92) shows that if

$$\frac{k_2}{m_2} = \omega^2 \tag{7.94}$$

then $X_1 = 0$; the steady-state response of the particle of mass $m_1$ is zero.

These form the theory on which a dynamic vibration absorber operates. The primary system is composed of the particle of mass $m_1$ attached to a spring of stiffness $k_1$ and acted on by a harmonic force of magnitude $F_0$ at a frequency $\omega$. A large steady-state response occurs when the primary system, by itself, is subject to an input with a frequency $\omega$ close to the natural frequency, $\omega_{11} = \sqrt{k_1/m_1}$. The auxiliary system, or the absorber, is composed of the particle of mass $m_2$, which is connected to the primary system through a spring of stiffness $k_2$. The resulting system has two degrees of freedom. The amplitudes of the steady-state responses of the primary system and the auxiliary system are given by Equations (7.92) and (7.93) respectively. Equation (7.92) shows that when the absorber is designed (or tuned) such that Equation (7.94) is satisfied the amplitude of the primary system is zero.

The term "absorber" is somewhat of a misnomer. The system is effective because the natural frequencies of the two-degree-of-freedom system are away from the natural frequency of the primary system. The natural frequencies are obtained by determining the poles of the transfer functions of Equations (7.90) and (7.91). The results are

$$\omega_{1,2} = \left[ \frac{(k_1 + k_2)m_2 + k_2 m_1 \pm \sqrt{[(k_1 + k_2)m_2 + k_2 m_1]^2 - 4 m_1 m_2 k_1 k_2}}{2 m_1 m_2} \right]^{1/2} \tag{7.95}$$

The absorber is said to be tuned to the frequency $\omega_{22} = \sqrt{k_2/m_2}$. When the primary system is subject to a harmonic input at the tuned frequency the steady-state response of the primary system is zero. The application of Newton's law to the free-body diagram of the absorber under these conditions shows that the steady-state amplitude of the absorber mass is

$$X_2 = -\frac{F_0}{k_2} \tag{7.96}$$

---

**Example 7.12**

A 500-kg machine is mounted on a foundation of stiffness $7.2 \times 10^6$ N/m. During operation it is subject to a harmonic force of magnitude 5,000 N at a frequency of 118 r/s. (a) Determine the steady-state amplitude of the machine under these conditions. (b) Specify the mass and stiffness of a dynamic vibration absorber that is tuned to 118 r/s and when added to the primary system has a steady-state amplitude of 5 mm. (c) Plot the frequency response of the primary system both with and without the absorber.

### Solution

(a) The natural frequency of the primary system is

$$\omega_{11} = \sqrt{\frac{k_1}{m_1}} = \sqrt{\frac{7.2 \times 10^6 \text{ N/m}}{500 \text{ kg}}} = 120 \text{ r/s} \qquad \text{(a)}$$

The steady-state amplitude of the primary system before an absorber is added is obtained using Equation (7.8) as

$$X_1 = \left| \frac{F_0}{m(\omega_{11}^2 - \omega^2)} \right|$$

$$= \left| \frac{5,000 \text{ N}}{(500 \text{ kg}) \left[ (120 \text{ rad/s})^2 - (118 \text{ rad/s})^2 \right]} \right| = 2.10 \text{ cm} \qquad \text{(b)}$$

(b) If the absorber is designed such that it is tuned to 118 r/s and the input frequency is 118 r/s then its steady-state amplitude is given by Equation (7.96). Requiring the steady-state amplitude to be less than 5 mm leads to

$$\frac{F_0}{k_2} < 5 \text{ mm}$$

$$k_2 > \frac{5,000 \text{ N}}{0.005 \text{ m}} = 1 \times 10^6 \text{ N/m} \qquad \text{(c)}$$

Tuning the absorber to 118 r/s requires

$$\frac{k_2}{m_2} = (118 \text{ r/s})^2$$

$$m_2 = \frac{1 \times 10^6 \text{ N/m}}{(118 \text{ r/s})^2} = 71.8 \text{ kg} \qquad \text{(d)}$$

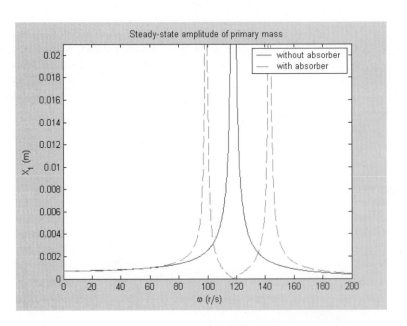

**FIG. 7.30** Contrast of the steady-state amplitude of a machine operating near resonance with and without the addition of a vibration absorber.

The minimum value of the stiffness is used in Equation (c) to determine the absorber of minimum mass.

(c) The substitution of the given and calculated values into Equation (7.92) leads to

$$X_1 = \frac{1.39 \times 10^5 - 10\omega^2}{\omega^4 - 3.03 \times 10^4\, \omega^2 + 2.01 \times 10^8} \qquad \text{(e)}$$

Equations (b) and (e) are plotted on the same graph in Figure 7.30.

Figure 7.30 shows that when the absorber is tuned to the input frequency the steady-state amplitude of the primary system is zero. The amplitude is large when the input frequency is away from the tuned frequency of the absorber. Figure 7.30 also illustrates that the lowest natural frequency of the two-degree-of-freedom system that occurs when the absorber is added is less than the tuned frequency. Thus a transient large amplitude response occurs as the input frequency builds to its steady-state value.

Configurations for dynamic vibration absorbers other than that of Figure 7.29 exist. It can be shown that when an auxiliary mass-spring system is tuned to the input frequency of a harmonic response, the steady-state amplitude of the particle in the system to which the absorber is attached is zero.

## 7.6.2 Higher-Order Filters

Passive filters that use only resistors, capacitors, and inductors do not have external energy sources. Active filters contain resistors and capacitors and amplifiers. Filters are often referred to by order, which is equal to the order of their transfer function. Higher-order filters provide greater design flexibility in achieving the required objectives.

The response of a filter is dependent on its transfer function. The polynomial in the numerator of the transfer function affects the frequency response, especially for small and large frequencies. For an $n$th-order filter the numerator of its transfer function can be written as

$$N(s) = a_n s^n + a_{n-1} s^{n-1} + \cdots a_1 s + a_0 \qquad (7.97)$$

The following can be deduced about the frequency response from the numerator of the transfer function. Let $V(\omega)$ represent the frequency response of the filter, the amplitude of the steady-state output as a function of the input frequency:

- $V(0) = 0$ if $a_0 = 0$ and $V(0) = C$, a nonzero constant, when $a_0 \neq 0$.
- $\lim_{\omega \to \infty} V(\omega) = 0$ if $a_n = 0$ and $\lim_{\omega \to \infty} V(\omega) = C$, a nonzero constant, when $a_n \neq 0$.
- When $a_n = 0$, the rate at which $V(\omega)$ approaches zero depends on the highest power of $s$ with a nonzero coefficient. $V(\omega)$ approaches zero fastest when $N(s) = a_0$.

These trends in $V$ are used to develop the necessary forms of the numerator of transfer functions for the different types of filters:

- A low-pass filter, which allows low-frequency signals to pass but filters high-frequency signals, requires $V(0) = C$ and $\lim_{\omega \to \infty} V(\omega) = 0$; thus its transfer function must have $a_0 \neq 0$ and $a_n = 0$.
- A high-pass filter, which allows high-frequency signals to pass but filters low frequency signals, must have $V(0) = 0$ and $\lim_{\omega \to \infty} V(\omega) = C$; thus its transfer function must have $a_0 = 0$ and $a_n \neq 0$.
- A band-pass filter, which allows signals in a band of frequencies to pass but filters signals outside the band, should have $V(0) = 0$ and $\lim_{\omega \to \infty} V(\omega) = 0$; thus its transfer function must have $a_0 = 0$ and $a_n = 0$.
- A band-reject filter, which filters signals in a bandwidth and allows signals outside the bandwidth to pass, should have $V(0) = C$ and $\lim_{\omega \to \infty} V(\omega) = C$; thus its transfer function must have $a_0 \neq 0$ and $a_n \neq 0$.

The numerator of a circuit's transfer function defines its potential use a filter. Properties of the filter such as bandwidth are determined by the location of the poles of the transfer function. The values of component parameters may be adjusted to place poles at the desired locations.

---

## Example 7.13

Consider the circuit of Figure 7.31. (a) Derive the transfer function $G(s) = V_2(s)/V_1(s)$. (b) State the potential use of the circuit as a filter. (c) Determine the poles of the transfer function. (d) Draw the Bode plots and frequency response curves for this circuit.

### Solution

(a) The application of KCL at nodes $A$ and $B$ and KVL around each of the loops, using the currents illustrated in Figure 7.31, leads to

$$v_1(t) - i_1 - \sqrt{2}\frac{di_1}{dt} - \sqrt{2}\int_0^t i_1 dt - \frac{1}{\sqrt{2}}\frac{d}{dt}(i_1 - i_2) = 0 \tag{a}$$

$$-\frac{1}{\sqrt{2}}\int_0^t (i_2 - i_3)dt + \frac{1}{\sqrt{2}}\frac{d}{dt}(i_1 - i_2) = 0 \tag{b}$$

$$-i_3 + \frac{1}{\sqrt{2}}\int_0^t (i_2 - i_3)dt = 0 \tag{c}$$

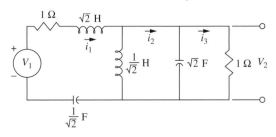

**FIG. 7.31** The circuit of Example 7.13 is an example of a band-bass filter.

Transforming Equations (a)–(c) into the Laplace domain leads to

$$\left(1 + \sqrt{2}s + \frac{1}{\sqrt{2}}s + \frac{\sqrt{2}}{s}\right)I_1(s) - \frac{1}{\sqrt{2}}sI_2(s) = V_1(s) \tag{d}$$

$$-\frac{1}{\sqrt{2}}sI_1(s) + \left(\frac{1}{\sqrt{2}}s + \frac{1}{\sqrt{2}s}\right)I_2(s) - \frac{1}{\sqrt{2}s}I_3(s) = 0 \tag{e}$$

$$-\frac{1}{\sqrt{2}s}I_2(s) + \left(1 + \frac{1}{\sqrt{2}s}\right)I_3(s) = 0 \tag{f}$$

Equations (d)–(f) are solved using Cramer's rule leading to

$$I_3(s) = \frac{\begin{vmatrix} 1 + \sqrt{2}s + \frac{1}{\sqrt{2}}s + \frac{\sqrt{2}}{s} & -\frac{1}{\sqrt{2}}s & V_1(s) \\ -\frac{1}{\sqrt{2}}s & \frac{1}{\sqrt{2}}s + \frac{1}{\sqrt{2}s} & 0 \\ 0 & -\frac{1}{\sqrt{2}s} & 0 \end{vmatrix}}{\begin{vmatrix} 1 + \sqrt{2}s + \frac{1}{\sqrt{2}}s + \frac{\sqrt{2}}{s} & -\frac{1}{\sqrt{2}}s & 0 \\ -\frac{1}{\sqrt{2}}s & \frac{1}{\sqrt{2}}s + \frac{1}{\sqrt{2}s} & -\frac{1}{\sqrt{2}s} \\ 0 & -\frac{1}{\sqrt{2}s} & 1 + \frac{1}{\sqrt{2}s} \end{vmatrix}}$$

$$= \frac{0.5s^2}{s^4 + \sqrt{2}s^3 + 3s^2 + \sqrt{2}s + 1} V_1(s) \tag{g}$$

Noting that $V_2 = (1)I_3$ leads to

$$G(s) = \frac{0.5s^2}{s^4 + \sqrt{2}s^3 + 3s^2 + \sqrt{2}s + 1} \tag{h}$$

(b) The filter is fourth order. Since the numerator has both $a_4 = 0$ and $a_0 = 0$ the filter has a potential application as a band-pass filter.

(c) The **MATLAB** work session used to determine the poles of $G(s)$ and to draw the Bode plot for the system is illustrated in Figure 7.32.

(d) The sinusoidal transfer function is determined from Equation (h) as

$$G(j\omega) = \frac{-0.5\omega^2(1 - 3\omega^2 + \omega^4) + j0.5\sqrt{2}\omega^2(\omega - \omega^3)}{(1 - 3\omega^2 + \omega^4)^2 + 2(\omega - \omega^3)^2} \tag{i}$$

The steady-state amplitude is determined using Equation (i) as

$$V_1|G(j\omega)| = \frac{V_1\omega^2}{\sqrt{((1 - 3\omega^2 + \omega^4)^2 + 2(\omega - \omega^3)^2}} \tag{j}$$

```
>> N=[0.5 0 0];
>> D=[1 2^0.5 3 2^0.5 1]

D =

   1.0000   1.4142   3.0000   1.4142   1.0000

>> N=[0.5 0 0];
>> D=[1 2^0.5 3 2^0.5 1];
>> [r,p,k]=residue(N,D)

r =

  0.0536 - 0.2454i
  0.0536 + 0.2454i
 -0.0536 + 0.1082i
 -0.0536 - 0.1082i

p =

 -0.4776 + 1.3612i
 -0.4776 - 1.3612i
 -0.2295 + 0.6541i
 -0.2295 - 0.6541i

k =

  []

>> bode(N,D)
>>
```

(a)

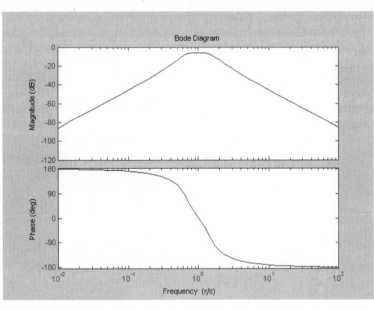

(b)

**FIG. 7.32** (a) The MATLAB work session used to determine the poles of the transfer function and develop the Bode plot of the filter of Example 7.13; (b) the Bode diagram shows that the circuit may be used as a band-pass filter.

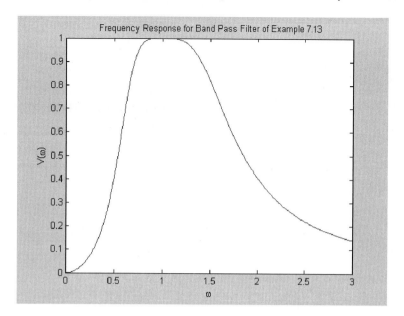

**FIG. 7.33** The flat nature of the frequency response curve near $\omega = 1$ illustrates the use of the circuit of Example 7.13 as a band-pass filter.

The frequency response is illustrated in Figure 7.33. Both the Bode diagram and the frequency response curves show the nature of this band-pass filter.

## 7.7 RESPONSE DUE TO PERIODIC INPUT

A system input $F(t)$ is periodic if there exists a value of $T$, called the period, such that $F(t + T) = F(t)$ for all $t$. The fundamental frequency of a periodic input is defined as

$$\omega_1 = \frac{2\pi}{T} \tag{7.98}$$

Examples of periodic inputs are given in Figure 7.34.

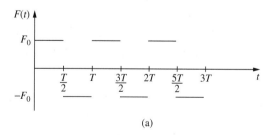

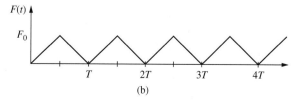

**FIG. 7.34** Example of periodic functions: (a) square wave; (b) triangular wave.

A piecewise continuous periodic function has a **Fourier series** representation given by

$$F(t) = \frac{a_0}{2} + \sum_{k=1}^{\infty} [a_k \cos(\omega_k t) + b_k \sin(\omega_k t)] \qquad (7.99)$$

where the **Fourier coefficients** are

$$a_k = \frac{2}{T} \int_0^T F(t) \cos(\omega_k t) dt \qquad k = 0, 1, 2, \ldots \qquad (7.100)$$

$$b_k = \frac{2}{T} \int_0^T F(t) \sin(\omega_k t) dt \qquad k = 1, 2, \ldots \qquad (7.101)$$

and the harmonic frequencies are

$$\omega_k = k\omega_1 = \frac{2\pi k}{T} \qquad (7.102)$$

The Fourier series representation of Equation 7.99 converges point-wise to $F(t)$ at all values of $t$ where $F$ is continuous. If $F(t)$ has a jump discontinuity at $t^*$ the Fourier series converges to $\frac{1}{2}[F(t^{*+}) + F(t^{*-})]$. Figure 7.35 illustrates the convergence of the Fourier series representation of the square wave input of Figure 7.34(a).

The Laplace transform of a periodic function can be obtained using its Fourier series representation

$$\mathcal{L}\{F(t)\} = \frac{1}{2}\mathcal{L}\{a_0\} + \mathcal{L}\left\{ \sum_{k=1}^{\infty} [a_k \cos(\omega_k t) + b_k \sin(\omega_k t)] \right\} \qquad (7.103)$$

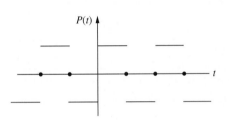

$P(t)$

$t$

**FIG. 7.35** The convergence of Fourier series representation for a square wave.

Since the Fourier series converges, the property of linearity of the transform may be applied leading to

$$F(s) = \frac{a_0}{2s} + \sum_{k=1}^{\infty} \left( a_k \frac{s}{s^2 + \omega_k^2} + b_k \frac{\omega_k}{s^2 + \omega_k^2} \right) \qquad (7.104)$$

If a periodic input is applied to a system whose transfer function is $G(s)$ and whose output is $x(t)$, then

$$X(s) = \left[ \frac{a_0}{2s} + \sum_{k=1}^{\infty} \left( a_k \frac{s}{s^2 + \omega_k^2} + b_k \frac{\omega_k}{s^2 + \omega_k^2} \right) \right] G(s) \qquad (7.105)$$

A procedure similar to that used in Section 7.2 to derive the steady-state response due to a sinusoidal input is applied to derive the steady-state response of the system as

$$x(t) = \frac{a_0}{2} x_s(t) + \sum_{k=1}^{\infty} |G(j\omega_k)|[a_k \cos(\omega t + \phi_k) + b_k \sin(\omega t + \phi_k)]$$

$$(7.106)$$

where

$$\phi_k = \tan^{-1}\left\{\frac{\text{Im}[G(j\omega_k)]}{\text{Re}[G(j\omega_k)]}\right\} \tag{7.107}$$

and $x_u(t)$ is the steady-state response of the system due to a unit step input.

## Example 7.14

The voltage input to the fourth-order band-pass filter of Example 7.13 is the periodic waveform of Figure 7.36. Determine the output voltage from the filter.

### Solution

The voltage signal of Figure 7.36 is periodic of period $T = 6\pi$ s. The definition of the voltage over one period is

$$v_1(t) = \begin{cases} 200 \text{ V} & 0 < t < 3\pi \text{ s} \\ 0 & 3\pi \text{ s} < t < 6\pi \text{ s} \end{cases} \tag{a}$$

The fundamental frequency of the voltage is

$$\omega_1 = \frac{2\pi}{T} = \frac{1}{3} \tag{b}$$

The Fourier coefficients are calculated using Equations (7.100) and (7.101) as

$$a_0 = \frac{2}{6\pi}\int_0^{6\pi} F(t)\,dt$$

$$= \frac{2}{6\pi}\int_0^{3\pi} (200)\,dt = 200 \tag{c}$$

$$a_k = \frac{2}{6\pi}\int_0^{3\pi} 200\cos\frac{k}{3}t\,dt$$

$$= \frac{200}{k\pi}\sin\frac{k}{3}t\Big|_{t=0}^{t=3\pi} = 0 \tag{d}$$

$$b_k = \frac{2}{6\pi}\int_0^{3\pi} 200\sin\frac{k}{3}t\,dt$$

$$= -\frac{200}{k\pi}\cos\frac{k}{3}t\Big|_{t=0}^{t=3\pi}$$

$$= \frac{200}{k\pi}(1 - \cos k\pi)$$

$$= \begin{cases} 0 & k = 2,4,6,\ldots \\ \dfrac{400}{k\pi} & k = 1,3,5,\ldots \end{cases} \tag{e}$$

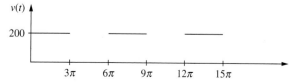

**FIG. 7.36** The periodic input for Example 7.14.

Thus the Fourier series representation for $F(t)$ is obtained by using Equations (c)–(e) in Equation (7.99) leading to

$$F(t) = 100 + \sum_{k=1,2,3}^{\infty} \frac{400}{k\pi} \sin\left(\frac{k}{3}t\right) \tag{f}$$

The sinusoidal transfer function obtained in Equation (j) of Example 7.13 leads to

$$|G(j\omega_k)| = \frac{\omega_k^2}{\sqrt{\left(1 - 3\omega_k^2 + \omega_k^4\right)^2 + 2\left(\omega_k - \omega_k^3\right)^2}} \tag{g}$$

$$\phi_k = \tan^{-1}\left(\frac{\sqrt{2}\left(\omega_k - \omega_k^3\right)}{\left(1 - 3\omega_k^2 + \omega_k^4\right)}\right) \tag{h}$$

The final value theorem is used to determine the steady-state response of the system due to a unit step input

$$v_u = \lim_{s \to 0} G(s)$$

$$= \lim_{s \to 0} \frac{0.5s^2}{s^4 + \sqrt{2}s^3 + 3s^2 + \sqrt{2}s + 1} = 0 \tag{i}$$

The steady-state response is obtained using Equation (7.105) as

$$v_2(t) = \frac{400}{\pi} \sum_{k=1,3,5}^{\infty} \frac{|G(j\omega_k)|}{k} \sin\left(\frac{k}{3}t + \phi\right) \tag{j}$$

Plots of the Fourier series representation for $F(t)$, evaluated using an upper limit on the summation of 50 terms and of the steady-state response, both obtained using MATLAB, are shown in Figure 7.37.

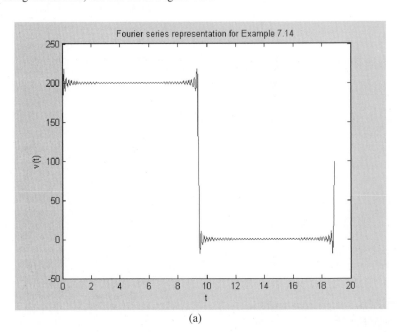

**FIG. 7.37** (a) The Fourier series representation of the voltage input of Figure 7.36; (b) the steady-state voltage across the terminals of the filter for the periodic input of Figure 7.36.

(a)

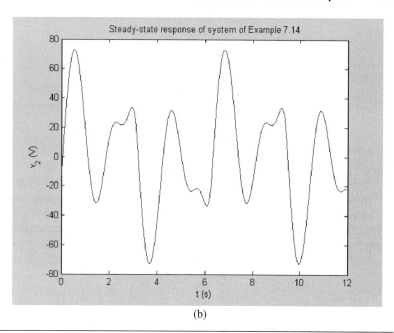

Steady-state response of system of Example 7.14

**FIG. 7.37** (*Continued*)

(b)

## 7.8  FURTHER EXAMPLES

Aspects of frequency response for several examples considered in Chapters 2–6 are considered in the following problems. MATLAB is used to draw Bode diagrams and Nyquist diagrams as well as to aid in calculations.

### Example 7.15

An *LR* circuit with $L = 0.25$ H and $R = 1$ kΩ has a voltage source $v(t) = 100 \sin(1000t)$ V. Determine its steady-state response.

**Solution**

The time constant for the first-order circuit is $T = L/R = 0.25$ H/1,000 Ω = $2.5 \times 10^{-4}$ s. The steady-state amplitude is determined using Equation (7.61) as

$$I = \frac{V_0}{RT}\sqrt{\frac{1}{\omega^2 + \dfrac{1}{T^2}}}$$

$$= \frac{100 \text{ V}}{(1{,}000 \ \Omega)(2.5 \times 10^{-4}) \text{ s}}\sqrt{\frac{1}{(1{,}000 \text{ r/s})^2 + \dfrac{1}{\left(2.5 \times 10^{-4} \text{ s}\right)^2}}}$$

$$= 9.70 \text{ mA} \tag{a}$$

The phase is determined using Equation (7.62) as

$$\phi = -\tan^{-1}(\omega T) = -\tan^{-1}\left[(1,000 \text{ r/s})(2.5 \times 10^{-4} \text{ s})\right] = -0.245 \text{ rad} \qquad \text{(b)}$$

Thus the steady-state response is

$$i(t) = 9.70 \sin(1,000t - 0.245) \text{ mA} \qquad \text{(c)}$$

---

## Example 7.16

Consider the thermal system of Example 4.14. (a) Determine the maximum and minimum temperatures in the room during the day. (b) At what time does the temperature in the room reach its maximum if the outside temperature is maximum at 3 p.m.?

### Solution

Defining $\theta = T - 80$ F in Equation (j) of Example 4.14 and rearranging leads to

$$2.86\frac{d\theta}{dt} + \theta = 10\sin\left(\frac{\pi t}{12}\right) \qquad \text{(a)}$$

where $t$ is measured in hours. The time constant for this first-order system is $T = 2.86$ hr and the input frequency is $\pi/12$ r/s. The nondimensional frequency is $\beta = \omega T = 0.748$. The steady-state response is of the form $\theta(t) = \Theta \sin[(\pi/12)t + \phi]$, where $\Theta = (10 \text{ F})\sqrt{1/(1+\beta^2)} = 8.01$ F and $\phi = -\tan(0.748) = -0.642$ rad. Thus the steady-state response for the room temperature is

$$\theta(t) = 8.01\sin\left(\frac{\pi}{12}t - 0.642\right)\text{F} \qquad \text{(b)}$$

(a) The maximum and minimum room temperatures are $T_{\max} = 80 \text{ F} + 8.01 \text{ F} = 88.0$ F and $T_{\min} = 80 \text{ F} - 8.01 \text{ F} = 72.0$ F.
(b) Since the maximum temperature occurs at 3 p.m. this must correspond to a value of $t$ such that $\pi t/12 = \pi/2$ or $t = 6$ hr. The maximum room temperature occurs when $(\pi/12)t - 0.642 = \pi/2$ or $t = 8.45$ hr. Thus the maximum temperature in the room occurs 2.45 hr after 3 p.m. or at 5:27 p.m. This is obviously not a very accurate result. However, the model assumes all properties are lumped and the temperature in the room is uniform. A more accurate model considers a variation in temperature across the room at any instant.

---

## Example 7.17

In Example 7.16 it is shown that the maximum temperature in the room of Example 4.14 is 88.0 F. What is the minimum value of $R_{eq}$ such that the maximum temperature in the room is 85 F assuming all other parameters are as given in Example 4.14?

**Solution**

In terms of the notation of Example 7.16, for the maximum temperature of the room to be at most 85 F, the steady-state amplitude must be limited to 5 F, $\Theta \leq 5$ F, which leads to

$$5 \text{ F} \leq (10 \text{ F}) \sqrt{\frac{1}{1+\beta^2}} \tag{a}$$

Equation (a) is rearranged to

$$\beta \geq \sqrt{3} \tag{b}$$

Equation (j) of Example 4.14 can be written in terms of the equivalent resistance as

$$99.2 R_{eq} \frac{d\theta}{dt} + \theta = 10 \sin\left(\frac{\pi t}{12}\right) \tag{c}$$

The time constant and the input frequency are determined from Equation (c) as $T = 99.2 R_{eq}$ hr and $\omega = \pi/12$ r/hr respectively. Note that $\beta = \omega T = 99.2 \pi R_{eq}/12 = 25.8 R_{eq}$.

The substitution of this result into Equation (b) gives

$$25.8 R_{eq} \geq \sqrt{3}$$

$$R_{eq} \geq 6.71 \times 10^2 \text{ hr} \cdot \text{F/Btu} \tag{d}$$

---

**Example 7.18**

Consider a mass-spring-viscous damper system designed such that the system is critically damped with a natural frequency of 1 r/s. (a) Qualitatively discuss the Bode diagram. Determine its low-frequency and high-frequency asymptotes. Discuss the behavior of the phase. (b) Determine the features of the system's Nyquist diagram. Determine its low- and high-frequency limits, and its intercepts. (c) Use MATLAB to plot the Bode diagram and the Nyquist diagrams.

**Solution**

The transfer function for a critically damped mechanical system with a natural frequency of 1 r/s is

$$G(s) = \frac{1}{(s+1)^2} \tag{a}$$

The sinusoidal transfer function for the system is obtained as

$$G(j\omega) = \frac{1 - \omega^2 - j2\omega}{(\omega^2 + 1)^2} \tag{b}$$

The real and imaginary parts of the transfer function and its magnitude are

$$\text{Re}[G(j\omega)] = \frac{1 - \omega^2}{(\omega^2 + 1)^2} \tag{c}$$

$$\text{Im}[G(j\omega)] = \frac{-2\omega}{(\omega^2 + 1)^2} \tag{d}$$

$$|G(j\omega)| = \frac{1}{1 + \omega^2} \tag{e}$$

The phase for the system is

$$\phi = -\tan^{-1}\left(\frac{-2\omega}{1 - \omega^2}\right) \tag{f}$$

(a) The Bode diagram is a plot of

$$L(\omega) = 20\log|G(j\omega)| = 20\log\frac{1}{1 + \omega^2} = -20\log\left(1 + \omega^2\right) \tag{g}$$

versus $\log \omega$. The low-frequency asymptote is obtained as

$$L(0) = -20\log(1) = 0 \text{ dB} \tag{h}$$

The high-frequency asymptote is obtained as

$$\lim_{\omega \to \infty} L(\omega) = \lim_{\omega \to \infty} -20\log(100 + \omega^2) = -20\log\omega^2 = -40\log\omega \tag{i}$$

Thus the low-frequency asymptote is a horizontal line $L(\omega) = 0$ dB while the high-frequency asymptote is a line of slope $-40$ dB.

Form Equation (f) it is noted that the phase is zero for $\omega = 0$ and decreases with increasing $\omega$. When $\omega = 1$ r/s, $\phi = -\pi/2$. For large $\omega$, $\tan\phi \to 0$, but both the numerator and denominator are negative, thus $\phi = -\pi$ for large $\omega$.

(b) The Nyquist diagram is a plot of $\text{Re}|G(j\omega)|$, given by Equation (c) on the horizontal axis vs. $\text{Im}|G(j\omega)|$, given by Equation (d) on the vertical axis. The diagram is plotted with $\omega$ as a parameter for $-\infty < \omega < \infty$. Equations (c) and (d) leads to the following:

- The Nyquist diagram intercepts the horizontal axis only for $\omega = 0$ when $\text{Re}[G(j\omega)] = 1$.
- The Nyquist diagram intercepts the vertical axis only for $\omega = \pm 1$. When $\omega = 1, \text{Im}[G(j\omega)] = -0.5$. When $\omega = -1 \; \text{Im}[G(j\omega)] = 0.5$.
- As $\omega \to \pm\infty$, $\text{Re}|G(j\omega)| \to 0$ and $\text{Im}|G(j\omega)| \to 0$.
- As $\omega$ increases from 0, Equation (d) shows that $\text{Im}|G(j\omega)|$ grows negatively while $\text{Re}|G(j\omega)|$ decreases from 1. This indicates that the Nyquist plot is traversed clockwise as $\omega$ increases.

(c) The MATLAB-generated Bode diagrams and Nyquist plot for this system are illustrated in Figures 7.38 and 7.39.

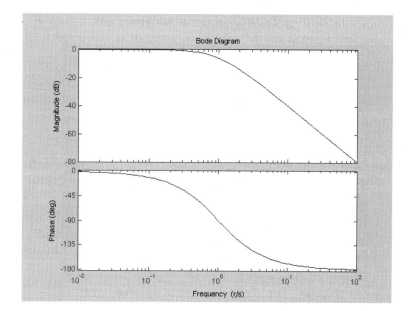

**FIG. 7.38** The Bode diagram for the system of Example 7.18. The low-frequency asymptote of 0 dB and the high-frequency asymptote of $-40 \log \omega$ are clearly illustrated.

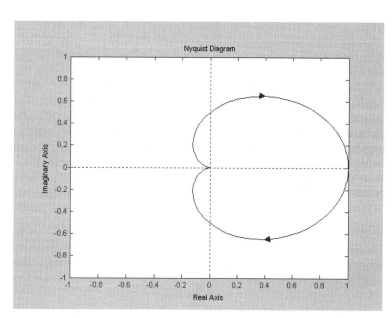

**FIG. 7.39** The Nyquist diagram for the system of Example 7.18.

**Example 7.19**

The centrifuge of Example 2.26 has parameters $m = 99.5$ kg, $k = 1 \times 10^6$ N/m, $c = 2000$ N·s/m, $m_0 = 0.5$ kg, $e = 0.1$ m. (a) Determine the steady-state response for $\omega = 200$ r/s. (b) Determine the frequency response for the system.

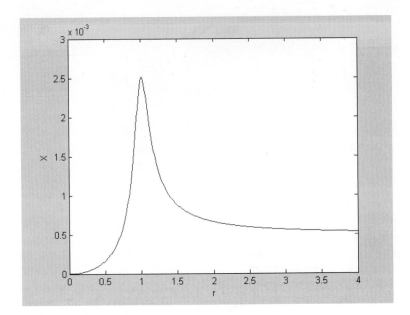

**FIG. 7.40** The frequency response of the centrifuge of Example 7.19.

**Solution**

The substitution of the given values into Equation (e) of Example 2.25 leads to

$$100\ddot{y} + 2{,}000\dot{y} + 1 \times 10^6 y = 0.05\omega^2 \sin(\omega t) \tag{a}$$

The natural frequency and damping ratio for the system are obtained as $\omega_n = \sqrt{1 \times 10^6 / 100} = 100$ r/s and $\zeta = (2000 \text{ N} \cdot \text{s/m})/2(100 \text{ kg})(100 \text{ r/s}) = 0.1$. The steady-state amplitude and phase for the system for this system are obtained using Equations (7.70) and (7.71) with $F_0 = 0.05\omega^2$. The application of Equation (7.70) leads to

$$X = \frac{0.05\omega^2}{100\omega_n^2} \frac{1}{\sqrt{(1 - r^2)^2 + (0.2r)^2}} \tag{b}$$

Noting that $r = \omega/\omega_n$, Equation (b) is rewritten as

$$X = \frac{5 \times 10^{-4} r^2}{\sqrt{(1 - r^2)^2 + (0.2r)^2}} \tag{c}$$

Equation (c) is graphed in Figure 7.40.

**Example 7.20**

For the circuit of Figure 7.41, (a) determine the steady-state response for $v_2(t)$ when $v_1(t) > 100 \sin(10t)$ V; (b) draw the Bode plot for the system and comment on the use of this circuit as a filter; and (c) qualitatively discuss and then plot the Nyquist diagram for this system.

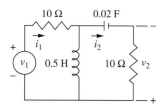

**FIG. 7.41**  The circuit of Example 7.20.

**Solution**

Let $i_1(t)$ and $i_2(t)$ be the nodal currents. The application of KVL to each loop leads to

$$0.5\frac{di_1}{dt} - 0.5\frac{di_2}{dt} + 10i_1 = v_1(t) \tag{a}$$

$$-0.5\frac{di_1}{dt} + 0.5\frac{di_2}{dt} + 10i_2 + 50\int_0^t i_2 dt = 0 \tag{b}$$

Taking the Laplace transforms of Equations (a) and (b) leads to

$$(0.5s + 10)I_1(s) - 0.5sI_2(s) = V_1(s) \tag{c}$$

$$-0.5sI_1(s) + \left(0.5s + 10 + \frac{50}{s}\right)I_2(s) = 0 \tag{d}$$

Cramers' rule is used to solve Equations (c) and (d) for $I_2(s)$ leading to

$$I_2(s) = \frac{0.05s^2 V_1(s)}{s^2 + 2.5s + 50} \tag{e}$$

Noting that $v_2 = 10i_2$ the transfer function $G(s) = V_2(s)/V_1(s)$ is determined as

$$G(s) = \frac{0.5s^2}{s^2 + 2.5s + 50} \tag{f}$$

The sinusoidal transfer function for this circuit is

$$G(j\omega) = \frac{0.5(j\omega)^2}{(j\omega)^2 + 2.5(j\omega) + 50}$$

$$= \frac{-0.5\omega^2}{(50 - \omega^2) + 2.5j\omega} \tag{g}$$

For $\omega = 10$ Equation (g) becomes

$$G(j10) = \frac{-50}{-50 + 25j}$$

$$= \frac{-50(-50 - 25j)}{3125}$$

$$= 0.8 + 0.4j \tag{h}$$

It is apparent from Equation (h) that $|G(j10)| = 0.894$ and $\phi = \tan^{-1}(0.4/0.8) = 0.464$ r. Thus the steady-state response is

$$v_2(t) = 89.4 \sin(10t + 0.464) \text{ V} \tag{i}$$

(b) Equation (g) is used to show that

$$\text{Re}[G(j\omega)] = \frac{-0.5\omega^2(50 - \omega^2)}{(50 - \omega^2)^2 + (2.5\omega)^2} \tag{j}$$

$$\text{Im}[G(j\omega)] = \frac{-1.25\omega^3}{(50 - \omega^2)^2 + (2.5\omega)^2} \tag{k}$$

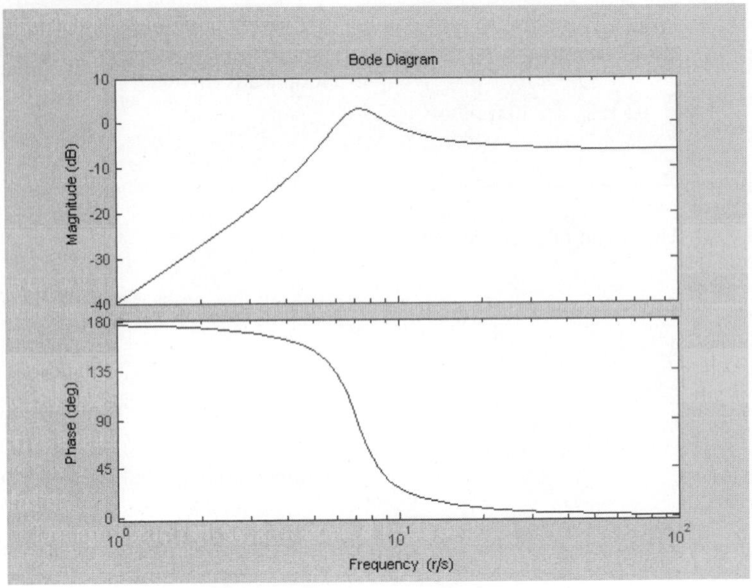

FIG. **7.42** The Bode diagram for the circuit of Example 7.20. This diagram shows that the circuit can be used as a high-pass filter.

The magnitude and phase of the sinusoidal transfer function are

$$|G(j\omega)| = \frac{0.5\omega^2}{\sqrt{(50 - \omega^2)^2 + (2.5\omega)^2}} \tag{1}$$

$$\phi = \tan^{-1}\left(\frac{2.5\omega}{50 - \omega^2}\right) \tag{m}$$

Equation (l) is used to obtain

$$L(\omega) = 20\log\left[\frac{0.5\omega^2}{\sqrt{(50 - \omega^2)^2 + (2.5\omega)^2}}\right] \tag{n}$$

The low-frequency asymptote is $20\log(0.5\omega^2/50) = 20\log(0.01\omega^2) = -40 + 40\log(\omega)$.

The high frequency asymptote on the Bode diagram is $20\log(0.5) = -6.02$. The MATLAB-generated Bode diagram for the circuit, shown in Figure 7.42, confirms these asymptotes.

The denominator of the transfer function is second order. Since the numerator is also second order, but with the coefficient of its constant term equal to zero, the circuit could be used as a high-pass filter. This is confirmed by the Bode diagram.

(c) In drawing the Nyquist plot it is noted that $\text{Re}[G(j\omega)] = 0$ for $\omega = 0$, $\pm\sqrt{50}$. Thus the Nyquist plot intercepts the imaginary axis at $\text{Im}[G(0)] = 0$, $\text{Im}[G(j\sqrt{50})] = -1.41$, and $\text{Im}[G(-j\sqrt{50})] = 1.41$. It is further noted that $\text{Im}[G(j\omega)] = 0$ only for $\omega = 0$ and as $\omega \to \pm\infty$. Thus the Nyquist plot intercepts the real axis only at $\text{Re}[G(j\omega)] = 0$ and $\text{Re}[\lim_{\omega \to \pm\infty} G(j\omega)] = 0.5$. The MATLAB-generated Nyqusit diagram of Figure 7.43 confirms these points.

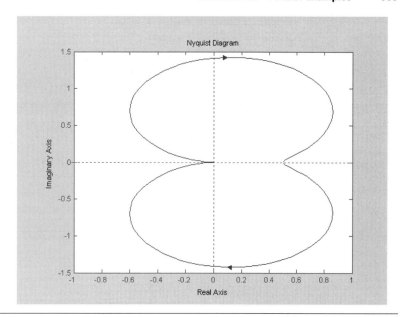

**FIG. 7.43** The Nyquist diagram for the circuit of Example 7.20.

**Example 7.21**

Determine the steady-state response of the system of Figure 7.44.

**Solution**

The differential equation governing the motion of the system is

$$40\ddot{x} + 960\dot{x} + 1.44 \times 10^5 x = 960y(t) \tag{a}$$

where $y(t)$ is the prescribed displacement of the base to which the viscous damper is attached. The transfer function $G(s) = X(s)/Y(s)$ is obtained from Equation (a) as

$$G(s) = \frac{24s}{s^2 + 24s + 3{,}600} \tag{b}$$

The input to the system has an amplitude of 0.002 m and a frequency of 80 r/s. The steady-state response is obtained from the sinusoidal transfer function

$$
\begin{aligned}
G(80j) &= \frac{24(80j)}{(80j)^2 + 24(60j) + 3{,}600} \\
&= \frac{1{,}920j}{-2{,}800 + 1{,}920j} \cdot \frac{1{,}920j(-2{,}800 - 1{,}920j)}{-2{,}800 + 1{,}920j} \\
&= \frac{2.68 \times 10^6 - 5.38 \times 10^6}{1.15 \times 10^7} = 0.233 - 0.468j
\end{aligned} \tag{c}
$$

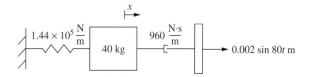

**FIG. 7.44** The system of Example 7.21.

The magnitude and phase of $G(80j)$ are $|G(80j)| = \sqrt{(0.233)^2 + (-0.468)^2} = 0.523$ and $\phi = \tan^{-1}(-0.468/0.233) = -1.11$ rad. Thus the steady-state response due to the sinusoidal base motion is

$$x(t) = (0.002)(0.523)\sin(80t - 1.11)$$
$$= 1.05 \times 10^{-3} \sin(80t - 1.11)\,\text{m} \qquad \text{(d)}$$

## Example 7.22

Show that when $\sqrt{k_2/m_2} = \omega$ the steady-state amplitude of the middle of the bar $X_1$ of the system in Figure 7.45 is zero; thus the auxiliary mass-spring system acts as a vibration absorber. Determine the steady-state amplitudes of the angular rotation of the bar $\Theta$ and the absorber $X_2$. Assume $m_2 = 5$ kg in all calculations.

### Solution

The free-body diagrams of the three-degree-of-freedom system at an arbitrary instant, assuming small $\theta$, are shown in Figure 7.46. The small angle assumption also implies that gravity cancels with static spring forces and thus neither are shown on the free-body diagrams. The application of conservation laws to the free-body diagrams leads to

$$\begin{bmatrix} m_1 & 0 & 0 \\ 0 & I & 0 \\ 0 & 0 & m_2 \end{bmatrix} \begin{bmatrix} \ddot{x}_1 \\ \ddot{\theta} \\ \ddot{x}_2 \end{bmatrix} + \begin{bmatrix} 2k_1 & 0 & -k_2 \\ 0 & 2k_1\ell^2 & 0 \\ -k_2 & 0 & k_2 \end{bmatrix} \begin{bmatrix} x_1 \\ \theta \\ x_2 \end{bmatrix} = \begin{bmatrix} F(t) \\ F(t)\ell \\ 0 \end{bmatrix} \qquad \text{(a)}$$

Taking the Laplace transform of Equation (a), substituting in given numerical values and assuming $m_2 = 5$ kg and $k_2 = (5\,\text{kg})(60\,\text{r/s})^2 = 1.8 \times 10^4$ N/m leads to

$$\begin{bmatrix} 50s^2 + 2 \times 10^5 & 0 & -1.8 \times 10^4 \\ 0 & 2.8s^2 + 7.2 \times 10^4 & 0 \\ -1.8 \times 10^4 & 0 & 5s^2 + 1.8 \times 10^4 \end{bmatrix} \begin{bmatrix} X_1 \\ \Theta \\ X_2 \end{bmatrix} = \begin{bmatrix} 1 \\ 0.3 \\ 0 \end{bmatrix} F(s)$$

$$\text{(b)}$$

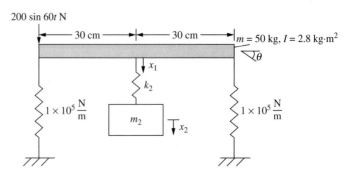

**FIG. 7.45** The system of Example 7.22.

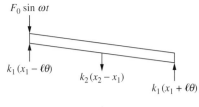

$F_0 \sin \omega t$

$k_1(x_1 - \ell\theta)$

$k_2(x_2 - x_1)$

$k_1(x_1 + \ell\theta)$

$k_2(x_2 - x_1)$

**FIG. 7.46** The free-body diagrams of a primary system and auxilary mass at an arbitrary instant.

Cramer's rule is used to determine the system's transfer functions as

$$G_1(s) = \frac{X_1(s)}{F(s)} = \frac{5s^2 + 1.8 \times 10^4}{250s^4 + 1.90 \times 10^6 s^2 + 3.28 \times 10^9} \tag{c}$$

$$G_2(s) = \frac{\Theta(s)}{F(s)} = \frac{0.3}{2.8s^2 + 7.2 \times 10^4} \tag{d}$$

$$G_3(s) = \frac{X_2(s)}{F(s)} = \frac{1.8 \times 10^4}{250s^4 + 1.90 \times 10^6 s^2 + 3.28 \times 10^9} \tag{e}$$

The steady-state amplitudes are obtained from the sinusoidal transfer functions as

$$X_1 = 200|G(60j)| = \left| \frac{200[5(60j)^2 + 1.8 \times 10^4]}{250(60j)^4 + 1.96 \times 10^6(60j)^2 + 3.28 \times 10^9} \right| = 0 \tag{f}$$

$$\Theta = 200|G_2(60j)| = \left| \frac{200(.3)}{2.8(60j)^2 + 7.2 \times 10^4} \right| = 9.70 \times 10^{-4} \text{ rad} \tag{g}$$

$$X_2 = 200|G_3(60j)| = \left| \frac{200(1.8 \times 10^4)}{250(60j)^4 + 1.90 \times 10^6(60j)^2 + 3.28 \times 10^9} \right| = 11.3 \text{ mm} \tag{h}$$

---

## Example 7.23

Contrast the Bode plots and Nyquist diagrams for the concentration perturbation $C_{A2p}$ of the system of Example 6.29 for the case where there is no time delay and for the case where the system has a time delay of 10 s.

### Solution

The appropriate transfer function is determined from Equation (k) of Example 6.29, which is repeated below as

$$G(s) = \frac{4.43 \times 10^{-7} e^{-10s}}{(s + 3.97 \times 10^{-3})(s + 3.20 \times 10^{-3})} \tag{a}$$

The sinusoidal transfer function obtained for the system without the time delay is

$$G(j\omega) = \frac{4.43 \times 10^{-7}(1.27 \times 10^{-5} - \omega^2 - j7.17 \times 10^{-3}\,\omega)}{(1.27 \times 10^{-5} - \omega^2)^2 + (7.17 \times 10^{-3}\,\omega)^2} \tag{b}$$

The real and imaginary parts of the transfer function are

$$\text{Re}[G(j\omega)] = \frac{4.43 \times 10^{-7}(1.27 \times 10^{-5} - \omega^2)}{(1.27 \times 10^{-5} - \omega^2)^2 + (7.17 \times 10^{-3}\,\omega)^2} \tag{c}$$

$$\text{Im}[G(j\omega)] = \frac{3.18 \times 10^9\,\omega}{(1.27 \times 10^{-5} - \omega^2)^2 + (7.17 \times 10^{-3}\,\omega)^2} \tag{d}$$

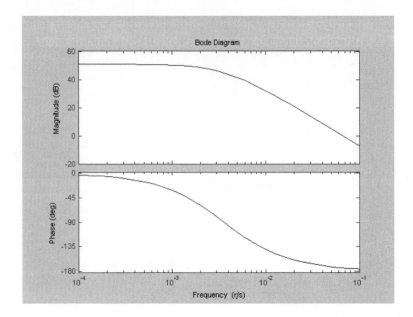

**FIG. 7.47** The Bode diagram for the system of Examples 6.29 and 7.23 without time delay.

The magnitude and phase of the sinusoidal transfer function are

$$|G(j\omega)| = \frac{4.43 \times 10^{-7}}{\sqrt{(1.27 \times 10^{-5} - \omega^2)^2 + (7.17 \times 10^{-3}\,\omega)^2}} \tag{e}$$

$$\phi = \tan^{-1}\left(\frac{-7.17 \times 10^{-3}\,\omega}{1.27 \times 10^{-5}\,\omega^2}\right) \tag{f}$$

The MATLAB-generated Bode diagram and Nyquist plots for the system without the time delay are given in Figures 7.47 and 7.48 respectively.

The sinusoidal transfer function for the system with the time delay is

$$G(j\omega) = \frac{4.43 \times 10^{-7} e^{-j10\omega}\left(1.27 \times 10^{-5} - \omega^2 - j7.17 \times 10^{-3}\,\omega\right)}{(1.27 \times 10^{-5} - \omega^2)^2 + (7.17 \times 10^{-3}\,\omega)^2} \tag{g}$$

Noting that $e^{-j10\omega} = \cos(10\omega) - j\sin(10\omega)$ Equation (g) can be rewritten as

$$\mathrm{Re}[G(j\omega)] = \frac{4.43 \times 10^{-7}\left[\cos(10\omega)(1.27 \times 10^{-5} - \omega^2) + 7.17 \times 10^{-3}\omega\sin(10\omega)\right]}{(1.27 \times 10^{-5} - \omega^2)^2 + (7.17 \times 10^{-3}\omega)^2}$$

$$\tag{h}$$

$$\mathrm{Im}[G(j\omega)] = \frac{4.43 \times 10^{-7}\left[7.17 \times 10^{-3}\omega\cos(10\omega) + (1.27 \times 10^{-5} - \omega^2)\sin(10\omega)\right]}{(1.27 \times 10^{-5} - \omega^2)^2 + (7.17 \times 10^{-3}\omega)^2}$$

$$\tag{i}$$

The magnitude of the sinusoidal transfer function is not affected by the time delay and is given by Equation (e). The phase is affected by the time delay and

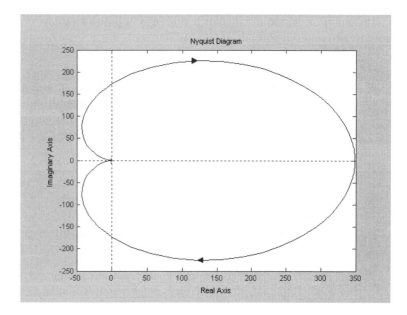

**FIG. 7.48** The Nyquist diagram for the system of Examples 6.29 and 7.23 without time delay.

is determined as

$$\phi = \tan^{-1}\left(\frac{7.17 \times 10^{-3}\omega \cos\left(10\omega\right) + (1.27 \times 10^{-5} - \omega^2)\sin\left(10\omega\right)}{\cos\left(10\omega\right)(1.27 \times 10^{-5} - \omega^2) + 7.17 \times 10^{-3}\omega \sin\left(10\omega\right)}\right) \qquad \text{(j)}$$

Figure 7.49 illustrates the steady-state phase as a function of the frequency. Note that the denominator of the argument of the inverse tangent function of

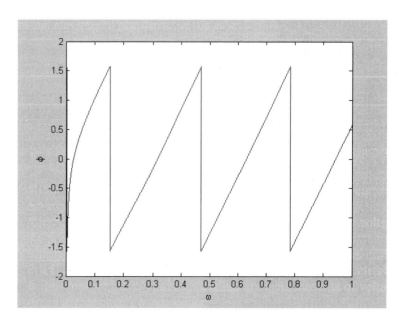

**FIG. 7.49** The phase portion of plot of the Bode diagram with time delay.

Equation (j) is zero when

$$\tan(10\omega) = -\frac{1.27 \times 10^{-5} - \omega^2}{7.17 \times 10^{-3}\omega} \tag{k}$$

Equation (k) has an infinite number of solutions, for which $\phi = \pi/2$.

The distance from the origin to a point on the Nyqusit diagram is the magnitude of the sinusoidal transfer function for the appropriate value of $\omega$. Since the magnitude of the sinusoidal transfer function is unaffected by the time delay, the Nyquist diagram is also unaffected.

## 7.9 SUMMARY

### 7.9.1 Chapter Highlights

- Chapter 7 completes the study of system response, studying the steady-state response of dynamic systems.
- The undamped mechanical system is a special case. When the input frequency coincides with the natural frequency, a resonance condition occurs. Resonance is characterized by a response that grows without bound.
- The sinusoidal transfer function is developed from the transfer function. The steady-state response is developed using the sinusoidal transfer function.
- The steady-state response is characterized by an amplitude and a phase.
- A system's frequency response describes how its amplitude and phase change with input frequency.
- Bode plots show the frequency response using logarithmic scales.
- Bode plots can be constructed for products of transfer functions by adding the individual Bode plots.
- Nyquist diagrams are polar plots of the sinusoidal transfer function.
- The frequency response of all first-order systems is the same in a nondimensional form.
- The frequency response of second-order and higher-order systems is highly dependent on the numerator of the transfer function.
- Frequency responses for series *LRC* circuits, second-order mechanical systems with force input, and second-order mechanical systems with motion input are considered.
- Electric circuits can be designed to be filters. The type of filter depends on the numerator of the transfer function.
- Vibration isolators are used to eliminate steady-state vibrations of a particle in a mechanical system.
- Periodic inputs have a Fourier series representation, which is used to develop the steady-state response.
- The steady-state response due to a general periodic input is an infinite series of harmonic terms whose frequencies are integer multiples of the fundamental frequency.

## 7.9.2 Important Equations

- General form of periodic input
$$F(t) = F_0 \sin{(\omega t + \psi)} \tag{7.1}$$

- General form of the steady-state response
$$x(t) = X \sin{(\omega t + \psi)} \tag{7.2}$$

- Resonance for undamped mechanical systems
$$x(t) = \frac{F_0}{2m\omega_n^2}[\sin{(\omega_n t)} - \omega_n t \cos{(\omega_n t)}] \tag{7.12}$$

- Steady-state response using the sinusoidal transfer function
$$x_s(t) = F_0|G(j\omega)| \sin{(\omega t + \phi)} \tag{7.31}$$

$$\phi = \tan^{-1}\left(\frac{\mathrm{Im}[G(j\omega)]}{\mathrm{Re}[G(j\omega)]}\right) \tag{7.27}$$

- Frequency response of a series $LRC$ circuit
$$M = \frac{L\omega_n I}{V_0} = \frac{r}{\sqrt{(1-r^2)^2 + (2\zeta r)^2}} \tag{7.35}$$

- Bode diagram
$$L(\omega) = 20 \log|G(j\omega)| \tag{7.38}$$

- Frequency response for first-order systems
$$M = \sqrt{\frac{1}{\beta^2 + 1}} \tag{7.63}$$

$$\phi = -\tan^{-1}(\beta) \tag{7.64}$$

- Frequency response for second-order mechanical systems with force input
$$\frac{m\omega_n^2 X}{F_0} = \frac{1}{\sqrt{(1-r^2)^2 + (2\zeta r)^2}} \tag{7.70}$$

$$\phi = \tan^{-1}\left(\frac{2\zeta r}{r^2 - 1}\right) \tag{7.71}$$

- Frequency response for mechanical systems with motion input
$$T = \sqrt{\frac{1 + (2\zeta r)^2}{(1-r^2) + (2\zeta r)^2}} \tag{7.82}$$

$$\omega_n^2 \Gamma = \omega_n^2 r^2 \sqrt{\frac{1 + (2\zeta r)^2}{(1-r^2) + (2\zeta r)^2}} \tag{7.83}$$

$$\Lambda = \frac{r^2}{\sqrt{(1-r^2)^2 + (2\zeta r)^2}} \tag{7.84}$$

- Fourier series representation of periodic function

$$F(t) = \frac{a_0}{2} + \sum_{k=1}^{\infty} \left[ a_k \cos\left(\omega_k t\right) + b_k \sin\left(\omega_k t\right) \right]$$    (7.99)

- Fourier coefficients

$$a_k = \frac{2}{T} \int_0^T F(t) \cos\left(\omega_k t\right) dt$$    (7.101a)

$$b_k = \frac{2}{T} \int_0^T F(t) \sin\left(\omega_k t\right) dt$$    (7.101b)

# PROBLEMS

**7.1** Determine, for the series *RL* circuit of Figure P7.1, (a) the sinusoidal transfer function when $\omega = 800$ rad/s, (b) the steady-state current through the circuit when $\omega = 800$ rad/s.

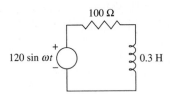

**FIG. P7.1**

**7.2** Determine, for the two-mesh circuit of Figure P7.2, (a) the sinusoidal transfer function $G(s) = I_m(s)/V_1(s)$ when the frequency of the sinusoidal voltage source is 2.1 kHz, (b) the steady-state current through the 10-Ω resistor when the frequency of the sinusoidal voltage source is 2.1 kHz.

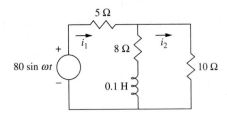

**FIG. P7.2**

**7.3** Determine the steady-state response of the spring-viscous damper system of Figure P7.3.

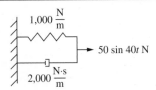

**FIG. P7.3**

**7.4** Determine the steady-state response of the system of Figure P7.4 when $\omega = 80$ rad/s.

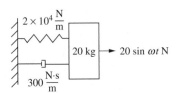

**FIG. P7.4**

**7.5** Determine the steady-state response for the absolute displacement of the system of Figure P7.5.

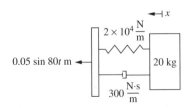

**FIG. P7.5**

**7.6** The thin disk of Example 2.17 is subject to a time-dependent moment of the form $M(t) = 20.5 \sin(200t)$ N·m. Determine the steady-state amplitude of angular oscillation of the disk using the following system parameters: $m_1 = 10$ kg, $m_2 = 15$ kg, $r = 20$ cm, $I = 0.1$ kg.m$^2$, $k_1 = k_2 = 1 \times 10^4$ N/m, and $c_1 = c_2 = 2.5 \times 10^3$ N·s/m.

**7.7** Determine the steady-state current through the inductor in the circuit of Figure P7.7.

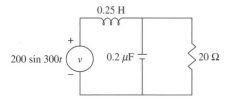

**FIG. P7.7**

**7.8** The two-tank system of Figure P.7.8 is in the steady-state shown when the exit pressure is 2.5 kPa. Determine the steady-state liquid levels in the tanks of the system of Figure P7.8 when the exit pressure varies sinusoidally according to $p(t) = 2.5 + 0.6 \sin(50t)$ kPa.

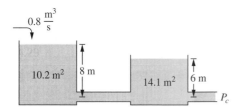

**FIG. P7.8**

**7.9** Determine the steady-state response for the acceleration of the mass in the system of Figure P7.9.

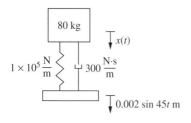

**FIG. P7.9**

**7.10** Determine the steady-state response for $x_1(t)$ for the system of Figure P7.10.

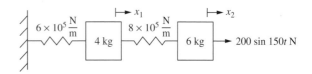

**FIG. P7.10**

**7.11** Draw a Bode diagram for a series *LRC* circuit with $\zeta = 1.2$ and $\omega_n = 500$ rad/s, and discuss its low-frequency and high-frequency asymptotes.

**7.12** Draw a Bode diagram for the system of Figure P7.4. Discuss the low-frequency and high-frequency asymptotes.

**7.13** Draw a Bode diagram for an undamped mechanical mass-spring system of natural frequency 100 rad/s.

**7.14** Draw a Bode diagram for the vehicle of Problem 6.30 when it traverses a road whose contour is sinusoidal, as illustrated in Figure P7.14. Note that the sinusoidal road contour provides a sinusoidal input to the suspension system in which the frequency is proportional to vehicle speed.

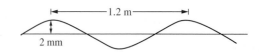

**FIG. P7.14**

**7.15** The transfer function for a dynamic system is

$$G(s) = \frac{s+3}{s^2 + 10s + 50} \qquad (a)$$

(a) Determine the low-frequency and high-frequency asymptotes for its Bode diagram. (b) Qualitatively discuss the system's Nyquist diagram by determining (i) the points corresponding to $\omega = 0$, $\pm\infty$, (ii) its axes intercepts, and (iii) the quadrants in which the Nyquist diagram exists. (c) Use MATLAB to draw the Bode plot and Nyquist diagram for the system.

**7.16** Repeat Problem 7.15 with a system transfer function of

$$G(s) = \frac{s^2 + 2s + 4}{s^3 + 5s^2 + 6s + 10} \qquad (a)$$

**7.17** Draw a Bode diagram for the time delay system whose transfer function is

$$G(s) = \frac{(2s+3)e^{-3s}}{s(s+6)(s+3)(s+2)} \quad \text{(a)}$$

Determine the low- and high-frequency asymptotes.

**7.18** Draw a Bode diagram for a system whose transfer function is

$$G(s) = \frac{s^2 + 3s + 4}{s(s^2 + 5s + 12)} \quad \text{(b)}$$

**7.19** Without plotting, determine the salient features of the Nyqusit diagram for the system of Problem 7.17.

**7.20** Without plotting, determine the salient features of the Nyquist diagram for the system of Problem 7.18.

**7.21** The transfer function for a circuit being proposed as a filter is

$$G(s) = \frac{s^3 + 2s + 6}{s^3 + 10s^2 + 20s + 100} \quad \text{(a)}$$

(a) From the transfer function discuss the type of filter for which this circuit may be suitable. (b) Use MATLAB to draw the Bode diagram for this transfer function and use the Bode diagram to discuss its suitability for use as a filter.

**7.22** A 500-kg machine rests on a foundation of stiffness $5 \times 10^6$ N/m. During operation it is subject to a harmonic force of magnitude 150 N at a frequency of 110 r/s. (a) What is the steady-state response of the machine assuming the system is undamped? (b) What is the minimum stiffness of a dynamic vibration absorber that could be added to the machine to eliminate steady-state vibrations when the machine operates at 100 r/s if the steady-state amplitude of the absorber is to be limited to 1 cm? (c) What is the steady-state amplitude of the machine with the absorber of part (b) in place when a 10-kg component is removed from the machine?

**7.23** The machine of Figure P7.23 has an unbalanced rotating component. The component, which rotates at a constant speed $\omega$, is of mass $m_0$ whose center is located at distance $e$ from the axis of rotation. The unbalanced rotating component induces steady-state vibrations of the machine. The differential equation governing the displacement of the machine from its equilibrium position is

$$m\ddot{x} + c\dot{x} + kx = m_0 e \omega^2 \sin(\omega t) \quad \text{(a)}$$

Show that the steady-state amplitude and phase for the displacement of the machine are

$$X = \frac{m_0 e}{m} \frac{r^2}{\sqrt{(1-r^2) + (2\zeta r)^2}} \quad \text{(b)}$$

$$\phi = \tan^{-1}\left(\frac{2\zeta r}{r^2 - 1}\right) \quad \text{(c)}$$

where $r = \omega / \omega_n$.

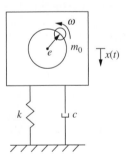

**FIG. P7.23**

# 8

# Feedback Control Systems

The previous chapters have focused on the modeling and response of dynamic systems. The transient response of a first-order system is characterized by the system's time constant. The transient response of second-order systems is characterized by the system's natural frequency and damping ratio. The steady-state response due to a harmonic input is characterized by the steady-state amplitude and phase and is graphically represented by frequency response curves, Bode diagrams, and Nyquist plots.

It is often necessary to modify the transient response of a system by decreasing its rise time or settling time or to eliminate oscillations of an underdamped system. This can be done by use of an actuator, a device that acts on the system to modify its behavior. Actuators themselves are dynamic systems with transfer functions. An actuator can be a mechanical system, an electrical system, a hydraulic system, or a combination of systems.

An actuator, by itself, is generally not sufficient to modify a dynamic system to achieve a desired response. Often actuators are placed in combination with dynamic systems in a feedback control loop. The purpose of this chapter is to introduce feedback control systems and to provide examples of their analysis and how they can be used to achieve the desired response of a dynamic system.

The chapter begins with a survey of block diagrams that provides graphical models of feedback control systems in terms of transfer functions of system components. Common control strategies are introduced and their effects on first- and second-order systems examined. Examples are presented. This chapter is simply an introduction to feedback control systems, which in itself is an exhaustive topic.

## 8.1 BLOCK DIAGRAMS

A block **diagram provides** a schematic representation of system components and their input and output. Block diagrams are drawn in the transform domain or the $s$ domain. The transfer function for a system component

**FIG. 8.1** The representation of a system component in a block diagram using its transfer function.

$$A(s) \quad\boxed{\quad G(s) \quad}\quad C(s) = A(s)G(s)$$

**FIG. 8.2** Series components.

$$A(s) \quad\boxed{\quad G_1(s) \quad}\quad A(s)G_1(s) \quad\boxed{\quad G_2(s) \quad}\quad C(s) = A(s)G_1(s)G_2(s)$$

$G(s)$ is represented by a block, as illustrated in Figure 8.1. The input to the system component is $A(s)$. Recalling that the transfer function is the ratio of the transform of the output to the transform of the input, the transform of the component's output is

$$C(s) = G(s)A(s) \tag{8.1}$$

### 8.1.1 Block Diagram Algebra

Block diagrams show the relationship between the transfer functions of system components. Two systems, represented by block diagrams, are taken to be equivalent if, given the same input, they have the same output. Block diagram algebra is a tool used to simplify the block diagram of a complex system by replacing it with a simpler but equivalent block diagram in which the equivalent system has a transfer function obtained using Equation (8.1).

Figure 8.2 illustrates two components in **series** where the output from the first component is fed as input to the second component. Thus the output from the second series component is

$$C(s) = G_2(s)[G_1(s)A(s)] = [G_1(s)G_2(s)]A(s) \tag{8.2}$$

Equation (8.2) shows that the equivalent transfer function for system components in series is the product of the transfer functions of the individual components.

A **branch point** in a block diagram illustrates output from one component used as input for multiple system components. The output $C(s)$ from Figure 8.3 is fed as input to both components represented by transfer functions $G_2(s)$ and $G_3(s)$.

A **summing point** also called a summing junction, in a block diagram represents a junction where two or more transfer functions are added to or subtracted from one another. The output from the summing point illustrated in Figure 8.4 is $G_1(s) - G_2(s)$.

**Parallel system components** are illustrated in Figure 8.5 where both components have the same input fed from the branch point and their outputs summed at the summing point. If the input to the components

$$C(s) \quad C(s) \quad\boxed{\quad G_2(s) \quad}$$
$$G_3(s)C(s) \quad\boxed{\quad G_3(s) \quad}\quad C(s)$$

**FIG. 8.3** A branch point in a block diagram.

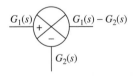

**FIG. 8.4** A summing point or summing junction in a block diagram.

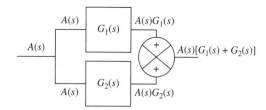

**FIG. 8.5** Parallel components.

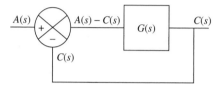

**FIG. 8.6** A system with closed-loop feedback.

is $A(s)$ then the output from the parallel combination is

$$C(s) = G_1(s)A(s) + G_2(s)A(s)$$

$$= [G_1(s) + G_2(s)]A(s) \qquad (8.3)$$

Equation (8.3) shows that two parallel system components can be replaced by a single component whose transfer function is the sum of the component transfer functions.

In the **feedback loop** of Figure 8.6 the branch point sends the output to a summing point where it is subtracted from the loop's input. Let $A(s)$ be the input to the feedback loop, let $C(s)$ be its output, and let $G(s)$ be the transfer function of the system component. The input to the component is $A(s) - C(s)$. Thus the output from the component $C(s)$ is

$$C(s) = [A(s) - C(s)]G(s) \qquad (8.4)$$

Rearranging Equation (8.4) leads to

$$H(s) = \frac{C(s)}{A(s)} = \frac{G(s)}{1 + G(s)} \qquad (8.5)$$

## Example 8.1

Determine the system transfer function $G(s) = C(s)/A(s)$ for the feedback loop of Figure 8.7(a).

### Solution

The input and output from system components is illustrated on the block diagram of Figure 8.7(b). The input to the component whose transfer function is $G_1(s)$ is $A(s) - G_2(s)C(s)$ while its output is $C(s)$. Thus using the definition of a transfer function

$$C(s) = G_1(s)[A(s) - G_2(s)C(s)] \qquad (a)$$

Rearranging Equation (a) leads to

$$H(s) = \frac{C(s)}{A(s)} = \frac{G_1(s)}{1 + G_1(s)G_2(s)} \qquad (b)$$

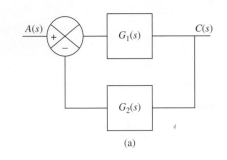

(a)

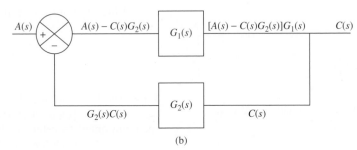

FIG. **8.7** (a) The system of Example 8.1; (b) a block diagram illustrating input and output for each system component.

(b)

The transfer function

$$H(s) = \frac{G_1(s)}{1 + G_1(s)G_2(s)} \tag{8.6}$$

is called the **equivalent open-loop transfer function** or the loop transfer function. The system of Figure 8.6 is a special case of the system of Figure 8.7 in which $G_2(s) = 1$. The equivalent open-loop transfer function is the transfer function of a system used in an open-loop system that leads to the same system response as the system when placed in the feedback loop.

**Example 8.2**

The transfer function for a system component is

$$G(s) = \frac{1}{s+2} \tag{a}$$

Determine the transfer function for the system when this component is placed in a feedback loop (a) as in Figure 8.6 and (b) as in Figure 8.7 (a) with $G_2(s) = 0.5$.

**Solution**

(a) The transfer function of the system when the component is placed in a feedback loop as in Figure 8.6 is obtained by applying Equation (8.5).

$$\frac{C(s)}{A(s)} = \frac{\dfrac{1}{s+2}}{1 + \dfrac{1}{s+2}}$$

$$= \frac{1}{s+3} \tag{b}$$

(b) The transfer function of the system when the component is placed in a feed-back loop as in Figure 8.7 is obtained by applying Equation (8.6)

$$\frac{C(s)}{A(s)} = \frac{\dfrac{1}{s+2}}{1 + \dfrac{0.5}{s+2}}$$

$$= \frac{1}{s+2.5} \tag{c}$$

**TABLE 8.1 BLOCK DIAGRAM EQUIVALENCIES**

| Name | Block Diagram | Equivalent Block Diagram |
|---|---|---|
| Series components | | |
| Parallel components | | |
| Feedback loop | | |
| Relocation of branch point 1 | | |
| Relocation of branch point 2 | | |
| Summing junction | | |

Block diagram algebra is used to show that a block diagram in the right column of Table 8.1 is equivalent to the corresponding block diagram in the left column. These equivalencies can be used to reduce complex block diagrams and eventually obtain an equivalent transfer function.

## Example 8.3

Use block diagram algebra to determine an equivalent open-loop transfer function for the system of Figure 8.8.

### Solution

The block diagram of Figure 8.8 is systematically reduced using the block diagram algebra of Table 8.1. The transfer functions $G_5$ and $G_6$ are in parallel, and the combination is replaced by a single transfer function $G_5 + G_6$. The transfer functions $G_1$ and $G_2$ are in series, and the combination is replaced by a single transfer function $G_1 G_2$. The summing junction for the feedback loop including $G_4$ is moved before the summing junction for the feedback loop including $G_1$ and $G_2$. These simplifications are reflected in Figure 8.9(a). The unity feedback loop involving $G_1$ and $G_2$ is in series with $G_3$. This is reflected in Figure 8.9(b), which also includes the elimination of this feedback loop. The resulting transfer function from elimination of the upper feedback loop is

$$\hat{G}(s) = \frac{\dfrac{G_1 G_2 G_3}{1 - G_1 G_2}}{1 + \left(\dfrac{G_4}{G_1 G_2}\right)\left(\dfrac{G_1 G_2 G_3}{1 - G_1 G_2}\right)}$$

$$= \frac{G_1 G_2 G_3}{1 - G_1 G_2 + G_1 G_2 G_3 G_4} \tag{a}$$

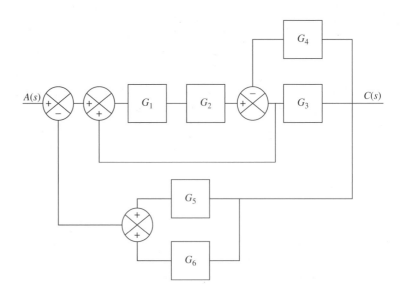

**FIG. 8.8** The block diagram of the system of Example 8.3.

which is reflected in Figure 8.9(c). The last feedback loop is replaced by noting that the closed-loop transfer function is

$$H(s) = \frac{\hat{G}}{1 + (G_5 + G_6)\hat{G}}$$

$$= \frac{G_1 G_2 G_3}{1 - G_1 G_2 + G_1 G_2 G_3 G_4 + (G_5 + G_6)G_1 G_2 G_3} \tag{b}$$

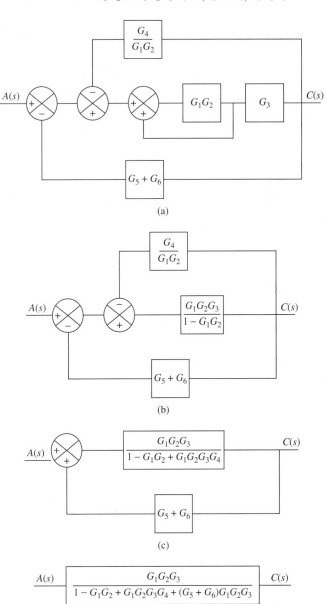

(a)

(b)

(c)

(d)

**FIG. 8.9** The steps in the reduction in the block diagram of Figure 8.8 to an equivalent open-loop transfer function.

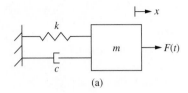

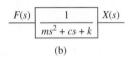

**FIG. 8.10** (a) A mass-spring viscous damper system; (b) the block diagram representation of the system.

## 8.1.2 Block Diagram Modeling of Dynamic Systems

Block diagrams can be drawn to illustrate the relationship, in the transform domain, between the input and output of a dynamic system. Consider the mass-spring-viscous damper system of Figure 8.10(a). The transfer function relationship between the input and output is

$$X(s) = G(s)F(s) = \frac{1}{ms^2 + cs + k}F(s) \qquad (8.7)$$

The block diagram representation of this system is illustrated in Figure 8.10(b).

A cam and follower system is used to provide a prescribed motion input to the base of the mass-spring-viscous damper system of Figure 8.11(a). The force developed in the spring and viscous damper mechanism is

$$F = k(y - x) + c(\dot{y} - \dot{x}) \qquad (8.8)$$

which in turn drives the motion of the mass

$$F = m\ddot{x} \qquad (8.9)$$

Taking the Laplace transforms of Equations (8.8) and (8.9) leads to

$$F = (cs + k)(Y - X) \qquad (8.10a)$$

$$X = \frac{1}{ms^2}F \qquad (8.10b)$$

Since the transform of the force is dependent on the difference between the transforms of the input and output, the block diagram for the system, shown in Figure 8.11(b), contains a feedback loop. The system transfer function is obtained using Equation (8.5) as

$$\frac{X(s)}{F(s)} = \frac{\dfrac{cs + k}{ms^2}}{1 + \dfrac{cs + k}{ms^2}}$$

$$= \frac{cs + k}{ms^2 + cs + k} \qquad (8.11)$$

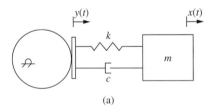

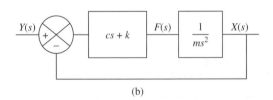

**FIG. 8.11** (a) The mechanical system with motion input; (b) the block diagram of the system includes a feedback loop.

**Example 8.4**

Draw the block diagram for an armature-controlled dc motor. The system input is the voltage in the armature circuit and the output is the angular velocity of the motor. Use the block diagram to determine the system transfer function.

**Solution**

The differential equations for the current in the armature circuit and the angular velocity of the motor for an armature-controlled motor are Equations (a) and (b) of Example 3.23, as follows.

$$L_a \frac{di_a}{dt} + R_a i_a + K_b \omega = v_a \tag{a}$$

$$J \frac{d\omega}{dt} + c_t \omega - K_1 i_a = 0 \tag{b}$$

Taking Laplace transforms of Equations (a) and (b) leads to

$$I_a(s) = \frac{V(s) - K_b \Omega(s)}{L_a s + R_a} \tag{c}$$

$$\Omega(s) = \frac{K_1 I_a(s)}{Js + c_t} \tag{d}$$

Equation (d) shows that the input for the transfer function for the angular velocity is the transform of the armature current, while Equation (c) shows that the input for the transfer function of the armature current is $V(s) - K_b \Omega(s)$. This implies that the angular velocity output is multiplied by $K_b$ in a feedback loop and then brought to a summing junction, where it is subtracted from the input voltage and the difference used as input for the armature current. The appropriate block diagram for the system is illustrated in Figure 8.12. The application of Equation (8.6) leads to the transfer function

$$\frac{\Omega(s)}{V_a(s)} = \frac{\dfrac{K_1}{(L_a s + R_a)(Js + c_t)}}{1 + \dfrac{K_1 K_b}{(L_a s + R_a)(Js + c_t)}}$$

$$= \frac{K_1}{(L_a s + R_a)(Js + c_t) + K_1 K_b} \tag{e}$$

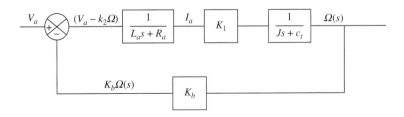

**FIG. 8.12** The block diagram for an armature-controlled motor.

**Example 8.5**

The differential equations for the flow rate perturbations in a three-tank liquid-level problem due to a perturbation in the inlet flow to the first tank are

$$A_1 \frac{dh_1}{dt} + \frac{1}{R_1} h_1 - \frac{1}{R_1} h_2 = q_i(t) \tag{a}$$

$$A_2 \frac{dh_2}{dt} - \frac{1}{R_1} h_1 + \left( \frac{1}{R_1} + \frac{1}{R_2} \right) h_2 - \frac{1}{R_2} h_3 = 0 \tag{b}$$

$$A_3 \frac{dh_3}{dt} - \frac{1}{R_2} h_2 + \left( \frac{1}{R_2} + \frac{1}{R_3} \right) h_3 = 0 \tag{c}$$

Draw a block diagram for the system.

### Solution

Taking the Laplace transforms of Equations (a)–(c) and rearranging leads to

$$H_1(s) = \frac{1}{A_1 s + \dfrac{1}{R_1}} \left[ Q_i + \frac{1}{R_1} H_2(s) \right] \tag{d}$$

$$H_2(s) = \frac{1}{A_2 s + \dfrac{1}{R_1} + \dfrac{1}{R_2}} \left[ \frac{1}{R_1} H_1(s) + \frac{1}{R_2} H_3(s) \right] \tag{e}$$

$$H_3(s) = \frac{1}{A_3 s + \dfrac{1}{R_2} + \dfrac{1}{R_3}} \left[ \frac{1}{R_2} H_2(s) \right] \tag{f}$$

Equations (d)–(f) are used to develop the block diagram of Figure 8.13.

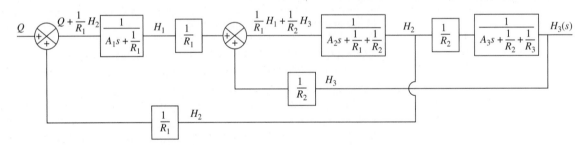

**FIG. 8.13** The block diagram model for a three-tank liquid-level system.

**Example 8.6**

The differential equations governing the concentration perturbations in a CSTR with a two-way reaction due to a perturbation in inlet flow rate are

$$V \frac{dC_A}{dt} + (q_s + k_1 V) C_A - k_2 V C_B = q C_{Ai} \tag{a}$$

$$V \frac{dC_B}{dt} - k_1 V C_A + (q_s + k_2 V) C_B = q C_{Bi} \tag{b}$$

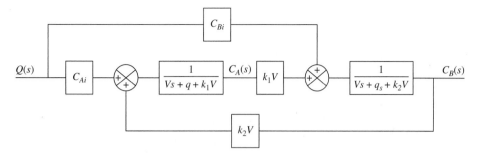

**FIG. 8.14** The block diagram for the CSTR system of Example 8.6.

Draw a block diagram for this system where $q$ is the system input.

**Solution**

Taking the Laplace transforms of Equations (a) and (b) leads to

$$C_A(s) = \frac{C_{Ai}Q(s)}{Vs + q_s + k_1 V} + \frac{k_2 V C_B(s)}{Vs + q_s + k_1 V} \tag{c}$$

$$C_B(s) = \frac{C_{Bi}Q(s)}{Vs + q_s + k_2 V} + \frac{k_1 V C_A(s)}{Vs + q_s + k_2 V} \tag{d}$$

An appropriate block diagram is illustrated in Figure 8.14.

## 8.2 USING SIMULINK IN BLOCK DIAGRAM MODELING

SIMULINK is a dynamic simulation software package developed in conjunction with MATLAB. SIMULINK has many capabilities, including modeling nonlinear or discrete time systems, which are not covered. The focus in this study is on building models and running dynamic simulations using SIMULINK.

SIMULINK models are developed using block diagrams that may be developed using either the transfer function method discussed in Section 8.1 or the state-space method introduced in Chapter 9. Dynamic simulations are run using SIMULINK, which employs numerical integration software from MATLAB. The numerical methods used are discussed in Chapter 9.

The user is referred to the SIMULINK help function to access SIMULINK from the MATLAB command window, access the SIMULINK Library Browser, and open a model. The model is built by clicking on and dragging the appropriate icons from the Library Browser into the model workspace.

As an example consider the mass-spring-viscous damper system of Figure 8.10(a) with $m = 1$, $k = 25$, and $c = 6$, whose block diagram is illustrated in Figure 8.10(b). The SIMULINK model is developed by reproducing the block diagram in the model workspace. The following

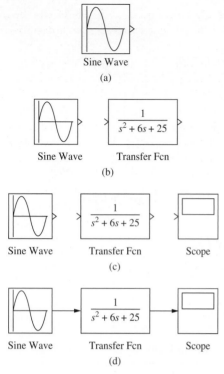

Sine Wave

(a)

Sine Wave         Transfer Fcn

(b)

Sine Wave         Transfer Fcn         Scope

(c)

Sine Wave         Transfer Fcn         Scope

(d)

**FIG. 8.15** (a) Sine Wave; (b) Sine Wave and system Transfer Fcn; (c) Scope added; (d) icons connected.

steps are illustrated sequentially in Figure 8.15. The model is built first using the default sine source provided by SIMULINK and then revised for a general sinusoidal excitation. Numerical values are assumed as $m = 1$ kg, $k = 25$ N/m, and $c = 6$ N·s/m.

- Click on the Sine Wave icon from the sources menu of the Library Browser and drag it into the model workspace [Figure 8.15(a)].
- Click on the Transfer Fcn icon from the continuous menu of the Library Browser and drag it into the model workspace, placing it to the right of the Sine Wave icon.
- Once in the model workspace click on the transfer function icon to bring up the form for Block Parameters. A transfer function is defined in SIMULINK as it is in MATLAB; enter the coefficients of the polynomials in the numerator and denominator in a vector. For this system leave the 1 for the numerator and enter 1 6 25 inside the brackets for the polynomial in the denominator [Figure 8.15(b)].
- Click on the Scope icon from the sinks menu of the Library Browser. The Scope allows the display of the simulation results. Drag the Scope icon into the model workspace and place it to the right of the Transfer Fcn icon [Figure 8.15(c)].
- Connect the icons as instructed by SIMULINK help; connect the Sine Wave to the Transfer Fcn and connect the Transfer Fcn to the Scope [Figure 8.15(d)].

The model is complete. A simulation is run by clicking on Simulation and then Start from the model workspace toolbar. The simulation results are viewed by clicking on the Scope icon. A box titled Scope appears, showing a plot of the displacement as a function of time. The

**FIG. 8.16** SIMULINK simulation of response of mass-spring-viscous damper system.

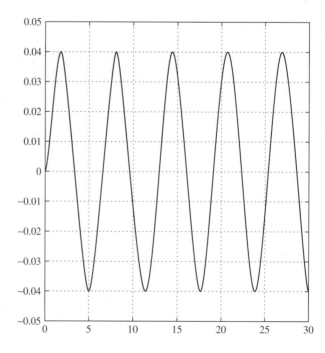

default limits on the vertical axis are from $-5$ to $+5$. These can be adjusted by right-clicking on an axis in the Scope window. An appropriate scale for this simulation is from $-.05$ to $+.05$. The default on the time scale is 10 s. The time for the simulation to run can be changed by clicking on Simulation and then Simulation Parameters from the toolbar in the model workspace. The simulation obtained from the scope with these modifications is shown in Figure 8.16.

Figure 8.17 shows a SIMULINK model with two revisions. The Sine Wave generator has been modified by clicking on its icon, and changing the frequency in the Block Parameters window to 5 r/s. The Scope has been replaced by a block, which sends the output to the MATLAB workspace. This is obtained from the source menu from the Library Browser. The To workspace icon is clicked and dragged to the model workspace. In the Block Parameters box that appears when the icon is clicked, the Variable Name was changed to x and the Save Format was selected as structure with time. When the simulation is run, instead of a graphical representation of the response as with a scope, the numerical values calculated are returned to the MATLAB workspace under a structure named "x." The "with time" means that the corresponding values of time are also sent to the MATLAB workspace. This allows a customized plot to be drawn.

The data is accessed in MATLAB by the commands

$$t = x(1).\text{time}$$

$$x1 = x(1).\text{signals}.\text{values}$$

The first command stores the time values in the vector t, while the second command stores the displacement values in the vector x1. A customized plot of the response generated from the simulation data is shown in Figure 8.18.

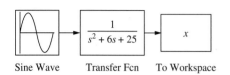

**FIG. 8.17** SIMULINK model of mass-spring-viscous damper system with numerical values of response sent to MATLAB workspace.

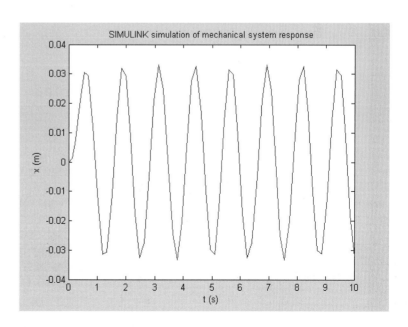

**FIG. 8.18** The customized plot of the SIMULINK simulation.

**Example 8.7**

Develop SIMULINK models and run simulations for (a) the armature-controlled dc motor of Example 8.4 with $J = 1.8$ kg·m$^2$, $c_t = 2.15$ N·m·s/r, $L_a = 0.15$ H, $R_a = 100$ Ω, $K_1 = 120$ N·m/A, and $K_2 = 1.67$ V·s/r with a unit step input for the armature voltage; (b) the three-tank liquid-level system of Example 8.5 with $A_1 = A_2 = A_3 = 10$ m$^2$, $R_1 = 5$ s$^2$/m, $R_2 = 10$ s$^2$/m, and $R_3 = 5$ s$^2$/m when the first tank has a unit step increase in inlet flow rate; and (c) the CSTR with the two-way reaction of Example 8.6 with $V = 1.5 \times 10^{-3}$ m$^3$, $q_s = 1.5 \times 10^{-6}$ m$^3$/s, $C_{Ai} = 0.25$ mol/L, $C_{Bi} = 0.15$ mol/L, $k_1 = 2.0 \times 10^{-3}$s$^{-1}$, and $k_2 = 1.0 \times 10^{-3}$s$^{-1}$ when the inlet flow rate has a perturbation of $5 \times 10^{-7}$ m$^3$/s.

(a) The SIMULINK model for the armature-controlled dc servomotor is shown in Figure 8.19. The following is noted regarding this model:

- The model has two sinks (outputs). The sinks are located appropriately in the block diagram to send numerical values for the armature circuit current and the motor speed to the MATLAB workspace.

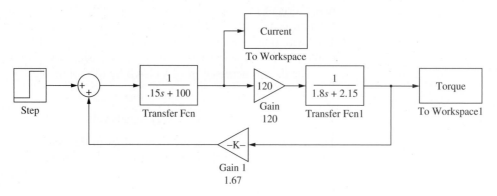

**FIG. 8.19** SIMULINK model of armature controlled dc motor.

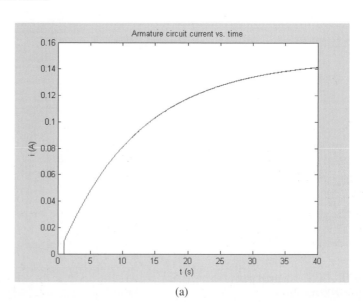

**FIG. 8.20** The MATLAB-generated graphs of SIMULINK results for the armature-controlled dc motor of Example 8.7(a). (a) current (b) angular velocity.

(a)

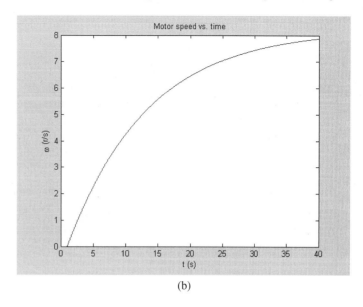

**FIG. 8.20** (*Continued*).

(b)

- The block diagram uses the Gain block from the Mathematical Functions menu of the SIMULINK Library Browser. The gain block simply multiplies the signal in the line by the value of the gain.
- The block diagram contains a feedback loop. The lines for the feedback loop and to the current sink are drawn by right-clicking on the point where the line is to start and drag to the end.

The output plotted from the MATLAB workspace is illustrated in Figure 8.20. It shows the armature current approaching slightly more than 0.14 A and the motor speed approaching 8 r/s.

(b) The SIMULINK model for the three-tank liquid-level system is illustrated in Figure 8.21. This model uses two feedback loops and three scopes. The system response is shown in Figure 8.22.

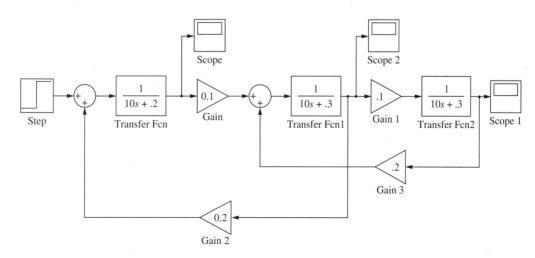

**FIG. 8.21** SIMULINK model of three-tank liquid level system developed using block diagram of Figure 8.13.

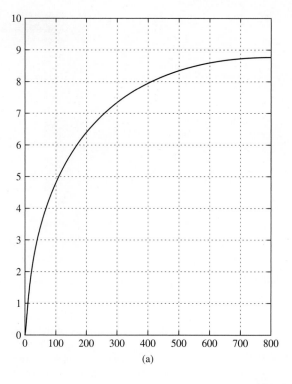

(a)

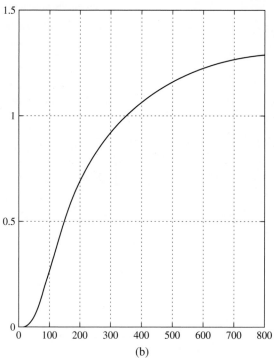

(b)

**FIG. 8.22** (a) $h_1(t)$ from scope; (b) $h_2(t)$ from scope 2; (c) $h_3(t)$ from scope 1.

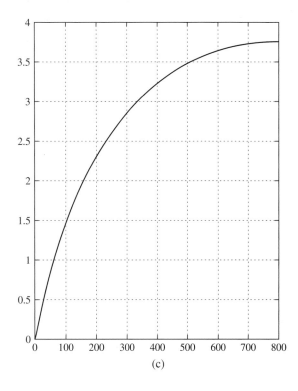

**FIG. 8.22** (*Continued*).

(c) The SIMULINK model for the CSTR is shown in Figure 8.23. The following is noted regarding this model:

- The step input is multiplied by a gain equal to the flow rate perturbation.
- The input is fed into two places in the block diagram.

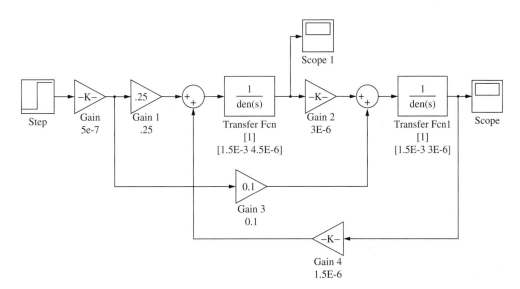

**FIG. 8.23** SIMULINK model of CSTR developed using block diagram of Figure 8.14.

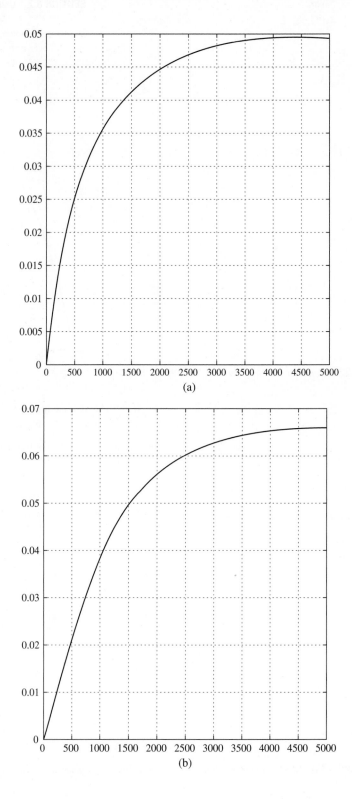

**FIG. 8.24**  $C_A(t)$ from scope1;
(b) $C_B(t)$ from scope.

- The transfer functions blocks are annotated below the block. The annotation shows the numerator polynomial and the denominator polynomial, using vector representations for their coefficients.

The scope traces when the simulation is run are shown in Figure 8.24.

# 8.3 FEEDBACK CONTROL

A dynamic system that is to be controlled in a feedback control system is called a **plant**. A plant's transfer function $G(s)$, is an **open-loop transfer function**. The output from the plant is sent through a feedback loop to the summing point, as illustrated in the block diagram of Figure 8.25. The output $C(s)$ is compared with the system input $A(s)$. An actuator, which has the transfer function $G_a(s)$, performs a control function on the difference, the result of which is sent to the plant. The function $G_a(s)$ represents the combined action of actuators and controllers. The resulting equivalent **open-loop transfer function** is

$$H(s) = \frac{C(s)}{A(s)} \tag{8.12}$$

The input to the controller is $A(s)-C(s)$ while its output is $B(s) = [A(s)-C(s)]G_a(s)$. The plant's output is $C(s) = B(s)G(s) = [A(s)-C(s)]G_a(s)\,G(s)$. Equation (8.12) is applied to determine the closed-loop transfer function as

$$H(s) = \frac{G_a(s)G(s)}{1 + G_a(s)G(s)} \tag{8.13}$$

The controller in the control system of Figure 8.25 is placed in the feedforward portion of the loop, but it could alternatively be placed in the feedback portion of the loop, as illustrated in Figure 8.26(a). In this latter case the closed-loop transfer function is determined as

$$H(s) = \frac{G(s)}{1 + G_a(s)G(s)} \tag{8.14}$$

There are many combinations possible for configuration of control systems. This study is not intended to be exhaustive; thus, to introduce the effect of actuators and controllers on the closed-loop transfer functions and system response, only systems with controllers placed in the feedforward loop are studied.

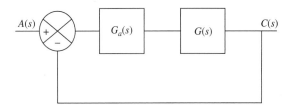

**FIG. 8.25** The feedback control system with a controller in the feedforward part of the loop.

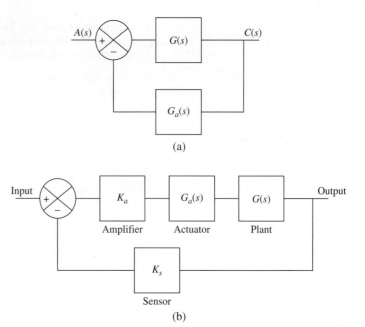

**FIG. 8.26** (a) The control system with a controller in the feedback portion of loop; (b) the feedback control loop with an amplifier and sensor.

The feedback loop of Figure 8.25 is an idealized loop that is used for discussion purposes. A more practical feedback loop is illustrated in Figure 8.26(b). All transfer functions feeding into a summing point must have the same units. Often the system output $C(s)$ does not have the same units as the input $A(s)$. A sensor converts the plant output into a form that can be compared with the input, including consistent units. Often the error signal is small and must be amplified to feed into the actuator.

### 8.3.1 Proportional Control

A controller applies **proportional control** when its transfer function is a constant

$$G_a(s) = K_p \tag{8.15}$$

with $K_p$ as the proportional gain. The substitution of Equation (8.15) into Equation (8.13) leads to

$$H(s) = \frac{K_p G(s)}{1 + K_p G(s)} \tag{8.16}$$

The operational amplifier circuit of Example 3.29 and Figure 3.46 is an example of a proportional controller. For this active circuit, the output voltage $v_2$ is related to the input voltage $v_1$ by $v_2 = (R_2/R_1)v_1$. Thus its actuator transfer function is $G_a(s) = V_2(s)/V_1(s) = R_2/R_1$, which shows that it is a proportional controller with a proportional gain $K_p = R_2/R_1$.

The transfer function for the hydraulic servomotor of Example 4.23 and Figure 4.35 is determined as $G_a(s) = Y(s)/X(s) = \hat{C}b/[\hat{C}a + A_p (a+b)s] = b/a/\{1+(A_p/\hat{C})[1+(b/a)]s\}$. In practical application $(A_p/\hat{C})(1 + (b/a))s << 1$ and can be neglected, in which case the transfer function is approximated as $G_a(s) = b/a = K_p$. Thus a hydraulic servomotor in which the piston and the spool valve are connected via a walking beam can serve as a proportional controller with a proportional gain of $b/a$.

## 8.3.2 Integral Control and Proportional Plus Integral (PI) Control

The actuator transfer function for an integral controller with integral gain $K_i$ is

$$G_a(s) = \frac{K_i}{s} \qquad (8.17)$$

The time-dependent output of an integral controlled for an input $E(t)$ is

$$B(t) = K_i \int_0^t E(t)dt \qquad (8.18)$$

Substitution of Equation (8.17) into Equation (8.13) leads to

$$H(s) = \frac{\dfrac{K_i}{s} G(s)}{1 + \dfrac{K_i}{s} G(s)}$$

$$= \frac{K_i G(s)}{s + K_i G(s)} \qquad (8.19)$$

The operational amplifier circuit of Figure 8.27 may be used as an integral controller. The first operational amplifier functions as an integrator and inverter as in Figure 3.28. The second amplifier serves to invert the sign and change the gain. The relation between the input voltage $v_1$ and the output voltage $v_2$ is $v_2 = R_3/R_1R_2C \int_0^t v_1 dt$. The actuator transfer function for the circuit is $G_a(s) = V_2(s)/V_1(s) = R_3/R_1R_2Cs$.

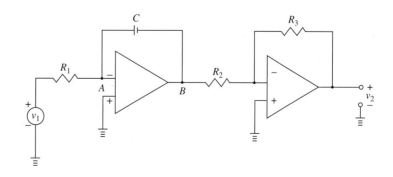

**FIG. 8.27** An operational amplifier circuit that can be used as an integral controller.

Thus the amplifier circuit may serve as an integral controller with an integral gain of $K_i = R_3/R_1 R_2 C$.

The transfer function of an actuator with proportional plus integral (PI) control is

$$G_a(s) = K_p + \frac{K_i}{s} \tag{8.20}$$

Substitution into Equation (8.13) leads to

$$H(s) = \frac{\left(K_p + \dfrac{K_i}{s}\right) G(s)}{1 + \left(K_p + \dfrac{K_i}{s}\right) G(s)}$$

$$= \frac{(K_i + s K_p) G(s)}{s + (K_i + s K_p) G(s)} \tag{8.21}$$

## 8.3.4 Derivative Control and Proportional Plus Derivative (PD) Control

The transfer function for an actuator with derivative control is

$$G_a(s) = K_d s \tag{8.22}$$

where $K_d$ is the derivative gain. The substitution of Equation (8.22) into Equation (8.13) leads to

$$H(s) = \frac{K_d s G(s)}{1 + K_d s G(s)} \tag{8.23}$$

A dashpot is a mechanical device that delivers proportional control. Derivative control responds to the rate of change of the signal error, not the error itself. Thus derivative control is used in conjunction with proportional control and/or integral control.

The transfer function for an actuator with proportional plus derivative (PD) control is

$$G_a(s) = K_p + K_d s \tag{8.24}$$

The substitution of Equation (8.24) into Equation (8.13) leads to

$$H(s) = \frac{(K_p + K_d s) G(s)}{1 + (K_p + s K_d) G(s)} \tag{8.25}$$

## 8.3.5 Proportional Plus Integral Plus Derivative (PID) Control

The transfer function for an actuator with proportional plus integral plus derivative control (PID) control is

$$G_a(s) = K_p + K_d s + \frac{K_i}{s} \tag{8.26}$$

If $E(t)$ is the input to a PID controller then its output is

$$B(t) = K_p E(t) + K_d \frac{dE}{dt} + K_i \int_0^t E(t)dt \qquad (8.27)$$

Substitution of Equation (8.26) into Equation (8.13) leads to

$$H(s) = \frac{\left(K_p + K_d s + \dfrac{K_i}{s}\right) G(s)}{1 + \left(K_p + K_d s + \dfrac{K_i}{s}\right) G(s)}$$

$$= \frac{\left(K_d s^2 + K_p s + K_i\right) G(s)}{s + \left(K_d s^2 + K_p s + K_i\right) G(s)} \qquad (828)$$

## Example 8.8

Show that the operational amplifier circuit of Example 3.33 and Figure 3.51 can serve as an electronic PID controller and identify $K_p$, $K_i$, and $K_d$.

### Solution

The relation between the output potential $v_2$ and the input potential $v_1$ is given in Equation (f) of Example 3.33, repeated here

$$v_2 = \frac{C_1 R_2 R_4}{R_3} \frac{dv_1}{dt} + \left(\frac{R_2 R_4}{R_1 R_3} + \frac{C_1 R_4}{C_2 R_3}\right) v_1 + \frac{R_4}{C_2 R_1 R_3} \int_0^t v_1 dt \qquad (a)$$

Taking the Laplace transform of Equation (a) leads to

$$V_2(s) = \frac{C_1 R_2 R_4}{R_3} s V_1(s) + \left(\frac{R_2 R_4}{R_1 R_3} + \frac{C_1 R_4}{C_2 R_3}\right) V_1(s) + \frac{R_4}{C_2 R_1 R_3} \frac{V_1(s)}{s} \qquad (b)$$

The controller s transfer function is determined from Equation (b) as

$$G(s) = \frac{V_2(s)}{V_1(s)} = \left(\frac{R_2 R_4}{R_1 R_3} + \frac{C_1 R_4}{C_2 R_3}\right) + \frac{C_1 R_2 R_4}{R_3} s + \frac{R_4}{C_2 R_1 R_3} \frac{1}{s} \qquad (c)$$

Equation (c) is the transfer function of a PID controller, Equation (8.26) with

$$K_p = \frac{R_2 R_4}{R_1 R_3} + \frac{C_1 R_4}{C_2 R_3} \qquad (d)$$

$$K_i = \frac{R_4}{C_2 R_1 R_3} \qquad (e)$$

$$K_d = \frac{C_1 R_2 R_4}{R_3} \qquad (f)$$

## 8.3.6 Error and Offset

The **error** $e(t)$ in a feedback control system is defined as the difference between the input and the output. The transfer function for the error is obtained using Equations (8.12) and (8.13) as

$$E(s) = A(s) - C(s)$$

$$= \frac{A(s)}{1 + G_a(s)G(s)} \tag{8.29}$$

The **offset** $\eta$ is defined as the difference between the input and the output at steady state. The offset for a stable system is obtained by applying the final value theorem to Equation (8.29)

$$\eta = \lim_{s \to 0} sE(s)$$

$$= \lim_{s \to 0} \frac{sA(s)}{1 + G_a(s)G(s)} \tag{8.30}$$

## 8.3.7 Response Due to Unit Step Input

The response of the feedback control system due to a unit step input is obtained by using $A(s) = 1/s$ in Equation (8.12) and use of Equation (8.13) leading to

$$C(s) = \frac{G_a(s)G(s)}{s[1 + G_a(s)G(s)]} \tag{8.31}$$

For a specific plant and specific controller Equation (8.31) can be inverted.

The offset of a feedback control system when subject to a unit step input is obtained from Equation (8.30) as

$$\eta = \lim_{s \to 0} \frac{1}{1 + G_a(s)G(s)} \tag{8.32}$$

For a system with a proportional controller with $G_a(s) = K_p$, Equation (8.32) leads to

$$\eta = \frac{1}{1 + K_p G(0)} \tag{8.33}$$

Equation (8.33) illustrates that a system with only proportional control has a nonzero offset. The steady-state response differs from the input. The error for a system with only proportional control cannot be zero because proportional control simply multiplies the instantaneous error signal by a constant. If the instantaneous error were zero the output from the controller, which is the input to the plant, would be zero and the plant would not operate.

The application of Equation (8.32) to a system with integral control whose transfer function is of the form of Equation (8.19) leads to

$$\eta = \lim_{s \to 0} \frac{1}{1 + \dfrac{K_i}{s} G(s)} = 0 \tag{8.34}$$

Thus a system with integral control has no offset. Since the integral controller integrates the time history of the input, an instantaneous error of zero does not lead to a zero transfer function as input to the plant. A similar analysis shows that the offset is also zero for a PI controller and a PID controller.

The application of Equation (8.32) to a system with a differential controller with the closed-loop transfer function of Equation (8.23) leads to

$$\eta = \lim_{s \to 0} \frac{1}{1 + K_d s G(s)} = 1 \tag{8.35}$$

Thus derivative control leads to an offset that is independent of the derivative gain when subject to a step input.

Equations (8.33)–(8.35) show that when a proportional controller, a differential controller, or a PD controller is used as an actuator in the feedforward loop of a feedback control system, the final steady state is altered from that if the system were open loop. When an integral, PI, or PID controller is used, the presence of the integrator eliminates the offset.

## 8.4 FEEDBACK CONTROL OF FIRST-ORDER PLANTS

The standard form of the transfer function for a first-order plant is

$$G(s) = \frac{K}{s + \dfrac{1}{T}} \tag{8.36}$$

where $K$ is a constant and $T$ is the system's time constant. The transient behavior of first-order systems is considered in Section 6.4. The response of a first-order plant when placed in a feedback loop with feedforward control is considered in this section.

The closed-loop transfer function for a first-order system with a proportional controller is obtained by substituting Equation (8.36) into Equation (8.16) leading to

$$H(s) = \frac{K_p \dfrac{K}{s + \dfrac{1}{T}}}{1 + K_p \dfrac{K}{s + \dfrac{1}{T}}}$$

$$= \frac{K_p K}{s + \dfrac{1}{T} + K_p K} \tag{8.37}$$

The transfer function of Equation (8.37) is that of a first-order system with a time constant of

$$\hat{T} = \frac{T}{1 + K_p K T} \tag{8.38}$$

and a multiplicative constant of $K_p K$. For $K_p > 0$, $\hat{T} < T$.

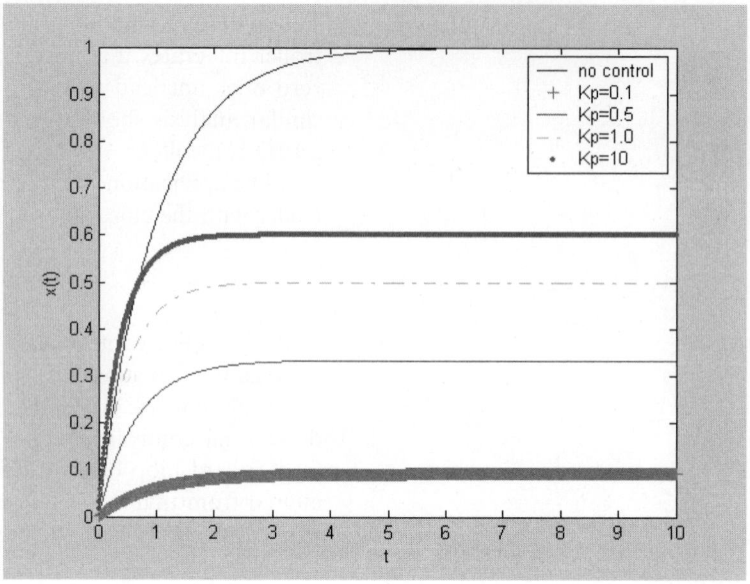

**FIG. 8.28** The use of a proportional controller with a first-order plant leads to offset. The offset is smaller for controllers with larger gains. The settling time is less than the settling time for a plant without a controller and is smaller for controllers with larger gains.

The output from a closed-loop system with a first-order plant actuated by a proportional controller and subject to a unit step input $A(s) = 1/s$ is

$$C(s) = \frac{K_p K}{s\left(s + \dfrac{1}{T} + K_p K\right)} \tag{8.39}$$

The response of a first-order plant to a step input is compared with the open-closed loop response of a first-order system with a proportional controller in Figure 8.28. The final value of the open-loop system is $KT$ where the final value of the closed-loop system is obtained using the final value theorem as

$$\lim_{t \to \infty} c(t) = \lim_{t \to \infty} sC(s)$$

$$= \frac{K_p K}{\dfrac{1}{T} + K_p K}$$

$$= K_p K \hat{T} \tag{8.40}$$

The error for the system with a unit step input is

$$E(s) = \frac{s + \dfrac{1}{T}}{s\left(s + \dfrac{1}{\hat{T}}\right)} \tag{8.41}$$

The offset is obtained using the final value theorem as

$$\eta = \lim_{s \to 0} sE(s)$$

$$= \lim_{s \to 0} \frac{s + \dfrac{1}{T}}{s + \dfrac{1}{\hat{T}}}$$

$$= \frac{\hat{T}}{T} \tag{8.42}$$

The response of the first-order system with a proportional controller due to a unit step input is obtained by inverting Equation (8.39) resulting in

$$c(t) = K_p K \hat{T}\left(1 - e^{-t/\hat{T}}\right) \tag{8.43}$$

The 2 percent settling time for the closed-loop system with the proportional controller is $t_s = 3.92\hat{T}$. The benchmark times such as 2 percent setting time are less for a first-order plant when placed in a feedback control loop with proportional controller. The benchmark times decrease with increasing values of proportional gain. However, the use of proportional control alone leads to a nonzero offset.

The closed-loop transfer function of a first-order system with an integral controller with $G_a(s) = K_i/s$ is obtained by substituting Equation (8.36) into Equation (8.19) resulting in

$$H(s) = \frac{\dfrac{K_i K}{s\left(s + \dfrac{1}{T}\right)}}{1 + \dfrac{K_i K}{s\left(s + \dfrac{1}{T}\right)}}$$

$$= \frac{K_i K}{s^2 + \dfrac{1}{T}s + K_i K} \tag{8.44}$$

Equation (8.44) is the transfer function of a second-order system with natural frequency

$$\omega_n = \sqrt{KK_i} \tag{8.45}$$

and damping ratio

$$\zeta = \frac{1}{2T\sqrt{K_i K}} \tag{8.46}$$

The larger the value of the integral gain is, the smaller is the value of the damping ratio and the larger the value of the natural frequency of the system. If $K_i > 1/4KT^2$ is underdamped.

The transform of the error when an integral controller is used for a first-order plant is

$$E(s) = A(s) \frac{s^2 + \dfrac{1}{T}}{s^2 + \dfrac{1}{T}s + K_i K} \tag{8.47}$$

The transform of the response due to a unit step input is

$$C(s) = \frac{K_i K}{s\left(s^2 + \dfrac{1}{T}s + K_i K\right)} \tag{8.48}$$

As noted in Section 8.3 the offset when an integral controller is used on a first-order plant is $\eta = 0$.

The closed-loop response of a first-order system with an integral controller due to a unit step input is obtained using Table 6.5. If the closed-loop system is underdamped the response is

$$x(t) = \frac{K_i K}{\omega_n^2}\left[1 - e^{-\zeta \omega_n t}\cos(\omega_d t) + \frac{\zeta \omega_n}{\omega_d}e^{-\zeta \omega_n t}\sin(\omega_d t)\right] \tag{8.49}$$

The use of an integral controller with a first-order plant leads to a second-order system. The system is underdamped if the integral gain is large enough. The use of the integral controller eliminates the offset that occurs when a proportional controller is used. The natural frequency of the resulting second-order system increases as the integral gain increases while the damping ratio decreases. The increase in natural frequency decreases the benchmark times, but the decrease in damping ratio increases the benchmark times. However, the use of an integral controller allows only one parameter, the damping ratio or the natural frequency, to be specified. The output from closed-loop system with an integral controller and first-order plant with a unit step input is illustrated in Figure 8.29.

The use of a proportional plus integral controller in a closed-loop system with a first-order plant leads to a closed loop transfer function of

$$H(s) = \frac{K(K_p s + K_i)}{s^2 + \left(KK_p + \dfrac{1}{T}\right)s + KK_i} \tag{8.50}$$

The closed-loop system is a second-order system with the same natural frequency as a system with just integral control, given by Equation (8.45) and damping ratio

$$\zeta = \frac{K_p}{2}\sqrt{\frac{K}{K_i}} + \frac{1}{2T\sqrt{KK_i}} \tag{8.51}$$

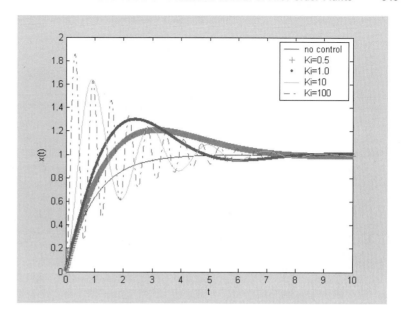

**FIG. 8.29** The step response of a first-order plant in a feedback loop with an integral controller in the feedforward part of the loop. There is no offset with an integral controller, but the system is underdamped for larger $K_i$.

The use of proportional plus integral control for a system with a step input leads to no offset. The damping ratio of the system is larger than that when just integral control is used.

The use of a PI controller with a first-order plant provides more flexibility than does the use of an integral controller. An integral controller has only one parameter, the integral gain, which can be chosen to control the system response. Thus only the natural frequency or the damping ratio can be specified when using an integral controller. When a PI controller is used, two parameters (the proportional gain and the integral gain) may be chosen. The integral gain may be chosen to specify the system's natural frequency and the proportional gain is then chosen to specify the system's damping ratio.

The use of a PD controller in a closed-loop system with a first-order plant leads to

$$H(s) = \frac{KK_d s + KK_p}{(1 + KK_d)s + \dfrac{1}{T} + KK_p} \tag{8.52}$$

As noted in Section 8.3 the addition of derivative control to a proportional controller does not affect the offset.

The transform of the response of a first-order system with closed-loop PD control due to a unit step input is

$$C(s) = \frac{KK_p}{\dfrac{1}{T} + KK_p} \frac{1}{s} - \frac{\left(KK_p - \dfrac{KK_d}{T}\right)\tilde{T}}{(1 + KK_d)^2} \frac{1}{s + \dfrac{1}{\tilde{T}}} \tag{8.53}$$

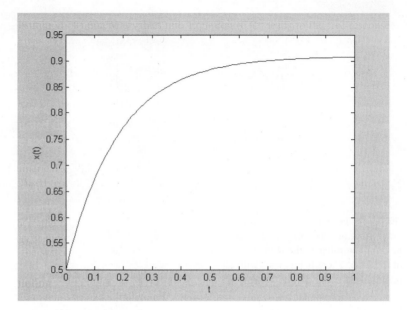

**FIG. 8.30** The use of a proportional plus derivative controller to a first-order plant with a step input leads to a discontinuity

where the time constant is

$$\tilde{T} = \frac{1 + KK_d}{\dfrac{1}{T} + KK_p} \qquad (8.54)$$

Figure 8.30 illustrates the application of a proportional plus derivative controller to a first-order plant with a transfer function $G(s) = 1/(s+1)$ with $K_p = 10$ and $K_d = 1$ when the system is subject to a unit step input. The addition of the derivative term does not affect the offset caused by the proportional controller, but does lead to a discontinuity in the response at $t = 0$. This discontinuity is caused by differentiating a unit step input that is discontinuous at $t = 0$. Mathematically the initial value theorem is used to show that the response of a first order system at $t = 0$ due to a unit step is $x_s(0) = KK_d/(1 + KK_d)$. If the input is continuous at $t = 0$ the response is continuous. The addition of the derivative term causes the system to respond more quickly and reduces the rise time.

The closed-loop transfer function of a proportional plus integral plus derivative controller used with a first-order plant is

$$H(s) = \frac{K\left(K_d s^2 + K_p s + K_i\right)}{(1 + KK_d)s^2 + \left(KK_p + \dfrac{1}{T}\right)s + KK_i} \qquad (8.55)$$

The closed-loop system for a first-order plant with a PID controller is a second-order system. The natural frequency and damping ratio of the closed-loop system are obtained from the denominator of Equation (8.55). Comparing Equation (8.55) to the closed-loop transfer function of a first-order plant with an integral controller shows that an effect of adding damping is to decrease the natural frequency and that an effect of adding a proportional term is to increase the damping ratio. An increase in the value of the integral gain leads to an increase in natural frequency and a decrease in damping ratio.

The numerator of the closed-loop transfer function is important in the rate of response due to the controller. When derivative control is present the numerator of the closed-loop transfer function for a first-order plant is of the same order as the denominator. Thus the step response of the closed-loop system is discontinuous at $t = 0$. The rate at which the response occurs is affected by the numerator of $H(s)$. When derivative control is present the numerator is of a higher order. This leads to a decrease in the benchmark times for the transient response of a second-order system.

Figure 8.31 illustrates the use of PID control on a first-order plant with a transfer function $G(s) = 1/(s + 1)$ with a unit step input with $K_p = 10$ and $K_d = 1$. The step response is discontinuous with $x_s(0) = KK_d/(1 + KK_d)$. The addition of the integral control eliminates the offset.

This section provides an introduction to the closed-loop response of dynamic systems with various controllers. Several principles are evident and are addressed in subsequent sections. A proportional controller is the basic controller, but its use leads to error and offset in the closed-loop response. This offset is eliminated by adding integral control. However, the use of integral control increases the order of the system and is destabilizing. While the closed-loop response of a first-order plant is inherently stable an increase in integral gain leads to a decrease in the damping ratio, which itself is destabilizing. Derivative control is used to enhance sensitivity and increase the rate of response.

The properties of the closed-loop system of a first-order plant acted on by various controllers in the feedforward portion of the loop are summarized in Table 8.2.

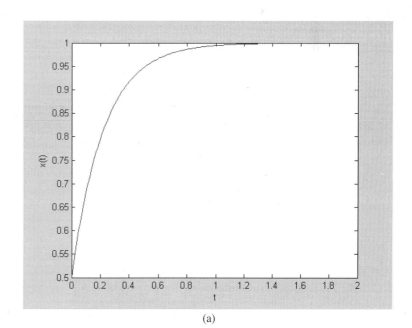

**FIG. 8.31** The step response of a first-order plant with PID control in the feedforward loop with $K = 1$, $T = 1$, $K_p = 10$, $K_d = 1$: (a) $K_i = 9$, the system is overdamped; (b) $K_i = 100$, the system is underdamped. (*Continued*)

(a)

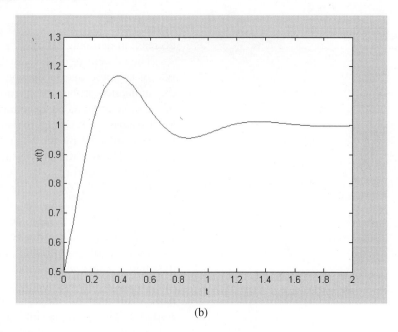

**FIG. 8.31** (*Continued*)

(b)

**TABLE 8.2 SUMMARY OF CLOSED-LOOP RESPONSE OF FIRST-ORDER PLANTS, $G(S)=K/[S + (1/T)]$ WITH FEEDFORWARD ACTUATOR $G_a(S)$**

| Controller | $G_a(s)$ | Order of Closed-Loop System | Closed-Loop Transfer Function | Closed-Loop System Parameters | Offset |
|---|---|---|---|---|---|
| Proportional | $K_p$ | first | Equation (8.37) | $\hat{T} = \dfrac{T}{1 + K_p K T}$ | $\dfrac{\hat{T}}{T}$ |
| Integral | $\dfrac{k_i}{s}$ | second | Equation (8.44) | $\hat{\omega}_n = \sqrt{KK_i}$ | 0 |
| | | | | $\hat{\zeta} = \dfrac{1}{2T\sqrt{KK_i}}$ | |
| PD | $K_p + K_d s$ | first | Equation (8.52) | $\hat{T} = \dfrac{1 + KK_d}{\dfrac{1}{T} + KK_p}$ | $\dfrac{1}{1 + K_p K T}$ |
| PI | $K_p + \dfrac{K_i}{s}$ | second | Equation (8.50) | $\hat{\omega}_n = \sqrt{KK_i}$ | 0 |
| | | | | $\hat{\zeta} = \dfrac{K_p}{2}\sqrt{\dfrac{K}{K_i}} + \dfrac{1}{2T\sqrt{KK_i}}$ | |
| PID | $K_p + K_d s + \dfrac{K_i}{s}$ | second | Equation (8.55) | $\hat{\omega} = \sqrt{\dfrac{KK_i}{1 + KK_d}}$ | 0 |
| | | | | $\hat{\zeta} = \dfrac{KK_p + \dfrac{1}{T}}{2\sqrt{KK_i(1 + KK_d)}}$ | |

Note: $\hat{T}$ is the time constant of a first-order closed-loop system, and $\hat{\omega}_n$ and $\hat{\zeta}$ are the natural frequency and damping ratio of a second-order closed-loop system.

**Example 8.9**

A tank of cross-sectional area 20 m² has a steady-state level of 12 m for an incoming flow rate of 1.5 m³/s. A closed-loop control system is being designed for the tank to improve its response when the flow rate has a step change. In all cases provide the closed-loop transfer function of the system. (a) What is the required gain of a proportional controller such that the 2 percent settling time of the system is 180 s? (b) What is the required gain of an integral controller such that the response is underdamped with a damping ratio of 0.8? (c) If a proportional plus integral controller is used with the integral gain as determined in part (b), what is the required proportional gain such that the system is critically damped? (d) For the output to be compared to the input after the feedback loop, a sensor is placed in the system to convert the input from perturbation of flow rate to its resultant final value of perturbation in liquid level. This is accomplished by multiplying the perturbation flow rate by the resistance $R$. Plot the output for the system when this sensor is used with each of the controllers specified in parts (a)–(c) when the perturbation in flow rate is 0.02 m³/s.

**Solution**

The resistance of the system at the steady state is calculated as

$$R = 2\frac{H}{Q}$$

$$= 2\frac{12\,\text{m}}{1.5\,\text{m}^3/\text{s}} = 16\ \text{s/m}^2 \tag{a}$$

The linearized equation for the perturbation in the level of the tank is

$$A\frac{dh}{dt} + \frac{1}{R}h = q(t) \tag{b}$$

The transfer function for the system is

$$G(s) = \frac{1}{A\left(s + \dfrac{1}{T}\right)} \tag{c}$$

where the time constant is

$$T = RA$$

$$= \left(16\,\text{s/m}^2\right)\left(20\ \text{m}^2\right) = 320\ \text{s} \tag{d}$$

The constant $K$ in Equation (8.36) for this system is

$$K = \frac{1}{A} = 0.05\ \text{m}^{-2} \tag{e}$$

(a) The 2 percent settling time for the closed-loop system with a proportional controller is $3.92\hat{T}$ where $\hat{T}$ is the time constant for the system. Requiring the settling time to be 180 s leads to

$$\hat{T} = \frac{t_s}{3.92} = \frac{180\ \text{s}}{3.92} = 45.92\ \text{s} \tag{f}$$

Equation (8.38) is rearranged to solve for the proportional gain in terms of the time constant of the plant and the time constant of the closed-loop system as

$$K_p = \frac{T - \hat{T}}{KT\hat{T}} \tag{g}$$

Substituting calculated values into Equation (g) leads to

$$K_p = \frac{(320\,\text{s}) - (45.92\,\text{s})}{(0.05\,\text{m}^{-2})(320\,\text{s})(45.92\,\text{s})} = 0.373\,\text{m}^2/\text{s} \tag{h}$$

The equivalent closed-loop transfer function for this system is obtained using Equation (8.37) as

$$H(s) = \frac{KK_p}{s + \dfrac{1}{\hat{T}}}$$

$$= \frac{(0.05)(0.373)}{s + \dfrac{1}{45.92}}$$

$$= \frac{1.87 \times 10^{-2}}{s + 2.18 \times 10^{-2}} \tag{i}$$

(b) The damping ratio for a feedback system with an integral controller is given by Equation (8.46). The integral gain is determined in terms of the damping ratio as

$$K_i = \frac{1}{4KT^2\zeta^2} \tag{j}$$

Requiring the damping ratio to be 0.8 leads to

$$K_i = \frac{1}{4(0.05\,\text{m}^{-2})(320\,\text{s})^2(0.8)^2} = 7.63 \times 10^{-5}\,\frac{\text{m}^2}{\text{s}^2} \tag{k}$$

The transfer function when this controller is used is obtained using Equation (8.44) as

$$H(s) = \frac{(0.05)(7.63 \times 10^{-5})}{s^2 + \dfrac{1}{320}s + (0.05)(7.63 \times 10^{-5})}$$

$$= \frac{3.82 \times 10^{-6}}{s^2 + 3.13 \times 10^{-3}s + 3.82 \times 10^{-6}} \tag{l}$$

(c) The damping ratio for a proportional plus integral controller used with a first-order plant is given by Equation (8.51). Rearranging to solve for the proportional gain leads to

$$K_p = 2\sqrt{\frac{K_i}{K}}\left(\zeta - \frac{1}{2T\sqrt{KK_i}}\right) \tag{m}$$

Noting that since the integral controller of part (b) is designed for a damping ratio of 0.8, requiring a damping ratio of 1 when a PI controller is used leads to

$$K_p = 2\sqrt{\frac{7.63 \times 10^{-5}}{0.05}}(1 - 0.8)$$

$$= 1.57 \times 10^{-2} \tag{n}$$

The transfer function when a PI controller is used is given by Equation (8.50), which for this system becomes

$$H(s) = \frac{(0.05)\left(1.57 \times 10^{-2}s + 7.63 \times 10^{-5}\right)}{s^2 + \left[\dfrac{1}{320} + (0.05)(1.57 \times 10^{-2})\right]s + (0.05)(7.63 \times 10^{-5})}$$

$$= \frac{7.85 \times 10^{-4}s + 3.82 \times 10^{-6}}{s^2 + 3.91 \times 10^{-3}s + 3.82 \times 10^{-6}} \tag{o}$$

(d) The use of the sensor before the feedback control loop leads to an input of

$$A(t) = qRu(t)$$

$$= \left(0.02 \,\frac{\text{m}^3}{\text{s}}\right)\left(16 \,\frac{\text{s}}{\text{m}^2}\right)u(t) = 0.32u(t) \tag{p}$$

The responses of the system using the feedback controllers specified in parts (a)–(c) are shown in Figure 8.32. The offset when using the proportional controller is apparent.

**FIG. 8.32** The response of the tank of Example 8.9 with feedback control using (a) the proportional controller of part (a), (b) the integral controller of part (b), and (c) the proportional plus integral controller of part (c). (*Continued*)

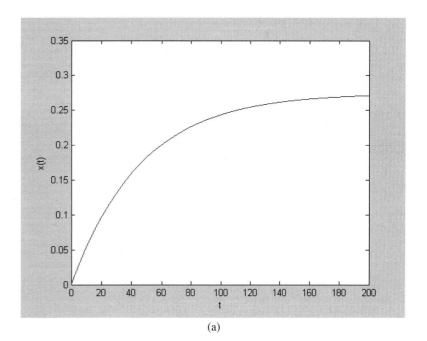

(a)

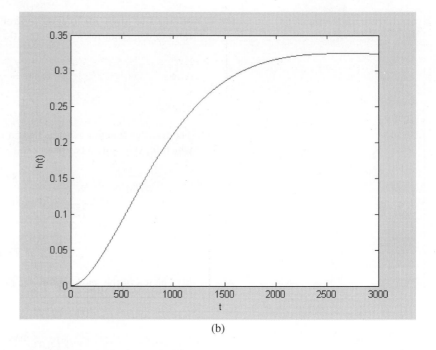

(b)

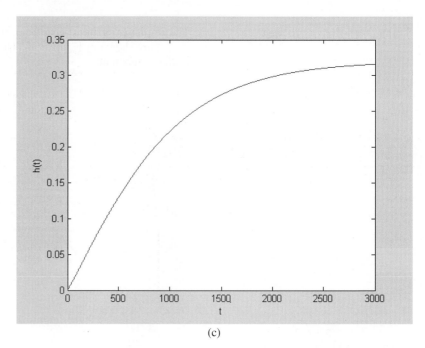

**FIG. 8.32**  (*Continued*)

(c)

**Example 8.10**

Develop a SIMULINK model for a PID controller to be used as an actuator in the feedforward part of a feedback control loop for the system of Example 8.9.

**Solution**

The SIMULINK model is illustrated in Figure 8.33. The model is developed such that the gains can be changed independently. The model uses the Integrator block and the Derivative block from the continuous menu of the SIMULINK Library Browser.

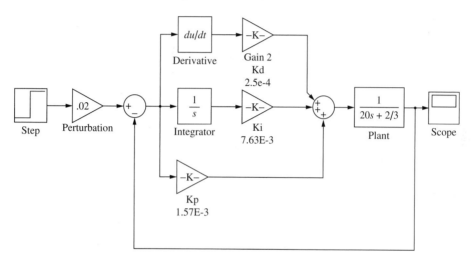

**FIG. 8.33** The SIMULINK model for Example 8.10; a first-order plant in a feedback loop with a PID controller.

## 8.5 CONTROL OF SECOND-ORDER PLANTS

The transfer function of a second-order system is of the form

$$G(s) = \frac{As + B}{s^2 + 2\zeta\omega_n s + \omega_n^2} \tag{8.56}$$

where $\zeta$ is the damping ratio for the system, $\omega_n$ is its natural frequency, and $A$ and $B$ are system constants. The use of a proportional controller on a second-order plant in the feedback control loop of Figure 8.25 leads to a closed-loop transfer function of

$$H(s) = \frac{K_p(As + B)}{s^2 + (2\zeta\omega_n + K_pA)s + K_pB + \omega_n^2} \tag{8.57}$$

Equation (8.57) is the transfer function of a second-order system of natural frequency

$$\hat{\omega}_n = \sqrt{\omega_n^2 + K_pB} \tag{8.58}$$

and damping ratio

$$\hat{\zeta} = \frac{\zeta + \dfrac{K_p A}{2\omega_n}}{\sqrt{1 + \dfrac{K_p B}{\omega_n^2}}} \tag{8.59}$$

The transfer function for the response of a second-order plant with proportional feedback control when subject to a unit step input is

$$C(s) = \frac{K_p(As + B)}{s\left(s^2 + 2\hat{\zeta}\hat{\omega}_n + \hat{\omega}_n^2\right)} \tag{8.60}$$

The transform of the error due to a unit step input is

$$E(s) = \frac{1}{s}[1 - H(s)]$$

$$= \frac{s^2 + 2\zeta\omega_n s + \omega_n^2}{s\left(s^2 + 2\hat{\zeta}\hat{\omega}_n s + \hat{\omega}_n^2\right)} \tag{8.61}$$

The offset is evaluated as

$$\lim_{t \to \infty} e(t) = \lim_{s \to 0} sE(s)$$

$$= \left(\frac{\omega_n}{\hat{\omega}_n}\right)^2$$

$$= \frac{1}{1 + \dfrac{K_p B}{\omega_n^2}} \tag{8.62}$$

## Example 8.11

Plot the response of a second-order plant with $A = 0$, $B = 1$, $\zeta = 1.5$, and $\omega_n = 10$ when subject to a unit step input and placed in a feedback loop with a proportional controller for various values of the nondimensional parameter $\kappa = K_p B/\omega_n^2$.

### Solution

For the given values, Equations (8.58) and (8.59) are used to determine the natural frequency and damping ratio of the closed-loop system as

$$\hat{\omega}_n = 10\sqrt{1 + \kappa} \tag{a}$$

$$\hat{\zeta} = \frac{1.5}{\sqrt{1 + \kappa}} \tag{b}$$

Equation (b) shows that with $A = 0$ the damping ratio for the closed-loop system is less than the damping ratio for the plant itself. The closed-loop system is overdamped ($\hat{\zeta} > 1$) when $\kappa < 1.25$, is critically damped for $\kappa = 1.25$, and is overdamped for $\kappa > 1.25$. The response of the system in each case is determined using Table 6.6. Plots for different values of $\kappa$ are given in Figure 8.34.

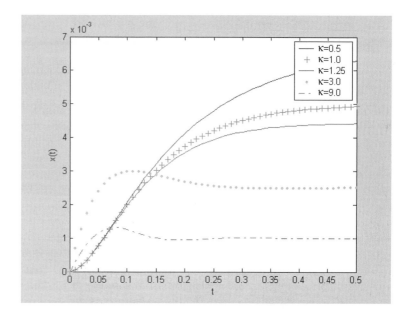

**FIG. 8.34** The response of second-order plant subject to unit step input when placed in a feedback loop with a proportional controller. The natural frequency and damping ratio of the closed-loop system vary with the proportional gain. The offset is also dependent on the gain.

The use of an integral controller in a feedback loop with a second-order plant leads to a closed-loop transfer function of

$$H(s) = \frac{K_i(As + B)}{s^3 + 2\zeta\omega_n s^2 + (\omega_n^2 + K_i A)s + K_i B} \quad (8.63)$$

The closed loop transfer function of Equation (8.63) is that of a third-order system. The system is stable if each of its three poles has negative real parts. The application of Routh's stability criterion to a third-order polynomial of the form

$$p(s) = s^3 + a_2 s^2 + a_1 s + a_0 \quad (8.64)$$

shows that all the roots of $p(s)$ have negative real parts if and only if

$$a_1 a_2 > a_0 \quad (8.65)$$

The application of Equation (8.65) to the transfer function of Equation (8.63) leads a stability criterion of

$$2\zeta\omega_n(\omega_n^2 + K_i A) > K_i B$$

$$K_i < \frac{2\zeta\omega_n^3}{B - 2\zeta\omega_n A} \quad (8.66)$$

If the integral gain $K_i$ satisfies Equation (8.66) the closed-loop system is stable. If $K_i$ is equal to the value on the right-hand side of Equation (8.66) then one pole has a real part of zero and the system is neutrally stable.

In this case the response due to a unit step input is oscillatory with constant amplitude. If $K_i$ is greater than that value on the right-hand side of Equation (8.63) then the system is unstable and the response due to a unit step input grows without bound.

The use of a PI controller in a closed-loop feedback system with a second-order plant leads to a closed-loop transfer function of the form

$$H(s) = \frac{K_p A s^2 + (K_i A + K_p B)s + K_i B}{s^3 + (2\zeta\omega_n + K_p A)s^2 + (\omega_n^2 + K_p B + K_i A)s + K_i B} \tag{8.67}$$

Equation (8.67) is the transfer function for a third-order system. The application of Equation (8.65) to the transfer function of Equation (8.67) leads to a stability criterion of

$$(2\zeta\omega_n + K_p A)(\omega_n^2 + K_p B + K_i A) > K_i B \tag{8.68}$$

Equation (8.68) can be rearranged to

$$ABK_p^2 + (2\zeta\omega_n B + \omega_n^2 A + K_i A)K_p + 2\zeta\omega_n^3 + 2\zeta\omega_n K_i A - K_i AB > 0 \tag{8.69}$$

If $A = 0$, Equation (8.69) shows that if $K_i$ satisfies Equation (8.66), then Equation (8.69) is satisfied for all positive values of $K_p$. If $K_i$ does not satisfy Equation (8.66) then a value of $K_p$ can be obtained to render the system stable. Thus the addition of proportional control enhances the stability of the system. The same result is obtained if $A$ and $B$ have the same sign. If $B = 0$ then $A$, $K_i$ and $K_p$ all being of the same sign is sufficient for stability.

---

## Example 8.12

Consider a second-order plant with $A = 0$, $B = 1$, $\zeta = 1.5$, and $\omega_n = 10$. (a) If the plant is placed in a feedback loop with an integral controller, for what values of the integral gain is the system stable? (b) Determine and plot the output from the feedback loop for various values of the integral gain when the system is subject to a unit step input. (c) If the plant is placed in a feedback loop with a PI controller with an integral gain of $K_i = 4,000$, or what values of the proportional gain constant is the system stable? (d) Determine and plot the output from the feedback loop for various values of the proportional gain when the system is subject to a unit step input.

### Solution

(a) The application of Equation (8.66) to the specified system leads to a stability criterion of

$$K_i < \frac{2(1.5)(10)^3}{(1)} = 3,000 \tag{a}$$

(b) The equivalent closed-loop transfer function for the system, in terms of the integral gain, is obtained using Equation (8.63) as

$$H(s) = \frac{K_i}{s^3 + 30s^2 + 100s + K_i} \tag{b}$$

**TABLE 8.3 POLES FOR EXAMPLE 8.12 IN TERMS OF THE INTEGRAL GAIN**

| $K_i$ | $s_1$ | $s_2$ | $s_3$ |
|---|---|---|---|
| 1 | $-26.35$ | $-3.81$ | $-0.010$ |
| 50 | $-26.26$ | $-3.13$ | $-0.609$ |
| 88.6621 | $-26.33$ | $-1.835$ | $-1.835$ |
| 100 | $-26.35$ | $-1.853 + 0.680\,j$ | $-1.853 - 0.680\,j$ |
| 500 | $-26.98$ | $-1.510 + 4.031\,j$ | $-1.510 - 4.031\,j$ |
| 1,000 | $-27.69$ | $-1.535 + 5.877\,j$ | $-1.535 - 5.877\,j$ |
| 2,000 | $-28.94$ | $-0.533 + 8.297\,j$ | $-0.533 - 8.927\,j$ |
| 2,500 | $-29.48$ | $-0.258 + 9.205\,j$ | $-0.258 - 9.205\,j$ |
| 3,000 | $-30.00$ | $10\,j$ | $-10\,j$ |
| 4,000 | $-30.95$ | $0.473 + 11.36\,j$ | $-0.473 - 11.36\,j$ |

The poles of the transfer function for various values of $K_i$ are given in Table 8.3. For $K_i < 88.6621$ all poles of Equation (b) real, distinct, and negative. For $K_i = 88.6621$ the transfer function has two real negative poles as $-1.835$ is repeated. For $88.6621 < K_i < 3,000$ the transfer function has one real negative pole and two complex conjugate poles with negative real parts. For $K_i = 3,000$ the transfer function has one real negative pole and two purely complex poles. For $K_i > 3,000$ the transfer function has one real negative pole and two complex conjugate poles with positive real parts.

The transform of the output when the input is a unit step input is

$$C(s) = \frac{1}{s} H(s)$$

$$= \frac{K_i}{s(s^3 + 30s^2 + 100s + K_i)} \tag{c}$$

For $K_i < 88.6621$ the appropriate partial fraction decomposition of Equation (c) is

$$C(s) = \frac{B_0}{s} + \frac{B_1}{s - s_1} + \frac{B_2}{s - s_2} + \frac{B_3}{s - s_3} \tag{d}$$

where the residues are evaluated as

$$B_0 = 1 \tag{e}$$

$$B_i = \lim_{s \to s_i} \frac{(s - s_i)K_i}{s(s^3 + 30s^2 + 100s + K_i)} \tag{f}$$

The inversion of Equation (d) leads to

$$c(t) = 1 + B_1 e^{s_1 t} + B_2 e^{s_2 t} + B_3 e^{s_3 t} \tag{g}$$

For $K_i = 88.6621$ the appropriate partial fraction decomposition of Equation (c) is

$$C(s) = \frac{1}{s} - \frac{5.61 \times 10^{-3}}{s + 26.33} - \frac{0.994}{s + 1.835} - \frac{1.973}{(s + 1.835)^2} \tag{h}$$

Inversion of Equation (h) leads to

$$c(t) = 1 - 5.61 \times 10^{-3} e^{-26.33t} - 0.994 e^{-1.835t} - 1.973t e^{-1.835t} \tag{i}$$

For $88.6621 < K_i < 3{,}000$ the appropriate partial fraction decomposition of Equation (c) is

$$C(s) = \frac{1}{s} + \frac{B_1}{s - s_1} + \frac{D_1 s + D_2}{s^2 - 2s_{2r}s + (s_{2r}^2 + s_{2j}^2)} \tag{j}$$

where $s_2 = s_{2r} + j s_{2j}$, $B_1$ is determined using Equation (f), and $D_1$ and $D_2$ are obtained using the usual methods for quadratic factors. The inversion of Equation (j) leads to

$$c(t) = 1 + B_1 e^{s_1 t} + D_1 e^{2s_{2r}t} \cos\left[\sqrt{s_{2r}^2 + s_{2j}^2}\, t\right] + \frac{D_2 + 2s_{2r}D_1}{\sqrt{s_{2r}^2 + s_{2j}^2}} e^{2s_{2r}t} \sin\left[\sqrt{s_{2r}^2 + s_{2j}^2}\, t\right] \tag{k}$$

For $K_i = 3{,}000$ the appropriate partial fraction decomposition of Equation (c) is

$$C(s) = \frac{1}{s} - \frac{0.1}{s + 30} + \frac{0.9s - 3}{s^2 + 100} \tag{l}$$

Inversion of Equation (l) leads to

$$c(t) = 1 - 0.1 e^{-30t} + 0.9 \cos(10t) - 3 \sin(10t) \tag{m}$$

The partial fraction decomposition and the response of the system for $K_i > 3{,}000$ are similar to those given by Equations (j) and (k) with the exception that $s_{2r}$ is positive rather than negative.

Responses for several values of $K_i$ are plotted in Figure 8.35.

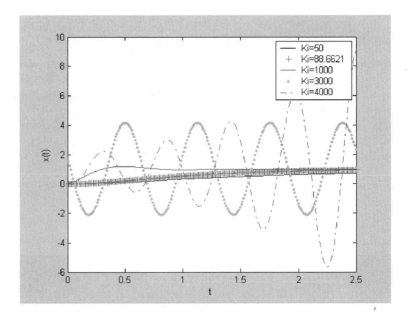

**FIG. 8.35** The step response of second-order plant with an integral controller. For low values of the integral gain the system approaches the final value asymptotically. As the integral gain increases the response becomes oscillatory. When the integral gain reaches a critical value the response becomes unstable and a further increase leads to unbounded growth.

**TABLE 8.4 POLES FOR SYSTEM OF EXAMPLE 8.12 FOR VARIOUS VALUES OF THE PROPORTIONAL GAIN**

| $K_p$ | $s_1$ | $s_2$ | $s_3$ |
|---|---|---|---|
| 0 | $-30.95$ | $0.473 + 11.36j$ | $0.473 - 11.35j$ |
| 10 | $-30.67$ | $0.333 + 11.42j$ | $0.333 - 11.42j$ |
| 33.33 | $-30$ | $11.55j$ | $-11.55j$ |
| 50 | $-29.51$ | $-0.245 + 11.64j$ | $-0.245 - 11.64j$ |
| 100 | $-27.96$ | $-1.018 + 11.92j$ | $-1.018 - 11.92j$ |
| 500 | $-10$ | $-10 + 17.32j$ | $-10 - 17.32j$ |
| 1,000 | $-4.018$ | $-12.99 + 28.76j$ | $-12.99 - 28.76j$ |
| 2,000 | $-1.956$ | $-14.02 + 42.99j$ | $-14.02 - 14.99j$ |
| 10,000 | $-0.397$ | $-14.80 + 99.34j$ | $-14.80 - 99.34j$ |

(c) An integral controller with $K_i = 4,000$ leads to an unstable closed-loop system. If proportional control is added, Equation (8.69) shows that, to have a stable system

$$K_p > \frac{K_i - 2\zeta\omega_n^3}{2\zeta\omega_n} = \frac{4,000 - 2(1.5)(10)^3}{2(1.5)(10)}$$

$$K_p > \frac{100}{3} \tag{n}$$

(d) The closed-loop transfer function for the system in terms of the proportional gain is obtained using Equation (8.67) as

$$H(s) = \frac{K_p s + 4,000}{s^3 + 30s^2 + (100 + K_p)s + 4,000} \tag{o}$$

Table 8.4 presents the poles of the transfer function of Equation (o) for several values of the proportional gain. For all values of $K_p$ the transfer function has one real and negative pole and two complex conjugate poles. For $0 \le K_p < 100/3$ the complex poles have a positive real part and thus the closed-loop system is unstable. For $K_p = 100/3$ the complex poles have a real part equal to zero and the closed-loop system is neutrally stable. For $K_p > 100/3$ the complex poles have negative real parts and the system is stable. It is also noted that when $K_p = 500$ the real part of the complex poles equals the value of the real pole. For proportional gains greater than 500 the absolute value of the real part of the complex poles is larger than the absolute value of the real pole. For these values of the proportional gain the settling time is dominated by the real pole.

The transform of the system output when it is subject to a unit step input is

$$C(s) = \frac{K_p s + 4,000}{s\left[s^3 + 30s^2 + (100 + K_p)s + 4,000\right]} \tag{p}$$

For all values of $K_p$ the appropriate partial fraction decomposition of Equation (p) is

$$C(s) = \frac{1}{s} + \frac{B_1}{s - s_1} + \frac{D_1 s + D_2}{s^2 + 2s_{2r}s + s_{2r}^2 + s_{2j}^2} \tag{q}$$

The system response is of the same form as Equation (j) and is plotted for several values of $K_p$ in Figure 8.36.

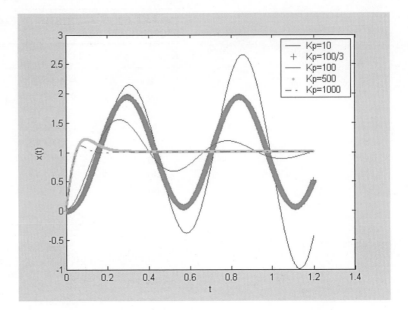

**FIG. 8.36** The step response of the second-order plant of Example 8.12 with a PI controller with $K_i = 4,000$ and various values of $K_p$. The system is unstable with only the integral gain. As the proportional gain increases the system becomes stable.

The use of a proportional plus derivative controller in a feedback loop with a second-order plant leads to a closed-loop transfer function of the form

$$H(s) = \frac{K_d A s^2 + (K_p A + K_d B)s + K_p B}{(1 + K_d A)s^2 + (K_p A + K_d B + 2\zeta\omega_n)s + K_p B + \omega_n^2}$$ (8.70)

Consider first the case when $A = 0$ and Equation (8.70) reduces to

$$H(s) = \frac{K_d B s + K_p B}{s^2 + (K_d B + 2\zeta\omega_n)s + K_p B + \omega_n^2}$$ (8.71)

Equation (8.71) is the transfer function of a second-order system. The natural frequency is unaffected by the derivative gain is the same as that for a system with only proportional control and is given by Equation (8.58) as

$$\hat{\omega}_n = \sqrt{\omega_n^2 + K_p B}$$ (8.58)

The damping ratio of the closed-loop system is

$$\hat{\zeta} = \frac{K_d B + 2\zeta\omega_n}{2\hat{\omega}_n}$$ (8.72)

Equation (8.72) shows that the use of derivative control results in an increase in the system's damping ratio from that if only proportional control is used.

For a nonzero value of $A$, Equation (8.70) represents the transfer function for a second-order system, but with the numerator of the same order as the denominator. The response of such a system due to a unit step input is discontinuous at $t = 0$. The natural frequency and damping ratio for this system are

$$\hat{\omega}_n = \sqrt{\frac{K_p B + \omega_n^2}{1 + K_d A}} \tag{8.73}$$

$$\hat{\zeta} = \frac{K_d B + 2\zeta\omega_n}{2\hat{\omega}_n(1 + K_d A)} \tag{8.74}$$

---

**Example 8.13**

(a) Consider a second-order plant with $A = 0, B = 1, \zeta = 1.5$, and $\omega_n = 10$. The plant is placed in a feedback loop with a PD controller with $K_p = 50$. Determine and plot the response of the system for various values of $K_d$ when it is subject to a unit step input. (b) Repeat part (a) with $K_p = 500$. (c) Repeat part (a) for a second-order plant with $A = 1, B = 1, \zeta = 1.5$, and $\omega_n = 10$.

**Solution**

(a) The response of a system with a proportional controller with $K_p = 50$ is overdamped. When derivative control is introduced in the controller the response continues to be overdamped, as illustrated in Figure 8.37(a). The system is quicker to respond to the disturbance for higher values of the derivative gain but has a larger settling time. Overshoot occurs for larger values of $K_d$ (b). The response of a system with a proportional controller with $K_p = 500$ is underdamped. The system is underdamped for $K_d < 18.98$. As in part (a) the system is quicker to respond for

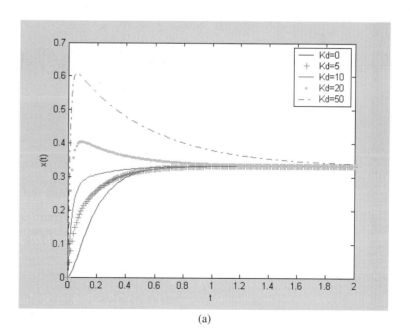

**FIG. 8.37** The step response of a second-order plant of Example 8.13 in a feedback loop with a PD controller for various values of the derivative gain with (a) $K_p = 50$ in which the system is overdamped without a derivative gain and (b) $K_p = 500$ in which the system is underdamped without a derivative gain. (*Continued*)

(a)

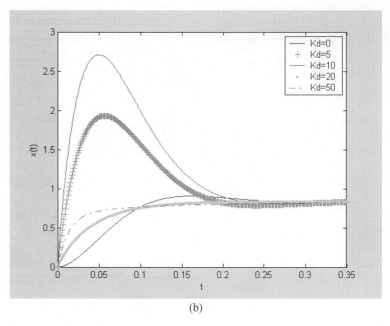

(b)

larger values of the derivative gain, but the settling time is also longer. When the system is overdamped, overshoot occurs. These responses are illustrated in Figure 8.37(b). (c) For $K_p = 50$ the natural frequency and damping ratio are determined using Equations (8.58) and (8.72) as

$$\hat{\omega}_n = \sqrt{\frac{150}{1 + K_d}} \tag{a}$$

$$\hat{\zeta} = \frac{K_d + 30}{2\sqrt{150(1 + K_d)}} \tag{b}$$

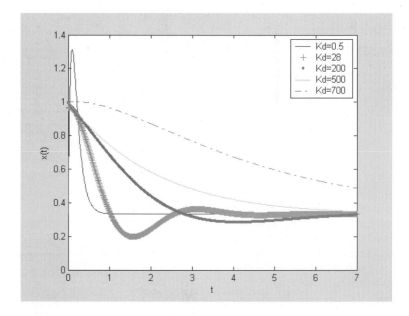

**FIG. 8.38** The step response of a second-order plant whose transfer function is $G(s) = (s + 1)/(s^2 + 30s + 100)$ placed in a feedback loop with a PD controller with a constant proportional gain and varying values of derivative gain.

The natural frequency decreases as the derivative gain increases. Analysis using Equation (b) reveals that the system is underdamped when $0.556 < K_d < 539.4$, critically damped when $K_d = 0.556$ or $K_d = 539.4$ and overdamped otherwise. The damping ratio has a minimum value of 0.44 when $K_d = 28$. The responses for several values of $K_d$ are plotted in Figure 8.38. The response of each system is discontinuous at $t = 0$ with a value of $x_s(0) = K_d/(1 + K_d)$. The addition of a derivative term does not affect the offset. For this system the response is slower for larger values of $K_d$.

The closed-loop transfer function when a PID controller is used with a second-order plant in a feedback loop is

$$H(s) = \frac{K_d A s^3 + (K_p A + K_d B)s^2 + (K_p B + K_i A)s + K_i B}{(1 + K_d A)s^3 + (2\zeta\omega_n + K_p A + K_d B)s^2 + (\omega_n^2 + K_p B + K_i A)s + K_i B}$$

(8.75)

Equation (8.75) is that of a third-order system. The application of Routh's criterion leads to a stability condition of the form

$$(2\zeta\omega_n + K_p A + K_d B)(\omega_n^2 + K_p B + K_i A) > (1 + K_d A)K_i B \qquad (8.76)$$

If $A = 0$, Equation (8.76) reduces to

$$(2\zeta\omega_n + K_d B)(\omega_n^2 + K_p B) > K_i B \qquad (8.77)$$

Comparison between Equations (8.68) and (8.77) shows that the addition of a derivative term to a PI controller enhances the stability of the system with a second-order plant. If $A$ is other than zero no general conclusions can be drawn regarding the effect of the derivative gain on stability. Also, as with the PD controller, since for $A \neq 0$ the order of the polynomial in the numerator of $H(s)$ is the same as the order of the polynomial in the denominator, the response of the system due to a unit step input is discontinuous at $t = 0$.

## Example 8.14

Consider a second-order plant with $A = 0$, $B = 1$, $\zeta = 1.5$, and $\omega_n = 10$. The plant is placed in a feedback loop with a PID controller with $K_i = 4,000$ and $K_p = 10$. Plot the response of the system when subject to a unit step input for various values of the derivative gain.

### Solution

The transfer function for this system in terms of the derivative gain is

$$H(s) = \frac{K_d s^2 + 10s + 4,000}{s^3 + (30 + K_d)s^2 + 110s + 4,000} \qquad (a)$$

Example 8.13 shows that a second-order system with actuation by a PI controller with $K_i = 4,000$ and $K_p = 10$ is unstable; $H(s)$ has a root with a positive real part. The application of Equation (8.77) to the denominator of Equation (a) leads to a stability criteria of $K_d > 6.37$ for the application of a PID controller

**TABLE 8.5 POLES OF $H(s) = (K_d s^2 + 10s + 4,000)/(s^3 + (30 + K_d)s^2 + 110s + 4,000)$ FOR VARIOUS VALUES OF $K_d$**

| $K_d$ | $s_1$ | $s_2$ | $s_3$ |
|---|---|---|---|
| 0 | −30.67 | $0.337 + 11.42j$ | $0.337 − 11.42j$ |
| 5 | −35,12 | $0.0559 + 10.67j$ | $0.559 + 10.67j$ |
| 6.360 | −36.36 | $10.49j$ | $−10.49j$ |
| 10 | −37.76 | $−0.118 + 10.03j$ | $−0.118 − 10.03j$ |
| 20 | −49.41 | $−0.294 + 8.99j$ | $−0.294 − 8.99j$ |
| 100 | −129.4 | $−0.3101 + 5.56j$ | $−0.3101 − 5.56j$ |

with the same values of the proportional and integral gains. Table 8.5 gives the poles of $H(s)$ for several values of $K_d$. Figure 8.39 shows MATLAB-generated plots of the step response of the system for several values of $K_d$.

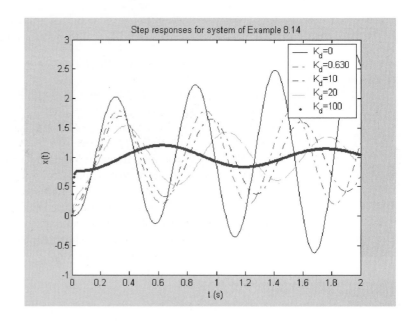

**FIG. 8.39** The step responses of system of Example 8.14 for various values of $K_d$. The response for $K_d = 0$ is unstable and grows without bound. The response for $K_d = 6.360$ is neutrally stable. The responses for $K_d > 6.360$ are stable.

## 8.6 CONTROL SYSTEM DESIGN

Sections 8.3–8.5 illustrate the basics of control systems. The types of control strategies are introduced in Section 8.3 and their use illustrated on first-order plants in Section 8.4 and second-order plants in Section 8.5. These applications illustrate the effect of the controllers on the closed-loop response. General concepts discovered include the following:

- Response parameters such as time constants, damping ratios, and natural frequencies are dependent on controller gains.

- Proportional controllers often lead to desirable response characteristics but also to error and offset.
- Integral control is used to eliminate error and offset.
- Integral control is destabilizing and its use may lead to an unstable closed-loop response for second-order and higher plants.
- Derivative control is used to enhance the transient response characteristics.
- Derivative control adds damping to the closed-loop system and thus allows higher values of proportional and integral gains.

Control system design involves the appropriate selection of gains to achieve the desired transient response or frequency response of the closed-loop system. This section is an introduction to common techniques used in control system design.

## 8.6.1 Design Using Root-Locus Diagrams

An important consideration in control system design is stability. Obviously it is required that the closed-loop system is absolutely stable, but transient response is also dependent on relative stability. Absolute stability requires that all poles lie in the left half of the complex plane while relative stability requires that all poles lie to the left of a specified horizontal line.

The location of the poles in a closed-loop system depends on the values of the controller gains. One goal of a designer is to determine controller gains to achieve the desired level of stability. Some control systems are adaptive, in which the gains may vary during the operation of the system. Thus the designer must be aware of pole location as the gains vary.

The root locus method, introduced in Section 6.3 and presented in detail in Appendix D, is used to determine the location of the poles with the variation of a parameter, usually one of the gains. Suppose the poles of the transfer function are the roots of an equation of the form of Equation (6.32)

$$Q(s) + KR(s) = 0 \qquad (6.32)$$

The root locus is a complex plane plot of the roots of Equation (6.32) with the variation of $K$. If $Q(s)$ is of $n$th order the plot has $n$ branches, one for each root. The construction of the root locus plot is described in Appendix D and the MATLAB commands to generate a root-locus plot are given in Section 6.3.

The root-locus plot and its features are illustrated using the closed-loop transfer function of Example 8.12

$$H(s) = \frac{K_i}{s^3 + 30s^2 + 100s + K_i} \qquad (8.78)$$

Equation (8.78) results from the use of an integral controller of gain $K_i$ on a second-order plant of damping ratio 1.5 and natural frequency 10.

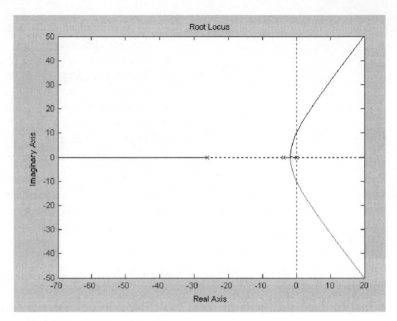

**FIG. 8.40** MATLAB-generated root-locus plot for the system of Example 8.12

The denominator of Equation (8.78) is put into the form of Equation (6.32) with

$$Q(s) = s^3 + 30s^2 + 100s \qquad (8.79a)$$

$$R(s) = 1 \qquad (8.79b)$$

The MATLAB commands to generate the root locus plot for Equation (8.78) are

$$Q = [1\ 30\ 100\ 0];$$

$$R = [1];$$

$$\text{rlocus (R,Q)}$$

The resulting root-locus plot is given in Figure 8.40. Since $Q(s)$ is third order, the root-locus diagram has three branches. Also since $Q(s)$ is third order there are either one real pole and two complex conjugate poles or three real poles. The root-locus diagram clearly shows the path of one pole that is real for all values of $K$. The paths of the other two poles are along the real axis for some values of $K$. A value of K exists such that the two paths meet, corresponding to a real root of multiplicity 2, and then the paths diverge from the real axis. The direction of the paths is obtained by examining Equation (8.79). The poles for $K = 0$ correspond to the roots of $Q(s)$, which are easily determined as 0, −3.82, and −26.2. The roots −3.82 and −26.2 correspond to the poles of the plant. The root $s = 0$ is a pole of the root-locus diagram, but not a pole of the plant. This pole occurs only for $K = 0$ and is introduced when Equation (8.78) is derived from Equation (8.63) by the multiplication of both numerator and denominator by zero. Thus, as $K$ increases the root that remains real

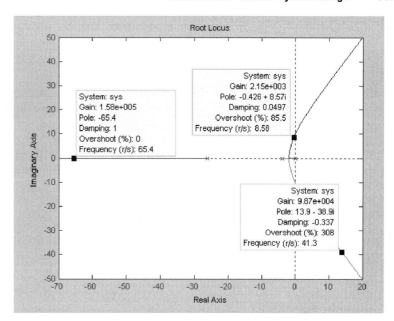

FIG. **8.41** The MATLAB-generated root locus diagram, annotated to show the properties of points on root-locus paths.

for all values of $K$ moves along the real axis in the negative direction. The root that starts at $s = 0$ for $K = 0$ moves along the real axis in the negative direction and the root that starts at $s = -3.82$ moves along the real axis in the positive direction until they meet at the breakaway point. Then both diverge from the real axis and become complex conjugates. The system remains stable until these roots cross the imaginary axis at the crossover frequency.

The **dominant root** is the root with the largest real part. This root is clearly along the path starting from $s = 0$. This root is dominant because it leads to the greatest term in the system response.

Properties of the roots at a point on any of the paths may be obtained by using a mouse to click on that point, as illustrated in Figure 8.41. For example the point chosen on the path that exclusively follows the real axis is at $s = -65.4$ and corresponds to a value of $K = 1.58 \times 10^5$. All roots that lie on the real axis are at least critically damped and are assigned a damping ratio of 1. Now consider the point chosen corresponding to $s = -0.426 + 8.57j$. This point corresponds to a value of $K = 2.15 \times 10^3$ (recall from Example 8.12 that the system is stable for $K < 3,000$). Complex poles are represented in terms of damping ratio $\zeta$ and frequency $\omega$ corresponding the definitions of damping ratio and damped natural frequency of a second-order system

$$s = -\zeta\omega \pm j\omega\sqrt{1 - \zeta^2} \qquad (8.80)$$

Thus for this point $\zeta\omega = -0.426$ and $\omega\sqrt{1 - \zeta^2} = 8.57$. These equations are solved simultaneously yielding $\zeta = 0.0497$ and $\omega = 8.58$. The third point chosen is at $s = 13.9 - 38.9j$ and corresponds to an unstable system with a gain of $K = 9.87 \times 10^4$. The damping ratio and frequency for this point are calculated using Equation (8.80), but since the real part of $s$ is positive, the damping ratio is negative.

The percent overshoot is related to the damping ratio by

$$\eta = \left( e^{-\zeta\pi/\sqrt{1-\zeta^2}} \right) \tag{8.81}$$

A general point in the complex $s$ plane is of the form $s = s_r + js_j$. Each point in the upper half of the complex plane also represents a unique pair of damping ratio and natural frequency related to $s$ by Equation (8.80). Rearranging Equation (8.80) to solve for $\zeta$ and $\omega$ leads to

$$s_r^2 + s_j^2 = \omega^2$$
$$\omega = |s| \tag{8.82}$$

and

$$\phi = \tan^{-1}\left(\frac{s_j}{s_r}\right) = \frac{\sqrt{1-\zeta^2}}{-\zeta} \tag{8.83}$$

Thus curves for constant $\omega$ are circles in the s plane centered at the origin. The curves of the constant $\zeta$ correspond to the lines of the constant $\phi$. A system with only one parameter to choose can be designed such that

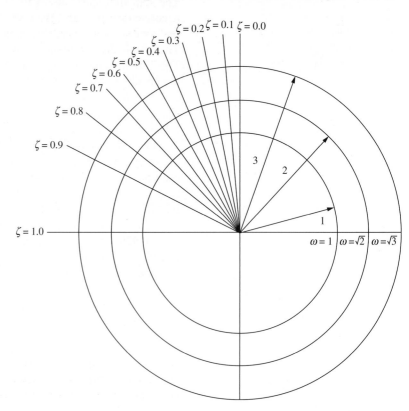

FIG. 8.42 Constant $\omega$ curves are circles in complex plane while constant $j$ curves are lines in second quadrant.

the dominant root corresponds to a specific value of the damping ratio or a specific value of the frequency, not both. The system of Figure 8.40 can be designed such that the dominant pole has any damping ratio between 0 and 1. However, a value of $K$ cannot be found leading to the dominant root having larger natural frequencies because the radius of the circle of constant $\omega$ may fall outside the range for which the system is stable. The representation of the $s$ plane in terms of $\zeta$ and $\omega$ is illustrated in Figure 8.42. In this representation the positive imaginary axis corresponds to $\zeta = 0$ while the negative real axis corresponds to $\zeta = 1$.

## Example 8.15

A second-order plant with

$$G(s) = \frac{1}{s^2 + 30s + 200} \tag{a}$$

is placed in a feedback control loop with a PI controller. Use the root-locus plot as well as some algebraic methods to answer the following: (a) If $K_p = 25$ describe the properties of the root-locus diagram, drawn using $K_i$ as the parameter. Identify any poles, breakaway points, and crossover frequencies. For what values of $K_i$ is the closed-loop response stable? (b) If $K_p = 25$ what is the required integral gain such that the dominant pole of the closed-loop system has an overshoot of 25 percent? What is the corresponding frequency of the dominant pole? (c) Plot the step response of the system when the controller of part (b) is used. (d) If $K_i = 2{,}000$ use MATLAB to draw the root-locus diagram using $K_p$ as the parameter. Describe how the poles of the transfer function change as the proportional gain varies. What is the minimum overshoot that can be attained? Discuss how the dominant pole varies as $K_p$ changes and its effect on the system response.

### Solution

The closed-loop transfer function when the plant of Equation (a) is used with a PI controller is obtained from Equation (8.67) as

$$H(s) = \frac{K_p s + K_i}{s^3 + 30s^2 + (200 + K_p)s + K_i} \tag{b}$$

(a) Setting $K_p = 25$ leads to

$$\frac{25s + K_i}{s^3 + 30s^2 + 225s + K_i} \tag{c}$$

MATLAB is used to generate the root-locus plot of Figure 8.43 using

$$Q = [1 \ 30 \ 225 \ 0]$$

$$R = [1]$$

The poles of the diagram are the roots of $Q(s) = s^3 + 30s^2 + 225s$. $Q(s)$ has a root $s = 0$ and a double root at $s = -15$. There are no zeroes for this diagram. Since $Q(s)$ is third order, the root-locus diagram has three branches. One branch originates at $s = -15$ and $K_i$ moves along the real axis in the negative direction. Since there are no zeroes all branches terminate at infinity. A second branch

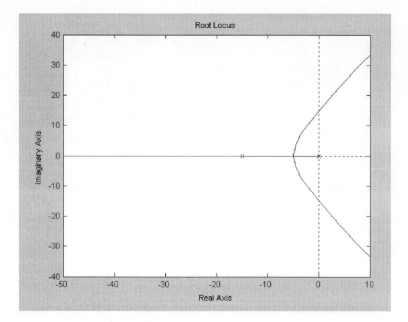

**FIG. 8.43** The MATLAB-generated root-locus diagram for the closed-loop system of Example 8.15 with $K_p = 25$.

originates at $s = -15$ and moves along the real axis in the positive direction for increasing $K$ until it reaches the breakaway point. The third branch originates at the pole $s = 0$ and moves along the real axis in the negative direction for increasing $K$ until it meets the second branch at the breakaway point.

Clicking the mouse as close as possible to the breakaway point shows that it occurs at $s = -5.14$ corresponding to $K = 500$. The exact values of the breakaway point and corresponding gain are obtained using Equation (D.11), which when applied to this system leads to

$$R\frac{dQ}{ds} - Q\frac{dR}{ds} = 0$$

$$(1)(3s^2 + 60s + 225) - (s^3 + 30s^2 + 225s) = 0$$

$$3s^2 + 60s + 225 = 0 \tag{d}$$

The solutions of Equation (d) are $s = -5$ and $s = -15$. The latter root corresponds to the breakin point where the two branches originate. Thus the breakaway point is $s = -5$. The integral gain is obtained from the magnitude criterion, Equation (D.3) that must be satisfied at all points on a branch of the root-locus diagram. To this end

$$K_i = \left|\frac{Q(-5)}{R(-5)}\right|$$

$$K_i = \left|(-5)^3 + 30(-5)^2 + 225(-5)\right| = 500 \tag{e}$$

The crossover frequency is the frequency that corresponds to where the branch crosses the imaginary axis. The crossover frequency and the corresponding gain are determined from substituting $s = j\omega$ in Equation (6.32)

$$(j\omega)^3 + 30(j\omega)^2 + 225(j\omega) + K = 0$$

$$(225\omega - \omega^3) + j(K - 30\omega^2) = 0 \tag{f}$$

Setting the real part of the right-hand side of Equation (f) to zero leads to a crossover frequency of $\omega = 15$. Then setting the imaginary part to zero leads to $K_i = 6,750$. Thus the system is stable for all $K_i < 6,750$.

(b) The required damping ratio for a 25% overshoot is obtained from Equation (8.81) as $\zeta = 0.404$. Equation (8.83) gives the phase of the point on the branch of the root locus as $\phi = 1.986$ r $= 113.8°$.

Clicking on the appropriate branch of the root-locus curve of Figure 8.43 reveals that for $K_i = 1.63 \times 10^3$, the damping ratio is 0.397 corresponding to $s = -3.32 + 7.67j$ and a frequency of 8.36 r/s. The exact values of the gain and frequency for $\zeta = 0.404$ are obtained by noting that the point on the branch of the root locus is of the form of Equation (8.80)

$$s = -\zeta\omega + j\omega\sqrt{1 - \zeta^2}$$

$$= -0.404\omega + 0.915\omega j \tag{g}$$

The substitution of Equation (g) into Equation (6.32) written for this example leads to

$$s^3 + 30s^2 + 225s + K = 0$$

$$(-0.674\omega^3 - 20.22\omega^2 - 90.9\omega + K) + j(-0.739\omega^3 - 22.18\omega^2 + 205.8\omega) = 0 \tag{h}$$

Setting the imaginary part of the right-hand side of Equation (g) to zero leads to $\omega = 7.43$ r/s. Setting the real part of the right-hand side to zero leads to $K = 2.07 \times 10^3$.

(c) The closed-loop transfer function is

$$H(s) = \frac{25s + 2.07 \times 10^3}{s^3 + 30s^2 + 225s + 2.07 \times 10^3} \tag{i}$$

The following MATLAB commands are used to determine the step response of Figure 8.44.

$$N = [25\ 2.07E3]$$

$$D = [1\ 30\ 225\ 2.07E3]$$

$$\text{step}(N,D)$$

(d) For $K_i = 2,000$ the closed-loop transfer function is

$$H(s) = \frac{K_ps + 2,000}{s^3 + 30s^2 + (200 + K_p)s + 2,000} \tag{j}$$

The following MATLAB commands are used to generate the root-locus diagram

$$Q = [1\ 30\ 200\ 2000]$$

$$R = [1\ 0]$$

$$\text{rlocus}(R,Q)$$

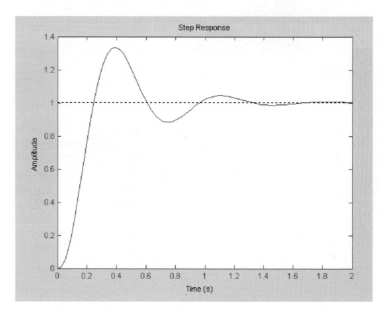

**FIG. 8.44** The step response of the system of Example 8.15(b). The PI controller is designed for a 25% overshoot.

The resulting root-locus diagram is shown in Figure 8.45. The following is known or can be determined from the diagram:

- The diagram has three poles: $-25.32$, $-2.39 \pm 8.58j$. Branches originate from these points.
- The diagram has one zero, $s = 0$. The branch entirely along the real axis terminates at this zero.
- The system is stable for all values of $K_p$.
- There is no value of $K_p$ such that the system has three real roots.
- The overshoot is approximately 40 percent for $K_p = 0$ and decreases to a minimum of 5 percent for $K_p = 233$.

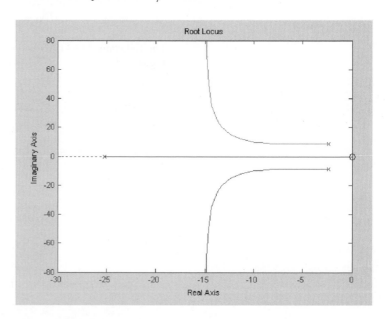

**FIG. 8.45** The root-locus diagram for part (d) of Example 8.15.

- Because the system has a zero at $s = 0$ the dominant pole is initially the complex pole, but as the proportional gain increases the real part of this pole decreases while the value of the real pole increases and becomes dominant for $K_p = 200$.

The step responses of the system with $K_p = 100$ and $K_p = 500$ are shown in Figures 8.46(a) and 8.46(b) respectively.

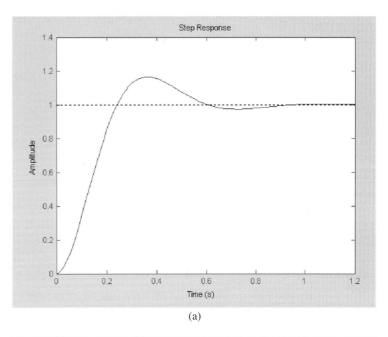

(a)

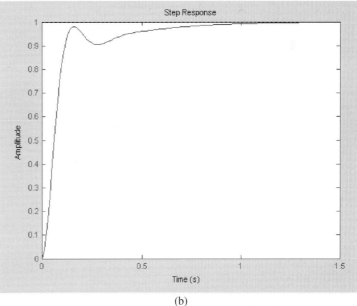

**FIG. 8.46** The step responses of the system of Example 8.15(d) with (a) $K_p = 100$ and (b) $K_p = 500$.

(b)

Concepts of the root-locus plot may be used in designing control systems without necessarily having to draw the root-locus diagram.

A compensator is a control strategy with a transfer function of the form

$$G_c(s) = K_c \frac{s+a}{s+b} \tag{8.84}$$

The compensator is placed in series with a plant in the feed forward part of a feedback loop such that the equivalent closed-loop transfer function is of the form

$$H(s) = \frac{K_c \dfrac{s+a}{s+b} G(s)}{1 + K_c \dfrac{s+a}{s+b} G(s)} \tag{8.85}$$

The compensator can be designed such that the damping ratio and frequency of the dominant poles of the closed-loop transfer function are specified.

## Example 8.16

A plant has a transfer function

$$G(s) = \frac{1}{s(s+3)} \tag{a}$$

Design a compensator with $b = 3$ such that the dominant poles of the closed-loop transfer function have a damping ratio of 0.25 and a frequency 10 r/s.

### Solution

Substitution of Equation (a) into Equation (8.85) and simplifying leads to

$$H(s) = \frac{K_c(s+a)}{s(s+5)(s+3) + K_c(s+a)} \tag{b}$$

The denominator of Equation (b) is written in the form of Equation (6.32) with

$$Q(s) = s(s+5)(s+3) \tag{c}$$
$$R(s) = s+a \tag{d}$$

For $\zeta = 0.25$ and $\omega = 10$, the dominant pole, using Equation (8.80) is located at

$$s = -(0.25)(10) + j(10)\sqrt{1 - (0.25)^2}$$
$$= -2.5 + 9.683j \tag{e}$$

For this point to be on a branch of the root-locus diagram the angle criterion must be satisfied

$$\phi_{s+a} - \phi_s - \phi_{s+5} - \phi_{s+3} = (2n+1)\pi \tag{f}$$

for some integer $n$. The angles are calculated as

$$\phi_s = \tan^{-1}\left(\frac{9.683}{-2.5}\right) = 1.819$$

$$\phi_{s+5} = \tan^{-1}\left(\frac{9.863}{2.5}\right) = 1.3226$$

$$\phi_{s+3} = \tan^{-1}\left(\frac{9.863}{0.5}\right) = 1.5301 \tag{g}$$

Substituting equation (g) into Equation (f) and solving for $\phi_{s+a}$ leads to

$$\phi_{s+a} = (2n+1)\pi + 1.819 + 1.3326 + 1.5201$$

$$= (2n+1)\pi + 4.6717 \tag{h}$$

Choosing $n = -1$ in Equation (h) leads to $\phi_{s+a} = 1.5301$. The imaginary part of $s + a$ must be 9.863. The value of $a$ is obtained from

$$-2.5 + a = \frac{9.863}{\tan(1.5301)} = 0.416$$

$$a = 2.916 \tag{i}$$

The necessary value of $K_c$ is obtained using the magnitude criterion

$$K_c = \left|\frac{Q(-2.5 + 9.863j)}{R(-2.5 + 9.863j)}\right|$$

$$K_c = 98.9 \tag{j}$$

## 8.6.2 Ziegler-Nichols Tuning Rules

The design of a PID controller requires the selection of three parameters: the proportional gain $K_p$, the integral gain $K_i$, and the derivative gain $K_d$. It may be difficult to track the effect of changing each parameter on the transient response of a system. It may also be difficult to choose a set of gains such that transient response specifications are met. The design of a controller to achieve specific transient response specifications is often called tuning the controller.

Many tuning rules for controllers have been suggested. Perhaps the most famous and widely used are the Ziegler-Nichols tuning rules. Using these rules, a controller is designed such that the step response of the closed loop system has a 25 percent overshoot. The rules are based on two methods, only the second of which, the more widely applicable set, is given here.

The second Ziegler-Nichols method involves determining the gain of a proportional controller that corresponds to the system's crossover frequency. The method is not applicable if the system does not have a crossover frequency. Let $\omega_c$ be the crossover frequency and let $K_c$ be the proportional gain at the crossover frequency. The tuning rules suggest values of the gains

**TABLE 8.6 ZIEGLER-NICHOLS TUNING RULES (SECOND METHOD)**

| Controller | $K_p$ | $K_i$ | $K_d$ |
|---|---|---|---|
| Proportional | $0.5K_c$ | | |
| PI | $0.45K_c$ | $\dfrac{4\pi}{3\omega_c}$ | |
| PID | $0.6K_c$ | $\dfrac{\pi}{\omega_c}$ | $\dfrac{\pi}{4\omega_c}$ |

for a proportional controller, a PI controller, and a PID controller in terms of $\omega_c$ and $K_c$. These rules are presented in Table 8.6.

## Example 8.17

A PID controller is to be used with a plant whose transfer function is

$$G(s) = \frac{1}{s(s^2 + 4s + 10)} \tag{a}$$

Design a PID controller using the Ziegler-Nichols tuning rules. Show the step response of the controlled system.

### Solution

Application of the Ziegler-Nichols tuning rules first requires determination of the crossover frequency and the corresponding gain when a proportional controller is used. The closed-loop transfer function of the plant subject to proportional control is

$$H(s) = \frac{K_p}{s^3 + 40s^2 + 10s + K_p} \tag{b}$$

The MATLAB commands used to generate a root-locus diagram for this system are

$$Q = [1\ 40\ 10\ 0]$$

$$R = [1]$$

$$\text{rlocus(R,Q)}$$

The resulting root-locus diagram is presented in Figure 8.47. Clicking the mouse near the branch crossing the imaginary axis brings up a box that provides the crossover frequency as $\omega_c = 3.17$ r/s with a corresponding gain of $K_c = 40.1$. The PID controller is designed according to the Ziegler-Nichols tuning rules provided in Table 8.6.

$$K_p = 0.6K_c = 24.06 \tag{c}$$

$$K_i = \frac{\pi}{\omega_c} = 0.991 \tag{d}$$

$$K_d = \frac{\pi}{4\omega_c} = 0.248 \tag{e}$$

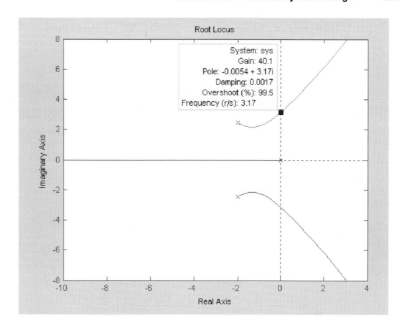

FIG. **8.47** The root-locus diagram for the system of Example 8.17 with a proportional controller. The crossover frequency is 3.17 r/s with a corresponding gain of 40.1.

The closed-loop transfer function for the plant under the action of a PID controller whose gains are given in Equations (c)–(e) is

$$H(s) = \frac{0.248s^2 + 24.06s + 0.991}{s^4 + 4s^3 + 10.248s^2 + 34.06s + 0.991} \tag{f}$$

The step response of the resulting closed-loop system is given in Figure 8.48. Note that the overshoot is 20 percent. Parameters may be fine-tuned to give a faster response or otherwise improve on the response.

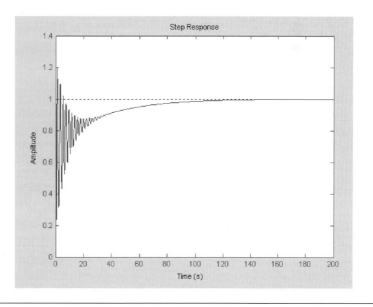

FIG. **8.48** The step response of the system of Example 8.17 with a PID controller designed using Ziegler-Nichols tuning rules. The maximum overshoot is 20 percent.

## 8.7 FURTHER EXAMPLES

The following examples illustrate the development of block diagrams from transfer functions and differential equations, the determination of the closed-loop response and its properties, and the stability of closed-loop systems. The actuator transfer functions are illustrative only. Applications are often more complex than illustrated here. These examples are intended to illustrate methods of analysis.

### Example 8.18

(a) Draw a block diagram for the two-tank CSTR problem of Example 6.29 that includes the time delay. Include only components $A$ and $B$ in both tanks in the block diagram. (b) Develop a SIMULINK model for this system. Run a simulation with $V_1 = V_2 = 1.5 \times 10^{-3}$ m³, $q = 1.5 \times 10^{-6}$ m³/s, $C_{Ai} = 0.25u(t)$ mol/L, $k_1 = 2.0 \times 10^{-3}$ s⁻¹, $k_2 = 1.0 \times 10^{-3}$ s⁻¹, and $\tau = 0.8$ s.

**Solution**

(a) Taking the Laplace transforms of Equations (a), (b), (d), and (e) of Example 6.29 and rearranging gives

$$\bar{C}_{A1p} = \frac{q\bar{C}_{A1i}}{V_1 s + q + k_1 V_1} \tag{a}$$

$$\bar{C}_{B1}(s) = \frac{k_1 \bar{C}_{A1}(s)}{V_1 s + q} \tag{b}$$

$$\bar{C}_{A2}(s) = \frac{q\bar{C}_{A1}(s)e^{-\tau s}}{V_2 s + q + k_1 V_2} \tag{c}$$

$$\bar{C}_{B2}(s) = \frac{1}{V_2 s + q + k_2 V_2}\left(q\bar{C}_{B1}(s)e^{-\tau s} + k_1 V_2 \bar{C}_{A2}(s)\right) \tag{d}$$

Equations (a)–(d) are used to develop the block diagram of Figure 8.49.

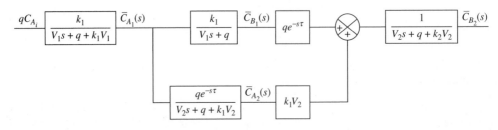

**FIG. 8.49** The block diagram for the two-tank CSTR with time delay.

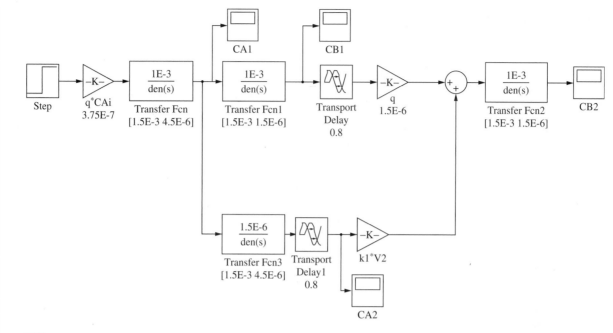

**FIG. 8.50** The SIMULINK model of the block diagram of Figure 8.49.

(b) The SIMULINK model of the block diagram of Figure 8.49 is shown in Figure 8.50. The concentrations of species *A* and *B*, obtained from the SIMULINK model are illustrated in Figures 8.51a and b respectively. The model includes a time delay. The Time Delay block is inserted into the SIMULINK model from the Continuous Systems menu of the SIMULINK Library Browser.

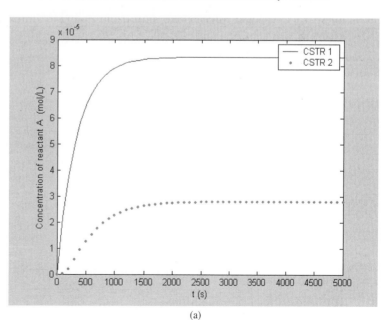

**FIG. 8.51** The results from running the SIMULINK simulation of the double CSTR of Example 8.18 (a) species *A*, (b) species *B*. (*Continued*)

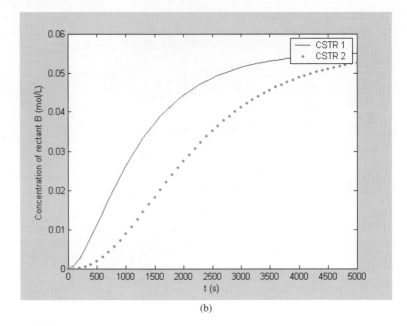

**FIG. 8.51** (*Continued*)

(b)

---

## Example 8.19

(a) Draw a block diagram for the mechanical system of Figure 8.52. (b) Develop a SIMULINK model for the system. Run a simulation with $F(t) = 30\sin(100t)$ N.

### Solution
Differential equations governing the response of the system are derived by applying Newton's second law to free-body diagrams of the particles drawn at an arbitrary instant. The resulting equations are

$$\ddot{x} + 3\dot{x}_1 - 2\dot{x}_2 + 300x_1 - 200x_2 = F(t) \tag{a}$$

$$2\ddot{x}_2 - 2\dot{x}_1 + 3\dot{x}_2 - \dot{x}_3 - 200x_1 + 500x_2 - 300x_3 = 0 \tag{b}$$

$$2\ddot{x}_3 - \dot{x}_2 + \dot{x}_3 - 300x_2 + 300x_3 = 0 \tag{c}$$

Taking the Laplace transforms of Equations (a)–(c) and rearranging leads to

$$X_1(s) = \frac{1}{s^2 + 3s + 300}[F(s) + (2s + 200)X_2(s)] \tag{d}$$

**FIG. 8.52** The mechanical system of Example 8.19.

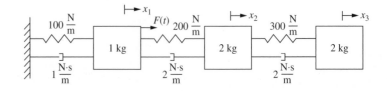

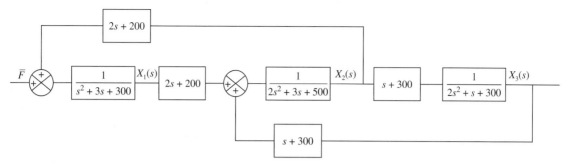

**FIG. 8.53** The block diagram for the system of Example 8.19.

$$X_2(s) = \frac{1}{2s^2 + 3s + 500}[(2s + 200)X_1(s) + (s + 300)X_3(s)] \tag{e}$$

$$X_3(s) = \frac{(s + 300)X_2(s)}{2s^2 + s + 300} \tag{f}$$

(a) Equations (d)–(f) are used to guide the development of the block diagram of Figure 8.53. (b) The SIMULINK model for the system with a sinusoidal input is shown in Figure 8.54. The simulation results are shown in Figure 8.55.

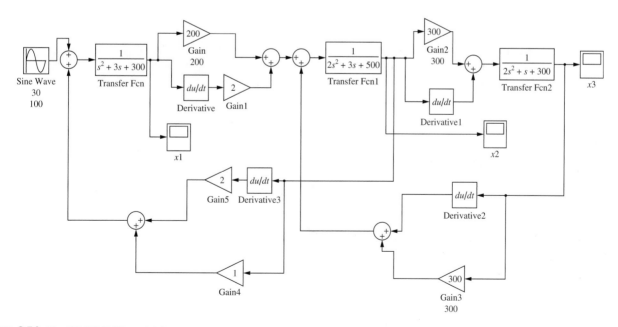

**FIG. 8.54** The SIMULINK model for the system of Example 8.19 with a sinusoidal input.

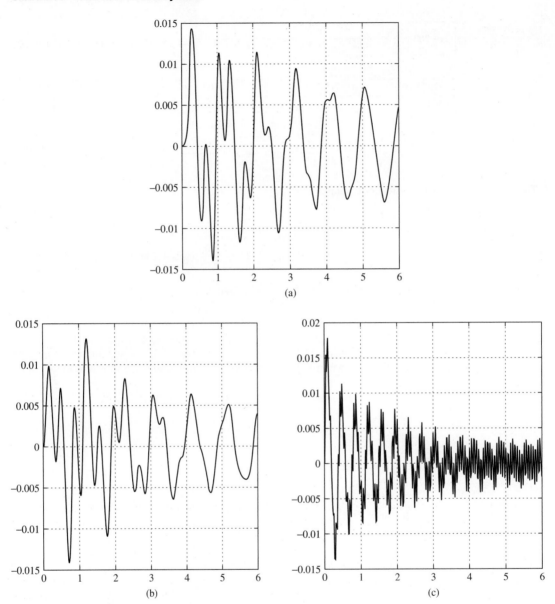

**FIG. 8.55** Simulation results from SIMULINK model: (a) $x_1(t)$; (b) $x_2(t)$; (c) $x_3(t)$.

**Example 8.20**

A single-tank liquid level system has an area of 2.0 m$^2$ and a resistance of 5.0 s/m$^2$. (a) What is the required gain of a proportional controller in the feedforward loop such that the system's 2 percent settling time for a step input is 1.0 s? (b) What is the steady-state perturbation in liquid level when this controller is used and the system has a unit step change in flow rate?

## Solution

The transfer function for the singe-tank liquid-level problem is obtained from Equation (4.65) as

$$G(s) = \frac{1}{As + \frac{1}{R}} = \frac{\frac{1}{A}}{s + \frac{1}{T}} \tag{a}$$

Where the first-order time constant is $T = RA$. The substitution of known values leads to

$$G(s) = \frac{0.5}{s + 0.1} \tag{b}$$

and $T = 10.0$ s.

When a proportional controller is used in the feedforward loop for the first-order system the closed-loop transfer function is first order with a time constant given by Equation (8.37). The application of Equation (8.38) for the transfer function of Equation (b) leads to

$$\hat{T} = \frac{10}{1 + K_p(0.5)(10)} = \frac{10}{1 + 5K_p} \tag{c}$$

Noting that the 2 percent settling time for this first-order system is $3.92\,\hat{T}$ and requiring this to be 1 s leads to

$$3.92\left(\frac{10}{1 + 5K_p}\right) = 1 \tag{d}$$

Equation (d) is solved giving $K_p = 7.64$ and then $\hat{T} = 0.255$ s. Using this value of the proportional gain in Equation (8.37) leads to the closed-loop transfer function of

$$H(s) = \frac{3.82}{s + 3.92} \tag{e}$$

The step responses of the system without a controller and with a proportional controller are illustrated in Figure 8.56. The offset when the controller is used is calculated using Equation (8.40) as $\hat{T}/T = 0.0255$. The magnitude of the responses shown in Figure 8.56 do not compare because the units on each response are different. The uncontrolled response has meters (m) as units, whereas the controlled response must have the same units as the input (flow rate) which are cubic meters per second (m$^3$/s). The proportional gain must have units that allow the output to be compared with the input. In this case, $K_p = 7.64$ m$^2$/s. It is clear from Figure 8.56 that the use of the controller significantly affects the dynamic response of the system, reducing the settling time and the rise time.

(b) The steady-state liquid level when this controller is used is obtained using the final value theorem applied to equation (e) as

$$h_s = \lim_{s \to 0} H(s)$$

$$= \frac{3.82}{3.92}$$

$$= 0.974 \text{ m} \tag{f}$$

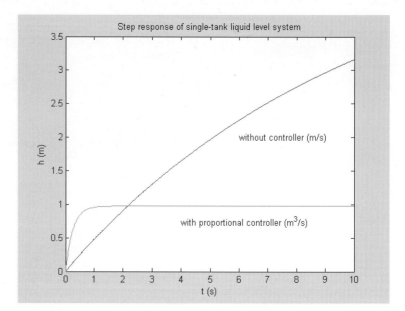

**FIG. 8.56** The step response of the first-order system with a proportional controller has offset from that of the response of the system without a controller.

---

**Example 8.21**

Figure 8.57 shows the block diagram for the perturbation concentrations for the CSTR of Example 6.31 due to a change in the concentration of the inlet stream of a reactant $A$ when a PI controller is used in the feedforwrd loop to control $C_A(t)$. (a) What are the required gains of the PI controller such that the system is critically damped with a natural frequency of 2.0 r/s? (b) What is the resulting transfer function $G_B(s) = \bar{C}_B(s)/\bar{C}_{Ai}(s)$ when this controller is used? (c) Determine the step response for $C_B(t)$ when this controller is used.

**Solution**

(a) The natural frequency of the closed-loop system when a PI controller is used on a first-order system is given by Equation (8.44). Requiring the natural frequency to be 2.0 r/s leads to

$$2.0 = \sqrt{(0.387)K_i}$$

$$K_i = 10.3 \tag{a}$$

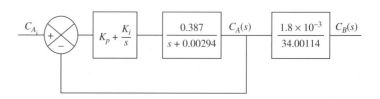

**FIG. 8.57** The block diagram for the system of Example 8.21, a CSTR with a PI controller to control the perturbation concentration of reactant $A$.

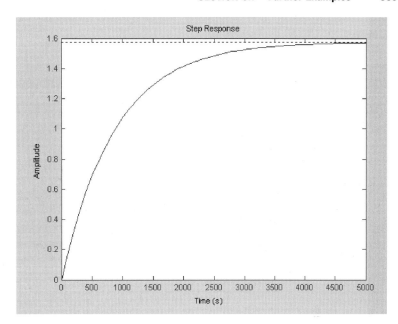

**FIG. 8.58** The step response for the CSTR with a PI controller of Example 8.21.

The damping ratio of the closed-loop system is given by Equation (8.51). Setting the damping ratio to 1, requiring the system to be critically damped, leads to

$$1 = \frac{K_p}{2}\sqrt{\frac{0.387}{10.3}} + \frac{1}{2\left(\dfrac{1}{0.00294}\right)(2.0)}$$

$$K_p = 10.3 \tag{b}$$

(b) The closed-loop transfer function for $\bar{C}_{Ai}(s)$ is obtained using Equation (8.50) as

$$H(s) = \frac{(0.387)(10.3s + 10.3)}{s^2 + 2s + 4}$$

$$= \frac{3.99s + 3.99}{s^2 + 2s + 4} \tag{c}$$

The transfer function for $C_2(t)$ is

$$G(s) = \left(\frac{1.8 \times 10^{-3}}{s + 0.00114}\right)\left(\frac{3.99 + 3.99s}{s^2 + 2s + 4}\right)$$

$$= \frac{7.18 \times 10^{-3}(s + 1)}{(s + 0.00114)(s^2 + 2s + 4)} \tag{d}$$

(c) The MATLAB-generated plot for the step response is given in Figure 8.58.

**Example 8.22**

A PID controller with $K_p = 50.0$ kJ/s·C, $K_i = 35.0$ MJ/s²·C, and $K_d = 6.5$ MJ/C is to be used in the feedforward loop to provide control to the thermal system of Example 4.15. When this controller is used, determine (a) the closed-loop transfer function, (b) the system's natural frequency and damping ratio (b) a MATLAB-generated plot of the step response, and (d) the initial value of the step response.

### Solution

The open-loop transfer function for the system is obtained using Equation (i) of Example 4.15 as

$$\bar{\Theta}(s) = \frac{5.95 \times 10^{-9}}{s + 1.22 \times 10^{-5}} \tag{a}$$

(a) The closed-loop transfer function for a first-order system using a PID controller is given by Equation (8.55), which for this system becomes

$$H(s) = \frac{5.95 \times 10^{-9}(6.5 \times 10^6 s^2 + 5.0 \times 10^6 s + 3.5 \times 10^6)}{[1 + (5.95 \times 10^{-9})(6.5 \times 10^6)s^2] + [(5.95 \times 10^{-9})(5.0 \times 10^6) + 1.22 \times 10^{-5}]s + 5.95 \times 10^{-9}3.5 \times 10^6}$$

$$= \frac{3.87 \times 10^{-2}s^2 + 2.98 \times 10^{-2}s + 2.08 \times 10^{-2}}{1.00387s^2 + 2.98 \times 10^{-2}s + 2.08 \times 10^{-2}} \tag{b}$$

(b) Equation (b) is that of a second-order system of natural frequency $\omega_n = \sqrt{2.08 \times 10^{-2}/1.00387} = 0.144$ r/s and damping ratio $\zeta = 2.98 \times 10^{-2}/2(1.00387)(0.144) = 0.103$.

(c) A MATLAB-generated plot of the step response is given in Figure 8.59. The amplitude of the response has units of Joules per second (J/s). Since an integral term is used in the controller the response has no offset (d). Due to the derivative term in the controller the step response is discontinuous at $t = 0$ with a discontinuity

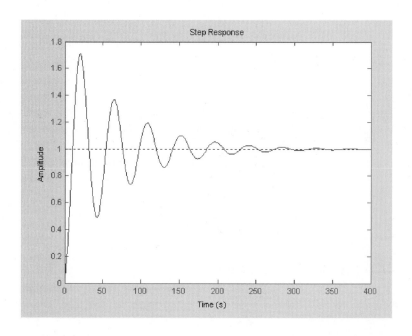

**FIG. 8.59** The step response of the system with a PID controller used on the tank of Example 8.22. The amplitude has units of (J/s).

given by the initial value theorem of

$$X_s(0) = \lim_{s \to \infty} H(s) = \frac{3.87 \times 10^{-2}}{1.00387} = 3.86 \times 10^{-2} \qquad (c)$$

## Example 8.23

Figure 8.60 shows the block diagram of a two-tank liquid-level system in which a proportional controller is used in the feedforward loop and the level in the second tank is feedback to the controller. (a) Determine the transfer of the closed-loop system. (b) Determine the natural frequency and damping ratio of the closed-loop system. (c) Determine the step response for both the first and second tanks. (d) Determine the steady-state liquid-level perturbation in each tank when this controller is used and the system has a unit step change in input flow rate.

### Solution

The transfer function for the system is

$$G(s) = \frac{0.05}{s^2 + 0.3s + 0.01} \qquad (a)$$

Equation (a) is the transfer function of a second-order system and is of the form of Equation (8.56) with $A = 0, B = 0.05, \omega_n = 0.141$ r/s, and $\zeta = 1.06$. (a) The closed-loop transfer function for a second-order system under the action of a proportional controller is given in Equation (8.57). The application to this system leads to

$$H(s) = \frac{2.0(0.05)}{s^2 + 0.3s + 2.0(0.05) + 0.01}$$

$$= \frac{0.1}{s^2 + 0.3s + 0.11} \qquad (b)$$

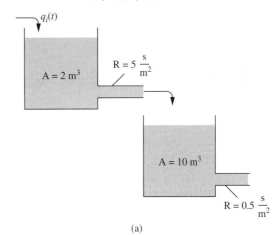

(a)

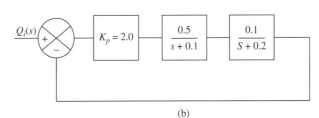

(b)

**FIG. 8.60** (a) The two-tank liquid-level system of Example 8.23; (b) the block diagram for the system with a proportional controller.

(b) Equation (b) is the transfer function of a second-order system with $\omega_n = \sqrt{0.12} = 0.346$ r/s and $\zeta = 0.3/2(0.346) = 0.433$.

(c) The closed-loop transform of the response for the first tank $H_1(s)$ is obtained using the block diagram and definitions of transfer functions. Noting that $H_1(s)$, the output from $G_1(s)$ is

$$H_1(s) = [Q_i(s) - H_2(s)]K_p G_1(s) \tag{c}$$

Since $G_1(s)$ and $G_2(s)$ are in series

$$H_2(s) = G_2(s)H_1(s) \tag{d}$$

Eliminating $H_2(s)$ between Equations (c) and (d) leads to

$$H_1(s) = \frac{K_p G_1(s)Q_i(s)}{1 + K_p G_1(s)G_2(s)} \tag{e}$$

The substitution of known values into Equation (e) leads to

$$H_1(s) = \frac{2.0\left(\dfrac{0.5}{s+0.1}\right)Q_i(s)}{1 + 2.0\left(\dfrac{0.5}{s+0.1}\right)\left(\dfrac{0.1}{s+0.2}\right)}$$

$$= \frac{\dfrac{1}{s+0.1}Q_i(s)}{\dfrac{s^2 + 0.3s + 0.12}{(s+0.2)(s+0.1)}}$$

$$= \frac{(s+0.2)Q_i(s)}{s^2 + 0.3s + 0.12} \tag{f}$$

MATLAB-generated plots for the step responses of $H_1(s)$ and $H_2(s)$ are given in Figure 8.61. Note that the overshoot for $h_1(t)$ is much greater than the overshoot for $h_2(t)$, due to the presence of the $s$ term in the numerator for $H_1(s)$.

The steady-state liquid levels are obtained using the final value theorem as

$$h_1 = \lim_{s \to \infty} H_1(s)$$

$$= \frac{0.2}{0.12}$$

$$= 1.67 \tag{g}$$

$$h_2 = \lim_{s \to \infty} H_2(s)$$

$$= \lim_{s \to \infty} G_2(s)H_1(s)$$

$$= 1.67 \lim_{s \to \infty}\left(\frac{0.1}{s+0.2}\right)$$

$$= 0.833 \tag{h}$$

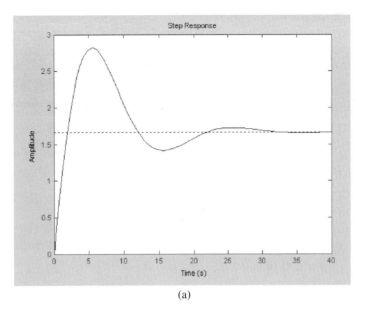

(a)

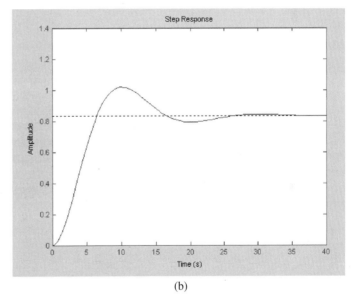

(b)

**FIG. 8.61** The step responses for the two-tank liquid-level problem of Example 8.23 with proportional control: (a) $h_1(t)$; (b) $h_2(t)$.

---

**Example 8.24**

The controller used for the system of Example 8.23 is being modified to a PI controller with $K_p = 2$. (a) For what values of the integral gain $K_i$ will the system be stable? (b) Use MATLAB to plot the impulsive response of the system for one value of $K_i$ for which the closed-loop system is stable and one value of $K_i$ for which the system is unstable.

## Solution

The open-loop transfer function is given by Equation (a) of Example 8.23. The closed-loop transfer function for a second-order system with a PI controller is given in Equation (8.67), which when applied to this system leads to

$$H(s) = \frac{0.2s + 0.1K_i}{s^3 + 0.3s^2 + 0.12s + 0.1K_i} \tag{a}$$

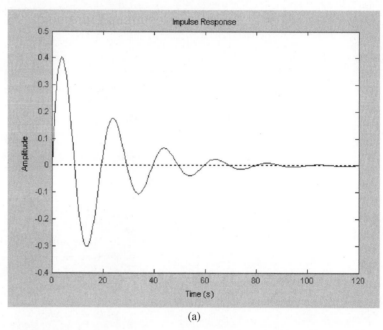

(a)

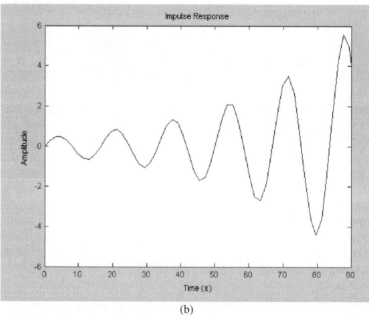

(b)

**FIG. 8.62** The closed-loop impulsive response of the two-tank liquid-level system of Example 8.24 when a PI controller is used with $K_p = 2.0\,\text{m}^2/\text{s}$: (a) $K_i = 0.2\,\text{m}^2/\text{s}^2$; (b) $K_i = 0.5\,\text{m}^2/\text{s}^2$.

(a) The system is stable when all poles of the closed-loop transfer function have negative real parts. Application of Routh's criterion applied to the third-order denominator, as developed in Equation (8.65), leads to a stability criterion of

$$(0.3)(0.12) > 0.1K_i$$

$$K_i < 0.36 \tag{b}$$

(b) The MATLAB-generated plots are shown in Figure 8.62.

---

**Example 8.25**

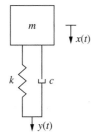

**FIG. 8.63** The system for Example 8.25.

An electronic PID controller is being designed to provide closed-loop feedforward control to a simplified vehicle suspension system, as illustrated in Figure 8.63. The open-loop transfer function for the system is

$$G(s) = \frac{2\zeta\omega_n + \omega_n^2}{s^2 + 2\zeta\omega_n s + \omega_n^2} \tag{a}$$

The controller is being designed for a system with $k = 1 \times 10^5$ N/m, $c = 3{,}000$ N·s/m, and an empty vehicle mass of 400 kg.

(a) A PID controller is to be used with $K_p = 2.0$ and $K_d = 1.5$. For what values of $K_i$ is the system stable?

(b) For what values of $K_i$ are all poles of the transfer function real?

(c) Compare the closed-loop and open-loop responses when subject to a step input for $K_i = 1.0$ and $K_i = 5.0$.

(a) The natural frequency and damping ratio of the open-loop system are

$$\omega_n = \sqrt{\frac{1 \times 10^5 \text{ N/m}}{400 \text{ kg}}} = 15.8 \text{ r/s} \quad \text{and} \quad \zeta = \frac{3{,}000 \text{ N·s/m}}{2(400 \text{ kg})(15.8 \text{ r/s})} = 0.237.$$

Equation (a) is of the form of Equation (8.56), the transfer function of a second-order system with $A = 2\zeta\omega_n = 7.49$ and $B = \omega_n^2 = 250.0$. Applying Equation (8.75), the closed-loop transfer function of a second-order plant in a feedback loop with a PID controller in the feedforward part of the loop leads to

$$H(s) = \frac{1.5(7.49)s^3 + [2.0(7.49) + 1.5(250)]s^2 + [2.0(250) + 7.49K_i]s + 250K_i}{[1 + 1.5(7.49)]s^3 + [7.49 + 2.0(7.49) + 1.5(250)]s^2 + [250 + 2.0(250) + 7.49K_i]s + 250K_i}$$

$$= \frac{11.24s^3 + 390.0s^2 + (500 + 7.49K_i)s + 250K_i}{12.24s^3 + 397.5s^2 + (750 + 7.49K_i)s + 250K_i} \tag{b}$$

The application of Routh's criterion to the denominator of $H(s)$ leads to a stability criterion of

$$(397.5)(750 + 7.49K_i) > 12.24(250K_i) \tag{c}$$

Equation (c) is rearranged leading to $K_i < 3602.3$.

(b) A trial and error solution using the MATLAB script of Figure 8.64 is used to determine that the poles of $H(s)$ are all real for $K_i < 1.504$.

(c) It is clear from the MATLAB-generated plots of Figure 8.65 that the use of this PID controller does not enhance the response of the system to a step change. The settling times are longer with the controller than without the controller.

```
Ki=input ('Input value of integral gain   ')
num=[11.24 390 500+7.49*Ki 250*Ki]
den=[12.24 397.5 750+7.49*Ki 250*Ki]
[r,p,k]=residue(num,den)
```

**FIG. 8.64** The MATLAB script used to iterate on values of $K$ such that all poles of the closed-loop transfer function are real. The script is run changing the values of $K_i$ and inspecting the poles obtained from the residue function.

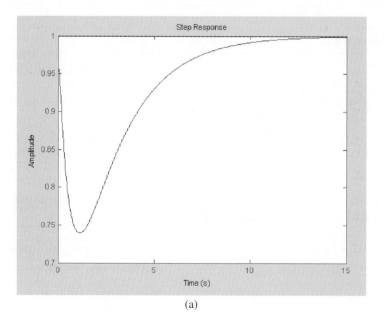

(a)

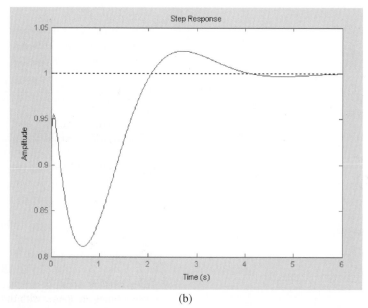

**FIG. 8.65** The step responses of Example 8.25: (a) PID controller with $K_i = 1$; (b) a PID controller with $K_i = 5$; (c) no controller.

(b)

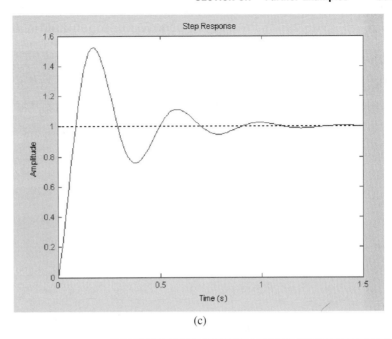

FIG. **8.65** (*Continued*)                                                                 (c)

---

**Example 8.26**

The transfer function for the concentration of reactant $A$ in the third reactor of a three-CSTR system in which the reactors are identical and in which the same reaction occurs in each, that is

$$G(s) = \frac{q^3}{(Vs + q + kV)^3}$$

$$= \left( \frac{\dfrac{q}{V}}{s + k + \dfrac{q}{V}} \right)^3 \qquad \text{(a)}$$

where the flow rate is $q = 1.2 \times 10^{-6}$ m$^3$/s, the volume of each reactor is $V = 6.25 \times 10^{-4}$ m$^2$, and the rate of reaction in each reactor is $k = 2.5 \times 10^{-3}$ s$^{-1}$. (a) Determine the closed-loop transfer function when the reactors are placed in a feedback loop with (i) a proportional controller of proportional gain $K_p$, (ii) a PI controller with proportional gain $K_p$ and integral gain $K_i$, and (iii) a PD controller with proportional gain $K_p$ and differential gain $K_d$. (b) Discuss the stability of the closed-loop response when each controller is used.

(a) The substitution of the given values into Equation (a) leads to

$$G(s) = \frac{7.08 \times 10^{-9}}{(s + 4.42 \times 10^{-3})^3} \qquad \text{(b)}$$

The equivalent closed-loop transfer function is derived from the open-loop transfer function using Equation (8.14). (i) For a proportional controller $G_a(s) = K_p$.

The application of Equation (8.14) leads to

$$H(s) = \cfrac{K_p \left[ \cfrac{7.09 \times 10^{-9}}{(s + 4.42 \times 10^{-3})^3} \right]}{1 + K_p \left[ \cfrac{7.09 \times 10^{-9}}{(s + 4.42 \times 10^{-3})^3} \right]}$$

$$= \frac{7.09 \times 10^{-9} K_p}{s^3 + 1.33 \times 10^{-2} s^2 + 5.86 \times 10^{-5} s + 8.64 \times 10^{-8} + 7.09 \times 10^{-9} K_p} \quad \text{(c)}$$

(ii) For a PI controller $G_a(s) = K_p + \dfrac{K_i}{s}$. The application of Equation (8.14) leads to

$$H(s) = \cfrac{\left( K_p + \cfrac{K_i}{s} \right) \left[ \cfrac{7.09 \times 10^{-9}}{(s + 4.42 \times 10^{-3})^3} \right]}{1 + \left( K_p + \cfrac{K_i}{s} \right) \left[ \cfrac{7.09 \times 10^{-9}}{(s + 4.42 \times 10^{-3})^3} \right]}$$

$$= \frac{7.09 \times 10^{-9} (K_p s + K_i)}{s^4 + 1.33 \times 10^{-2} s^3 + 5.86 \times 10^{-5} s^2 + (8.64 \times 10^{-8} + 7.09 \times 10^{-9} K_p) s + 7.09 \times 10^{-9} K_i}$$

$$\text{(d)}$$

(iii) For a PD controller $G_a(s) = K_p + K_d s$. The application of Equation (8.14) leads to

$$H(s) = \cfrac{(K_p + K_d s) \left[ \cfrac{7.09 \times 10^{-9}}{(s + 4.42 \times 10^{-3})^3} \right]}{1 + (K_p + K_d s) \left[ \cfrac{7.09 \times 10^{-9}}{(s + 4.42 \times 10^{-3})^3} \right]}$$

$$= \frac{7.09 \times 10^{-9} (K_p + K_d s)}{s^3 + 1.33 \times 10^{-2} s^2 + (5.86 \times 10^{-5} + 7.09 \times 10^{-9} K_d) s + 8.64 \times 10^{-8} + 7.09 \times 10^{-9} K_p}$$

$$\text{(e)}$$

(b) The closed-loop system is stable when all poles of the transfer function have negative real parts. (i) The poles of the closed-loop transfer function when the proportional controller is used are roots of

$$s^3 + 1.33 \times 10^{-2} s^2 + 5.86 \times 10^{-5} s + 8.64 \times 10^{-8} + 7.09 \times 10^{-9} K_p = 0 \quad \text{(f)}$$

Routh's criterion applied to a third-order transfer function as applied in Equation (8.65) leads to the stability criterion

$$(1.33 \times 10^{-2})(5.86 \times 10^{-5}) > 8.64 \times 10^{-8} + 7.09 \times 10^{-9} K_p \quad \text{(g)}$$

Equation (g) is rearranged to $K_p < 97.7$.
(ii) The poles of the closed-loop transfer function when the PI controller is used as the roots of

$$s^4 + 1.33 \times 10^{-2} s^3 + 5.86 \times 10^{-5} s^2 + (8.64 \times 10^{-8} + 7.09 \times 10^{-9} K_p) s + 7.09 \times 10^{-9} K_i$$

$$\text{(h)}$$

The use of Routh's stability criterion is discussed in Section 6.3. The matrix for this system developed using Equation (6.12) is

$$
\mathbf{B} = \begin{bmatrix} 1 & 5.86 \times 10^{-5} & 7.09 \times 10^{-9} K_i \\ 1.33 \times 10^{-2} & 8.64 \times 10^{-8} + 7.09 \times 10^{-9} K_p & 0 \\ 5.21 \times 10^{-5} - 5.33 \times 10^{-7} K_p & 7.09 \times 10^{-9} K_i & 0 \\ \dfrac{4.50 \times 10^{-12} - 3.23 \times 10^{-13} K_p + 3.78 \times 10^{-15} K_p^2 - 9.43 \times 10^{-11} K_i}{5.21 \times 10^{-5} - 5.33 \times 10^{-7} K_p} & 0 & 0 \\ 7.09 \times 10^{-9} K_i & 0 & 0 \end{bmatrix}
$$

(i)

The system is stable if and only if all elements of the first column of **B** are positive. Requiring $b_{3,1} > 0$ requires $K_p < 97.7$. Requiring $b_{5,1} > 0$ requires $K_i > 0$. Noting that if $b_{3,1} > 0$ then the denominator of $b_{4,1}$ is positive, thus if $K_p < 97.7$ then the closed-loop response is stable if

$$4.50 \times 10^{-12} - 3.23 \times 10^{-13} K_p + 3.78 \times 10^{-15} K_p^2 - 9.43 \times 10^{-11} K_i > 0 \quad \text{(j)}$$

Equation (j) represents a parabola when plotted in the $(K_i, K_p)$ plane, as illustrated in Figure 8.66. Stable and unstable regions are indicated in the figure with the parabola and the line $K_p = 97.7$ as boundaries of the regions between stability and instability.

(iii) The poles of the closed-loop transfer function when the PD controller is used are the roots of

$$s^3 + 1.33 \times 10^{-2} s^2 + (5.86 \times 10^{-5} + 7.09 \times 10^{-9} K_d)s$$
$$+ 8.64 \times 10^{-8} + 7.09 \times 10^{-9} K_p = 0 \quad \text{(k)}$$

Applying Routh's stability criterion in the same manner as for the proportional controller leads to a stability criterion of

$$(1.33 \times 10^{-2})(5.86 \times 10^{-5} + 7.09 \times 10^{-9} K_d) > 8.64 \times 10^{-8} + 7.09 \times 10^{-9} K_p \quad \text{(l)}$$

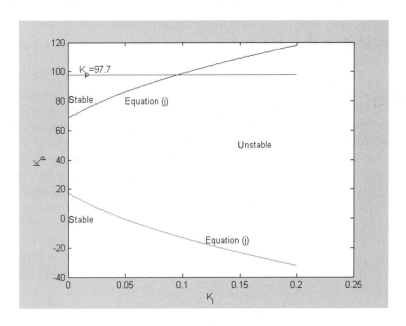

**FIG. 8.66** The stability regions for the PI controller of Example 8.26.

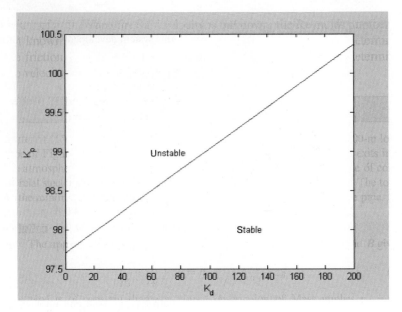

**FIG. 8.67** The stability regions for use of the PD controller on a three-CSTR system of Example 8.26.

Equation (l) is rearranged to

$$97.7 + 1.33 \times 10^{-2} K_d > K_p \tag{m}$$

Equation (m) shows that if $K_p < 97.7$ then the system is stable for any positive value of $K_d$. The larger the value of $K_d$ the larger the maximum value of $K_p$ for stability. Equation (m) is plotted in Figure 8.67, illustrating the regions of stability and instability.

## 8.8 SUMMARY

### 8.8.1 Chapter Highlights

- Block diagrams provide a schematic representation of a dynamic system and its input and output.
- Block diagrams are developed in the transform domain in terms of transfer functions.
- Block diagrams are used to illustrate feedback in dynamic systems.
- SIMULINK is a MATLAB package that simulates dynamic systems using block diagrams. SIMULINK uses numerical methods to run simulations for dynamic systems from block diagrams.
- Feedback control is used to modify the response of dynamic systems.
- The dynamic system response of a feedback control system is governed by the closed-loop transfer function.
- Actuators are placed in feedback control systems to modify the closed-loop transfer function and enhance system response. This study concentrates on actuators placed in the feedforward portion of the loop.

- Types of controllers are proportional, integral, derivative, proportional plus integral (PI), proportional plus derivative (PD), and proportional plus integral plus derivative (PID).
- Offset occurs with the use of a proportional controller, but is eliminated when used in conjunction with an integral controller.
- Proportional and/or derivative control of a first-order plant leads to a closed-loop transfer function for a first-order system.
- A controller with an integral gain used with a first-order plant leads to a closed-loop transfer function for a second-order system.
- Proportional and/or derivative control of a second-order plant leads to a closed-loop transfer function for a second-order system.
- A controller with an integral gain used with a second-order plant leads to a closed-loop transfer function for a third-order system.
- An integral controller used with a second-order plant could lead to an unstable system response.
- Root-locus diagrams are valuable tools in control system design because they are used to determine the placement of poles as a value of a gain change.
- Lines of constant angle in the root-locus plane correspond to lines of constant damping ratio. Constant frequency curves are circles in the root-locus plane.
- The dominant pole, the pole with the largest real part, often controls the transient behavior of a system.
- Tuning rules provide guidelines for choosing gains for a PI or PID controller.
- Ziegler-Nichols tuning, summarized in Table 8.6, offers the best known tuning rules. The controller gains are chosen to provide 25 percent overshoot in the step response.

## 8.8.2  Important Equations

- Definition of the transfer function in terms of input and output

$$C(s) = G(s)A(s) \tag{8.1}$$

- Closed-loop transfer function for a feedback loop

$$\frac{C(s)}{A(s)} = \frac{G(s)}{1 + G(s)} \tag{8.5}$$

- Closed-loop transfer function for a plant with an actuator in the feedforward portion of the loop

$$H(s) = \frac{G_a(s)G(s)}{1 + G_a(s)G(s)} \tag{8.13}$$

- Proportional control

$$G(s) = K_p \tag{8.15}$$

- Integral control

$$G_a(s) = \frac{K_i}{s} \tag{8.17}$$

- Derivative control

$$G_a(s) = K_d s \tag{8.22}$$

- PID control

$$G_a(s) = K_p + K_d s + \frac{K_i}{s} \tag{8.26}$$

- Offset for unit step input

$$\eta = \lim_{s \to 0} \frac{1}{1 + G_a(s)G(s)} \tag{8.32}$$

- Proportional control of a first-order plant

$$H(s) = \frac{K_p K}{s + \dfrac{1}{T} + K_p K} \tag{8.37}$$

- Integral control of a first-order plant

$$H(s) = \frac{K_i K}{s^2 + \dfrac{1}{T} s + K_i K} \tag{8.44}$$

- PID control of a first-order plant

$$H(s) = \frac{K(K_d s^2 + K_p s + K_i)}{(1 + KK_d)s^2 + \left(KK_p + \dfrac{1}{T}\right)s + KK_i} \tag{8.55}$$

- Proportional control of a second-order plant

$$H(s) = \frac{K_p(As + B)}{s^2 + (2\zeta\omega_n + K_p A)s + K_p B + \omega_n^2} \tag{8.57}$$

- Integral control of a second-order plant

$$H(s) = \frac{K_i(As + B)}{s^3 + 2\zeta\omega_n s^2 + (\omega_n^2 + K_i A)s + K_i B} \tag{8.63}$$

- PID control of a second-order plant

$$H(s) = \frac{K_d As^3 + (K_p A + K_d B)s^2 + (K_p B + K_i A)s + K_i B}{(1 + K_d A)s^3 + (2\zeta\omega_n + K_p A + K_d B)s^2 + (\omega_n^2 + K_p B + K_i A)s + K_i B} \tag{8.75}$$

- Root-locus diagram

$$Q(s) + KR(s) = 0 \tag{6.32}$$

# PROBLEMS

**8.1** The differential equations for the perturbations in concentration of components $A$ and $B$ in a CSTR due to a perturbation in concentration of the reactant in the inlet stream are

$$V\frac{dC_A}{dt} + (k+qV)C_A = qC_{Ai} \qquad \text{(a)}$$

$$V\frac{dC_B}{dt} - kC_A + qC_B = 0 \qquad \text{(b)}$$

Draw a block diagram for the system.

**8.2** The differential equations for the perturbations in the concentrations of components $A$ and $B$ in a CSTR with a two-way reaction due to perturbation in flow rate of the inlet stream are

$$0.2\frac{dC_A}{dt} + 1.2C_A - 0.4C_B = 0.2q \qquad \text{(a)}$$

$$0.2\frac{dC_B}{dt} - 0.4C_A + 0.9C_B = 0.1q \qquad \text{(b)}$$

Draw a block diagram that models these differential equations.

**8.3** Draw a block diagram that models the differential equations for the temperatures of the oil and water in the counterflow heat exchanger of Example 6.41

**8.4** Draw a block diagram that models the concentrations of the drug in the plasma and tissue in the two-compartment model of Example 4.30.

**8.5** Use block diagram algebra to simplify the block diagram of Figure P8.5.

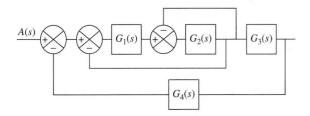

FIG. **P8.5**

**8.6** Use block diagram algebra to simplify the block diagram of Figure P8.6.

**8.7** Determine the equivalent closed-loop transfer function for the system of Figure P8.7.

**8.8** Determine the closed-loop transfer function $H(s) = C(s)/A(s)$ for the feedback system of Figure P8.8.

**8.9** Determine the closed-loop transfer function $H(s) = C(s)/A(s)$ for the feedback system of Figure P8.9.

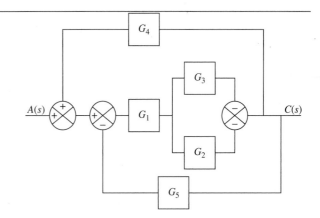

FIG. **P8.6**

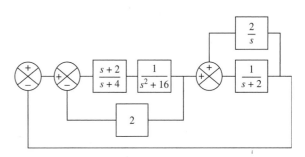

FIG. **P8.7**

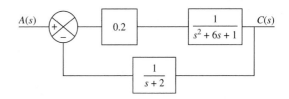

FIG. **P8.8**

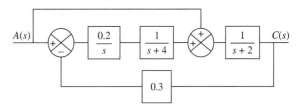

FIG. **P8.9**

**8.10** Determine the closed-loop transfer function $H(s) = C(s)/A(s)$ for the feedback system of Figure P8.10.

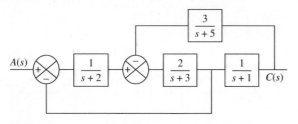

**FIG. P8.10**

**8.11** What is the minimum gain of a proportional controller required to reduce the 2 percent settling time of the thermal system of Example 4.15 by 30 percent? What is the eventual steady-state temperature when this proportional controller is used?

**8.12** What is the value of the gain of an integral controller that is required such that when used on the thermal system of Example 4.15 the closed-loop response has a damping ratio of 2.5?

**8.13** A PI controller with $K_p = 10$ and $K_i = 5.2$ is used with a plant whose transfer function is

$$G(s) = \frac{0.2}{s+5} \qquad (a)$$

(a) Determine the closed-loop transfer function $H(s)$. (b) What are the natural frequency and damping ratio of the closed-loop system? (c) Determine the impulsive response of the closed-loop system. (d) Determine the step response of the closed-loop system.

**8.14** A tank of cross-sectional area 1.2 m² exits into a pipe whose resistance is 6.2 s/m². For each of following controllers determine (i) the closed-loop transfer function $H(s)$, (ii) important system parameters such as time constant, or natural frequency and damping ratio, (iii) the impulsive response of the closed-loop system, and (iv) the offset when subject to a unit step input:

(a) Proportional controller of gain 0.2
(b) Integral controller of gain 1.5
(c) PI controller with $K_p = 0.2$ and $K_i = 1.5$
(d) PD controller with $K_p = 0.2$ and $K_d = 1.5$
(e) PID controller with $K_p = 0.2$, $K_i = 1.5$, and $K_d = 1.5$.

**8.15** Two identical tanks of cross section 1.2 m² are connected to pipes of identical resistances $R_1 = R_2 = 6.2$ s/m², as illustrated in Figure P8.15(a).

(a) Determine the open-loop transfer functions for the system defined as $G_1(s) = H_1(s)/Q(s)$ and $G_2(s) = H_2(s)/Q(s)$.

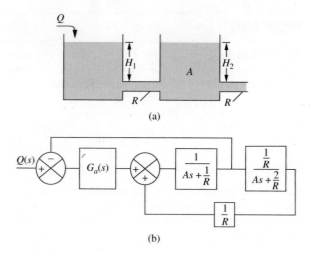

**FIG. P8.15**

Parts (b)–(f) refer to the closed-loop feedback system of Figure P8.15(b) in which the control system uses output from the first tank to compare with the input. For each of the following actuators determine (i) the closed-loop transfer functions pertaining to the liquid levels in both tanks, (ii) the poles of the transfer function, (iii) whether the closed-loop system is stable, (iv) the impulsive response, and (v) the step response:

(b) Proportional controller of gain 0.2
(c) Integral controller of gain 1.5
(d) PI controller with $K_p = 0.2$ and $K_i = 1.5$
(e) PD controller with $K_p = 0.2$ and $K_d = 1.5$
(f) PID controller with $K_p = 0.2$, $K_i = 1.5$, and $K_d = 1.5$.

**8.16** Repeat Problem 8.15, but the actuator is in a control system in which the output from the second tank is compared with the input, as in Figure P8.16.

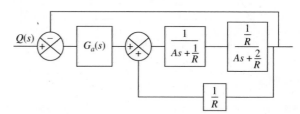

**FIG. P8.16**

Problems 8.17–8.21 refer to a two-stage CSTR with identical reactors and one-way reactions. The transfer function for the concentration of component $A$ in the second stage of the CSTR is

$$\bar{C}_{A2}(s) = \frac{0.2}{(s+0.4)^2} \qquad (a)$$

A feedback control system is being designed in which the actuator is placed in the feedforward part of the loop.

**8.17** What is the value of the gain of a proportional controller such that the closed-loop response has a damping ratio of 0.8?

**8.18** For what values of the gain of an integral controller is the closed-loop system is stable?

**8.19** For what values of the proportional gain is the closed-loop system stable when a PI controller with $K_i = 1$ is used as the actuator?

**8.20** For what values of the integral gain is the closed-loop system stable when a PI controller with $K_p = 2$ is used as the actuator?

**8.21** For what values of $K_d$ is the closed-loop system stable when a PID controller with $K_i = 1$ and $K_p = 2$ is used as the actuator?

Problems 8.22–8.24 refer to a three-stage CSTR with identical reactors and one-way reactions. The open-loop transfer function for the concentration of the reactant in the third stage is

$$\bar{C}_{A3}(s) = \frac{0.1}{(s + .2)^3} \qquad (a)$$

A feedback control loop is being designed in which the actuator is placed in the feedforward part of the loop.

**8.22** Consider a proportional controller of gain $K_p$ as the actuator. (a) Determine the closed-loop transfer function. (b) For what values of $K_p$ is the closed-loop system stable? (c) Use MATLAB to determine the impulsive response of the closed-loop system with $K_p = 1$.

**8.23** Consider an integral controller of gain $K_i$ as the actuator. (a) Determine the closed-loop transfer function. (b) For what values of $K_i$ is the closed-loop system stable? (c) Use MATLAB to determine the step response of the closed-loop system with $K_i = 0.5$.

**8.24** Consider a PID controller as the actuator. (a) Determine the closed-loop transfer function. (b) For what values of $K_d$ is the closed-loop system stable if $K_i = 0.5$ and $K_p = 1.5$. (c) Use MATLAB to determine the step response of the system with $K_i = 0.5$, $K_p = 1.5$, and $K_d = 0.5$.

Problems 8.25–8.28 refer to the simplified model of a vehicle suspension system of Figure P8.25. The transfer function for the system is

$$G(s) = \frac{X(s)}{Y(s)} = \frac{2\zeta\omega_n s + \omega_n^2}{s^2 + 2\zeta\omega_n s + \omega_n^2} \qquad (a)$$

where $\zeta$ is the system's damping ratio and $\omega_n$ is its natural frequency. A golf cart of empty mass 500 kg has a suspension

system such that when empty its natural frequency is 20 r/s and has a damping ratio of 0.7. An electronic controller is being designed to improve the performance of the suspension system.

**8.25** (a) Can a proportional controller be designed such that the system is critically damped when empty? If so, what is the value of the proportional gain? (b) Consider the situation in which the golf cart is carrying golfers and their clubs. Assume the maximum added mass is 250 kg. Determine the necessary value of proportional gain as a function of the added mass such that the system is always critically damped.

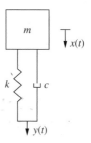

**FIG. P8.25**

**8.26** Determine the effect of the additional mass from passengers and clubs (from 0 to 250 kg) on the impulsive response of the closed-loop system when (a) a proportional controller is used, (b) an integral controller is used, (c) a PD controller is used.

**8.27** Suppose an integral controller of integral gain 1.3 is used. (a) Determine the step response of the closed-loop system. (b) Determine the 2 percent settling time. (c) Determine the 10–90 percent rise time.

**8.28** Repeat Problem 8.27 but assume a differential controller of differential gain 1.2 is used.

**8.29** Develop a SIMULINK model for the system of Problem 8.2.

**8.30** Develop a SIMULINK model for the system of Example 8.24 with a PID controller. Run the simulation for the system with the gains specified in Example 8.25. Then study the effect of changing each gain while holding the other two at the specified values. Describe the effects of changing the gains on the step response.

**8.31** Develop a SIMULINK model for the three-CSTR system of Problem 8.24. Set up the model so that the response for the reactant concentration in each tank can be monitored.

**8.32** Use the construction method outlined in Appendix D to construct the root-locus diagram for the closed-loop transfer function

$$H(s) = \frac{K}{s^3 + 20s + 80 + K} \tag{a}$$

The diagram must show all zeroes, poles, breakaway points, the crossover frequency, and asymptotes.

**8.33** What is the value of $K$ that leads to the dominant pole of the system of Problem 8.32 having a damping ratio of 0.3? What is the corresponding frequency?

**8.34** A second-order plant with a transfer function

$$G(s) = \frac{1}{s^2 + 20s + 80} \tag{b}$$

is placed in a feedback loop with a PI controller with $K_p = 30$. Use the construction method outlined in Appendix D to construct the root-locus diagram for the closed-loop transfer function. The diagram must show all zeroes, poles, breakaway and break-in points, the crossover frequency, and asymptotes.

**8.35** What is the largest value of the gain for the system of Problem 8.34 such that the dominant pole is less than –2. What are the corresponding damping ratio and frequency?

**8.36** The second-order plant of Problem 8.34 is placed in a feedback loop with a PID controller with $K_p = 30$ and $K_i = 50$. Use the methods of Appendix D to construct the root-locus diagram using the derivative gain as the parameter.

**8.37** A second-order plant whose transfer function is

$$G(s) = \frac{2}{(s + 1.2)(s + 1.8)} \tag{a}$$

is placed in a feedback loop with a PI controller with $K_p = 6$. (a) Use MATLAB to draw the root-locus diagram for the closed-loop system. (b) Use the root-locus diagram to determine the value of $K$ such that the dominant pole has the minimum possible overshoot. (c) What are the damping ratio and frequency corresponding to the minimum overshoot?

**8.38** Use MATLAB to draw the root-locus diagram for the system of Problem 8.37 with a PI controller for several values of $K_p$: 0, 2.0, 4.0, 8.0, 10.0, and 20.0. (The integral gain is still the parameter that varies in each root-locus diagram.) Make a chart of the minimum possible overshoot for each value of the proportional gain and the corresponding integral gain. Discuss how the minimum possible overshoot varies with proportional gain.

**8.39** A third-order pant of transfer function

$$G(s) = \frac{s + 2}{s^3 + 3s^2 + 5s + 10} \tag{a}$$

(a) Use MATLAB to plot the root-locus of the closed-loop transfer function when the plant is placed in a feedback loop with a proportional controller. (b) Repeat part (a) when the plant is placed in a feedback loop with a PI controller with $K_p = 25$. (c) Repeat part (a) when the plant is placed in a feedback loop with a PID controller with $K_p = 25$ and $K_i = 100$. (d) In each case discuss how to choose the parameter on which the root-locus diagram is drawn. Discuss the advantages and disadvantages of using each type of controller.

**8.40** A third-order plant of transfer function

$$G(s) = \frac{s + 1}{(s + 2)(s + 3)(s + 4)} \tag{b}$$

is placed in a feedback loop with a PID controller. Use an iterative or trial and error process by selectively varying the values of the gains, using MATLAB to draw root-locus diagrams, to design the controller. Considerations in the design should include absolute and relative stability, maximum overshoot, frequency of response, and settling time.

**8.41** A second-order plant has an open-loop transfer function of

$$G(s) = \frac{1}{s(s + 1)(s + 2)} \tag{a}$$

(a) Draw the root-locus diagram when the plant is placed in a feedback loop with a proportional controller. What is the minimum overshoot that can be attained using a proportional controller? (b) Instead of using a proportional controller it is proposed to place the plant in a feedback loop with a compensator whose transfer function is of the form

$$G_c(s) = K_c \frac{s + a}{s + b} \tag{b}$$

Design the compensator by choosing appropriate values of $K_c$, $a$, and $b$ such that the maximum overshoot of the step response is 10 percent and the frequency is 1.5 r/s.
(c) Plot and compare the step response of the system with and without the compensator.

**8.42** A plant has a transfer function

$$G(s) = \frac{1}{s(s^2 + 5s + 20)} \tag{c}$$

Use the Ziegler-Nichols tuning rules to design (a) a proportional controller, (b) a PI controller, and (c) a PID controller for the plant.

# 9

# State-Space Methods

Methods for determining the transient and forced response of dynamic systems described in Chapters 6 and 7 become difficult to manage for complex systems, including systems with multiple inputs and/or large number of dependent variables. The algebra in applying these methods becomes intractable. These methods are also applicable only for linear systems and are best suited for systems whose properties remain constant. However, as with any topic in engineering and science, one needs to understand simple systems before studying complex systems. A study of second-order systems leads to the understanding of concepts of natural frequency, damping ratio, steady-state amplitude, and phase. Such concepts are generalized in the study of higher-order systems. A good knowledge of the modeling and response of linear systems is necessary before studying the behavior of nonlinear systems. Knowledge of the response of time invariant systems is required before examining time-varying systems.

The rapid development of high-powered digital computers and sophisticated software packages such as MATLAB allow the application of a more general method for determining the response of dynamic systems. This method, called the state-space method, is applicable to all dynamic systems and is described in this chapter. The state-space method is introduced by example in Section 9.1 and then generalized in Section 9.2. Subsequent sections explore its application to linear and nonlinear systems.

The concept of states is first introduced in Section 2.9 where a state is related to a component that stores energy. The number of states is equal to the number of independent energy storage components in the system. State variables are often defined in terms of an energy storage component. This chapter shows how the system response is determined using state variables.

## 9.1 AN EXAMPLE IN THE STATE SPACE

Recall that the differential equation governing the motion of the mechanical system of Figure 9.1 is

$$m\ddot{x} + c\dot{x} + kx = F(t) \qquad (9.1)$$

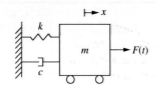

**FIG. 9.1** The mechanical system composed of mass, spring, and viscous damper with a force input.

Define **state-space variables** according to

$$y_1 = x \tag{9.2a}$$
$$y_2 = \dot{x} \tag{9.2b}$$

Using the state-space variables of Equation (9.2), Equation (9.1) can be rewritten as two first-order differential equations as

$$\dot{y}_1 = y_2 \tag{9.3a}$$
$$\dot{y}_2 = \frac{1}{m}F(t) - \frac{c}{m}y_2 - \frac{k}{m}y_1 \tag{9.3b}$$

Equations (9.3) are summarized in matrix form as

$$\begin{bmatrix} \dot{y}_1 \\ \dot{y}_2 \end{bmatrix} = \begin{bmatrix} 0 & 1 \\ -\dfrac{k}{m} & -\dfrac{c}{m} \end{bmatrix} \begin{bmatrix} y_1 \\ y_2 \end{bmatrix} + \begin{bmatrix} 0 \\ \dfrac{1}{m} \end{bmatrix} F(t) \tag{9.4}$$

Equation (9.4) can be written as

$$\dot{\mathbf{y}} = \mathbf{A}\mathbf{y} + \mathbf{B}F(t) \tag{9.5}$$

where **y** is called the state vector, **A** is the state matrix, and **B** is the input vector.

The input to the mechanical system is $F(t)$ and its output is $x(t)$. A block diagram of the state-space modeling for the system is illustrated in Figure 9.2. The system input is multiplied by $1/m$, and then is added to $-(k/m)y_1 - (c/m)y_2$ coming from feedback loops at a summing junction. The sum is integrated leading to $y_2(t)$, which is fed back to the summing junction. This result is integrated again leading to $y_1(t)$, which is fed back to the summing junction as well as sent as system output. The block diagram of Figure 9.2 is equivalent to the block diagrams introduced in Section 8.1 if the integrators are replaced by $1/s$.

The mass-spring-viscous damper system has two independent storage elements. The spring stores potential energy, $V = \frac{1}{2}kx^2$. The inertia element stores kinetic energy $T = \frac{1}{2}m\dot{x}^2$. The system is a second-order system because there are two independent storage elements. The state variables defined in Equations (9.2) are directly related to the potential energy and kinetic energy.

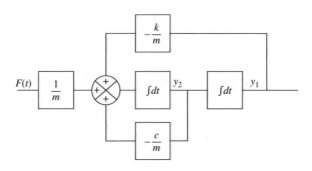

**FIG. 9.2** A block diagram of the state-space model for the system of Figure 9.1.

There are many methods available to determine the response of Equation (9.5) for a given $F(t)$. Given the appropriate initial conditions, Equation (9.4) is amenable to solution by single-step numerical methods such as the Runge-Kutta methods. The general analytical solution is the sum of the homogeneous solution, the solution obtained when $F(t) = 0$, and the particular solution, the solution specific to $F(t)$. Many methods are available for both.

First consider the solution obtained using the Laplace transform method. To this end define

$$\mathbf{Y}(s) = \begin{bmatrix} Y_1(s) \\ Y_2(s) \end{bmatrix} \tag{9.6a}$$

$$\mathbf{y}(0) = \begin{bmatrix} y_1(0) \\ y_2(0) \end{bmatrix} \tag{9.6b}$$

Taking the Laplace transform of Equation (9.5) using the initial conditions of Equation (9.6b) leads to

$$s\mathbf{Y}(s) = \mathbf{A}\mathbf{Y}(s) + \mathbf{y}(0) + \mathbf{B}F(s) \tag{9.7}$$

Equation (9.7) is rearranged to

$$(s\mathbf{I} - \mathbf{A})\mathbf{Y}(s) = \mathbf{y}(0) + \mathbf{B}F(s) \tag{9.8}$$

where $\mathbf{I}$ is the $2 \times 2$ identity matrix. The solution of Equation (9.8) is obtained as

$$\mathbf{Y}(s) = (s\mathbf{I} - \mathbf{A})^{-1}\mathbf{y}(0) + (s\mathbf{I} - \mathbf{A})^{-1}\mathbf{B}F(s) \tag{9.9}$$

Define

$$\mathbf{G}(t) = \mathcal{L}^{-1}\{(s\mathbf{I} - \mathbf{A})^{-1}\} \tag{9.10}$$

The convolution property is used to invert Equation (9.9) leading to

$$\mathbf{y}(t) = \mathbf{G}(t)\mathbf{y}(0) + \int_0^t \mathbf{G}(\tau)\mathbf{B}F(t - \tau)d\tau \tag{9.11}$$

**Example 9.1**

Use the state-space method to determine and plot the response of a second-order mechanical system with $m = 1\,\text{kg}$, $c = 5\,\text{N·s/m}$, and $k = 6\,\text{N/m}$ subject to $F(t) = 3e^{-2t}\,\text{N}$. The systems' initial conditions are $x(0) = 0$ and $\dot{x}(0) = 1\,\text{m/s}$.

**Solution**

Substitution of the given values into Equation (9.4) leads to

$$\begin{bmatrix} \dot{y}_1 \\ \dot{y}_2 \end{bmatrix} = \begin{bmatrix} 0 & 1 \\ -6 & -5 \end{bmatrix} \begin{bmatrix} y_1 \\ y_2 \end{bmatrix} + \begin{bmatrix} 0 \\ 1 \end{bmatrix} 3e^{-2t} \tag{a}$$

with

$$\mathbf{y}(0) = \begin{bmatrix} 0 \\ 1 \end{bmatrix}$$
(b)

The matrix $(s\mathbf{I} - \mathbf{A})$ is obtained as

$$s\mathbf{I} - \mathbf{A} = s\begin{bmatrix} 1 & 0 \\ 0 & 1 \end{bmatrix} - \begin{bmatrix} 0 & 1 \\ -6 & -5 \end{bmatrix} = \begin{bmatrix} s & -1 \\ 6 & s+5 \end{bmatrix}$$
(c)

The inverse matrix is obtained as

$$(s\mathbf{I} - \mathbf{A})^{-1} = \frac{1}{s^2 + 5s + 6}\begin{bmatrix} s+5 & 1 \\ -6 & s \end{bmatrix}$$
(d)

Noting that $s^2 + 6s + 5 = (s+3)(s+2)$ partial fraction decomposition leads to

$$(s\mathbf{I} - \mathbf{A})^{-1} = \begin{bmatrix} \dfrac{-2}{s+3} + \dfrac{3}{s+2} & \dfrac{1}{s+3} - \dfrac{1}{s+2} \\ -\dfrac{6}{s+3} + \dfrac{6}{s+2} & \dfrac{3}{s+3} - \dfrac{2}{s+2} \end{bmatrix}$$
(e)

The application of Equation (9.10) leads to

$$\mathbf{G}(t) = \begin{bmatrix} -2e^{-3t} + 3e^{-2t} & e^{-3t} - e^{-2t} \\ -6e^{-3t} + 6e^{-2t} & 3e^{-3t} - 2e^{-2t} \end{bmatrix}$$
(f)

Equation (9.11) becomes

$$\begin{bmatrix} y_1(t) \\ y_2(t) \end{bmatrix} = \begin{bmatrix} -2e^{-3t} + 3e^{-2t} & e^{-3t} - e^{-2t} \\ -6e^{-3t} + 6e^{-2t} & 3e^{-3t} - 2e^{-2t} \end{bmatrix}\begin{bmatrix} 0 \\ 1 \end{bmatrix}$$

$$+ \int_0^t \begin{bmatrix} -2e^{-3\tau} + 3e^{-2\tau} & e^{-3\tau} - e^{-2\tau} \\ -6e^{-3\tau} + 6e^{-2\tau} & 3e^{-3\tau} - 2e^{-2\tau} \end{bmatrix}\begin{bmatrix} 0 \\ 1 \end{bmatrix} 3e^{-3(t-\tau)}\,d\tau$$
(g)

The evaluation of the convolution integral leads to

$$\begin{bmatrix} y_1(t) \\ y_2(t) \end{bmatrix} = \begin{bmatrix} e^{-3t} - e^{-2t} \\ 3e^{-3t} - 2e^{-2t} \end{bmatrix} + \begin{bmatrix} 3te^{-3t} + 3e^{-3t} - 3e^{-2t} \\ 9te^{-3t} + 6e^{-3t} - 6e^{-2t} \end{bmatrix}$$

$$= \begin{bmatrix} 3te^{-3t} + 4e^{-3t} - 4e^{-2t} \\ 9te^{-3t} + 9e^{-3t} - 8e^{-2t} \end{bmatrix}$$
(h)

The time-dependent response of the system is illustrated in Figure 9.3. The plot illustrates that the response satisfies the initial conditions of Equation (b). The state-plane plot with $y_1$ on the horizontal axis and $y_2$ on the vertical axis is illustrated in Figure 9.4. The path in the state-plane is called a **trajectory**, which is plotted using $t$ as a parameter. The trajectory in the state-plane begins at $(0, 1)$ corresponding to the initial conditions at $t = 0$. The trajectory is continuous and approaches $(0, 0)$ for large $t$.

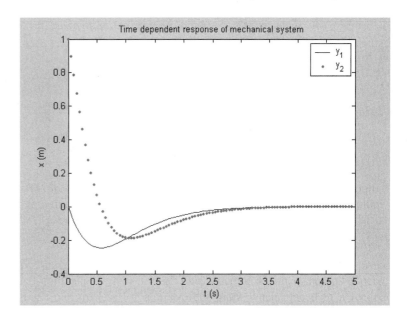

**FIG. 9.3** The response of the mechanical system of Example 9.1 is determined using state-space methods.

SIMULINK may be used to develop state-space models of dynamic systems. The SIMULINK model for the system of Example 9.1 with $F(t) = 0$ is illustrated in Figure 9.5. The following is noted regarding this model:

- An initial condition of $y_2(0) = 1$ is specified using the Block Properties box for the integrator. This specifies that the variable in the block diagram after the integration has an initial value of 1.
- An XY graph, generated from the Sinks menu of the Library Browser, is used to generate a state-plane plot. The upper input to

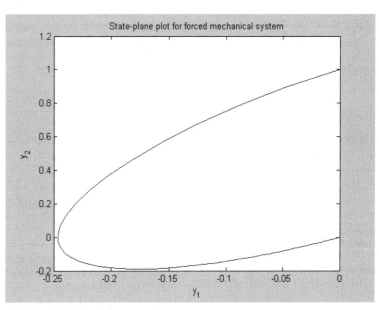

**FIG. 9.4** The state-plane plot of the forced response for the system of Example 9.1. The trajectory originates at (0, 1) corresponding to the system's initial conditions, is continuous, and approaches (0, 0) corresponding to the final state of the system.

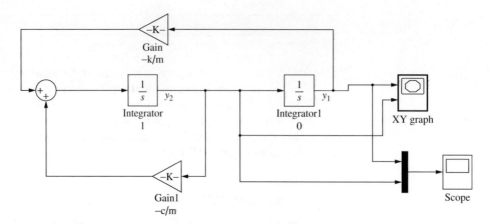

**FIG. 9.5** The SIMULINK model for the system in Example 9.1 with $F(t) = 0$.

the box is $y_1$, which is plotted on the $x$ axis while the lower input is $y_2$, which is plotted on the $y$ axis.

- A Mux from the Signal Routing menu of the Library Browser is used to send two signals to the same scope.
- The gains are set to $-c/m$ and $-k/m$ where $m$, $c$, and $k$ are the current values in the MATLAB workspace.

The scope output, showing the time-dependent responses, is shown in Figure 9.6.

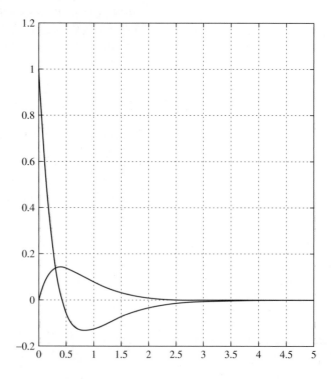

**FIG. 9.6** The scope output from a SIMULINK model for Example 9.1 showing time-dependent responses for $y_1(t)$ with an initial condition of 1 and $y_2(t)$ with an initial condition of 0.

## 9.2 GENERAL STATE-SPACE MODELING

The second-order differential equation for the system described in Section 9.1 was replaced by two first-order differential equations. This alternate formulation, called the **state-space formulation**, is generalized in this section. The system of Section 9.1, which has one input $F(t)$ and one output $x(t)$ is called a single-input single-output (SISO) system. As illustrated the state-space formulation can also be applied to multiple-input multiple-output (MIMO) systems.

### 9.2.1 Basic Concepts

Consider a dynamic system with the prescribed inputs $u_1(t)$, $u_2(t)$, ..., $u_k(t)$ and desired outputs $x_1(t), x_2(t), \ldots, x_n(t)$. The **state** of the system is defined by three sets of information: (1) A set of variables, called state variables, $y_1(t)$, $y_2(t)$, ..., $y_m(t)$ such that $m$ is the smallest number of variables needed to completely define the behavior of the system at an arbitrary time $t$; (2) the initial values of the state variables; and (3) the system input. The dynamic system is modeled such that the state variables are determined from the system input by a set of equations written in matrix form as

$$\dot{\mathbf{y}} = \mathbf{Ay} + \mathbf{Bu} \tag{9.12}$$

where $\mathbf{A}$ is a $m \times m$ matrix called the state matrix and $\mathbf{B}$ is a $m \times k$ matrix called the input matrix. The output variables are related to the state variables and the inputs through

$$\mathbf{x} = \mathbf{Cy} + \mathbf{Du} \tag{9.13}$$

where $\mathbf{C}$ is an $n \times m$ matrix called the output matrix and $\mathbf{D}$ is an $n \times k$ matrix called the transmission matrix.

The state variables are chosen for convenience and practicality. For a system in which the output variables are identical to the dependent variables used in modeling the system, the state variables are often chosen as the output variables and all derivatives necessary to completely specify the state at any time, given appropriate initial conditions. Such is the case for the second-order mechanical system of Section 9.1, for which the state-space representation is given by Equation (9.4). For this system the number of inputs $k$ is 1, the number of outputs $m$ is one, and the number of state variables $m$ is 2. The input vector $F$ is thus a $1 \times 1$ matrix or a scalar, $\mathbf{u} = [F]$. The output vector $\mathbf{x}$ is also a scalar $\mathbf{x} = [x(t)]$. The state vector is of size $2 \times 1$ defined by

$$\mathbf{y} = \begin{bmatrix} y_1(t) \\ y_2(t) \end{bmatrix} \tag{9.14}$$

Comparison of Equation (9.4) with Equation (9.12) shows that the $2 \times 2$ state matrix is

$$\mathbf{A} = \begin{bmatrix} 0 & 1 \\ -\dfrac{k}{m} & -\dfrac{c}{m} \end{bmatrix} \tag{9.15}$$

and the $2 \times 1$ output matrix is

$$\mathbf{B} = \begin{bmatrix} 0 \\ \dfrac{1}{m} \end{bmatrix} \tag{9.16}$$

The relationship between the state variables, the input variables, and the output variables is

$$x = \begin{bmatrix} 1 & 0 \end{bmatrix} \begin{bmatrix} y_1 \\ \dot{y}_1 \end{bmatrix} + [0][F(t)] \tag{9.17}$$

Comparison of Equation (9.16) with Equation (9.13) shows that the $1 \times 2$ output matrix is $\mathbf{C} = \begin{bmatrix} 1 & 0 \end{bmatrix}$ and the $1 \times 1$ transmission matrix is $\mathbf{D} = [0]$.

The state-space formulation of mathematical models is further illustrated in the following examples.

---

**Example 9.2**

The differential equations governing the perturbations of liquid levels due to perturbations in inlet flow rates derived as Equations (u) and (v) of Example 4.9 are

$$A_1 \frac{dh_1}{dt} + \frac{1}{R_1} h_1 - \frac{1}{R_1} h_2 = q_{i1}(t) \tag{a}$$

$$A_2 \frac{dh_2}{dt} - \frac{1}{R_1} h_1 + \left( \frac{1}{R_1} + \frac{1}{R_2} \right) h_2 = q_{i2}(t) \tag{b}$$

Develop the state-space formulation for the system.

**Solution**

The state variables can be defined as the output variables $y_1 = h_1$ and $y_2 = h_2(t)$. Equations (a) and (b) can be written in state-space form as

$$\begin{bmatrix} \dot{y}_1 \\ \dot{y}_2 \end{bmatrix} = \begin{bmatrix} -\dfrac{1}{R_1 A_1} & \dfrac{1}{R_1 A_1} \\ \dfrac{1}{R_1 A_2} & -\left( \dfrac{1}{R_1 A_2} + \dfrac{1}{R_2 A_2} \right) \end{bmatrix} \begin{bmatrix} y_1 \\ y_2 \end{bmatrix} + \begin{bmatrix} \dfrac{1}{A_1} & 0 \\ 0 & \dfrac{1}{A_2} \end{bmatrix} \begin{bmatrix} q_{i1}(t) \\ q_{i2}(t) \end{bmatrix} \tag{c}$$

Equation (c) is of the form of Equation (9.12). The appropriate form of Equation (9.13) is

$$\begin{bmatrix} h_1 \\ h_2 \end{bmatrix} = \begin{bmatrix} 1 & 0 \\ 0 & 1 \end{bmatrix} \begin{bmatrix} y_1 \\ y_2 \end{bmatrix} + \begin{bmatrix} 0 & 0 \\ 0 & 0 \end{bmatrix} \begin{bmatrix} q_{i1}(t) \\ q_{i2}(t) \end{bmatrix} \tag{d}$$

---

The system of Example 9.2 has two independent energy storage components. Each tank stores potential energy in the form of head. The chosen state variables are the head perturbations in the tanks.

## Example 9.3

Determine the state-space model for the armature-controlled dc motor of Example 3.23.

### Solution

The inputs to the system are the armature voltage $v_a(t)$ and an external torque applied to the shaft $T$. The system outputs are the armature current $i_a$ and the angular velocity of the armature $\omega$. The differential equations derived in Example 3.23 are

$$L_a \frac{di_a}{dt} + R_a i_a + K_b \omega = v_a \tag{a}$$

$$J \frac{d\omega}{dt} + c_t \omega - K_a i_a = T \tag{b}$$

The state variables are defined by

$$y_1 = i_a \tag{c}$$

$$y_2 = \omega \tag{d}$$

Equations (a) and (b) are rewritten in terms of the state variables defined in Equations (c) and (d) as

$$\frac{dy_1}{dt} = -\frac{R_a}{L_a} y_1 - \frac{K_b}{L_a} y_2 + \frac{1}{L_a} v_a(t) \tag{e}$$

$$\frac{dy_2}{dt} = \frac{K_a}{J} y_1 - \frac{c_t}{J} y_2 + \frac{1}{J} T(t) \tag{f}$$

The state and input matrices are determined from Equations (e) and (f) as

$$\mathbf{A} = \begin{bmatrix} -\dfrac{R_a}{L_a} & -\dfrac{K_b}{L_a} \\ \dfrac{K_a}{J} & -\dfrac{c_t}{J} \end{bmatrix} \tag{g}$$

$$\mathbf{B} = \begin{bmatrix} \dfrac{1}{L_a} & 0 \\ 0 & \dfrac{1}{J} \end{bmatrix} \tag{h}$$

The output and transmission matrices are obtained from Equations (c) and (d) as

$$\mathbf{C} = \begin{bmatrix} 1 & 0 \\ 0 & 1 \end{bmatrix} \tag{i}$$

$$\mathbf{D} = \begin{bmatrix} 0 & 0 \\ 0 & 0 \end{bmatrix} \tag{j}$$

The system of Example 9.3 has two independent storage components: the rotor storing kinetic energy and the inductor in the armature circuit storing energy in the presence of a magnetic field. The stored energy in the inductor is $\frac{1}{2}Li^2 = \frac{1}{2}Ly_1^2$ while the kinetic energy of the shaft is $\frac{1}{2}J\omega^2 = \frac{1}{2}Jy_2^2$. Thus the chosen state variables are directly related to the energy in each storage component.

---

## Example 9.4

Determine the state-space model for the series $LRC$ circuit of Example 3.10. The voltage delivered by the source $v(t)$ is the input and the current through the circuit $i(t)$ is the ouptput.

### Solution

The integrodifferential equation obtained in Example 3.10 for the current through the series $LRC$ circuit is

$$L\frac{di}{dt} + Ri + \frac{1}{C}\int_0^t i\,dt = v(t) \qquad \text{(a)}$$

The current is related to the charge $q$ by $i = dq/dt$. Using current as the dependent variable in Equation (a) leads to

$$L\frac{d^2q}{dt^2} + R\frac{dq}{dt} + \frac{1}{C}q = v(t) \qquad \text{(b)}$$

The state variables are chosen as

$$y_1 = q \qquad \text{(c)}$$

$$y_2 = i = \frac{dq}{dt} \qquad \text{(d)}$$

Equation (b) is rewritten in terms of state variables as

$$\frac{dy_2}{dt} = -\frac{1}{LC}y_1 - \frac{R}{L}y_2 + \frac{1}{L}v(t) \qquad \text{(e)}$$

Equations (d) and (e) are summarized in matrix form as

$$\begin{bmatrix} \dot{y}_1 \\ \dot{y}_2 \end{bmatrix} = \begin{bmatrix} 0 & 1 \\ -\dfrac{1}{LC} & -\dfrac{R}{L} \end{bmatrix} \begin{bmatrix} y_1 \\ y_2 \end{bmatrix} + \begin{bmatrix} 0 \\ \dfrac{1}{L} \end{bmatrix} v(t) \qquad \text{(f)}$$

From Equation (f) it is apparent that the state and input matrices are

$$\mathbf{A} = \begin{bmatrix} 0 & 1 \\ -\dfrac{1}{LC} & -\dfrac{R}{L} \end{bmatrix} \qquad \text{(g)}$$

$$\mathbf{B} = \begin{bmatrix} 0 \\ \dfrac{1}{L} \end{bmatrix} \qquad \text{(h)}$$

The relations between the output variables and state variables are summarized as

$$[i(t)] = [0 \quad 1]\begin{bmatrix} y_1 \\ y_2 \end{bmatrix} \tag{i}$$

The output and transmission matrices are obtained from Equation (i) as $\mathbf{C} = [0 \quad 1]$ and $\mathbf{D} = [0]$.

The system of Example 9.4 has two energy storage components. The energy stored in the electric field of the capacitor is $\frac{1}{2}q^2/C = \frac{1}{2}y_1^2/C$, while the energy stored in the magnetic field of the inductor is $\frac{1}{2}Li^2 = \frac{1}{2}Ly_2 2$. The chosen state variables are directly related to the stored energies.

## Example 9.5

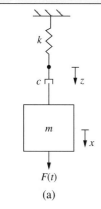

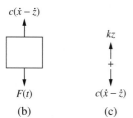

**FIG. 9.7** (a) The system of Example 9.5; (b) and (c) the free-body diagrams of the system of Example 9.6 at arbitary.

Derive a state-space model for the system of Figure 9.7(a). The input to the system is $F(t)$ and the system output is $x(t)$.

### Solution

Let $z(t)$ be the displacement of the joint between the spring and the viscous damper. The application of Newton's second law to the free-body diagram of the block at an arbitrary instant, as shown in Figure 9.7(b), gives

$$F(t) - c(\dot{x} - \dot{z}) = m\ddot{x} \tag{a}$$

The application of Newton's second law to the free-body diagram of the joint between the viscous damper and the mass, as illustrated in Figure 9.7(c), noting that the mass of the joint is zero leads to

$$-c(\dot{x} - \dot{z}) + kz = 0 \tag{b}$$

Define state variables as

$$y_1 = x \tag{c}$$

$$y_2 = \dot{x} \tag{d}$$

$$y_3 = z \tag{e}$$

The use of state variables in Equations (a) and (b) gives

$$m\dot{y}_2 - c\dot{y}_3 = -cy_2 + F(t) \tag{f}$$

$$c\dot{y}_3 = cy_2 - ky_3 \tag{g}$$

Equations (d), (f), and (g) are summarized in a matrix form as

$$\begin{bmatrix} 1 & 0 & 0 \\ 0 & m & -c \\ 0 & 0 & c \end{bmatrix}\begin{bmatrix} \dot{y}_1 \\ \dot{y}_2 \\ \dot{y}_3 \end{bmatrix} = \begin{bmatrix} 0 & 1 & 0 \\ 0 & -c & 0 \\ 0 & c & -k \end{bmatrix}\begin{bmatrix} y_1 \\ y_2 \\ y_3 \end{bmatrix} + \begin{bmatrix} 0 \\ 1 \\ 0 \end{bmatrix}F(t) \tag{h}$$

Noting that

$$\begin{bmatrix} 1 & 0 & 0 \\ 0 & m & -c \\ 0 & 0 & c \end{bmatrix}^{-1} = \begin{bmatrix} 1 & 0 & 0 \\ 0 & \dfrac{1}{m} & \dfrac{1}{m} \\ 0 & 0 & \dfrac{1}{c} \end{bmatrix} \quad \text{(i)}$$

premultiplication of Equation (h) by the inverse of the $3 \times 3$ matrix on its left-hand side leads to

$$\begin{bmatrix} \dot{y}_1 \\ \dot{y}_2 \\ \dot{y}_3 \end{bmatrix} = \begin{bmatrix} 1 & 0 & 0 \\ 0 & \dfrac{1}{m} & \dfrac{1}{m} \\ 0 & 0 & \dfrac{1}{c} \end{bmatrix} \begin{bmatrix} 0 & 1 & 0 \\ 0 & -c & 0 \\ 0 & c & -k \end{bmatrix} \begin{bmatrix} y_1 \\ y_2 \\ y_3 \end{bmatrix} + \begin{bmatrix} 1 & 0 & 0 \\ 0 & \dfrac{1}{m} & \dfrac{1}{m} \\ 0 & 0 & \dfrac{1}{c} \end{bmatrix} \begin{bmatrix} 0 \\ 1 \\ 0 \end{bmatrix} F(t)$$

$$= \begin{bmatrix} 0 & 1 & 0 \\ 0 & 0 & -\dfrac{k}{m} \\ 0 & 1 & -\dfrac{k}{c} \end{bmatrix} \begin{bmatrix} y_1 \\ y_2 \\ y_3 \end{bmatrix} + \begin{bmatrix} 0 \\ \dfrac{1}{m} \\ 0 \end{bmatrix} F(t) \quad \text{(j)}$$

The $3 \times 3$ state matrix and the $3 \times 1$ input matrix are determined directly from Equation (j).

The $1 \times 3$ output matrix is $\mathbf{C} = \begin{bmatrix} 1 & 0 & 0 \end{bmatrix}$ and the $1 \times 1$ transmission matrix is $\mathbf{D} = [0]$.

Consider an alternate choice of state variables as

$$w_1 = z - x \quad \text{(k)}$$
$$w_2 = \dot{x} \quad \text{(l)}$$
$$w_3 = z \quad \text{(m)}$$

Noting from Equations (k) and (m) that

$$\dot{w}_3 = \dot{w}_1 + w_2 \quad \text{(n)}$$

Equations (a), (b), and (n) are used to write

$$\begin{bmatrix} c & 0 & 0 \\ 0 & m & c \\ -1 & 0 & 1 \end{bmatrix} \begin{bmatrix} \dot{w}_1 \\ \dot{w}_2 \\ \dot{w}_3 \end{bmatrix} + \begin{bmatrix} 0 & 0 & -k \\ 0 & 0 & 0 \\ 0 & -1 & 0 \end{bmatrix} \begin{bmatrix} w_1 \\ w_2 \\ w_3 \end{bmatrix} = \begin{bmatrix} 0 \\ 1 \\ 0 \end{bmatrix} F(t) \quad \text{(o)}$$

Multiplication by the inverse of the first coefficient matrix leads to

$$\begin{bmatrix} \dot{w}_1 \\ \dot{w}_2 \\ \dot{w}_3 \end{bmatrix} = \begin{bmatrix} 0 & 0 & \dfrac{1}{ck} \\ 0 & -\dfrac{c}{m} & -\dfrac{1}{mk} \\ 0 & 1 & \dfrac{1}{ck} \end{bmatrix} \begin{bmatrix} w_1 \\ w_2 \\ w_3 \end{bmatrix} + \begin{bmatrix} 0 \\ \dfrac{1}{m} \\ 0 \end{bmatrix} F(t) \quad \text{(p)}$$

The output and transmission matrices are defined by

$$[x] = \begin{bmatrix} -1 & 0 & 1 \end{bmatrix} \begin{bmatrix} w_1 \\ w_2 \\ w_3 \end{bmatrix} + [0][F(t)] \tag{q}$$

Example 9.5 has several interesting features. First, it is used to illustrate that the choice of state variables is not unique. Second, it seems to contradict the previous assertion that each state corresponds to an independent energy storage component. There are only two independent energy storage components: kinetic energy stored by the inertia element and potential energy stored by the spring. Yet three seemingly independent state variables are used. However, a close examination of Equations (j) and (p) leads to a different conclusion. The examination of Equation (j) shows that the second and third equations have no dependence on $y_1$. These equations may be solved independently and then $y_1$ calculated using the first equation. This suggests that only two equations are independent. However, the third state variable is necessary because of the form of the output matrix of Equation (q). The form of Equation (9.13) does not allow for the output to depend on derivatives of the state variables. Thus the third state variable is necessary, not because of energy storage, but for specifying the output. If, for example, $z$ is the chosen output, specification of $y_1$ or $w_1$ is not necessary.

In each of the previous examples the transmission matrix is a zero matrix of an appropriate size. A nonzero transmission matrix occurs for systems in mechanical systems with motion input and other systems in which the time derivative of the input appears explicitly in the system's mathematical model. The right-hand side of Equation (9.12) does not allow for the explicit appearance of derivatives of input variables in the state-space formulation of the differential equations. For mechanical systems with motion input and similar systems the state variables are defined using input variables as well as output variables.

## Example 9.6

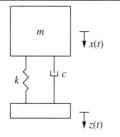

**FIG. 9.8** The system of Example 9.6. The block is attached through a spring and viscous damper, which is parallel to a base with a prescribed displacement $z(t)$.

The input to the mechanical system of Figure 9.8 is the prescribed displacement $z(t)$ of the base. The mathematical model for the absolute displacement of the block $x(t)$ is

$$m\ddot{x} + c\dot{x} + kx = c\dot{z} + kz \tag{a}$$

Determine a state-space model for this system.

### Solution

Defining state variables as in the previous examples leads to a state-space formulation in which the vector of derivatives of the state variables is dependent on the derivative of the input function, which is not allowed in the model described by Equations (9.12) and (9.13). A remedy to this problem is to define the state variables in terms of both the input and output variables such that Equation (a) can be written in terms of the derivative of a state variable, the state variables themselves, and the input variable $z(t)$. To this end define

$$y_1 = x \tag{b}$$

$$y_2 = \dot{x} + \mu z \tag{c}$$

where $\mu$ is to be chosen to achieve the desired state-space model. Equation (c) is rearranged and then differentiated leading to

$$\dot{x} = y_2 - \mu z \tag{d}$$

$$\ddot{x} = \dot{y}_2 - \mu \dot{z} \tag{e}$$

The substitution of Equations (c), (d), and (e) into Equation (a) leads to

$$m(\dot{y}_2 - \mu\dot{z}) + c(y_2 - \mu z) + ky_1 = c\dot{z} + kz \tag{f}$$

The time derivative of the input $\dot{z}(t)$ is eliminated from Equation (f) by choosing

$$\mu = -\frac{c}{m} \tag{g}$$

Use of Equation (g) in Equation (f) leads to

$$\dot{y}_2 = -\frac{k}{m}y_1 - \frac{c}{m}y_2 + \left(\frac{k}{m} - \frac{c^2}{m^2}\right)z \tag{h}$$

Equations (b) and (h) are used to write the state-space formulation as

$$\begin{bmatrix} \dot{y}_1 \\ \dot{y}_2 \end{bmatrix} = \begin{bmatrix} 0 & 1 \\ -\dfrac{k}{m} & -\dfrac{c}{m} \end{bmatrix} \begin{bmatrix} y_1 \\ y_2 \end{bmatrix} + \begin{bmatrix} \dfrac{c}{m} \\ \dfrac{k}{m} - \dfrac{c^2}{m^2} \end{bmatrix} [z(t)] \tag{i}$$

---

### Example 9.7

The mathematical model of the system of Figure 9.8, obtained using the displacement of the block relative to its base $w(t) = z(t) - x(t)$ as the output variable, is

$$m\ddot{w} + c\dot{w} + kw = -m\ddot{z} \tag{a}$$

Derive a state-space model for the system.

### Solution

The use of state variables defined by Equations (b) and (c) of Example 9.6 does not lead to a state-space model that does not involve time derivatives of the input. Since the second derivative of the input appears explicitly in the mathematical model it is necessary to define state variables as

$$y_1 = w + vz \tag{b}$$

$$y_2 = \dot{y}_1 + \mu z \tag{c}$$

where $v$ and $\mu$ are to be chosen to develop an appropriate state-space model. Equations (b) and (c) are rearranged and differentiated leading to

$$y_2 = \dot{w} + v\dot{z} + \mu z \tag{d}$$
$$\dot{y}_2 = \ddot{w} + v\ddot{z} + \mu\dot{z} \tag{e}$$

Equations (d) and (e) are rearranged to

$$\dot{w} = y_2 - v\dot{z} - \mu z \tag{f}$$
$$\ddot{w} = \dot{y}_2 - v\ddot{z} - \mu\dot{z} \tag{g}$$

The substitution of Equations (b), (f), and (g) into Equation (a) leads to

$$m(\dot{y}_2 - v\ddot{z} - \mu\dot{z}) + c(y_2 - v\dot{z} - \mu z) + k(y_1 - vz) = -m\ddot{z} \tag{h}$$

Derivatives of the input are eliminated from Equation (h) by choosing

$$v = 1 \tag{i}$$
$$-\mu m - cv = 0 \Rightarrow \mu = -\frac{c}{m} \tag{j}$$

The use of Equations (i) and (j) in Equation (h) leads to

$$\dot{y}_2 = -\frac{k}{m}y_1 - \frac{c}{m}y_2 + \left(\frac{k}{m} - \frac{c^2}{m^2}\right)z \tag{k}$$

The state-space model obtained from Equations (b), (c), and (k) as

$$\begin{bmatrix} \dot{y}_1 \\ \dot{y}_2 \end{bmatrix} = \begin{bmatrix} 0 & 1 \\ -\dfrac{k}{m} & -\dfrac{c}{m} \end{bmatrix}\begin{bmatrix} y_1 \\ y_2 \end{bmatrix} + \begin{bmatrix} -\dfrac{c}{m} \\ \dfrac{k}{m} - \dfrac{c^2}{m^2} \end{bmatrix}[z(t)] \tag{l}$$

$$[w(t)] = \begin{bmatrix} 1 & 0 \end{bmatrix}\begin{bmatrix} y_1 \\ y_2 \end{bmatrix} + [-1][z(t)] \tag{m}$$

Thus $\mathbf{C} = \begin{bmatrix} 1 & 0 \end{bmatrix}$ and $\mathbf{D} = [-1]$.

## 9.2.2 Multidegree-of-Freedom Mechanical Systems

The number of degrees of freedom required to model a mechanical system is the minimum number of dependent variables necessary to completely describe the motion of every particle with mass in the system. The system of Example 9.1 is a one-degree-of-freedom system. The system of Example 9.6 has only one degree of freedom, even though $x$ and $z$ are not kinematically related. Knowledge of $x$ is sufficient to specify the motion of every particle in the system with mass. The dependent variable $z$ specifies the displacement of a point in the system that does not have inertia. It might be appropriate to specify the system as having one and a half degrees of freedom.

The differential equations governing a linear $n$-degree-of-freedom mechanical system using $x_1, x_2, \ldots, x_n$ as dependent variables can be written in the general form of

$$\mathbf{M}\ddot{\mathbf{x}} + \mathbf{C_d}\dot{\mathbf{x}} + \mathbf{K}\mathbf{x} = \mathbf{F} \tag{9.18}$$

where $\mathbf{x}$ is a $n \times 1$ vector of the dependent variables, $\mathbf{M}$ is an $n \times n$ mass matrix, $\mathbf{C_d}$ is an $n \times n$ damping matrix, $\mathbf{K}$ is the $n \times n$ stiffness matrix, and $\mathbf{F}$ is the $n \times 1$ input vector. The distinction made in the previous paragraph between degrees of freedom and dependent variables ensures that the mass matrix is nonsingular and its inverse exists. The input vector may be written in the form

$$\mathbf{F} = \mathbf{Q}\mathbf{u} \tag{9.19}$$

where $\mathbf{u}$ is the $k \times 1$ vector of system inputs and $\mathbf{Q}$ is an $n \times k$ matrix.

A general state-space formulation for systems described by equations of the form of Equation (9.18) is obtained by defining two $n \times 1$ vectors of state variables $\mathbf{y}_1$ and $\mathbf{y}_2$ by

$$\begin{aligned} y_{1,i} &= x_i & i &= 1, 2, \ldots, n \\ y_{2,i} &= \dot{x}_i & i &= 1, 2, \ldots, n \end{aligned} \tag{9.20}$$

It is clear from Equation (9.20) that

$$\dot{\mathbf{y}}_1 = \mathbf{y}_2 \tag{9.21}$$

Furthermore, Equation (9.18) can be written as

$$\mathbf{M}\dot{\mathbf{y}}_2 = -\mathbf{C}_d\mathbf{y}_2 - \mathbf{K}\mathbf{y}_1 + \mathbf{Q}\mathbf{u} \tag{9.22}$$

Since it is guaranteed that $\mathbf{M}$ is nonsingular, Equation (9.22) can be premultiplied by $\mathbf{M}^{-1}$ leading to

$$\dot{\mathbf{y}}_2 = -\mathbf{M}^{-1}\mathbf{C}_d\mathbf{y}_2 - \mathbf{M}^{-1}\mathbf{K}\mathbf{y}_1 + \mathbf{M}^{-1}\mathbf{Q}\mathbf{u} \tag{9.23}$$

A $2n \times 1$ state vector $\mathbf{y}$ is obtained by augmentation of $\mathbf{y}_1$ and $\mathbf{y}_2$ such that

$$\mathbf{y} = \begin{bmatrix} \mathbf{y}_1 \\ \cdots \\ \mathbf{y}_2 \end{bmatrix} \qquad y_i = \begin{cases} x_i & i = 1, 2, \ldots, n \\ \dot{x}_i & i = n+1, n+2 \ldots, 2n \end{cases} \tag{9.24}$$

Using the definition of $\mathbf{y}$ from Equation (9.24), Equations (9.21) and (9.23) can be summarized in the form of Equation (9.12) where $\mathbf{A}$ is a $4n \times 4n$ matrix defined by augmenting four $n \times n$ matrices such that

$$\mathbf{A} = \begin{bmatrix} \mathbf{0}_n & \vdots & \mathbf{I} \\ \cdots & \vdots & \cdots \\ -\mathbf{M}^{-1}\mathbf{K} & \vdots & -\mathbf{M}^{-1}\mathbf{C}_d \end{bmatrix} \tag{9.25}$$

and $\mathbf{B}$ is the $2n \times k$ matrix defined by

$$\mathbf{B} = \begin{bmatrix} \mathbf{0}_k \\ \cdots \\ \mathbf{M}^{-1}\mathbf{Q} \end{bmatrix} \tag{9.26}$$

where $\mathbf{I}$ is the $n \times n$ identity matrix, $\mathbf{0}_n$ is the $n \times n$ zero matrix, and $\mathbf{0}_k$ is the $n \times k$ zero matrix. The $n \times 2n$ output and the $n \times k$ transmission matrices for this formulation are

$$\mathbf{C} = \begin{bmatrix} \mathbf{I} & \vdots & \mathbf{0}_n \end{bmatrix} \tag{9.27}$$

$$\mathbf{D} = \mathbf{0}_k \tag{9.28}$$

---

## Example 9.8

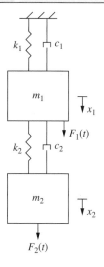

**FIG. 9.9** The two-degree-of-freedom system of Example 9.8.

Determine a state-space model for the system of Figure 9.9.

**Solution**

The differential equations governing the motion of the system of Figure 9.9 are

$$m_1 \ddot{x}_1 + (c_1 + c_2)\dot{x}_1 - c_2 \dot{x}_2 + (k_1 + k_2)x_1 - k_2 x_2 = F_1(t) \tag{a}$$

$$m_2 \ddot{x}_2 - c_2 \dot{x}_1 + c_2 \dot{x}_2 - k_2 x_1 + k_2 x_2 = F_2(t) \tag{b}$$

The inputs are $F_1(t)$ and $F_2(t)$. The output variables are $x_1(t)$ and $x_2(t)$. Equations (a) and (b) are written in the form of Equations (9.18) and (9.19) as

$$\begin{bmatrix} m_1 & 0 \\ 0 & m_2 \end{bmatrix} \begin{bmatrix} \ddot{x}_1 \\ \ddot{x}_2 \end{bmatrix} + \begin{bmatrix} c_1 + c_2 & -c_2 \\ -c_2 & c_2 \end{bmatrix} \begin{bmatrix} \dot{x}_1 \\ \dot{x}_2 \end{bmatrix} + \begin{bmatrix} k_1 + k_2 & -k_2 \\ -k_2 & k_2 \end{bmatrix} \begin{bmatrix} x_1 \\ x_2 \end{bmatrix} = \begin{bmatrix} 1 & 0 \\ 0 & 1 \end{bmatrix} \begin{bmatrix} F_1(t) \\ F_2(t) \end{bmatrix} \tag{c}$$

The mass, damping, and stiffness matrices are identified from Equation (c). State variables are defined as

$$\begin{bmatrix} y_1 \\ y_2 \\ y_3 \\ y_4 \end{bmatrix} = \begin{bmatrix} x_1 \\ x_2 \\ \dot{x}_1 \\ \dot{x}_2 \end{bmatrix} \tag{d}$$

Since the mass matrix is a diagonal matrix its inverse is a diagonal matrix with the reciprocals of the corresponding elements along its diagonal. The state-space formulation is obtained using Equations (9.25) and (9.26) as

$$\begin{bmatrix} \dot{y}_1 \\ \dot{y}_2 \\ \dot{y}_3 \\ \dot{y}_4 \end{bmatrix} = \begin{bmatrix} 0 & 0 & 1 & 0 \\ 0 & 0 & 0 & 1 \\ -\dfrac{k_1 + k_2}{m_1} & \dfrac{k_2}{m_2} & -\dfrac{c_1 + c_2}{m_1} & \dfrac{c_2}{m_2} \\ \dfrac{k_2}{m_1} & -\dfrac{k_2}{m_2} & \dfrac{c_2}{m_1} & -\dfrac{c_2}{m_2} \end{bmatrix} \begin{bmatrix} y_1 \\ y_2 \\ y_3 \\ y_4 \end{bmatrix} + \begin{bmatrix} 0 & 0 \\ 0 & 0 \\ \dfrac{1}{m_1} & 0 \\ 0 & \dfrac{1}{m_2} \end{bmatrix} \begin{bmatrix} F_1(t) \\ F_2(t) \end{bmatrix} \tag{e}$$

The output and transmission matrices are given by Equations (9.27) and (9.28) respectively.

## Example 9.9

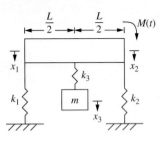

FIG. 9.10 The three-diagram-of freedom system of Example 9.9.

Derive a state-space model for the three-degree-of-freedom mechanical system of Figure 9.10. The bar is uniform of mass $m$ and centroidal moment of inertia $I$. The system inputs are the applied moment $M$ and the applied force $F$. The system outputs are the displacements $x_1$, $x_2$, and $x_3$ as illustrated in Figure 9.10.

### Solution

A mathematical model for the system can be derived either by applying the basic conservation laws for rigid bodies undergoing planar motion to the free-body diagrams of the bar and the block at an arbitrary instant or by using Lagrange's equations. Using either method the resulting differential equations are summarized in matrix form as

$$
\begin{bmatrix}
\dfrac{m}{4}+\dfrac{I}{L^2} & \dfrac{m}{4}-\dfrac{I}{L^2} & 0 \\[2mm]
\dfrac{m}{4}-\dfrac{I}{L^2} & \dfrac{m}{4}+\dfrac{I}{L^2} & 0 \\[2mm]
0 & 0 & m_1
\end{bmatrix}
\begin{bmatrix} \ddot{x}_1 \\ \ddot{x}_2 \\ \ddot{x}_3 \end{bmatrix}
+
\begin{bmatrix}
-0 & 0 & -\dfrac{c}{4} \\[2mm]
0 & -0 & -\dfrac{c}{4} \\[2mm]
-\dfrac{c}{4} & -\dfrac{c}{4} & c
\end{bmatrix}
\begin{bmatrix} \dot{x}_1 \\ \dot{x}_2 \\ \dot{x}_3 \end{bmatrix}
$$

$$
+
\begin{bmatrix}
k_1+\dfrac{k_3}{4} & 0 & -\dfrac{k_3}{4} \\[2mm]
0 & k_2+\dfrac{k_3}{4} & -\dfrac{k_3}{4} \\[2mm]
-\dfrac{k_3}{4} & -\dfrac{k_3}{4} & k_3
\end{bmatrix}
\begin{bmatrix} x_1 \\ x_2 \\ x_3 \end{bmatrix}
=
\begin{bmatrix} -\dfrac{M(t)}{L} \\[2mm] \dfrac{M(t)}{L} \\[2mm] F(t) \end{bmatrix}
\qquad\text{(a)}
$$

Equation (a) is of the form of Equation (9.18) with

$$
\mathbf{M} =
\begin{bmatrix}
\dfrac{m}{4}+\dfrac{I}{L^2} & \dfrac{m}{4}-\dfrac{I}{L^2} & 0 \\[2mm]
\dfrac{m}{4}-\dfrac{I}{L^2} & \dfrac{m}{4}+\dfrac{I}{L^2} & 0 \\[2mm]
0 & 0 & m_1
\end{bmatrix}
\qquad\text{(b)}
$$

$$
\mathbf{C}_d =
\begin{bmatrix}
-0 & 0 & -\dfrac{c}{4} \\[2mm]
0 & -0 & -\dfrac{c}{4} \\[2mm]
-\dfrac{c}{4} & -\dfrac{c}{4} & c
\end{bmatrix}
\qquad\text{(c)}
$$

$$
\mathbf{K} =
\begin{bmatrix}
k_1+\dfrac{k_3}{4} & 0 & -\dfrac{k_3}{4} \\[2mm]
0 & k_2+\dfrac{k_3}{4} & -\dfrac{k_3}{4} \\[2mm]
-\dfrac{k_3}{4} & -\dfrac{k_3}{4} & k_3
\end{bmatrix}
\qquad\text{(d)}
$$

$$\mathbf{Q} = \begin{bmatrix} -\dfrac{1}{L} & 0 \\ \dfrac{1}{L} & 0 \\ 0 & 1 \end{bmatrix} \tag{e}$$

The inverse of the mass matrix is determined using the symbolic capabilities of MATLAB as

$$\mathbf{M}^{-1} = \begin{bmatrix} \dfrac{L^2}{4I} + \dfrac{1}{m} & -\dfrac{L^2}{4I} + \dfrac{1}{m} & 0 \\ -\dfrac{L^2}{4I} + \dfrac{1}{m} & \dfrac{L^2}{4I} + \dfrac{1}{m} & 0 \\ 0 & 0 & \dfrac{1}{m_1} \end{bmatrix} \tag{f}$$

Equations (c), (d), and (f) are used in Equation (9.25) leading to

$$\mathbf{A} = \begin{bmatrix} 0 & 0 & 0 & 1 & 0 & 0 \\ 0 & 0 & 0 & 0 & 1 & 0 \\ 0 & 0 & 0 & 0 & 0 & 1 \\ -\dfrac{(4I+mL^2)(4k_1+k_3)}{16mI} & -\dfrac{(4I-mL^2)(4k_2+k_3)}{16mI} & \dfrac{k_3}{2m} & 0 & 0 & \dfrac{c}{2m} \\ -\dfrac{(4I-mL^2)(4k_1+k_3)}{16mI} & -\dfrac{(4I+mL^2)(4k_2+k_3)}{16mI} & \dfrac{k_3}{2m} & 0 & 0 & \dfrac{c}{2m} \\ \dfrac{k_3}{4m_1} & \dfrac{k_3}{4m_1} & -\dfrac{k_3}{m_1} & \dfrac{c}{4m} & \dfrac{c}{4m} & -\dfrac{c}{m_1} \end{bmatrix} \tag{g}$$

$$\mathbf{B} = \begin{bmatrix} 0 & 0 \\ 0 & 0 \\ 0 & 0 \\ -\dfrac{L}{2I} & 0 \\ \dfrac{L}{2I} & 0 \\ 0 & \dfrac{1}{m_1} \end{bmatrix} \tag{h}$$

The output and transition matrices for the system are

$$\mathbf{C} = \begin{bmatrix} 1 & 0 & 0 & 0 & 0 & 0 \\ 0 & 1 & 0 & 0 & 0 & 0 \\ 0 & 0 & 1 & 0 & 0 & 0 \end{bmatrix} \tag{i}$$

$$\mathbf{D} = \begin{bmatrix} 0 & 0 \\ 0 & 0 \\ 0 & 0 \end{bmatrix} \tag{j}$$

## 9.3 STATE-SPACE SOLUTIONS FOR FREE RESPONSE

The free response of a system is its response due to initial conditions when the input is zero. The state-space problem for the free response of a system with $n$ state variables is

$$\dot{\mathbf{y}} = \mathbf{A}\mathbf{y} \tag{9.29}$$

$$\mathbf{y}(0) = \mathbf{y}_0 \tag{9.30}$$

### 9.3.1 Laplace Transform Solution

Let $\mathbf{Y}(s)$ be the vector of Laplace transforms of the state variables. The Laplace transform of each equation constituting the system of Equation (9.29) is taken and the initial conditions of Equation (9.30) applied. The resulting equations are summarized in matrix form as

$$s\mathbf{Y}(s) - \mathbf{y}_0 = \mathbf{A}\mathbf{Y}(s) \tag{9.31}$$

where $\mathbf{Y}(s) = \mathcal{L}\{\mathbf{y}(t)\}$. Equation (9.31) is rearranged as

$$(s\mathbf{I} - \mathbf{A})\mathbf{Y}(s) = \mathbf{y}_0 \tag{9.32}$$

where $\mathbf{I}$ represents the identity matrix. Equation (9.32) is premultiplied by $(s\mathbf{I} - \mathbf{A})^{-1}$ leading to

$$\mathbf{Y}(s) = (s\mathbf{I} - \mathbf{A})^{-1}\mathbf{y}_0 \tag{9.33}$$

The solution of Equations (9.29) and (9.30) is written as

$$\mathbf{y} = \mathcal{L}^{-1}\{(s\mathbf{I} - \mathbf{A})^{-1}\}\mathbf{y}_0 \tag{9.34}$$

The inverse Laplace transform of Equation (9.34) is performed for all elements of $(s\mathbf{I} - \mathbf{A})^{-1}$.

---

**Example 9.10**

The parameter values for the mechanical system of Example 9.5 are $m = 10$ kg, $k = 1 \times 10^5$ N/m, and $c = 400$ N·s/m. Determine the free response of the system when $x(0) = 0$ m, $\dot{x}(0) = 0.1$ m/s, and $z(0) = 0$.

**Solution**

The substitution of the given parameter values into Equation (j) of Example 9.5 with $F(t) = 0$ leads to

$$\begin{bmatrix} \dot{y}_1 \\ \dot{y}_2 \\ \dot{y}_3 \end{bmatrix} = \begin{bmatrix} 0 & 1 & 0 \\ 0 & 0 & -10{,}000 \\ 0 & 1 & -250 \end{bmatrix} \begin{bmatrix} y_1 \\ y_2 \\ y_3 \end{bmatrix} \tag{a}$$

The initial condition vector is

$$\mathbf{y}_0 = \begin{bmatrix} 0 \\ 0.1 \\ 0 \end{bmatrix} \tag{b}$$

The matrix $s\mathbf{I} - \mathbf{A}$ is evaluated as

$$s\mathbf{I} - \mathbf{A} = \begin{bmatrix} s & 0 & 0 \\ 0 & s & 0 \\ 0 & 0 & s \end{bmatrix} - \begin{bmatrix} 0 & 1 & 0 \\ 0 & 0 & 10{,}000 \\ 0 & 1 & -250 \end{bmatrix} = \begin{bmatrix} s & -1 & 0 \\ 0 & s & 10{,}000 \\ 0 & -1 & s+250 \end{bmatrix} \tag{c}$$

The determinant of the matrix is

$$|s\mathbf{I} - \mathbf{A}| = s(s^2 + 250s + 10{,}000) \tag{d}$$

The roots of the determinant are $s = -200, -50, 0$. The inverse of $s\mathbf{I} - \mathbf{A}$ is determined using either the methods of Appendix B or the symbolic computational features of MATLAB as

$$(s\mathbf{I} - \mathbf{A})^{-1} = \frac{1}{s(s^2 + 250s + 10{,}000)} \begin{bmatrix} s^2 + 250s + 10{,}000 & s+250 & -10{,}000 \\ 0 & s(s+250) & -10{,}000s \\ 0 & s & 1 \end{bmatrix} \tag{e}$$

The inverse Laplace transform is determined as

$$\mathcal{L}\{(s\mathbf{I} - \mathbf{A})^{-1}\}^{-1} = \begin{bmatrix} 1 & \dfrac{1}{40} + \dfrac{1}{600}e^{-200t} + \dfrac{2}{75}e^{-50t} & -1 - \dfrac{1}{3}e^{-200t} + \dfrac{4}{3}e^{-50t} \\[3mm] 0 & \dfrac{4}{3}e^{-50t} - \dfrac{1}{3}e^{-200t} & -\dfrac{200}{3}e^{-50t} + \dfrac{200}{3}e^{-50t} \\[3mm] 0 & \dfrac{1}{150}e^{-50t} - \dfrac{1}{150}e^{-200t} & -\dfrac{1}{3}e^{-50t} + \dfrac{4}{3}e^{-200t} \end{bmatrix} \tag{f}$$

The state variables are determined using Equation (9.34) and the initial condition vector of Equation (b) as

$$\begin{bmatrix} y_1 \\ y_2 \\ y_3 \end{bmatrix} = 0.1 \begin{bmatrix} \dfrac{1}{40} + \dfrac{1}{600}e^{-200t} + \dfrac{2}{75}e^{-50t} \\[3mm] \dfrac{4}{3}e^{-50t} - \dfrac{1}{3}e^{-200t} \\[3mm] \dfrac{1}{150}e^{-50t} - \dfrac{1}{150}e^{-200t} \end{bmatrix} \tag{g}$$

## 9.3.2 Exponential Solution

An alternate form of the free response can be obtained by assuming an exponential solution of Equations (9.29) and (9.30) of the form

$$\mathbf{y} = \mathbf{Y}e^{\lambda t} \tag{9.35}$$

where $\mathbf{Y}$ is a vector of constants and $\lambda$ is a parameter, both of which are to be determined through the solution process. Substitution of Equation (9.35) into Equation (9.29) leads to

$$\lambda \mathbf{Y} e^{\lambda t} = \mathbf{A} \mathbf{Y} e^{\lambda t} \tag{9.36}$$

Equation (9.36) is rearranged to

$$(\mathbf{A} - \lambda \mathbf{I}) \mathbf{Y} e^{\lambda t} = 0$$
$$(\mathbf{A} - \lambda \mathbf{I}) \mathbf{Y} = 0 \tag{9.37}$$

A nontrivial solution of the system of equations represented by Equation (9.37) occurs if and only if

$$|\mathbf{A} - \lambda \mathbf{I}| = 0 \tag{9.38}$$

The evaluation of the determinant of Equation (9.38) leads to a polynomial equation in $\lambda$, the roots of which are the eigenvalues of $\mathbf{A}$. For each $\lambda$ that is an eigenvalue of $\mathbf{A}$, a nontrivial solution of Equation (9.37) is called an eigenvector of $\mathbf{A}$ corresponding to the eigenvalue $\lambda$. An eigenvector is not unique because any nonzero multiple of an eigenvector is also a nontrivial solution of Equation (9.38).

The eigenvalues of a system with $n$ state variables are computed by finding the roots of the $n$th-order polynomial generated by the evaluation of the determinant of Equation (9.38). The coefficients of the polynomial are all real. Thus, if complex eigenvalues occur, they occur in complex conjugate pairs. Let $\lambda_1, \lambda_2, \ldots, \lambda_n$ be the eigenvalues of $\mathbf{A}$ ordered such that $|\lambda_1| \le |\lambda_2| \le |\lambda_3| \le \cdots \le |\lambda_n|$. Eigenvectors $\mathbf{Y}_1, \mathbf{Y}_2, \ldots, \mathbf{Y}_n$ are defined such that $\mathbf{Y}_k$ corresponds to the eigenvalue $\lambda_k$.

If the eigenvalues are distinct then $n$ linearly independent solutions of Equation (9.29) in the form of Equation (9.35) exist. The general solution is a linear combination of such solutions. Thus it can be written

$$\mathbf{y} = \sum_{k=1}^{n} W_k \mathbf{Y}_k e^{\lambda_k t} \tag{9.39}$$

where $W_1, W_2, \ldots, W_n$ are arbitrary constants.

The $n \times n$ matrix $\mathbf{P}(t)$ is defined such that its $k$th column is $e^{\lambda_k t} \mathbf{Y}_k$. Equation (9.38) is rewritten using the $\mathbf{P}$ matrix as

$$\mathbf{y} = \mathbf{P}(t) \mathbf{W} \tag{9.40}$$

where $\mathbf{W}$ is the $n \times 1$ vector whose $k$th element is $W_k$. The application of Equation (9.30) to Equation (9.40) leads to

$$\mathbf{y}_0 = \mathbf{P}(0) \mathbf{W} \tag{9.41}$$

The matrix $\mathbf{P}(0)$ is a matrix whose columns are the eigenvectors of $\mathbf{A}$, which must be linearly independent. Thus $\mathbf{P}(0)$ cannot be singular and its inverse exists. Premultiplying Equation (9.41) by $\mathbf{P}(0)^{-1}$ leads to

$$\mathbf{W} = \{\mathbf{P}(0)\}^{-1} \mathbf{y}_0 \tag{9.42}$$

Substitution for $\mathbf{W}$ from Equation (9.42) into Equation (9.40) gives

$$\mathbf{y} = \mathbf{P}(t)\{\mathbf{P}(0)\}^{-1}\mathbf{y}_0 \tag{9.43}$$

## Example 9.11

Use Equation (9.43) to determine the response of the system of Examples 9.5 and 9.10.

### Solution

The matrix $\mathbf{A}$ is obtained in Equation (a) of Example 9.10 as

$$\mathbf{A} = \begin{bmatrix} 0 & 1 & 0 \\ 0 & 0 & -10{,}000 \\ 0 & 1 & -250 \end{bmatrix} \tag{a}$$

The eigenvalues of $\mathbf{A}$ are obtained using Equation (9.38)

$$|\mathbf{A} - \lambda\mathbf{I}| = 0 = \begin{vmatrix} -\lambda & 1 & 0 \\ 0 & -\lambda & -10{,}000 \\ 0 & 1 & -250 - \lambda \end{vmatrix} \tag{b}$$

The determinant is evaluated using expansion by its first column

$$-\lambda[-\lambda(-250 - \lambda) - (1)(-10{,}000)] = 0$$
$$\lambda(\lambda^2 + 250\lambda + 10{,}000) = 0 \tag{c}$$

The solutions of Equation (c) are

$$\lambda_1 = 0 \qquad \lambda_2 = -50 \qquad \lambda_3 = -200 \tag{d}$$

Eigenvectors of $\mathbf{A}$ are determined as

$$\mathbf{Y}_1 = \begin{bmatrix} 1 \\ 0 \\ 0 \end{bmatrix} \qquad \mathbf{Y}_2 = \begin{bmatrix} 1 \\ -50 \\ -\dfrac{1}{4} \end{bmatrix} \qquad \mathbf{Y}_3 = \begin{bmatrix} 1 \\ -200 \\ -4 \end{bmatrix} \tag{e}$$

The matrix $\mathbf{P}$ is written as

$$\mathbf{P} = \begin{bmatrix} 1 & e^{-50t} & e^{-200t} \\ 0 & -50e^{-50t} & -200e^{-200t} \\ 0 & -\dfrac{1}{4}e^{-50t} & -4e^{-200t} \end{bmatrix} \tag{f}$$

From which $\mathbf{P}(0)$ is obtained as

$$\mathbf{P}(0) = \begin{bmatrix} 1 & 1 & 1 \\ 0 & -50 & -200 \\ 0 & -\dfrac{1}{4} & -4 \end{bmatrix} \tag{g}$$

and then

$$
\mathbf{P}(0)^{-1} = \begin{bmatrix} 1 & \dfrac{1}{40} & -1 \\[2mm] 0 & -\dfrac{2}{75} & \dfrac{4}{3} \\[2mm] 0 & \dfrac{1}{600} & -\dfrac{1}{3} \end{bmatrix}
\tag{h}
$$

The response is obtained using Equation (9.43) as

$$
\mathbf{y} = \begin{bmatrix} 1 & e^{-50t} & e^{-200t} \\[1mm] 0 & -50e^{-50t} & -200e^{-200t} \\[1mm] 0 & -\dfrac{1}{4}e^{-50t} & -4e^{-200t} \end{bmatrix} \begin{bmatrix} 1 & \dfrac{1}{40} & -1 \\[2mm] 0 & -\dfrac{2}{75} & \dfrac{4}{3} \\[2mm] 0 & \dfrac{1}{600} & -\dfrac{1}{3} \end{bmatrix} \begin{bmatrix} 0 \\[1mm] 0.1 \\[1mm] 0 \end{bmatrix}
$$

$$
\mathbf{y} = 0.1 \begin{bmatrix} \dfrac{1}{40} + \dfrac{1}{600}e^{-200t} + \dfrac{2}{75}e^{-50t} \\[3mm] \dfrac{4}{3}e^{-50t} - \dfrac{1}{3}e^{-200t} \\[3mm] \dfrac{1}{150}e^{-50t} - \dfrac{1}{150}e^{-200t} \end{bmatrix}
\tag{i}
$$

### 9.3.3 General Description of Free Response

As expected, the Laplace transform method applied in Example 9.10 and the exponential solution applied in Example 9.11 lead to the same free response for a system. Since the two methods must lead to identical results, the comparison of Equations (9.34) and (9.43) makes it apparent that

$$
\mathcal{L}^{-1}\left\{(s\mathbf{I} - \mathbf{A})^{-1}\right\} = \mathbf{P}(t)\mathbf{P}(0)^{-1}
\tag{9.44}
$$

Both of these equations can be formulated as

$$
\mathbf{y} = \Phi(t)\mathbf{y}_0
\tag{9.45}
$$

where $\Phi(t)$ is called the **state transition matrix** and is determined as either of the matrices in Equation (9.44).

It is shown in Section 9.5 that the transfer functions for the state variables are proportional, by a matrix multiplication, to $(s\mathbf{I} - \mathbf{A})^{-1}$, which can be obtained from

$$
(s\mathbf{I} - \mathbf{A})^{-1} = \frac{1}{|s\mathbf{I} - \mathbf{A}|}\,\mathrm{adj}(\mathbf{A})
\tag{9.46}
$$

where $\mathrm{adj}(\mathbf{A})$ is the adjoint matrix of $\mathbf{A}$. Equation (9.46) implies that the poles of the transfer functions are the values of $s$ such that $|s\mathbf{I} - \mathbf{A}| = 0$. Thus the transfer functions have the same number of poles as state variables. As noted in Section 6.3 the system is stable only if all poles have negative real parts. The resulting mathematical form of the free response corresponding to a pole $s_i$ is proportional to $e^{s_i t}$. The general form of the free response is a linear combination of such a response for all poles. Complex poles lead to trigonometric terms in the response.

The eigenvalues of **A** are calculated as the values of $\lambda$ such that $|\mathbf{A} - \lambda \mathbf{I}| = 0$. Thus the eigenvalues of **A** are identical to the poles of the system's transfer function. The system is stable only if all eigenvalues have negative real parts. The free-response is as just discussed and in Chapter 6.

---

**Example 9.12**

Determine the state transition matrix for the *LRC* circuit of Example 9.4 when (a) $L = 0.1$ H, $R = 1$ k$\Omega$, $C = 0.5$ µF; (b) $L = 0.1$ H, $R = 1$ k$\Omega$, $C = 0.4$ µF; and (c) $L = 0.1$ H, $R = 1$ k$\Omega$, $C = 0.1$ µF.

**Solution**

(a) The substitution of the given values into Equation (g) of Example 9.4 leads to

$$\mathbf{A} = \begin{bmatrix} 0 & 1 \\ -2 \times 10^7 & -1 \times 10^4 \end{bmatrix} \tag{a}$$

The eigenvalues of **A** are determined from

$$\begin{vmatrix} -\lambda & 1 \\ -2 \times 10^7 & -1 \text{x} 10^4 - \lambda \end{vmatrix} = 0$$

$$(-\lambda)(-1 \times 10^4 - \lambda) - (1)(-2 \times 10^7) = 0$$

$$\lambda^2 + 1 \times 10^4 \lambda + 2 10^7 = 0 \tag{b}$$

The solutions of Equation (b) are $\lambda = -8.87 \times 10^3$, $-1.13 \times 10^3$. The corresponding eigenvectors are

$$\mathbf{Y}_1 = \begin{bmatrix} 1 \\ -8.87 \times 10^3 \end{bmatrix} \qquad \mathbf{Y}_2 = \begin{bmatrix} 1 \\ -1.13 \times 10^3 \end{bmatrix} \tag{c}$$

The matrices **P** and $\mathbf{P}(0)^{-1}$ are consequently determined as

$$\mathbf{P} = \begin{bmatrix} e^{-8.87 \times 10^3 t} & e^{-1.13 \times 10^3} \\ -8.87 \times 10^3 e^{-8.87 \times 10^3 t} & -1.13 \times 10^3 e^{-1.13 \times 10^3 t} \end{bmatrix} \tag{d}$$

$$\mathbf{P}(0)^{-1} = \begin{bmatrix} -0.146 & -1.29 \times 10^{-4} \\ -1.146 & 1.29 \times 10^{-4} \end{bmatrix} \tag{e}$$

The state transition matrix is

$$\Phi(t) = \mathbf{P}\mathbf{P}(0)^{-1} = \begin{bmatrix} -0.146 & -1.29 \text{x} 10^{-4} \\ 1.23 \times 10^3 & -1.146 \end{bmatrix} e^{-8.87 \times 10^3 t}$$

$$+ \begin{bmatrix} -0.146 & -1.29 \times 10^{-4} \\ 1.65 \times 10^2 & -0.146 \end{bmatrix} e^{-1.13 \times 10^3 t} \tag{f}$$

The state transition matrix of Equation (f) is an example of a state transition matrix for a second-order overdamped system.

(b) The matrix $\mathbf{A}$ obtained using these parameters is

$$\mathbf{A} = \begin{bmatrix} 0 & 1 \\ -2.5 \times 10^7 & -1 \times 10^4 \end{bmatrix} \tag{g}$$

A procedure similar to that of part (a) is used to obtain the eigenvalues of $\mathbf{A}$ as $\lambda = -5 \times 10^3, -5 \times 10^3$. The system has repeated eigenvectors. The exponential solution may be modified to obtain a response in the case of repeated eigenvectors, but it is beyond the scope of this study. In this case the state transition matrix is determined as

$$\Phi(t) = \mathcal{L}^{-1}\left\{ (s\mathbf{I} - \mathbf{A})^{-1} \right\} \tag{h}$$

Use of MATLAB to perform symbolic computations leads to

$$(s\mathbf{I} - \mathbf{A})^{-1} = \begin{bmatrix} \dfrac{s + 1 \times 10^4}{(s + 5,000)^2} & \dfrac{1}{(s + 5,000)^2} \\[3mm] -\dfrac{2.5 \times 10^7}{(s + 5,000)^2} & \dfrac{s}{(s + 5,000)^2} \end{bmatrix} \tag{i}$$

$$\Phi(t) = \begin{bmatrix} 1 + 5,000t & t \\ -2.5 \times 10^7 t & 1 - 5,000t \end{bmatrix} e^{-5,000t} \tag{j}$$

The state transition matrix of Equation (j) is an example of a state transition matrix for a second-order system with critical damping.

(c) The matrix $\mathbf{A}$ obtained using the given parameters is

$$\mathbf{A} = \begin{bmatrix} 0 & 1 \\ -1 \times 10^8 & -1 \times 10^4 \end{bmatrix} \tag{k}$$

The eigenvalues of $\mathbf{A}$ are obtained as $\lambda_{1,2} = -5 \times 10^3 \pm j8.66 \times 10^3$. The eigenvectors of $\mathbf{A}$ are determined as

$$\mathbf{Y}_1 = \begin{bmatrix} 1 \\ -5 \times 10^3 + j8.66 \times 10^3 \end{bmatrix} \quad \mathbf{Y}_2 = \begin{bmatrix} 1 \\ -5 \times 10^3 - j8.66 \times 10^3 \end{bmatrix} \tag{l}$$

The use of MATLAB for computation leads to

$$\Phi(t) = \begin{bmatrix} (0.5 - 0.289j)e^{\lambda_1 t} + (0.5 + 0.289j)e^{\lambda_2 t} & 5.78 \times 10^{-5}\left(e^{\lambda_1 t} + e^{\lambda_2 t}\right) \\ 5.77 \times 10^{-3}(e^{\lambda_1 t} - e^{-\lambda_2 t}) & (0.5 + 0.289j)e^{\lambda_1 t} + (0.5 - 0.289j)e^{\lambda_2 t} \end{bmatrix} \tag{m}$$

Equation (m) is further simplified by noting that $\lambda_2 = \bar{\lambda}_1$, the sum of two complex conjugates is twice the real part of one, the difference of two complex conjugates is twice the imaginary part of one, $\text{Re}[e^{jbt}] = \cos(bt)$ and $\text{Im}[e^{jbt}] = \sin(bt)$. The application of these simplifications leads to

$$\Phi(t) = \begin{bmatrix} \cos(8.66 \times 10^3 t) + 0.578\sin(8.66 \times 10^3 t) & 1.16 \times 10^{-4}\cos(8.66 \times 10^3 t) \\ 1.16 \times 10^{-2}\sin(8.66 \times 10^3 t) & \sin(8.66 \times 10^3 t) - 0.578\cos(8.66 \times 10^3 t) \end{bmatrix} e^{-500t} \tag{n}$$

## 9.4 STATE-SPACE ANALYSIS OF RESPONSE DUE TO INPUTS

The general form of the equations for the state variables when the system is subject to nonzero inputs is Equation (9.12)

$$\dot{\mathbf{y}} = \mathbf{A}\mathbf{y} + \mathbf{B}\mathbf{u} \tag{9.12}$$

Several convenient methods of solution exist for Equation (9.12). The application of the Laplace transform method leads to a solution in terms of integrals that can often be easily evaluated. State-space formulation is convenient for the application of self-starting numerical methods such as the Runge-Kutta methods.

### 9.4.1 Laplace Transform Solution

Application of the Laplace transform to Equation (9.12), including the application of initial conditions of the form $\mathbf{y}(0) = \mathbf{y}_0$ leads to

$$s\mathbf{Y}(s) - \mathbf{y}_0 = \mathbf{A}\mathbf{Y}(s) + \mathbf{B}\mathbf{U}(s) \tag{9.47}$$

Solving Equation (9.47) for $\mathbf{Y}(s)$ gives

$$\mathbf{Y}(s) = (s\mathbf{I} - \mathbf{A})^{-1}\mathbf{y}_0 + (s\mathbf{I} - \mathbf{A})^{-1}\mathbf{B}\mathbf{U}(s) \tag{9.48}$$

Equation (9.48) shows that the matrix of transfer functions for the state variables is

$$\mathbf{G}(s) = (s\mathbf{I} - \mathbf{A})^{-1}\mathbf{B} \tag{9.49}$$

Equation (9.49) shows that the poles of the transfer functions are the roots of $|(s\mathbf{I} - \mathbf{A})^{-1}|$ and that the matrix of impulsive responses is

$$\mathbf{y}(t) = \mathcal{L}^{-1}\{(s\mathbf{I} - \mathbf{A})^{-1}\}\mathbf{B} \tag{9.50}$$

Recalling that the state transition matrix is defined as $\Phi(t) = \mathcal{L}^{-1}\{(s\mathbf{I} - \mathbf{A})^{-1}\}$, Equation (9.50) is rewritten as

$$\mathbf{y}(t) = \Phi(t)\mathbf{B} \tag{9.51}$$

Taking the inverse transform of Equation (9.48) leads to

$$\mathbf{y}(t) = \Phi(t)\mathbf{y}_0 + \mathcal{L}^{-1}\{(s\mathbf{I} - \mathbf{A})^{-1}\mathbf{B}\mathbf{U}(s)\} \tag{9.52}$$

The convolution property of the transforms written in its inverse form and applied to matrices of transforms that are multiplicatively compatible is

$$\mathcal{L}^{-1}\{\mathbf{F}(s)\mathbf{G}(s)\} = \mathbf{f}(t) * \mathbf{g}(t) = \int_0^t \mathbf{f}(t - \tau)\mathbf{g}(\tau)d\tau \tag{9.53}$$

The application of Equation (9.53) to Equation (9.52) and using the definition of the state transition matrix leads to

$$\mathbf{y}(t) = \Phi(t)\mathbf{y}_0 + \int_0^t \Phi(t - \tau)\mathbf{B}\mathbf{u}(\tau)d\tau \tag{9.54}$$

Equation (9.54) provides the general response of the state variables to any form of system input.

**Example 9.13**

Determine the impulsive response of the series $LRC$ circuit of Example 9.4 for each set of circuit parameters of Example 9.12.

**Solution**

Noting that for each set of circuit parameters of Example 9.12 $L = 0.1$ H, the input matrix determined in Example 9.4 is written as

$$\mathbf{B} = \begin{bmatrix} 0 \\ 10 \end{bmatrix} \tag{a}$$

(a) The state transition matrix determined in Example 9.12(a)

$$\Phi(t) = \begin{bmatrix} -0.146 & -1.29 \times 10^{-4} \\ 1.23 \times 10^3 & -1.146 \end{bmatrix} e^{-8.87 \times 10^3 t}$$

$$+ \begin{bmatrix} -0.146 & -1.29 \times 10^{-4} \\ 1.65 \times 10^2 & -0.146 \end{bmatrix} e^{-1.13 \times 10^3 t} \tag{b}$$

Equation (9.51) is used to determine the impulsive response of this circuit as

$$\mathbf{y}(t) = \Phi(t)\mathbf{B} = \begin{bmatrix} -1.29 \times 10^{-3} \left( e^{-8.87 \times 10^3 t} + e^{-1.13 \times 10^3 t} \right) \\ 11.46 e^{-8.897 \times 10^3 t} - 1.46 e^{-1.13 \times 10^3 t} \end{bmatrix} \tag{c}$$

Noting from Equation (i) of Example 9.4 that the $i(t) = y_2(t)$ gives the current in the circuit due to an impulsive voltage is

$$i(t) = 11.46 e^{-8.897 \times 10^3 t} - 1.46 e^{-1.13 \times 10^3 t} \text{A} \tag{d}$$

(b) The state transition matrix determined in Example 9.12(b) is

$$\Phi(t) = \begin{bmatrix} 1 + 5,000t & t \\ -2.5 \times 10^7 t & 1 - 5,000t \end{bmatrix} e^{-5,000t} \tag{e}$$

The impulsive response for this circuit is

$$\mathbf{y}(t) = \begin{bmatrix} 10te^{-5,000t} \\ 10(1 - 5,000t)e^{-5,000t} \end{bmatrix} \tag{f}$$

(c) The state transition matrix determined in Example 9.12(c) is

$$\Phi(t) = \begin{bmatrix} \cos(8.66 \times 10^3 t) + 0.578 \sin(8.66 \times 10^3 t) & 1.16 \times 10^{-4} \cos(8.66 \times 10^3 t) \\ 1.16 \times 10^{-2} \sin(8.66 \times 10^3 t) & \sin(8.66 \times 10^3 t) - 0.578 \cos(8.66 \times 10^3 t) \end{bmatrix} e^{-500t} \tag{g}$$

The impulsive response is

$$\mathbf{y}(t) = \begin{bmatrix} 1.16 \times 10^{-3} e^{-500t} \cos(8.66 \times 10^3 t) \\ 10 e^{-500t} \left( \sin(8.66 \times 10^3 t) - 0.578 \cos(8.66 \times 10^3 t) \right) \end{bmatrix} \tag{h}$$

## Example 9.14

Determine the response of the two-tank liquid-level system of Examples 4.8 and 9.2 when $A_1 = 10.9$ m$^2$, $A_2 = 15.8$ m$^2$, $R_1 = 6.5$ s/m$^2$, $R_2 = 3.8$ s/m$^2$, and the perturbation in flow rates are $q_{i1} = -0.2u(t)$ m$^3$/s and $q_{i2} = 0.2u(t)$ m$^3$/s.

### Solution

Substituting the given parameters into Equation (c) of Example 9.2 leads to the state-space model of

$$\begin{bmatrix} \dot{y}_1 \\ \dot{y}_2 \end{bmatrix} = \begin{bmatrix} -0.0141 & 0.0141 \\ 0.00974 & -0.0264 \end{bmatrix} \begin{bmatrix} y_1 \\ y_2 \end{bmatrix} + \begin{bmatrix} 0.0917 & 0 \\ 0 & 0.0633 \end{bmatrix} \begin{bmatrix} -0.2u(t) \\ 0.2u(t) \end{bmatrix} \quad \text{(a)}$$

The eigenvalues of the state matrix are determined as

$$|\mathbf{A} - \lambda \mathbf{I}| = 0$$

$$\begin{vmatrix} -0.0141 - \lambda & 0.0141 \\ 0.00974 & -0.0264 - \lambda \end{vmatrix} = 0$$

$$(-0.0141 - \lambda)(-0.0264 - \lambda) - (0.00974)(0.0141) = 0$$

$$\lambda^2 + 0.0405\lambda + 1.37 \times 10^{-4} = 0$$

$$\lambda = -0.0368, -0.0186 \quad \text{(b)}$$

The eigenvectors are calculated as

$$\mathbf{Y}_1 = \begin{bmatrix} 1 \\ 1.61 \end{bmatrix} \qquad \mathbf{Y}_2 = \begin{bmatrix} 1 \\ 0.319 \end{bmatrix} \quad \text{(c)}$$

The state transition matrix is determined as

$$\Phi(t) = \begin{bmatrix} -0.2471e^{-0.0368t} + 1.2471e^{-0.0186t} & 0.775e^{-0.0368t} - 0.775e^{-0.0186t} \\ -0.398e^{-0.0368t} + 0.398e^{-0.0186t} & 1.247e^{-0.0368t} - 0.247e^{-0.0186t} \end{bmatrix} \quad \text{(d)}$$

Noting that $\mathbf{y}_0 = 0$, the application of Equation (9.54) leads to

$$\begin{bmatrix} y_1 \\ y_2 \end{bmatrix} = \int_0^t \begin{bmatrix} -0.2471e^{-0.0368(t-\tau)} + 1.2471e^{-0.0186(t-\tau)} & 0.775e^{-0.0368(t-\tau)} - 0.775e^{-0.0186(t-\tau)} \\ -0.398e^{-0.0368(t-\tau)} + 0.398e^{-0.0186(t-\tau)} & 1.247e^{-0.0368(t-\tau)} - 0.247e^{-0.0186(t-\tau)} \end{bmatrix}$$

$$\times \begin{bmatrix} 0.0917 & 0 \\ 0 & 0.0633 \end{bmatrix} \begin{bmatrix} -.2u(\tau) \\ 0.2u(\tau) \end{bmatrix} d\tau \quad \text{(e)}$$

The simplification of Equation (e) leads to

$$\begin{bmatrix} y_1 \\ y_2 \end{bmatrix} = \begin{bmatrix} \int_0^t \left( 0.0143e^{-0.0368(t-\tau)} - 0.0327e^{-0.0186(t-\tau)} \right) u(\tau) d\tau \\ \int_0^t \left( 0.0231e^{-0.0368(t-\tau)} - 0.0104e^{-0.0186(t-\tau)} \right) u(\tau) d\tau \end{bmatrix} \quad \text{(f)}$$

The evaluation of integrals in Equation (f) leads to

$$\begin{bmatrix} y_1 \\ y_2 \end{bmatrix} = \begin{bmatrix} 1.370 - 0.387e^{-0.0368t} + 1.76e^{-0.0186t} \\ -0.0414 - 0.601e^{-0.0368t} + 0.560e^{-0.0186t} \end{bmatrix} u(t) \quad \text{(g)}$$

**Example 9.15**

Determine the steady-state response of the system of Examples 9.5 and 9.10 when the system is subject to an input of $F(t) = 100 \sin(50t)$ N.

**Solution**

The initial conditions are irrelevant for the steady-state response. Thus it is assumed that $\mathbf{y}_0 = 0$. The state transition matrix is given in Equation (f) of Example 9.10. $\mathbf{B}$ and $\mathbf{u}$ are determined from Equation (j) of Example 9.5 as

$$\mathbf{B} = \begin{bmatrix} 0 \\ 0.1 \\ 0 \end{bmatrix} \qquad \mathbf{u} = [100 \sin(50t)] \qquad \text{(a)}$$

The application of the convolution integral solution, Equation (9.54) leads to

$$\begin{bmatrix} y_1 \\ y_2 \\ y_3 \end{bmatrix} = \int_0^t \begin{bmatrix} 1 & \frac{1}{40} + \frac{1}{600}e^{-200(t-\tau)} + \frac{2}{75}e^{-50(t-\tau)} & -1 - \frac{1}{3}e^{-200(t-\tau)} + \frac{4}{3}e^{-50(t-\tau)} \\ 0 & \frac{4}{3}e^{-50(t-\tau)} - \frac{1}{3}e^{-200(t-\tau)} & -\frac{200}{3}e^{-50(t-\tau)} + \frac{200}{3}e^{-50(t-\tau)} \\ 0 & \frac{1}{150}e^{-50(t-\tau)} - \frac{1}{150}e^{-200(t-\tau)} & -\frac{1}{3}e^{-50(t-\tau)} + \frac{4}{3}e^{-200(t-\tau)} \end{bmatrix}$$

$$\times \begin{bmatrix} 0 \\ 0.1 \\ 0 \end{bmatrix} 100 \sin(50\tau)d\tau \qquad \text{(b)}$$

The simplification of Equation (b) leads to

$$\begin{bmatrix} y_1 \\ y_2 \\ y_3 \end{bmatrix} = \begin{bmatrix} \int_0^t 10\left(\frac{1}{40} + \frac{1}{600}e^{-200(t-\tau)} + \frac{2}{75}e^{-50(t-\tau)}\right)\sin(50\tau)d\tau \\ \int_0^t 10\left(\frac{4}{3}e^{-50(t-\tau)} - \frac{1}{3}e^{-200(t-\tau)}\right)\sin(50\tau)d\tau \\ \int_0^t 10\left(\frac{1}{150}e^{-50(t-\tau)} - \frac{1}{150}e^{-200(t-\tau)}\right)\sin(50\tau)d\tau \end{bmatrix} \qquad \text{(c)}$$

It is noted in Example 9.5 that the system output is $x = y_1$. Thus from Equation (c)

$$x(t) = \int_0^t 10\left(\frac{1}{40} + \frac{1}{600}e^{-200(t-\tau)} + \frac{2}{75}e^{-50(t-\tau)}\right)\sin(50\tau)d\tau$$

$$= \frac{1}{4}\int_0^t \sin(50\tau)d\tau + \frac{1}{60}e^{-200t}\int_0^t e^{200\tau}\sin(50\tau)d\tau + \frac{4}{15}e^{-50t}\int_0^t e^{50\tau}\sin(50\tau)d\tau$$

$$\text{(d)}$$

The evaluation of the integrals in Equation (d) leads to

$$x(t) = 0.08[1 - \cos(50t)] + 2.74 \times 10^{-3}\sin(50t) + 1.96 \times 10^{-5}e^{-200t}[1 - \cos(50t)]$$

$$+ 2.67 \times 10^{-3}e^{-50t}[1 - \cos(50t)] \qquad \text{(e)}$$

The transient response is included in Equation (e). The steady-state response is obtained by taking the limit of Equation (e) as $t \to \infty$ leading to

$$x_{ss}(t) = 0.08[1 - \cos{(50t)}] + 2.74 \times 10^{-3}\sin{(50t)} \tag{f}$$

The use of trigonometric identities allows Equation (f) to be rewritten in the form of Equation (7.2) as

$$x_{ss}(t) = 0.08 + 8.00 \times 10^{-2}\sin{(50t - 1.54)} \tag{g}$$

## 9.4.2 Numerical Solutions

The state-space formulation of a mathematical model is convenient for the application of numerical methods to determine system response. Recall, from Chapter 1, that the possible objectives for mathematical modeling of a dynamic system are system analysis, system design, and system synthesis. The numerical simulation of dynamic system response can aid in achieving each of these objectives. The numerical computations involving complicated equations are often required for system analysis and system design. It is illustrated in Chapters 6 and 7 that the algebra necessary for the analysis of higher-order systems, systems of more than second order, is tedious and lengthy. Algebraic mistakes are easy to make and difficult to detect. Numerical simulation provides a convenient alternative.

Closed-form solutions for the response of dynamic systems are handy when easily obtained. Numerical solutions are still only approximate solutions and error is introduced in every computation. Thus, especially for long times, closed-form solutions are more accurate. As will be illustrated in Example 9.16, the application of state-space methods leads to the determination of the total solution from which a steady-state response, if desired, can be extracted. Numerical solutions also determine transient behavior before a steady state is reached. System design requires analysis of the response of the system in terms of unspecified values of system parameters. Indeed, part of the design is to determine the appropriate values of these parameters. An understanding of the qualitative changes in the response of a system as a parameter changes is easier to obtain when a closed-form solution exists.

This is why knowledge of both transfer function methods and state-space methods is desirable. Transfer function methods allow analysis of the system in terms of system parameters, which is useful in system design. The use of transfer function methods, as shown in Chapters 6 and 7, allows for a qualitative description of both transient and steady-state system responses. These in turn allow, as shown in Chapter 8, for an understanding of how control systems are used to control and modify system response. However, the algebra leading to the determination of the response is complicated and tedious. State-space methods provide a fast and accurate determination of system response and allow for a quicker response from control systems.

Numerical methods for the direct integration of state-space equations are self-starting methods; that is, only the vector of initial conditions, the matrices $\mathbf{A}$ and $\mathbf{B}$, and the input are required to numerically determine system response. Numerical methods approximate the system response at discrete times: $0 < t_1 < t_2 < \cdots < t_{k-1} < t_k < t_{k+1} < \cdots$. It is assumed that the difference between each discrete time is a constant $h$, called the **step size**, defined such that

$$t_k - t_{k-1} = h \qquad k = 1, 2, \ldots \tag{9.55}$$

The numerical approximation for $\mathbf{y}(t_k)$ is defined as $\mathbf{y}_k$. Self-starting numerical methods use a defined algorithm to approximate $\mathbf{y}_1$ using $\mathbf{y}_0$. The same algorithm is applied to approximate $\mathbf{y}_2$ using the approximation for $\mathbf{y}_1$. The algorithm is successively applied, calculating $\mathbf{y}_{k+1}$ using the approximation for $\mathbf{y}_k$ until it is terminated for a predetermined value of $k$.

Since numerical methods are approximate methods there is a difference between the approximate solution $\mathbf{y}_k$ and the exact solution $\mathbf{y}(t_k)$. This difference is called the **error** of the approximation. An error, called the **local error**, occurs each time the algorithm is applied to compute $\mathbf{y}_{k+1}$ from $\mathbf{y}_k$. The local error propagates; that is, the total error in the approximation for $\mathbf{y}_{k+1}$ is the sum of the local errors from each previous step.

The error induced in a numerical approximation is determined by deriving the numerical algorithm from a Taylor series expansion for the exact solution $\mathbf{y}(t + h)$ about $\mathbf{y}(t)$. As expected the error is strongly dependent on the step size $h$. The local error is usually determined to be proportional to an integer power of the step size

$$E = O(h^n) \tag{9.56}$$

Equation (9.56) is read as "the error is of the order of $h^n$." The equation illustrates that the error for such algorithms is reduced by decreasing the step size.

The most popular self-starting numerical methods are called **Runge-Kutta methods**. For a system with only one state variable a Runge-Kutta algorithm is of the form

$$y_{k+1} = y_k + \phi(y_k, t_k h)h \tag{9.57}$$

where $\phi(y_k, t_k, h)$ is a function determined from an appropriate Taylor series expansion. It can generally be shown that an $n$th-order Runge-Kutta method has a local error of $O(h^{n+1})$ and a global error of $O(h^n)$.

The simplest of the Runge-Kutta methods is **Euler's method**, a first-order method, that, when applied to a system of the form of Equation (9.12), leads to

$$\mathbf{y}_k = \mathbf{y}_{k-1} + [\mathbf{A}\mathbf{y}_{k-1} + \mathbf{B}\mathbf{u}(t_{k-1})]h \tag{9.58}$$

The local error for Euler's method is $O(h^2)$ while the global error is $O(h)$.

## Example 9.16

Develop a MATLAB M-file that uses Euler's method to approximate the solution to Example 9.15.

### Solution

The state-space equations for the system are those of Equation (j) of Example 9.5 using the values given in the problem statements of Examples 9.10 and 9.15. The result is

$$\begin{bmatrix} \dot{y}_1 \\ \dot{y}_2 \\ \dot{y}_3 \end{bmatrix} = \begin{bmatrix} 0 & 1 & 0 \\ 0 & 0 & -10{,}000 \\ 0 & 1 & -250 \end{bmatrix} \begin{bmatrix} y_1 \\ y_2 \\ y_3 \end{bmatrix} + \begin{bmatrix} 0 \\ 0.1 \\ 0 \end{bmatrix} 100\sin(50t) \qquad \text{(a)}$$

The application of Equation (9.58) to Equation (a) leads to

$$\begin{bmatrix} y_{1,k+1} \\ y_{2,k+1} \\ y_{3,k+1} \end{bmatrix} = \begin{bmatrix} y_{1,k} \\ y_{2,k} \\ y_{3,k} \end{bmatrix} + \left\{ \begin{bmatrix} 0 & 1 & 0 \\ 0 & 0 & -10{,}000 \\ 0 & 1 & -250 \end{bmatrix} \begin{bmatrix} y_{1,k} \\ y_{2,k} \\ y_{3,k} \end{bmatrix} + \begin{bmatrix} 0 \\ 10\sin(50t_k) \\ 0 \end{bmatrix} \right\} h \qquad \text{(b)}$$

```
% Example 9.16
% Euler's Method
%
% State matrix
A=[0 1 0; 0 0 -10000;0 1 -250]
% Input matrix
B=[0; 1; 0]
% Initial conditions
y(1,1)=0;
y(2,1)=0;
y(3,1)=0;
% Input step size and final time
h=input('Input step size   ');
tf=input('Input final time for computation  ')
t=0;
ta(1)=0;
k=1;
while t < tf
% Definition of state-vector at time t(k)
  C=[y(1,k);y(2,k);y(3,k)];
% Calculation of changes in state variables between t(k) and t(k+1)
  YN=C+h*(A*C+B*10*sin(50*t));
  k=k+1;
% Definition of state variables at t(k+1)
  y(1,k)=YN(1);
  y(2,k)=YN(2);
  y(3,k)=YN(3);
  t=t+h;
  ta(k)=t;
end
% System output is y(1)
plot(ta,y(1,:))
xlabel('t (s)')
ylabel('x (m)')
%
% End of Example9_16.m
```

**FIG. 9.11**  The script of Example9_16.m.

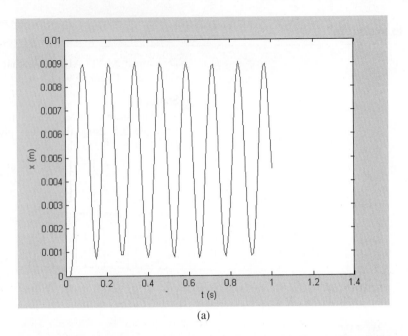

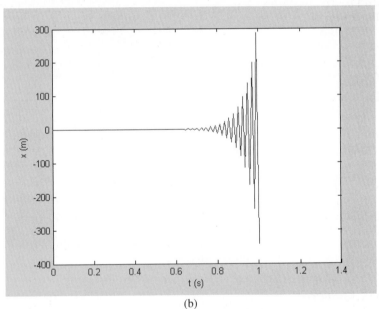

**FIG. 9.12** The integration of the state-space equations of Example 9.16 using Euler's method: (a) $h = 0.01$, (b) $h = 0.011$.

A MATLAB M-file used to approximate the response of this system is given in Figure 9.11. The file allows user input for values of $h$ and $t_\infty$, the final value of $t$ for which the response is approximated. Figure 9.12 illustrates the response obtained from the execution of Example9_16.m for several values of $h$. Figure 9.12(a) shows that the numerical method is stable for $h = 0.01$ while Figure 9.12(b) shows that the numerical method is unstable for $h = 0.011$.

The most popular Runge-Kutta method is the classic fourth-order Runge-Kutta method. Its application to approximate the solution of Equation (9.12) at discrete values of time leads to

$$\mathbf{y}_{k+1} = \mathbf{y}_k + \frac{h}{6}(\mathbf{b}_1 + 2\mathbf{b}_2 + 2\mathbf{b}_3 + \mathbf{b}_4) \tag{9.59}$$

where

$$\mathbf{b}_1 = \mathbf{A}\mathbf{y}_k + \mathbf{B}\mathbf{u}(t_k)$$

$$\mathbf{b}_2 = \mathbf{A}\left(\mathbf{y}_k + \frac{h}{2}\mathbf{b}_1\right) + \mathbf{B}\mathbf{u}\left(t_k + \frac{h}{2}\right)$$

$$\mathbf{b}_3 = \mathbf{A}\left(\mathbf{y}_k + \frac{h}{2}\mathbf{b}_2\right) + \mathbf{B}\mathbf{u}\left(t_k + \frac{h}{2}\right)$$

$$\mathbf{b}_4 = \mathbf{A}(\mathbf{y}_k + h\mathbf{b}_3) + \mathbf{B}\mathbf{u}(t_{k+1}) \tag{9.60}$$

The local error for the numerical algorithm described by Equations (9.59) and (9.60) is $O(h^5)$ while the global error is $O(h^4)$.

---

**Example 9.17**

Show the calculations for the first step of the application of the classical fourth-order Runge-Kutta method to the system of Example 9.15. Use $h = 0.004$ s.

**Solution**

The state-space formulation of the differential equations for this system is given in Equation (a) of Example 9.16. The initial conditions are all zero, thus $\mathbf{y}_0 = 0$. The application of Equation (9.60) using the forms of $\mathbf{A}$ and $\mathbf{B}$ determined in Example 9.15 leads to

$$\mathbf{b}_1 = \begin{bmatrix} 0 & 1 & 0 \\ 0 & 0 & -10{,}000 \\ 0 & 1 & -250 \end{bmatrix}\begin{bmatrix} 0 \\ 0 \\ 0 \end{bmatrix} + \begin{bmatrix} 0 \\ 0.1 \\ 0 \end{bmatrix} 100 \sin[(50)(0)] = \begin{bmatrix} 0 \\ 0 \\ 0 \end{bmatrix} \tag{a}$$

$$\mathbf{b}_2 = \begin{bmatrix} 0 & 1 & 0 \\ 0 & 0 & -10{,}000 \\ 0 & 1 & -250 \end{bmatrix}\left\{\begin{bmatrix} 0 \\ 0 \\ 0 \end{bmatrix} + \frac{0.004}{2}\begin{bmatrix} 0 \\ 0 \\ 0 \end{bmatrix}\right\} + \begin{bmatrix} 0 \\ 0.1 \\ 0 \end{bmatrix} 100 \sin[(50)(0.002)] = 0 \begin{bmatrix} 0 \\ 0.998 \\ 0 \end{bmatrix} \tag{b}$$

$$\mathbf{b}_3 = \begin{bmatrix} 0 & 1 & 0 \\ 0 & 0 & -10{,}000 \\ 0 & 1 & -250 \end{bmatrix}\left\{\begin{bmatrix} 0 \\ 0 \\ 0 \end{bmatrix} + \frac{0.004}{2}\begin{bmatrix} 0 \\ 0.998 \\ 0 \end{bmatrix}\right\} + \begin{bmatrix} 0 \\ 0.1 \\ 0 \end{bmatrix} 100 \sin[(50)(0.002)]$$

$$= \begin{bmatrix} 0.00200 \\ 0.998 \\ 0 \end{bmatrix} \tag{c}$$

$$\mathbf{b}_4 = \begin{bmatrix} 0 & 1 & 0 \\ 0 & 0 & -10{,}000 \\ 0 & 1 & -250 \end{bmatrix} \left\{ \begin{bmatrix} 0 \\ 0 \\ 0 \end{bmatrix} + 0.004 \begin{bmatrix} 0.00200 \\ 0.998 \\ 0 \end{bmatrix} \right\}$$

$$+ \begin{bmatrix} 0 \\ 0.1 \\ 0 \end{bmatrix} 100 \sin\left[(50)(0.004)\right] = \begin{bmatrix} 0.0040 \\ 1.593 \\ 0.0040 \end{bmatrix} \tag{d}$$

The approximation $\mathbf{y}_1$ is obtained using Equation (9.59)

$$\begin{bmatrix} y_{1,1} \\ y_{2,1} \\ y_{3,1} \end{bmatrix} = \frac{0.004}{6} \left\{ \begin{bmatrix} 0 \\ 0 \\ 0 \end{bmatrix} + 2\begin{bmatrix} 0 \\ 0.998 \\ 0 \end{bmatrix} + 2\begin{bmatrix} 0.00200 \\ 0.998 \\ 0 \end{bmatrix} + \begin{bmatrix} 0.0040 \\ 1.593 \\ 0.0040 \end{bmatrix} \right\}$$

$$= \begin{bmatrix} 5.33 \times 10^{-6} \\ 3.68 \times 10^{-3} \\ 2.67 \times 10^{-6} \end{bmatrix} \tag{e}$$

## 9.4.3 Use of MATLAB Program ode45.m

MATLAB has several M-files in its library that use Runge-Kutta methods to numerically integrate a system of differential equations. The most widely applicable of these files is ode45.m, which uses a fourth-order Runge-Kutta method. The basic format for using ode45.m to solve a system of $n$ equations of the form $\dot{y}_i = f_i(y_1, y_2, \ldots, y_n, t)$ is

$$[\text{t,y}] = \text{ode45(f,ts,y0)}$$

where

- t is a vector of times at which the numerical response is returned;
- y is a two-dimensional vector of responses (the second subscript identifies the variable and the first subscript corresponds to the time; for example $y(10,2)$ is the numerical approximation for $y_2$ at $t_{10}$);
- f is the name of the user-supplied function subprogram that supplies the functions $f_1, f_2, \ldots f_n$; the appropriate form for the function statement is

$$\text{function dy} = \text{f(t,y)}$$

where dy is the name of the vector that returns the values of $f_i$ $i = 1, 2, \ldots, n$;
- ts is a vector of two elements, ts $= [\text{t0,tf}]$ where t0 is the initial value of time and tf is the final value of the time at which the numerical approximation to the response is to be calculated;
- y0 is the vector of initial conditions.

---

**Example 9.18**

Write a MATLAB M-file that uses ode45.m to develop a numerical approximation to the solution of Example 9.15 for $0 < t < 1$ s.

**Solution**

The script for Example9_18.m is presented in Figure 9.13, which also presents the script of the function subprogram required for its execution.

```
% Example 9.18
% Use of ode45.m
%
% This program uses ode45 to solve a system of 3 ODEs of the form
%      dy_i/dt=f_i(y_1,y_2,y_3,t) i=1,2,3
%
% Input initial conditions
disp('Input initial conditions')
for k=1:3
  str=['y0(',num2str(k),')=   '];
  y1(k)=input(str);
end
% The initial condition vector must be a row vector
y0=[y1(1) y1(1) y1(1)]
% Input final time
tf=input('Input final time for numerical calculation  ');
ts=[0 tf];
% Calling ode45.m
%   t=vector of times at which solution is calculated
%   y=matrix containing system response
%   'f'=name of user supplied function which provides right-hand side
%       of differential equations. The format for the function is
%             function dy=f(t,y)
%       where dy is the vector of derivatives to be returned to ode45
%
[t,y]=ode45(@f,ts,y0)
plot(t,y(:,1))
xlabel('t (s)')
ylabel('x (m)')
title('ode45 solution for Example 9.18')
%
% End of Example9_18.m
```

(a)

```
% Function which provides right-hand sides of differential equations
%
%    dy is vector of derivatives to be returned to ode45
%    t is a scalar value of time for which the derivatives are to be
%       evaluated, it is supplied by ode45
%    y is the vector of values of numerical approximations at the
%       previous time step, it is supplied by ode45
%
```

**FIG. 9.13** (a) The script of Example9_18.m; (b) the user-supplied function required for the execution of Example9_18.m. *(Continued)*

```
function dy=f(t,y)
% The right-hand side is of the form Ab+c[10sin(50t)]
A=[0 1 0;0 0 -10000;0 1 -250];
b=[y(1);y(2);y(3)];
c=[0;1;0];
dy=A*b+c*10*sin(50*t);
```

(b)

**FIG. 9.13**  (*Continued*)

The plot of $x(t)$ generated from the execution of Example9_18.m is shown in Figure 9.14. The plot is very similar to that generated from the execution of Example9_16.m.

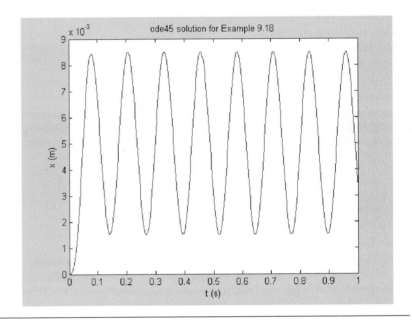

**FIG. 9.14** The numerical solution to Example 9.18 obtained from the execution of Example9_18.m, which uses MATLAB file ode45.m for numerical integration.

## 9.5 RELATIONSHIP BETWEEN TRANSFER FUNCTIONS AND STATE-SPACE MODELS

The transfer function $G(s) = X(s)/Y(s)$ for a system with a single input $y(t)$ and a single output $x(t)$ is unique. However, the state-space formulation of the system can take many forms and is thus not unique. Thus it should be possible to determine a unique transfer function from a state-space formulation and to determine a state-space formulation from a transfer function.

Consider first the state-space formulation of an $n$th-order system developed in the form of Equations (9.12) and (9.13) where $x_1$, $x_2$, ..., $x_n$ ares the state variables. For the purpose of this discussion assume, without loss of generality, that $x_n$ is the system output for the purposes of transfer function modeling. Also assume that the system has a single input; thus the input $f(t)$ is a $1 \times 1$ matrix and **B** is an $n \times 1$ matrix.

Taking the Laplace transform of Equation (9.12), assuming all initial conditions are zero, leads to

$$s\mathbf{Y}(s) = \mathbf{A}\mathbf{Y}(s) + \mathbf{B}F(s) \qquad (9.61)$$

Equation (9.61) can be rearranged as

$$(s\mathbf{I} - \mathbf{A})\mathbf{Y}(s) = \mathbf{B}F(s) \qquad (9.62)$$

Cramer's rule is used to solve for $Y_n(s)$ as

$$Y_n = \frac{|\mathbf{V}_n|}{|s\mathbf{I} - \mathbf{A}|} U(s) \qquad (9.63)$$

where $\mathbf{V}_n$ is $s\mathbf{I} - \mathbf{A}$ with its last column replaced by **B**.

Since $G(s)$ is unique it is clear that, for any state-space formulation, the transfer function is obtained from Equation (9.63) as

$$G(s) = \frac{|\mathbf{V}_n|}{|s\mathbf{I} - \mathbf{A}|} \qquad (9.64)$$

Equation (9.64) also suggests a method to determine a state-space model from the transfer function. The denominator of the transfer function is an $n$th-order polynomial in $s$, which can be used to construct a matrix whose determinant is the denominator. Then the $n \times 1$ vector **B** can be chosen to adjust the numerator to the proper form. To this end assume that the denominator of $G(s)$ is a polynomial of the form

$$D(s) = s^n + a_{n-1}s^{n-1} + a_{n-2}s^{n-2} + \cdots + a_1 s + a_0 \qquad (9.65)$$

Consider the matrix

$$\mathbf{A} = \begin{bmatrix} 0 & 1 & 0 & 0 & \cdots & 0 \\ 0 & 0 & 1 & 0 & \cdots & 0 \\ 0 & 0 & 0 & 1 & \cdots & 0 \\ \vdots & \vdots & \vdots & \vdots & \ddots & \vdots \\ -a_0 & -a_1 & -a_2 & -a_3 & \cdots & -a_{n-1} \end{bmatrix} \qquad (9.66)$$

Evaluation of $|s\mathbf{I} - \mathbf{A}|$ by row expansion using the last row shows that $|s\mathbf{I} - \mathbf{A}| = D(s)$. For example consider a third-order system, using a $3 \times 3$ matrix of the form of Equation (9.64)

$$|s\mathbf{I} - \mathbf{A}| = \begin{vmatrix} s & -1 & 0 \\ 0 & s & -1 \\ a_0 & a_1 & s + a_2 \end{vmatrix}$$

$$= a_0 \begin{vmatrix} -1 & 0 \\ s & -1 \end{vmatrix} - a_1 \begin{bmatrix} s & 0 \\ 0 & -1 \end{bmatrix} + (s + a_2) \begin{vmatrix} s & -1 \\ 0 & s \end{vmatrix}$$

$$= a_0(1) - a_1(-s) + (s + a_2)(s^2)$$

$$= s^3 + a_2 s^2 + a_1 s + a_0 \qquad (9.67)$$

## Example 9.19

Use Equation (i) of Example 9.6 to derive the transfer function for the system.

### Solution

The system output $x(t)$ is equal to the state variable $y_1$. Thus the transfer function is

$$G(s) = \frac{|\mathbf{V}_1|}{|s\mathbf{I} - \mathbf{A}|}$$

$$= \frac{\begin{vmatrix} \dfrac{c}{m} & -1 \\ \dfrac{k}{m} - \dfrac{c^2}{m^2} & s + \dfrac{c}{m} \end{vmatrix}}{\begin{vmatrix} s & -1 \\ \dfrac{k}{m} & s + \dfrac{c}{m} \end{vmatrix}}$$

$$= \frac{\dfrac{c}{m}\left(s + \dfrac{c}{m}\right) + \dfrac{k}{m} - \dfrac{c^2}{m^2}}{s\left(s + \dfrac{c}{m}\right) + \left(\dfrac{k}{m}\right)}$$

$$= \frac{cs + k}{ms^2 + cs + m} \tag{a}$$

## Example 9.20

The transfer function of a third-order system is

$$G(s) = \frac{2s^2 + 5s + 6}{s^3 + 3s^2 + 12s + 60} \tag{a}$$

Determine a state-space formulation for this system.

### Solution

The state matrix is obtained according to Equation (9.6b) as

$$\mathbf{A} = \begin{bmatrix} 0 & 1 & 0 \\ 0 & 0 & 1 \\ -60 & -12 & -3 \end{bmatrix} \tag{b}$$

Equation (9.64) and Equation (a) imply

$$|\mathbf{V}_3| = 2s^2 + 5s + 6$$

$$\begin{vmatrix} s & -1 & b_1 \\ 0 & s & b_2 \\ 60 & 12 & b_3 \end{vmatrix} = 2s^2 + 5s + 6$$

$$b_3 s^2 - 60 b_2 - 60 s b_1 - 12 s b_2 = 2s^2 + 5s + 6 \tag{c}$$

Equation (c) must be true for all values of $s$. Thus coefficients of like powers of $s$ are equated leading to

$$-60b_2 = 6 \tag{d}$$

$$-60b_1 - 12b_2 = 5 \tag{e}$$

$$b_3 = 2 \tag{f}$$

The solution of Equations (d)–(f) is $b_1 = -0.0633$, $b_2 = -0.1$, $b_3 = 2$. Thus a state-space formulation is

$$\begin{bmatrix} \dot{y}_1 \\ \dot{y}_2 \\ \dot{y}_3 \end{bmatrix} = \begin{bmatrix} 0 & 1 & 0 \\ 0 & 0 & 1 \\ -60 & -12 & -3 \end{bmatrix} \begin{bmatrix} y_1 \\ y_2 \\ y_3 \end{bmatrix} + \begin{bmatrix} -0.0633 \\ -0.1 \\ 2 \end{bmatrix} f(t) \tag{g}$$

## 9.6 MATLAB AND SIMULINK MODELING IN THE STATE-SPACE

The use of MATLAB to determine the transient response of a dynamic system from its transfer function is introduced in Section 6.2. MATLAB-generated Bode plots and Nyquist diagrams from a system's transfer function are illustrated in Chapter 7. The development of SIMULINK models from block diagrams of systems in terms of transfer functions is introduced in Section 8.1. It is shown in this chapter that state-space modeling is an alternative for transfer function modeling. Section 9.5 shows that the system's transfer function can be determined from a state-space model and that a state-space model may be determined from the transfer function.

### 9.6.1 MATLAB

MATLAB and SIMULINK may also be used for state-space modeling of dynamic systems. A general state-space formulation for a linear system is of the form of Equations (9.12) and (9.13). These equations are used to define MATLAB and SIMULINK models in the state space. The matrices **A**, **B**, **C**, and **D** define the state-space model. For a given system these matrices can be entered into the MATLAB workspace. A state-space model for the system is then defined by the statement

$$\text{system} = \text{ss(A,B,C,D)}$$

where ss is a MATLAB command that defines the state space and stores its attributes as the variable system. Figure 9.15 provides a copy of the workspace when a state-space model of Example 9.1 is entered with $m = 1$ kg, $c = 5$ N·s/m, and $k = 6$ N/m. The command

$$\text{step(system)}$$

generates the plot of the step response illustrated in Figure 9.16. Similarly the command

$$\text{impulse(system)}$$

generates the impulsive response of the system.

```
>> A=[0 1;-6 -5];
>> B=[0;1];
>> C=[1 0];
>> D=[0];
>> system=ss(A,B,C,D)

a =
      x1 x2
  x1  0  1
  x2 -6 -5


b =
      u1
  x1  0
  x2  1


c =
      x1 x2
  y1  1  0


d =
      u1
  y1  0

Continuous-time model.
>>
```

**FIG. 9.15** The MATLAB workspace for entering the state-space model of Example 9.1.

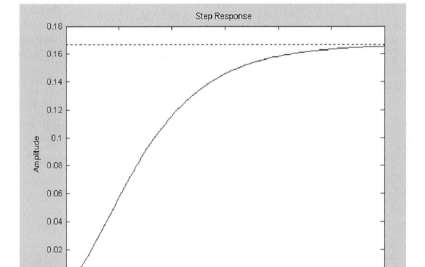

**FIG. 9.16** The step response for the system of Example 9.1, obtained using a state-space model.

The response of a linear system due to a time-dependent input is obtained using the 'lsim' command. The times at which the simulation is to occur are defined in a vector, say t. The values of the input at these times is stored in a vector, say F. Then the command

$$lsim(system, F, t)$$

calculates a numerical simulation for the response of the system at the times specified in the vector t and plots the response. The MATLAB code necessary to run a simulation of the current mechanical system due to an input of the form $F(t) = 3e^{-3t}$ and the resulting simulation are illustrated in Figure 9.17. The 'lsim' command leads to a plot that compares the response to the input. It is difficult to determine much detail of the response from Figure 9.17. Thus a customized plot of the response is desirable and can be obtained by using the command

$$x = lsim(system, F, t)$$

which, instead of plotting the response, stores the numerical values of the response in the vector x. Then, a customized plot, such as that illustrated in Figure 9.18, can be developed.

```
>> t=0.:0.1:3;
>> F=3*exp(-3*t);
>> lsim(system,F,t)
>>
```

(a)

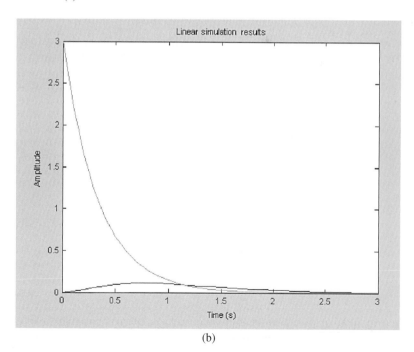

**FIG. 9.17** (a) The MATLAB commands to generate the simulation of the mechanical system with $F(t) = 3e^{-3t}$; (b) the resulting plot comparing the response to the input.

(b)

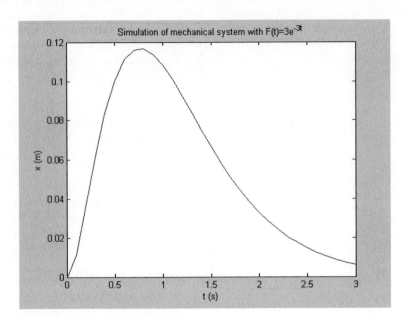

**FIG. 9.18** The customized plot developed using data generated from 'y=lsim(system,F,t)'.

It is shown in Section 9.5 that a state-space model can be used to determine the system's transfer function and that the transfer function may be used to determine a state-space model. MATLAB can perform both tasks. Given a state-space model, defined as system, the transfer function is determined using the command

$$[N,D] = ss2tf(A,B,C,D)$$

where N and D are the vectors of coefficients defining the numerator and denominator of the system's transfer function. The application of this command to the current mechanical system is shown in Figure 9.19.

```
>> [N,D]=ss2tf(A,B,C,D)

N =

    0      0   1.0000


D =

1.0000   5.0000   6.0000

>> tf(N,D)

Transfer function:
        1
    -------------
    s^2 + 5 s + 6

>>
```

**FIG. 9.19** Illustration of the development of the transfer function from the state-space model using MATLAB.

Conversely, given the transfer function of a system in the form of the coefficient vectors for the numerator and denominator, a state-space model is obtained using the command

$$[A,B,C,D] = tf2ss(N,D)$$

The application of this command to the transfer function of Example 9.20 is illustrated in Figure 9.20. Recall that a state-space formulation is not unique. The formulation provided in Figure 9.20 is an alternative to the formulation provided in Example 9.20.

```
>> [N,D]=ss2tf(A,B,C,D)

N =

       0     0   1.0000


D =

   1.0000   5.0000   6.0000

>> tf(N,D)

Transfer function:
      1
------------
s^2 + 5 s + 6

>> clear
>> N=[2 5 6];
>> D=[1 3 12 60];
>> [A,B,C,D]=tf2ss(N,D)

A =

    -3  -12  -60
     1    0    0
     0    1    0


B =

     1
     0
     0


C =

     2    5    6


D =

     0
>>
```

**FIG. 9.20** Illustration of the development of a state-space model from the system's transfer function.

---

**Example 9.21**

Consider the two-degree-of-freedom mechanical system of Example 9.8 with $m_1 = 1$ kg, $m_2 = 0.5$ kg, $c_1 = c_2 = 2$ N·s/m, $k_1 = 20$ N/m, and $k_2 = 10$ N/m. (a) Develop a state-space model for the system in MATLAB. (b) Use the state-space model to determine the impulsive responses for the system. (c) Use the state-space model to determine the system response if $F_1(t) = 10 \sin(10t)$ and $F_2(t) = 5 \sin(5t)$. (d) Use the state-space model to determine the transfer functions for the system.

```
% Example 9.21
%
% Definition of parameters
m1=1;
m2=0.5;
c1=2;
c2=2;
k1=20;
k2=10;
% Definition of state-space matrices
A=[0 0 1 0;0 0 0 1;-(k1+k2)/m1 k2/m2 -(c1+c2)/m1 c2/m2;k2/m1 -k2/m2 c2/m1 -c2/m2];
B=[0 0;0 0;1/m1 0;0 1/m2];
C=[1 0 0 0;0 1 0 0];
D=[0 0; 0 0];
% State-sapce formulation
system=ss(A,B,C,D)
% Impulsive response
impulse(system)
% Forced responses
% Defining time range for calculation of forced response
t=0:0.02:4;
% Defining system inputs
F1=1*sin(10*t);
F2=0.5*sin(5*t);
F=[F1;F2];
figure
% Numerical simulation of forced response
lsim(system,F,t)
y=lsim(system,F,t);
figure
plot(t,y(:,1),'-',t,y(:,2),'--')
xlabel('t (s)')
ylabel('displacement (m)')
legend('x_1','x_2')
% transfer function
[N1,D1]=ss2tf(A,B,C,D,1)
[N2,D2]=ss2tf(A,B,C,D,2)
%
% End of Example9_21.m
```

**FIG. 9.21** The script of Example9_21.m.

```
> Example9_21

a =
    x1  x2  x3  x4
  x1  0   0   1   0
  x2  0   0   0   1
  x3  -30  20  -4   4
  x4  10  -20   2  -4


b =
   u1 u2
  x1  0  0
  x2  0  0
  x3  1  0
  x4  0  2


c =
   x1 x2 x3 x4
  y1  1  0  0  0
  y2  0  1  0  0


d =
   u1 u2
  y1  0  0
  y2  0  0

Continuous-time model.

N1 =

   0  -0.0000   1.0000   4.0000  20.0000
   0  -0.0000  -0.0000   2.0000  10.0000


D1 =

 1.0000   8.0000  58.0000 120.0000 400.0000


N2 =

   0  -0.0000  -0.0000   8.0000  40.0000
   0  -0.0000   2.0000   8.0000  60.0000


D2 =

 1.0000   8.0000  58.0000 120.0000 400.0000

>>
```

(a)

**FIG. 9.22** (a) The workspace output from the execution of Example9_21.m; (b) the impulsive responses; (c) the forced responses plotted directly from 'lsim'; (d) the customized plots of the forced response.   (*Continued*)

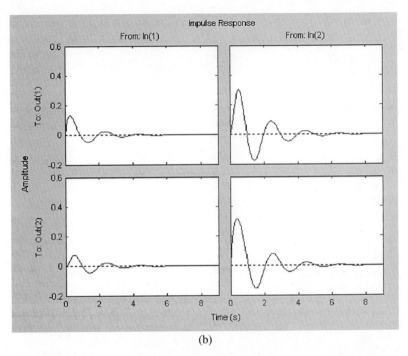

(b)

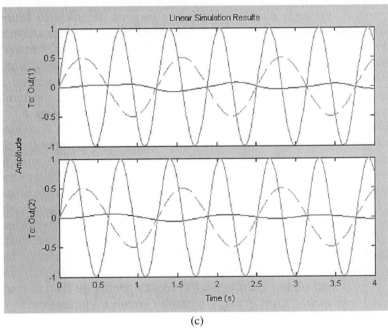

(c)

**FIG. 9.22**  (*Continued*)

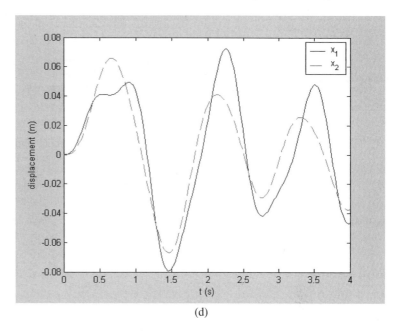

**FIG. 9.22**  (*Continued*)

(d)

**Solution**

The script for the MATLAB program Example9_21.m is shown in Figure 9.21, while the output from its execution is shown in Figure 9.22. This system has two inputs $F_1(t)$ and $F_2(t)$ and two outputs $x_1(t)$ and $x_2(t)$. This leads to the following modifications in syntax for the state-space modeling from that of a SISO system:

- The impulsive response consists of four plots. Each plot corresponds to one input and one output. For example the plot in the top left of Figure 9.22(b) is the response for $x_1$ due to $F_1 = \delta(t)$ and $F_2 = 0$.
- The variable defining the input for 'lsim' must be a vector with two columns. Application of 'lsim' leads to two plots, one for each output variable.
- There are four transfer functions, $G_{i,j}(s) = X_i(s)/F_j(s)$. The use of 'ss2tf' for multiple output systems requires an additional argument defining which row of the transfer function matrix to determine. From the output in Figure 9.22(a) it is determined that

$$G_{1,1} = \frac{s^2 + 4s + 20}{s^4 + 8s^3 + 58s^2 + 120s + 400} \tag{a}$$

$$G_{2,2} = \frac{2s^2 + 8s + 60}{s^4 + 8s^3 + 58s^2 + 120s + 400} \tag{b}$$

### 9.6.2  SIMULINK

There are several methods of applying SIMULINK to simulate the response of a dynamic system from its state-space formulation. It is illustrated in Section 9.1 how a dynamic model is built in SIMULINK from the block diagram of the state-space formulation of the system. This is further illustrated in the following example.

**Example 9.22**

For the system of Example 9.21 (a) draw a block diagram of the system from its state-space formulation and (b) develop and run a SIMULINK model for the system.

**Solution**

(a) A block diagram based on the state-space model of Example 9.8, using the numerical values of Example 9.21, is illustrated in Figure 9.23. A SIMULINK model based on the block diagram of Figure 9.23 is illustrated in Figure 9.24. The output from the scope when the simulation is run is shown in Figure 9.25.

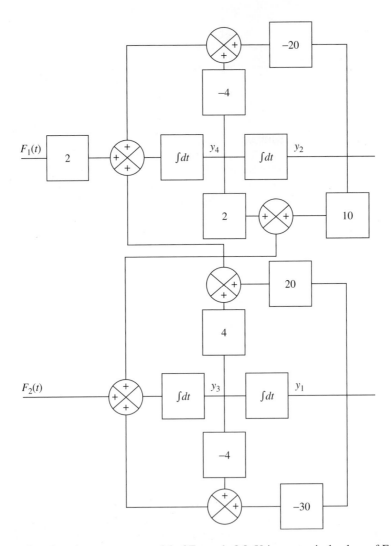

**FIG. 9.23**  The block diagram based on the state-space model of Example 9.8. Using numerical values of Example 9.21.

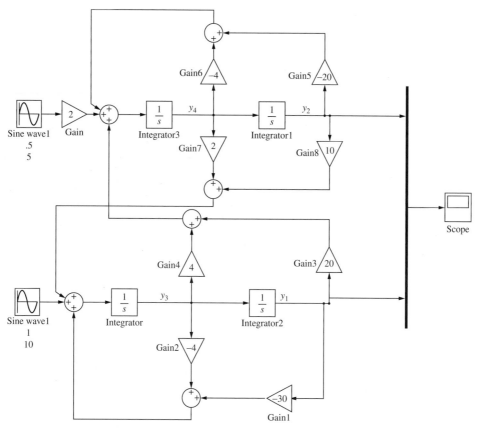

**FIG. 9.24** A SIMULINK model based on the block diagram of Figure 9.23.

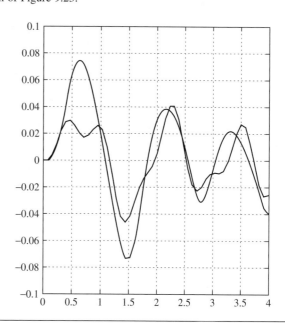

**FIG. 9.25** The output from the scope when the simulation is run.

A block diagram is not necessary for a SIMULINK model of a linear system formulated in a state space. SIMULINK has a State Space block in the Continuous Systems menu of the Library Browser. The State Space block can be directly used to define the state matrices.

---

### Example 9.23

Use the State Space block to develop a SIMULINK model for the system of Examples 9.21 and 9.22.

### Solution

The SIMULINK model using the state-space block is illustrated in Figure 9.26. The model has two input and two outputs. Two sine waves are routed through a 'Mux' to vectorize the input. The scope trace, shown in Figure 9.27 plots the response for both output variables.

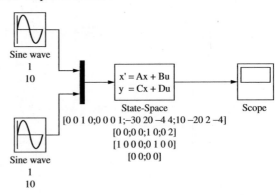

**FIG. 9.26** The SIMULINK model using the State Space block.

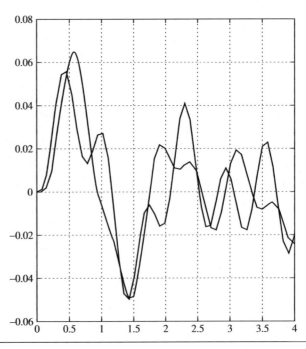

**FIG. 9.27** The scope trace for the responses for both output variables.

**Example 9.24**

Consider the two-tank liquid-level system of Example 4.8 with $A_1 = A_2 = 2.0\,\text{m}^3$, $R_1 = R_2 = 5\dfrac{\text{s}}{\text{m}^2}$ (a) Develop a state-space model for the system. (b) Use the State Space block to develop a SIMULINK model for the system. (c) Develop a SIMULINK model for the closed-loop system that occurs when a PID controller is used in the feedforward portion of a feedback loop where the controller controls the level in the second tank. Run the simulation for the following proposed controllers: (i) $K_i = 0$, $K_p = 1$, $K_d = 0$; (ii) $K_i = 0.1, K_p = 1$, $K_d = 0$; and (iii) $K_i = 0, K_p = 1$, $K_d = 0.25$.

**Solution**

The differential equations governing the liquid levels in the tanks due to a perturbation in inlet flow rate are

$$A_1 \frac{dh_1}{dt} + \frac{1}{R_1} h_1 - \frac{1}{R_1} h_2 = q_i \tag{a}$$

$$A_2 \frac{dh_2}{dt} - \frac{1}{R_1} h_1 + \left(\frac{1}{R_1} + \frac{1}{R_2}\right) h_2 = 0 \tag{b}$$

The state variables are $y_1 = h_1$ and $y_2 = h_2$. If the liquid levels in both tanks are considered as output the appropriate state-space model is

$$\mathbf{A} = \begin{bmatrix} -\dfrac{1}{A_1 R_1} & \dfrac{1}{A_1 R_1} \\ \dfrac{1}{A_2 R_1} & -\dfrac{1}{A_2}\left(\dfrac{1}{R_1} + \dfrac{1}{R_2}\right) \end{bmatrix} \tag{c}$$

$$\mathbf{B} = \begin{bmatrix} \dfrac{1}{A_1} \\ 0 \end{bmatrix} \tag{d}$$

$$\mathbf{C} = \begin{bmatrix} 1 & 0 \\ 0 & 1 \end{bmatrix} \tag{e}$$

$$\mathbf{D} = \begin{bmatrix} 0 \\ 0 \end{bmatrix} \tag{f}$$

(b) The SIMULINK model is shown in Figure 9.28, while the scope trace obtained using the parameters specified in Example 8.17 is shown in Figure 9.29. (c) Since the control system monitors $h_2(t)$ the state-space model for the liquid-level system is modified such that $h_2(t)$ is the only output. Thus the output matrix and transmission matrix are modified to

$$\mathbf{C} = \begin{bmatrix} 0 & 1 \end{bmatrix} \tag{g}$$

$$\mathbf{D} = \begin{bmatrix} 0 \end{bmatrix} \tag{h}$$

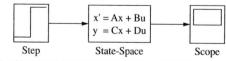

Step            State-Space            Scope

[−1/(R1*A1) 1/(R1*A1);1/(R1*A2) −1/A2*(1/R1 + 1/R2)]

[1/A1;0]

[1 0;0 1]

[0;0]

**FIG. 9.28** The SIMULINK model for the system of Example 9.24.

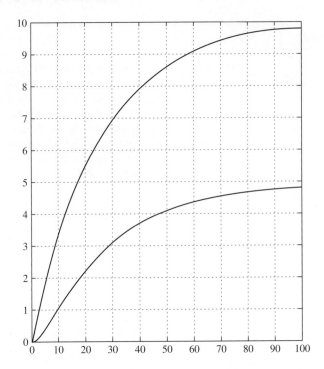

**FIG. 9.29** The scope trace using the parameters specified in Example 8.17.

The SIMULINK model with a feedback loop and a PID controller in the feedforward part of the loop is shown in Figure 9.30. This SIMULINK model uses the State Space box for the open-loop system. The scope traces for the various controller parameters are given in Figure 9.31. The offset is clear in Figure 9.31(a), in which a proportional controller is used.

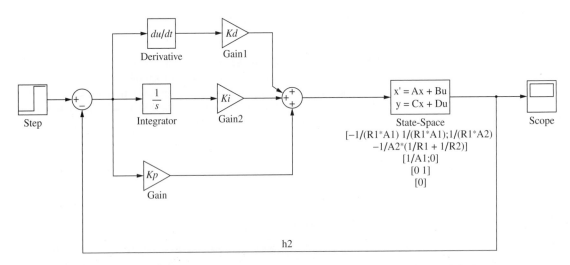

**FIG. 9.30** The SIMULINK model with a feedback loop and PID controller in the feedforward part of the loop.

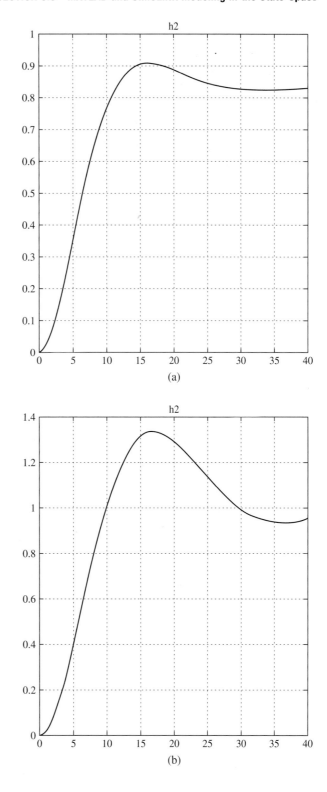

**FIG. 9.31** The scope traces for the various controller parameters: (a) Proportional; (b) PI; and (c) PID.  (*Continued*)

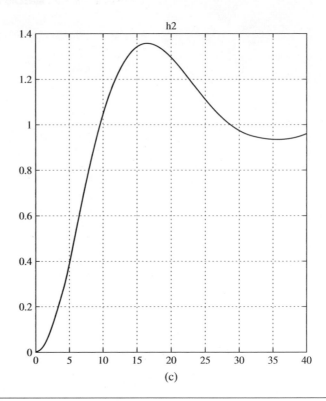

**FIG. 9.31** (*Continued*)

(c)

## 9.7 NONLINEAR SYSTEMS AND SYSTEMS WITH VARIABLE COEFFICIENTS

State-space methods are convenient for the formulation and numerical solution of mathematical models for systems that cannot be modeled using transfer functions. The application of the Laplace transform to the mathematical model of a dynamic system leads to the development of a system transfer function for systems with linear differential equations with constant coefficients. The success of the transform method depends on the application of two very important transform properties: linearity of the transform and transforms of derivatives.

Nonautonomous systems include systems in which parameters are time dependent and thus coefficients of differential equations are time dependent. For a linear system with constant coefficients the linearity property is used to write the transform of a term such as $c\dot{x}$ as $\mathcal{L}\{c\dot{x}\} = c\mathcal{L}\{\dot{x}\}$. This implies that the operations of multiplication by a scalar and taking the Laplace transform are interchangeable. Then the property of transforms of derivatives is used leading to $\mathcal{L}\{c\dot{x}\} = csX(s)$. However, if the coefficient is time dependent then the operations of multiplication by the coefficient and taking the transform cannot be interchanged. No general Laplace transform property exists for the transform of the multiplication of two time-dependent functions. An exception occurs if the coefficient is a polynomial in $t$. In this case the property $\mathcal{L}\{t^n f(t)\} = (-1)^n d^n F(s)/ds^n$ can be applied. However, this is of little help because its application to a differential equation leads to a differential equation, usually nonlinear, to solve for $X(s)$. Thus it is generally not possible to use the transfer function approach for a system with variable coefficients.

Examples of dynamic systems with variable coefficients are mechanical systems with variable mass, such as rockets and rocket sleds; circuits that are in a time-varying electric field leading to a time-dependent inductance; and chemical systems in which reaction rates may be functions of time due to the presence of catalysts. The state-space model for linear systems with variable coefficients is developed as described in Section 9.2 with the exception that some of the elements of the state matrix $\mathbf{A}$, defined in Equation (9.12), are time dependent. In certain cases some elements of the input matrix $\mathbf{B}$ may be time dependent.

The methods of Section 9.3 used to determine the free response and of Section 9.4.1 used to determine the forced response are not applicable for systems with variable coefficients. The differential equations are usually solved using a numerical method, such as the Runge-Kutta methods described in Section 9.4.2.

Nonlinear systems have mathematical models resulting in nonlinear differential equations. Nonlinear terms prevent the linearity property from being used and thus the application of the transform method does not lead to a transfer function.

Assumptions have been made throughout this book that when applied allowed nonlinear terms to be neglected or allowed nonlinear differential equations to be approximated by linear differential equations. Examples of the former include assuming a linear force-displacement relation in a spring, neglecting forms of friction other than viscous damping, and assuming that the variation of specific heat with temperature is negligible. An example of the latter is the small angle or small displacement assumption used to linearize mathematical models of mechanical systems. Also included in the latter are the linearization of nonlinear differential equations for perturbations from equilibrium that were made in liquid-level problems by using piping system's resistances at the original steady state or in nonisothermal CSTR systems in which the rate of reaction is dependent on temperature through the Arrhenius equation.

Linearized equations are often sufficient to meet the objectives of the modeling process. However, nonlinear systems have behaviors that cannot be predicted using linear models and that may be significant when large perturbations from equilibrium or steady state occur. Examples of nonlinear system behavior that cannot be predicted using linear models include the following:

- System parameters depend on initial conditions. The free response of a linear first-order system is characterized by the time constant, which is independent of initial conditions. The free-response of a second-order system is characterized by the period and damping ratio, which are independent of initial conditions. A nonlinear first-order system may still have exponential behavior, but the time constant may depend on initial conditions.

- Resonances occur for linear systems only when an input frequency coincides with a natural frequency. In addition to these primary resonances subharmonic and superharmonic resonances occur in nonlinear systems at frequencies different from the system's natural frequencies.
- The frequency response of a linear system is continuous whereas the frequency response of a nonlinear system may have discrete jumps at certain frequencies.
- The amplitude of the response at a fixed frequency is proportional to the amplitude of excitation. For certain multidegree-of-freedom nonlinear systems a mode may become saturated in that its amplitude remains constant as the amplitude of the input is increased.
- A linear system has one stable equilibrium position. A nonlinear system may have multiple equilibrium positions, some of which may be unstable.

The differential equations for a state-space model for a nonlinear system cannot be written in the form of Equation (9.12) because the state matrix could not be formulated. The differential equations for an $n$th-order system with state variables $y_1$, $y_2$, ..., $y_n$ and system inputs $u_1$, $u_2$, ..., $u_k$ are formulated as

$$
\begin{aligned}
\dot{y}_1 &= f_1(y_1, y_2, \ldots, y_n, u_1, u_2, \ldots, u_k, t) \\
\dot{y}_2 &= f_2(y_1, y_2, \ldots, y_n, u_1, u_2, \ldots, u_k, t) \\
&\vdots \\
\dot{y}_n &= f_n(y_1, y_2, \ldots, y_n, u_1, u_2, \ldots, u_k, t)
\end{aligned}
\tag{9.66}
$$

The Runge-Kutta method described in Section 9.4 is developed for a linear system and thus cannot be applied to develop a numerical solution for Equation (9.68). However, a more general form of a fourth-order Runge-Kutta method exists and can be applied. This method is employed in the MATLAB command 'ode45' whose use is illustrated in examples.

Applications in this study are limited to second-order systems. For such systems a graphical representation of the solution, called the phase plane plot or the state plane plot, is used to provide useful insights into the behavior of the system. A phase plane plot for a second-order system is a plot of the variation of one state variable with the other, say with $y_1$ on the horizontal axis and $y_2$ on the vertical axis.

The formulation of the state-space model for nonlinear systems, the application of ode45 to solve the differential equations, and the development and analysis of the phase plane are illustrated in the following examples.

**Example 9.25**

The nonlinear differential equation governing the free response of a simple pendulum of mass $m$ and length $\ell$ is

$$\ddot{\theta} + \omega_n^2 \sin\theta = 0 \tag{a}$$

where $\omega_n = \sqrt{g/\ell}$ is the natural frequency of a linearized system model. (a) Develop a state-space formulation for the system. (b) Write a MATLAB M-file using the command 'ode45' to determine a numerical solution for the system and to plot the time-dependent response and the phase plane. (c) Use the MATLAB M-file to determine the numerical solution with $\omega_n = 1$ r/s and for initial conditions $\dot{\theta}(0) = 0$ and (i) $\theta(0) = \pi/6$, (ii) $\theta(0) = \pi/3$, and (iii) $\theta(0) = \pi/2$. Compare and contrast the solutions.

**Solution**

(a) Using state-space variables defined as $y_1 = \theta$ and $y_2 = \dot{\theta}$ the state-space formulation of Equation (a) is

$$\begin{aligned} \dot{y}_1 &= y_2 \\ \dot{y}_2 &= \sin(y_1) \end{aligned} \tag{b}$$

(b) The script of Example9_25.m, which uses ode45 to solve Equation (b), is shown in Figure 9.32. The program plots the time-dependent response of the system as well as the state plane. Plots generated from the execution of Example9_25.m are given in Figure 9.33. The comparison of Figures 9.33(b) and 9.33(c) shows that the frequency and period of a nonlinear system is dependent on initial conditions.

```
% Example9.25
% Nonlinear response of simple pendulum
% Initial conditions
theta0=input('input initial angular displacement in rad   ')
y0=[theta0 0];
tf=30;
% Vector of derivatives is provided from user-supplied file pend.m
[t,y]=ode45(@pend,[0 tf],y0);
% Plot time response
figure
plot(t,y(:,1))
xlabel('t (s)')
ylabel('\theta (rad)')
str=['Pendulum response for \theta_0=',num2str(theta0),'rad']
title(str)
figure
% State plane plot
plot(y(:,1),y(:,2))
xlabel('\theta (rad)')
ylabel('\omega (rad/s)')
str1=['State plane for pendulum with \theta_0=',num2str(theta0),'rad']
title(str1)
%
% End of Example9_25.m
```
(a)

**FIG. 9.32** (a) The script of the file Example9_25.m; (b) the script of pend.m required for Example9_25.m. (*Continued*)

```
% User supplied program to generate vector of derivatives for dynamic
%       response of simple pendulum
function dy=pend(t,y)
dy1=y(2);
dy2=sin(y(1));
dy=[dy1;dy2];
```

(b)

**FIG. 9.32**  (*Continued*)

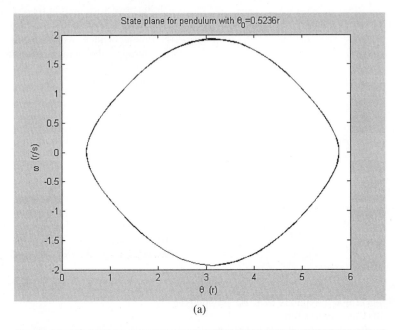

(a)

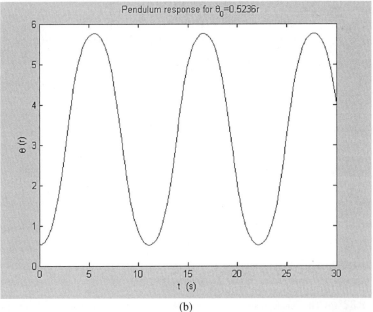

**FIG. 9.33**  Plots obtained from the
execution of Example9_25.m: (a) the
state plane for the pendulum with
$\theta_0 = 0.5236$ r; (b) the pendulum response
for $\theta_0 = 0.5236$ r; and (c) the pendulum
response for $\theta_0 = 1.5708$ r.

(b)

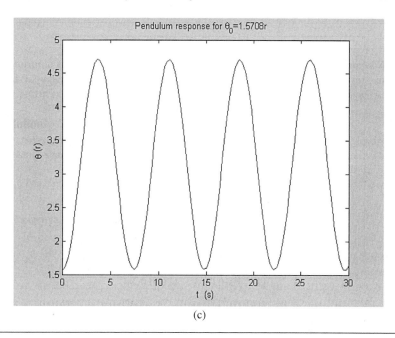

**FIG. 9.33** (*Continued*)    (c)

SIMULINK is a valuable tool in modeling nonlinear systems. SIMULINK has blocks that allow nonlinear manipulation of a signal.

## Example 9.26

Use SIMULINK to model the nonlinear pendulum of Example 9.25.

### Solution

The SIMULINK model for the nonlinear system is shown in Figure 9.34. The model has a Trigonometric Function block, chosen from the Continuous

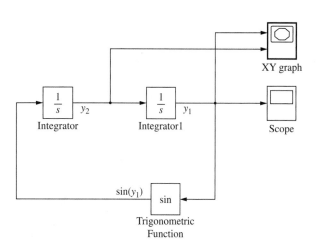

**FIG. 9.34** The SIMULINK model for the nonlinear system of Example 9.26.

System menu of the Library Browser. From the Block Properties menu a choice of Sin was made. This block mathematically takes the sine of the value of the signal that enters the block. Thus the signal leaving the block is $\sin(y_1)$. The initial condition is specified at the second integrator. The trace of the scope is identical to the response obtained using 'ode45'.

---

**Example 9.27**

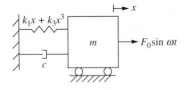

**FIG. 9.35** The system with spring with cubic nonlinearity of Example 9.27.

The spring in the system of Figure 9.35 is a nonlinear spring. Its force-displacement relation is

$$F = k_1 x + k_3 x^3 \tag{a}$$

The spring is said to have a cubic nonlinearity. The differential equation governing the motion of the system is

$$m\ddot{x} + c\dot{x} + k_1 x + k_3 x^3 = F_0 \sin(\omega t) \tag{b}$$

If $m = 1$ kg, $c = 0.08$ N·s/m, $k_1 = 1$ N/m, $k_3 = 0.1$ N/m$^3$, $F_0 = 1$ N, and $\omega = 1$ r/s, Equation (b) becomes

$$\ddot{x} + 0.08\dot{x} + x + 0.1x^3 = \sin(t) \tag{c}$$

Equation (c) is a form of Duffing's equation. Use SIMULINK to develop and run a state-space simulation of Equation (c).

**Solution**

Defining $y_1 = x$ and $y_2 = \dot{x}$, the state-space formulation of Equation (c) is

$$y_1 = y_2 \tag{d}$$
$$\dot{y}_2 = \sin(t) - y_2 - y_1 - y_1^3 \tag{e}$$

The SIMULINK model for Equations (d) and (e) is illustrated in Figure 9.36. The model uses the Polynomial block from the Math Operations menu of the

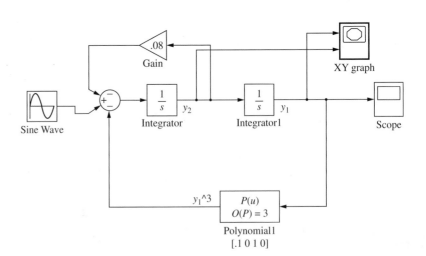

**FIG. 9.36** The SIMULINK model for the system of Example 9.27.

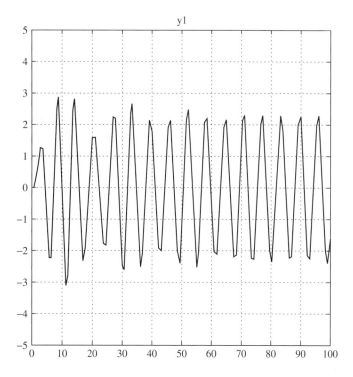

y1

**FIG. 9.37** The time-dependent response and the state plane generated by the simulation of Example 9.27.

Library Browser. In the Block Properties box the coefficients of the polynomial are entered as $[1\ 0\ 0.1\ 0]$ corresponding to a polynomial of the form $P(u) = u + 0.1u^3$. Thus since $y_1$ is the signal entering the block, the signal leaving the block has a value of $y_1 + 0.1y_1^3$. The time-dependent response and the state plane generated from running a simulation are shown in Figure 9.37.

## 9.8 FURTHER EXAMPLES

### Example 9.28

Consider the two-compartment model of the pharmokinetic system of Examples 4.30 and 6.35(b). (a) Derive a state-space model for the system. (b) Determine the state transition matrix for the system. (c) Use the state transition matrix to derive the time-dependent tissue and plasma concentrations.

**Solution**

(a) Defining state-space variables by $y_1 = C_p$ and $y_2 = C_t$, Equations (g) and (h) of Example 4.30 are written in state-space form as

$$\begin{bmatrix} \dot{y}_1 \\ \dot{y}_2 \end{bmatrix} = \begin{bmatrix} -(k_e + k_1) & \dfrac{V_t}{V_p}k_2 \\ \dfrac{V_p}{V_t}k_1 & -k_2 \end{bmatrix} \begin{bmatrix} y_1 \\ y_2 \end{bmatrix} + \begin{bmatrix} \dfrac{1}{V_p} \\ 0 \end{bmatrix} I(t) \tag{a}$$

Using the numerical values of Example 6.35(b) in Equation (a) leads to

$$
\begin{bmatrix} \dot{y}_1 \\ \dot{y}_2 \end{bmatrix} = \begin{bmatrix} -0.307 & 0.0133 \\ 2.02 & -0.133 \end{bmatrix} \begin{bmatrix} y_1 \\ y_2 \end{bmatrix} + \begin{bmatrix} 0.025 \\ 0 \end{bmatrix} I(t) \tag{b}
$$

(b) The state transition matrix is determined as $\Phi(t) = \mathcal{L}^{-1}\{(s\mathbf{I} - \mathbf{A})^{-1}\}$. To this end,

$$
(s\mathbf{I} - \mathbf{A})^{-1} = \begin{bmatrix} s + 0.307 & -0.0133 \\ -2.02 & s + 0.133 \end{bmatrix}^{-1}
$$

$$
= \begin{bmatrix} \dfrac{s + 0.133}{s^2 + 0.440s + 0.0140} & \dfrac{0.0133}{s^2 + 0.880s + 0.00140} \\[3mm] \dfrac{2.02}{s^2 + 0.440s + 0.0140} & \dfrac{s + 0.307}{s^2 + 0.440s + 0.0140} \end{bmatrix} \tag{c}
$$

Taking the inverse Laplace transform of Equation (c), noting that $s^2 + 0.4405 + 0.0140 = (s + 0.4054)(s + 0.0345)$ leads to

$$
\Phi(t) = \begin{bmatrix} 0.7345 & -0.03585 \\ -5.446 & 0.2655 \end{bmatrix} e^{-0.4054t} + \begin{bmatrix} 0.2655 & 0.03585 \\ 5.446 & 0.7345 \end{bmatrix} e^{-0.0345t} \tag{d}
$$

(c) The time-dependent infusion specified from Example 6.35 is $I(t) = 0.302[1 - u(t - 12)]$ mg/hr. The application of Equation (9.54) to determine the system response leads to

$$
\begin{bmatrix} y_1 \\ y_2 \end{bmatrix} = \int_0^t \left\{ \begin{bmatrix} 0.7345 & -0.03585 \\ -5.446 & 0.2655 \end{bmatrix} e^{-0.4054(t-\tau)} + \begin{bmatrix} 0.2655 & 0.03585 \\ 5.446 & 0.7345 \end{bmatrix} e^{-0.0345(t-\tau)} \right\}
$$

$$
\times \begin{bmatrix} 0.025 \\ 0 \end{bmatrix} 0.302[1 - u(\tau - 12)]d\tau \tag{e}
$$

Evaluation of the integrals in Equation (e) leads to

$$
y_1(t) = 0.0137[1 - e^{-0.4054t}]u(t) - 0.0137[1 - e^{-0.4054(t-12)}]u(t - 12)
$$

$$
+ 0.0581[1 - e^{-0.0345t}]u(t) - 0.0581[1 - e^{-0.0345(t-12)}]u(t - 12) \tag{f}
$$

$$
y_2(t) = -0.1014[1 - e^{-0.4054t}]u(t) - 0.1014[1 - e^{-0.4054(t-12)}]u(t - 12)
$$

$$
+ 1.9181[1 - e^{-0.0345t}]u(t) - 1.918[1 - e^{-0.0345(t-12)}]u(t - 12) \tag{g}
$$

**Example 9.29**

Consider the three-degree-of-freedom mechanical system of Figure 9.38. (a) Determine a state-space model for the system. (b) Determine the eigenvalues of the state matrix and discuss what they imply about the free response of the system. (c) Use MATLAB to determine the impulsive response for the system.

**Solution**

The differential equations governing the motion of the system are

$$
\begin{bmatrix} 20 & 0 & 0 \\ 0 & 30 & 0 \\ 0 & 0 & 20 \end{bmatrix} \begin{bmatrix} \ddot{x}_1 \\ \ddot{x}_2 \\ \ddot{x}_3 \end{bmatrix} + \begin{bmatrix} 3,000 & -3,000 & 0 \\ -3,000 & 7,500 & -4,500 \\ 0 & -4,500 & 4,500 \end{bmatrix} \begin{bmatrix} \dot{x}_1 \\ \dot{x}_2 \\ \dot{x}_3 \end{bmatrix}
$$

$$
+ \begin{bmatrix} 4 \times 10^5 & -4 \times 10^5 & 0 \\ -4 \times 10^5 & 1 \times 10^6 & -6 \times 10^5 \\ 0 & -6 \times 10^5 & 6 \times 10^5 \end{bmatrix} \begin{bmatrix} x_1 \\ x_2 \\ x_3 \end{bmatrix} = \begin{bmatrix} 0 \\ 0 \\ F(t) \end{bmatrix}
\tag{a}
$$

(a) State variables for this system are $y_1 = x_1$, $y_2 = x_2$, $y_3 = x_3$, $y_4 = \dot{x}_1$, $y_5 = \dot{x}_2$, $y_6 = \dot{x}_3$. The resulting state-space formulation is obtained using Equations (9.25) and (9.26) as

$$
\begin{bmatrix} \dot{y}_1 \\ \dot{y}_2 \\ \dot{y}_3 \\ \dot{y}_4 \\ \dot{y}_5 \\ \dot{y}_6 \end{bmatrix} = \begin{bmatrix} 0 & 0 & 0 & 1 & 0 & 0 \\ 0 & 0 & 0 & 0 & 1 & 0 \\ 0 & 0 & 0 & 0 & 0 & 1 \\ -2 \times 10^4 & 2 \times 10^4 & 0 & -150 & 150 & 0 \\ 1.33 \times 10^4 & -3.33 \times 10^4 & 2 \times 10^4 & 100 & -250 & 150 \\ 0 & 3 \times 10^4 & -3 \times 10^4 & 0 & 225 & -225 \end{bmatrix} \begin{bmatrix} y_1 \\ y_2 \\ y_3 \\ y_4 \\ y_5 \\ y_6 \end{bmatrix} + \begin{bmatrix} 0 \\ 0 \\ 0 \\ 0 \\ 0 \\ 0.05 \end{bmatrix} F(t)
\tag{b}
$$

(b) The MATLAB 'eig' command is used to determine the eigenvalues of the state matrix as $0, 0, -87.5 \pm j125.6, -225 \pm j96.6$. The eigenvalues of the state matrix are also the poles of the transfer function. Since 0 is a double pole of the transfer function, $s^2$ is a factor of the denominator of the transfer function. Thus, a partial fraction decomposition of the transfer function contains $1/s$ and $1/s^2$ terms. When inverted the free response contains terms proportional to 1 and $t$. The additional terms of the free response are proportional to $e^{-87.5t}\sin(125.6t)$ and $e^{-225t}\sin(96.6t)$.

(c) The MATLAB-generated plots for the impulsive response of this unrestrained system are shown in Figure 9.39. Since the system is unrestrained, the application of an impulse imparts an initial momentum to the system. The system itself satisfies conservation of linear momentum and thus continues in motion indefinitely. While viscous damping dissipates energy associated with the sinusoidal portions of the respone it does not dissipate energy from the systems rigid-body motion which occurs because the system is unrestrained.

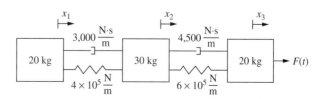

**FIG. 9.38** The system of Example 9.29.

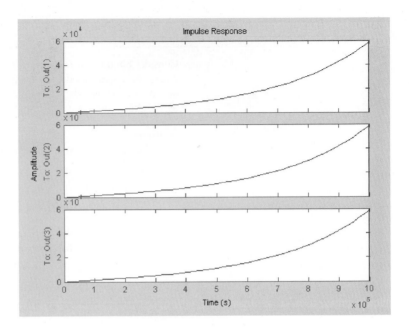

**FIG. 9.39** The impulsive response for the system of Example 9.29. Since the system is unrestrained an impulse sets the system in motion indefintieily.

---

## Example 9.30

(a) Consider the two-loop circuit of Example 7.20. (a) Determine a state-space model for the circuit where the system input is $v_1(t)$ and the system output is $v_2(t) = 10i_2(t)$. (b) Develop a SIMULINK model for this system. Run a simulation for $V_0 = 120$ **V** and $\omega = 10$ r/ s

### Solution

The integrodifferential equations for the circuit are Equations (a) and (b) of Example 7.20 repeated here

$$0.5\left(\frac{di_1}{dt} - \frac{di_2}{dt}\right) + 10i_1 = v_1(t) \tag{a}$$

$$-0.5\left(\frac{di_1}{dt} - \frac{di_2}{dt}\right) + 10i_2 + 50\int_0^t i_2 dt = 0 \tag{b}$$

Define state variables as

$$y_1 = i_1 - i_2 \tag{c}$$

$$y_2 = q_2 = \int_0^t i_2 dt \tag{d}$$

Equations (a) and (b) are written in terms of the defined state variables as

$$0.5\dot{y}_1 + 10(y_1 + \dot{y}_2) = v_1(t) \tag{e}$$

$$-0.5\dot{y}_1 + 10\dot{y}_2 + 50y_2 = 0 \tag{f}$$

Subtracting Equation (f) from Equation (e) leads to

$$\dot{y}_1 = v_1(t) - 10y_1 + 50y_2 \tag{g}$$

Adding Equation (e) to Equation (f) leads to

$$\dot{y}_2 = \frac{1}{20}[v_1(t) - 10y_1 - 50y_2] \tag{h}$$

Equations (g) and (h) are summarized in the form of Equation (9.12) as

$$\begin{bmatrix} \dot{y}_1 \\ \dot{y}_2 \end{bmatrix} = \begin{bmatrix} -10 & 50 \\ -0.5 & -2.5 \end{bmatrix} \begin{bmatrix} y_1 \\ y_2 \end{bmatrix} + \begin{bmatrix} 1 \\ 0.05 \end{bmatrix} v(t) \tag{i}$$

The substitution of Equations (c) and (g) into Equation (a) leads to

$$0.5[v_1(t) - 10y_1 - 50y_2] + 10i_1 = v(t)$$
$$i_1 = 0.05v_1(t) + 0.5y_1 - 2.5y_2 \tag{j}$$

Noting from Equation (c) that $i_2 = i_1 - y_1$, use of Equation (j) leads to

$$i_2 = 0.05v_1 - 0.5y_1 - 2.5y_2 \tag{k}$$

Equation (j) is used to determine the appropriate formulation of Equation (9.13) for this state-space model as

$$10i_2 = [-5 - 25] \begin{bmatrix} y_1 \\ y_2 \end{bmatrix} + 0.5v(t) \tag{l}$$

Equations (i) and (l) form the state-space model for the system.
(b) The SIMULINK model developed using the State Space block is illustrated in Figure 9.40 and the requested simulation in Figure 9.41.

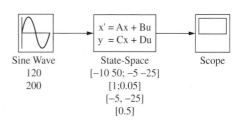

Sine Wave    State-Space    Scope
120       [−10 50; −5 −25]
200          [1;0.05]
             [−5, −25]
              [0.5]

**FIG. 9.40** The SIMULINK model using the State Space block for Example 9.30.

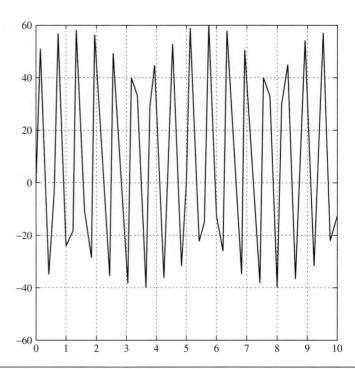

**FIG. 9.41** The simulation for Example 9.30.

---

**Example 9.31**

Develop a state-space model for a second-order system of natural frequency $\omega_n$ and damping ratio $\zeta$ when placed in a feedback control system with a PI controller in the feedforward part of the loop.

**Solution**

The closed-loop transfer function of a second-order system is given by Equation (8.67), repeated here

$$H(s) = \frac{K_p A s^2 + (K_i A + K_p B)s + K_i B}{s^3 + (2\zeta\omega_n + K_p A)s^2 + (\omega_n^2 + K_p B + K_i A)s + K_i B} \tag{a}$$

where the open-loop transfer function is $G(s) = (As + B)/(s^2 + 2\zeta\omega_n s + \omega_n^2)$.

The state matrix is obtained using Equation (9.64) as

$$\mathbf{A} = \begin{bmatrix} 0 & 1 & 0 \\ 0 & 0 & 1 \\ -K_i B & -(\omega_n^2 + K_p B + K_i A) & -(2\zeta\omega_n + K_p A) \end{bmatrix} \tag{b}$$

The input matrix is assumed of the form $\mathbf{B} = [\, b_1 \quad b_2 \quad b_3 \,]^T$. Its elements are obtained using the method illustrated in Example 9.20 to this end

$$|\mathbf{V}_3| = K_p A s^2 + (K_i A + K_p B)s + K_i B$$

$$\begin{vmatrix} s & -1 & b_1 \\ 0 & s & b_2 \\ K_i B & \omega_n^2 + K_p B + K_i A & b_3 \end{vmatrix} = K_p A s^2 + (K_i A + K_p B)s + K_i B$$

$$b_3 s^2 - K_i B b_2 - K_i B b_1 s - (\omega_n^2 + K_p B + K_i A)b_2 s = K_p A s^2 + (K_i A + K_p B)s + K_i B$$

(c)

Equating coefficients of like powers of $s$ in Equation (c) leads to

$$\mathbf{B} = \begin{bmatrix} \dfrac{\omega_n^2}{K_i B} \\ -1 \\ K_p A \end{bmatrix}$$

(d)

## 9.9 SUMMARY

### 9.9.1 Chapter Highlights

- State-space modeling is an alternative to dynamic system modeling with transfer functions.
- State-space formulations can be developed for linear and nonlinear systems.
- The state space model of a linear system is defined using four matrices: the state matrix, the input matrix, the output matrix, and the transmission matrix.
- The free-response using the state-space formulation can be determined using either the Laplace transform method or matrix methods that involve determining the eigenvalues of the state matrix.
- The free response can be written in terms of the state transition matrix.
- The convolution integral is used to determine the forced response.
- Numerical methods such as Euler's method or Runge-Kutta methods are often employed to determine the forced response.
- The state-space formulation may be used to determine a system's transfer function, and the system's transfer function, may be used to determine a state-space model.
- The state-space method can be used for MATLAB modeling of dynamic systems. Given the state-space formulation MATLAB can be used to determine the system's impulsive response, step response, and response due to any system input.

- SIMULINK may be used to develop state-space models. A SIMULINK model can be developed from a block diagram of the state-space model, or the State Space block can be used to define the state space.
- SIMULINK is well suited for nonlinear problems formulated in the state space.

### 9.9.2 Important Equations

- Definition of the state space

$$\dot{\mathbf{y}} = \mathbf{A}\mathbf{y} + \mathbf{B}\mathbf{u} \tag{9.12}$$

$$\mathbf{x} = \mathbf{C}\mathbf{y} + \mathbf{D}\mathbf{u} \tag{9.13}$$

- State transition matrix

$$\boldsymbol{\Phi}(t) = \mathcal{L}^{-1}\big\{(s\mathbf{I} - \mathbf{A})^{-1}\big\} \tag{9.44}$$

- Free response

$$\mathbf{y} = \boldsymbol{\Phi}(t)\mathbf{y}_0 \tag{9.45}$$

- Convolution integral for the forced response

$$\mathbf{y}(t) = \boldsymbol{\Phi}(t)\mathbf{y}_0 + \int_0^t \boldsymbol{\Phi}(t - \tau)\mathbf{B}\mathbf{u}(\tau)d\tau \tag{9.54}$$

## PROBLEMS

Problems 9.1–9.9 require the determination of a state-space model from the governing differential equations. For each problem specify the state matrix, the input matrix, the output matrix, and the transmission matrix.

**9.1** Determine a state-space model for a system governed by

$$\dddot{x} + 3\ddot{x} + 16\dot{x} + 25x = F(t) \tag{a}$$

**9.2** Determine a state-space model for a system governed by

$$0.2\frac{dy_1}{dt} + 2y_1 - y_2 = F_1(t) \tag{a}$$

$$0.25\frac{dy_2}{dt} - y_1 + 3y_2 = F_2(t) \tag{b}$$

**9.3** Determine a state-space model for a system governed by

$$2\frac{dy_1}{dt} + 3\frac{dy_2}{dt} + 4y_1 - 2y_2 = F(t) \tag{a}$$

$$3\frac{dy_1}{dt} + 2\frac{dy_2}{dt} - 2y_1 + 4y_2 = 0 \tag{b}$$

**9.4** Derive a state-space model for the electrical circuit of Example 3.27, that includes the effect of mutual inductance.

**9.5** Derive a state-space model for the operational amplifier circuit of Example 3.19 assuming the input sources are independent.

**9.6** Derive a state-space model for a three-degree-of-freedom mechanical system whose governing differential equations are

$$\begin{bmatrix} 2 & 0 & 0 \\ 0 & 4 & 0 \\ 0 & 0 & 3 \end{bmatrix}\begin{bmatrix} \ddot{x}_1 \\ \ddot{x}_2 \\ \ddot{x}_3 \end{bmatrix} + \begin{bmatrix} 2 & -2 & 0 \\ -2 & 2 & 0 \\ 0 & 0 & 0 \end{bmatrix}\begin{bmatrix} \dot{x}_1 \\ \dot{x}_2 \\ \dot{x}_3 \end{bmatrix}$$

$$+ \begin{bmatrix} 3 & -2 & 0 \\ -2 & 5 & -3 \\ 0 & -3 & 3 \end{bmatrix}\begin{bmatrix} x_1 \\ x_2 \\ x_3 \end{bmatrix} = \begin{bmatrix} 0 \\ 0 \\ F(t) \end{bmatrix} \tag{a}$$

**9.7** Derive a state-space model for the electric circuit of Example 3.14 if $L_1 = 0.2$ H, $L_2 = 0.5$ H, $R_1 = 1.0$ kΩ, $R_2 = 1.5$ kΩ, $R_3 = 2.0$ Ω, and $C = 0.25$ μF.

**9.8** Determine a state-space model for the system of Example 2.28.

**9.9** Determine a state-space model for the system of Figure P9.9.

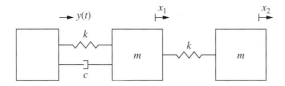

**FIG. P9.9**

**9.10** Determine the state transition matrix for the system of Problem 9.1.

**9.11** Determine the state transition matrix for the system of Problem 9.2.

**9.12** Determine the state transition matrix for the system of Problem 9.5.

**9.13** Determine the state-transition matrix for the system of Problem 9.9.

**9.14** Determine the free response of the system of Problem 9.1 if $x(0) = 1$, $\dot{x}(0) = 0$, and $\ddot{x}(0) = 0$.

**9.15** Determine the free response of the system of Problem 9.2 if $y_1(0) = 0$ and $y_2(0) = 0.5$.

**9.16** The capacitor in the circuit of Figure P9.16 has an initial charge of 1.2 mC. The switch is closed at $t = 0$, allowing the capacitor to discharge. (a) Determine a state-space model for the system where the output is the charge on the capacitor. (b) Use the state-space model to determine the time-dependent charge in the capacitor.

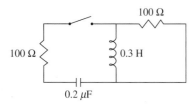

**FIG. P9.16**

**9.17** Two identical railroad cars, as illustrated in Figure P9.17, are being connected by moving one car with a velocity of 8 m/s toward a second car that is stationary. When connected, the cars have an elastic coupling of stiffness

$5 \times 10^7$ N/m. Use a state-space model to determine the displacements of each railroad car after coupling.

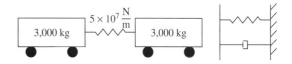

**FIG. P9.17**

**9.18** Use a state-space model to determine the time-dependent displacement of the railroad cars of Problem 9.17 if after the lead car travels 10 m it engages a bumper modeled as a spring of stiffness $2 \times 10^6$ N/m in parallel with a viscous damper of damping coefficient $1.2 \times 10^4$ N·s/m.

**9.19** Repeat Problem 9.18 assuming that the bumper is non-linear and exerts a force of

$$F = 2 \times 10^6 x_1 + 1.2 \times 10^4 \dot{x}_1^{1.6} \qquad (a)$$

on the car, which engages the bumper. (a) Write an M-file that uses MATLAB's 'ode45' to develop a numerical solution for the state-space equations. (b) Use SIMULINK to determine the response.

**9.20** Use state-space methods to determine the response of the system of Problem 9.1 if $F(t) = 0.5 \sin(20t)$.

**9.21** Use state-space methods to determine the response of the system of Problem 9.1 if $F(t) = 0.5[u(t) - u(t - 3)]$.

**9.22** Use state-space methods to determine the response of the system of Problem 9.2 if $F_1(t) = 0.2e^{-0.2t}$ and $F_2(t) = 0.5e^{-0.2t}$.

**9.23** Use state-space methods to determine the response of the system of Problem 9.5 if $v_1(t) = 120 \sin(800t)$ V.

**9.24** Use state-space methods to determine the liquid-level perturbations in the three-tank system of Figure P9.24 when the flow rate into the first tank is suddenly increased by 0.2 m³/s.

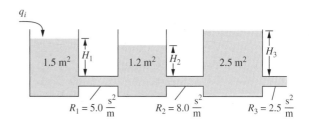

**FIG. P9.24**

**9.25** Use state-space methods to determine the response of a one-degree-of-freedom system of mass 10 kg, stiffness $4 \times 10^5$ N/m, and damping coefficient 200 N·s/m when subject to a force $F(t) = 100 \sin(200t)$ N.

**9.26** Use state-space methods to determine the response of the system of Problem 9.25 if the mass varies according to $m(t) = 10e^{-0.1t}$ kg. Write an M-file that uses 'ode45' to numerically solve the resulting equations.

**9.27** Use state-space methods to determine the response of the system of Problem 6.30(b). Use MATLAB's 'ode45' to solve the resulting equations.

**9.28** Write a MATLAB M-file that determines the response of the nonlinear pendulum equation $\ddot{\theta} + \omega_n^2 \sin\theta = 0$ for $\omega_n = 10$ r/s with initial conditions of $\theta(0) = \theta_0$ and $\dot{\theta}(0) = 0$. Run the program for several values of $\theta_0$ and discuss the dependence of the period on this initial condition.

**9.29** Write a MATALAB M-file that uses state-space methods to determine the response of the CSTR system of Example 4.17 (a) using the linearized model of part (c) of that example, and (b) using the nonlinear model of part (a).

**9.30** Determine the transfer function $G(s) = X(s)/F(s)$ for the system of Problem 9.1 from its state-space formulation.

**9.31** Determine the transfer functions $G_{1,1}(s) = Y_1(s)/\bar{F}_1(s)$ and $G_{2,1}(s) = Y_2(s)/F_1(s)$ for the system of Example 9.3 from its state-space formulation.

**9.32** The transfer function for a system is

$$G(s) = \frac{2s + 3}{s^2 + 6s + 25} \qquad \text{(a)}$$

Determine a state-space formulation for it.

**9.33** The transfer function for a system is

$$G(s) = \frac{3s^2 + 2s + 5}{s^4 + 3s^3 + 10s^2 + 20s + 50} \qquad \text{(b)}$$

Determine a state-space formulation for it.

**9.34** Determine a state-space formulation for the feedback control system of Figure P9.34. Develop a SIMULINK model for the system. Run the model with a step input.

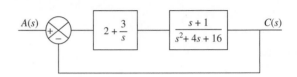

**FIG. P9.34**

**9.35** Determine a state-space formulation for the feedback control system of Figure P9.35.

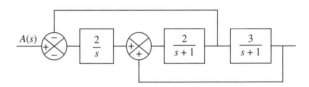

**FIG. P9.35**

**9.36** Develop a SIMULINK model for system of Figure P9.35 from its block diagram.

**9.37** Develop a SIMULINK model for the system of Figure P9.24 from its state-space model.

**9.38** Use state-space methods to determine the response of the heat exchanger in Example 6.41.

# Appendix A

## Complex Algebra

A **complex number** is of the form

$$c = a + jb \tag{A.1}$$

where

$$j = \sqrt{-1} \tag{A.2}$$

and $a$ and $b$ are real numbers. The number $a$ is called the **real part** of $c$, $a = \text{Re}[c]$, and $b$ is called the **imaginary part** of $c$, $b = \text{Im}[c]$. A number whose imaginary part is zero is a real number. A number whose real part is zero and whose imaginary part is not zero is called an imaginary number.

Powers of $j$ are calculated using Equation (A.2)

$$j^2 = \left(\sqrt{-1}\right)\left(\sqrt{-1}\right) = -1 \tag{A.3}$$

$$j^3 = j^2\left(\sqrt{-1}\right) = -j \tag{A.4}$$

$$j^4 = jj^3 = j(-j) = -j^2 = 1 \tag{A.5}$$

The sum of two complex numbers is also a complex number. The real part of the sum is the sum of the real parts of the numbers. The imaginary part of the sum is the sum of the imaginary parts.

$$(a_1 + jb_1) + (a_2 + jb_2) = (a_1 + a_2) + j(b_1 + b_2) \tag{A.6}$$

The distributive property is used to obtain the product of two complex numbers

$$\begin{aligned}(a_1 + jb_1)(a_2 + jb_2) &= a_1a_2 + ja_1b_2 + jb_1a_2 + j^2b_1b_2 \\ &= (a_1a_2 - b_1b_2) + j(a_1b_2 + b_1a_2)\end{aligned} \tag{A.7}$$

The **complex conjugate** of a number is a complex number such that the real parts of both numbers are the same and the imaginary parts are negatives of one another. The complex conjugate of $c$ is denoted by $\bar{c}$ and is given by

$$\bar{c} = a - jb \tag{A.8}$$

The use of Equation (A.7) shows that the product of a number and its complex conjugate is a real number given by

$$(c)(\bar{c}) = a^2 + b^2 \tag{A.9}$$

The quotient of two complex numbers is a complex number and can be written in the form of Equation (A.1) by multiplying both the numerator and denominator of the quotient by the complex conjugate of the denominator.

---

## Example A.1

Write the complex number

$$c = \frac{3 + j2}{3 - j4} \tag{a}$$

in the form $c = a + jb$.

### Solution

Multiplying the numerator and denominator of Equation (a) by $3 + j4$ leads to

$$c = \frac{(3 + j2)(3 + j4)}{(3 - j4)(3 + 4j)} \tag{b}$$

Equation (A.7) is used to evaluate the numerator of Equation (b), while Equation (A.9) is used for the denominator leading to

$$c = \frac{[(3)(3) - (2)(4)] + j[(3)(4) + (2)(3)]}{(3)^2 + (4)^2}$$

$$= \frac{1}{25} + j\frac{18}{25} \tag{c}$$

---

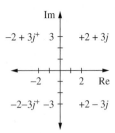

**FIG. A.1** Illustration of complex numbers as points in the complex plane.

A complex number can be represented geometrically as a point in the complex plane, as illustrated in Figure A.1. The horizontal axis for the complex plane is the real axis and the vertical axis is the imaginary axis.

The polar form of a complex number, illustrated in Figure A.2, is a representation in terms of a magnitude and a phase angle. The **magnitude** is the distance of the point in the complex plane from the origin,

$$|c| = \sqrt{a^2 + b^2} \tag{A.10}$$

From Equations (A.9) and (A.10) it is seen that

$$|c| = (c\bar{c})^{1/2} \tag{A.11}$$

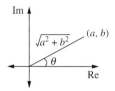

**FIG. A.2** The polar representation of a complex number.

The phase angle $\theta$ is the angle, measured in radians, made by a line drawn from the origin to the point with the real axis. The phase angle is measured positive counterclockwise. The polar form of a complex number is

$$c = |c|e^{j\theta} \tag{A.12}$$

**Euler's identity** states that

$$e^{j\theta} = \cos(\theta) + j\sin(\theta) \tag{A.13}$$

Use of Equation (A.13) in Equation (A.12) gives

$$c = |c|\cos(\theta) + j|c|\sin(\theta) \tag{A.14}$$

Equation (A.10), in conjunction with the following equation, obtained from Equation (A.14) is used to convert a complex number to polar form

$$\theta = \tan^{-1}\left\{\frac{\text{Im}[c]}{\text{Re}[c]}\right\} \tag{A.15}$$

When applying Equation (A.15) it is important to remember that $0 \leq \theta < 2\pi$. Thus the value of the inverse tangent must be assigned depending on the quadrant in which $c$ exists.

---

**Example A.2**

Convert the complex number $c = -3 + j4$ into polar form.

**Solution**

The magnitude of $c$, obtained using Equation (A.10) is

$$|c| = \sqrt{(-3)^2 + (4)^2} = 5 \tag{a}$$

The phase angle for $c$ is determined using Equation (A.15)

$$\theta = \tan^{-1}\left(\frac{4}{-3}\right) \tag{b}$$

The inverse tangent of Equation (b) must be evaluated corresponding to an angle in the second quadrant. The appropriate value is

$$\theta = 2.21 \text{ r} \tag{c}$$

Using the numerical values from Equations (a) and (c) in Equation (A.12) leads to

$$c = 5e^{j2.21} \tag{d}$$

# Appendix B

## Matrix Algebra

## B.1 DEFINITIONS

A **matrix** is a collection of elements arranged in rows and columns. A matrix of $m$ rows and $n$ columns is said to be of size $m \times n$. The number of rows equals the number of columns in a square matrix. Throughout this text the name of a matrix is written using bold type and is identified by its elements arranged appropriately in rows and columns enclosed by square brackets. An example of a $3 \times 3$ matrix is

$$\mathbf{A} = \begin{bmatrix} 1 & 2 & -1 \\ 3 & 2 & 3 \\ 0 & 2 & -2 \end{bmatrix} \tag{B.1}$$

An element of a matrix is identified according to where it resides in the matrix. The element in the $i$th row and $j$th column of matrix $\mathbf{A}$ is written as $a_{i,j}$. For example for the matrix of Equation (B.1), $a_{3,2} = 2$.

Useful definitions regarding matrices are as follows:

- A **column vector** is a matrix with only one column, whereas a **row vector** is a matrix with only one row.
- The **diagonal elements** of a square matrix $\mathbf{A}$ are those elements $a_{i,j}$ such that $i = j$.
- The **off-diagonal elements** of a square matrix are those elements $a_{i,j}$ such that $i \neq j$.
- A **diagonal** matrix is a square matrix whose off-diagonal terms are all zero.
- The $n \times n$ **identity matrix I** is a diagonal matrix such that $a_{i,i} = 1$ for $i = 1, 2, \ldots, n$.
- The **transpose** of a matrix $\mathbf{A}$, written $\mathbf{A}^T$ is obtained by interchanging the rows and columns of $\mathbf{A}$. Thus if $\mathbf{B} = \mathbf{A}^T$ then $b_{i,j} = a_{j,i}$. The transpose of a column vector is a row vector and vice versa.
- A square matrix is **symmetric** if $\mathbf{A}^T = \mathbf{A}$.

## B.2  Matrix Arithmetic

The **sum** of two matrices $\mathbf{A}$ and $\mathbf{B}$, both of size $m \times n$, is the $m \times n$ matrix $\mathbf{C} = \mathbf{A} + \mathbf{B}$ whose elements are calculated as

$$c_{i,j} = a_{i,j} + b_{i,j} \qquad i = 1, 2, \ldots, m \quad j = 1, 2, \ldots, n \qquad \text{(B.2)}$$

**Matrix multiplication** may be defined for two matrices, written as $\mathbf{C} = \mathbf{AB}$, if the number of columns of $\mathbf{A}$ is equal to the number of rows of $\mathbf{B}$. If $\mathbf{A}$ is of size $m \times n$ and $\mathbf{B}$ is of size $n \times p$ then $\mathbf{C} = \mathbf{AB}$ is a $m \times k$ matrix such that

$$c_{i,j} = \sum_{k=1}^{n} a_{i,k} \, b_{k,j} \qquad i = 1, 2, \ldots, m \quad j = 1, 2, \ldots, K_p \qquad \text{(B.3)}$$

Matrix multiplication is not commutative: in general $\mathbf{AB} \neq \mathbf{BA}$.

## B.3  Determinants

A determinant of a square matrix is a scalar quantity with great significance and is often used in computations involving matrices. The formal definition of the determinant is beyond the scope of this study, but it is a linear combination of permutations of elements of the matrix. Each permutation in the evaluation of the determinant of an $n \times n$ matrix is a product of $n$ terms, with one element from each column and one element from each row. One permutation is a product of the diagonal elements.

The determinant of a $2 \times 2$ matrix is simply the product of the diagonal elements minus the product of the off-diagonal elements. If

$$\mathbf{A} = \begin{bmatrix} a_{1,1} & a_{1,2} \\ a_{2,1} & a_{2,2} \end{bmatrix} \quad \text{then} \quad \det(\mathbf{A}) = |\mathbf{A}| = a_{1,1}a_{2,2} - a_{1,2}\,a_{2,1} \qquad \text{(B.4)}$$

The determinant of a larger matrix can be computed using a row or column expansion. For example a row expansion of the determinant of an $n \times n$ matrix $\mathbf{A}$ using the $i$th row is

$$\det(\mathbf{A}) = \sum_{j=1}^{n} (-1)^{i+j} a_{i,j} \det\left(\mathbf{M}_{i,j}\right) \qquad \text{(B.5)}$$

where $\mathbf{M}_{i,j}$ is the $(n-1) \times (n-1)$ matrix obtained from $\mathbf{A}$ by deleting its $i$th row and $j$th column. Thus the determinant of an $n \times n$ matrix can be written as a sum of $n$ determinants of $(n-1) \times (n-1)$ matrices. Row or column expansions can be employed to evaluate the determinant of each of the $(n-1) \times (n-1)$ matrices as a sum of $n-1$ determinants of $(n-2) \times (n-2)$ matrices. This process can be continued until $2 \times 2$ matrices are obtained whose determinants are evaluated using Equation (B.4).

**Example B.1**

Evaluate the determinant of the $3 \times 3$ matrix $\mathbf{A}$ of Equation (B.1) by row expansion using the second row.

**Solution**

Evaluation of the determinant of $\mathbf{A}$ using Equation (B.5) with $i = 2$ leads to

$$\det(\mathbf{A}) = -3 \begin{vmatrix} 2 & -1 \\ 2 & -2 \end{vmatrix} + 2 \begin{vmatrix} 1 & -1 \\ 0 & -2 \end{vmatrix} - 3 \begin{vmatrix} 1 & 2 \\ 0 & 2 \end{vmatrix} \tag{a}$$

The $2 \times 2$ determinants are evaluated using Equation (B.4) leading to

$$\det(\mathbf{A}) = -3[(2)(-2) - (-1)(2)] + 2[(1)(-2) - (-1)(0)] - 3[(1)(2) - (2)(0)] = -4 \tag{b}$$

## B.4  Matrix Inverse

The inverse of a square matrix $\mathbf{A}$, written $\mathbf{A}^{-1}$, is the matrix such that

$$\mathbf{A}\mathbf{A}^{-1} = \mathbf{A}^{-1}\mathbf{A} = \mathbf{I} \tag{B.6}$$

$\mathbf{A}^{-1}$ exists if $\det(\mathbf{A}) \neq 0$. Such a matrix is said to be nonsingular.
    If $\mathbf{B} = \mathbf{A}^{-1}$ then

$$b_{i,j} = (-1)^{i+j} \frac{\det(\mathbf{M}_{j,i})}{\det(\mathbf{A})} \tag{B.7}$$

The inverse of a diagonal matrix is a diagonal matrix with the reciprocals of the diagonal elements along the diagonal.

**Example B.2**

Determine $\mathbf{A}^{-1}$ for the matrix defined by Equation (B.1).

**Solution**

From Example B.1 $\det(\mathbf{A}) = -4$. The application of Equation (B.7) leads to

$$\mathbf{A}^{-1} = -\frac{1}{4} \begin{bmatrix} \begin{vmatrix} 2 & 3 \\ 2 & -2 \end{vmatrix} & -\begin{vmatrix} 2 & -1 \\ 2 & -2 \end{vmatrix} & \begin{vmatrix} 2 & -1 \\ 2 & 3 \end{vmatrix} \\ -\begin{vmatrix} 3 & 3 \\ 0 & -2 \end{vmatrix} & \begin{vmatrix} 1 & -1 \\ 0 & -2 \end{vmatrix} & -\begin{vmatrix} 1 & -1 \\ 3 & 3 \end{vmatrix} \\ \begin{vmatrix} 3 & 2 \\ 0 & 2 \end{vmatrix} & -\begin{vmatrix} 1 & 2 \\ 0 & 2 \end{vmatrix} & \begin{vmatrix} 1 & 2 \\ 3 & 2 \end{vmatrix} \end{bmatrix}$$

$$= -\frac{1}{4} \begin{bmatrix} -10 & 2 & 8 \\ 6 & -2 & -6 \\ -6 & -2 & -4 \end{bmatrix} = -\begin{bmatrix} \frac{5}{2} & -\frac{1}{2} & -2 \\ \frac{3}{2} & \frac{1}{2} & \frac{3}{2} \\ -\frac{3}{2} & \frac{1}{2} & 1 \end{bmatrix} \tag{a}$$

## B.5 Systems of Equations

A system of $n$ simultaneous equations in terms of $n$ unknowns, $x_1$, $x_2, \cdots, x_n$ is of the form

$$a_{1,1}x_1 + a_{1,2}x_2 + \cdots a_{1,n}x_n = b_1$$
$$a_{2,1}x_1 + a_{2,2}x_2 + \cdots + a_{2,n}x_n = b_2$$
$$\vdots$$
$$a_{n,1}x_1 + a_{n,2}x_2 + \cdots a_{n,n}x_n = b_n \tag{B.8}$$

Equation (B.7) can be summarized in matrix form as

$$\mathbf{Ax} = \mathbf{b} \tag{B.9}$$

where

$$\mathbf{A} = \begin{bmatrix} a_{1,1} & a_{2,1} & \cdots & a_{n,1} \\ a_{2,1} & a_{2,2} & \cdots & a_{2,n} \\ \vdots & \vdots & \cdots & \vdots \\ a_{n,1} & a_{n,2} & \cdots & a_{n,n} \end{bmatrix} \quad \mathbf{x} = \begin{bmatrix} x_1 \\ x_2 \\ \vdots \\ x_n \end{bmatrix} \quad \mathbf{b} = \begin{bmatrix} b_1 \\ b_2 \\ \vdots \\ b_n \end{bmatrix} \tag{B.10}$$

A solution to Equation (B.7) is obtained by premultiplying both sides of Equation (B.9) by $\mathbf{A}^{-1}$, which leads to

$$\mathbf{x} = \mathbf{A}^{-1}\mathbf{b} \tag{B.11}$$

Equation (B.11) is the solution of Equation (B.8) if $\mathbf{A}^{-1}$ exists, that is, if $\mathbf{A}$ is nonsingular. If $\mathbf{A}$ is a singular matrix $[\det(\mathbf{A}) = 0]$ then a solution of Equation (B.8) exists only for certain forms of $\mathbf{b}$. However, when a solution does exist it is not unique.

**Example B.3**

Use the matrix inverse method to determine the solution of the system of equations

$$x_1 + 2x_2 - x_3 = 4$$
$$3x_1 + 2x_2 + 3x_3 = -2 \tag{a}$$
$$2x_2 - 2x_3 = 5$$

**Solution**

The system of equations represented by Equation (a) has a matrix formulation of the form of Equation (B.9) with the matrix $\mathbf{A}$ of Equation (B.1) and $\mathbf{b} = \begin{bmatrix} 4 & -2 & 5 \end{bmatrix}^T$. The inverse of $\mathbf{A}$ is given in Equation (a) of Example B.2. The application of Equation (B.11) leads to

$$\begin{bmatrix} x_1 \\ x_2 \\ x_3 \end{bmatrix} = \begin{bmatrix} \frac{5}{2} & -\frac{1}{2} & -2 \\ -\frac{3}{2} & \frac{1}{2} & \frac{3}{2} \\ -\frac{3}{2} & \frac{1}{2} & 1 \end{bmatrix} \begin{bmatrix} 4 \\ -2 \\ 5 \end{bmatrix} = \begin{bmatrix} 1 \\ \frac{1}{2} \\ -2 \end{bmatrix} \tag{b}$$

## B.6 Cramer's Rule

Cramer's rule provides an alternate method for the solution of a set of simultaneous equations formulated in the matrix form of Equation (B.9). Cramer's rule is used to determine each element of the solution vector individually, rather than simultaneously.

Consider a set of simultaneous equations formulated as in Equation (B.9). Let $\mathbf{V}_k$ be the matrix formed by replacing the $k$th column of $\mathbf{A}$ with the vector $\mathbf{b}$. That is

$$(V_k)_{i,j} = \begin{cases} a_{i,j} & j \neq k \\ b_i & j = k \end{cases} \tag{B.12}$$

Cramer's rule is used to determine $x_k$ as

$$x_k = \frac{\det(\mathbf{V}_k)}{\det(\mathbf{A})} \qquad k = 1, 2, \ldots, n \tag{B.13}$$

---

**Example B.4**

Use Cramer's rule to determine $x_1$, an element of the solution vector of Example B.3.

**Solution**

Recalling that $\det(\mathbf{A}) = -4$, Cramer's rule, Equation (B.13), is applied giving

$$x_1 = \frac{\det(\mathbf{V}_1)}{\det(\mathbf{A})}$$

$$= \frac{\begin{vmatrix} 4 & 2 & -1 \\ -2 & 2 & 3 \\ 5 & 2 & -2 \end{vmatrix}}{-4} \tag{a}$$

The determinant in the numerator of Equation (a) is evaluated using a first-row expansion and Equations (B.3) and (B.4)

$$x_1 = -\frac{1}{4}\left[ 4\begin{vmatrix} 2 & 3 \\ 2 & -2 \end{vmatrix} - 2\begin{vmatrix} -2 & 3 \\ 5 & -2 \end{vmatrix} + (-1)\begin{vmatrix} -2 & 2 \\ 5 & 2 \end{vmatrix} \right] \tag{b}$$

$$= -\frac{1}{4}[4(-10) - 2(-11) - (-14)] = 1$$

---

## B.7 Eigenvalues and Eigenvectors

Consider the system of $n$ equations represented in matrix form as

$$\mathbf{A}\mathbf{x} = \lambda \mathbf{x} \tag{B.14}$$

where $\lambda$ is a parameter. Equation (B.14) can be rewritten as

$$(\mathbf{A} - \lambda \mathbf{I}) = \mathbf{0} \tag{B.15}$$

The application of Cramer's rule to determine any element of the solution vector of Equation (B.15) leads to

$$x_k = \frac{\det(\mathbf{V}_k)}{\det(\mathbf{A} - \lambda\mathbf{I})} \qquad k = 1, 2, \ldots, n \tag{B.16}$$

Since $\mathbf{V}_k$ is defined as in Equation (B.12), its $k$th column is the vector on the right-hand side of Equation (B.15), the zero vector. However, it has been noted that the determinant of a matrix with a column of zeroes is itself zero. Thus Equation (B.16) becomes

$$x_k = \frac{0}{\det(\mathbf{A} - \lambda\mathbf{I})} \qquad k = 1, 2, \ldots, n \tag{B.17}$$

It is clear from Equation (B.14) that $\mathbf{x} = \mathbf{0}$, the **trivial solution**, is a solution of the set of equations. Equation (B.17) shows that the trivial solution is the only solution unless

$$\det(\mathbf{A} - \lambda\mathbf{I}) = 0 \tag{B.18}$$

Equation (B.18) is satisfied only for certain values of $\lambda$, called the eigenvalues of $\mathbf{A}$.

The **eigenvalues** of a square matrix $\mathbf{A}$ are the values of the parameter $\lambda$ such that $\mathbf{A}\mathbf{x} = \lambda\mathbf{x}$ has nontrivial solutions. The **eigenvectors** of $\mathbf{A}$ are the corresponding nontrivial solutions. An eigenvector $\mathbf{y}$ corresponding to an eigenvalue $\lambda$ $\mathbf{A}\mathbf{y} = \lambda\mathbf{y})$ is not unique because any nonzero multiple of the eigenvector $c\mathbf{y}$ with $c \neq 0$ is also an eigenvector corresponding to the eigenvalue $\lambda$, $\mathbf{A}(c\mathbf{y}) = \lambda(c\mathbf{y})$.

Equation (B.18) provides a method for calculating the eigenvalues of $\mathbf{A}$. The elements of the matrix $\mathbf{A} - \lambda\mathbf{I}$ are the same as the elements of $\mathbf{A}$ except for the diagonal elements, which are $a_{i,i} - \lambda$. Recalling that one permutation in the evaluation of the determinant is the product of the diagonal elements, it becomes clear that the evaluation of $\det(\mathbf{A} - \lambda\mathbf{I})$ is an $n$th-order polynomial in $\lambda$. Since the eigenvalues are determined using Equation (B.18), they are thus the roots of the polynomial. Hence an $n \times n$ matrix has $n$ eigenvalues $\lambda_1, \lambda_2, \ldots, \lambda_n$. If all elements of $\mathbf{A}$ are real complex eigenvalues occur in complex conjugate pairs. The eigenvectors are then calculated using Equation (B.15).

## Example B.5

Determine the eigenvalues and eigenvectors of the matrix $\mathbf{A}$ defined in Equation (B.1).

### Solution

The eigenvalues of $\mathbf{A}$ are determined through the application of Equation (B.18) leading to

$$\begin{vmatrix} 1 - \lambda & 2 & -1 \\ 3 & 2 - \lambda & 3 \\ 0 & 2 & -2 - \lambda \end{vmatrix} = 0 \tag{a}$$

The evaluation of the determinant in Equation (a) by column expansion using the first column leads to

$$(1-\lambda)\begin{vmatrix} 2-\lambda & 3 \\ 2 & -2-\lambda \end{vmatrix} - 3\begin{vmatrix} 2 & -1 \\ 2 & -2-\lambda \end{vmatrix} + 0\begin{vmatrix} 2 & -1 \\ 2-\lambda & 3 \end{vmatrix} = 0$$

$$(1-\lambda)[(2-\lambda)(-2-\lambda) - (3)(2)] - 3[(2)(-2-\lambda) - (-1)(2)] = 0$$

$$-\lambda^3 + \lambda^2 + 16\lambda - 4 = 0$$

(b)

The solutions of Equation (b) are $\lambda_1 = -3.664, \lambda_2 = 0.2471, \lambda_3 = 4.417$. The eigenvectors are obtained by solving

$$\begin{bmatrix} 1-\lambda & 2 & -1 \\ 3 & 2-\lambda & 3 \\ 0 & 2 & -2-\lambda \end{bmatrix}\begin{bmatrix} y_1 \\ y_2 \\ y_3 \end{bmatrix} = \begin{bmatrix} 0 \\ 0 \\ 0 \end{bmatrix}$$

(c)

with values of $\lambda$ corresponding to each of the eigenvalues. For $\lambda = -3.664$ this leads to

$$\begin{bmatrix} 4.664 & 2 & -1 \\ 3 & 5.664 & 3 \\ 0 & 2 & 1.664 \end{bmatrix}\begin{bmatrix} y_1 \\ y_2 \\ y_3 \end{bmatrix} = \begin{bmatrix} 0 \\ 0 \\ 0 \end{bmatrix}$$

(d)

The equations represented by the second and third rows of Equation (d) are used to obtain $y_1 = 0.571y_3$ and $y_2 = -0.832y_3$. The first equation does not yield additional information and $y_3$ may be taken as any nonzero constant. Thus the eigenvector corresponding to the eigenvalue $\lambda_1 = -3.664$ is $\mathbf{y}_1 = c_1[0.571 \quad -0.832 \quad 1]^T$. In a similar fashion the eigenvectors corresponding to the remaining eigenvalues are $\mathbf{y}_2 = c_2[-1.66 \quad 1.12 \quad 1]^T$ and $\mathbf{y}_3 = c_3[3.21 \quad 3.59 \quad 1]^T$.

# Appendix C
## MATLAB

MATLAB is a special computer programming language designed for engineering and scientific computation. The name "MATLAB" is contracted from MATRIX LABORATORY. In this context MATLAB considers all data as an array or a matrix. Scalars are considered a matrix with only one row and one column.

MATLAB can be used as a programming language or as an interactive computational worksheet. In either case the MATLAB work environment must be open. Executable statements are entered into the work environment. The environment allocates memory for variables and follows instructions provided by executable statements. Programs developed to be executed using MATLAB are stored in M-files, which have an extension .m. These files are executed by simply typing the name of the file in the MATLAB work environment. Any data created and assigned a variable name is stored in the work environment and may be recalled and used during a work session.

This appendix provides a summary of MATLAB commands and features used in the text. It is not exhaustive, nor does it present concepts of programming. Proficiency in programming is not required to use MATLAB. The uses of the commands presented are functional but may not be the most efficient commands to achieve the objective. MATLAB has a useful HELP feature, which the reader is encouraged to use.

MATLAB has many toolboxes that provide programs and allow special executable commands specific to the toolbox. The Symbolic Math Toolbox is included with the student edition of MATLAB and SIMULINK, and it is available as an add-on for the professional version. The Control System Toolbox is available as an add-on for both the student version and the professional edition, but at a reduced price for the student version. The text also uses SIMULINK, a MATLAB-based modeling and simulation package.

The tables of commands and features presented in this appendix are by no means comprehensive. They are limited to the commands and features used in this text.

**TABLE C.1 MATLAB OPERATORS**

| Operator | Name | Sample Scalar Calculation | Sample Matrix Calculation |
|---|---|---|---|
| + | Addition | $3 + 4 = 7$ | [1 2;3 4] + [5 6;7 8] = [6 8;10 12] |
| − | Subtraction | $3 - 4 = -1$ | [1 2;3 4] −[5 6;7 8] = [−4 − 4; −4 − 4] |
| * | Multiplication | $3{*}4 = 12$ | [1 2;3 4]*[5 6;7 8] = [19 22;43 50] |
| / | Division | $3/4 = 0.75$ | |
| ^ | Exponentiation | $3{^{\wedge}}4 = 81$ | [1 2;3 4]^2 = [7 10; 15 22] |
| .* | Term-by-term matrix multiplication | | [1 2;3 4].*[5 6;7 8] = [5 12;21 24] |
| .^ | Term-by-term matrix exponentiation | | [1 2;3 4].^2 = [1 9;4 16] |
| ' | Matrix transpose | | [1 2;3 4]' = [1 3;2 4] |
| = | Assignment | A = 1 | A = [1 2;3 4] |

## C.1 MATLAB BASICS

MATLAB stores all data in matrices. A matrix is entered in MATLAB using an assignment statement with the elements of the matrix enclosed in square brackets ([]). The elements are entered row by row with each element in the row separated by a space. Rows are separated by a semicolon (;). While MATLAB considers a scalar as a matrix with one row and one column, the assignment of a scalar to a variable does not require enclosing the scalar in square brackets. Individual elements of a matrix may be defined or referenced by using subscripts to indicate the row and column in the matrix where the element resides. Since the physical use of subscripts is not possible the subscripts are identified in parentheses after the variable name For example "A(1,2)" refers to the element residing in the first row and second column of the matrix **A**.

Operators are used to represent mathematical calculations. Table C.1 gives MATLAB operators and sample results of the operation. Table C.2 presents the hierarchy of operations. Some of the available MATLAB functions are listed in Table C.3. Characters with special significance are listed in Table C.4.

**TABLE C.2 HIERARCHY OF OPERATIONS FROM HIGHEST TO LOWEST**

1. Grouping (functions and parentheses)
2. Exponentiation
3. Multiplication and division from left to right
4. Addition and subtraction from left to right

**TABLE C.3 MATLAB FUNCTIONS**

| Function | Name | Mathematical Equation | Examples | Comments |
|---|---|---|---|---|
| abs(x) | Absolute value | $\lvert x \rvert$ | abs($-4$) = 4<br>abs ($-3 + 4$j) = 5 | Absolute value of a real number or magnitude of a complex number |
| angle(x) | Angle | $\tan^{-1}\left\{\dfrac{\text{Im}[x]}{\text{Re}[x]}\right\}$ | angle(2-j)=2.679 | Result in radians between 0 and $2\pi$. |
| atan(x) | Inverse tangent | $\tan^{-1}(x)$ | atan(1) = 0.7854 | Result in radians between $-\pi/2$ and $\pi/2$ |
| atan2(y,x) | Four-quadrant inverse tangent | $\tan^{1}\left(\dfrac{y}{x}\right)$ | atan2(2, $-1$) = 2.0344 | Result in radians between 0 and $2\pi$ |
| ceil(x) | Ceiling function | Smallest integer greater than or equal to $x$ | ceil(6.11) = 7 | |
| conj(x) | Complex conjugate | $\bar{x}$ | conj($-3 + 4$j) = $-3 - 4$j | |
| cos(x) | Cosine | $\cos(x)$ | cos(pi) = $-1$ | Argument must be in radians |
| exp(x) | e to the x | $e^x$ | exp(2) = 7.3891 | |
| floor(x) | Floor function | Largest integer less than or equal to $x$ | floor(3.15) = 3 | |
| imag(x) | Imaginary part | Im $[x]$ | imag($-3 + 4$j) = 4 | |
| log(x) | Natural logarithm | $\ln(x)$ | log(10) = 2.3026 | |
| log10(x) | Base 10 logarithm | $\log(x)$ | log10(10) = 1 | |
| max(A) | Maximum value of elements of the vector A | $\max\limits_{i=1\ldots n} A_i$ | max([2 1 0 5]) = 5 | |
| min(A) | Minimum value of elements in the vector A | $\min\limits_{i=1\ldots n} A_i$ | min([1 2 -1 -2])=-2 | |
| real(x) | Real part | Re$[x]$ | real($-3 + 4$j) = $-3$ | Either $i$ or $j$ may be used to represent $\sqrt{-1}$ |
| sin(x) | Sine | $\sin(x)$ | sin(pi) = 0 | Argument must be in radians; pi is MATLAB-defined variable = 3.14157... |
| tan(x) | Tangent | $\tan(x)$ | tan(pi/4) = 1 | Argument must be in radians and cannot be an odd multiple of pi/2 |

**TABLE C.4 CHARACTERS WITH SPECIAL SIGNIFICANCE**

| Character | Significance | Example | Comments |
|---|---|---|---|
| % | Comment statement | % Calculate natural frequency | A comment statement is a nonexecutable statement used to document a program. The character must preede a comment. |
| ; | Suppress printing | a=3*x^2; | The presence of the semicolon at the end of an assignment statement suppresses the printing of the result. |
| ; | Row separation | [1 2;3 4] | When used in defining a matrix the semicolon indicates the end of one row and the beginning of the next. |
| ' ' | Literal printing | disp('The natural frequency is') | Text enclosed by single quotes will be printed as is. |
| : | Vector generation | G = 0:2:20 | A method of generating a vector of elements that constitute an arithmetic sequence. The example generates a vector G such that $G(1) = 0$, $G(2) = 2$, $G(3) = 4$, ..., $G(11) = 20$. The colons separate the first element, the constant difference between elements, and the maximum value of the last element. |
| ... | Statement continuation | $X = 3 + 6...$ $+ 4 + 7$ | Allows for the continuation of lengthy equations on multiple lines. |

MATLAB allows the definition of vectors whose elements constitute an arithmetic sequence through the following command

$$name = i:j:k$$

where name is the vector name, i is the value of name(1), j is the constant difference between elements of name, $j = name(m+1)-name(m)$, and k is the largest (or smallest if j is negative) possible value of an element of name.

If a vector, say t, is defined, a second vector, say x, whose elements are defined from the elements of t using a common definition, a statement of the form

$$x = f(t)$$

will define the vector x for a specified function f. Term-by-term calculations involving matrix elements must be defined using a period (.) before an operator. For example if t is a column vector the statement

$$x = t*t$$

generates an error "inner matrix dimensions must agree." This statement attempts to determine the product of the $1 \times n$ vector with the $1 \times n$ vector. However, it the statement

$$x = t.*t$$

is used, a vector x is generated such that $x_i = t_i^2$.

### Example C.1

A $3 \times 3$ matrix is defined as

$$\mathbf{A} = \begin{bmatrix} 4 & -1 & 2 \\ 3 & 6 & 5 \\ 2 & -1 & 0 \end{bmatrix} \tag{a}$$

(a) Write the command to enter the matrix into MATLAB with the variable name **A**.
(b) Write a MATLAB statement that calculates the product $(a_{3,2})(a_{1,3})$ and assigns it to the variable $q$.

#### Solution

(a) The assignment statement to enter the matrix **A** into MATLAB's memory is

$$A = [4\ \text{-}1\ 2;3\ 6\ 5;2\ \text{-}1\ 0]$$

(b) The assignment statement to calculate the desired product is

$$q = A(3,2)*A(1,3)$$

### Example C.2

Write a MATLAB statement that evaluates the algebraic expression

$$b = \frac{3 + 4\sin\left(x^{2y+5}\right)}{2 - 4\cos\left(\dfrac{x-2}{3+2x}\right)} \tag{a}$$

using the minimum number of parentheses.

#### Solution

Using the operators of Table C.1, the hierarchal rules of Table C.2, and the functions of Table C.3 the MATLAB assignment statement for Equation (a) using the fewest number of parentheses is

$$b = (3 +4*\sin(x^{\wedge}(2*y + 5)))/(2 - 4*\cos((x - 2)/(3 + 2*x)))$$

### Example C.3

Write MATLAB statements that define a vector of values $t_i = 2 + 0.02 * i$ for $i = 1, 2, \ldots, 100$ and a vector of values $x_i = e^{-2t_i}[\cos(4t_i)]^2$

#### Solution

It is noted that $t_{100} = 4$. The appropriate MATLAB statements are

$$t = 1:.02:4;$$
$$x = \exp(-2*t).*\cos(4*t).^{\wedge}2;$$

Note that since $\exp(-2*t)$ generates a vector and $\cos(4*t)$ generates a vector, the .* operator must be used for term-by-term multiplication. Similarly the .^2 operator must be used to square the individual elements of a vector.

## C.2 Plotting and Annotating Graphs

The development of graphs is very easy using MATLAB. A graph can be developed directly in the MATLAB work environment or from the execution of a program. A plot developed during the execution of a program may be annotated using statements in the program or from the workspace after execution of the program.

A two-dimensional plot in MATLAB is generated by plotting two vectors, using one as the abscissa and the other as the ordinate. For example the command

$$\text{plot(t,x)}$$

constructs a graph plotting the elements of t on the horizontal axis and the elements of x on the vertical axis. The vectors t and x must be of the same length. The plot is scaled automatically.

The commands

$$t = 0:0.05:10;$$
$$x = \sin(2*t);$$
$$\text{plot(t,x)}$$

generate vectors t and x and then plots x as a function of t. The resulting graph is illustrated in Figure C.1.

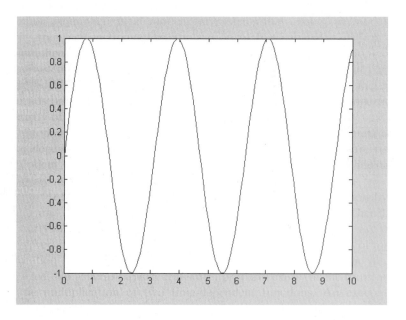

**FIG. C.1** The MATLAB-plot generated using 'plot(t,x)'.

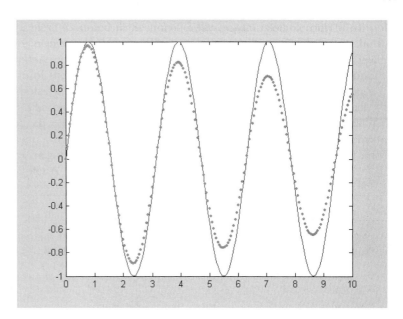

**FIG. C.2** Two graphs on one set of axes, generated in MATLAB.

More than one graph can be displayed on one set of axes. The commands

$$t = 0{:}0.05{:}10;$$
$$x = \sin(2{*}t);$$
$$y = \exp(-.05{*}t).{*}\sin(2{*}t);$$
$$\text{plot}(t,x,\text{`-'},t,y,\text{`.'})$$

generate vectors t, x, and y and plots x vs. t and y vs. t on the same set of axes. The resulting plot is shown in Figure C.2. The plot of x is continuously connected while the values of y in the generated vector are connected by a period (.), as instructed by the plot command.

The plot can be annotated with labels on the axes, a title, a legend, and specifically placed text. The plot in Figure C.2 can be annotated using the following commands

title('Comparison of undamped and damped responses')

xlabel('time (s)')

ylabel('displacement (mm)')

legend('undamped response','damped response\zeta =0.05')

text(6,0,'y(t) = e^-.^0^5^tsin(2t)')

text(1.2,.8,'x(t) = sin(2t)')

Use of these commands leads to the graph of Figure C.3. The commands are explained in Table C.5. These commands contain features that allow exponentiation and Greek letters in the displayed text. Some of the available features are summarized in Table C.6. In each case the trigger for the feature is used in the string inside the single quotes.

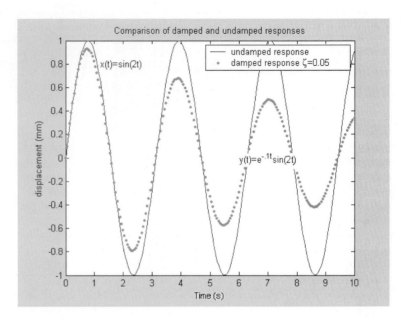

**FIG. C.3** An annotated graph.

## C.3 Programming Commands

MATLAB is a programming language. As such it has all the aspects of a programming language, including the assignment of variables, input and output functions, conditional statements, and user-defined functions.

**TABLE C.5 COMMANDS FOR ANNOTATION OF GRAPHS**

| Command | Purpose | Comments |
|---|---|---|
| grid | To add grid lines to the graph | |
| legend(' ',' ',...,' ') | To add a legend to the graph | Legend appears in a box in the upper right-hand corner. The number of arguments is equal to the number of graphs on the set of axes. The arguments provide the annotation for the legend for the graphs in order of listing in the plot statement. |
| text(x,y,' ') | To add text to the graph at a specified location | The left of the text begins at the location (x,y) on the graph where x and y are coordinates measured along the horizontal and vertical axes of the graph. |
| title(' ') | To add a title to the graph | The title appears centered above the graph, outside the graphics box. |
| xlabel(' ') | To add a label to the horizontal axis | The label appears centered below the horizontal axis. |
| ylabel(' ') | To add a label to the vertical axis | The label appears centered to the left of vertical axis. |

**TABLE C.6 TRIGGERS FOR SPECIAL FEATURES IN ANNOTATING GRAPHS**

| Feature | Syntax Example | Comments |
|---|---|---|
| Superscript | ^t | The t appears as a superscript in the printed text. If more than one character is to appear as a superscript the caret (^) must be repeated after each character. For example ^2^t leads to a superscript of $^{2t}$. |
| Subscript | _3 | The 3 appears as a subscript in the printed text. If more than one character is to appear as a subscript the underscore(_) must be repeated after each character. For example _3_,_2 leads to a subscript of $_{3,2}$. |
| Lowercase Greek letter | \alpha | The Greek letter $\alpha$ is displayed. The name of the Greek letter to be displayed is spelled out in lowercase after the backslash. |
| Uppercase Greek letter | \Omega | The uppercase Greek letter $\Omega$ is displayed. The name of the Greek letter to be displayed is spelled out with the first letter in uppercase and the remaining letters in lowercase. |

A program consists of executable statements placed in a specific order designed to achieve a desired objective. The programs used in this text are modest and the commands used only rudimentary, but the programs use all these aspects of programming.

**C.3.1 Input and Output**   "Output" refers to how the program communicates its results. The programs in this text are all designed to output directly into the MATLAB worksheet. Not only can statements be written to physically display the value of variables in the workspace, but once a value of a variable is calculated in the program it is stored in memory and can be recalled and used from the workspace.

By default the workspace displays, in the workspace, the result of calculations from every assignment statement immediately upon completion of the execution of the statement. MATLAB also displays the name of the variable being displayed. This output may be suppressed by ending the statement with a semicolon (;). The value of a variable may be displayed in the workspace just by entering the name of the variable. This can be done from a statement in a program or directly from the workspace.

In lieu of the default output of results, MATLAB allows the display of more information regarding the variable through use of the 'disp' command. Suppose a program calculates a steady-state amplitude X in units of meters. Instead of simply displaying the variable name, the following 'disp' statement displays more information regarding the variable

disp('The steady-state amplitude in m is')

X

**TABLE C.7 LOGICAL AND RELATIONAL OPERATORS**

| Symbol | Name | Example | Result of Example |
|--------|------|---------|-------------------|
| ~ | Not | ~(6>4) | False |
| == | Is equal to | 3 == 2 + 1 | True |
| ~= | Is not equal to | 2~ = 2 + 1 | True |
| < | Less than | 4 < −1 | False |
| <= | Less than or equal to | 4 <= 4 | True |
| > | Greater than | 4 > 1 | True |
| >= | Greater than or equal to | 4 >= 6 | False |
| & | And | 4 < 3& 4 > −1 | False |
| ! | Or | 4 < 3 ! 4 > − 1 | True |

"Input" refers to the entering of data into the program. The programs in this text allow only for data to be entered interactively from the workspace using an input statement. Suppose it is desired to enter the value of a damping ratio and assign it to the variable z. The following input statement may be used

z = input ('Please enter the value of the damping ratio')

The user is then prompted to provide a value for z to be entered in the workspace.

**C.3.2 Conditional Statements**    Programs often decide how to proceed based on specified conditions. If a condition is satisfied the program is to do one thing; if the condition is not satisfied the program is to do something else. Program control is exerted when the program decides which step is the next to be executed. The program control structure commonly used in this text is an 'if' statement. The basic structure of an 'if' statement is

if conditional statement
        statements to be executed
end

The conditional statement in an 'if' statement is a logical statement that is either true or false. The conditional statement is written using logical and relational operators that are summarized in Table C.7. If the conditional statement is true then the statements following the conditional are executed in sequence. If the conditional is false program control transfers to the 'end' statement. The 'end' statement is nonexecutable and serves as a place marker.

An expanded structure for the 'if' statement occurs when an 'else' clause is added to permit the execution of statements when the conditional is false.

>
>if conditional statement
>
>>statements to be executed
>
>else
>
>>statements to be executed
>
>end

In this structure if the conditional is true the immediate statements are executed until the program reaches the 'else' when control is transferred to the 'end' statement. If the conditional is false control shifts to the 'else' statement.

When a false conditional sends the program control to an 'else' which would be followed by another conditional an alternative is to use an 'elseif' statement

>if conditional 1
>
>>statements to be executed
>
>elseif conditional 2
>
>>statements to be executed
>
>else
>
>>statements to be executed
>
>end

If conditional 1 is false then program control shifts to the 'elseif' statement. If conditional 2 is true then the statements following the 'elseif' are executed. If conditional 2 is false then program control flows to the 'else' statement. As many 'elseif' statements as necessary may be used in a conditional.

---

**Example C.4**

A numerical value of a variable 'b' has been determined previously in the program. Write MATLAB conditional statements for the following. (a) If b is less than 20 then its value is to be doubled. (b) If b is less than 20 its value is to be doubled, otherwise it is to tripled. (c) If b is less than 20 its value is to be doubled, if it is between 20 and 100 inclusive it is to be tripled, otherwise it is to be quadrupled.

(a) The appropriate conditional is

>if b < 20
>
>>b = 2*b;
>
>end

(b) The use of an else is required and the appropriate conditional is

$$
\begin{aligned}
&\text{if } b < 20 \\
&\qquad b = 2.*b; \\
&\text{else} \\
&\qquad b = 3.*b; \\
&\text{end}
\end{aligned}
$$

(c) The use of an 'elseif' is convenient. One possible conditional is

$$
\begin{aligned}
&\text{if } b < 20 \\
&\qquad b = 2.*b; \\
&\text{elseif } b <= 100 \\
&\qquad b = 3.*b; \\
&\text{else} \\
&\qquad b = 4.*b; \\
&\text{end}
\end{aligned}
$$

**C.3.3 Looping**   Looping allows statements to be executed multiple times. In some cases it is convenient to use a 'while' loop, which has the structure

$$
\begin{aligned}
&\text{while conditional statement} \\
&\qquad \text{statements to be executed} \\
&\text{end}
\end{aligned}
$$

If the conditional statement is true the immediately subsequent statements are executed. When the 'end' statement is reached, program control returns to the 'while' statement. The relational expression is again checked for validity. Once the relational expression is not true program control is shifted to the 'end' statement and the loop ends.

A 'for' loop is executed a specified number of times. It has the structure

$$
\begin{aligned}
&\text{for var} = i{:}j{:}k \\
&\qquad \text{statements} \\
&\text{end}
\end{aligned}
$$

where var is a variable name used as a counter in the loop, i is the beginning value of var, j is an incremental value, and k is the largest possible value for var. The counter is initially set to i. The following statements are executed. When 'end' is reached program control returns to the 'for' statement. The value of var is incremented by j. Before proceeding the new value of var is compared with k. If var is greater than k control is shifted to the 'end' statement and the loop is exited. Otherwise the following steps are again executed and the loop continues as described.

**C.3.4 User-Defined Functions** The modeling of a dynamic system focuses on being able to determine the system output given its input, which is time dependent. MATLAB is used as a tool in the determination of system response. MATLAB programs, M-files, are written in support of this effort. A programmer, developing a MATLAB code to determine the output of the dynamic system, should have as an objective that the program should be able to determine the response for any input, but the program itself does not need to be modified when the system input changes. A program to determine system response for an arbitrary input should have available a function that for any value of time provides the input value(s).

The preceding objective is attained through the use of user-defined functions, which are functions defined outside a program and which the program may call whenever necessary. A user-defined function is written to return a value or values to the calling program. Information is transferred from the calling program to the user-defined function. The user-defined function uses the information to calculate the desired results, which are then transferred to the calling program. The syntax for such a user-defined function is

$$\text{function } [x,y,z] = \text{name}(a,b,c)$$

$$\text{assignment statements to}$$

$$\text{calculate x, y, and z}$$

The calling program sends the values of a, b, and c, the arguments of name, to the user-defined function, which uses them to calculate x, y, and z. The values of x, y, and z are then returned to the calling program.

For example suppose that execution of an M-file requires values of two forces $F = 10\sin(2t)$ and $G = 20e^{-2t}$ at an arbitrary time $t$. The name of the function is force. An appropriate form for the user-defined function is

$$\text{function } [F,G] = \text{force}(t)$$
$$F = 10*\sin(2*t);$$
$$G = 20*\exp(-2*t);$$

The user-defined function must be called from the M-file. The syntax for calling this function is

$$[A,B] = \text{force(time)}$$

MATLAB searches for force as the name of a default function or a user-defined function. The name "force" must correspond to the name of the user-defined function. The value of time is sent to the user-defined function where it is used as t. Upon return the value calculated in the user-defined function as F is assigned to A in the calling program, and the value calculated in the user-defined function as G is assigned to B in the calling program.

**TABLE C.8 SYMBOLIC MATH TOOLBOX COMMANDS USED IN TEXT**

| Command | Syntax | Comments |
|---|---|---|
| syms | syms s t | Declares s and t as symbolic variables. |
| sym | sym('x') | Creates a symbolic representation of the value assigned to x. |
| laplace | laplace(F) | Takes the Laplace transform of function F defined in terms of a symbolic variable t; default is to return a function of symbolic variable s; both t and s must be declared as symbolic variables using syms. |
| ilaplace | ilaplace(G) | Take the inverse Laplace transform of function G defined in terms of a symbolic variable s; default its to return a function of symbolic variable t. |
| simplify | simplify(F) | Algebraically simplifies the symbolic expression for F using a variety of techniques. |
| expand | expand(F) | Algebraically expands products of polynomials contain in F. |
| subs | subs(F,y) | If F is a function of a symbolic variable, say x, and y is a numerical value or an algebraic expression, subs evaluates F when y is substituted for x. |
| vpa | vpa(F,n) | Variable precision arithmetic computes a symbolic expression F to a decimal expression with n significant digits. |

## C.4 SYMBOLIC MATH TOOLBOX

The Symbolic Toolbox performs symbolic algebra, linear algebra, and calculus. The toolbox is used in this text for the determination of Laplace transforms, inverse Laplace transforms, and transfer functions for multiple output systems. Table C.8 summarizes the commands from the Symbolic Math Toolbox used in this regard.

## C.5 CONTROL SYSTEM TOOLBOX

The Control System Toolbox performs calculations and develops diagrams used in the analysis and design of control systems. The functions from the Control System Toolbox are used in this text to determine poles and residues, define transfer functions, develop root-locus diagrams, and draw Bode and Nyquist diagrams.

The use of these functions is limited to transfer functions defined as the ratio of two polynomials $G(s) = N(s)/D(s)$. However, the Symbolic Math Toolbox is not required to use these functions because the coefficients defining the polynomials in the numerator and denominator of the transfer function are entered as vectors. An $n$th-order polynomial is entered as a row vector of $n$ elements. The first element in the vector is the coefficient of $s^n$. The command 'tf(N,D)' is used to define the transfer function.

## Example C.5

Use MATLAB to define the transfer function

$$G(s) = \frac{3s^2 + 2s + 1}{s^3 + 5s^2 + 6s + 12}$$

```
> N=[3 2 1];
>> D=[1 5 6 12];
>> G=tf(N,D)

Transfer function:
  3 s^2 + 2 s + 1
----------------------
s^3 + 5 s^2 + 6 s + 12
```

**FIG. C.4** The MATLAB worksheet for the development of the transfer function of Example C.5.

### Solution

The MATLAB worksheet for entering $N(s)$, $D(s)$, and defining $G(s)$ is shown in Figure C.4.

The development of the transfer function using the function 'tf' is not often necessary. Control System Toolbox functions requiring the transfer function can also use the definitions of $N(s)$ and $D(s)$. For example the function to plot the Bode diagram can be written as either

$$bode(G)$$

where G is the transfer function or

$$bode(N,D)$$

Some useful Control System Toolbox functions are described in Table C.9.

**TABLE C.9 SELECTED CONTROL SYSTEM TOOLBOX FUNCTIONS**

| Function | Use | Syntax | Comments |
|---|---|---|---|
| tf | Develop the transfer function from the numerator and denominator polynomials N(s) and D(s) respectively | G = tf(N,D) | MATLAB shows G as a function of s, but s is not defined as a symbolic variable and G is not a function of a symbolic variable. |
| residue | Determine the poles and residues of a transfer function | [r,p,k] = residue(N,D) | r = vector of residues<br>p = vector of poles<br>k =vector of coefficients obtained by dividing N by D. If the order of N is less than order of D then k is a null vector. |
| impulse | Impulsive response of system | impulse(G)<br>impulse(N,D) | Output is a plot of impulsive response. |
| step | Step response of system | step(G)<br>step(N,D) | Output is a plot of step respsonse. |
| bode | Bode diagram (magnitude and phase) | bode(tf)<br>bode(N,D) | |
| nyquist | Nyquist diagram | nyquist(tf)<br>nyquist(N,D) | |
| rlocus | Root-locus diagram | rlocus(R,Q) | Root-locus diagram is created for the equation Q(s) + KR(s) = 0. |

# Appendix D

## Construction of Root-Locus Diagrams

Root-locus diagrams are introduced in Section 6.3 to illustrate the effect of changing a parameter on the stability of a system. The diagrams are used in Section 8.6 in discussing pole placement when designing a controller. While root-locus diagrams are easily developed using MATLAB it is instructive to understand how the diagrams can be developed analytically and sketched by hand. Such knowledge leads to a better understanding of how the poles vary with a changing parameter.

## D.1 DEFINITIONS

A **root-locus diagram** is a diagram showing the loci of the roots of the equation

$$Q(s) + KR(s) \tag{D.1}$$

for all $K$ such that $0 \le K < \infty$ where $Q(s)$ and $R(s)$ are polynomials in $s$ with the order of $R(s)$ less than or equal to the order of $Q(s)$. A root-locus diagram is drawn in the complex $s$ plane. One locus represents the variation of one root of Equation (D.1) with the variation of $K$. If $Q$ is an $n$th-order polynomial in $s$ there are $n$ loci with each locus called a **branch** of the diagram.

Construction of a root-locus diagram begins by determining the roots of Equation (D.1) for $K = 0$, the roots of $Q(s)$ called the **poles** of the diagram and the roots of Equation (D.1) as $K$ becomes unbounded, the roots of $R(s)$, called the **zeros** of the diagram. If a branch of the diagram corresponds to a real root for $K = 0$, but for some value of $K$ its root becomes complex, the point on the real axis where the locus departs from the real axis is called the **breakaway point**. If a branch of the diagram corresponds to a complex root for $K = 0$, but for some value of $K$ its root becomes real, the point on the real axis where the locus joins the

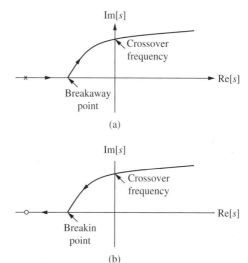

**FIG. D.1** Illustration of the crossorver frequency, breakaway point, and breakin point.

real axis is called the **breakin point**. The imaginary component of the root when a locus crosses the real axis is called the **crossover** frequency for the locus. The **angle of departure** from a complex pole is the angle of the tangent to the root-locus curve at a complex pole. The **angle of arrival** at a complex zero is the angle of the tangent to the root-locus diagram at a complex zero. These terms are illustrated in Figure D.1.

## D.2 ANGLE AND MAGNITUDE CRITERIA

Equation (D.1) may be rearranged as

$$K \frac{R(s)}{Q(s)} = -1 \tag{D.2}$$

Recall that $R(s)$ and $Q(s)$ are functions of a complex variable $s$ and that $K$ is a real positive number. The number $-1$ has a representation of $-1 + j0$ in the complex plan and a polar representation of $e^{i(2n+1)\pi}$ for any $n = 0, \pm 1, \pm 2, \ldots$ . Two complex numbers are equal if and only if they have the same magnitude and the same phase. Thus Equation (D.2) leads to

$$K \left| \frac{R(s)}{Q(s)} \right| = 1 \tag{D.3}$$

and

$$\phi = (2n + 1)\pi \qquad n = 0, \pm 1, \pm 2, \ldots \tag{D.4}$$

where $\phi$ is the phase of $KR(s)/Q(s)$. Equation (D.3) is the **magnitude criterion** that must be satisfied at every point on all branches of a root-locus plot. Equation (D.4) is the **angle criterion** that must be satisfied at every point on all branches of a root-locus plot. Since the phase of a product of complex numbers is the sum of the phases, Equation (D.4) results in

$$\phi_K + \phi_R - \phi_Q = (2n+1)\pi \qquad n = 0,\ \pm 1,\ \pm 2,\ \ldots \qquad \text{(D.5)}$$

where $\phi_K, \phi_R$, and $\phi_Q$ are the phases of $K$, $R(s)$, and $Q(s)$ respectively. Since $K$ is a positive real number $\phi_K = 0$ and Equation (D.4) becomes

$$\phi_R - \phi_Q = (2n+1)\pi \qquad n = 0,\ \pm 1,\ \pm 2,\ \ldots \qquad \text{(D.6)}$$

Equations (D.3) and (D.6) are essential for the analytic development of a root-locus diagram.

## D.3 CONSTRUCTION GUIDELINES

A set of guidelines may be developed from Equations (D.1)–(D.6) that facilitate the construction of a root-locus diagram. In the following it is noted that $Q(s)$ may be factored into $p$ linear factors of the form $s - s_{q,i}$ for $i = 1,\ 2,\ \ldots,\ p$ where $s_{q,i}$ are the poles of the diagram and $p$ is the total number of poles. In addition $R(s)$ may be factored into $z$ linear factors of the form $s - s_{r,j}$ for $j = 1,\ 2,\ \ldots,\ z$ where $s_{r,j}$ are the zeroes of the diagram and $z$ is the total number of zeroes. Furthermore it is noted that $\phi_R = \sum_{j=1}^{z} \phi_{r,j}$ where $\phi_{r,z}$ is the phase of the liner factor $s_{r,j}$. A similar expression can be written for $\phi_Q$.

The following develops some guidelines for developing a root-locus diagram. A summary of how these guidelines are used to sketch a root-locus diagram is presented afterward.

**1.** Since complex roots occur in complex conjugate pairs, a root-locus diagram is symmetric with respect to the real axis. Thus the diagram may be constructed in the upper half plane and then reflected about the real axis.

**2.** When $K = 0$ the roots of Equation (D.1) are the poles of the diagram. Thus the branches begin at the poles.

**3.** When $K = \infty$, Equation (D.2) implies that either $R(s) = 0$ or $Q(s) = \infty$. Thus a branch either ends at a zero or approaches infinity.

**4.** The angle condition is used to determine which parts of the real axis are on any branch of the root locus. This is perhaps best explained through consideration of $Q(s) = s(s + a)(s + b)$ where $a$ and $b$ are positive real numbers with $a < b$ and $R(s) = 1$. Then $\phi_R = 0$ and $\phi_Q = \phi_s + \phi_{s+a} + \phi_{s+b}$. The angle criterion [Equation (D.6)] then requires

$$-\phi_s - \phi_{s+a} - \phi_{s+b} = (2n+1)\pi \qquad \text{(D.7)}$$

for some integer value of $n$. Consider first a real value of $s > 0$. Then $\phi_s = \phi_{s+a} = \phi_{s+b}$, which leads to $-\phi_s - \phi_{s+a} - \phi_{s+b} = 0$ implying that Equation (D.7) is not satisfied. Thus no point lying on the positive real axis may be on the root-locus diagram for this system. Now consider a value of $s$ on the negative real axis with $|s| < a$. Then $\phi_s = -\pi$ and $\phi_{s+a} = \phi_{s+b} = 0$, which leads to $-\phi_s - \phi_{s+a} - \phi_{s+b} = -\pi$ and Equation (D.7) is satisfied with $n = -1$. Thus the region of the real axis $-a \le s \le 0$ is part of some branch of the root locus. (Note that since $s = 0$ and $s = -a$ are poles of the diagram, both lie on the root locus corresponding to $K = 0$).

Now consider a value of $s$ on the negative real axis with $a < |s| < b$. For this case $\phi_s = \pi$, $\phi_{s+a} = \pi$, and $\phi_{s+b} = 0$, which leads to $-\phi_s - \phi_{s+a} - \phi_{s+b} = -2\pi$ and thus Equation (D.7) cannot be satisfied. Thus, no point lying on the negative real axis such that $-b < s < -a$ is part of any branch of this system's root locus.

Finally consider a point on the negative real axis such that $|s| > b$. For this case $\phi_s = \phi_{s+a} = \phi_{s+b} = \pi$, which leads to $-\phi_s - \phi_{s+a} - \phi_{s+b} = -3\pi$ and Equation (D.7) is satisfied with $n = -2$. Thus the region of the negative real axis defined by $-\infty < s \le -b$ is on some branch of this system's root locus diagram.

The polynomial $Q(s)$ can always be factored into a product of linear and quadratic factors. The preceding discussion shows that for a value of $s$ on the real axis the phase of a linear factor is either 0 or $\pi$. Any quadratic factor which cannot be factored into two linear factors with real coefficients may be written in the form $s^2 + 2\zeta\omega s + \omega^2$ with $\zeta < 1$. If $s$ is a positive real number the quadratic factor clearly has a positive value and thus its phase is zero. If $s$ is a negative real number of the form $s = -\alpha$ then the quadratic factor is $\alpha^2 - 2\zeta\omega\alpha + \omega^2$. Since $\zeta < 1$

$$\alpha^2 - 2\zeta\omega\alpha + \omega^2 > \alpha^2 - 2\alpha\omega + \omega^2$$
$$= (\alpha - \omega)^2 \qquad\qquad (\text{D.8})$$
$$> 0$$

Equation (D.8) shows that the evaluation of a quadratic factor for negative real values of $s$ always leads to a positive value and thus $\phi = 0$. This implies that quadratic factors of $Q(s)$ and $R(s)$ do not have any effect as to whether a portion of the real axis is on a branch of the root-locus diagram. Thus only the factors of $Q(s)$ corresponding to real poles and the factors of $R(s)$ corresponding to real zeros need to be considered in determining the parts of the real axis that lie on a branch of the root-locus diagram.

**5.** The analysis of the location of branches of the root locus on the real axis is supplemented by the knowledge that, while separate branches of the root-locus diagram may intersect (corresponding to repeated roots), two branches may not have the same locus over a finite region of the complex plane. Thus if separate branches begin at two real poles and approach each other over a range of the real axis the preceding analysis shows as they eventually intersect at a point from which each locus leaves the real axis. The point at which the branches intersect is a breakaway point. Since the branch of the root locus follows increasing $K$, the break-away point corresponds to the maximum value of $K$ for which the root corresponding to that branch has an imaginary part equal to zero. For values of $s$ on the real axis Equation (D.1) can be interpreted as an equation to determine $K$ as a function of a real independent variable $s$. In this context $K(s)$ reaches a maximum value for a value of $s$ corresponding to the breakaway point. Thus the breakaway point can be determined as the value of $s$ such that

$$\frac{dK}{ds} = 0 \qquad\qquad (D.9)$$

The differentiation of Equation (D.1) with respect to $s$ leads to

$$\frac{dQ}{ds} + K\frac{dR}{ds} + R\frac{dK}{ds} = 0$$

$$\frac{dK}{ds} = -\frac{1}{R}\left(\frac{dQ}{ds} + K\frac{dR}{ds}\right) \qquad (D.10)$$

$$= -\frac{1}{R}\left(\frac{dQ}{ds} - \frac{Q}{R}\frac{dR}{ds}\right)$$

Thus the numerical values of the breakaway points are obtained by solving

$$R\frac{dQ}{ds} - Q\frac{dR}{ds} = 0 \qquad\qquad (D.11)$$

Only the values of $s$ obtained by solving Equation (D.11), which the analysis of step 4 has shown to be located on a branch of the root locus, are breakaway points.

An analogous argument is used to show that any breakin points can be determined using Equations (D.9) and (D.10). In the case of a breakin point the corresponding value of $K$ corresponds to the minimum value for which the branch has an imaginary part equal to zero.

It can be shown that tangent of a branch of the root locus at a breakaway point or a breakin point is vertical because the branch leaves or enters the real axis.

**6.** It is noted in guideline 2 that as $s \to \infty$ the branch of the root locus diagram approaches either a zero of the diagram or infinity. To draw the root-locus diagram accurately it is desirable to know the asymptotic behavior of those branches which approach infinity. Consider two factors of $Q(s)$ of the form $s - s_{q,m}$ and $s - s_{q,n}$. As $s \to \infty$ the specific values of the poles are much smaller than $s$ and both factors are approximated simply by $s$. This implies that as $s \to \infty$ in the vicinity of the asymptote of a branch of the root locus the angles of all factors of $Q(s)$ corresponding to poles are the same. It can also be argued that the angles of all factors of $R(s)$ corresponding to the zeroes are equal to this angle;

call it $\phi_a$. Thus an asymptote to a branch of the root-locus diagram is a straight line of angle $\phi_a$. The angle criteria [Equation (D.6)] then requires

$$z\phi_a - p\phi_a = (2n+1)\pi \qquad n = 0,\ \pm 1,\ \pm 2$$

$$\phi_a = \frac{(2n+1)\pi}{z - p} \qquad n = 0,\ \pm 1,\ \pm 2 \tag{D.12}$$

Equation (D.12) is used to determine the slopes of the asymptotes to the branches of the root locus. The number of asymptotes is the number of unique values of $\phi_a$ determined by Equation (D.12).

It can be shown that the asymptotes all intersect at a point on the real axis at the point

$$s = \frac{1}{p - z}\left(\sum_{j=1}^{p} s_{q,j} - \sum_{i=1}^{z} s_{r,i}\right) \tag{D.13}$$

**7.** The crossover frequencies are found by letting $s = j\omega$ in Equation (D.1). The real and imaginary parts of the left-hand side of the equation are identically set to zero. These equations are used to solve for the crossover frequency $\omega$ and the corresponding value of $K$.

**8.** The angle of departure from a complex pole or the angle of arrival at a complex zero is obtained using the angle criterion; Equation (D.6).

**9.** The shapes of the branches of the root locus are sketched by filling in several points on the branches. This may be done by trying to locate the point on the branch with a specific imaginary part, chosen between zero and the crossover frequency. A value of $s$ with this imaginary part is guessed and the angle criterion checked. If the left-hand side of Equation (D.6) exceeds $\pi$ the sum of the angles is decreased by choosing a value of $s$ with a greater real part and vice versa. This iterative process is continued until the angle criterion is satisfied.

## D.4 SUMMARY OF CONSTRUCTION STEPS

The following steps may be followed in developing a root-locus diagram.

**1.** The locations of the origin of the branches of the root-locus diagram are the poles of the diagram, the roots of $Q(s)$.
**2.** The locations of the ends of the branches of the root-locus diagram are the zeroes of the diagrams, the roots of $R(s)$, or at infinity.
**3.** The portions of the real axis that are on any branch of the root-locus diagram are determined using the ideas of guideline 4.
**4.** The breakaway or breakin points are obtained by determining the values of $s$ that satisfy Equation (D.10) and have been demonstrated in step 3 to be part of a branch.
**5.** The equations of the asymptotes to branches that approach $\infty$ as $K$ increases are obtained using Equations (D.12) and (D.13).

6. The crossover frequency and its corresponding value of $K$ are obtained by substituting $s = j\omega$ into Equation (D.1), breaking the resulting equation into its real and imaginary parts and solving the resulting equations for $K$ and $\omega$.

7. The angle of departure for any complex pole and the angle of arrival for any complex zero is obtained using the angle criterion.

8. Several points on a root-locus branch and their corresponding gains are determined as described in step 9 by using the angle criterion.

## D.5 EXAMPLES

### Example D.1

Construct a root-locus diagram for

$$s^3 + 6s^2 + 11s + 6 + K = 0 \tag{a}$$

**Solution**

Equation (a) is written in the form of Equation (D.1) by defining

$$Q(s) = s^3 + 6s^2 + 11s + 6 \tag{b}$$

$$R(s) = 1 \tag{c}$$

The three poles of the diagram, the roots of $Q(s)$ are determined as $-1$, $-2$, and $-3$. There are no zeroes of the diagram because $R(s)$ has no roots. Since $R(s)$ is third order there are three branches of the root-locus diagram. The steps in developing the root-locus are described below.

1. The branches all begin on the real axis at $s = -1$, $s = -2$, and $s = -3$, as illustrated in Figure D.2(a).

2. Since $R(s)$ has no roots, all branches approach infinity as $K$ approaches infinity.

3. Consider a real value of $s > -1$. Then $\phi_{s+1} = \phi_{s+2} = \phi_{s+3} = 0$. Thus the angle criterion is not satisfied and all points on the real axis such that $s > -1$ are not on any branch of the root-locus diagram. Next consider a real value of $s$ such that $-2 < s < -1$. Then $\phi_{s+1} = \pi$ and $\phi_{s+2} = \phi_{s+3} = 0$. Thus the angle criterion is satisfied and all points in the range $-2 \leq s \leq -1$ are on some branch of the root-locus diagram. Now consider a real value of $s$ such that $-3 < s < -2$. Then $\phi_{s+1} = \phi_{s+2} = \pi$ and $\phi_{s+3} = 0$. Thus the angle criterion is not satisfied and all points on the real axis such that $-3 < s < -2$ are not on any branch of the root-locus diagram. Finally consider a real value with $s < -3$. Then $\phi_{s+1} = \phi_{s+2} = \phi_{s+3} = \pi$ and the angle criterion is satisfied. Thus all real values of $s$ such that $s \leq -3$ are on a branch of the root-locus diagram. The results of this step are illustrated in Figure D.2(b).

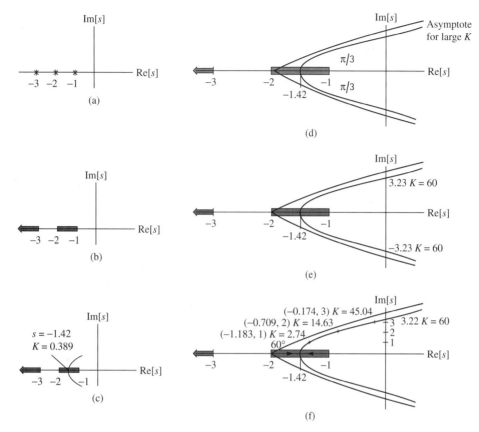

**FIG. D.2** The development of the root-locus diagram for Example D.2: (a) pole location; (b) regions on real axis that are part of a branch; (c) the breakaway point occurs at $s = -1.42$ with $K = 0.389$; (d) asymptotes for large $K$ pass through $\mathrm{Re}(s) = -2$ and make angles of $\pm \frac{\pi}{3}$ with Re axis; (e) the crossover point ocurs for $K = 60$ with $\omega = \mathrm{Im}(s) = \pm 3.23$; (f) three additional points are determined to fill in locus.

4.  The breakaway point for the branches originating at $s = -1$ and $s = -2$ is obtained by solving Equation (D.11), which for this problem is

$$R\frac{dQ}{ds} - Q\frac{dR}{ds} = 0$$

$$(1)\left(3s^2 + 12s + 11\right) - \left(s^3 + 6s^2 + 11s + 6\right)(0) = 0 \qquad \text{(d)}$$

$$3s^2 + 12s + 11 = 0$$

The solutions to Equation (d) are $s = -1.42$ and $s = -2.58$. Since $s = -1.42$ is in the range of real values of $s$ that are on a root-locus branch, as established in step 3, $s = -1.42$ is a breakaway point. Since $s = -2.58$ is not within this range, it is not a breakaway point. The tangent to the root-locus branches at the breakaway point is vertical, as illustrated in Figure D.2(c).

The value of $K$ at the breakaway point is obtained using the magnitude criterion [Equation (D.3)], $K = |Q(-1.42)/R(-1.42)| = 0.389$.

5.  There are three poles and no zeroes. Thus the angle of the asymptotes to the branches the approach infinity is determined using Equation (D.12)

$$\phi_a = \frac{(2n+1)\pi}{0-3} = -(2n+1)\frac{\pi}{3} \qquad \text{(e)}$$

For $n = 0$, Equation (e) gives $\phi_a = -\pi/3$. For $n = 1$, Equation (e) gives $\phi_a = -\pi$ and for $n = -1$, Equation (e) gives $\phi_a = \pi/3$. The asymptote with $\phi_a = -\pi$ corresponds to the branch that lies entirely on the negative real axis. The angles of the asymptotes to the other branches are $\pm\pi/3$.

The value of $s$ where these asymptotes meet on the real axis is determined using Equation (D.13) as

$$s = \frac{1}{3-0}[(-1) + (-2) + (-3)] = -2 \tag{f}$$

The asymptotes are illustrated in Figure D.2(d).

6. Substituting $s = j\omega$ into Equation (a) leads to

$$(j\omega)^3 + 6(j\omega)^2 + 11(j\omega) + 6 + K = 0$$
$$6 + K - 6\omega^2 + j(11\omega - \omega^3) = 0 \tag{g}$$

Equating the imaginary part of Equation (g) to zero leads to a crossover frequency of $\omega = \sqrt{11} = 3.32$. Then equating the real part of Equation (g) to zero gives the corresponding gain as $K = 60$. The crossover frequency is illustrated in Figure D.2(e).

7. Since there are no complex poles or zeroes there are no angles of departure or angles of arrival to calculate.

8. Three points on the branch in the third quadrant of the complex plane are determined. It is known that for this branch in the third quadrant $-1.42 \le s_r \le 0$ and $0 \le s_j \le 3.32$. First consider the point with $s_j = 3$. Guess $s_r = -0.5$. Checking the angle criterion, $\phi_{s+1} = \tan^{-1}(3/0.5) = 1.405$, $\phi_{s+2} = \tan^{-1}(3/1.5) = 1.107$, $\phi_{s+3} = \tan^{-1}(3/2.5) = 0.876$, and $\phi_{s+1} + \phi_{s+2} + \phi_{s+3} = 3.388$. The angle criterion is not met. Since the sum of the angles is greater than $\pi$ the guessed value of $s_r$ is increased to $-0.1$. This leads to $\phi_{s+1} + \phi_{s+2} + \phi_{s+3} = 3.088$. Since this value is less than $\pi$ the guess for $s_r$ is decreased to $-0.2$, leading to $\phi_{s+1} + \phi_{s+2} + \phi_{s+3} = 3.160$. This iteration continues until it is established that for $s_r = -0.174$, $\phi_{s+1} + \phi_{s+2} + \phi_{s+3} = 3.141$. The corresponding value of $K$ is found from the magnitude criterion [Equation (D.3)]

$$K = \left| \frac{Q(-0.174 + 3j)}{R(-0.174 + 3j)} \right| = 45.04 \tag{h}$$

Two additional points are determined in the same fashion. These points are used to construct the final version of the root-locus diagram, shown in Figure D.2(f).

## Example D.2

Construct the root-locus diagram for

$$s^3 + 2s^2 + (5 + K)s + 2K = 0 \tag{a}$$

### Solution

Equation (a) is written in the form of Equation (D.1) by defining

$$Q(s) = s^3 + 2s^2 + 5s \tag{b}$$
$$R(s) = s + 2 \tag{c}$$

The root-locus diagram is constructed by following the steps in the summary.

1.  The poles of the diagram are obtained by setting $Q(s) = 0$ leading to $s = 0, -1 \pm 2j$. The three branches originate from these points.
2.  The zeroes of the diagram are obtained by setting $R(s) = 0$ leading to $s = -2$. Thus a branch terminates either at $s = -2$ or at infinity.
3.  The angle criterion for determining points on the root-locus diagram is

$$\phi_{s+2} - \phi_s - \phi_{-1+2j} - \phi_{-1-2j} = (2n+1)\pi \qquad n = 0, \pm 1, \pm 2, \ldots \qquad \text{(d)}$$

For a point on the real axis $\phi_{-1-2j} = \phi_{-1+2j}$ and thus Equation (d) reduces to

$$\phi_{s+2} - \phi_s = (2n+1)\pi \qquad n = 0, \pm 1, \pm 2, \ldots \qquad \text{(e)}$$

Equation (d) is satisfied for real $s$ only when $-2 \le s \le 0$.

4.  A breakaway point or breakin point is obtained by solving Equation (D.11)

$$R\frac{dQ}{ds} - Q\frac{dR}{ds} = 0$$

$$(s+2)(3s^2 + 4s + 5) - (s^3 + 2s^2 + 5s)(1) = 0 \qquad \text{(f)}$$

$$s^3 + 4s^2 + 4s + 5 = 0$$

The solutions of Equation (f) are $s = -3.24$ and $s = -0.379 \pm 1.18j$. The complex roots are excluded because the equation is valid only for real $s$. The root $s = -3.24$ cannot be a breakaway point because it is not in the range $-2 \le s \le 0$, which has been determined as the range of real values of $s$ lying on a branch of the root-locus diagram. Thus there are no breakaway points or breakin points for this root-locus diagram.

5.  There are three poles and one zero. The angles of the asymptotes are determined using Equation (D.12)

$$\phi_a = \frac{(2n+1)\pi}{1-3} \qquad n = 0, \pm 1, \pm 2, \ldots$$

$$\phi_a = -\frac{\pi}{2}(2n+1) \qquad n = 0, \pm 1, \pm 2, \ldots \qquad \text{(g)}$$

Equation (g) leads to unique values of $\phi_a = \pi/2$ and $\phi_a = -\pi/2$. There are only two asymptotes because the third branch terminates at the zero, $s = -2$.

The point of intersection of the asymptotes with the real axis is obtained using Equation (D.13)

$$s = \frac{1}{3-1}[(0) + (-1+2j) + (-1-2j) - (-2)] = 0 \qquad \text{(h)}$$

Equations (g) and (h) imply that the asymptotes are the positive and negative imaginary axes.

6.  From the previous steps it is clear that the branches do not cross the imaginary axis. One branch begins at $s = 0$ and terminates at $s = 2$. The other branches originate at $s = -1 \pm 2j$ and asymptotically approach the imaginary axis. Thus there is no crossover frequency for this example.

7.  The application of the angle criterion to determine the angle of departure at the complex pole $-1 + 2j$ is illustrated in Figure D.3(a). The angle criterion

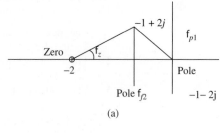

(a)

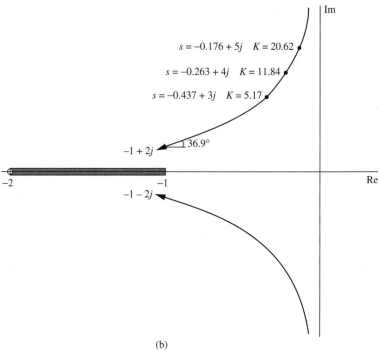

**FIG. D.3** (a) Illustration of the determination of angle of departure; (b) the root-locus diagram for Example D.2.

(b)

leads to

$$\phi_d = \pi + \phi_z - \phi_{p1} - \phi_{p2}$$

$$= \pi + 1.107 - 2.0344 - \frac{\pi}{2} \tag{i}$$

$$= 0.6435 \, r \, (36.9°)$$

8. Three points are established on the branch of the root locus that exists in the third quadrant. The real part of a point on this branch satisfies $-1 \leq s_r < 0$ while the imaginary part satisfies $s_j \geq 2$. Three points and their corresponding gains determined using the method illustrated in step 8 of Example D.1 are

$$s = -0.437 + 3j \quad K = 5.17$$
$$s = -0.263 + 4j \quad K = 11.84 \tag{j}$$
$$s = -0.176 + 5j \quad K = 20.62$$

The root-locus diagram for this example is sketched in Figure D.3(b).

# References

Antoniou, A. 1993. *Digital filters—analysis, desigln and applications,* 2nd ed. New York, NY: McGraw-Hill.

Bedford, A. M., and Fowler, W. 2004. *Engineering mechanics—dynamics,* 4th ed. Upper Saddle River, NJ: Pearson Prentice Hall.

Chapman, S. T. 2002. *MATLAB programming for engineers.* Australia: Brooks-Cole.

Chapra, S. C, and Canale, R. P. 2006. *Numerical methods for engineers,* 5th ed. Boston, MA: McGraw-Hill.

Churchill, R. V. 1972. *Operational mathematics,* 3rd ed. New York: McGraw-Hill.

Fournier, R. L. 1999. *Basic transport phenomena for biomedical engineering.* Philadelphia, PA: Taylor and Francis.

Irwin, J. D. 1991. *Basic engineering circuit analysis,* 6th ed. Englewood Cliffs, NJ: Prentice Hall.

John, J. A., and Haberman, W. L. 1988. *Introduction to fluid mechanics,* 3rd ed. Englewood Cliffs, NJ: Prentice Hall.

Kelly, S. G. 2000. *Fundamentals of mechanical vibrations.* 2nd ed. Boston, MA: McGraw-Hill.

———. 1996. *Schaum's outline for mechanical vibrations.* New York: McGraw-Hill.

Luyben, W. L. 1973. *Process modeling, simulation, and control for chemical engineers.* New York: McGraw-Hill.

Marshall, J. E., Goreek, H., Walton, K., and Korytowski, A. 1992. *Time delay systems, stability and performance criteria with applications.* New York: Ellis Horwood.

MathWorks. 2006. *Using MATLAB Version 6.* Natick, MA.

Nahvi, M., and Edminister, J. 2003. *Schaum's outline of electric circuits,* 4th ed. New York: McGraw-Hill.

Newton, C. M. 1984. "Modern Model-Based Methods and Software for Pharmokinetics Research," in C. Nicolin, ed. *Modeling and advances in biomedicine.* Singapore: World Scientific.

Ogata, K. O. 2004. *System dynamics,* 4th ed. Upper Saddle River, N.J.: Pearson Prentice Hall.

Palm, W. J. 2001. *Introduction to MATLAB for engineers.* Boston, MA: McGraw-Hill.

———. 1987. *Modeling analysis and control of dynamic systems.* New York: John Wiley and Sons.

Pelesko, J. A. 2003 and Bernstein, D. H. *Modeling MEMS and NEMS.* Boca Raton, FL: Chapman and Hall/ CRC.

Raven, F. 1995. *Automatic control engineering,* 5th ed. New York: McGraw-Hill.

Smith, J. M. 1981. *Chemical engineering kinetics,* 3rd ed. New York: McGraw-Hill.

Trietly, H. 1986. *Transducers in mechanical and electronic design.* New York: Marcel Dekker.

Umez-ERonini, E. *System dynamics and control.* Pacific Grove, CA: Brooks Cole.

Watton, J. 1989. *Fluid power systems.* New York: Prentice Hall.

White, F. M. 1999. *Fluid mechanics,* 4th ed. Boston: McGraw-Hill.

# Index

Acceleration vector, particles, 51
Algebra, 516–521, 673–675, 676–682
 block diagrams, 516–521
 complex, 673–675
 complex conjugates, 673–674
 complex numbers, 673
 Euler's identity, 675
 magnitude, 674–675
 matrix, 676–682
Amplification, 479–480
 motion input, 479–480
 range of, 480
Amplifier, electrical, 143
Amplitude, 371, 435
 second-order systems, 371
 steady-state, 435
Angle and magnitude criteria, root-locus
  diagrams, 699–700
Angle of arrival, root-locus diagrams, 699
Angle of departure, root-locus diagrams, 699
Angular velocity, rigid bodies, 52
Assumptions, mathematical modeling, 10–11
Asymptotes, Bode diagrams, 450–451

Band-pass filter, 482, 490
Band-reject filter, 482, 490
Bandwidth, defined, 461
Beating, frequency response, 437
Bernoulli's equation, 204–207
 control volume analysis, 204–207
 extended, 206
 gauge pressure, 205
 head, 206
 head loss, 206
 manometer equation, 206
 streamline, 204–205
Biological systems, 244–245
Block diagrams, 515–525, 525–533
 algebra, 516–521
 branch point, 516

equivalencies, 519
equivalent open-loop transfer function, 578
feedback control systems, 515–525,
  525–533
feedback loop, 517
modeling of dynamic systems, 522–525,
  525–533
parallel system components, 516–517
representation of, 515–516
series, components in, 516
SIMULINK, using for modeling of,
  525–533
summing point, 516
Bode diagrams, 450–462, 463–467
 asymptotes, 450–451
 bandwidth, 461
 construction of, 450–451
 corner (breakpoint) frequency, 461
 cutoff frequency, 461
 frequency response using, 450–462
 MATLAB for development of plots,
  463–467
 minimum phase transfer function, 462
 nonminimum phase transfer function, 462
 parameters, 461–462
 transfer functions, 452–453, 453–461
Body force, free-body diagrams, 81
Branch, root-locus diagrams, 689
Branch point, block diagrams, 516
Breakaway point, root-locus diagrams, 698
Breakin point, root-locus diagrams, 699
Brute force methods, 299–300

Capacitors, 138–141
 capacitance, 138
 electrical, 138–141
 farads, 138
 parallel plate capacity, 138
Center of mass, rigid bodies, 53
Change of scale, Laplace transforms, 286

Characters with special significance,
  MATLAB, 686
Charge, electrical, 134
Chemical systems, 198–200, 200–207,
  240–244, 245–268
 continuous stirred tank reactors (CSTR),
  240–244
 defined, 199
 examples of mathematical modeling,
  245–261
 introduction to, 198–200
 lumped parameter assumptions, 199–200
Circuits, 133, 136–144, 144–146, 146–151,
  151–156
 active elements, 133, 137
 capacitors, 138–141
 components, 136–144, 146–151
 current source, 143
 diagrams, 144
 equivalent equations, 149
 inductors, 141–143
 Kirchoff's laws, 144–146
 loop, 144
 Maxwell's equations, 144
 mechanical systems and, 143–144
 modeling of, 151–156
 node, 144
 Ohm's law, 136–137
 open, 143
 operational amplifier, 143
 parallel combinations, 148–149
 parallel components, 147
 passive elements, 133, 136
 reduction, 146–151
 resistors, 136–138
 series combinations, 147–148
 series components, 147
 short, 143
 switch, 143
 voltage source, 143

Closed-loop transfer function, 533
Colebrook equation, 209
Components, 146–149, 516–517
    parallel, 146–147, 148–149, 516–517
    series, 146–148, 516
Compressible flow, 214–215
Conditional statement, MATLAB, 692–693
Conduction, thermal systems, 233–234
Conservation of Charge, 144
Conservation of Energy, 144
Conservation of Mass, 201–202
    continuity equation, 202
    control volume analysis, 201–202
    incompressible fluids, 201–202
    volumetric flow rate, 201
Continuous stirred tank reactors (CSTR),
    240–244
Continuous system, defined, 60
Continuous time system, defined, 3
Control System Toolbox, MATLAB,
    696–697
Control systems, 3, 44–45, 515–600
    derivative, 536
    design of, 562–580
    feedback, 3, 515–600
    integral, 535–536
    proportional plus derivative (PD), 536
    proportional plus integral (PI), 535–536
    proportional plus integral plus derivative
        (PID), 536–537, 573–575
    proportional, 534–535
    root-locus diagrams, design of using,
        563–573
    types of, 44–45
    use of, 3
    Ziegler-Nichols tuning rules, 573–575
Control volume analysis, 200–207
    Bernoulli's equation, 204–207
    conservation of mass, 201–202
    continuity equation, 202
    defined, 200–201
    energy equation, 202–204
    internal energy, 202–203
    shaft work, 203
    specific energy, 203
    system, compared to, 200–201
    volumetric flow rate, 201
Convection, thermal systems, 234–235
Convolution integral, 345
Convolution of two functions, Laplace
    transforms, 286
Corner (breakpoint) frequency, defined,
    461
Coulomb damping, 72–74
    defined, 72–73
    kinetic coefficient of friction, 72
    static coefficient of friction, 72
Coupling field, electromechanical systems,
    172
Cramer's rule, 680
Crossover frequency, root-locus
    diagrams, 699

Current, electrical, 135, 143, 145
    defined, 135
    Kirchoff's current law (KCL), 145
    source, 143
Cutoff frequency, defined, 461
Cycles, units of, 9

D'Alembert's principle, 95–101
    particles, 95
    planar motion, 95–101
    rigid bodies, 95–101
Damped natural frequency, second-order
    systems, 373–374
Damped period, second-order systems, 373
Damping, 68–74
    coefficient, 69
    Coulomb, 72–74
    defined, 68
    hysteretic, 74
    viscous, 68–72
Deformable bodies, 51, 58–59
Degrees of freedom, 59–61, 101–103,
    103–104
    continuous system, 60
    defined, 60
    discrete system, 60
    distributed parameter system, 60
    multidegree systems, 103–104
    natural frequency, 101
    particle motion, 59–60
    single systems, 101–103
    standard form of differential equation,
        102
    viscous damping ratio, 102
Degrees, units of, 8
Derivative control, 536
Derivatives, Laplace transform of, 282–284
Determinants, matrices, 677–678
Diagrams, 81–87, 144, 209–210, 357–360,
    448–450, 450–462, 462–463, 463–467,
    515–525, 525–533, 563–573, 698–708
    block, 515–525, 525–533
    Bode, 450–462
    circuit, 144
    feedback control systems, 515–525, 525–
        533, 563–573
    free-body (FBD), 81–87
    frequency response, 450–462, 462–463
    frequency response curves, 448–450
    MATLAB use of for development of,
        463–467
    Moody, 209–210
    Nyquist, 462–463
    root-locus diagrams, 357–360, 563–573,
        698–708
    SIMULINK use in modeling, 525–533
    transient analysis, 357–360
Differential equations, 15–20, 102, 307–317
    integrodifferential, 313–315
    Laplace transform solutions of, 307–317
    linearization of, 15–20

MATLAB, use of for solution of,
    315–317
perturbation variables, 18–20
small angle approximation, 17
standard form of, 102
systems of, 310–313
systems with one dependant variable,
    307–310
Taylor series expansion, 15–16, 18
Differentiation of Laplace transforms, 286
Dimensions, 3–9
    basic, examples of, 4–5
    defined, 3–4
    derived, 7
    electrical systems, 5
    physical quantities, 6
    principle of dimensional homogeneity, 6
    units and, 3–9
Dirac delta function, 22
Direct current (dc) servomotors, 176–179
Discrete system, defined, 60
Discrete time system, defined, 3
Distributed parameter system, defined, 60
Dot notation, 2
Dynamic systems, 1–49, 522–525, 525–533
    block diagram modeling of, 522–525,
        525–533
    control systems, 3, 44–45
    defined, 1
    equilibrium, 13–15, 25–27
    introduction to, 1–3
    mathematical modeling, 1–2, 9–12
    plant, 533
    SIMULINK in block diagram modeling,
        525–533
    stability, 25–27
    system response, 1, 12–15, 25–27
Dynamic vibration absorbers, frequency
    response of, 486–489

Eigenvalues and eigenvectors, 680–682
Electrical systems, 133–197. See also
        Electromechanical systems
    active circuit elements, 133, 137
    capacitors, 138–141
    charge, 134
    circuit components, 136–144, 146–151
    circuit reduction, 146–151
    current, 135
    electrical field, 134
    electrical potential, 135
    electromechanical systems, 172–182
    energy methods, 156–157
    examples of mathematical modeling,
        182–189
    inductors, 141–143
    introduction to, 133–134
    Kirchoff's laws, 144–146
    mathematical modeling process for,
        133–134
    mechanical system analogies, 156–166
    modeling of circuits, 151–156

motion input, 163–165
multiple-loop circuits, 161–163
Ohm's law, 136–137
operational amplifiers, 143, 166–172
passive circuit elements, 133, 136
power, 136
resistors, 136–138
single-loop circuits, 157–159, 159–161
states, 165–166
voltage source, 143
Electromechanical systems, 172–182
coupling field, 172
defined, 172
direct current (dc) servomotors, 176–179
general theory of, 173–174
magnetic fields, 172–173
microelectromechanical systems (MEMS),
179–181
nanoelectromechanical systems (NEMS),
179–181
transducer, 172
Energy equation, 202–204
control volume analysis, 202–204
internal energy, 202–203
shaft work, 203
specific energy, 203
Energy methods, 104–114
energy storage, 110–111
equivalent systems, 105–107
Lagrange's equations, 111–112
mechanical systems, 104–114
principle of work and energy, 104–105
states and order of systems, 112–114
English system of units, 6, 8
Equation of state, pneumatic systems, 223
Equations, 2–3, 15–20, 104–105, 111–112,
144, 149, 200–207, 209, 307–317, 679.
*See also* Algebra; Newton's laws
Bernoulli's, 204–207
Colebrook, 209
conservation of mass, 201–202
continuity, 202
control volume analysis, 200–207
differential, 15–20, 307–317
dot notation, 2
energy, 202–204
equivalence, circuit components, 149
extended Bernoulli's, 206
integrodifferential, 313–315
Lagrange's, 111–112
linear, 2
linearization of differential, 15–20
manometer, 206
matrix systems of, 679
Maxwell's, 144
nonlinear, 2–3
principle of work and energy, 104–105
volumetric flow rate, 201
Equilibrium, 13–15, 25–27
impulsive response, 26
neutrally stable, 25–26
stability of, 25–27

system response of, 13–15
unstable, 25, 27
Equivalence equations, circuit components,
149
Equivalent open-loop transfer function,
578
Error, 538, 632
approximation, 632
feedback control, 538
local, 632
Euler's identity, 675
Euler's method, 632
Explicit assumptions, mathematical
modeling, 10–11
Exponential order, Laplace transform, 271
External force, 75–78

Faraday's law, 141
Farads, units of, 138
Feedback control systems, 3, 515–600
block diagrams, 515–525, 525–533
closed-loop transfer function, 533
configuration of, 533–539
derivative control, 536
design of, 562–580
error, 538
examples of mathematical modeling for,
576–594
feedback loop, 534
feedforward loop, 533
first-order plants, 539–551
integral control, 535–536
introduction to, 515
offset, 538
open-loop transfer function, 533
plant, 533
proportional control, 534–535
proportional plus derivative (PD) control,
536
proportional plus integral (PI) control,
535–536
proportional plus integral plus derivative
(PID) control, 536–537, 573–575
response to unit step input, 538–539
root-locus diagrams, design of using,
563–573
second-order plants, 551–562
SIMULINK, using for block diagram
modeling of, 525–533
use of, 3
Ziegler-Nichols tuning rules, 573–575
Feedback, mathematical modeling, 10
Feedback loop, 517, 534
Feedforward loop, 533
Filters, 482–485, 489–493
band-pass, 482, 490
band-reject, 482, 490
frequency response of, 482–485, 489–493
higher-order systems, 489–493
high-pass, 482, 490
low-pass, 482, 490
second-order systems, 482–485

Final value theorem, Laplace transforms,
285
First-order plants, 539–551
closed-loop response, 546
feedback control of, 539–551
First-order systems, 12, 361–367, 467–472
defined, 12
free response, 362–364
frequency response of, 467–472
impulsive response, 365–366
ramp response, 367
step response, 366–367
time constant, 361
transient analysis of, 361–367
Fluid systems, 198–267. *See also* Hydraulic
systems; Pneumatic systems
Bernoulli's equation, 204–207
Colebrook equation, 209
compressible flow, 214–215
control volume analysis, 200–207
examples of mathematical modeling,
245–261
gauge pressure, 205
head loss, 206, 208
incompressible flow, 198, 201–202
introduction to, 198–200
liquid-level, modeling of, 215–222
lumped parameter assumptions, 199–200
mathematical modeling, 215–222
Moody diagram, 209–210
orifices, 213–214
pipe flow, 207–215
Reynolds number, 208
streamline, 204–205
volumetric flow rate, 201
Force, 75–80, 81–87
body, 81
external, 75–78
Fourier series representation, 79
free-body diagrams (FBD), 81–87
fundamental frequency, 79
gravity, 81–82
harmonic excitation, 80
impulsive, 78
mechanical systems, 75–80
periodic, 79–80
resultant moment, 82
step, 78–79
surface, 82
torque, 75–78
Forced response, 12, 627–628
defined, 12
Euler's method, 632
inputs, due to, 627–628
Laplace transform solution, 627–631
numerical solutions for, 631–636
Runge-Kutta methods, 632–636
state-space solutions for, 627–638
Force-displacement relations, springs,
62–64
Fourier coefficients, 494
Fourier series representation, 79, 494

Free-body diagrams (FBD), 81–87
  body force, 81
  gravity force, 81–82
  mechanical systems, 81–87
  resultant, 82
  resultant moment, 82
  surface forces, 82
Free response, 12, 343, 362–364, 370–379,
    620–626
  defined, 12
  description of, 624–626
  exponential solution for, 621–624
  first-order systems, 362–364
  Laplace transform solution for, 620–621
  second-order systems, 370–379
  state transition matrix, 624
  state-space solutions for, 620–626
  transient analysis using, 343, 362–364,
    370–379
Frequency, 9, 79, 101, 371–374, 448, 461
  corner (breakpoint), 461
  cutoff, 461
  damped natural, 373–374
  defined, 371
  fundamental, 79
  natural, 101, 372
  ratio, 448
  second-order systems, 371–374
  units of, 9
Frequency domain, defined, 435
Frequency response, 435–514
  Bode diagrams, 450–462
  curves, 448–450
  dynamic vibration absorbers, 486–489
  examples of mathematical modeling,
    497–510
  filters, 482–485, 489–493
  frequency domain, 435
  graphical representation of, 448–467
  higher-order systems, 485–493
  introduction to, 435
  MATLAB, use of, 463–467
  motion input, 478–482
  Nyquist diagrams, 462–463
  one-degree-of-freedom mechanical system,
    472–477
  periodic input, 493–497
  s domain, 435
  second-order systems, 472–485
  sinusoidal transfer function, 440–447
  steady-state amplitude, 435
  steady-state phase, 435
  steady-state response, 435
  time domain, 435
  undamped second-order systems,
    436–440
Friction elements, 68–74
  Coulomb damping, 72–74
  damping, 68
  damping coefficient, 69
  hysteretic damping, 74
  kinetic coefficient of friction, 72

  mechanical systems, 68–74
  static coefficient of friction, 72
  viscous damping, 68–72
Functions, MATLAB, 685, 695
Fundamental frequency, 79

Gauge pressure, 205
General response, defined, 12
Graphs, plotting and annotating, 688–690
Gravity force, free-body diagrams, 81–82

Harmonic excitation, 80
Head, defined, 206
Head loss, 206, 208
  Bernoulli's equation, 206
  major losses, 208
  minor losses, 208
  pipe flow, 208
Higher-order systems, 12, 394–303, 485–493
  defined, 12
  dynamic vibration absorbers, 486–489
  filters, 489–493
  frequency response of, 485–493
  multi-degree-of-freedom mechanical
    systems, 399–403
  transient analysis of, 394–403
High-pass filter, 482, 490
Hydraulic systems, 222–223, 226–232
  defined, 222–223
  servomotor, 228–232
Hysteretic damping, 74

Impedance matrix, 399–400
Implicit assumptions, 10, 50–51
  mathematical modeling, 10
  mechanical system modeling, 50–51
Impulse, defined, 78
Impulsive force, 78
Impulsive response, 26, 343, 365–366,
    379–383
  defined, 26
  determination of, 343
  first-order systems, 365–366
  transient analysis using, 343, 365–366,
    379–383
Inductors, 141–143
  electrical, 141–143
  Faraday's law, 141
  inductance, 131
Inertia, 51–59
  deformable bodies, 51, 58–59
  elements, 51–59
  moment of, 53–54, 55
  particles, 51–52
  polar moment of, 54
  products of, 54
  rigid bodies, 51, 52–58
  tensor, 53
Initial conditions, mathematical
    modeling, 11
Initial value theorem, Laplace transforms,
    286

Input, 74–81, 163–165, 340–341, 478–482,
    493–497, 538–539, 627–638, 691–692
  amplification, 479–480
  electrical systems, 163–165
  frequency response, 478–482, 493–497
  isolation, 479–480
  MATLAB, 691–692
  mechanical systems, 74–81
  motion, 163–165, 479–480
  multiple, 340–341
  nonzero, 627
  periodic, 493–497
  state-space analysis of response due to,
    627–638
  transfer functions, 340–341
  unit step, feedback control response to,
    538–539
Input matrix, 607
Integral control, 535–536
Integrals, Laplace transform of, 284–285
Integrodifferential equations, 313–315
Internal energy, defined, 202–203
Inverse, matrices, 678
Inversion, 271, 289–307
  brute force methods, 299–300
  inverse Laplace transform, 271
  Laplace transforms, 289–307
  linearity of inverse transform, 290
  MATLAB, use of, 301–307
  partial fraction decompositions, 292–300
  periodic functions, 300–301
  poles, 292, 294–298
  residues, 293, 299–300
  tables, use of, 290–292
Isentropic process, pneumatic systems, 223
Isolation, 479–480
  motion input, 479–480
  range of, 480
Isothermal process, pneumatic systems, 223

Kinetic coefficient of friction, 72
Kinetic energy, particles, 51
Kirchoff's laws, 144–146
  Conservation of Charge, 144
  Conservation of Energy, 144
  current law (KCL), 145
  Maxwell's equations, 144
  voltage law (KVL), 145

Lagrange's equations, 111–112
Laminar flow, 208
Laplace transforms, 269–335, 620–621,
    627–631
  defined, 271
  derivatives, 282–284
  differential equations, solutions of,
    307–317
  direct integration, 272–276
  examples of Mathematical modeling
    using, 317–331
  existence, 271
  exponential order, 271

final value theorem, 285
first shifting theorem, 280–281
free response, solution for, 620–621
integrals, 284–285
integrodifferential equations, solution of
    using, 313–315
introduction to, 269–270
inverse, 271
inversion of, 289–307
linearity of, 280
MATLAB, use of, 276–278, 301–307,
    315–317
partial fraction decompositions, 292–300
poles, 292, 294–298
properties of, 278–289
residues, 293, 299–300
response due to inputs, solution for,
    627–631
second shifting theorem, 281–282
state-space solutions, 620–621, 627–631
tables, 272–273, 290–292
transform pairs, 271, 272–278
Linear equations, defined, 2
Linear momentum, particles, 52
Linear system, defined, 2
Linearity of inverse transform, 290
Linearity of Laplace transforms, 280
Linearization of differential equations, 15–20
Logarithmic decrement, second-order
    systems, 375
Loop, circuit diagrams, 144
Looping, MATLAB, 694
Low-pass filter, 482, 490
Lumped parameter assumptions, 199–200

M-files, MATLAB, 27–28
Magnetic fields, 172–173
Magnitude, defined, 674–675
Manometer equation, 206
Mass, see Conservation of Mass
Mathematical modeling, 1–2, 9–12, 43–44,
    114–123, 133–134, 151–156, 182–189,
    199–200, 215–222, 236–239, 245–261,
    408–426, 497–510, 576–594, 663–669
    defined, 1–2
    dynamic systems, 9–12
    electric circuits, 151–156
    electrical systems, 133–134, 182–189
    explicit assumptions, 10–11
    feedback, 10
    feedback control systems, 576–594
    fluid systems, 215–222
    frequency response, 497–510
    implicit assumptions, 10, 133–134
    initial conditions, 11
    liquid-level systems, 215–222
    lumped parameter assumptions, 199–200
    mechanical systems, 114–123
    model formulation, 43–44
    parameters, 11
    procedure for, 9–12
    state-space analysis, 636–638

system analysis, 9
system design, 9
system synthesis, 9
thermal systems, 236–239
transient analysis, 408–426
transport systems, 245–261
variables, 11
MATLAB, 27–43, 276–278, 301–307,
    315–317, 346–349, 463–467, 525–533,
    636–638, 641–649, 649–656, 658–663,
    683–697
    basics of, 684–688
    block diagram modeling using
        SIMULINK, 525–533
    Bode plots, for development of, 463–467
    characters with special significance, 686
    conditional statement, 692–693
    Control System Toolbox, 696–697
    differential equations, 315–317
    examples of, 28–43
    functions, 685, 695
    graphs, plotting and annotating, 688–690
    input, 691–692
    introduction to, 683
    inversion of transforms, 301–307
    Laplace transforms, 276–278, 301–307,
        315–317
    looping, 694
    M-files, 27–28
    Nyquist diagrams, for development of,
        463–467
    operators, 684, 692
    output, 691–692
    program ode45.m, use of, 636–638,
        658–663
    programming commands, 690–696
    SIMULINK, 525–533, 649–656
    state-space analysis, 636–638
    state-space modeling, 641–649, 649–656
    Symbolic Math Toolbox, 696
    transform pairs, 276–278
    transient system response, 346–349
Matrices, 103–104, 399–400, 607, 624,
    676–682
    algebra, 676–682
    arithmetic, 677
    Cramer's rule, 680
    definitions for, 676
    determinants, 677–678
    diagonal, 676
    eigenvalues and eigenvectors, 680–682
    identity, 676
    impedance, 399–400
    input, 607
    inverse, 678
    multidegree-of-freedom systems,
        formation of, 103–104
    state transition, 624
    state, 607
    systems of equations for, 679
    transmission, 607
    transpose of, 676

Maxwell's equations, 144
Mechanical systems, 50–132, 143–144,
    156–172, 399–403, 472–477. See also
    Electromechanical systems
    Coulomb damping, 72–74
    D'Alembert's principle, 95–101
    damping, 68
    deformable bodies, 51, 58–59
    degrees of freedom, 59–61, 101–103,
        103–104
    electrical circuits and, 143–144
    electrical systems, analogies of, 156–172
    energy methods, 104–114, 156–157
    external force, 75–78
    force-displacement relations, 62–64
    free-body diagrams (FBD), 81–87
    frequency response of, 472–477
    friction elements, 68–74
    hysteretic damping, 74
    implicit assumptions for, 50–51
    impulsive force, 78
    inertia elements, 51–61
    input, 74–81
    introduction to, 50–51
    mathematical models for, 114–123
    motion input, 80–81, 163–165
    multidegree-of-freedom, 399–403
    multiple-loop circuits, 161–163
    Newton's laws, 87–101
    one-degree-of-freedom, 472–477
    particles, 51–52, 87–90
    periodic force, 79–80
    planar motion, 93–95
    rigid bodies, 51, 52–58, 90–91, 91–93,
        93–95
    rotational motion, 91–92
    single-loop circuits, 157–159, 159–161
    springs, 61–68
    state-space analysis, 113
    states, 113–114, 165–166
    static deflection, 67–68
    step force, 78–79
    transient analysis of, 399–403
    viscous damping, 68–72
Microelectromechanical systems (MEMS),
    179–181
Modeling, see Mathematical modeling;
    State-space
Moment of inertia, rigid bodies, 53–54, 55
Moody diagram, 209–210
Motion, 80–81, 87–101
    input, 80–81
    Newton's laws of, 87–101
    planar, 93–95
    rigid body, 90–91
    rotational, 91–92
    three-dimensional, 95
Motion input, 163–165, 478–482
    amplification, 479–480
    electrical systems, 163–165
    frequency response, 478–482
    isolation, 479–480

Motion input (*continued*)
  range of amplification, 480
  range of isolation, 480
  second-order systems, 478–482
Multidegree-of-freedom systems, 103–104,
    111–112, 399–403, 615–619
  impedance matrix, 399–400
  Lagrange's equations, 111–112
  matrix formation of, 103–104
  Newton's law, 103–104
  state-space modeling, 615–619
  transfer functions, 399–400
  transient analysis of, 399–403
  undamped systems, 400–401

Nanoelectromechanical systems (NEMS),
    179–181
Natural frequency, 101, 372, 373–374
  damped, 373–374
  defined, 101
  second-order systems, 372, 373–374
Newton's laws, 87–101, 103–104
  D'Alembert's principle, 95–101
  multidegree-of-freedom systems, 103–104
  particles, 87–90, 95
  planar motion, 93–95
  planar motion, 93–95, 95–101
  rigid body motion, 90–91
  rotational motion, 91–92
  rotational motion, 91–92
  second law, 87–88
  three-dimensional motion, 95
Node, circuit diagrams, 144
Nonlinear equations, defined, 2–3
Nonlinear system, 2–3, 656–663
  defined, 2–3
  MATLAB program ode45.m, 658–663
  state-space analysis of, 656–663
Numerical solutions, 631–636
  error of approximation, 632
  Euler's method, 632
  local error, 632
  Runge-Kutta methods, 632–636
  state-space analysis, 631–638
  step size, 632
Nyquist diagrams, 462–463, 463–467
  frequency response using, 462–463
  MATLAB for development of, 463–467

Offset, feedback control, 538
Ohm's law, 136–137
One-degree-of-freedom mechanical system,
    frequency response, 472–477
Open circuit, 143
Open-loop transfer function, 533
Operational amplifiers, 143, 166–172
Operators, MATLAB, 684, 692
Order, *see* System order
Orifices, pipe flow, 213–214
Output, 1, 340–341, 691–692
  MATLAB, 691–692
  multiple, 340–341

transfer functions, 340–341
system, 1
Overall heat transfer coefficient, 235

Parallel axis theorem, rigid bodies, 54
Parallel components, 146–147, 148–149,
    516–517
  block diagrams, 516–517
  circuits, 146–147, 148–149
  combinations, 148–149
  series and, 147
Parallel system components, block diagrams,
    516–517
Parameters, mathematical modeling, 11
Partial fraction decompositions, 292–300
Particles, 51–52, 59–60, 87–90, 95
  acceleration vector, 51
  D'Alembert's principle, 95
  defined, 51
  degrees of freedom, 59–60
  kinetic energy, 51
  linear momentum, 52
  Newton's second law and, 87–90
  position vector, 51
  velocity vector, 51
Peak time, defined, 344
Percentage overshoot, defined, 344
Period beating, frequency response, 437
Periodic force, 79–80
Periodic functions, 286, 300–301, 493–497
  Fourier coefficients, 494
  Fourier series, 494
  frequency response due to, 493–497
  input, 493–497
  inversion of transforms of, 300–301
  Laplace transform of, 286
Perturbation variables, 18–20
Phase, 371, 453
  amplitude, 435
  second-order systems, 371
Pipe flow, 207–215
  Colebrook equation, 209
  compressible, 214–215
  laminar, 208
  losses, 208–213
  Moody diagram, 209–210
  orifices, 213–214
  resistance, 211–213
  Reynolds number, 208
  turbulent, 208
Planar motion, 52–53, 93–95, 95–101
  D'Alembert's principle, 95–101
  defined, 52–53
  Newton's laws for, 93–95
  rigid bodies, 52–53, 93–95, 95–101
Plants, 533, 539–551, 551–562
  defined, 533
  feedback control of, 533, 539–551,
    551–562
  first-order, 539–551
  open-loop transfer function, 533
  second-order, 551–562

Pneumatic systems, 222–226
  defined, 222–223
  equation of state, 223
  isentropic process, 223
  isothermal process, 223
  polytropic process, 224
Polar moment of inertia, rigid bodies, 54
Poles, 292, 294–298, 342, 349–352, 357–360,
    689
  complex, 295–296
  defined, 294
  Laplace transform, 292, 294–298
  poles, 342, 349–350
  real distinct, 294–295
  repeated, 297–298
  root-locus diagram, 357–360, 689
  stability analysis, 349–352
  transient functions, 342, 349–352
Polytropic process, pneumatic systems, 224
Position vector, particles, 51
Power, electrical, 136
Principle of dimensional homogeneity, 6
Principle of work and energy, 104–105
Products of inertia, rigid bodies, 54
Programming commands, MATLAB,
    690–696
Proportional control, 534–535
Proportional plus derivative (PD) control, 536
Proportional plus integral (PI) control,
    535–536
Proportional plus integral plus derivative
    (PID) control, 536–537, 573–575

Radians, units of, 8
Radiation, thermal systems, 235–236
Ramp response, 345, 367
  determination of, 345
  first-order systems, 367
  transient analysis using, 345, 367
Reduction, circuits, 146–151
Residues, 293, 299–300
  brute force methods, 299–300
  defined, 293
  Laplace transform, 293, 299–300
Resistance, pipe flow, 211–213
Resistors, 136–138
Resonance, frequency response, 437
Response, *see* Forced response; Frequency
    response; System response; Transient
    response
Resultant, free-body diagrams, 82
Resultant moment, 82
Revolutions, units of, 9
Reynolds number, 208
Rigid bodies, 51, 52–58, 90–91, 91–93,
    93–95, 95–101
  angular velocity, 52
  center of mass, 53
  D'Alembert's principle, 95–101
  inertia tensor, 53
  moment of inertia, 53–54, 55
  motion, 90–91

Newton's law and, 90–91
Parallel axis theorem, 54
planar motion, 52–53, 93–95, 95–101
polar moment of inertia, 54
products of inertia, 54
rotational motion, 91–93
three-dimensional motion, 95
Rise time, 10–90 percent, 344
Root-locus diagrams, 357–360, 563–573, 698–708
analysis using, 357–360
angle and magnitude criteria, 699–700
angle of arrival, 699
angle of departure, 699
branch, 689
breakaway point, 698
breakin point, 699
construction of, 698–708
crossover frequency, 699
dominant root, 565
feedback control systems, design of using diagrams, 563–573
poles, 689
transient analysis, 357–360
Rotational motion, 91–92
Routh's method, 353–355
Runge-Kutta methods, 632–636

s domain, frequency response in, 435
Second-order plants, 551–562
feedback control of, 551–562
integral control, 554–556
proportional control, 557–559
Second-order systems, 12, 368–393, 436–440, 472–485
amplitude, 371
damped natural frequency, 373–374
damped period, 373
defined, 12
filters, 482–485
free response, 370–379
frequency, 371–374
frequency response of, 436–440, 472–485
impulsive response, 379–383
logarithmic decrement, 375
motion input, 478–482
natural frequency, 372
one-degree-of-freedom mechanical, 472–477
phase, 371
step response, 383–391
transient analysis of, 368–393
transient response, 391–393
undamped, 436–440
Series components, 146–148, 516
block diagrams, 516
circuits, 146–148
combinations, 147–148
parallel and, 147
Servomotors, 176–179, 228–232
direct current (dc), 176–179
hydraulic, 228–232

Settling time, 2 percent, 344
Shaft work, defined, 203
Shifting theorems, Laplace transforms, 280–282
first, 280–281
second, 281–282
Short circuit, 143
SI (System International) system of units, 8
SIMULINK, 525–533, 649–656
block diagram modeling using, 525–533
state-space, modeling in, 649–656
Single-degree-of-freedom systems, 101–103
Sinusoidal transfer function, 440–447
Small angle approximation, 17
Specific energy, defined, 203
Springs, 61–68
force-displacement relations, 62–64
mechanical systems, 61–68
parallel, in, 64
series, in, 64–65
static deflection, 67–68
stiffness, 63
unstretched length, 62
Stability, 25–27. *See also* Equilibrium
impulsive response, 26
system equilibrium, 25–27
Stability analysis, 349–360
general theory of, 349–352
poles of the transfer function, 349–352
relative stability, 355–357
root-locus analysis, 357–360
Routh's method, 353–355
Standard form of differential equation, 102
State-space, 113, 601–672
examples of mathematical modeling, 663–669
exponential solution, 621–624
formulation, 607–609
free response, solutions for, 620–626
general example of, 601–606
input matrix, 607
introduction to, 601
Laplace transform solution, 620–621, 627–631
MATLAB modeling in, 641–649
MATLAB program code ode45.m, use of, 636–638, 658–663
mechanical systems, analysis of, 113
methods, 601–672
modeling, 607–619
multidegree-of freedom mechanical systems, 615–619
nonlinear systems, analysis of, 656–663
numerical solutions, 631–636
response due to inputs, 627–638
SIMULINK modeling in, 649–656
state matrix, 607
state of system, 607
state transition matrix, 624
transfer functions and state-space models, relationship between, 638–641

transmission matrix, 607
variable coefficients, analysis of systems with, 656–663
variables, 602
States, 112–114, 165–166, 607
electrical systems, 165–166
matrix, 607
mechanical systems, 113–114, 165–166
order of systems and, 112–114
system, of, 607
Static coefficient of friction, 72
Static deflection, springs, 67–68
Steady-state amplitude, 435
Steady-state phase, 435
Steady-state response, 12, 435
defined, 12
determination of, 435
Step force, 78–79
Step response, 343–345, 366–367, 383–391, 538–539
10–90 percent rise time, 344
2 percent settling time, 344
determination of, 343–345
feedback control, 538–539
first-order systems, 366–367
peak time, 344
percentage overshoot, 344
transient analysis using, 343–345, 366–367, 383–391
unit step input, response to, 538–539
Step size, defined, 632
Streamline, defined, 204–205
Summing point, block diagrams, 516
Surface forces, free-body diagrams, 82
Switch, electrical, 143
Symbolic Math Toolbox, MATLAB, 696
System analysis, use of, 9
System design, use of, 9
System order, 12, 112–114, 342
defined, 342
poles, 342
response, 12
states and, 112–114
transfer functions, 342
zeros, 342
System response, 1, 12–15, 25–27, 44. *See also* Transient response
defined, 1
equilibrium, 13–15, 25–27
first-order system, 12
forced, 12
free, 12
general, 12
higher-order system, 12
impulsive, 26
order, 12
second-order system, 12
stability, 25–27
steady-state, 12
transient, 12
types of, 44
System synthesis, use of, 9

Systems, 1–49, 50–132, 133–197, 198–267, 361–367, 368–393, 394–403, 403–408
  biological, 244–245
  chemical, 198–200, 240–244
  components for, 1, 11
  continuous, 60
  continuous time, 3
  control volume analysis, 200–207
  control, 3, 44–45
  defined, 200
  differential equations, 15–20
  dimensions, 3–9
  discrete, 60
  discrete time, 3
  distributed parameter, 60
  dynamic, 1–49
  electrical, 133–197
  electromechanical, 172–182
  feedback control, 3
  first-order, 12, 361–367
  fluid, 198–267
  higher-order, 12, 394–403
  hydraulic, 222–223, 226–232
  impulsive response, 26
  input, 1
  introduction to, 1–49
  linear, 2
  linearization of differential equations, 15–20
  liquid-level, 215–222
  mathematical modeling, 1–2, 9–12, 43–44
  mathematical solutions and, 44
  MATLAB, 27–43
  mechanical, 50–132
  microelectromechanical (MEMS), 179–181
  model formulation, 43–44
  multidegree-of-freedom, 103–104, 111–112, 399–403
  nanoelectromechanical (NEMS), 179–181
  nonlinear, 2–3
  output, 1
  pneumatic, 222–226
  response, 1, 12–15, 44
  scope of study for, 43–45
  second-order, 12, 368–393
  single-degree-of-freedom, 101–103
  stability of, 25–27
  states and order of, 112–114
  static, 1
  thermal, 198–200, 232–239
  time constant, 361
  time delay, 403–408
  transient analysis of, 361–367, 368–393, 394–403
  undamped, 400–401
  unit impulse function, 20–23
  unit step function, 20–21, 23–25
  units, 3–9

Taylor series expansion, 15–16,18
Thermal systems, 198–200, 200–207, 232–239, 245–268
  conduction, 233–234
  control volume analysis, 200–207
  convection, 234–235
  defined, 199
  examples of mathematical modeling, 245–261
  introduction to, 198–200
  lumped parameter assumptions, 199–200
  mathematical modeling, 236–239
  overall heat transfer coefficient, 235
  radiation, 235–236
Time, 3, 336–434, 435. See also Transient analysis
  constant, 361
  continuous, 3
  delay, 403–408
  discrete, 3
  domain, 435
  domain response, 336–434
  peak, 344
  rise time, 10-90 percent, 344
  settling time, 2 percent, 344
  transient analysis and time domain response, 336–434
Torque, 75–78
Torsional viscous damper, 69
Transducer, electromechanical systems, 172
Transfer functions, 336–342, 342–349, 349–360, 440–447, 451–453, 453–461, 462, 516–521, 533, 638–641
  block diagram algebra determination of, 516–521
  Bode diagrams, 440–447, 451–453, 453–461, 462
  branch point, 516
  closed-loop, 533
  defined, 336–339
  equivalent open-loop transfer function, 518
  minimum phase, 462
  multiple inputs and outputs, 340–341
  nonminimum phase, 462
  open-loop, 533
  poles, 342, 349–352, 357–360
  products of, 451–453
  root-locus diagram, 357–360
  sinusoidal, 440–447
  stability analysis, 349–360
  state-space models, relationship between, 638–641
  summing point, 516
  system order, 342
  transient response specification, 342–349
  zeros, 342
Transform pairs, 271, 272–278
  defined, 271
  direct integration, 272–276
  MATLAB, use of, 276–278
  table of, 272–273
Transforms, see Laplace transforms

Transient analysis, 336–434
  convolution integral, 345
  examples of mathematical modeling, 408–426
  first-order systems, 361–367
  free response, 343, 362–364, 370–379
  higher-order systems, 394–403
  impulsive response, 343, 365–366, 379–383
  MATLAB, use of, 346–349
  multi-degree-of-freedom mechanical systems, 399–403
  ramp response, 345, 367
  relative stability, 355–357
  Root-Locus analysis, 357–360
  Routh's method, 353–355
  second-order systems, 368–393
  specifications of, 342–349
  stability analysis, 349–360
  step response, 343–345, 366–367, 383–391
  time delay, 403–408
  time domain response, and, 336–434
  transfer functions, 336–342
  transient response, 336, 342–349, 391–393
Transient response, 12, 336, 342–349, 391–393
  convolution integral, 345
  defined, 12, 336
  free response, 343, 362–364, 370–379
  impulsive response, 343, 365–366, 379–383
  MATLAB, use of, 346–349
  ramp response, 345, 367
  second-order systems, 391–393
  specification of, 342–349
  step response, 343–345, 366–367, 383–391
Transmission matrix, 607
Transport systems, see Chemical systems; Fluid systems; Thermal systems
Turbulent flow, 208

Undamped second-order systems, 436–440
  beating, 437
  frequency response of, 436–440
  period beating, 437
  resonance, 437
Unit impulse function, 20–23
Unit ramp function, 25
Unit step function, 20–21, 23–25
Units, 3–9
  cycles, 9
  degrees, 8
  derived, 7
  dimensions and, 3–9
  English system, 6, 8
  frequency, 9
  radians, 8
  revolutions, 9
  SI (System International) system, 8
  system of, 6
Unstretched length, springs, 62

Variables, 11, 18–20, 602, 656–663
    coefficients, state-space analysis of systems
        with, 656–663
    mathematical modeling, 11
    perturbation, 18–20
    state-space, 602
Velocity vector, particles, 51

Viscoelastic materials, 69
Viscous damping, 68–72, 102
    damping coefficient, 69
    defined, 68
    ratio, 102
    torsional viscous damper, 69
    viscoelastic materials, 69

Voltage, 143, 145
    Kirchoff's voltage law (KVL), 145
    source, 143
Volumetric flow rate, 201

Zeros, transient functions, 342
Ziegler-Nichols tuning rules, 573–575